Dr. F   ıs

AF835923

FROM CYCLOTRONS TO CYTOCHROMES
Essays in Molecular Biology and Chemistry

MARTIN D. KAMEN

FROM CYCLOTRONS TO CYTOCHROMES

Essays in Molecular Biology and Chemistry

Edited by

NATHAN O. KAPLAN

Department of Chemistry
School of Medicine
University of California, San Diego
La Jolla, California

ARTHUR ROBINSON

Linus Pauling Institute of Science and Medicine
Menlo Park, California

1982

ACADEMIC PRESS
A Subsidiary of Harcourt Brace Jovanovich, Publishers
New York London
Paris San Diego San Francisco São Paulo Sydney Tokyo Toronto

Academic Press Rapid Manuscript Reproduction

ACADEMIC PRESS, INC.
111 Fifth Avenue, New York, New York 10003

United Kingdom Edition published by
ACADEMIC PRESS, INC. (LONDON) LTD.
24/28 Oval Road, London NW1 7DX

Library of Congress Cataloging in Publication Data
Main entry under title:

From cyclotrons to cytochromes.

Symposium honoring Martin Kamen, 1978, University of California, San Diego, La Jolla, Calif.
Includes index.
1. Molecular biology--Addresses, essays, lectures.
2. Biological chemistry--Addresses, essays, lectures.
3. Kamen, Martin David, Date. I. Kaplan, Nathan Oram, Date. II. Robinson, Arthur, Date.
III. Kamen, Martin David, Date. IV. University of California, San Diego.
QH506.F75 574.8'8 82-1785
ISBN 0-12-397580-8 AACR2

PRINTED IN THE UNITED STATES OF AMERICA

82 83 84 85 9 8 7 6 5 4 3 2 1

Contents

CYTOCHROMES: CHEMISTRY AND STRUCTURE

PHOSPHORYLATION, OXIDATION, AND PHOTOSYNTHESIS

PROTEIN STRUCTURE

REGULATORY MECHANISMS AND METABOLIC MISCELLANY

Contributors

Numbers in parentheses indicate the pages on which the authors' contributions begin.

Ileana Agalidis (381), *Laboratoire de Photosynthèse, Centre National de la Recherche Scientifique, 91190 Gif-sur-Yvette, Essonne, France*

R. P. Ambler (263), *Department of Molecular Biology, University of Edinburgh, Edinburgh EH9 3J4, Scotland*

Herrick Baltcheffsky (447), *Department of Biochemistry, Arrhenius Laboratory, University of Stockholm, S10691, Stockholm, Sweden*

Margareta Baltscheffsky (347), *Department of Biochemistry, Arrhenius Laboratory, University of Stockholm, S10691, Stockholm, Sweden*

H. A. Barker (67), *Department of Biochemistry, University of California, Berkeley, Berkeley, California 94720*

A. A. Benson (59), *Department of Marine Biology, Scripps Institute of Oceanography, La Jolla, California 92093*

Sosamma J. Berger (587), *Department of Pharmacology, School of Medicine, Washington University, St. Louis, Missouri 63110*

William F. Bosron (469), *Department of Medicine and Biochemistry, Indiana University School of Medicine, Indianapolis, Indiana*

A. F. Brodie (417), *Department of Biochemistry, School of Medicine, University of Southern California, Los Angeles, California 90033*

Melvin Calvin (15), *Laboratory of Chemical Biodynamics, University of California, Berkeley, California 94720*

Joyce G. Carter (587), *Department of Pharmacology, School of Medicine, Washington University, St. Louis, Missouri 63110*

Young-Ling Chiang (145), *Department of Chemistry, State University of New York, Albany, New York 12222*

Richard I. Christopherson (611), *Department of Biochemistry, School of Medicine, University of North Carolina, Chapel Hill, North Carolina 27514*

Mildred Cohn (51), *Department of Biochemistry and Biophysics, University of Pennsylvania Medical School, Philadelphia, Pennsylvania 19104*

Sidney P. Colowick (561), *Department of Microbiology, Vanderbilt University, Nashville, Tennessee 37232*

Elizabeth J. Crawford (77), *Department of Biology, University of California, San Diego, La Jolla, California 92093*

Michael A. Cusanovich (215), *Department of Biochemistry, University of Arizona, Tucson, Arizona 85721*

Werner P. Dafeldecker (469), *Biophysics Research Laboratory, Department of Biological Chemistry, Harvard Medical School, Boston, Massachusetts*

Gerald W. DeVries (587), *Department of Pharmacology, School of Medicine, Washington University, St. Louis, Missouri 63110*

Kent Diveley (77), *Department of Biology, University of California, San Diego, La Jolla, California 92093*

Pierre Douzou (203), *Institut National de la Sante et de la Recherche Medicale, U128 INSERM-CNRS, 34033 Montpellier Cedex, France*

Gerhart Drews (355), *Institut für Biologie II, Albert Ludwigs Universität, Schlänzlestrasse 1, D7800 Freiburg, Federal Republic of Germany*

Karl M. Dus (231), *Edward A. Doisy Department of Biochemistry, St. Louis University Medical Center, St. Louis, Missouri 63104*

Fujio Egami (135), *Mitsubishi-Kasei Institute for Life Sciences, Tokyo 194, Japan*

Johannes Everse (599), *Cancer Center, University of California, San Diego, La Jolla, California 92093*

Gerald D. Fasman (455), *Department of Biochemistry, Brandeis University, Waltham, Massachusetts 02254*

James A. Ferrendelli (587), *Department of Pharmacology, School of Medicine, Washington University, St. Louis, Missouri 63110*

Torgeir Flatmark (431), *Department of Biochemistry, University of Bergen, N-5000 Bergen, Norway*

William A. Frey (469), *Biophysics Research Laboratory, Department of Biological Chemistry, Harvard Medical School, Boston, Massachusetts*

Morris Friedkin (77), *Department of Biology, University of California, San Diego, La Jolla, California 92093*

Keiko Fujii (251), *Department of Biology, Faculty of Science, Osaka University, Toyonaka, Osaka 560, Japan*

Yoshihiro Fukumori (251), *Department of Biology, Faculty of Science, Osaka University, Toyonaka, Osaka 560, Japan*

Howard Gest (305), *Photosynthetic Bacteria Group, Department of Biology, Indiana University, Bloomington, Indiana 47401*

L. Grossman (669), *Department of Biochemistry, The Johns Hopkins University School of Hygiene and Public Health, Baltimore, Maryland 21205*

W. Haseltine (669), *Department of Biochemistry, The Johns Hopkins University School of Hygiene and Public Health, Baltimore, Maryland 21205*

Donald R. Helinski (691), *Department of Biology, University of California, San Diego, La Jolla, California 92093*

Robin Hill (299), *Department of Biochemistry, University of Cambridge, Cambridge, CB2 1QW, England*

Takekazu Horio (367), *Institute for Protein Research, Osaka University, Osaka, Japan*

C. W. Huang (397), *Department of Chemistry, University of Southern California, University Park, Los Angeles, California 90007*

Manabu Inuzuka (691), *Department of Biology, University of California, San Diego, La Jolla, California 92093*

A. J. Jacobs (417), *Department of Biochemistry, School of Medicine, University of Southern California, Los Angeles, California 90033*

William P. Jencks (485), *Department of Biochemistry, Brandeis University, Waltham, Massachusetts 02254*

J. B. Jones (397), *Department of Chemistry, University of Southern California, University Park, Los Angeles, California 90007*

Mary Ellen Jones (611), *Department of Biochemistry, School of Medicine, University of North Carolina, Chapel Hill, North Carolina 27514*

Tomisaburo Kakuno (367), *Institute for Protein Research, Osaka University, Osaka, Japan*

Herman M. Kalckar (291), *Department of Chemistry, Boston University, Boston, Massachusetts 02215*

V. K. Kalra (417), *Department of Biochemistry, School of Medicine, University of Southern California, Los Angeles, California 90033*

Nathan O. Kaplan (3, 599), *Cancer Center, University of California, San Diego, La Jolla, California 92093*

Kumiko Kawaguchi (641), *Department of Biochemistry, Dental School, University of Hiroshima, Hiroshima, Japan 754*

Tsoo E. King (145), *Department of Chemistry, State University of New York, Albany, New York 12222*

Alexander M. Klibanov (509), *Department of Nutrition and Food Science, Massachusetts Institute of Technology, Cambridge, Massachusetts 02139*

Robert Kolter (691), *Department of Biology, University of California, San Diego, La Jolla, California 92093*

Joseph Kraut (193), *Department of Chemistry, University of California, San Diego, La Jolla, California 92093*

G. Kumar (417), *Department of Biochemistry, School of Medicine, University of Southern California, Los Angeles, California 90033*

J. Lavorel (323), *Laboratoire de Photosynthèse, Centre National de la Recherche Scientifique, 91190 Gif-sur-Yvette, Essonne, France*

S. H. Lee (417), *Department of Biochemistry, School of Medicine, University of Southern California, Los Angeles, California 90033*

Lawrence Levine (571), *Department of Biochemistry, Brandeis University, Waltham, Massachusetts 02254*

Choh Hao Li (519), *Hormone Research L1, University of California, San Francisco, San Francisco, California 94143*

Ting-kai Li (469), *Department of Medicine and Biochemistry, Indiana University School of Medicine, Indianapolis, Indiana*

K. Lindan (669), *Department of Biochemistry, The Johns Hopkins University School of Hygiene and Public Health, Baltimore, Maryland 21205*

Fritz Lipmann (283), *The Rockefeller University, New York, New York*

John M. Lowenstein (629), *Graduate Department of Biochemistry, Brandeis University, Waltham, Massachusetts, 02254*

Oliver H. Lowry (587), *Department of Pharmacology, School of Medicine, Washington University, St. Louis, Missouri 63110*

Arne Lundin (347), *Department of Biochemistry, Arrhenius Laboratory, University of Stockholm, S10691, Stockholm, Sweden*

C. E. McKenna (397), *Department of Chemistry, University of Southern California, University Park, Los Angeles, California 90007*

M. C. McKenna (397), *Department of Chemistry, University of Southern California, University Park, Los Angeles, California 90007*

Edwin M. McMillan (35), *Lawrence Berkeley Laboratory, University of California, Berkeley, Berkeley, California 94720*

T. E. Meyer (157), *Department of Chemistry, University of California, San Diego, La Jolla, California 92093*

W. Greg Miller (215), *Department of Biochemistry, University of Arizona, Tucson, Arizona 85721*

J. W. Newton (391), *Northern Regional Laboratory, United States Department of Agriculture, Peoria, Illinois 61604*

H. T. Nguyen (397), *Department of Chemistry, University of Southern California, University Park, Los Angeles, California 90007*

Nozomu Nishi (367), *Institute for Protein Research, Osaka University, Osaka, Japan*

Thomas L. Poulos (193), *Department of Chemistry, University of California, San Diego, La Jolla, California 92093*

H. Primakoff (45), *Department of Physics, University of Pennsylvania, Philadelphia, Pennsylvania 19104*

Françoise Reiss-Husson (381), *Laboratoire de Photosynthèse, Centre National de la Recherche Scientifique, 91190 Gif-sur-Yvette, Essonne, France*

S. Riazuddin (669), *Department of Biochemistry, The Johns Hopkins University School of Hygiene and Public Health, Baltimore, Maryland 21205*

Arthur Robinson (3, 727), *2251 Dick George Road, Cave Junction, Oregon 86523*

Demoy W. Schulz (587), *Department of Pharmacology, School of Medicine, Washington University, St. Louis, Missouri 63110*

David Shemin (117), *Department of Biochemistry and Molecular Biology, Northwestern University, Evanston, Illinois 60201*

S. J. Singer (11), *Department of Biology, University of California, San Diego, La Jolla, California 92093*

Gilbu Soe (367), *Institute for Protein Research, Osaka University, Osaka, Japan*

David M. Stalker (691), *Department of Biology, University of California, San Diego, La Jolla, California 92093*

P. J. Stephens (397), *Department of Chemistry, University of Southern California, Los Angeles, California 90007*

Jekisiel Szulmajster (529), *Laboratorie de Enzymologie, Centre National de la Recherche Scientifique, 91190 Gif-sur-Yvette, Essonne, France*

Shigehiko Taniguchi* (641), *Department of Biochemistry, Dental School, University of Hiroshima, Hiroshima, Japan 754*

Henri Temianka (9), *California Chamber Orchestra, Los Angeles, California*

Hugo Theorell (7), *Medicinska Nobelinstitutet, Karolinska Institutet, S-104 01 Stockholm, Sweden*

Frederick I. Tsuji (537), *Marine Biology Research Division, University of California, San Diego, La Jolla, California 92093*

Bert L. Vallee (469), *Biophysics Research Laboratory, Department of Biological Chemistry, Harvard Medical School, Boston, Massachusetts*

Nancy M. Weare (509), *Department of Nutrition and Food Science, Massachusetts Institute of Technology, Cambridge, Massachusetts 02139*

Fred C. Westall (705), *Salk Institute, San Diego, California 92112*

Isadore Wodinsky (77), *A. D. Little, Inc., Cambridge, Massachusetts 02142*

Frances C. Womack (561), *Department of Microbiology, Vanderbilt University, Nashville, Tennessee 37232*

Harland G. Wood (99), *Department of Biochemistry, Case Western Reserve University, Cleveland, Ohio 44106*

H. Tonie Wright† (727), *Department of Biochemical Sciences, Princeton University, Princeton, New Jersey 08544*

Tateo Yamanaka (251), *Department of Biology, Faculty of Science, Osaka University, Toyonaka, Osaka 560, Japan*

Jinpei Yamashita (367), *Institute for Protein Research, Osaka University, Osaka, Japan*

**Present address: Department of Biochemistry, Okayama University Dental School, 2-5-1, Skikata-cho, Okayama City, Japan 700.*

†Present address: Department of Biochemistry, Medical College of Virginia, Richmond, Virginia 23298.

Prologue

INTRODUCTION

Nathan O. Kaplan
Arthur Robinson

This volume is dedicated to Martin Kamen. Students, associates, colleagues, and friends of Kamen over the past 40 years have contributed to this volume.

Kamen played an important role in the development of biochemical sciences in the United States. He began his scientific career in nuclear physics and chemistry, receiving both his undergraduate and graduate training at the University of Chicago. His early interests led him, in 1936, to the Radiation Laboratory of the University of California at Berkeley. He played an important role in the exciting research which centered around the cyclotron in Berkeley. Even though he was young in years, he was recognized as one of the most able investigators in the studies of artificially produced radioactive elements.

His work with the cyclotron led him into contact with the late Samuel Rubin. Both Kamen and Rubin saw the potential use of radioactive isotopes in biological studies. They attracted the interest of a number of young faculty members (W. Z. Hassid and H. A. Barker), as well as a number of students who were to become collaborators in these pioneering studies.

 ISBN 0-12-397580-8

The use of stable isotopes (^{13}C, ^{15}N, and ^{2}H) as tracers were being applied by Schonhiemer, Rittenberg, and their colleagues at Columbia University; Rubin and Kamen initiated studies with ^{11}C, an isotope with a half-life of approximately 20 minutes. Despite the short half-life, they were able to begin studies on the mechanism of CO_2 fixation, particularly with the plant photosynthetic system. Despite the drawbacks inherent in the use of ^{11}C, they saw the potential of using radioactive carbon in resolving key biological problems.

Theoretical physicists had calculated that ^{14}C, if it existed, had an even shorter half-life than ^{11}C. Nevertheless, Kamen continued to search for a longer half-life carbon isotope. Because of his access to the cyclotron, Kamen was able to demonstrate that an isotope resembling ^{14}C could be produced. The identification of this isotope was the result of a brilliant series of experiments by both Kamen and Rubin. Although the isotope could be produced by the cyclotron in small quantities, it was evident that this isotope, if available in quantities, could play a major role in elucidating major biological problems. With the advent of atomic piles, ^{14}C became available in large amounts; the use of this isotope has become a tool in nearly all biomedical studies.

World War II interrupted the ^{14}C studies. Following the War, Kamen joined the faculty of Washington University. Here he pioneered studies on the metabolic, as well as the molecular, characteristics of the photosynthetic bacteria. His interest in bacterial cytochromes developed from these studies, and he has been a major figure in this area of research for many years.

In 1957, Kamen moved to Brandeis University, where he played a unique role in shaping the scientific and academic programs of this new University. Later, at the University of California at San Diego, he was also instrumental in developing the high standars for which this institution is now noted. During the time he was at USC, he was instrumental in developing biology as a modern molecular science.

Kamen has been an unusual member of the biochemical community. His interests lie not only in science, but in the arts and humanities. He served the Universities in many ways because of his extensive breadth of knowledge as well as his enthusiastic approach.

The essays in this volume are small expressions of the respect and admiration of the scientific community; they represent a token of thanks for all those who have had the good fortune to have been associated with him during his illustrious career.

We are grateful to The Linus Pauling Institute of Science and Medicine, National Science Foundation, The Kettering Foundation, UC San Diego Foundation, Hoffman LaRoche Inc.,

Academic Press Inc., Calbiochem-Behring Corp., P-L Biochemicals Inc., and the Graduate Department of Biochemistry, Brandeis University for support of the Symposium honoring Dr. Kamen, which was held in La Jolla, California. We would also like to thank Dr. Virginia Swanson Kamen for her help and suggestions in this endeavor.

MEDICINSKA NOBELINSTITUTET
Biokemiska Avdelningen
Laboratorium för Enzymforskning
KAROLINSKA INSTITUTET

Solnavägen 1
104 01 STOCKHOLM 60
Telefon 34 05 60

Stockholm, June 30, 1978.

My dear old friend Martin Kamen.

I happen to know that you will reach a matured age on August 27 this year and therefore grasp the occasion to send you my very best wishes and heartiest congratulations. Margit of course joins me.

Among the hundreds of biochemists I have met and collaborated with in my life you belong to a separate, higher class than most of the others, because you are a master not only in science - biochemistry - but also in art - music, and as you well know I do the little I can in the same directions in music promoted by Margit. You and I also had the opportunity of collaborationg in this field. Let me take this occasion to tell you that I am greatly thankful to the fate, that brought us together. All the best for the future.

Yours ever

Hugo Theorell

Hugo Theorell

 ISBN 0-12-397580-8

Everything I've Always Wanted to Know about Cytochromes
but Did Not Dare to Ask

A loving tribute to Martin Kamen on his 65th birthday,
from an ignorant musician
H. Temianka

I'm amazed how Martin copes
With all those million isotopes.
Do they crawl inside his fiddle
With Beethoven right in the middle?
And Bartok, is he more attractive
When Martin makes him radioactive?
Does the law of fundamental decay
Make out-of-tune playing okay?
Do ionization chamber sessions
Explain Bach's sexual transgressions?
Could better fiddles be created
If they were slightly tritiated?
Does Martin's work on transmutations
Relate to Schoenberg's modulations?
Is it deuterophenylalanine
That makes Mozart sound divine
And so miraculously erogenous
While Martin is endogenous?
Has Chaikovsky been apprised[1]
Of Chaikoff's dogs hepatectomized?
How strange that the one makes listeners quiver
While the other removes the fat in the liver.
It's just a difference in degree
But still it's all biology
Which sometimes is molecular
While music is so secular.
It's Martin's greatest contribution
That he alone found the solution
In welding musicology
To the science of biology.
Therefore, dear Martin Kamen,
Accept this tribute from a layman
Not in a class with Linus Pauling
But humbly proud of his own calling.

With deep affection and admiration from

[1]*Reference to the late I. L. Chaikoff, Professor of Physiology, University of California, Berkeley, California.*

ISBN 0-12-397580-8

SOME MEMORIES, SOME JOYS

S. J. Singer

I always think of Martin Kamen when memories flood back about the "good old days" when U.C.S.D. was at its beginning. Now, thinking fondly about Martin on the occasion of his 65th birthday celebration, I find myself remembering in a golden haze the way things were at U.C.S.D.* Such memories are strong, although I have never written them down before. This seems like an appropriate occasion to do so. The following free associations may mean little to any one who was not there at the time, and to such unfortunates, I apologize for this bit of nostalgia.

Martin and I came to La Jolla at the same time, in September, 1961 - he to the Department of Chemistry, and I to the fledgling Department of Biology. It was a marvelous time, almost incredibly so, in retrospect. U.C.S.D. was brimming with new ideas and energy - full of vigor and enthusiasm. We were going to be the Harvard of the West; none of us who were in-

**U.S.C.D., University of California at San Diego.*

 ISBN 0-12-397580-8

volved had any doubts about that. There were no cynics among us. They had elected to remain in their comfortable niches with the other old blues. Clearly, there were no clairvoyants among us, either.

As a matter of fact, we were an interesting group of people, the U.C.S.D. "pioneers." We were certainly not radicals, on the outside of the academic establishment. On the contrary, all were respectable and respected academics, coming here from the best institutions. But, along with this high order of professionalism, there were some special qualities. We were all restless, willing to venture, not a little visionary, and not a little naive. In a big-spirited place and time, we were all exhilarated, unleashed, and very happy.

In this company, Martin stood out. His Chicago-style staccato irreverence always kept things from becoming too solemn. His tremendous range of interests - from who played third for Chicago the first three weeks of September, 1942, to who was Ginastera, to picosecond light-induced reactions - always came to the rescue when the rest of us were at a loss. And Martin, who never thought small, helped keep our ambitions large.

When the new campus in La Jolla was first planned, it was to have been a science-oriented campus. Thus, the first faculty recruited to it were in Physics and Chemistry, followed a little later by Biology. By the time Martin and I arrived, the decision had been made that La Jolla was to be a general campus of the University of California - so the faculty in Humanities and Social Sciences were to be recruited immediately. Initially, this was to be done through committees in different disciplines, which were dominated by some of the Brahmins in these disciplines from the established sister campuses of the University. They would suggest someone to us, who would be brought in to meet everyone and to give a public lecture that we all religiously attended. After a while, it became painfully apparent to Martin, myself, Dave Bonner, and a few others, that the people who were being touted to us in the Humanities and Social Sciences were, by and large, just not of the same caliber as our scientists. We began to suspect that the Brahmins were being somewhat condescending to us; after all, what really important figures in the Humanities and Social Sciences would come to a campus whose faculty consisted only of scientists, and whose library was essentially nonexistent?

I remember one afternoon when a group of us barged in to Herb York's office (which was always open to anyone) and began to complain about this state of affairs, at which point someone in the room said, "Well, if you think you can do better, go ahead!" I must admit that I was first taken aback by the idea. Martin's reaction was, "Sure, why not?"

That was the beginning of a great year for many of us. The idea of scientists recruiting faculty in the Humanities and

Social Sciences no doubt strikes many as at least presumptuous, if not absolutely ridiculous. But all of us had come from the elite institutions where we had known very well some faculty in a variety of disciplines whose judgments we respected. It turned out that we did not have to rely on our own limited knowledge of an area that we needed to develop. What we did was to impose on our old friends at our former institutions to give us the best advice about professionals in an area, and then to appraise the human qualities of the people they recommended when they were brought here.

In this role, Martin was terrific. He knew a tremendous number of people in the Arts and Humanities, and he unhesitatingly imposed upon them. Marvelous people streamed in and out of La Jolla at that time. Martin was always exhorting us that what we really wanted was a "cross between Jesus Christ, Buddha, and Napoleon." He also suggested U.C.S.D. deserved some special recognition or award at that time for having single-handedly raised academic salaries at a number of northeastern institutions. It was a marvelous paradox that for many of us who had been in academic life at established universities for a long time, these first years at La Jolla gave us more of an understanding of what a university was all about than we had ever had before.

Nor were we without the cultural events we had left behind. Martin saw to that. From his acquaintances in the musical world, he selected the best chamber musicians to come through La Jolla in those early years. He also was a fount of other cultural activities. I remember once his acting out on our living room floor the important scenes from "Virgin Spring," taking all the roles in succession. We never did see that movie - it would have been anticlimax.

Another occasion I will never forget began one November day. Martin noticed a newspaper ad for a concert by Isaac Stern to be given in Escondido at the local high school. We were not going to miss that rare an event, so we got tickets and two carfuls of people together - Martin, and I remember Bill and Nao Belser, but I don't remember who else - and we tooled out to Escondido. The concert was held in the high school gymnasium, fitted out with metal folding chairs. The event looked like the high school graduation ceremony. An ancient Knabe was rolled out on the stage. Nothing daunted, Stern and his accompanist Alexander Zakin launched into a Brahms' sonata. I still remember that first strident note on the violin, the momentum of which was so great that he nearly fell off the stage. That was the kind of performance it was - absolutely electric. During the intermission, the audience was exhorted for about 15 minutes by a Chamber of Commerce spokesman to support the concert series. When Stern and Zakin came out to play again, Stern turned and addressed the audience. "Please, when you provide the money to support these concerts, could you also buy a better piano!"

After the concert was over, and we got up to leave, Martin casually asked if would we like to go backstage and meet Stern. That sounded fine, so we trooped up the steps. As we approached Stern, he spied Martin, and in great surprise, rushed to embrace him, leaving the rest of us a little stunned. There followed a great rush of questions - what was Martin doing in Escondido, etc. When this stream of questions was answered, it then turned out that Stern and Zakin had not eaten dinner and we were asked if we would join them for dinner. This was Escondido around 1964, 11:00 at night on a weekday. After some searching, we found an open restaurant in the local bowling alley. We had a marvelous dinner conversation with Martin and Stern for the next two hours, about their mutual friends, music, the new campus, everything. Finally around 1 or 2 AM, Stern and Zakin had to leave to drive back to Los Angeles on, as I remember, a night when the fog was thick (even for Escondido). A great glowing occasion. Martin made things like that happen.

So, Martin, with pleasant memories, much affection, and many thanks, Happy 65th Birthday to you and Virginia from Ruth and me.

SCIENCE, TECHNOLOGY, AND THE QUALITY OF LIFE

Melvin Calvin

Department of Chemistry
University of California
Berkeley, California 94720

I have separated this discussion into two components: science (understanding), on the one hand, and technology (invention), on the other. They are two quite different endeavors, two quite different human activities with different requirements. They are dependent on each other. Science arises from technology and technology arises from science. However, they have different social consequences and I want to treat them separately.

EFFECT OF SCIENCE ON WORLD CONCEPT

I will first discuss how science affects our view of ourselves and the world in which we live, which is really the basic question. Science contributes primary input to our understanding of the meaning of our lives, and I would like to illustrate that with a series of historical illustrations, up to and including the present.

ISBN 0-12-397580-8

Copernicus, Kepler

Three or four hundred years ago our concept of the universe was rather circumscribed - the earth was its center and man was the important creature on the earth and, therefore, in the universe. Copernicus and Kepler recognized that this was not so, that the earth was not the center of our universe, that the earth was near the edge, and that the sun was the center of our local universe. That concept initiated an enormous change in our view of the world. The sun became the center and the earth one of several objects moving around the sun. This was not only a tremendous change in our view of the world, it has had an enormous impact on human thinking. Gelileo's telescope, of course, confirmed this idea by actually viewing and seeing the various planets that were moving around the sun, allowing precise calculation of the planetary orbits. This ultimately led to the equations which describe the planetary motion, which had to be understood in more fundamental terms, and this, in turn, gave rise to Newton and his law of gravity and motion.

Darwin

A second step in the impact of science on our view of the world - and all of these things effect the quality of our lives today - was taken later, a little over 100 years ago, when Charles Darwin removed man from the center of things, and moved him to the edge of the stream of life, to being a component in the stream of living things. In other words, man was removed from the center and made part of the evolutionary stream of life, after Darwin. This even more profoundly changed our view of ourselves and the world in which we live. There is, even today, some slight argument about this.

Einstein

In the twentieth century, I cannot help but call your attention to the man who revised our view of the universe in a fundamental way. He gave microscopic and macroscopic explanations for Newton's laws of gravity and motion, and changed our whole view of how the universe is organized. I am speaking, of course, of Einstein, whose 100th birthday was celebrated on March 14, 1979. There has been a great deal written about him recently, not only of his scientific work but of his personal career as well.

Bear in mind that Einstein not only revised our view of the universe, but he invented (he made real) the quantum of energy. Until then, the packet of energy which we now call a "quantum" was a mathematical fiction used to help understand the distribution of energy emitted from a hot body. Planck invented the term "quantum of action," but it was a mathematical invention, a fiction that he needed in order to obtain the right shape for the curve. Einstein went much beyond that. He recognized the equivalence of matter and energy and invented the quantum as a unit of energy in 1905. The year 1905 was fantastic for the scientific world. It was the year of special relativity when the equivalence of matter and energy was stated with the famous equation: energy is mass times the square of the velocity of light, that is, the energy of a massive particle is equal to the mass times the square of the velocity of light. Also in 1905 there was the recognition of the way in which light ejects electrons out of metal. In describing quanta, the recognition came that light sometimes seems to be a stream of particles and sometimes behaves as waves. Newton struggled with that problem. He concluded that light was a wave because it was diffracted by a grating into colors and formed interference patterns, and thus must be wave motion. Einstein recognized the duality. In addition, he gave reality to the idea of a particle, a quantum, of light energy.

Einstein in recognizing the reality of the quantum which gave rise to the quantum theory of matter, and with his special relativity, realized the equivalence of matter and energy. Much later, in 1918 or there abouts, he devised the theory of general relativity, recognizing the equivalence of acceleration and gravity. That is, you cannot tell the difference between an accelerating body and a body pulled by mass in a gravitational field. This gave rise to Einstein's entire view of the universe: the two equivalence principles that he invented, one in 1905 and one in 1917 - 1918. The latter is easier to understand today because of our experience with spaceships which did not exist at the time of Einstein. Today, we can experiment in reality. A man has been in a spaceship and in a spaceship he is weightless. If a man is moving with the same velocity as the spaceship he can float around anywhere in the spaceship. However, if the man turns the rocket propulsion on the ship after he lets go of the spaceship, it catches up with him, as though he is falling to the bottom of the ship. The moment you let go of the earth, you fall, and that is what happens to the man in the accelerating spaceship. The moment the spaceship accelerates, the man falls to the bottom, in a direction opposite to the one in which the rockets are pushing.

The equivalence of matter and energy and the equivalence of acceleration and apparent gravity are the two major theories which stem from Einstein's major inventions of the quantum theory and the relativity view of the universe.

Modern Chemistry and Molecular Biology

In recent years life itself has been lifted out of its unique position and made part of the evolution of the physical universe. We, begin now with matter itself in the collapsing stars - giving rise to the planets, which collect on atmosphere, which give rise to organic molecules which, in turn, give rise to living things. There is a stream of evolution of the physical universe, which ultimately gives rise to living things. Thus, life itself, now, has been moved from the unique place which it occupied up until relatively recently.

All these effects are very profound. Scientific understanding of the way in which our world works has modified our concepts of ourselves and our place in the universe. This, in turn, has changed what our beliefs are and what our rules of conduct are. All this comes from our view of ourselves and the physical world in which we live. Science in its main function, understanding of how the universe works, has a profound effect on our behavioral standards.

TECHNOLOGY AND APPLIED SCIENCE

What has technology done for our lives in the past and what can we expect from it in the future? I cannot project what we will discover in science, because it is in the nature of science to be unpredictable, i.e., we cannot specify what we have not yet discovered. A scientist sets as his life goal the discovery of new ideas and concepts of how the world functions, but what the new principle may be is, by definition, impossible for him to predict ahead of time.

Technology, i.e., applied science, is subject to surmises of what it will do, and predictions of what direction it will take, because it depends on already known scientific principles. Six different technological areas are discussed here: three of them are old and three of them are new; one of them might even be considered as in the future.

Steam Engine/Internal Combustion Engine

These have had a profound effect on the form of our everyday lives in a physical sense. These two inventions have actually reduced human dependence on physical labor and have given rise to what we have called the industrial revolution which began in Great Britain and later spread to Western Europe and the United States. All of the human conflicts that arose from the displacement of human labor by machines (steam engine and internal combustion engine) and, more recently, such specific applications as the automobile - hardly need to be described. What we will do as the next step in this evolution, I do not know. It seems inevitable that we are going to run out of fossil fuel for the internal combustion engine, and that is a subject for later discussion. Perhaps it well be necessary to invent a new type of engine, as well as finding renewable sources of fuel.

Maxwell's Electromagnetic Theory

The primary discovery by Faraday of electromagnetic induction gave rise to the electric motor and electric generator and all of the electrical industry which came out of this discovery. When Faraday was first demonstrating magnetic induction at the Royal Institution in London, he had a galvanometer connected to a coil of wire. When he moved a magnet past the coil the galvanometer deflected. The Royal Institution is an old British organization of private individuals for the social and intellectual recognition of science; we do not have a similar organization in the United States. In England, the Lords and Ladies would go to the Royal Institution and watch various types of scientific demonstrations and would make contributions - largely of money; it was part of the social scene in London to show up at the Royal Institution. When Faraday was demonstrating his magnet, one of the government ministers (finance, I believe) asked Faraday of what use his demonstration might ultimately be. There are two answers which are attributed to Faraday in response to that query. One was: "What good is a newborn baby?" and the other answer that I have heard (and I don't know which one is true) - "One day you will tax it." Both of those answers express the nature of the situation even today.

Electromagnetic induction is only one of the consequences of Maxwell's electromagnetic theory and Faraday's discovery of induction. The other one is the discovery of radio waves by Hertz and the communication industry which followed.

The situation today with respect to world events is so different from what it was 50 years ago that it is difficult to imagine. When I was a boy, events were happening in Europe (the beginning of World War I) and we were not finding out about them until hours, or sometimes days, later. Today, this is impossible. There is nothing that can happen anywhere in the world that we do not know about within hours, and actually at times we see it happening! The world is a completely different place because of this, and it is so different that most people cannot imagine what it was like to have things happen in one part of the country (or world) and not know about it for days and weeks. The world is now a much smaller place; this has given rise to all sorts of political and social problems which still have yet to be solved.

Invention of Synthetic Chemicals

This is the third major technological breakthrough, covering all the synthetic materials including polymers, drugs, etc. A good many of the agricultural chemicals that we use are put together by chemists. In fact, probably a majority of the one-half million or more organic compounds that are known are man-made synthetic compounds. The most heavily used of these are the polymers and synthetic fibers. All of us are, in some way, in touch with synthetic chemicals right now. I would be very surprised if you do not have, somewhere on your person, a synthetic fiber of some kind - polyacryonitrile, polyester, polyamide. Much of the natural fibers (cotton, wool, silk), plant and animal fiber, have been displaced, for the most part, by synthetic fibers. More importantly, other materials, much wood and metal, have been replaced by plastic materials, particularly plastics made generally from petroleum raw materials. This has created a major revolution in western society, and is beginning to effect the rest of the world as well. The synthetic drugs, synthetic agricultural chemicals, and synthetic materials, all have had an enormous impact on the way we live. We could not live the way we do now without those synthetics, nor would we have the food or the clothing, things which we take for granted, just as we take communication and transportation for granted. Our daily activities are dependent on the existence of those synthetic materials, and the developing world is rapidly approaching the same state.

Microprocessor

This is a technology whose day has just begun. There has been an enormous explosion in the use of microprocessors - solid state devices that are the basis for the computers, watches, tape recorders, and other such devices. The revolution in the aggregation, storage, and recovery of information is just emerging. We are just now learning how to use this way of collecting, storing, and recalling information - information of all types, about people, things, subjects, etc. All this information will become accessible with these "gadgets" by telephone, home consoles, and various other devices which will make this possible. This revolution, which is only just beginning, will have many social consequences. The increase in the ease and availability of these devices for storing, transforming, and recalling information will change the way we behave. First, the whole question of individual personal privacy will arise. The "bytes" of information, accessible to almost anybody, will be very important pieces of information, but that will reduce our privacy. Everybody will be able to look into your life by punching a button and see everything you have done since the day you were born. Perhaps it could be coded so that the information will not be as accessible, but this may be very difficult. The computer people indicate that this is next to impossible, however, if you want to spend enough money, you may achieve it.

You take it for granted that you have your privacy today unless someone takes it from you. In the future, it will be the other way around - you will not have your privacy unless you buy it. Your bank accounts, telephone calls, everything will be recorded on a computer, and people can get access to it unless you fix it so they cannot. The whole situation will be reversed. Privacy is one thing we would like to retain, and this revolution is already beginning to show how difficult this situation may become. The ease of access to information of all types (financial, scientific, population, political) will be available so people will not have to make decisions on incomplete knowledge, as we now do. We make decisions today with only a fraction of the information we actually need, thus, depending on our judgment for decision making. To make a decision when you do not have enough information seems to be called "judgment." The time is coming when this will not be necessary. I do not know what this will do to our behavior, but I believe it will dramatically change it. In the future, when a boy meets a girl, he can find out everything he wants or needs to know about her by pushing a button and vice versa. Our entire behavior will change as a result of the ease of information accessibility. Our "judgment" will be better; our human interactions will change. It is hard for me to predict precisely what will happen, except that it will lead to problems.

There are other aspects of the information revolution which will have to be dealt with, and I have mentioned just two of them: the privacy question and the accessibility question. We will have accessible to us all types of information which we now have to fight for, or do not get at all, and our basis of judgment will be different. This will result in a new capacity to predict at this point. This must be thought about, since some of you will have to deal with it.

Applied Relativity (Nuclear Energy)

This refers to the equation $e = mc^2$, that is, nuclear energy. Since this is a reality, we must learn to deal with it and use it properly. We are in the process of finding ways to deal with the problems of nuclear energy (sometimes successfully and sometimes unsuccessfully), but the method of operation is still in flux. We have not really evolved methodology for dealing with nuclear energy. Unfortunately, this problem will not disappear, since it appears that we need nuclear energy.

There are some who believe that the only form of nuclear energy that we can safely use is fusion energy, with the fusion reactor located about 93 million miles from earth, i.e., the sun. There are a number of ways in which we may be able to usefully capture solar energy. Two of the most interesting ways, and perhaps most accessible, stem from our growing knowledge of the molecular basis of photosynthesis. The first of these applied molecular biology activities involves genetic engineering - how do you modify the genetic behavior of a cell, invent a new crop, that does not presently exist.

1. *Petroleum Plantations*

In the past 4 years we have been growing two species of *Euphorbia* plants for the purpose of harvesting the latex produced and extracting the oil from it. This work was undertaken as an extension of our knowledge of the basic process of photosynthesis with a specific application, i.e., development of renewable resources from plants as alternate sources of energy. We have used in this work two species: *Euphorbia lathyris* (Fig. 1), which is an annual, and *Euphorbia tirucalli*, a perennial. These plants, seen in Fig. 2, shows the "plantation" in southern California about 1 year ago; the *E. tirucalli* are in the foreground and the *E. lathyris* in the background. The euphorbias exude a white latex, which is an emulsion of oil in water; we separate the emulsion to obtain black oil. The latex is a complex mixture of hydrocarbon-like materials. The overall yield of hydrocarbon is roughly 8% by weight of the dry euphorbias. Since we can grow some of these species to maturity in one season (6 months or thereabouts), the amount of dry weight corresponds to 6 barrels of oil per acre per growing season. This yield is using wild seed, without selection and without plant breeding.

Furthermore, we have not yet had enough material to develop a true commercial extraction process. Therefore, it is a bit premature to make economic estimates. Nevertheless, this has been done at the request and demand of the Department of Energy, in spite of the fact that the data upon which these es-

Fig. 1. Petroleum Plantation, Southern California, Nov., 1978. Euphorbia tirucalli (front) and E. lathyris (rear).

Fig. 2. Euphorbia lathyris, Southern California, Nov., 1978.

timates were based was our own, and thus I am aware of how "flimsy" it is. The data does provide estimates of the order of $20 per barrel for processing costs and another $40 per barrel for capital costs for the actual processing plant, based only on a benchtop extraction plant. The cracking pattern of the material from the euphorbias, by the way, is very similar to that obtained from the naphtha; the latter product sells for more than $40 per barrel in the United States and $50 per barrel in Europe. The preliminary economic estimates for the euphorbia process are shown in Table I.

Obviously an improvement eitherin the oil yield or the oil composition from euphorbias will greatly reduce the crudely estimated costs. In the first instance, improvements can be achieved by simply selecting proper seed and plants. However, ultimately, we want to reduce these plants to single cell tissue cultures. By selecting the cells for their oil productivity, both in terms of amount and composition, we could recover whole plants with the agronomic characteristics of oil production and composition that we want from the single cells separated out of the culture. It is conceivable that we might be able to introduce into these cells the genetic mechanism of the enzymes which would produce precisely the kinds of materials we desire, with control mechanisms responsible to the climatic and agronomic conditions which we have available. We are, of course, limited to the use of land which is today not agriculturally productive, and that means "inventing" a new crop, which is where this discussion began. There is, at this point, the possibility of genetic engineering to solve our major energy problem.

*TABLE I. Comparison of Independent Evaluations of Oil from Euphorbia in 1978**

	Chevron (1000 barrels/day)	*Stanford Research Institute (1600 barrels/day)*
Feedstock		
Cost from farm	*11*	*30*
Credit for bagasse	-	*19*
Net feedstock cost	*11*	*11*
Other operating costs	*10*	*14*
Capital cost (amortization, 15% return, etc.)	*46*	*41*
Total production cost ($/barrel)	*67*	*66*

**During proofreading of this paper in 1981 it should be pointed out that the number has come down, to approximately $40/barrel.*

2. *Diesel Trees*

On a recent trip to Brazil, my wife and I encountered another possibility for hydrocarbon production, a tree (*Copaifera langsdorfii*). We were told that if a hole were drilled in this tree, oil would come out. (We had heard of this tree on a previous visit and did not believe the stories about the production of oil after tapping.) We went to the Ducke Forest in the Amazonian jungle to observe one of these trees and watch the production of oil after trapping. The tree itself is shown in Fig. 3 and it is possible to see the bung stuck in the hole from which the oil flowed in Fig. 4. The tree we observed was 1 meter in diameter and about 30 meters high, obviously a mature tree. The amount of oil from one hole in the tree is about 20 liters in 2 hours, which means about five gallons of oil in about 2 hours. The material that comes from the tree can be used as directly a diesel fuel. There is no processing, and no companies - just nature, from the tree immediately into the diesel tank.

Can these trees grow in the United States. I have been in communication with the U.S. Department of Agriculture, and we hope that we will be able to obtain some seeds, or seedlings, of this particular tree for experimental purposes. The plan is to try to cultivate this species in southern Florida and near the Gulf Coast of Texas. The reason we feel that these trees will be successful in these environments is that the USDA does have an experimental planting near Miami which contains *Hevea* rubber trees which grow under the same conditions as the "diesel tree" we saw in the Amazon jungle.

How long it will take to grow a tree of tapping age is a moot question. Keep in mind, however, that the original giant rubber trees from the wilds of Brazil were tapped for almost 70 years. The *Hevea* seeds were eventually taken from Brazil to London. They were germinated there and then transferred to Malaysia where they were planted. Now all rubber is plantation rubber, taking 8 years from germination to tapping for the rubber tree. Perhaps a similar pattern will evolve for the *Copaifera* tree, i.e., a relatively short period from germination to tapping.

3. *Synthetic Chloroplasts*

This is the second of the two methods of applying molecular biology, that is, understanding biological energy capture, conversion, and transfer in such a thorough manner that we would be able to produce by synthetic means those functions of the energy capturing system which would be most useful to us. This development has made some progress as well. At present we understand considerably more about the nature of the photo-

Fig. 3. *Copaifera langsdorfii*, Rio de Janeiro, Brazil.

Fig. 4. Copaifera langsdorfii, Ducke Forest, Manaus, Brazil, showing bung in trunk of tree.

synthetic process. We now believe that it consists of two quantum processes in succession, both involving electron transfer across a phase boundary. If this turns out to be correct, it should be possible to construct that double layer with the photosensitive material residing on both sides. A more thorough examination of the nature of chlorophyll reveals that it is surfactant dyestuff and it indeed behaves in this manner.

However, the synthetic dyestuffs for use in the laboratory would be much more stable than the natural material (chlorophyll) and would be constructed to suit the purpose. We have achieved this to some extent. In fact, we have made vesicles, which are hollow spheres of bilipid membranes made from phospholipids, upon which are situated photosensitizers (Fig. 5). We have been able to photoinduce an electron transfer across that bilipid membrane, in an uphill direction, moving the electrons from a lower to a higher energy level (Figs. 6 and 7). That higher energy level can eventually rise to molecular hydrogen or perhaps even some reduced carbon product (Fig. 8).

On the other side of the membrane, the donor system from which that electron is removed, must ultimately be water. Whether that water will then lead to oxygen evolution, or to the oxidation of some compound to a useful product is yet to be determined. We are at the moment concerned with the possibility of finding a catalyst which will generate molecular oxygen on the donor side of this membrane, having already succeeded in generating molecular hydrogen on the acceptor side. However, it may very well be that a much simpler way to solve this problem would be to use the nascent oxygen on the donor side to produce oxygenated materials that cannot be produced in any other way.

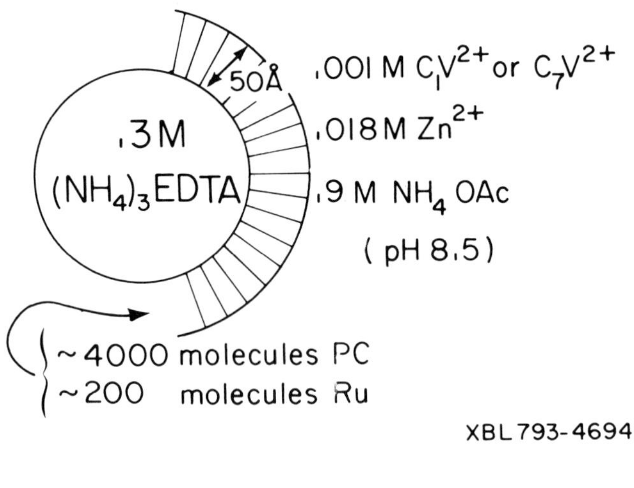

Fig. 5. Representation of vesicle cross section (W. E. Ford).

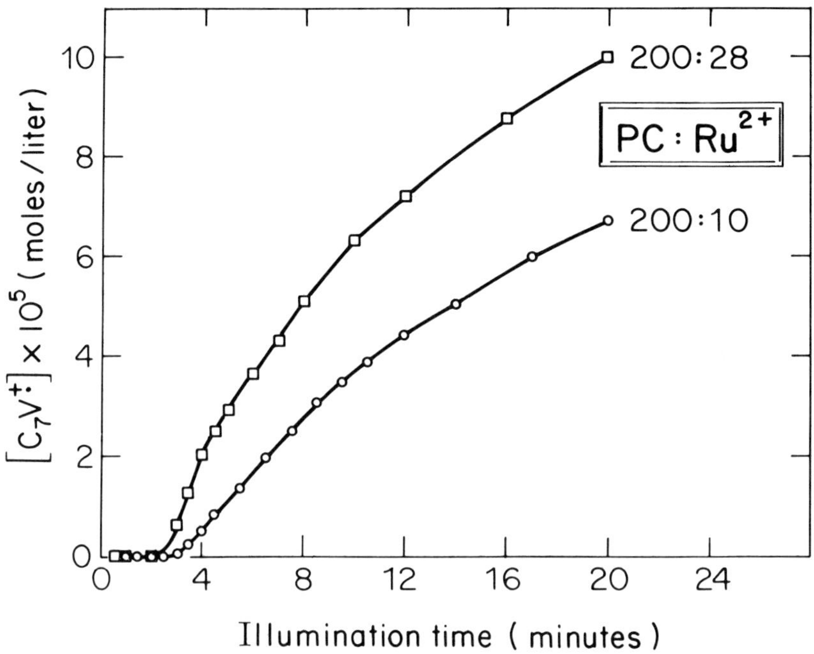

Fig. 6. Photoreduction kinetics (W. E. Ford).

Should this synthetic chloroplast device reach maturity it would be an additional component in the answer to the next phase of our human evolution, namely, eventually finding renewable energy sources for the maintenance and development of our life style. It is in this role that I think chemistry and chemists will play a major function in the construction of the society of the future.

CONCLUSION

Each of the problems outlined above is not only a problem but an opportunity. It is an opportunity to grasp the problem and solve it in a humanly acceptable way. Hard problems are what make life worthwhile, and the solution of hard problems gives real pleasure in living. I cannot refrain from a quote

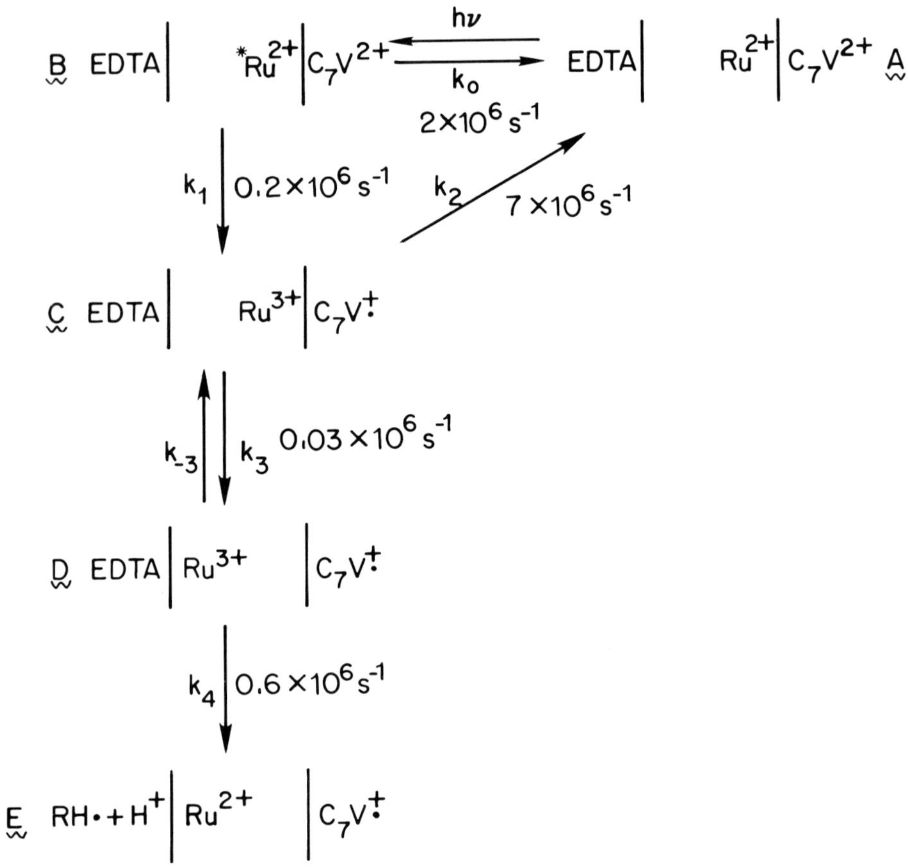

Fig. 7. Kinetic scheme for photoreduction (W. E. Ford).

by a biologist of broad experience [Thomas (1)]: "We need science, more and better science, not just for its technology, not for leisure, not even for health and longevity, but for the hope of wisdom which our kind of culture must acquire for its survival."

Recently I have been reading a good deal of Einsteinia, and I would like to conclude with a quotation that I had never heard before. Written in September, 1937, Einstein (2) noted: "... Our time is distinguished by wonderful achievements in the fields of scientific understanding and the technical applications of these insights. Who would not be cheered by this? But let us not forget that knowledge and skills alone cannot

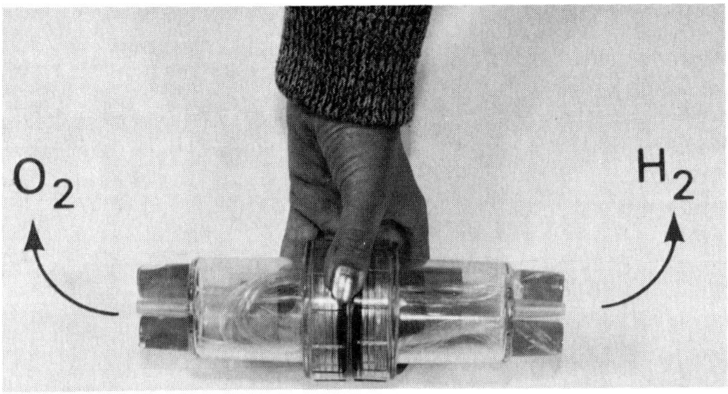

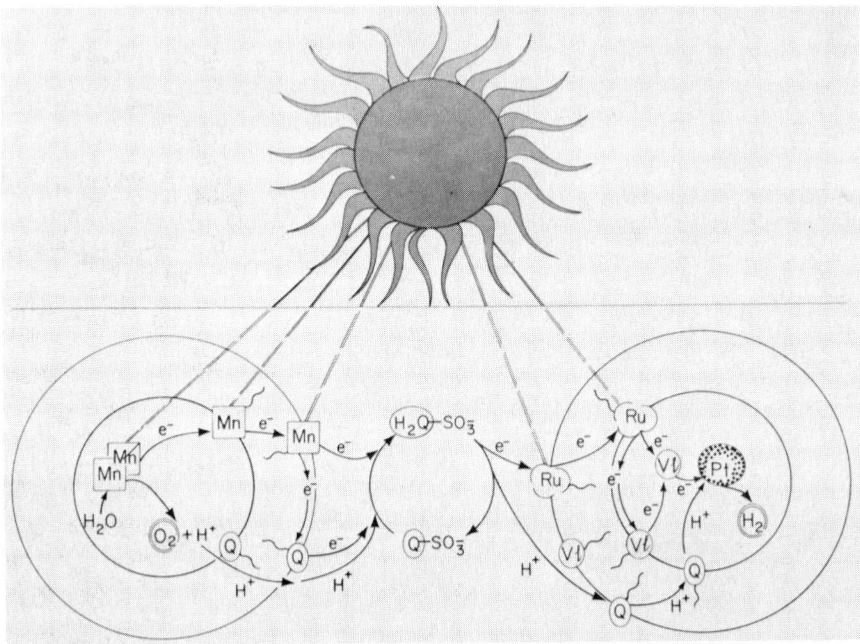

Fig. 8. Schematic for synthetic chloroplasts (Calvin, 1979).

lead humanity to a happy and dignified life. Humanity has every reason to place the proclaimer of high moral standards and values above the discoverers of objective truth. What humanity owes to personalities like Buddha, Moses and Jesus ranks, for me, higher than all the achievements of the inquiring and constructive mind."

REFERENCES

1. Thomas, Lewis (1979). "The Medusa and the Snail: More Notes of a Biology Watcher," p. 175. Viking Press, New York.
2. Dukas, Helen and Hoffman, Banesh, (eds.) (1979). "Albert Einstein: The Human Side," p. 70. Princeton University Press, Princeton, New Jersey.

Isotopes

ISOTOPES AT BERKELEY IN THE THIRTIES

Edwin M. McMillan

Lawrence Berkeley Laboratory
University of California
Berkeley, California

Frederick Soddy coined the word "isotope" in 1913 from Greek words meaning "the same place," intending it to be used only in cases where more than one kind of atom of a different mass could occupy the same place in the Periodic Table. He would no more have spoken of an element having only one isotope than we would use the word "sibling" in referring to an only child. Once a word comes into popular usage, however, its meaning tends to change. This has happened to "isotope," which is commonly used to mean a radioactive form of a normally stable element, and which is used in this sense here.

When the discovery of artificial radioactivity in Paris was announced early in 1934 there was a great deal of excitement at Lawrence's Radiation Laboratory in Berkeley. It was found that the deuteron beam of the cyclotron could produce activity of much greater strength than could be made using natural radioactive sources. When, a few months later, the neutron-induced activities were announced from Rome there again was a flurry of excitement and the intense neutron source given by deuterons on beryllium was used to repeat some of these obser-

 ISBN 0-12-397580-8

vations. There were, however, difficulties in spite of the strong sources at Berkeley. The deuteron beam was available only inside the cyclotron, and it was hard to avoid contamination of the targets and high background radiation. Methods of measurement were primitive and the earliest observations of artificial radioactivity at Berkeley were of a rather qualitative nature.

One important improvement was soon made. A thin aluminum window was put in the side of a reentrant brass tube that was already installed in the wall of the cyclotron chamber. The deuteron beam could pass through this window to strike a target inserted in the brass tube, and the target could then be removed and its activity studied away from the cyclotron with its large background of induced activity. It was desirable to have some material for a target holder that would not itself be activated, and I, as a new member of the laboratory staff, set about finding such a material. The first thing that I tried was platinum, which can be obtained in high purity and which has too high an atomic number for penetration by 2 MeV deuterons. For measurement of this induced activity I had a Lauritsen-type electroscope that I had made for observing gamma-rays emitted during bombardment, and which I installed in a tin can with a thin window for observing beta rays.

The surprising observation, first made on July 12, 1934, was that the platinum target exhibited an activity with a 2-minute half-life after deuteron bombardment! This could hardly be due to activation of the platinum, so I supposed that it was due to some component of the air between the window and the target, the active atoms being driven into the platinum by the recoil from the collisions in which they were formed. To verify this, I tried filling the brass tube with oxygen. No activity was found. With carbon dioxide in the tube the known 10-minute period from carbon plus deuterons was found, and with a vacuum in the tube the known 2 1/2 minute activity made in the aluminum window was able to recoil onto the target, where it was found. This established the fact that the 2-minute period was due to the bombardment of nitrogen.

At this point Stan Livingston joined me and we set up equipment to chemically identify the active substance. The target was now the air filling the brass tube. After the bombardment this air was transferred into a gas analysis apparatus by which it was found that the active material was an isotope of oxygen, and it was concluded that it was oxygen-15 formed by the reaction: nitrogen-14 plus deuteron gives oxygen-15 plus neutron. This was the first isotope identified at Berkeley which had not previously been found elsewhere. It was the first active isotope of oxygen to be discovered, and is still the longest-lived.

While Stan and I were working on oxygen-15, Ernest Lawrence

was bombarding various materials in the brass tube and found an activity induced in sodium which had a half-life of 15 hours, and which was remarkable in the large levels of activity that could be produced. It was, in fact, the same sodium-24 that had previously been discovered by Fermi and his group in Rome, but Lawrence was the first to emphasize its uses in many fields, including biology and medicine. This discovery received a lot of attention in the press which helped in obtaining support for nuclear research.

It is impossible to give a comprehensive account of isotope discoveries at Berkeley in the remaining years of the thirties. Sometimes the experimenters focused on cases of particular importance, like the discovery of helium-3 and hydrogen-3 by Alvarez and Cornog in 1939. In one case, in 1937, the target was bombarded in Berkeley and the chemical separations and activity measurements were done elsewhere. The target was a strip of molybdenum that formed part of the cyclotron dee which received a heavy bombardment, the place where it was sent was Palermo in Sicily, and the experimenters were Perrier and Segrè, who found not only a new isotope but a new chemical element, which they named "technetium" because it was the first artificially made element.

There were also systematic programs of investigation, of which the most impressive is that of Jack Livingood and Glenn Seaborg. They published some 20 papers over a time extending from 1936 into 1941, dealing with the activities produced by bombarding some 11 different elements. Jack made the measurements with a Lauritsen-type electroscope that he built himself, while Glenn made the chemical separations. Each measurement involved the timing of the motion of the electroscope fiber along an eyepiece scale with a stop watch; at that time, there were no automated methods. Each of the multitude of samples had to be followed for days or months to catch the longer decay periods. I remember being very impressed by the patience and dedication required by this work.

Another systematic program was that of Harold Walke, who came to Berkeley from England as a Commonwealth Fellow. He concentrated on the elements from potassium to chromium, and published 9 papers in the period 1937 - 1940, with various co-workers. Arthur Snell covered the range arsenic to rubidium, Ryokichi Sagane from Japan the range zirconium to molybdenum, and there were many others. Finally, I must mention a program that was systematic in the sense that the four neighboring elements iridium, platinum, gold, and mercury were bombarded with neutrons. However, only one paper, in 1937, resulted. Two of the authors of that paper, myself and Martin Kamen, are here today, the third, Sam Ruben, died in a laboratory accident involving phosgene a few years later. We reported 5 new isotopes, and I find on looking at a table of isotopes that these have all survived; in two cases the mass number has been changed, but the assignment of mass numbers was often a matter of guesswork in those days.

So far I have been discussing the discovery of isotopes; now I shall consider their uses. There are basically two ways in which radioactive isotopes are used One is as a source of radiation in such applications as radiography and radiotherapy where the sources are strong enough to blacken a photographic plate or produce a physiological effect, or in things like thickness gauges where much weaker sources can be used because all that is necessary is to activate a counter. The other basic type of application is as a tracer. The radioactive isotopes have the same (or nearly the same) chemical properties as their stable counterparts, and will follow the same reactions and go the same places in a chemical or physiological system. But their presence is made evident by their activity, which is easily measured by a counter. By the use of tracers the course of chemical reactions can be followed, even the complex reactions that comprise a living system, and the levels of activity required are so small that any direct effects of the radiation on the system are negligible.

Naturally occurring radioactive substances had already been used in both types of applications. Radium and radon were routinely used in radiotherapy, but were very expensive. Hevesy had pioneered the tracer methods, but the range of elements available was extremely limited. When Lawrence found radio-sodium he immediately realized the enormous possibilities opened up by the artificially made isotopes, with the cyclotron as a powerful way for making them. He promoted this idea both personally and in publications. He had always been interested in medicine and his younger brother John, who had become a physician, came to Berkeley to work with him. It is, therefore, natural that the first uses of isotopes at Berkeley were for medical purposes. There was also a program involving X-rays and neutrons for cancer therapy, but that is another story.

The first experimental use of isotopes in Berkeley was in 1936. Radio-sodium was administered to two human patients suffering from leukemia. The amounts of activity given were not very large, 13 and 53 mCi. In the two cases, no clinical effects were found, but also there were no harmful effects such as would have followed the administration of equivalent amounts of radium. As part of the experiment the activity in the blood, sweat, and urine was measured, and one of the stated conclusions at the end of the paper was: "A new method has been developed for studying the metabolism of sodium within the body. A similar line of investigation can be followed, using other artificially prepared radio-active elements."

This work was done by Dr. Joseph G. Hamilton, who had just received his M.D. from the University of California Medical School in San Francisco, and Dr. Robert S. Stone, a radiologist at that school. Dr. Hamilton had taken his B.S. in chemistry at Berkeley and was not primarily a clinician but became very interested in the use of isotopes as physiological tracers in humans and animals, which he made his life work until his death from leukemia in 1957, possibly as a result of the large amount of radioactive materials he had handled during his life. As I remember, he was not particularly cautious in handling them; he was aware of the danger but was impatient and eager to get on with the work. Shortly after the trial in the leukemia patients he made a tracer experiment in which three normal human subjects* were given by mouth small amounts of radium-sodium, about 1 mCi, and the rate of absorption into the bloodstream was followed by observing the counting rate of a Geiger counter held in the subject's hand, inside a lead box to shield it from the gamma-rays coming from his body. By 1938 this method had been used to study the rate of absorption of sodium, potassium, chlorine, bromine, and iodine in normal human subjects, and by 1939 Dr. Hamilton, along with Dr. Mayo H. Soley of the U.C. Hospital Thyroid Clinic, was engaged in a more sophisticated investigation of the absorption of iodine by the thyroid gland. Iodine, in contrast to the other elements named above, which become more or less uniformly distributed throughout the body, is specifically concentrated in the thyroid, and it is in cases like these that the tracer method shows its power. The concentration is, in fact, so great that the administration of radioactive iodine became a feasible method for the radiation treatment of thyroid diseases; a therapeutic dose could be given to the thyroid without appreciable whole-body radiation.

I find one physiological investigation done by Hamilton and G. A. Alles in 1938 particularly interesting. It had been known for a long time that potassium and rubidium have a specific stimulating effect on the heart muscle. It was also known that these elements have a small but easily measurable natural radioactivity, and many physiologists believed that the activity was responsible for the stimulating effect. Experiments using other naturally radioactive materials had been published that seemed to confirm this. Now, with both radio-potassium and radio-rubidium available from the cyclotron, a direct comparison could be made. It was found that activated potassium with 20,000,000 times the normal activity and rubidium with 70,000

**Waldo Cohn, one of these subjects, also took part in some of the work with radio-phosphorus at Berkeley.*

times gave the same stimulating effect on the hearts of experimental dogs as the ordinary elements, while radio-sodium with 60,000,000 times the activity gave no effect. That argument was decisively ended.

The other elements that I will mention are phosphorus, sulfur, iron, and carbon. Phosphorus is a very important element in biochemistry, and the isotope, phosphorus-32, has the convenient half-life of 2 weeks. Its use started in Berkeley in 1937, when Scott and Cook of the U.C. Medical School Division of Physiology started feeding it, in the form of phosphate, to growing chicks. They observed physiological effects in the form of a reduction in white blood cell counts. In another experiment they measured the amount excreted and the amount incorporated into various tissues. Later, in 1939, Dr. John Lawrence did a series of experiments with various co-workers in which radio-phosphorus was fed to normal mice and mice with various kinds of tumors, and to human leukemia patients. In these experiments, following a typical pattern, physiological effects of the radiation were observed, and the tracer aspect was carried out by measuring the activity in excreta and, in the case of the mice, in various tissues after the mice were killed. Unfortunately, radio-phosphorus did not turn out to be the "magic bullet" for curing leukemia, as had been hoped.

The work with sulfur-35, which has a half-life of nearly 3 months, was done at the California Institute of Technology starting in 1937, with the activated sulfur supplied from Berkeley. I was involved as an intermediate, preparing the targets for bombardment and sending the material to Pasadena, where Henry Borsook, the physiologist, Don Yost, the chemist, and some others did the experiments. One experiment was of a purely chemical nature. Elementary sulfur is very soluble in carbon bisulphide, and this solubility is useful in separating sulfur from other elements. The question is, does dissolved sulfur exchange with the combined sulfur in CS_2? To settle this question, active sulfur was dissolved in CS_2 and allowed to stand for several days. Some of the CS_2 was then distilled off and found to be inactive, showing that there had been no exchange. The other experiments were done with normal human subjects, observing the excretion of activity when the radio-sulfur was administered as sulfate or as thiamin, a sulfur-containing molecule also known as vitamin B_1. The activity level was so low that no radiation effects would be expected; these were purely tracer experiments.

The work with iron isotopes was also done at another institution, the University of Rochester, where the senior investigator was George H. Whipple, who had won the Nobel Prize in Physiology in Medicine in 1934 for his part in developing a treatment for pernicious anemia. His interest in iron was natural as this is an essential element in hemoglobin, the oxygen-carrying pigment in the blood. The initial experiments at Rochester in 1939 involved the feeding and injection of radioactive iron in dogs, with measurements of its excretion and absorption. Martin Kamen was one of the participants in these experiments, I believe in a way similar to mine in the sulfur work.

Carbon is the most important element in biochemistry. The long-lived carbon-14 is the most important tracer isotope. However, before its discovery by Kamen and Ruben in 1940 all that one had to work with was the 20-minute carbon-11, made by bombarding boron with deuterons. Twenty minutes, or a few times twenty minutes, is not long to carry out involved procedures. However, Martin Kamen, with Ruben and Hassid, started a study of photosynthesis using this isotope in 1938. The radio-carbon was converted to carbon dioxide which was passed into a vessel containing barley plants or the alga *Chlorella*, and after a period of up to an hour the plants were killed by boiling alcohol. The resulting extract was separated chemically into fractions of different composition, whose activities were measured with a Geiger counter. This was done with the plants both in the light and in the dark. Radioactive carbohydrates the end products of photosynthesis, were found, but the greater part of the activity went into other compounds which they were trying to identify. It was clear that carbohydrates were not the primary products made from the added carbon dioxide, but the actual mechanism of incorporation of carbon is far too complicated to be unraveled using a 20-minute isotope.

I have presented merely an outline of isotope uses at Berkeley in the thirties, without any attempt to be exhaustive, but I hope that I have conveyed some idea of what it was like then, when experimental facilities were primitive and the experimenters were feeling their way into the unknown.

APPENDIX

Table 1 displays a list of the isotopes discovered in Berkeley in the thirties (including the year 1940), the "pre-war" years in respect to the principal activities of the Radiation Laboratory. The table was compiled by noting the earliest references for each isotope given in the sixth edition (1967) of the "Table of Isotopes" by Lederer, Hollander, and Perlman, and in some cases by checking the references themselves. The mass numbers and isomer designations are not always those given in the original publications, but a clear identification of the half-life and element is implied. The number of entries in the table is 81, nearly one-third of the world total during the same years.

TABLE I. Isotopes Discovered in Berkeley 1934 - 1940.

Element	Z	A (m denotes isomer)	Discoverer and date
H	1	3	Alvarez and Cornog, 1939
He	2	3 (stable)	Alvarez and Cornog, 1939
C	6	14	Kamen and Ruben, 1940
O	8	15	McMillan and Livingston, 1934
Ar	18	41	Snell, 1936
K	19	38, 44	Walke et al., 1937
Ca	20	45	Walke et al., 1940
Sc	21	44m, 48, 49	Walke et al., 1937-40
V	23	48, 49	Walke et al., 1937-39
Cr	24	51	Walke et al., 1940
Mn	25	51, 52, 54	Livingood and Seaborg et al., 1937
Fe	26	55, 59	Livingood and Seaborg, 1938-39
Ni	28	57	Livingood and Seaborg, 1938
Zn	30	69m	Thornton, 1938
Ga	31	65, 72	Livingood and Seaborg, 1938-39
Ga	31	67	Alvarez, 1938
Ga	31	66	Mann, 1937
Ge	32	68, 69	Mann, 1938
As	33	78	Snell, 1937
Se	34	81m, 83	Snell, 1937
Se	34	81	Langsdorf, 1940
Br	35	78, 83	Snell, 1937
Kr	36	85m, 87	Snell, 1937
Kr	36	83m	Langsdorf, 1940
Rb	37	86, 88	Snell, 1937

Element	Z	A (m denotes isomer)	Discoverer and date
Tc	43	93, 99m	Seaborg, 1939
Tc	43	97m (new element)	Perrier and Segrè, 1937 (in Palermo)
Ru	44	103	Livingood, 1936
Ag	47	107m	Alvarez, 1940
Sn	50	113, 121, 123m, 125, 125m	Livingood and Seaborg, 193
Sb	51	117, 124, 125	Livingood and Seaborg, 1937-39
Sb	51	127, 129, 132	Albeson, 1939 (fission products)
Te	52	121m, 127m, 129m, 131, 131m	Seaborg, 1940
I	53	124, 130, 131	Livingood and Seaborg, 1938
I	53	132, 133m, 134	Abelson, 1939 (fission products)
I	53	135	Segrè and Wu, 1940 (fission product)
Xe	54	135	Segrè and Wu, 1940) (fission product)
Xe	54	133	Langsdorf, 1939
Ba	56	135m	Kalbtell, 1940
Ta	73	180m	Oldenberg, 1938
Ir	77	192ml	McMillan, Kamen and Ruben, 1937
Pt	78	195m	McMillan, Kamen and Ruben, 1937

Table I (continued)

Element	Z	A (m denotes isomer)	Discoverer and date
Pt	78	197	Cork and Lawrence, 1936
Au	79	196, 196m, 199	McMillan, Kamen and Ruben, 1937
At	85	211 (new element)	Corson, MacKenzie, and Segrè, 1940
Np	93	239 (new element)	McMillan and Abelson, 1940

WHY IS THE HALF-LIFE OF ^{14}C SO LONG

H. Primakoff

Department of Physics
University of Pennsylvania
Philadelphia, Pennsylvania 19104

INTRODUCTION

The discovery by Kamen and Rubin of the very long-lived beta-active ^{14}C nuclide ($^{14}C \rightarrow {}^{14}N + e^{-} + \bar{\nu}_e$) is of inestimable value for chemistry, biochemistry, biology, geology, and archaeology (^{14}C as a tracer and as a "clock"). In addition, the ^{14}C nuclide is of very considerable interest for nuclear physics. Indeed, because of its "abnormally" long half-life ($[t_{\frac{1}{2}}(^{14}C)]_{observed} = [5730 \pm 40]$yr), the ^{14}C nuclide demonstrates far more convincingly than other nuclides the participation in the nuclear beta-decay process of transiently present particles different from the nucleus' constituent nucleons [neutrons (n) and protons (p)]. Among these particles the ones of smallest

 ISBN 0-12-397580-8

mass, and so the ones most likely to be present in low-lying nuclear energy states, are the π-mesons or pions (π^+,π^-,π°) which mediate the nuclear forces between nucleons at relatively large internucleon distances in essentially the same way as photons mediate the electromagnetic forces between charged particles.

I shall try in this presentation to explain why the pionic contribution in the beta-decay of ^{14}C is so much more important than in the beta decay of other nuclei (e.g., the short-lived ^{11}C) and why this contribution inhibits the ^{14}C beta-decay rate by a very large factor.

ESTIMATION OF THE ^{14}C BETA-DECAY RATE ON THE BASIS OF THE BETA DECAY OF ITS CONSTITUENT NEUTRONS ONLY

The ^{14}C beta-decay rate has been estimated by various investigators supposing it to arise solely from the beta decay of one or another of the nucleus' 8 constituent neutrons ($n \to p + e^- + \bar{\nu}_e$). The results are extremely sensitive to the parameters characterizing the initial ^{14}C and final ^{14}N wave functions; if these wave functions are taken as

$$\Psi_{^{14}C;J=0} = C_{00}\Psi_{L=0,S=0} + C_{11}\Psi_{L=1,S=1}; \quad C_{00}^2 + C_{11}^2 = 1$$

$$\Psi_{^{14}N;J=1} = N_{01}\Psi_{L=0,S=1} + N_{10}\Psi_{L=1,S=0} + N_{21}\Psi_{L=2,S=1};$$

$$N_{01}^2 + N_{10}^2 + N_{21}^2 = 1$$

where L, S, and J ($J = |L \pm S|$ = "nuclear spin") are, respectively, orbital, spin, and total angular-momentum quantum numbers, the *neutronic* ^{14}C beta-decay amplitude works out to be proportional to the combination of parameters $(C_{00}N_{01} - C_{11}N_{10}/\sqrt{3})$

so that the neutronic beta-decay rate is proportional to $(C_{00}N_{01} - C_{11}N_{10/\sqrt{3}})^2$. Thus this rate has contributions only from transitions between ^{14}C and ^{14}N configurations of the same L value and, in addition, depends critically on how close $C_{00}N_{01}$ is to $C_{11}N_{10/\sqrt{3}}$; in particular, if N_{21} is close to 1, N_{01} and $N_{10/\sqrt{3}}$ must be both close to 0, and $C_{00}N_{01}$ cannot differ much from $C_{11}N_{10/\sqrt{3}}$ so that the neutronic ^{14}C beta-decay rate is small. Unfortunately, the values of the parameters C_{00}, C_{11}, N_{01}, N_{10}, N_{21} cannot be extracted unambiguously from the observed properties of the ^{14}C and ^{14}N nuclei, i.e., from the ^{14}N magnetic dipole moment and electric quadrupole moment and from the rates of radiative pionic reactions such as $\gamma + {}^{14}N \rightarrow \pi^+ + {}^{14}C$ and $\pi^- + {}^{14}N \rightarrow \gamma + {}^{14}C$. However, in spite of the uncertainty in the parameters, i.e., $C_{00} = 0.86 - 0.80$, $C_{11} = 0.51 - 0.60$, $N_{01} = 0.08 - 0.16$, $N_{10/\sqrt{3}} = 0.18 - 0.14$, $N_{21} = 0.95 - 0.96$, it appears difficult to predict a half-life for ^{14}C longer than about $250{,}000 \times t_{\frac{1}{2}}(n) = 5$ years (where we use $t_{\frac{1}{2}}(n)]_{observed} = 10.8$ min). Thus a discrepancy by a factor of about 1000 vis-à-vis the observed ^{14}C half-life remains to be explained. It should also be noted that the earliest calculations of the *neutronic* ^{14}C beta-decay amplitude in effect assumed $C_{00} = 1$, $C_{11} = 0$, $N_{01} = 1$, $N_{10} = 0$, $N_{21} = 0$ and, consequently, predicted a half-life for ^{14}C of about 1.3 days!!

ESTIMATION OF THE ^{14}C BETA-DECAY RATE ON THE BASIS OF THE PARTICIPATION IN THE BETA DECAY OF ITS TRANSIENTLY PRESENT PIONS AS WELL AS OF ITS CONSTITUENT NEUTRONS [see B. Goulard *et al.* (1977)]

In this calculation, the *pionic* ^{14}C beta-decay amplitude ($n_1 + n_2 \rightarrow p_1 + \pi^o + e^- + \bar{\nu}_e + n_2 \rightarrow p_1 + e^- + \bar{\nu}_e + n_2$ and $n_1 + n_2 \rightarrow n_1 + \pi^+ + e^- + \bar{\nu}_e + n_2 \rightarrow n_1 + e^- + \bar{\nu}_e + p_2$)

works out to have a magnitude *close* to the magnitude of the *neutronic* ^{14}C beta-decay amplitude (as determined by $|C_{00}N_{01} - C_{11}N_{10}/\sqrt{3}|$) and a sign *opposite* to the sign of the neutronic ^{14}C beta-decay amplitude (i.e., a sign opposite to $(C_{00}N_{01} - C_{11}N_{10}/\sqrt{3})$) for a range of permissible values of C_{00}, C_{11}, N_{01}, N_{10}, and N_{21}. Further, the magnitude and sign of the *pionic* ^{14}C beta-decay amplitude are quite insensitive to the values of C_{00}, C_{11}, N_{01}, N_{10} and depend largely on the relatively well determined value of N_{21} (N_{21} = 0.95 - 0.96). As a consequence of this near equality in magnitude and opposition in sign the *neutronic* and *pionic* ^{14}C beta-decay amplitudes *interfere destructively* and yield a {*neutronic* + *pionic*} ^{14}C beta-decay amplitude whose square gives a ^{14}C beta-decay rate which is much smaller than the above estimated neutronic ^{14}C beta-decay rate and which corresponds to a ^{14}C half-life of the order of several thousand years rather than several years. On the other hand, in beta-active nuclides different from ^{14}C, e.g., in ^{11}C ($^{11}C \rightarrow {}^{11}B + e^+ + \nu_e$), after correction for an energy release which is much bigger in ^{11}C than in ^{14}C, the *protonic* beta-decay amplitude is estimated as some 20 times larger numerically than the above estimated *neutronic* ^{14}C beta-decay amplitude while the accompanying *pionic* beta-decay amplitude is estimated as roughly of the same magnitude as in ^{14}C. Thus, because of their discrepant magnitudes, any destructive or constructive interference of the *protonic* and *pionic* ^{11}C beta-decay amplitudes hardly affects the {*protonic* + pionic} ^{11}C beta-decay amplitude and so can modify only slightly the short half-life of ^{11}C expected from the protonic contribution alone ($t_{\frac{1}{2}}(^{11}C)]_{observed}$ = 20.4 min).

CONCLUSION

^{14}C is remarkable in that its *neutronic* beta-decay amplitude is small and has a sign opposite to the sign of its comparably small *pionic* beta-decay amplitude; the resultant almost complete *destructive interference* of these two amplitudes is then responsible for the "abnormally" long half-life of ^{14}C. No such destructive interference of the *nucleonic* and *pionic* beta-decay amplitudes has so far been identified in any other beta-decay transition so that in this respect ^{14}C appears to be unique in the family of beta-active nuclides.

REFERENCE

Goulard, B., Lorazo, B., Primakoff, H., and Vergados, J. D. (1977). *Phys. Rev. C16*, 1999.

TWO ISOTOPES ARE BETTER THAN ONE

Mildred Cohn

Department of Biochemistry and Biophysics
University of Pennsylvania
School of Medicine
Philadelphia, PA 19104

Perhaps I can contribute a bit of pre-history on Kamen and Rubin's discovery of ^{14}C because I happened to be present during a relevant conversation between Ernest Lawrence and Harold Urey. It occurred on the lawn of the Bureau of Standards at the American Physical Society Meeting in the spring of 1938. I was talking with Professor Urey, whose laboratory I had left the preceding fall, when Professor Lawrence approached. Professor Urey introduced me to Professor Lawrence and informed him that I was now engaged in applying stable isotopes as tracers in biological systems. Furthermore, Urey pointed out with ill-suppressed glee that the only isotopes generally useful as tracers in biology were not the radioactive ones but rather the stable ones, namely, ^{2}H, ^{13}C, ^{15}N, and ^{18}O. Whereupon the normally voluble Lawrence had nothing to say in defense of radioactive isotopes. But I have been told that upon returning to Berkeley he informed his research team which included Martin Kamen, that they had to find a long-lived isotope that would be useful as a biological tracer. The rest is history. I doubt whether even Lawrence in his most optimistic moments, antici-

 ISBN 0-12-397580-8

pated the remarkable and universal usefulness that the discovery of carbon-14 by Kamen and Rubin has had for the study of biological systems.

The use of stable isotopes for the study of intermediary metabolism was pioneered by Schoenheimer and Rittenberg. With his characteristic enthusiasm, Schoenheimer, in 1937, pointed out to me the great advantage of using two isotopic tracers in a single metabolite to establish the concept of group transfer and to yield mechanistic details in the conversion of a labeled precursor to product. And, in fact, the very first isotopic tracer experiment in which I participated in Vincent du Vigneaud's laboratory in 1938 in collaboration with Schoenheimer and Rittenberg (1) involved the use of both ^{15}N and ^{2}H. It had been established by du Vigneaud that both the L and D isomers of the unnatural amino acid, phenylaminobutyric acid, when fed to a rat, were excreted as the L-acetyl derivative. By labeling the amino acid with ^{15}N and the body water of the rat with D_2O, it was possible to show that the α-H was lost in both cases but the ^{15}N was lost only for the D-amino acid. Consequently, only for the D-amino acid was the keto acid a possible intermediate.

In 1944, du Vigneaud's group (2) performed an elegant experiment with methionine synthesized with ^{34}S and ^{13}C in the γ and β positions, respectively, to study the pathway of the *in vivo* conversion of the essential amino acid, methionine, to cystine. An unequivocal demonstration of the utilization of the sulfur but the nonutilization of the carbon chain of methionine in its conversion to cystine by the rat was made possible by the doubly labeled metabolite.

By the time I arrived at Washington University School of Medicine in 1946, Martin Kamen was already there, and although I still planned to use stable isotopes as biochemical tracers, the availability and the advantages of the radioactive isotopes in most applications were clearly apparent. Martin was most generous in sharing his expertise and helping me to set up a radioactive isotope laboratory in Cori's Department of Biological Chemistry. I shall give a few illustrations of the use to which the double label approach was again used in their laboratory in the study of phosphate transfer reactions.

Shortly after Leloir and his co-workers (3) found in 1949 that glucose-1,6-diphosphate was a coenzyme in the conversion of glucose 1-phosphate to glucose 6-phosphate catalyzed by phosphoglucomutase, the equilibration by the enzyme of glucose 1-phosphate labeled with ^{32}P and ^{14}C with synthetic coenzyme, glucose 1,6-diphosphate, was investigated. An essential feature of the reaction mechanism was established since complete equilibration of both the phosphate and glucose moieties of the monophosphate and diphosphate ester occurred.

In further mechanistic studies of enzyme-catalyzed phosphoryl transfer reactions during the 1950's it proved to be most useful to use a radioactive tracer ^{32}P in combination with the stable isotope ^{18}O, particularly in oxidative phosphorylation. One example is an investigation of the mechanism of the phosphate - $H_2{}^{18}O$ exchange catalyzed by yeast inorganic pyrophosphatase (5). The question arose whether the exchange of phosphate - oxygen with water could be ascribed to the reversal of the overall hydrolysis reaction with the reversible formation of pyrophosphate. The rapid rate of oxygen exchange compared to the slower rate of reversal of pyrophosphate followed by labeling inorganic orthophosphate with ^{18}O and ^{32}P simultaneously clearly indicated the exchange could not be due to a reversal of the overall reaction.

In more recent years, I have made use of a property of the nucleus other than its mass or radioactive decay as a means of tracing atoms in enzymatic pathways, namely, its magnetic moment associated with its spin. Since isotopes of the same nucleus have different spins and, in some cases, such as carbon, only ^{13}C has a non-zero spin, again isotopes are of inestimable value for the solution of some problems of enzyme mechanism by NMR. One example is the solution of the stereochemistry of the enolase reaction, the dehydration of 2-P-glycerate to P-enolpyruvate, by NMR (6) which involved both deuterium and ^{13}C. I. A. Rose synthesized 3-monodeuterio-2-P-glycerate of known stereochemistry (3*R*) enzymatically and converted it to P-enolpyruvate. The problem was to establish the position of the deuterium in the vinyl group of the product. The proton NMR spectrum of P-enolpyruvate is shown in Fig. 1A, the two triplets arising from H_A and H_B, respectively, each coupled to ^{31}P. The assignment of the downfield peak as H_A, the proton trans to the phosphate group and cis to the carboxyl group was established from the proton spectrum of a sample of P-enolpyruvate, 60% 1-^{13}C and 40% 1-^{12}C shown in Fig. 1B. The assignment was possible since the trans ^{13}C-H coupling constant is larger than the cis. The proton NMR spectrum of deuterated P-enolpyruvate formed from (3*R*)2-P-glycerate-3-D by the enolase reaction is shown in Fig. 1C, and the peak due to H_A, is missing. Consequently H_A, deuterium now, is in the position trans to the phosphate group. Thus, the abstraction of the hydroxide group in the reaction must occur anti to the C-2 proton that is eliminated in the enolase reaction. The establishment of the stereochemistry of specifically deuterated P-enolpyruvate made possible the stereochemical analysis of additions to C-3 that occur in many enzymatic reactions.

In our current NMR studies a new multiple tracer is being exploited. An isotopic shift due to ^{18}O has been observed (7) in the ^{31}P spectrum as shown in Fig. 2 where the five isotopic species of inorganic phosphate labeled with 44% ^{18}O, i.e., the binominal distribution of $^{18}O_{4,3,2,1,0}$, are resolved. This ability to resolve the five species makes possible the simul-

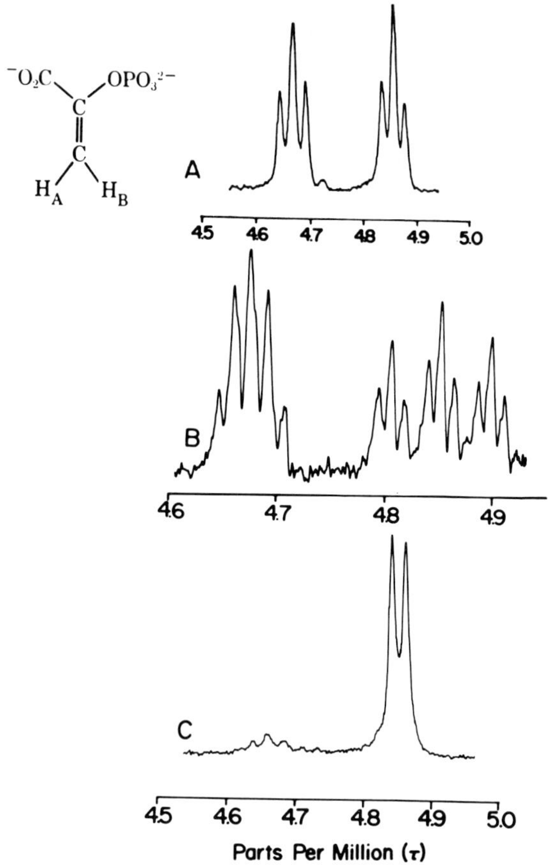

FIG. 1. Proton NMR spectra of P-enolpyruvate. (A) Vinyl protons of unlabeled P-enolpyruvate at 60 MHz; (B) same with ^{13}C (60%) in carboxyl group at 100 MHz; (C) deuterated P-enolpyruvate produced in the enolase reaction at 60 MHz.

taneous labeling of the oxygen and phosphorus of phosphates. In a substrate such as ATP with 3 phosphate groups, it is possible to synthesize the compound with different fractions of ^{18}O in the α, β, and γ phosphates, respectively, thereby labeling each phosphorus atom uniquely. The NMR spectroscopic technique permits continuous monitoring of a reaction and following simultaneously the fate of each phosphate moiety provided the oxygen does not exchange with solvent water. Alternatively, the fate of oxygen may be followed if phosphate-oxygen does exchange with water. ^{31}P NMR can now be used to determine the rate of phosphate-phosphate exchange in any reactant-product equilibrium as well as the site of bond cleav-

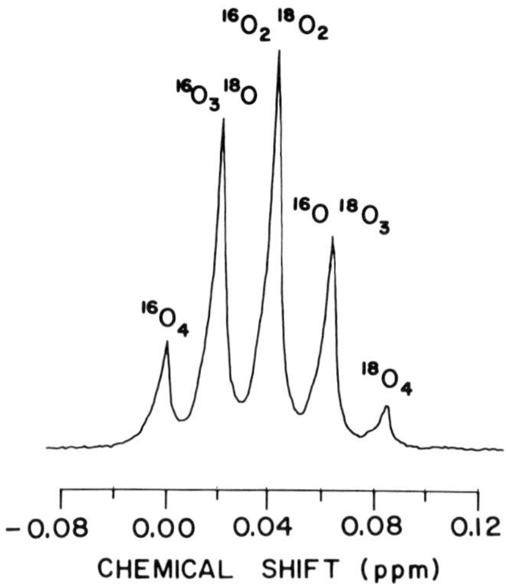

FIG. 2. ^{31}P NMR spectrum at 145.7 MHz of a randomized sample of inorganic phosphate ∿44% ^{18}O.

age. Such a reaction is illustrated by the $P_i(^{18}O) \rightleftarrows ADP(^{16}O)$ exchange catalyzed by polynucleotide phosphorylase (7). The spectrum after equilibrium is reached between 15 m*M* ADP and 10 m*M* P_i is shown in Fig. 3 with the ^{31}P spectra of P_i, the αP and βP resonances of ADP expanded. The initial P_i (93.4% ^{18}O) consisted of two species $^{16}O^{18}O_3$ and $^{18}O_4$ and at equilibrium a third species, $^{16}O_4$, has appeared. The αP and ADP consists of $^{16}O_4$ and $^{16}O_3{}^{18}O$; the single ^{18}O in the latter species arises from the α - β bridge oxygen proving that all four oxygens of P_i are incorporated and that the bond cleavage occurred between the αP and the bridge oxygen. This conclusion is confirmed by the fact that all four oxygens of P_i exchanged with all four oxygens of β-P.

The special property of an oxygen tracer of a phosphate, that it can reveal what the phosphate has been up to in a reaction sequence even though the intermediate cannot be observed, is exemplified in a comparative study of the phosphate-water (^{18}O) exchange catalyzed by native zinc alkaline phosphatase and by the enzyme in which cobalt has been substituted for zinc. The simplest exchange reaction scheme is:

$$E + P_i \underset{k_{-1}}{\overset{k_1}{\rightleftarrows}} E \cdot P_i \underset{k_{-2}}{\overset{k_2}{\rightleftarrows}} E\text{-}P + H_2O$$

where E·P is a noncovalent complex and E-P is the phosphorylated

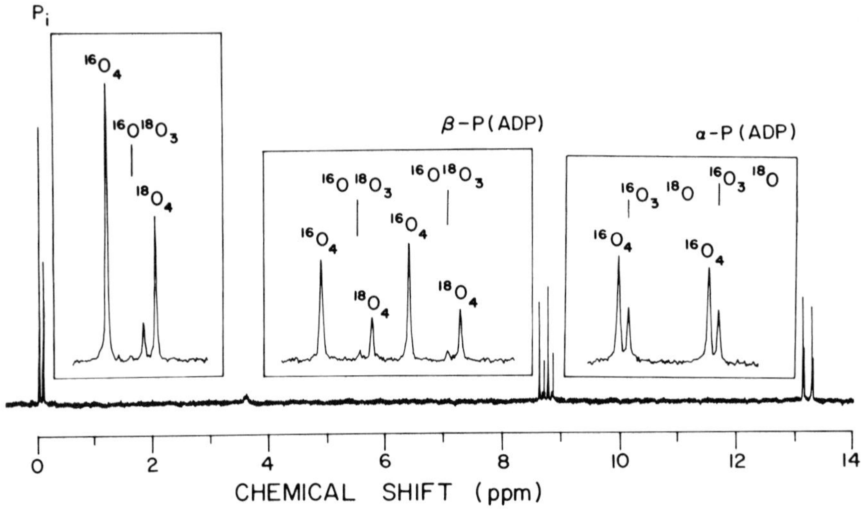

FIG. 3. ^{31}P NMR spectrum at 145.7 MHz of 10 mM P_i initially 93.4% ^{18}O and of 15 mM ADP, initially ^{16}O after equilibration by polynucleotide phosphorylase and termination of the exchange reaction by addition of EDTA.

enzyme; the reactions characterized by k_{-2} and k_{-1} describe the hydrolysis of phosphorylated enzyme during normal substrate turnover. Typical ^{31}P NMR spectra in the exchange reaction are shown in Fig. 4. The initial P_i (Fig. 4A) was enriched with ^{18}O (90.3%) and the distribution of ^{18}O species after partial exchange catalyzed by the native zinc enzyme is shown in Fig. 4B. The distribution of the species is consistent for a binomial distribution for 66.5% ^{18}O. On the other hand, a very nonrandom distribution is observed after partial exchange catalyzed by the cobalt enzyme in Fig. 4C, where the dashed line indicates the expected spectrum for a binomial distribution. The apparent anomalous distribution of species resulting from the cobalt enzyme catalyzed reaction can readily be explained by a consideration of the reaction scheme given above. From an expression introduced by Boyer *et al.* (9), the ratio $R = k_2/k_{-1}$ determines the average number of oxygen atoms exchanged on an enzyme-bound phosphate before it is released. Thus when $k_2 >> k_{-1}$ and $R >> 1$, all of the ^{18}O atoms in the P_i which is initially bound to the enzyme will be replaced by ^{16}O before P_i dissociates from E•P and, consequently, only $^{18}O_0$ ($^{16}O_4$) will appear in the bulk P_i in solution and be observed by ^{31}P NMR. On the other hand, when $R << 1$, a phosphate ion will exchange no more than one ^{18}O before dissociating and the distribution will remain, i.e., a binomial distribution will

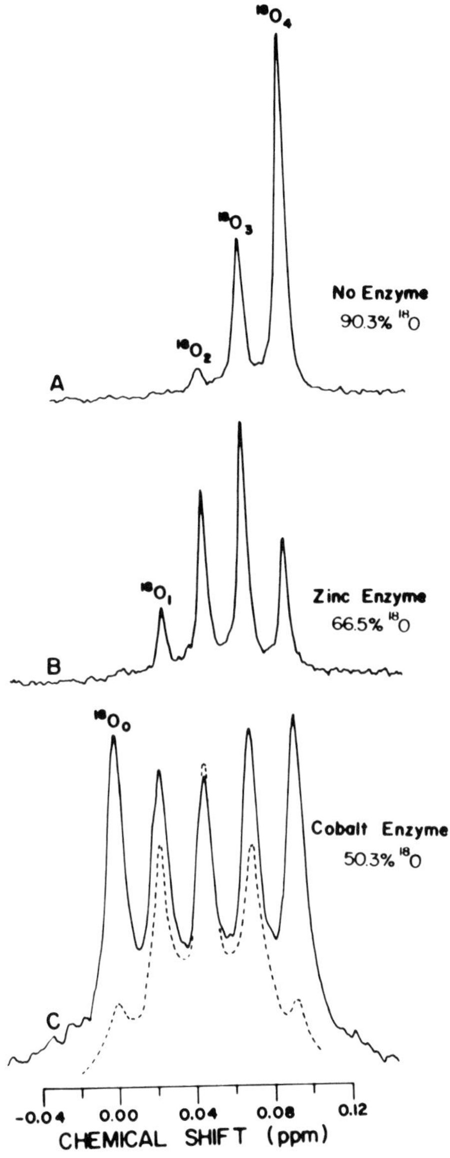

FIG. 4. ^{31}P NMR spectra at 145.7 MHz with the $^{18}O_0$ species as the reference point. (A) Initial $KH_2P^{18}O_4$, 90.3% ^{18}O; (B) same sample (66.5% ^{18}O) after partial exchange with $H_2{}^{16}O$ catalyzed by native (zinc) alkaline phosphatase; (C) same sample (51.3% ^{18}O) after partial exchange with $H_2{}^{16}O$ catalyzed by cobalt alkaline phosphatase. The solid line is the experimental spectrum and the dashed line represents the binomial distribution for 51.3% ^{18}O in phosphate for (C).

persist throughout the course of the exchange reaction as in the case of the reaction catalyzed by the zinc enzyme (Fig. 4B). In the case of the cobalt enzyme, the distribution of ^{18}O species is consistent with R = 3, so that an intermediate case exists in which a few phosphate - water exchanges occur before the phosphate dissociates from the enzyme. Thus the rate-determining step in the reaction is metal ion dependent and the ratio of rate constants can be determined quantitatively in this way.

In this brief presentation, I have attempted to give a few examples, from the great body of work which has developed since the early days of isotopic tracers, of the unique information which can be obtained by the use of nuclear properties of two elements simultaneously to probe metabolic pathways and enzyme mechanisms. In each case cited, it would have been very difficult, if not impossible, to answer the questions posed without the use of isotopes.

REFERENCES

1. du Vigneaud, V., Cohn, M., Brown, G. B., Irish, C. J., Schoenheimer, R., and Rittenberg, D. (1939). *J. Biol. Chem. 131*, 273-296.
2. du Vigneaud, V., Kilmer, G. W., Rachele, J. R., and Cohn, M. (1944). *J. Biol. Chem. 155*, 645-651.
3. Cardini, C. E., Paladini, A. C., Caputto, R., Leloir, L. F., and Trucco, R. E. (1949). *Arch. Biochem. Biophys. 22*, 87-100.
4. Sutherland, E. W., Cohn, M., Posternak, T., and Cori, C. F. (1949). *J. Biol. Chem. 180*, 1285-1295.
5. Cohn, M. (1958). *J. Biol. Chem. 230*, 369-379.
6. Cohn, M., Pearson, J. E., O'Connell, E. L., and Rose, I. A. (1970). *J. Am. Chem. Soc. 92*, 4095-4098.
7. Cohn, M. and Hu, A. (1978). *Proc. Nat. Acad. Sci. U.S. 75*, 200-203.
8. Bock, J. L. and Cohn, M. (1978). *J. Biol. Chem. 253*, 4082-4085.
9. Boyer, P. D., de Meis, L., Carvalho, M., and Hackney, D. D. (1977). *Biochemistry 16*, 136-140.

TRACERS IN BIOLOGY

A. A. Benson

Scripps Institution of Oceanography
University of California, San Diego
La Jolla, California

Biology made giant strides when Martin Kamen and Sam Ruben put carbon-11 to work. With the frantic energy of maniacs and uncompromising perseverance, they used their brief opportunity to the limit.

Established concepts were changed overnight (experiments in the Rat House usually began at 8:00 PM and ended after midnight). Carbon dioxide fixation, long thought to be a simply photochemical process, clearly involved a series of enzymic phosphorylations, reductions, carboxylations, and transformations. Reversibility of carboxylations and other reactions that had been unknown or hypothetical before became a reality. Carbon dioxide fixations by chemoautotrophic bacteria and by plants were found to be similar or closely related. Rates of reactions could be measured as one could discriminate between first reactions and those that followed. Intermediary metabolism became a reality, now that it could be observedw

As those concepts dawned on them, Kamen and Ruben became more and more impressed with the importance of the new energetics of metabolism emerging from the concepts of Lipmann and

 ISBN 0-12-397580-8

Kalckar. Their papers were followed with near reverence and became guide maps for the new developments with radiocarbon and the experiments Kamen and Ruben were planning.

Martin Kamen and Sam Ruben were a superb team. Both provided experimental skill, scientific background, and the fervor essential for innovation. They had what it took to make everything work. Martin built the cyclotron targets and made the combustion systems work. With this "aspirator" (Fig. 1) he collected the radioactive gases from his boron targets, and, glowing with positrons, he ran to the Rat House where Sam, with chemists and biologists, waited with counters clicking, at least all that had not broken down in the previous hour, and all hot plates on HIGH. It was the 1940 version of a "Space Launch"--all systems had to be GO.

The "Rat House" was a gray three-story square wooden structure with space for Bill Libby and Sam Ruben on the basement level. There was a classroom on the center level and some research space for Kamen and Ruben on the top floor. Laced with temporary wires and rubber tubing through holes drilled in the unpainted pine wood walls and vertical columns through all floors for some gas diffusion experiments, the place was eerie in the glare of the bare 200-Watt electric bulbs hanging from the ceilings. The Rat House was too plain to have been photographed. Sandwiched between the hillside and the old brick chemistry building, no photographer could get far enough away to see much of it except from atop the Campanile.

When carbon-11 experiments were scheduled, the building was "off limits" to Martin Kamen. He glowed with contamination from his cyclotron targets and could cause the backgrounds of their home-made Geiger tubes (Fig. 2) to climb. They could afford no deviation from their logarithmic decay charts for carbon-11, especially after five half-lives when the critical samples were to be counted. One of the marvelous aspects of the work they did was the apparent absence of any errors in measurement or deductions made from them - a real tribute to the excellence of work done against all sorts of experimental difficulties.

Kamen and Ruben placed great stress on planning. To get everything out of their precious five half-lives, every minute would count. I remember many planning sessions with Martin Kamen and Sam and graduate students Mary Belle Allen and Pete Yankwich. They treated each other's ideas with brutal respect, replete with the strong language of Cominsky Park, home of Martin's White Sox. From these, the experimental plan evolved; arguments suddenly ceased and, when everything worked, progress was made. At "game time," and all too often, there came the terse phone message from Martin, "Cyke's sick, Sam"; that ended the experiment until the cyclotron could be repaired and Martin

FIG. 1. "Aspirator" used for removing radioactive carbon-11 oxides from the boric acid cyclotron targets after deuteron bombardments. This 540-cc chamber was made about 1940 by Martin Kamen. He used it for carrying the highly radioactive gases to be used in every carbon-11 experiment with Sam Ruben and their collaborators in the Rat House.

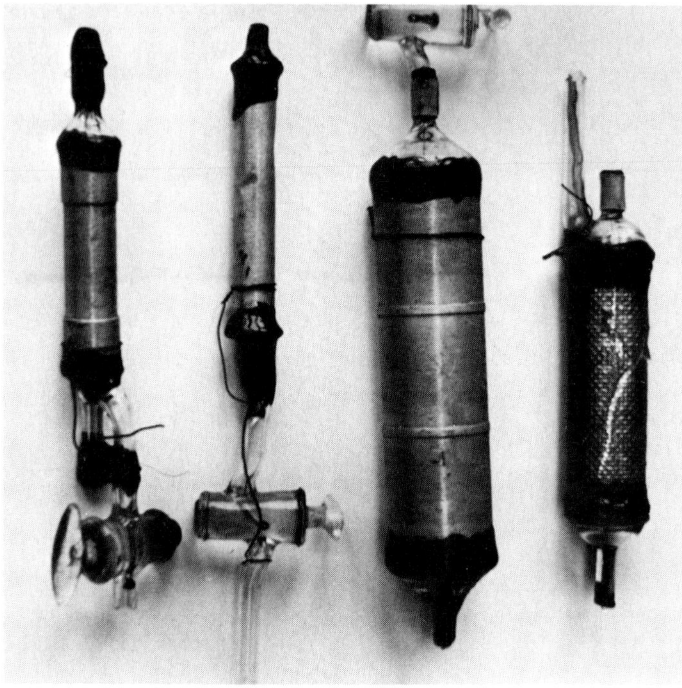

FIG. 2. Geiger-Mueller (G.-M.) tubes made by Martin Kamen and Sam Ruben used for measurement of carbon-11 samples in their studies of photosynthesis. The first three are thin-wall aluminum tubes for C-11. The fourth is an adaptation of the Libby Screen Wall Counter in which the cylindrical cathode of the G.-M. tube is replaced by a screen cylinder through which the soft beta radiation of C-14 can pass. This tube was probably used for experiments with C-14.

could schedule another bombardment of his targets.

With Martin Kamen, today we have Baird Hastings, who followed the Berkeley group doing similar experiments with the Harvard cyclotron. He, too, knows the excitement of following radioactive carbon dioxide through a living creature for the first time. Baird used the rat. Martin used *Chlorella*, bacteria, bacteria, barley, and a host of other organisms.

Besides overcoming technical obstacles, Martin Kamen and Sam Ruben built bridges from their skills as chemists and physicists to real biology. They did this elegantly. They recruited people who were the best in their fields - Hassid and

Barker, Doudoroff and van Niel - teamwork was an imperative requisite to success with carbon-11, and the teams were superb.

At this point enough was known of nuclear stability to predict properties of a long-lived carbon-14. If Kamen and Ruben had not gone so far as they had with carbon-11, and had so much trouble with its decay and need for longer biological experiments, they might not have been so strongly driven to find a long-lived radioisotope. With Lawrence's support and inspiring enthusiasm, Martin and Sam tried, using a number of nuclear reactions, to make carbon-14. They spent a year trying to make it from carbon-13, but with little success. The less likely reaction of bombarding nitrogen-14 with neutrons finally led to success and further difficulties. They had to adapt Libby's screen-wall Geiger counters to measure the soft radiation or count it as a gas. It was not easy to do biological experiments when sample measurements were so tedious. Martin had built dozens of stainless steel tanks to hold ammonium nitrate solution for neutron shielding around the 27-inch. After months of exposure to its neutron cloud, they bubbled nitrogen through them and harvested the carbon-14 in the form of carbon monoxide and dioxide. That was tedious, too, but it yielded useful quantities of the new radioisotope which was destined to revolutionize biology and medicine. With the first carbon-14, they could do longer, more detailed experiments. Figure 3 shows Sam Ruben holding their first big batch of C-14 in 200 mg $BaCO_3$. Martin Kamen and Sam Ruben invented carbon-14. I say "invented" because that is what they did. When you know it can be done and try every plausible trick to do it, that's invention! With this we did find the dark fixation product of CO_2 in algae and the first evidence of "C-4 photosynthesis." I had developed the solvent partition method for measuring pK_a for carboxylic acids and thereby of our dark $C^{14}O_2$ fixation product of *Chlorella*. The data fitted exactly the properties of succinic acid which we finally identified after the war.

Kamen and Ruben never really got a chance to exploit the attributes of their new isotope. It was tedious to measure and the amounts available were limited. They did, however, recognize its tremendous potential for progress in biology and medicine.

As the developments were becoming more and more exciting, the real difficulties began. Sam's time was consumed by a meteorological war project and the biological action of phosgene. Martin Kamen's time and peace-of-mind were increasingly beset by the irrationalities of Manhattan Project security (Fig. 4). This, and Sam's tragic accident brought to an end, for a time, a renaissance in biology. Not long after that, the situation had sunk so low that the inventor of radiocarbon and the engineer of classic voyages into the unknown reaches of

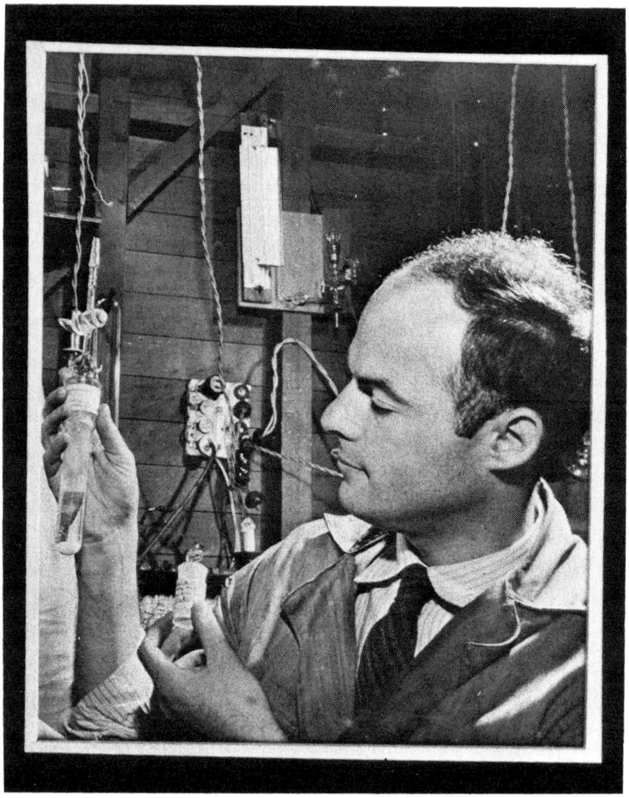

FIG. 3. Sam Ruben, Assistant Professor of Chemistry, University of California, 1938 - 1943, holds the first large sample of long-lived radiocarbon produced in collaboration with Martin Kamen. The "fine structure" of the Rat House may be seen in the background.

biology and physics was relegated to building Liberty ships in the Richmond yards. We are so fortunate that Martin's sense of humor and his marvelous ability to survive in a world of humans and, embarrassingly, sub-humans of those days, enabled him to successfully pass that stressful period and to emerge unscathed. After the war, the place was full of eager beavers anxious to go on where Martin and Sam had left off. For Martin Kamen, a new world of photosynthetic bacteria presented the real challenges. He had engendered a host of disciplines. All of them owe some part of their success to the legacy left by Martin Kamen and Sam Ruben.

The contributors will present first-hand reports of their

FIG. 4. The Old Radiation Laboratory seen from its North-west entrance. The security guards' station is at the main entrance in the center of this side of the building. The 37-inch room was in the center of the building.

classic developments made with isotopes. The tracer studies of Kamen, Barker, Hassid, and Ruben spread through this world and on to others.

RADIOACTIVE CARBON IN BACTERIAL METABOLISM: SOMETHING OLD AND SOMETHING NEW

H. A. Barker

Department of Biochemistry
University of California
Berkeley, California 94720

Martin Kamen must have been 25 when I first met him in 1938. I cannot now recall the exact circumstances of our first meeting, but the general setting is clear. Sam Ruben, an ambitious and dynamic young member of the Chemistry Department at Berkeley, had somehow become imbued with the idea of elucidating the then mysterious chemical process of photosynthesis by the use of carbon dioxide labeled with ^{11}C. Because of its short half-life, ^{11}C had to be prepared in the cyclotron just before it was to be used. Ernest Lawrence had assigned Kamen the job of preparing ^{11}C, as well as several other short-lived isotopes, for various biological applications. I first came to know Kamen as the man who delivered the radiocarbon.

I remember Kamen in this period as a slender young man with a thick head of dark hair, a frequently unshaven appearance, slender and finely shaped hands, a slouching walk, and a quiet and incredibly fast manner of speech.

So far as I can recall, Kamen seldom if ever delivered the ^{11}C on time. The 37-inch cyclotron was not available for the preparation of ^{11}C before 9 or 10 PM. Having been in use all

 ISBN 0-12-397580-8

day for other purposes it was evidently tired and frequently malfunctioned; several more hours often elapsed before Kamen and the cyclotron crew were able to nurse it along to provide a beam of adequate intensity. On a number of occasions the radiocarbon was not delivered at the "Rat House," the delapidated old wooden building where the experiments were done, until around midnight, and we did not finish the work and start home until dawn was emerging over the Berkeley hills.

The preparation of ^{11}C and other short-lived radioisotopes required the handling of cyclotron targets and adjacent parts that were highly radioactive. I seem to remember that Kamen wore gloves for some of these manipulations, although pictures of target preparation that have survived do not show this. As additional protection Kamen always wore an old and rather dirty lab coat. This coat or some other part of his clothing was so heavily contaminated with radioactive materials that he was not permitted in the lab when samples were being counted. But sometimes he would look in late in the evening to see how the experiment was going. As the counter heralded his approach, Ruben would emit a string of oaths and demand in no uncertain terms that Kamen go elsewhere - anywhere - until the counting was finished.

The study of photosynthesis, of course, required the use of some plants. Since Ruben had no background in botany, he soon persuaded a colleague of mine, Zev Hassid, to provide some expertise in this area. The first experiments on photosynthesis with ^{11}C were done with barley which was then a favorite experimental material for studies of plant nutrition and ion uptake. Barley, however, proved to be a rather inconvenient material for isotope studies. In discussing some of the problems with Hassid, I called his attention to the advantages of using unicellular algae, such as *Chlorella*, which Warburg had used extensively for studying photosynthesis. Since I had had experience in growing algae, I volunteered to provide some for a trial experiment. The trial was a success and so I was drawn into the Ruben-Kamen-Hassid group, initially as a supplier of *Chlorella*.

I also brought in some new ideas for the application of radioactive carbon dioxide to biological systems. I had had some experience with the biological formation of methane, which van Niel had postulated always resulted from the reduction of carbon dioxide, and I was also familiar with the Wieringa's demonstration of acetate synthesis from hydrogen and carbon dioxide by *Clostridium aceticum*, and with Wood and Werkman's presumptive evidence for the incorporation of carbon dioxide into succinic acid by propionic acid bacteria. When Ruben learned of these relatively simple microbial processes that utilized carbon dioxide, he was eager to collaborate in applying ^{11}C for their further elucidation. As a result I became a more active member of the group. A little later van Niel and some of his associates from the Hopkins Marine Station, including Stanley Carson and the late Jackson Foster, were drawn into the experimental work.

In looking back, I cannot refrain from mentioning the feeling of excitement in all the participants generated by the ^{11}C experiments. This resulted in part from the fact that the work involved a group effort requiring careful planning and rapid execution, in part from the almost instantaneous generation of results, and in part from the impression that we were opening up new frontiers of science. Nevertheless as the experiments progressed we soon came to recognize the serious limitations to experimental work imposed by the 21-minute half-life of ^{11}C and looked forward to the discovery of a radiocarbon isotope with a longer life.

The discovery of ^{14}C by Ruben and Kamen in 1940 (1) promised to revolutionize the use of radiocarbon as a tracer. However, the amount of ^{14}C initially prepared was very small and only one biological tracer experiment could be completed prior to December 1941 (2). This was a study of the incorporation of carbon dioxide into the carboxyl group of propionic acid during the fermentation of glycerol by propionic acid bacteria. With the entry of the United States into the war, all research of this type ceased at Berkeley and the major participants became involved in other endeavors. I was on sabbatical leave at the time. When I returned to Berkeley in the summer of 1942, Ruben was engaged in research on poison gases that soon resulted in his death, and Kamen was preoccupied with unmentionable research at the Radiation Lab that I presume was connected with the atomic bomb project.

I saw Kamen very infrequently during the next 2 years. Then, one day, in the fall of 1944 he came to Hassid's laboratory in a state of shock and bewilderment, and told us that he had just been summarily dismissed from his position in the Radiation Laboratory apparently because he had dinner in a San Francisco restaurant with a Russian consular official under circumstances that an overzealous FBI agent considered to be suspicious. This was a devastating blow to his career as a nuclear chemist and was made even more serious by the fact that he had recently purchased a fine old Italian viola and was still making payments thereon. Fortunately, there was no shortage of work in the bay area in those days and Kamen soon started a temporary career in one of the local shipyards. Most of his spare time during the next several months was spent in my lab where we applied ^{14}C to the investigation of several bacterial fermentations.

The source of the ^{14}C was some 600 liters of saturated ammonium nitrate solution which Kamen had prepared several years earlier and which had been exposed to stray neutron radiation from the 60-inch cyclotron for about a month. Tom Norris of the Chemistry Department and I spent about a week recovering the ^{14}C from this solution. It was a messy job that involved boiling aliquots of the solution in a 12-liter flask, passing the vapors through a condenser to remove water and over hot copper oxide to oxidize reduced forms of carbon; the CO_2 was then absorbed in alkali and precipitated as barium carbonate. Kamen and I got half the product, which amounted to 1.8 g of $BaCO_3$ having the relatively low specific radioactivity of 180 cpm/mg or 35 cpm/μmol, estimated on a rather inefficient counter.

I had no equipment for estimating radioactivity at that time, but Kamen was able, through his connections in the Radiation Lab, to obtain a binary scaling circuit and components for the construction of Geiger tubes. Kamen also introduced me to the techniques of making thin windows from sheets of mica and assembling Geiger tubes of acceptable quality. Before long we had all the tools for experiments with ^{14}C.

During the next few months we were able to carry out several simple but rather satisfactory quantitative tracer experiments on the utilization of carbon dioxide for the synthesis of acetic acid during the fermentation of glucose by *Clostridium thermoaceticum* and on the role of acetate in the synthesis of 4-carbon and 6-carbon fatty acids by *Clostridium kluyveri* (3). These experiments illustrated the extraordinary utility of ^{14}C as a tracer in biological systems and confirmed that the discovery of this isotope was one of the great scientific contributions of our time.

I always look back on my association with Martin during this period with great pleasure and satisfaction. The things I learned from him had an important influence on the direction of my career for a number of years thereafter.

NEW ENZYMATIC REACTIONS IN THE DEGRADATION OF LYSINE AND β-LYSINE

The second part of my paper is about some of our recent studies of the chemistry and enzymology of the degradation of lysine and related compounds by various bacteria. Most of the experiments have been done with ^{14}C-labeled compounds, which is now so common that it is hardly worth mentioning.

One of the processes we have studied is the conversion of lysine to acetic and butyric acids by extracts of certain anaerobic bacteria. The overall reaction was established some years ago by Thressa Stadtman (4). She showed that the carbon chain of lysine is cleaved in the course of the fermentation mainly between the second and third carbon atoms to form acetic acid from carbons-1 and -2 and butyric acid from carbons-3 to -6. The bacteria also have some capacity to cleave lysine between carbon atoms-4 and -5, but we shall not be concerned with this process.

We found some time ago that the degradation of lysine involves the enzymatic reactions shown in Fig. 1 (3). The amino group on carbon-1 of lysine is first moved to carbon-2 to form so-called β-lysine. The amino group in carbon-6 is then moved to carbon-5 to form 3,5-diaminohexanoate. This compound then undergoes a typical enzymatic oxidative deamination at carbon-3 to form a β-keto acid, 3-keto-5-aminohexanoate.

The nature of the reaction causing the decomposition of this β-keto acid puzzled us for a considerable time. We found that the utilization of the β-keto acid required the presence of acetyl-CoA and so first assumed that the β-keto acid was converted to its CoA thioester by a CoA transferase and then cleaved by reaction with CoA in a typical thiolase reaction to form acetyl-CoA and the CoA derivative of the acid with two fewer carbon atoms. However, we found that free CoA did not stimulate the reaction as expected and we were unable to detect formation of the postulated thioester of the β-keto acid. Later Yorifuji *et al.* (5) undertook the purification of the activity responsible for the β-keto acid cleavage in the presence of acetyl-CoA and obtained a single enzyme that catalyzed a readily reversible conversion of 3-keto-5-aminohexanoate and acetyl-CoA to L-3-aminobutyryl-CoA and acetoacetate (Reaction 1).

$$CH_3\underset{\displaystyle NH_3^+}{\underset{|}{C}}HCH_2COCH_2CO_2^- + CH_3CO\cdot CoA \rightleftharpoons CH_3\underset{\displaystyle NH_3^+}{\underset{|}{C}}HCH_2CO\cdot CoA + AcAcO^- \qquad (1)$$

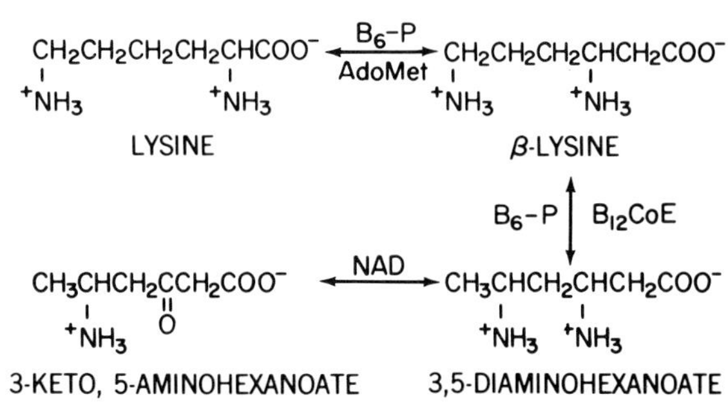

FIG. 1. Pathway of lysine degradation by Clostridium subterminale.

This was a novel type of reaction for β-keto acid cleavage and synthesis that had not been seen in other organisms.

We have recently found (7) that the same type of reaction participates in the degradation of L-β-lysine by an aerobic *Pseudomonas* species. β-Lysine utilization by this organism is initiated by acetylation of the amino group on carbon-6 (6) as shown in Fig. 2. Acetyl-CoA is the acetylating agent. The acetylated β-lysine undergoes a transamination reaction with α-ketoglutarate to form the β-keto acid, 3-keto-6-acetamidohexanoate. The further decomposition of this compound occurs only in the presence of acetyl-CoA. The enzyme responsible for this reaction has been partially purified (7) and shown to convert 3-keto-6-acetamidohexanoate and acetyl-CoA to 4-acetamidobutyryl-CoA and acetoacetate. The reaction is similar to that catalyzed by the clostridial β-keto acid cleavage enzyme, but the two enzymes differ in their substrate specificities and in other properties.

Since both aerobic and anaerobic bacteria catalyze this type of β-keto acid cleavage reaction, it seems probable that it will be found to occur in other organisms as well.

The further degradation of 4-acetamidobutyryl-CoA by extracts of *Pseudomonas* B4 also presents an interesting problem of enzymology which we have been able to define but not yet solve. The problem is the identification of the chemical steps involved in the conversion of 4-acetamidobutyryl-CoA to acetate, 4-aminobutyrate, and CoA. We believe that 4-aminobutyrate is an intermediate in the degradation of 4-acetamidobutyryl-CoA because the bacteria contain an active transaminase and an active succinic semialdehyde dehydrogenase that convert 4-aminobutyrate to succinate, which can be further metabolized in the usual way. We have investigated the degradation of 4-acetamido-

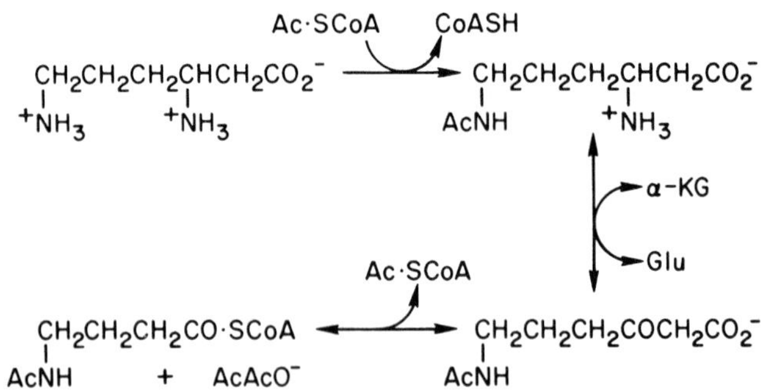

FIG. 2. Pathway of β-lysine degradation by a Pseudomonas sp.

butyryl-CoA with an enzyme preparation that was initially purified on the basis of its ability to form reduced coenzyme A from this compound, i.e., it functioned as a thioesterase. Later the enzyme was further purified on the basis of its ability to form acetate from 4-acetamidobutyryl-CoA. The acetate- and the CoA-forming activities always stay together and appear to be associated with a single protein. The other products of the reaction were found to be 4-aminobutyrate, 4-acetamidobutyrate, and the lactam of 4-aminobutyrate. The yields of these products have varied considerably with different enzyme preparations, but generally 30 to 40% of the substrate is recovered in 4-acetamidobutyrate and the remainder is found in 4-aminobutyrate and its lactam; the yield of acetate corresponds approximately to the total molar quantity of the latter compounds.

On the basis of these results there seemed to be several possible pathways for the formation of acetate and 4-aminobutyrate from 4-acetamidobutyryl-CoA as shown in Fig. 3. One possible pathway was the removal of CoA by the thiolesterase to form 4-acetamidobutyrate followed by the deacetylation of the latter compound. This pathway was excluded by the observation that the enzyme cannot deacetylate or otherwise metabolize 4-acetamidobutyrate. A second possibility was the deacetylation of the substrate prior to CoA removal. This would give 4-aminobutyryl-CoA which would rapidly cyclize by a nonenzymatic reaction to form the lactam of 4-aminobutyrate; the latter might then be hydrolyzed to 4-aminobutyrate. This pathway was eliminated by the finding that the lactam is not converted to 4-aminobutyrate. A third possibility was the direct

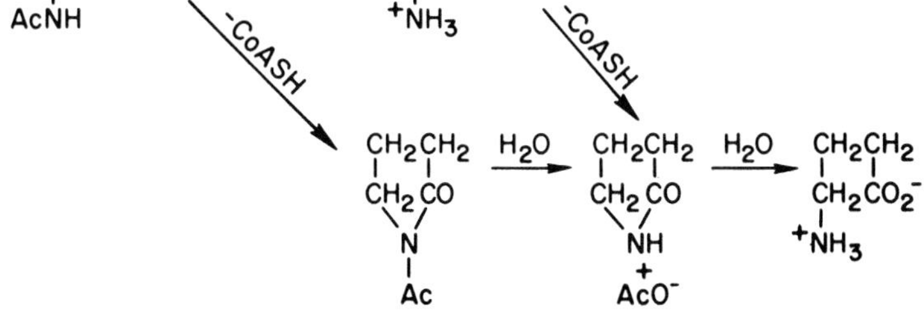

FIG. 3. Possible reactions for the conversion of 4-acetamidobutyryl-CoA to acetate and 4-aminobutyrate.

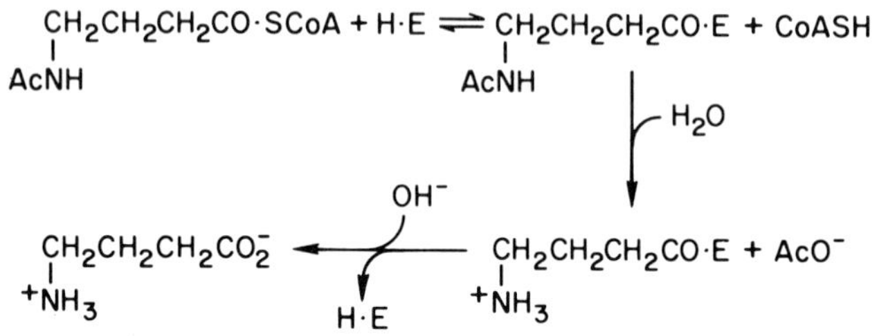

FIG. 4. Postulated mechanism for the enzymatic deacetylation of 4-acetamidobutyryl-CoA.

cyclization of the substrate to form the *N*-acetyl derivative of the lactam, a rather reactive compound that might be easily deacetylated to form the lactam of 4-aminobutyrate. However, this pathway was also eliminated by the finding that neither the lactam nor the acetylated lactam is utilized by the enzyme.

Since these lines of evidence eliminate most of the more obvious low molecular weight intermediates, it seems probable that the deacetylation of 4-acetamidobutyryl-CoA involves enzyme-bound intermediates of the type shown in Fig. 4. We postulate that the acyl moiety of 4-acetamidobutyryl-CoA is first transferred to the enzyme with the release of reduced coenzyme A. The acetyl group is then removed by a hydrolytic reaction. The presumed advantage of an enzyme-bound intermediate for the deacetylation step is that the 4-aminobutyryl moiety will be protected from rapid cyclization by some property of the enzyme, such as a strong affinity for the charged amino group. Therefore the hydrolytic conversion of the 4-aminobutyryl moiety to 4-aminobutyrate will presumably compete effectively with the cyclization reaction. According to this scheme, the considerable formation of 4-acetamidobutyrate always observed with our enzyme preparations would be attributable to a nonphysiological side reaction involving either the substrate or the first enzyme-bound intermediate.

At present this mechanism for the deacetylation reaction is only a theory. We hope to obtain some further evidence for or against it before long.[1]

[1] *[See added Note on separate page.]*

REFERENCES

1. Kamen, M. D. (1963). *J. Chem. Educat. 40*, 234-242.
2. Carson, S. F., Foster, J. W., Ruben, S., and Barker, H. A. (1941). *Proc. Nat. Acad. Sci. U.S. 27*, 229-235.
3. Barker, H. A. (1978). *Annu. Rev. Biochem. 47*, 1-33.
4. Stadtman, T. C. (1964). *Ann. N.Y. Acad. Sci. 112* (Art 2), 728-734.
5. Yorifuji, T., Jeng, I. M., and Barker, H. A. (1977). *J. Biol. Chem. 252*, 20-31.
6. Edmunds, H. N. and Barker, H. A. (1973). *Arch. Biochem. Biophys. 154*, 460-470.
7. Ohsugi, M., Kahn, J., Hensley, C., Chew, S., Bozler, G., Robertson, J. M., and Barker, H. A. (1980). *Arch. Biochem. Biophys. 203*, 437-452.
8. Ohsugi, M., Kahn, J., Hensley, C., Chew, S., and Barker, H. A. (1981). *J. Biol. Chem. 256*, 7642-7651.

Note Added in Proof

[1]Further research (8) has shown that the postulated mechanism for the conversion of 4-acetamidobutyryl-CoA to acetate, 4-aminobutyrate and CoASH is probably not entirely correct. The expected rapid exchange between CoASH and the CoA moiety of the 4-acetamidobutyryl-CoA analog, 3-acetamidopropionyl-CoA, does not occur. Furthermore, 4-aminobutyryl-CoA and 3-aminopropionyl-CoA accumulate under certain conditions during the degradation of 4-acetamidobutyryl-CoA and 3-acetamidopropionyl-CoA, respectively. This establishes that the acetyl group is removed before the CoA moiety. However, synthetic or enzymatically formed 4-aminobutyryl-CoA is converted exclusively to the lactam of 4-aminobutyrate by a nonenzymatic reaction. Consequently, we have postulated that 4-aminobutyrate is formed directly from an undefined 4-aminobutyryl-CoA-enzyme complex. The yield of 4-aminobutyrate from 4-acetamidobutyryl-CoA was shown to increase specifically with enzyme concentration. This suggests that the state of aggregation of the enzyme determines the identity of the product.

THE REEMERGENCE OF POSITRON-EMITTING ISOTOPES AS TRACERS OF METABOLISM: A NEW APPROACH FOR MEASURING CHANGES IN RATES OF CELLULAR PROLIFERATION *In Vivo*

Morris Friedkin
Elizabeth J. Crawford
Kent Diveley

Department of Biology
University of California
San Diego, La Jolla, California 92093

Isadore Wodinsky

A. D. Little, Inc.
Cambridge, Massachusetts 02142

When Martin Kamen turned to a search for a more practical isotope of carbon he had good reason to be disillusioned with the first tracer he had used so skillfully (1). Tired of the helter-skelter of chasing metabolic events with carbon-11, Kamen wanted a nuclide that would not demand a constant eye on the clock during preparation of labeled precursors. Carbon-11 had an alarming rate of decay: $t_{1/2}$ = 20 min. Kamen yearned for a more leisurely and careful approach to the biochemical operations required to isolate metabolic intermediates.

For biological experiments, the cyclotron had to be near at hand for generation of ^{11}CO and $^{11}CO_2$ from bombardment of boron-10 with deuterons (2). The pace was nerve-racking from the moment the ^{11}C oxides emerged from the glass train. Very little travel time was available between cyclotron and biological laboratory. One thing was certain: within 2 hours (6 half times) after a cyclotron run, 98% of the radioactivity would have decayed, a physical reality that severely limited the general usefulness of carbon-11.

Thus when Martin Kamen jammed his counter with the product

 ISBN 0-12-397580-8

of the $^{14}N(n,p)^{14}C$ reaction at the Berkeley cyclotron, a new era of biochemistry opened up. With a half-life of over 5000 years, carbon-14 provided ample time for less hectic experiments. The early but elegant research with carbon-11 soon became an almost forgotten step in the history of biochemical investigations with radioisotopes.

Within the last decade, however, interest in short-lived isotopes has revived. The cyclotron which played such an important role in Martin Kamen's early scientific career has again emerged as an important generator of carbon-11 and other positron-emitting isotopes. In the summer of 1978 the Second International Symposium on Radiopharmaceutical Chemistry held at St. Catharine's College, Oxford, England, featured many papers on the synthesis of compounds labeled with ^{11}C, ^{15}O, ^{13}N, and ^{18}F (Table 1). Why is there so much effort being expended to prepare these tracers despite the difficulties and high cost of the technology involved?

The renewed interest in carbon-11 and other short-lived nuclides (^{13}N, ^{15}O and ^{18}F) stems from the unique way in which they decay. In contrast to such isotopes as ^{32}P, ^{14}C, and ^{3}H, which emit electrons with widely varying energies, the isotope used by Kamen emits positrons. Based on theoretical aspects of quantum mechanisms the existence of the positron had been predicted by Dirac in 1932. C. D. Anderson discovered the positron, a particle which acted in cloud chambers as if it had the mass of an electron but exhibiting a charge equal but opposite to that of the electron.

Kamen's counters, however, could not detect the positrons emitted by carbon-11 but rather secondary gamma rays produced upon annihilation of the positrons.

POSITRON ANNIHILATION

The annihilation process which accompanies positron-electron interaction has been and continues to be a phenomenon of great importance in physics and in biology. It is impossible to conserve energy and momentum if only one gamma quantum is emitted during the annihilation of a pair (electron/positron). The simplest possible process starting from a pair with small relative velocity is the emission of two gamma-rays traveling in opposite directions. In this case the total momentum is zero. This is the most important mode of annihilation.

Although ^{11}C, ^{13}N, ^{15}O, and ^{18}F emit positrons which vary in velocity (Table 2), the energy released upon annihilation of a positron-electron pair is always the same for these nuclides: the emission of two 511 KeV photons 180° apart. This back-to-

TABLE 1

Compounds Labeled with Positron-Emitting Isotopes[a]

Class	*Abstract number*
Amino acids	
6-^{18}F-fluorotryptophan	A3
^{13}N-L-asparagine	42
^{13}N-L-glutamic acid	43
^{13}N-L-tyrosine	43
^{13}N-L-phenylalanine	43
^{18}F-fluorophenylalanine	A3
Carbohydrates	
^{11}C-2-deoxy-D-glucose	4
^{18}F-2-2-deoxy-2-fluoro-D-glucose	4
^{18}F-3-deoxy-3-fluoro-D-glucose	5
^{11}C-lactic acid	29
Nucleic acid derivatives	
5-^{18}F-5-fluoro-2'-deoxyuridine	A6
2'-^{18}F-2'-deoxyuridine	6
Fatty acids	
2-^{18}F-fluorostearic acid	A1
9,10-^{18}F-fluorostearic acid	A1
16-^{18}F-fluorohexadecanoic acid	A1
Steroids	
^{18}F-4-fluoroesterone	7
^{18}F-4-fluoroestradiol	7
Drugs	
^{11}C-imipramine	45
{N-^{11}C-3}-morphine	28
^{13}N-streptozotocin	44
^{11}C-hexamethonium	27
Miscellaneous	
$^{30}PH_3$ ($t_{1/2}$=2.5 min)	49
5-^{18}F-fluoro-3,4-dihydroxyphenylalanine	9
^{13}N-nitrous oxide	41
^{11}C-acetylene	41
^{11}C-hippuric acid	A21
$C^{15}O_2$, $C^{15}O$, $^{15}O_2$, $H^{15}O_2$, $^{13}N_2$, $^{13}NH_3$	25

[a] *Compounds described or referred to in reference (3). This listing although incomplete is presented to acquaint the reader with a sample of the intense activity in this field.*

TABLE 2[a]

Positron-Emitting Isotopes

	^{18}F	^{11}C	^{13}N	^{15}O
Maximum energy (MeV)	0.633	0.959	1.197	1.738
Most probable energy (MeV)	0.2025	0.326	0.432	0.696
Half-life (min)	109.7	20.3	10.0	2
Path length for electron of same maximum energy in water (cm)	0.239	0.498	0.535	0.822
Thickness of tissue within which 90% of positrons are stopped (cm)	0.09	0.11	0.14	0.18

[a] *Modified from Table 1 (4).*

back relationship may vary slightly if the positron-electron pair is not completely at rest when annihilation occurs (5). Depending on the radionuclide, the positrons are emitted with slightly different energies (Table 2). Once a positron is emitted its range is not a simple function of continuous and unidirectional slowing down. In fact the positron is finally annihilated after having followed a very tortuous path. This is due to elastic and inelastic scattering by atoms (4). The net effect is the formation of secondary gamma rays from sites only a few millimeters from the initial point of positron emission.

In contrast to other radionuclides such as ^{99m}Tc whose low-energy gamma radiation varies significantly with depth, the two 511 KeV photons which are created by the annihilation of a positron are not affected by the presence of tissue overlying the point of emission. Because of the unique nature of annihilation coincidental detection, mathematical reconstruction of cross-section images is possible.

The emission of two gamma-rays 180° apart is the basis of modern positron-emission transaxial tomography involving devices capable of detecting coincident gamma rays with complex arrays of detectors. Computers solve the problem of the initial site of positron emission (6, 7).

The inherent resolution of positron emission tomography depends on the geometry of the detectors arranged in such a way so as to record the maximum number of coincident photons in three dimensions. This is usually accomplished by constant rotation of either the positron-emitting object or the array of detectors (Fig. 1). Because the coincident 511 KeV gamma-rays are so powerful, many are not captured by crystal detectors now available. The efficiency could be markedly improved by development of devices that could register every double event within "sight" of the camera. Bismuth germanate has been considered as a potential scintillation detector instead of the activated sodium iodide crystals now generally used (7a).

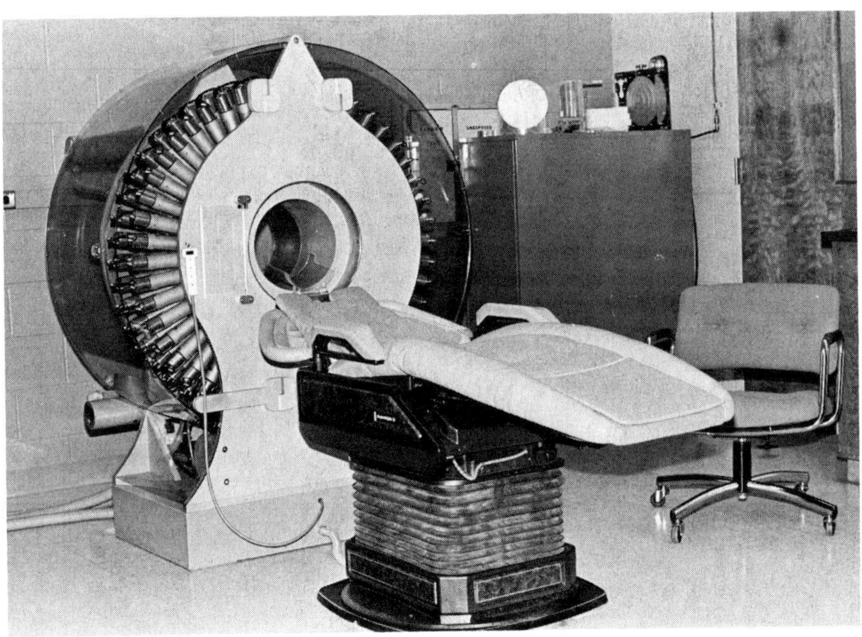

FIG. 1. Positron emission transverse tomograph (PETT V). This instrument was designed for visualization of the human head and experimental animals. We are greatly indebted to Dr. Michel M. Ter-Pogossian (Mallinckrodt Institute of Radiology, Washington University School of Medicine, St. Louis, Missouri) for this photograph (7a).

THE ADVANTAGE OF USING POSITRON-EMITTING ISOTOPES

The availability of tracers labeled with positron-emitting isotopes and the technical means of detecting gamma-rays produced by annihilation of positrons makes possible a great variety of *in vivo* studies of the kinetics of metabolism. This is a new situation in nuclear medicine which offers extraordinary opportunities for physiological imaging (8 - 13).

In the past, numerous radiopharmaceuticals have been used in nuclear medicine but what was missing in the long list of mostly inorganic ionic tracers were gamma-emitting organic compounds that could enter the mainstream of metabolic events. Isotopes such as ^{3}H, ^{14}C, and ^{15}N have played a critical role in unraveling intermediary metabolism. Without these isotopes, the modern era of biochemistry would have been greatly delayed, if not impossible.

With the advent of the positron emission tomograph the use of organic compounds labeled with positron-emitting isotopes is a rapidly developing field in nuclear medicine as shown in Table 1. Precursors labeled with ^{11}C, ^{15}O, ^{13}N, and ^{18}F emit gamma-rays as the result of positron annihilation. This makes possible the detection of *in vivo* metabolic events involving carbohydrates, amino acids, nucleic acids, and lipids with external counting devices. Although the half-lives of positron-emitting isotopes are exceedingly short (Table 2), rapid synthetic methods for their incorporation into important biological precursors have been developed.

POSSIBILITY OF MEASURING *In Vivo* CHANGES IN THE PROLIFERATIVE STATES OF TISSUES WITH POSITRON-EMITTING ISOTOPES

If a proliferative process is defined as the increase in cell number characterized by a concomitant increase in DNA, then any analytical method that assesses these parameters can be used as a probe of the rate of proliferation. This is an important concept in evaluating the efficacy of various anticancer agents whose main action must be to stop proliferation. Since an increase in mass is a direct consequence of proliferation, radiological approaches have been used predominantly to measure increases in cell number. However, mass changes can occur in nonproliferating tissues. Although the net accumulation of RNA and protein generally reflects the proliferative process there are situations in which cells synthesize much protein and RNA, yet proliferation as defined above does not occur.

It is for the above reasons that the incorporation of labeled thymidine into newly replicated DNA has become such a valuable measure of proliferation. Hughes *et al.* attempted to do this with ^{125}I-labeled iododeoxyuridine (14). We have previously investigated the use of carbon-11 labeled thymidine as a means of studying cellular proliferation with external detectors.

$^{11}CO_2$ produced at the Brookhaven 152-cm cyclotron was converted to formaldehyde which, in turn, was used for the enzymatic conversion of deoxyuridine 5'-phosphate to [^{11}C]thymidylate. Enzymatic treatment of the nucleotide with alkaline phosphatase gave [^{11}C]thymidine (15). Positron-emitting [^{11}C]thymidine and [^{11}C]thymidylate have been used in a variety of *in vivo* tracer studies of DNA synthesis in mice and tumor-bearing rabbits (15, 16).

INCREASED INTRACELLULAR PHOSPHORYLATION AND TRANSPORT OF URIDINE IN PROLIFERATING CELLS

When resting mammalian cells are stimulated to synthesize DNA by the action of various growth factors, the uptake of uridine is markedly elevated as the result of at least two major processes: (1) a rapid increase in the rate of intracellular phosphorylation and (2) a subsequent increase in the rate of transfer of the nucleoside across the plasma membrane by facilitated diffusion. The net effect of these processes is a marked difference in the rate of uridine uptake and metabolic trapping when quiescent cells are compared with proliferating cell populations (Table 3) (17 - 20).

Due to increased cellular uptake of labeled nucleoside it is difficult to assess absolute rates of RNA synthesis because of uncertainties in establishing the exact specific activity of intracellular UTP. It is reasonable to assume, however, that the proliferative process involves increased rates of synthesis of all forms of RNA (messenger, ribosomal, and transfer) and that the metabolic trapping of labeled uridine will reflect the interaction of all three parameters: increases in transport, phosphorylation, and utilization for newly synthesized RNA. It can be argued, therefore, that although the rate of RNA synthesis cannot be accurately assessed, the total uridine uptake in all forms (cold acid-soluble and hot acid-soluble) can be used as a reflection of increased cell proliferation.

TABLE 3[a]

Uridine Uptake in Quiescent, Serum-Stimulated, and Transformed Cells

Cell type	*Condition*	*Transport (pmoles/5 sec/ mg protein)*	*Intracellular phosphorylation (pmoles/10min/mg protein)*
Mouse embryo fibroblasts	Quiescent	0.30	64
	10% serum	0.29	127
	Log growth	0.62	383
Balb 3T3 fibroblasts	Quiescent	0.34	35
	20% serum	0.36	141
	Log growth	0.81	432
Transformed 3T3 cells			
SV40	Log growth	0.70	349
Polyoma	Log growth	0.66	401

[a] *Modified from Table 1 (20).*

THE POTENTIAL USEFULNESS OF [5-^{18}F]FLUOROURIDINE AS A PROBE FOR CHANGES IN PROLIFERATION

Our objective has been to take advantage of the phenomenon of increased intracellular phosphorylation and increased transport of uridine in rapidly dividing cells by preparing an analogue of uridine labeled with a positron-emitting isotope. Our premise was that the enhanced tissue uptake of such a precursor should be easily detected by positron emission tomography. Although ^{11}C-labeled uridine could serve such a purpose, we chose ^{18}F-labeled 5-fluorouridine as a more practical analogue for studying changes in proliferation:

1. Fluorouridine is utilized by cells in metabolic pathways very similar to that of uridine (21 - 23).
2. Fluorine-18 has a half-time of 110 min, a value five times longer than that of carbon-11, making possible more extended periods of physiological imaging.
3. A positron-emitting and reactive nuclide of fluorine, $^{18}F_2$, can be easily generated in the cyclotron by bombardment of neon gas with deuterons, ^{20}Ne(d,α)^{18}F, and then utilized for the synthesis of [5-^{18}F]fluorouracil (24).
4. Fluorouracil can be converted rapidly to fluorouridine and fluorouridylate by enzymatic procedures (see following sections).

Believing that the four parameters listed above provided a reasonable basis for assessing the *in vivo* rate of proliferation, we proceeded with preliminary tracer experiments in which we utilized carbon-14 labeled precursor, working out conditions for later studies with fluorine-18 in collaborative research with Dr. Alfred Wolf's group at the Brookhaven National Laboratories.

We describe here tracer experiments with [2-^{14}C]5-fluorouridylate (an *in vivo* precursor of [2-^{14}C]5-fluorouridine) in which two major changes in proliferative states have been examined, one involving a neoplastic infiltration of various tissues, and the other, an increased growth of normal cells. Mice were injected intracerebrally with L1210 leukemia cells, a procedure that leads to rapid infiltration of distant organs with neoplastic cells (25). In the other case, increased erythropoiesis in mouse spleen was triggered by chemically induced lysis of red cells (26). In each of the above cases the total tissue uptake of ^{14}C from a pulse of [2-^{14}C]5-fluorouridylate reflected increases in the rate of cellular proliferation. Furthermore, inhibition of growth by drugs such as methotrexate and actinomycin D resulted in decreased uptake of radioactivity as expected.

METHODS

Lactobacillus leichmanii cells (ATCC 7830) were cultured in ATCC medium 17 (1 liter of medium contains 100 gm of skim milk, 100 ml of neutralized tomato juice and 5 g of yeast extract). Bacterial suspensions thus obtained were used as inocula for large-scale growth in media with low concentrations of vitamin B_{12} (Difco B_{12} assay medium supplemented with 0.1 µg of vitamin B_{12} per liter). These growth conditions yielded bacteria which appeared as moderately elongated rods. These cells upon extraction gave maximum pyrophosphorylase activity. Overnight incubation of 14 liters of medium at 37°C, produced approximately 50 g of cells (wet weight) which after centrifugation were usually stored at -4°C for no longer than 1 to 2 days. The method for extraction and purification of the enzyme was a modification of previous procedures (27). A suspension of cells, 50 g plus 100 ml of 0.05 *M* phosphate buffer, pH 7.4, was sonicated twice for 2 min at 0°C in a Bronson W-350 sonicator and then centrifuged for 20 min at 30,000-60,000 *g*. The supernatant fluid was cooled and acidified to pH 4.5 with 0.2 *N* acetic acid. After 5 min the precipitate was collected by centrifugation and then taken up with 80 ml of 0.05 *M* phosphate buffer pH 7.4. This preparation usually contained about 10 mg of protein per milliliter catalyzing the conversion of fluorouracil to the nucleotide at a rate of 1 µmol per minute per milliliter of enzyme under conditions of the assay described below.

The Assay of UMP Pyrophosphorylase

To 100 µl of mixture consisting of 1 m*M* [2-^{14}C]5-fluorouracil (0.5 µCi/µmol), 2.5 m*M* pyrophosphorylribosylphosphate, 5 m*M* reduced glutathione, 2.5 m*M* $MgCl_2$, and 0.05 *M* $NaHCO_3$ (freshly prepared) were added 5-15 µl of enzyme. After 15 min at 37°C the incubation was terminated by brief immersion of the assay tubes in a boiling water bath. Five microliters of the heat-treated assay mixture was applied to Whatman No. 3 paper and then subjected to ascending chromatography with a mixture of acetonitrile and 0.1 *N* ammonium acetate (70:30). [2-^{14}C]5-Fluorouridylate that was taken to dryness by rapid these conditions, was determined by scintillation counting.

Enzymatic Synthesis of [2-^{14}C]5-fluorouridylate

A typical 10-ml preparation (100 times the assay mixture described above) contained 10 µmol of pyrophosphorylribosylphosphate and 0.5 ml of UMPpyrophosphorylase. After 30 min at 37°C the incubation mixture was diluted with 15 ml of H_2O and then applied to a column of DEAE-cellulose (1.2×12 cm). After a preliminary wash with 20 ml of H_2O and 5 ml of 0.25 M NH_4HCO_3, elution with 25 ml of 0.25 M NH_4HCO_3 yielded a solution of [2-^{14}C]5-fluorouridylate that was taken to dryness by rapid evaporation *in vacuo* in a Büchi Rotovap (bath 90°-100°C). The final product, representing 60-80% conversion of the pyrimidine base to nucleotide was dissolved in isotonic saline.

Cold and Hot Perchloric Acid Extractions

Tissues were frozen on solid CO_2 immediately after dissection and after weighing were homogenized with 10 volumes of ice-cold 3% $HClO_4$ and centrifuged. After two more suspensions with cold acid solution, same volume as above, the 3 extracts were counted separately, but summed for total cold acid-soluble radioactivity. Each residue was then subjected to treatment with 10 volumes of 3% $HClO_4$ for 20 min at 80°C. This hot perchloric acid extract contains labeled RNA and unlabeled DNA (28). In separate experiments separation of RNA and DNA by alkaline treatment (29) showed that [2-^{14}C]5-fluorouridylate does not label DNA.

RESULTS

Infiltration of Mouse Organs after Intracerebral Injection with L1210 Leukemia Cells

After intracerebral injection of mice with suspensions of L1210 leukemia cells, the brains became infiltrated diffusely with neoplastic cells and after several days distant organs were also involved. The total tissue uptake of ^{14}C, 3 hours after injection of [2-^{14}C]5-fluorouridylate, reflected the progressive state of leukemia cell proliferation in tissues such as spleen (Fig. 2), lung (Fig. 3), liver (Fig. 4), and brain (Fig. 5). These tissues showed significant increases of radioactivity after pulse labeling, consistent with previous histological evidence of infiltration with leukemic cells (25). Not only was the total tissue radioactivity increased, but also the incorporation into RNA (Figs. 2B-5B). The labeling of RNA was proportional in all cases to the total tissue uptakes.

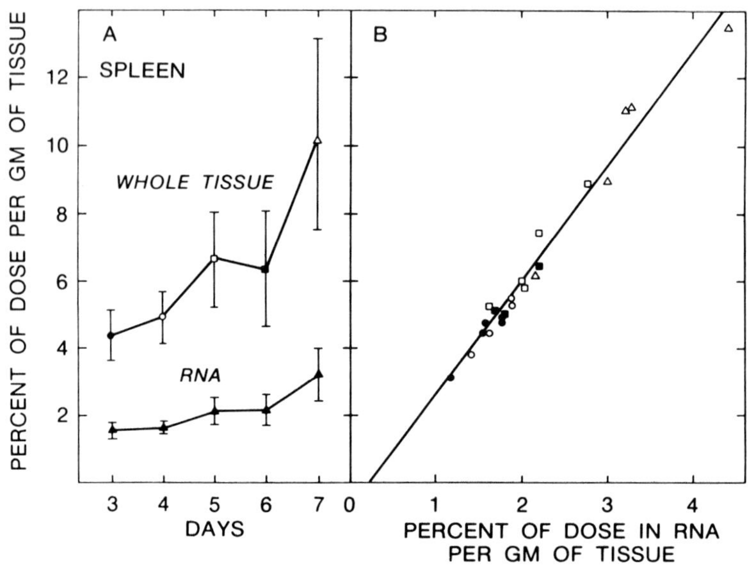

FIG. 2. The proliferation of leukemic cells in mouse spleen as revealed by uptake of radioactivity after a 3-hour pulse of [2-^{14}C]5-fluorouridylate. On day 0 CDF strain male mice (20 g) were injected intracerebrally with 1×10^4 murine leukemia cells (subline LE52 B07 of L1210 ascites) in a volume of 0.05 ml. Controls were injected with 0.05 ml of isotonic saline. The progress of the leukemic infiltration of the spleen was measured as follows: On days (as indicated in the figure), the leukemic mice were given an intraperitoneal injection of [2-^{14}C]5-fluorouridylate (1 μCi, 1 μmol) and killed 3 hours later. Spleens were frozen immediately after dissection, weighed, and treated with $HClO_4$, as described in Methods.

(A) The sum of the radioactivity in the cold and hot extracts represents the total tissue uptake (expressed in the figure as percentage of the injected dose per gram of tissue, wet weight). The radioactivity in the hot extracts is due to incorporation into RNA (expressed in the figure as percent of the injected dose in RNA per gram of tissue). The bars represent standard deviations.

(B) Panel shows the linear relationship between total tissue uptake and incorporation into RNA for all values obtained in (A). The symbol used in each case corresponds with symbols used for individual whole tissue uptakes in (A).

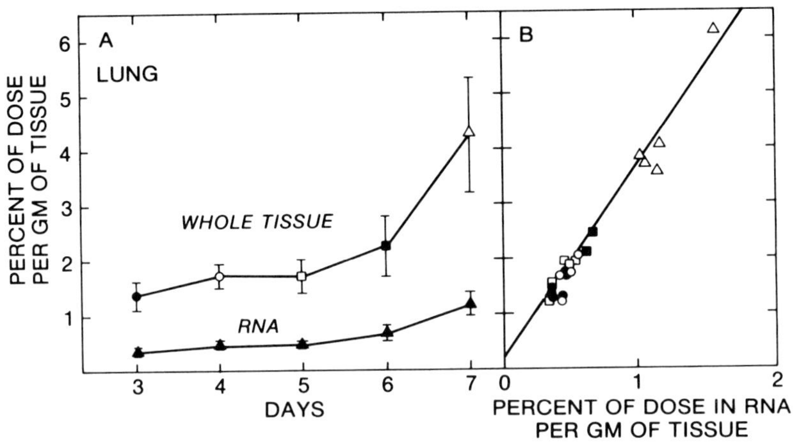

FIG. 3. The proliferation of leukemic cells in mouse lung as revealed by uptake of radioactivity after a 3-hour pulse with [2-^{14}C]5-fluorouridylate. See legend of Fig. 1 for description of protocol.

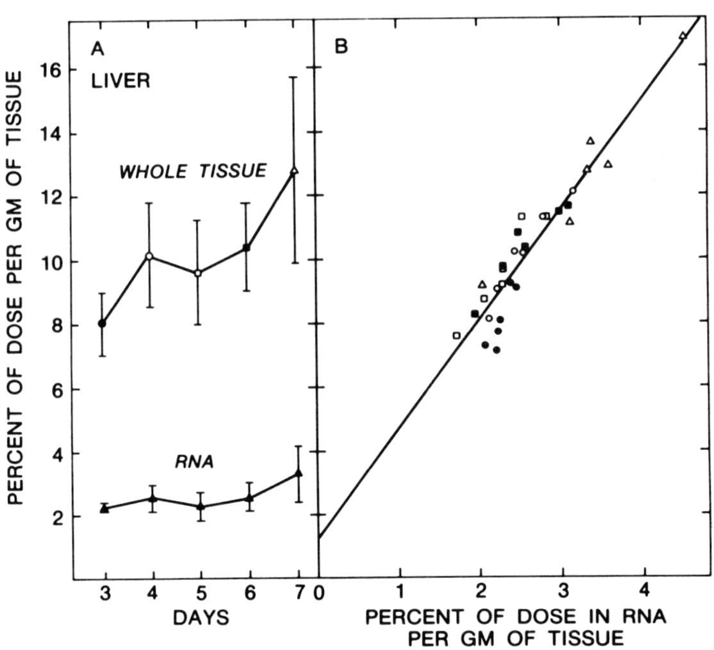

FIG. 4. The proliferation of leukemic cells in mouse liver as revealed by uptake of radioactivity after a 3-hour pulse with [2-^{14}C]5-fluorouridylate. See legend of Fig. 1 for description of protocol.

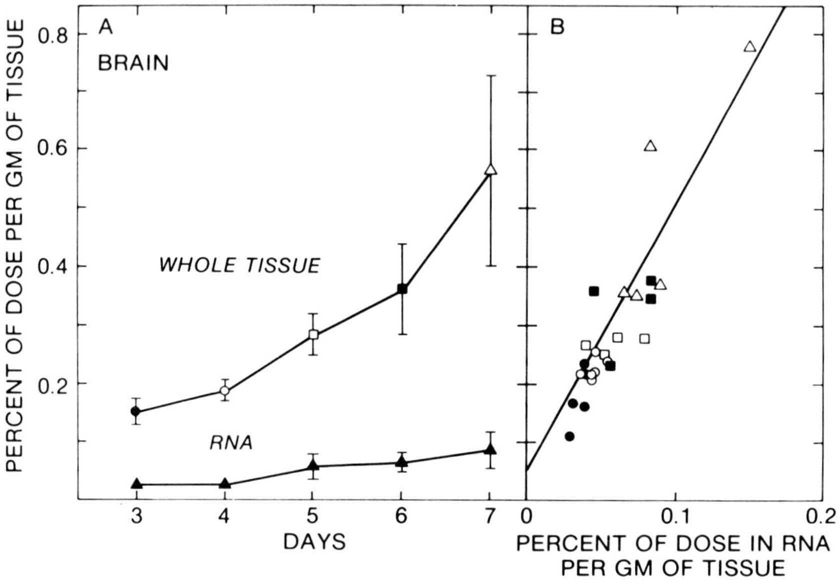

FIG. 5. The proliferation of leukemic cells in mouse brain as revealed by uptake of radioactivity after a 3-hour pulse with [2-^{14}C]5-fluorouridylate. See legend of Fig. 1 for description of protocol.

Uptake of Radioactivity in Leukemic Mice Treated with Methotrexate after a Pulse of [2-^{14}C]5-Fluorouridylate

The total tissue uptake of radioactivity after a pulse of [2-^{14}C]5-fluorouridylate can reflect not only increases of proliferation due to infiltration by leukemic cells but also a return of the tissue to normalcy after treatment of leukemic animals with methotrexate, a powerful antileukemic agent (30). As shown in Table 4, the spleens of control nonleukemic animals exhibited uptakes of 4.2 (percentage of the total dose remaining in 1 g of tissue 3 hours after the pulse). This value rose almost 3-fold to 11.5 in untreated leukemic mice and dropped back to normal in leukemic animals treated with methotrexate (4.2 in animals on a low methotrexate dosage regimen).

Similar patterns of uptake were observed in liver and lung, i.e., increased uptakes in animals with infiltrated tissue, and normal uptakes in leukemic mice treated with methotrexate. In all cases involving the experimental conditions described in Table 4, the uptake of radioactivity into RNA was proportional to the total tissue uptake.

TABLE 4

Total Radioactivity in Various Tissues of Normal and Leukemic Mice Treated with Methotrexate[a]

	Percent of dose per gm of tissue[b]					
	Control mice			*Leukemic mice*		
Tissues	*Untreated*	*Low methotrexate*[c]	*High methotrexate*[c]	*Untreated*	*Low methotrexate*[c]	*High methotrexate*[c]
Spleen	4.2 ± 0.66	4.4 ± 2.6	4.6 ± 0.47	11.5 ± 3.9	4.2 ± 0.98	3.3 ± 1.3
Liver	5.3 ± 1.7	7.3 ± 1.9	7.2 ± 2.3	10.6 ± 4.1	6.4 ± 1.5	5.0 ± 3.5
Lung	1.0 ± 0.22	1.3 ± 0.61	1.0 ± 0.89	4.9 ± 3.2	1.3 ± 0.43	0.84 ± 0.33
Brain	0.13 ± 0.068	0.21 ± 0.13	0.15 ± 0.16	0.39 ± 0.11	0.24 ± 0.61	0.17 ± 0.06
Intestine	4.8 ± 1.3	6.0 ± 0.66	5.5 ± 0.80	5.1 ± 1.4	6.8 ± 1.9	4.0 ± 1.3

[a] *On the seventh day of leukemic infiltration, mice were killed 3 hr after intraperitoneal injection of [2-^{14}C]5-fluorouridylate (1 μCi, 1 μmol).*

[b] *Mean ± standard deviation; 5-6 mice in each group.*

[c] *On days 3 and 5 after intracerebral injection of leukemia cells or saline as described in legend of Fig. 1 mice were administered subcutaneous solutions of neutralized methotrexate (4 mg per ml) for a final dosage of 20 μg per gm body weight. This is referred to as the "low methotrexate" protocol. A separate group of mice, leukemic and saline-injected controls, were treated once on day 4 with a neutralized solution of methotrexate (80 mg per ml) for a final dosage of 400 μg per gm body weight. This is referred to as the "high methotrexate" protocol.*

Uptake of Radioactivity in Erythropoietic Spleens after a Pulse of [2-^{14}C]5-fluorouridylate

For a period of several days after injection of mice with phenylhydrazine, the spleens enlarged and became the main organ of erythropoiesis (26). This is due to chemical lysis of circulating red cells. At various times after treatment with phenylhydrazine we injected the mice with [2-^{14}C]5-fluorouridylate. Three hours after injection of the tracer we measured total tissue uptake of radioactivity as well as the amount incorporated into RNA. At five days after lysis of erythrocytes, maximal incorporation was observed: 11% of the injected tracer was taken up per gram of spleen (Fig. 6, Expt. 7), a value at least twice that of normal control animals (Fig. 6, Expt. 1). Actinomycin D at levels known to inhibit RNA synthesis (31) markedly inhibited the total tissue uptake of ^{14}C as well as the incorporation into RNA (Fig. 6, Expt. 3). After 5 days the spleens returned to a more normal state as expected, and ^{14}C uptakes reflected the subsidence of proliferation. In all cases, from the period of increased erythropoiesis to the final state of relative quiescence, the uptakes into RNA were proportional to the total tissue uptakes in a surprising linear relationship (Fig. 6).

Comparison of [^{3}H]Fluorouridine with [^{3}H]Uridine Uptakes in Acid-Soluble Pools of Various Untransformed and Virus-Transformed Mouse Cells

Cultures of different mouse cell types were labeled with [^{3}H]fluorouridine and [^{3}H]uridine for 10 minutes (2 μCi per ml; 2 μ*M*) and then extracted into cold 5% trichloroacetic acid as described by Rozengurt *et al.* (Table 1 of Ref. 20). The ratio of incorporation of radioactivity into the acid soluble pools from [^{3}H]fluorouridine/[^{3}H]uridine was as follows: quiescent 3T3 cells, 0.82; quiescent 3T3 cells exposed to 10% bovine serum albumin for 1 hour, 0.90; log growth 3T3 cells 1.0; log growth 3T6 cells, 0.83; SV40-3T3 cells, 1.3; and mouse embryo fibroblasts, 1.3. These results show that the 5-fluoro analog of uridine is utilized as well as uridine by mouse cells in culture and supports the notion that fluorouridine can serve as an accurate probe of cell proliferation. We wish to thank Dr. E. Rozengurt for providing the facilities and cell cultures that made these comparative uptake studies possible.

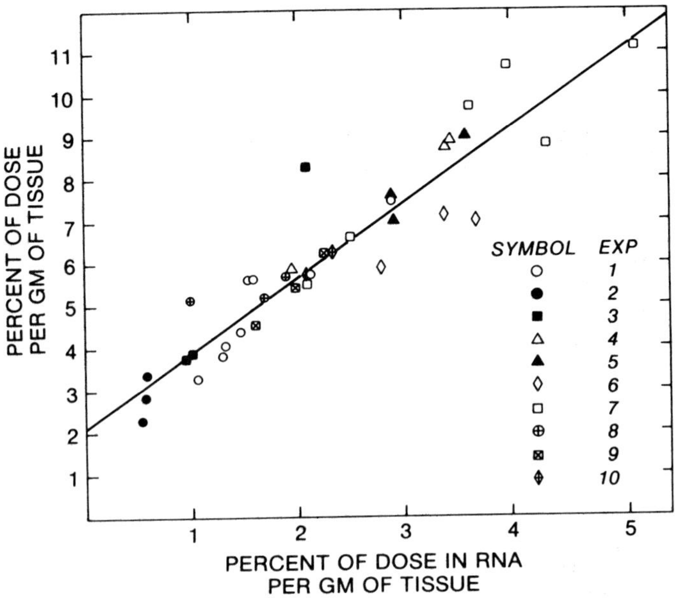

FIG. 6. The proliferation of erythropoietic cells in mouse spleen as revealed by uptake of radioactivity after a pulse of [2-^{14}C]5-fluorouridylate. Increased erythropoiesis was produced in mice by the method of Spivak et al. (26). Mice were injected intraperitoneally with 40 µg of phenylhydrazine-HCl per g (freshly prepared and neutralized solution containing 10 mg/ml) at 9 AM and 5 PM on day 1 and at 5 PM on day 2. Decreased rates of tissue RNA synthesis were obtained by subcutaneous injection of actinomycin D, 800 µg/kg, 2.7 hours prior to injection of [2-^{14}C]5-fluorouridylate (31).

Expt. 1, normal mice, no phenylhydrazine treatment (PT). Expt. 2, no PT; actinomycin D injected 2.7 hours prior to injection of pulse of [2-^{14}C]5-fluorouridylate. Expt. 3, PT 120 hours prior to, and actinomycin D 2.7 hours prior to pulse. Expt. 4, PT 49 hours prior to pulse. Expt. 5, PT 69 hours prior to pulse. Expt. 6, PT 94 hours prior to pulse. Expt. 7, PT 120 hours prior to pulse. Expt. 8, PT 142 hours prior to pulse. Expt. 9, PT 266 hours prior to pulse. Expt. 10, PT 190 hours prior to pulse. In all cases mice were killed 3 hours after an intraperitoneal injection of [2-^{14}C]5-fluorouridylate (1 µCi, 1 µmol). See Methods for details of analysis of tissue following dissection.

DISCUSSION

Our original premise was that the uptake of fluorouridine, an analog of uridine, should be elevated in tissues undergoing increased proliferation because the phosphorylation of intracellular uridine as well as nucleoside transport is known to be higher in exponentially growing cells than in quiescent cells. This has been demonstrated in mouse tissues infiltrated with leukemic cells, as well as in mouse spleen, the major site of erythropoiesis after red cell lysis. In these investigations we have employed the fluoronucleotide as a precursor. As first shown by Heidelberger's laboratory (32), the major advantage of using a ribosyl derivative of 5-fluorouracil in contrast to the free base (33-35) is the greater efficiency of intracellular uptake when the nucleoside is used. We have found that the fluoronucleotide is also a most effective precursor. Although a fraction of [2-^{14}C]5-fluorouridylate may actually enter cells as such, it is probable that most of the fluoronucleotide undergoes *in vivo* dephosphorylation to the fluoronucleoside prior to entry into cells. We used the fluoronucleotide for most of our studies since an extra preparative step with alkaline phosphatase could be eliminated. Our premise is that the cellular uptake of radioactivity from injected [5-^{18}F]fluorouridine, a positron-emitting tracer, should also reflect different rates of *in vivo* proliferation in neoplasms, hematologic tissues, and other organs undergoing rapid growth changes. We conclude, therefore, that the positron emission tomograph could be used as an external detector of different rates of cellular growth by revealing the uptake of [5-^{18}F]fluorouridine in localized tissue areas. Furthermore, it should be possible to detect changes in the rate of proliferation in tissues of the same animal by repeated injections (i.e. before and after a potentially active anticancer agent) because of the rapid decay of fluorine-18. Initial studies with carbon-14 labeled fluorouridine and fluorouridylate, such as those described in this report, have prompted the preparation and use of [5-^{18}F]fluorouridylate and [5-^{18}F]fluorouridine in collaborative studies with Dr. Al Wolf's group (36) at Brookhaven National Laboratory. Further animal studies are now in progress with the fluorine-18 label compounds (see Addendum).

It is important to emphasize that 5-fluorouridine has powerful antimetabolic properties (37, 22, 23). We propose to employ as a tracer of proliferative processes, a fluorinated analogue of uridine that is paradoxially antiproliferative on its own account. However, when used as the ^{18}F-fluoronucleoside, tracer amounts of very high specific activity can be prepared and used at levels that do not produce pharmacological effects. The group at Brookhaven National Laboratory has recently developed a method for the rapid synthesis of [5-^{18}F]fluorouridine that does not involve the use of enzymes (38).

Our studies are very similar in concept to the approach of Houghton and Taylor who used the relationship between total tissue radioactivity and labeled DNA after injection of [^{3}H]thymidine into tumor bearing animals to assess the rate of DNA synthesis (39). They refer to this correlation as fractional incorporation (FI). Since the FI is independent of the precursor concentration in tissues these investigators suggest that FI may provide a valuable index of DNA synthesis.

CONCLUSIONS

Isotope studies of physiological processes normally require clean separations of labeled intermediates. If the dynamic events of *in vivo* metabolism are to be followed by external detectors of gamma-emitting isotopes, how then can unique biochemical conversions be recognized by an external camera that "sees" all labeled compounds no matter what their form may be? Only in rare cases such as pulmonary investigations can volatile radioactive gases such as O_2, N_2, CO, and CO_2 be easily analyzed. We believe the studies described herein provide an answer at least for one important cellular process: proliferation.

Although the approach described herein provides an extraordinary opportunity to observe normal and pathological biochemical pathways with external detectors, the methodology involves a complex and expensive technology. First a cyclotron is required for generation of the positron-emitting isotopes. Because of the short lives of these radionuclides, facilities for both the rapid synthesis and immediate utilization of organic compounds labeled with these isotopes must be near at hand. Finally, the external detection instrumentation of the positron emission tomograph involves sophisticated electronic components and a committed on-line computer for the final imaging process. Obviously only a few research centers can afford to support such costly biological research. Nevertheless, the promise of this unique union between biology and physics continues to attract a growing number of interdisciplinary scientists.

ACKNOWLEDGMENT

This investigation (M.F.) was supported in part by a grant from the National Cancer Institute USPHS CA11449.

Research carried out (I.W.) was supported by contract NOI-CM-53765 with the Division of Cancer Treatment, National Cancer Institute, DHEW.

REFERENCES

1. Kamen, M. D. (1957). "Isotopic Tracers in Biology," pp. 293-299. Academic Press, New York.
2. Ruben, S., Kamen, M. D., and Hassid, W. Z. (1940). *J. Amer. Chem. Soc. 62,* 3443-3455.
3. Abstr., 2nd *Intern. Symp. on Radiopharmaceut. Chem.,* St. Catherine's College, Oxford, July 3-7, 1978.
4. Cho, Z. H., Chan, J. K., Ericksson, L., Singh, M., Graham, S., MacDonald, N. S., and Yano, Y. (1975). *J. Nucl. Med. 16,* 1174-1176.
5. Muehllehner, G. (1976). *J. Nucl. Med. 17,* 757.
6. Brownell, G. L., Burnham, C. A., Wilensky, S., Aronow, S., Kazemi, H., and Strieder, D. (1969). *Proc. Symp. Med. Radioisotope Scintigraphy Salzburg,* Aug. 6-15, 1968, pp. 163-174. Vienna IAEA.
7. Ter-Pogossian, M. M., Phelps, M. E., Hoffman, E. J., and Mullani, N. A. (1975). *Radiology 114,* 89-98.
7a. Ter-Pogossian, M. M., Mullani, N. A., Hood, J. T., Higgins, C. S., and Ficke, D. C. (1978). *J. Computer Assist. Tomogr. 2,* 539-544.
7b. Cho, Z. H. and Farukhi, M. R. (1977). *J. Nucl. Med. 18,* 840-844.
8. Raichle, M. E., Larson, K. B., Phelps, M. E., Grubb, R. L., and Welch, M. J. (1975). *Amer. J. Physiol. 228,* 1936-1948.
9. Gallagher, B. M., Ansari, A., Atkins, H., Casella, V., Christman, D. R., Fowler, J. S., Ido, T., MacGregor, R. R., Som, P., Wan, C. N., Wolf, A. P., Kuhl, D. E., and Reivich, M. (1977). *J. Nucl. Med. 18,* 990-996.
10. Phelps, M. E., Hoffman, E. J., Selin, C., Huang, S. C., Robinson, G., MacDonald, N., Schelbert, H., and Kuhl, D. E. (1978). *J. Nucl. Med. 19,* 1311-1319.
11. Gallagher, B. M., Fowler, J. A., Gutterman, N. I., Gutterson, N. I., MacGregor, R. R., Wan, C. N., and Wolf, A. P. (1978). *J. Nucl. Med. 19,* 1154-1161.
12. Spolter, L., Cohen, M. B., Chang, C. C., and MacDonald, N. S. (1979). *J. Nucl. Med. 20,* 662.

13. Hübner, K. F., Andrews, G. A., Buonocore, E., Hayes, R. L., Washburn, L. C., Collmann, I. R., and Gibbs, W. D. (1979). *J. Nucl. Med. 20,* 507-513.
14. Hughes, W. L., Commerford, S. L., and Gitlin, S. (1964). *Fed. Proc. 23,* 640-648.
15. Christman, D., Crawford, E. J., Friedkin, M., and Wolf, A. P. (1972). *Proc. Natl. Acad. Sci. U.S. 69,* 988-992.
16. Crawford, E. J., Christman, D., Atkins, H., Friedkin, M., and Wolf, A. P. (1978). *Intern. J. Nucl. Med. Biol. 5,* 61-69.
17. Stambrook, P. J., Sisken, J. F., and Ebert, J. D. (1973). *J. Cell. Physiol. 82,* 267-276.
18. Plagemann, P. G. W. and Richey, D. P. (1974). *Biochim. Biophys. Acta 344,* 263-305.
19. Quinlan, D. C. and Hochstadt, J. (1974). *Proc. Natl. Acad. Sci. U.S. 71,* 5000-5003.
20. Rozengurt, E., Mierzejewski, K., and Wigglesworth, N. (1978). *J. Cell. Physiol. 97,* 241-252.
21. Heidelberger, C. (1965). *Progr. Nucleic Acid Res. Mol. Biol. 4,* 1-50.
22. Mandel, H. G. (1969). *Progr. Mol. Subcell. Biol. 1,* 82-135.
23. Holstege, A., Herrmann, B., and Keppler, D. O. R. (1978). *FEBS Lett. 95,* 361-365.
24. Fowler, J. S., Finn, R. D., Lambrecht, R. M., and Wolf, A. P. (1973). *J. Nucl. Med. 14,* 63-64.
25. Thomas, L. B., Chirigos, M. A., Humphreys, S. R., and Goldin, A. (1962). *J. Nat. Cancer Inst. 28,* 1355-1389.
26. Spivak, J. L., Marmor, J., and Dickerman, H. W. (1972). *J. Lab. Clin. Med. 79,* 526-540.
27. Crawford, I., Kornberg, A., and Simms, E. S. (1957). *J. Biol. Chem. 226,* 1093-1101.
28. Schneider, W. C. (1945). *J. Biol. Chem. 161,* 293-303.
29. Schmidt, G. and Thannhauser, S. J. (1945). *J. Biol. Chem. 161,* 83-89.
30. Humphreys, S. R., Thomas, L. B., Chirigos, M. A., Goldin, A., Crawford, E. J., and Friedkin, M. (1962). *Nature (London) 195,* 453-455.
31. Schwartz, H. S., Sodergren, J. E., and Ambaye, R. Y. (1968). *Cancer Res. 28,* 192-197.
32. Harbers, H., Chaudhuri, N. K., and Heidelberger, C. (1959). *J. Biol. Chem. 234,* 1255-1262.
33. Levin, V. A. and Chadwick, M. (1972). *J. Nat. Cancer Inst. 49,* 1577-1584.
34. Chadwick, M. and Rogers, W. I. (1972). *Cancer Res. 32,* 1045-1056.
35. Shani, J., Wolf, W., Schlesinger, T., Atkins, H. L., Bradley-Moore, P. R., Casella, V., Fowler, J. S., Greenberg, D., Ido, T., Lambrecht, R. M., MacGregor, R.,

Mantescu, C., Neirinckx, R., Som, P., and Wolf, A. P. (1978). *Intern. J. Nucl. Med. Biol. 5,* 19-28.
36. Crawford, E. J., Friedkin, M., Fowler, J., Gallagher, B., MacGregor, R., Wolf, A., Wodinsky, I., and Goldin, A. (1978). *J. Nucl. Med. 19,* 702.
37. Wilkinson, D. S. and Crumley, J. (1976). *Cancer Res. 36,* 4032-4038.
38. Shiue, C.-Y., Fowler, J. S., MacGregor, R. R., and Wolf, A. P. (1979). *In* "Radiopharmaceuticals II : Proceedings of the Second International Symposium on Radiopharmaceuticals," Society of Nuclear Medicine, 259-264.
39. Houghton, P. J. and Taylor, D. M. (1977). *Brit. J. Cancer 35,* 68/77.
40. Crawford, E. J., Friedkin, M., Wolf, A. P., Fowler, J. S., Gallagher, B. M., Lambrecht, R. M., MacGregor, R. R., Shiue, C. Y., Wodinsky, I., and Goldin, A. (1982). *Adv. in Enzyme Regulation 20,* in press.

ADDENDUM (26 October 1981)

A paper describing the use of [^{18}F]5-fluorouridine and [^{18}F]5-fluorouridylate at Brookhaven National Laboratory is now in press (40). Proliferative changes in normal and tumor-bearing mice and rabbits are followed with the ^{18}F-containing probe.

FROM CO_2 TO ACETATE

Harland G. Wood

Department of Biochemistry
Case Western Reserve University
Cleveland, Ohio 44106

Martin Kamen caught my interest early in my scientific career and has held it ever since. He is a close friend, always exciting and, as a scientist, he's tops in my book. We have had some memorable times together. I remember one in particular. I was at a Cold Spring Harbor Symposium in 1948 on one of my first meetings with Martin. A fairly large group of us gathered one evening for some drinks and as time slipped by, we began singing songs. After a while, we had exhausted the old favorites and things began to slow down. In my childhood, our family had often gathered around the piano on a Sunday evening and dad and mother Wood would lead us in some Methodist hymns. Thus, I was quite familiar with these hymns of the so-called shouting Methodist. To pick things up, I started to sing a hymn and before long, we had a pretty good revival meeting underway. Again, however, with the passing of time, we were calling up hymns which were less familiar and I and the others fumbled for the words. On those occasions, to my utter amazement, Martin would chime in with the words loud and clear, taking over the lead. He knew every hymn in the book.

 ISBN 0-12-397580-8

It was then I really began to appreciate the talents of my young Jewish friend. His scope is amazing, not just in science, but in all walks of life from the batting averages of baseball players to politics and the arts.

Martin entered my scientific life with a bang in 1940. In 1936, Werkman and I (1) found while I was a graduate student at Iowa State University that when the propionic acid bacteria ferment glycol, there is a net uptake of CO_2 rather than production of CO_2. This was very exciting because it showed that the ability of living forms to utilize CO_2 was not confined to autotrophs but was a property of heterotrophs as well. The propionic acid bacteria are true heterotrophs requiring organic compounds such as glucose, amino acids, and vitamins for growth. At that time, little if anything was known about the mechanism of the utilization of CO_2 including that by the autotrophs and we were eager to learn how such synthesis took place. We obtained evidence that the reaction might be by addition of CO_2 to a 3-carbon compound because the net CO_2 utilized equalled the succinate formed in the fermentation of glycerol. Since we had been able to isolate pyruvate from fermentations by the propionic acid bacteria, we considered that the CO_2 was converted to succinate as follows (2):

$$CO_2 + CH_3{\cdot}CO{\cdot}COOH \rightleftharpoons HOOC{\cdot}CH_2{\cdot}CO{\cdot}COOH \qquad (1)$$

$$HOOC{\cdot}CH_2{\cdot}CO{\cdot}COOH + 4H \rightleftharpoons HOOC{\cdot}CH_2{\cdot}CH_2COOH + H_2O \qquad (2)$$

In 1938, there was little we could do to directly verify this postulate. The enzymological approach with bacteria was in its infancy. For example, the preparation of extracts from bacteria which were capable of glycolytic reactions by use of the Booth-Green mill (3) was described in 1938 and by grinding the cells with pulverized glass in 1940 (4). We were, therefore, elated when carbon isotopes became available and the way was opened to study this problem by tracer techniques.

A.O.C. Nier, a physicist at the University of Minnesota, had concentrated ^{13}C by thermal diffusion using methane and we initiated collaborative studies with him. We were soon able to show by mass analysis that $^{13}CO_2$ is converted exclusively to the carboxyl groups of succinate during the fermentation of glycerol by the propionic acid bacteria and we also obtained the same distribution in succinate formed by *E. coli* fermenting pyruvate (5).

These results were very gratifying, since they fit in well with our earlier proposals. Then, Kamen and his co-workers, really gave us a shock. We were amazed that Carson, Foster, Ruben, and Kamen (6) had found using $^{11}CO_2$ that the ^{11}C is converted to every position of propionic acid during the fermentation. They degraded the labeled propionate using alkaline permanganate and found that roughly three times as much label was recovered in oxalate, as in the carbonate. They recognized that this degradation might not be reliable and therefore degraded the propionate by dry distillation of the barium salt and obtained diethyl ketone and barium carbonate. If there was a uniform distribution of ^{11}C in the propionate the yield should be 83.4 and 16.6%, respectively. They found 88% of the ^{11}C in the ketone and 12% in the barium carbonate. In addition, radioactive CH_3I was obtained from the $(C_2H_5)_2CO$ by oxidation with I_2 in alkaline solution. They stated, "In view of these results it seems reasonable to conclude that all the carbons in the propionic acid are labelled, although by the latter method it is not possible to distinguish between the α and β carbons.... It seems that an appreciable fraction of the propionic acid has been synthesized from C^*O_2 rather than from a simple transformation of glycerol without degradation of the carbon skeleton. In other words, CO_2 is reduced to propionic acid with an organic compound (in this case glycerol) acting as the ultimate reducing agent. The reduction of CO_2 by all living cells (photosynthetic and non-photosynthetic) for synthesis of cellular constituents or excretory products is strongly supported by the present findings."

I was really angry with myself because we had the ^{13}C-propionic acid from our fermentations in the freezer and I had not even examined it for ^{13}C. The California group had captured the prize since it appeared they had shown that there was total synthesis of propionate from CO_2 by the propionic acid bacteria. Their results were very convincing but they had not determined how much of the tracer was in the α and β carbons of propionate, respectively. If there was total synthesis of propionate from CO_2, the tracer should be equal in both positions. We decided, therefore, to degrade our ^{13}C-propionate by a procedure which would give us this information. The degradation was accomplished as shown in reactions (3) to (6).

$$CH_3{\cdot}CH_2{\cdot}COOH \xrightarrow{\text{P + Br then } H_2O} CH_3{\cdot}CHBr{\cdot}COOH \tag{3}$$

$$CH_3{\cdot}CHBr{\cdot}COOH \xrightarrow{AgOH} CH_3{\cdot}CHOH{\cdot}COOH \tag{4}$$

$$CH_3{\cdot}CHOH{\cdot}COOH \xrightarrow{KMnO_4} CH_3CHO + CO_2 \tag{5}$$

$$CH_3CHO \xrightarrow{I_2 \text{ (alkali)}} CH_3I + HCOONa \tag{6}$$

We were simply dumbfounded when we found there was no excess ^{13}C in the iodoform or formate; it was all in the CO_2 from the carboxyl group. Shortly thereafter, Carson *et al.* (8) using $^{14}CO_2$, reported they were in error in the previous studies and that the ^{14}C was confined to the carboxyl group.

I am quite sure Martin Kamen had little to do with the chemical procedures and counting of these experiments. As I understand it, his job was to generate the ^{11}C (half-life ~20 min) and run with it to where the experiment was to be done. In fact, he was so radioactive himself, they could not trace carbon when he was around. They could only trace Martin. You can imagine that as a result of these events the name of Martin Kamen remained indelibly in my mind.

Of more interest to this story is the fact that Martin Kamen was not to be denied and did turn the tables. With H. A. Barker he showed that certain heterotrophic bacteria do, in fact, totally synthesize an organic compound from CO_2. They (9) showed that *Clostridium thermoaceticum* incorporates $^{14}CO_2$ into both carbons of acetate during fermentations of glucose. They proposed the following reaction to describe the overall process.

$$C_6H_{12}O_6 + 2H_2O \longrightarrow 2CH_3COOH + 2CO_2 + 8H \quad (7)$$

$$2CO_2 + 8H \longrightarrow CH_3COOH + 2H_2O \quad (8)$$

Sum: $$C_6H_{12}O_6 \longrightarrow 3CH_3COOH \quad (9)$$

Clostridium thermoaceticum was originally isolated by Fontaine *et al.* (10) who had shown that acetate is the only product of the fermentation. Thus, if glucose was fermented by the usual cleavage to two C_3 compounds and the C_3 compounds were cleaved to C_2 and C_1 compounds, clearly, the C_1 compounds were utilized and converted to acetate.

The proposal that *C. thermoaceticum* as well as some other organisms, which Barker and his co-workers had studied, could totally synthesize acetate from CO_2 was of immense interest to me because heterotrophs in general can only carry out carboxylation reactions with CO_2 but cannot totally synthesize multicarbon compounds from CO_2. But the question remained - do they really make acetate in which both carbons of a single molecule of acetate come from CO_2 or do they only make part of the acetate molecules with carbon from CO_2 in the carboxyl group and another part with the carbon in the methyl group? If they really totally synthesized acetate, it meant they had properties resembling, in some respects, that of an autotroph. If so, perhaps something could be learned about autotrophism by studying these bacteria. At that time, we knew nothing about the total synthesis of compounds from CO_2; the Calvin cycle of photosynthesis had not yet been discovered.

There was a way to answer this question and we undertook to find the answer (11). The procedure consisted of fermenting glucose in the presence of $^{13}CO_2$ and then isolating the acetate and, so to speak, weigh the molecules in a mass spectrometer. If molecules occurred with 2-g atoms greater mass ($^{13}CH_3{}^{13}COOH$) than the molecules that contained only ^{12}C ($^{12}CH_3{}^{12}COOH$), then acetate was being totally synthesized from CO_2. Also, the molecules which were made containing only one ^{13}C could be measured. By comparing the relative proportion of each mass, one could estimate the proportion of the molecules made by total synthesis from CO_2. What we found is shown in Table I. One-third of the acetate molecules were formed by total synthesis from CO_2 as predicted by Barker and Kamen. Another third were formed with CO_2 in only one position and a third were formed from the ^{12}C-carbon of the glucose. We found by degradation of the acetate that the carboxyl group contained twice as much excess ^{13}C as the methyl group, so it was clear that the molecules with a mass one greater than that of ^{12}C-acetate had their ^{13}C in the carboxyl group. We obtained evidence that these molecules probably arose by secondary exchange of $^{13}CO_2$ with unlabeled acetate formed from the glucose. Thus, we found that the proposal of H. A. Barker and Martin Kamen indeed was correct. We will see, however, that the total synthesis of acetate from CO_2 is indirect and that the CO_2 enters the carboxyl group of acetate via an indirect route through exchange of CO_2 with the carboxyl group of pyruvate.

We and others undertook to determine the mechanism by which this synthesis was accomplished. For some time, however, we were not successful. By pulse-labeling, which Calvin and his group used so successfully in delving into the pathway of photosynthesis, we made no progress. No intermediate compounds were detected that gave a clue to the mechanism. Formate and the carboxyl of lactate acquired ^{14}C from $^{14}CO_2$, but that information was of little help in deciphering the mechanism.

The first real breakthrough came when Poston, Kuratomi, and Stadtman (12) in 1964, found that $[^{14}CH_3\text{-Co}]B_{12}$ is converted to the methyl of group of acetate in the fermentation of pyruvate by extracts of *C. thermoaceticum*. Now, knowing what to look for, we (13) found by pulse-labeling with $^{14}CO_2$ that the Co methyl of the corrinoids of *C. thermoaceticum* became rapidly labeled. It then became clear why we had not found labeled intermediates in our previous experiments. The key intermediate compounds were enzyme bound and were not free in solution.

TABLE I

Types of Acetate Formed by *C. thermoacetICUM*

	Percentage formed from CO_2	
	CH_3	COOH
$^{13}CH_3-^{13}COOH$	33	33
$^{12}CH_3-^{13}COOH$	0	33
$^{12}CH_3-^{12}COOH$	0	0

The broad outline of the pathway of acetate synthesis began to take form (14). It was proposed as shown in Fig. 1 that the pyruvate is converted to acetate and CO_2 and the CO_2 acts as the oxidant being converted to formate, which in turn combines with tetrahydrofolate (THF) forming formyltetrahydrofolate. Through a series of steps, the formyltetrahydrofolate is converted to CH_3-THF. Then there is a transmethylation of the methyl to the cobalt of a corrinoid enzyme forming a methyl corrinoid, much as occurs in methionine synthesis. The methyl corrinoid is then carboxylated and converted to acetate. In this scheme, the intermediate product is shown as a carboxymethyl corrinoid but this portion of the sequence still remains in doubt (15). There is considerable evidence which supports the other portions of the scheme. It had previously been shown

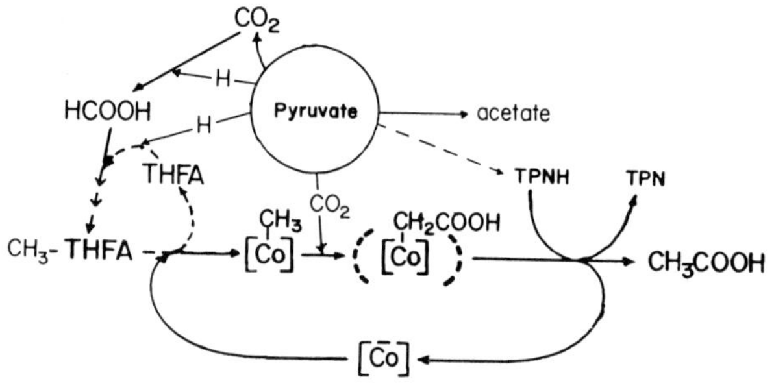

FIG. 1. Outline proposed for the total synthesis of acetate from CO_2 by C. thermoaceticum [Ljungdahl et al. (14)]. CH_3-THF is N-5-methyltetrahydrofolate and [Co] signifies a corrinoid enzyme.

that [^{14}C]formate is converted to the methyl group of acetate more rapidly than is $^{14}CO_2$. Furthermore, Parker *et al.* (16) in pulse-labeling experiments with $^{14}CO_2$, showed that the methyl of CH_3-THF acquires high activity in 15 seconds in accord with a precursor role. In addition, Ghambeer *et al.* (17) showed that $^{14}CH_3$-THF is converted to [2-^{14}C]CH_2COOH by extracts of *C. thermoaceticum* in fermentations of pyruvate. However, attempts to strengthen the evidence by isolation of the enzymes which convert [$^{14}CH_3$-Co]B_{12} or $^{14}CH_3$-THF to acetate were not successful, although Poston *et al.* (18) were able to obtain two protein fractions which together with ferredoxin catalyzed the formation of [2-^{14}C]CH_3COOH from [$^{14}CH_3$-Co]B_{12} during fermentations of pyruvate and Ljungdahl and his collaborators isolated most of the enzymes involved in the conversion of formate to CH_3-THF.

The reason for the failures became apparent, when it was discovered that pyruvate serves as the carboxylating agent of the methyl corrinoid enzyme and not CO_2. Previously it had been considered that pyruvate was required to generate ATP and to serve as a reducing agent for the reactions. Therefore, ATP and a TPNH- or DPNH-generating system were added and attempts were made to find an enzyme catalyzing the conversion of CO_2 and $^{14}CH_3$-THF to acetate in the absence of pyruvate.

During the course of the studies with crude extracts, we were surprised to find that the rate of synthesis of acetate from CH_3-THF and pyruvate was not effected when CO_2 was carefully excluded. This finding was not compatible with a mechanism involving CO_2 fixation. We, thus, were forced to the conclusion that the carboxylation of the CH_3-corrinoid was not with free CO_2 but it might be via transcarboxylation from pyruvate. We knew that there is a rapid exchange of CO_2 with the carboxyl of pyruvate. Thus, if the carboxylation were by transcarboxylation, CO_2 would occur in both positions of the acetate just as if the carboxylation was directly with CO_2. We, therefore, conducted kinetic tracer experiments using $^{14}CO_2$ and cold α-ketobutyrate and compared these results with the reverse situation, that is, with cold CO_2 and [1-^{14}C]α-ketobutyrate (19). If there was transcarboxylation, the reaction should occur as follows via the corrinoid enzyme:

$$CH_3{\cdot}CH_2\ CO{\cdot}^{14}COOH + CH_3THF \longrightarrow CH_3{\cdot}CH_2{\cdot}COOH + CH_3{\cdot}^{14}COOH + THF \qquad (10)$$

Thus, cold propionate and ^{14}C acetate would result. We used α-ketobutyrate rather than pyruvate so as to eliminate formation of cold acetate from pyruvate since we wished to determine the ^{14}C in the acetate as compared to the $^{14}CO_2$ and the ^{14}C of the carboxyl of the keto acid. The results are shown

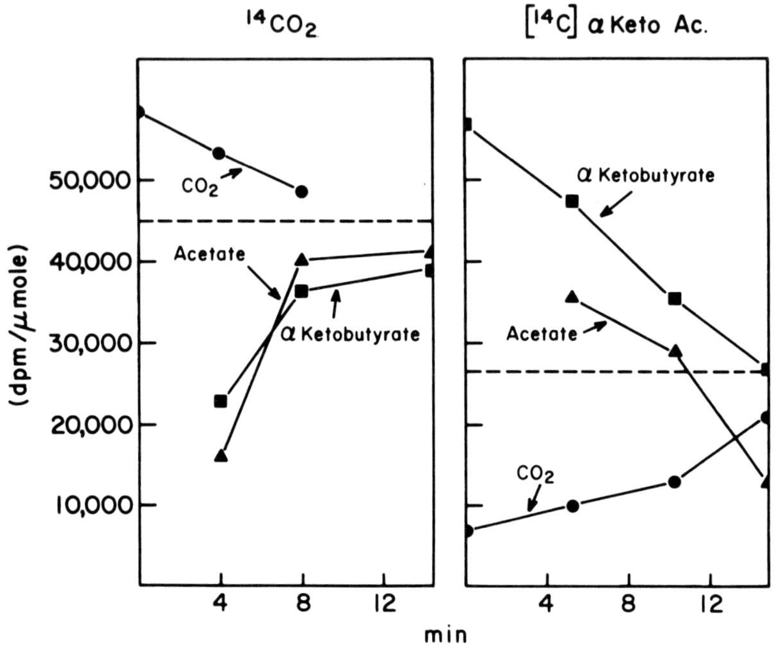

FIG. 2. Changes in the specific radioactivity of CO_2, the carboxyl of α-ketobutyrate, and the acetate formed in kinetic experiments with CH_3-THF in the presence of either $^{14}CO_2$ and cold α-ketobutyrate or [1-^{14}C]α-ketobutyrate and cold CO_2. From Schulman et al. (19).

in Fig. 2. The specific radioactivity of CO_2, of the α-ketobutyrate and of the acetate are plotted against time. It is seen that the specific activity of the acetate is correlated with that of the α-ketobutyrate rather than that of the CO_2. Thus, when $^{14}CO_2$ was used together with cold α-ketobutyrate (left panel of Fig. 2), with time, the cold α-ketobutyrate acquired radioactivity because of the exchange reaction with $^{14}CO_2$ and at the same time the specific activity of the $^{14}CO_2$ decreased because of this exchange. It is seen that the specific activity of the acetate increased parallel with that of the α-ketobutyrate. The dashed line is the theoretical specific radioactivity that the CO_2 and α-ketobutyrate would attain when they are at isotopic equilibrium.

The panel on the right of Fig. 2 shows the reverse situation, that is when [1-^{14}C]α-ketobutyrate and cold CO_2 were used. In this case, the specific radioactivity of the acetate started high and decreased as the specific activity of the α-ketobutyrate decreased due to exchange with the cold CO_2. Clearly, the specific activity of the acetate is directly related to the activity of carboxyl group of the α-ketobutyrate and not with that of the CO_2. Thus, it was clear that CO_2 is not the direct source of the carboxyl of the acetate.

In view of the above results, it was necessary to modify the scheme representing the fermentation (Fig. 3). An exchange reaction has been inserted in which a C_2 intermediate is formed (shown in brackets). This C_2 compound combines with $^{14}CO_2$ to give carboxyl labeled pyruvate. One molecule of this pyruvate then is converted to acetate and CO_2. The CO_2 is reduced to formate then to formyl-THF and to CH_3·THF. Via transmethylation, the methyl is transferred to the cobalt of the corrinoid enzyme. The second molecule of pyruvate is utilized in the transcarboxylation to the $^{14}CH_3$-corrinoid enzyme and via this transcarboxylation the second and third molecules of acetate are formed; the one contains ^{14}C from $^{14}CO_2$ in both of its carbons.

With this information in hand, we again undertook isolation of the corrinoid enzyme, which after all is the unique component

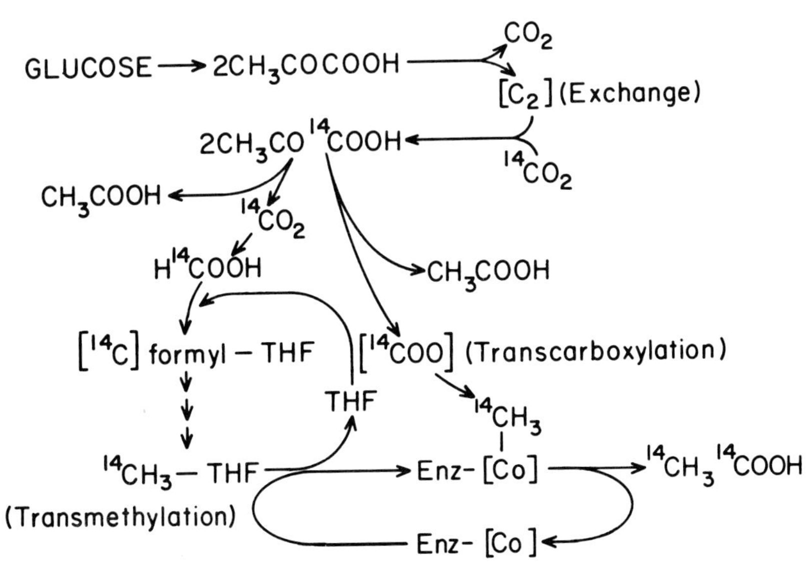

FIG. 3. Formation of doubly labeled acetate by exchange of $^{14}CO_2$ with the carboxyl of pyruvate and by transcarboxylation.

of the system. A major obstacle in the isolation of this corrinoid enzyme has been the lack of a specific assay for the enzyme. Even if pyruvate and $^{14}CH_3$-THF are used as substrates to simplify the system, several enzymes are involved and some means is required to supply these supplementary non-corrinoid enzymes so that the rate of the reaction can be made dependent on the concentration of the corrinoid enzyme which is to be assayed. This requirement has been accomplished by using a propylated crude extract of *C. thermoaceticum* to supply these supplementary enzymes. A specific property of corrinoid enzymes is that they react with propyl iodide by combination of the propyl group with the cobalt. Propyl iodide has been found to inhibit corrinoid enzymes involved in methionine (20, 21), methane (22), and acetate synthesis (17). It is characteristic of the propylated corrinoid enzymes that they are reactivated by light through cleavage of the propyl-Co bond and Ghambeer *et al.* (17) have demonstrated that propylated extracts of *C. thermoaceticum* are reactivated by light. We have established that the non-corrinoid enzymes are not inhibited by propyl iodide and that the propylated extract can serve as a source of the required supplementary enzymes. The assay system is as follows:

$$CH_2 \cdot CO \cdot COOH + {}^{14}CH_3THF \xrightarrow[\text{2. Cofactors; 3. Purified corrinoid enzyme}]{\text{1. Non-corrinoid enzymes of propylated extract}} {}^{14}CH_3COOH + CH_3COOH + THF \quad (10a)$$

Cofactors which have been found essential are thiamin pyrophosphate, *S*-adenosylmethionine, ATP, and ferrous iron. Addition of dithiothreitol is beneficial. Thiamin pyrophosphate may be required for the transcarboxylation from pyruvate and *S*-adenosylmethionine is known to be required as a component of transmethylation systems for methionine synthesis.

Using this assay system we have recently purified the corrinoid enzyme from *C. thermoaceticum* (23). The enzyme is sensitive to light, inactivated by oxidation in air, and, is in general, difficult to handle. My graduate student, Francine Welty, who isolated this enzyme was literally working in the dark a good part of the time. The enzyme was purified using a folate affinity column, DEAE Bio-Gel columns and by ultrafiltration.

The ultracentrifugation pattern of the enzyme before and after ultrafiltration is shown in Fig. 4. The enzyme is re-

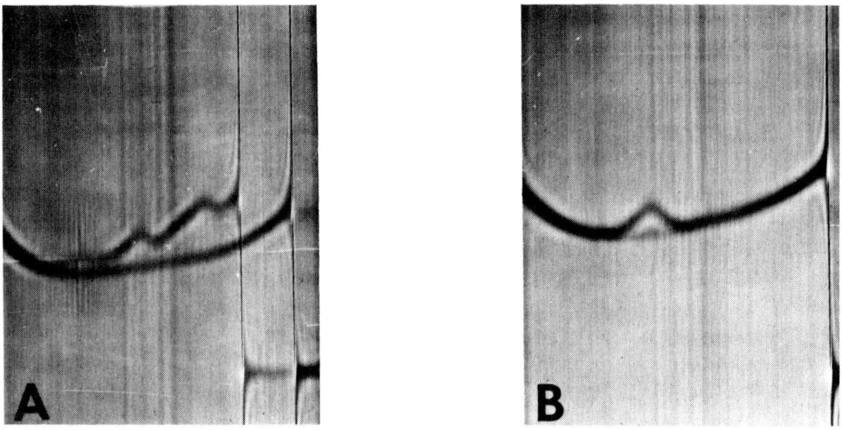

FIG. 4. Sedimentation velocity centrifugation of the cornoid enzyme. A. Preparation from the DEAE-Bio-Gel column (specific activity 54 nmol/min/mg protein). B. Preparation after filtration on an Amicon XM-300A membrane (specific activity 106 nmol/min/mg protein). Welty and Wood (23).

tained by the membrane and the smaller proteins are removed. After filtration, only a single peak was observed with a sedimentation coefficient of 10.5 S. By sedimentation equilibrium, the molecular weight of this protein was found to be 158,000.

On gel electrophoresis in sodium dodecyl sulfate (Fig. 5), only a single band was observed corresponding to a mol. wt. of 40,000. Thus, the enzyme appears to be a tetramer made up of four subunits.

The tetrameric structure was confirmed by electron microscopy (Fig. 6). Examples of the tetrameric structure are enclosed in squares. There is dissociation of the enzyme during preparation of the carbon grids giving rise to dimeric and monomeric structures. The distance from the center of one subunit to the center of an adjoining subunit of many of the tetramers was 45 - 50 Å which is in accord with a mol. wt. of 40,000 to 50,000 and agrees well with the mol. wt. determined by SDS electrophoresis.

The question remained whether the purified corrinoid protein catalyzed the transmethylation from CH_3-THF or was simply an acceptor of the methyl group. If the corrinoid enzyme catalyzed the transmethylation, the reaction shown in the legend of Fig. 7 should occur, i.e., the corrinoid enzyme should be methylated and in the absence of the transcarboxylase of the propylated extract, it would not be converted to acetate. The corrinoid enzyme was incubated with $^{14}CH_3$-THF with the

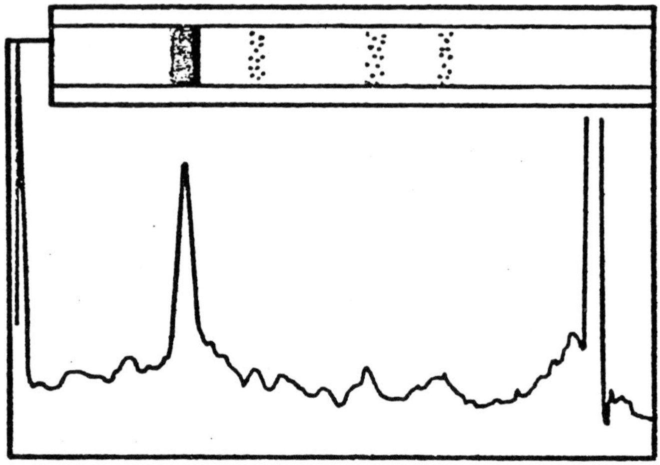

FIG. 5. Polyacrylamide gel electrophoresis of corrinoid enzyme with SDS and densitometry of the stained gel. Molecular weight of the subunit in SDS is ∿40,000 (23).

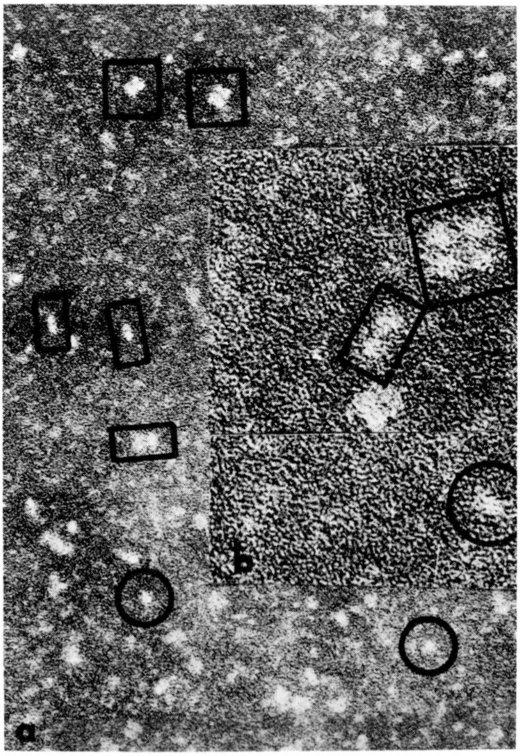

FIG. 6. Electron micrograph of the corrinoid methyltransferase. Magnification 300,000 X, inset 1,000,000 X. □ tetramer, ▭ dimer, ○ monomer. From Welty and Wood (23).

propylated extract omitted and the mixture was then passed through a Sephadex G-25 column to separate the protein from the excess CH_3-THF. The solid line of Fig. 7 is the reading at 280 mm and the dashed line, the radioactivity. The first peak represents the protein. There also is some 280 reading associated with the second peak, because the tetrahydrofolate also absorbs at 280. Clearly, there is radioactivity associated with the protein peak. Knowing the amount of protein and the specific radioactivity of the $^{14}CH_3$-THF, it was possible to calculate how many methyl groups were bound per mole of enzyme. The value was appropriately four, which agrees with the fact that the enzyme is a tetramer and probably contains four corrinoids.

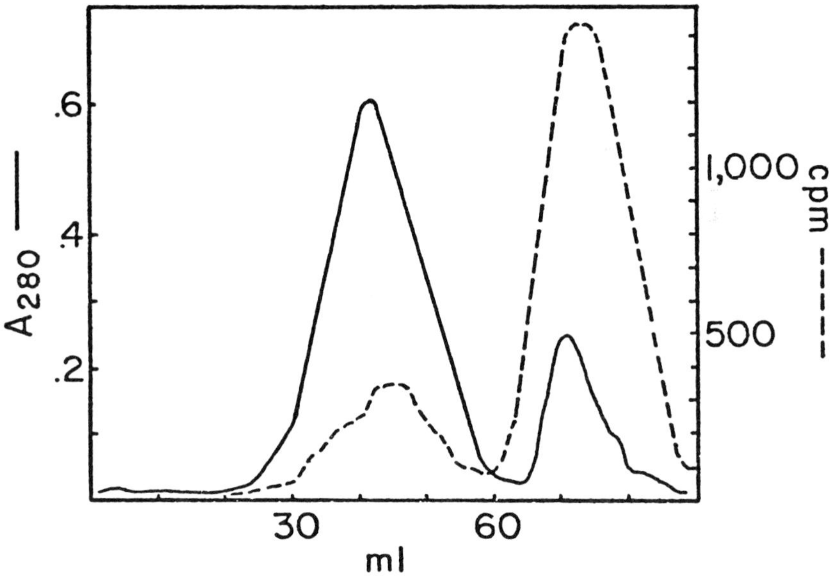

Fig. 7. Evidence that the corrinoid enzyme in the absence of the propylated extract catalyzes the reaction $^{14}CH_3$-THF + Enz-[Co] → THF + Enz-[$\overset{CH_3}{|}$Co]. See the text for description. From Welty and Wood (23).

This brings me to the end of the summary of our studies with *C. thermoaceticum*. The unique part of this pathway is the transmethylation and the transcarboxylation leading to formation of the acetate. With the purified corrinoid in hand, we can now turn our attention to purifying the transcarboxylase. The same assay system can be used. The purified corrinoid enzyme will be added in excess and the propylated extract will be fractionated in an attempt to purify and isolate the transcarboxylase. When this is accomplished, it should be possible to delve into the details of the chemistry of this remarkable organometallo reaction and determine whether it is a Grignard type reaction with formation of an acetyl-cobalt as an intermediate.

We started out by considering that this mechanism involved a total synthesis of acetate from CO_2 and might be an example of an autotrophic mechanism which has been retained by heterotrophs. This idea lost some of its strength when it was found that the carboxylation involves a transcarboxylation and does not occur by a direct fixation of CO_2. However, Ralph Wolfe of the University of Illinois has recently isolated a bacterium which utilizes CO_2 and hydrogen to form acetate. It also will grow heterotrophically with fructose and form $^{14}CH_3{}^{14}COOH$ in the presence of $^{14}CO_2$. Preliminary studies by Tanner *et al.* (24) indicate that this organism carries out the synthesis by the same reactions used by *C. thermoaceticum,* i.e., acetate is formed from pyruvate and CH_3-THF. It is attractive to consider that this organism might use reactions (11) to (16) rather than the Calvin cycle for its chemoautotrophism.

$$CO_2 + H_2 \longrightarrow HCOOH \quad (11)$$

$$HCOOH + THF \longrightarrow formyl\text{-}THF + H_2O \quad (12)$$

$$formyl\text{-}THF + 2H_2 \longrightarrow CH_3\text{-}THF + H_2O \quad (13)$$

$$CH_3\text{-}THF + [Co]\text{-}Enz \longrightarrow \overset{CH_3}{\overset{|}{[Co]}}\text{-}Enz + THF \quad (14)$$

$$H_2 + CO_2 + acetyl\text{-}CoA \longrightarrow pyruvate + CoA \quad (15)$$

$$Pyruvate + \overset{CH_3}{\overset{|}{[Co]}}\text{-}Enz \longrightarrow acetyl\text{-}CoA + acetate + [Co]\text{-}Enz \quad (16)$$

$$2CO_2 + 4H_2 \longrightarrow acetate + 2H_2O$$

I find it attractive to think that nature has developed more than one route for autotrophic growth.

Incidentally, Ralph Wolfe, named this newly isolated species which synthesizes acetate *Acetebacterium woodii*. I am indeed honored and flattered to join a long list of distinguished scientists whose names have been made immortal through bacterial nomenclature.

Finally, I would like to recognize my fine collaborators.

Kenneth Lentz showed that formate is a better precursor of the methyl group of acetate than is CO_2. He also was involved in the tracer studies with different types of labeled glucose, which showed that the sugar probably is metabolized by the Emlden-Meyerhof Pathway.

Lars Ljungdahl and I have had a long association, both as a graduate student and, subsequently, as a staff member in our department. Since moving to the University of Georgia, he has continued studies with *C. thermoaceticum* and *C. formicoaceticum*. The latter grows on fructose and has a similar mechanism for synthesis of acetate from CO_2. A large part of the work presented here was done in collaboration with him.

Eckart Irion's expert help led to the isolation of the corrinoids from the pulse labeling experiments and the proof that methyl corrinoids are intermediates in the synthesis of acetate.

Albert Sun was involved in the isolation of formyltetrahydrofolate synthetase from *C. thermoaceticum*.

Raj Ghambeer and Marvin Schulman demonstrated that extracts of *C. thermoaceticum* convert CH_3-THF to the methyl group of acetate and they also were involved in the transcarboxylation studies.

Donald Parker was responsible for demonstrating by pulse labeling that CH_3-THF is an intermediate in the synthesis of acetate. He also took part in the studies using trideuteromethyl-THF and showing that a carboxymethyl corrinoid probably is not intermediate in the synthesis of acetate.

Francine Welty accomplished the purification of the corrinoid enzyme.

Finally, Lew Waber, now at Johns Hopkins, has studied another organism, *Clostridium acidiurici*, which ferments purines and synthesizes doubly labeled acetate. He has shown that it produces acetate from CO_2 by an entirely different mechanism. This is a whole story in itself which time does not permit me to develop (25).

REFERENCES

1. Wood, H. G. and Werkman, C. H. (1936). *Biochem. J. 30*, 48.
2. Wood, H. G. and Werkman, C. H. (1938). *Biochem. J. 32*, 201.
3. Booth, J. H. and Green, D. E. (1938). *Biochem. J. 32*, 855.
4. Wiggert, W. P. and Werkman, C. H. (1940). *Iowa State College J. Sci. 14*, 179.
5. Wood, H. G., Werkman, C. H., Hemingway, A., and Nier, A. O. (1940). *J. Biol. Chem. 135*, 789; (1941) *139*, 377.
6. Carson, S. F., Foster, J. W., Ruben, S., and Kamen, M. D. (1940). *Science 92*.
7. Wood, H. G., Werkman, C. H., Hemingway, A., and Nier, A. O. (1941). *Proc. Soc. Exp. Biol. Med. 46*, 313.
8. Carson, S. F., Foster, J. W., Ruben, S., and Barker, H. A. (1941). *Proc. Natl. Acad. Sci. U.S. 27*, 229.
9. Barker, H. A. and Kamen, M. D. (1945). *Proc. Natl. Acad. Sci. U.S. 31*, 219.
10. Fontaine, F. E., Peterson, W. H., McCoy, E., and Johnson, M. J. (1942). *J. Bacteriol. 43*, 701.
11. Wood, H. G. (1952). *J. Biol. Chem. 194*, 905.
12. Poston, J. M., Kuratomi, K., and Stadtman, E. (1964). *Ann. N.Y. Acad. Sci. 112*, 804.
13. Ljungdahl, L., Irion, E., and Wood, H. G. (1965). *Biochem. 4*, 2771.
14. Ljungdahl, L., Irion, E., and Wood, H. G. (1966). *Fed. Proc. 25*, 1642.
15. Parker, D. J., Wood, H. G., Ghambeer, R. K., and Ljungdahl, L. G. (1972). *Biochemistry 11*, 3074.
16. Parker, D. J., Wu, Tsai-Feng, and Wood, H. G. (1971). *J. Bacteriol. 108*, 770.
17. Ghambeer, R. K., Wood, H. G., Schulman, M., and Ljungdahl, L. (1971). *Arch. Biochem. Biophys. 143*, 471.
18. Poston, J. M., Kuratomi, K., and Stadtman, E. R. (1966). *J. Biol. Chem. 241*, 4209.
19. Schulman, M., Parker, D., Ljungdahl, L. G., and Wood, H. G. (1972). *J. Bacteriol. 109*, 633.
20. Brot, N. and Weissbach, H. (1965). *J. Biol. Chem. 240*, 3064.
21. Taylor, R. T., Whitefield, C., and Weissbach, H. (1968). *Arch. Biochem. Biophys. 125*, 240.
22. Wood, J. M. and Wolfe, R. S. (1966). *Biochemistry 5*, 3598.
23. Welty, F. K. and Wood, H. G. (1978). *J. Biol. Chem.*, in press.

24. Tanner, R. S., Wolfe, R. S., and Ljungdahl, L. J. (1978). *J. Bacteriol. 109,* 626.
25. Waber, L. and Wood, H. G. (1979). *J. Bacteriol. 140,* 468.

ADDENDUM

Since this chapter was written, there have been exciting new developments in this area. Drake, H. L., Hu, S.-I., and Wood, H. G. (*J. Biol. Chem.,* in press) have isolated from *C. thermoaceticum* phosphotransacetylase, methyltransferase, pyruvate-ferredoxin oxidoreductase, and ferredoxin which, together with a fifth fraction containing 3 or 4 proteins, converts CH_3THF and pyruvate to acetylphosphate. When CoA is present in substrate amounts and phosphotransacetylase is omitted, the product is acetyl-CoA. One of the components of the fifth fraction is a nickel enzyme, carbon monoxide dehydrogenase (Drake, H. L., Hu, S.-I., and Wood, H. G. (1980) *J. Biol. Chem. 255,* 7174-7180). It has now been found (Hu, S.-I., Drake, H. L., and Wood, H. G., *J. Bacteriol.,* in press) that methyltransferase and the impure fraction containing CO dehydrogenase catalyzes the formation of acetyl-CoA from CO, CH_3THF, and CoA. CO dehydrogenase catalyzes the following reaction

$$CO + H_2O \longrightarrow [HCOOH] \longrightarrow CO_2 + 2H^+ + 2e$$

where [HCOOH] represents an unidentified formyl derivative. Formate per se is not an intermediate. The same formyl derivative apparently is formed from the carboxyl group of pyruvate by the pyruvate-ferredoxin oxidoreductase. Thus, either CO or pyruvate can serve as a source of the carboxyl in the formation of acetyl-CoA with CH_3THF.

These observations are of interest in relation to autotrophic bacteria which grow on CO as the sole source of carbon. Hu, Drake, and Wood have proposed that this type of autotrophism may occur by the following reactions [Eqs. (1a - 5a)]:

$$2\ CO + 2H_2O \longrightarrow 2[HCOOH] \quad (1a)$$

$$2\ CO + 2H_2O \longrightarrow 2\ CO_2 + 4H + 4e \quad (2a)$$

$$THF + [HCOOH] + 4H + 4e \longrightarrow CH_3THF + 2H_2O \quad (3a)$$

$$HSCoA + CH_3THF + [HCOOH] \longrightarrow CH_3COSCoA + THF + H_2O \quad (4a)$$

$$HSCoA + 4CO + H_2O \longrightarrow CH_3COSCoA + 2CO_2 \quad (5a)$$

For this growth, the conversion of CO to CO_2 is the source of the low-potential electron required for reductions and the conversion of acetyl-CoA to acetyl phosphate by phosphotransacetylase and then to acetate by acetate kinase the source of the ATP required for the anabolic processes.

$$CH_3COSCoA + H_3PO_4 \longrightarrow CH_3COPO_4H_2 + HSCoA \quad (6a)$$

$$CH_3COPO_4H_2 + ADP \longrightarrow CH_3COOH + ATP \quad (7a)$$

Accordingly, *C. thermoaceticum* should grow with CO as the source of carbon since it apparently has the necessary enzymatic equipment. I am informed by J. G. Zeikus (University of Wisconsin) that he has succeeded in obtaining such growth with *C. thermoaceticum*.

Autotrophic organisms which grow on CO_2 and H_2 may use the same pathway. In this case, H_2 via hydrogenase is the source of electrons and the formyl derivative is formed by formic dehydrogenase. In this way, reactions 15 and 16 (see text) are bypassed since a formyl derivative formed from CO_2 replaces pyruvate as the carboxylating agent forming acetyl-CoA from the CH_3THF via CH_3-[Co]-Enz (reaction 14, see text).

FROM GLYCINE TO HEME

David Shemin

Department of Biochemistry
and Molecular Biology
Northwestern University
Evanston, Illinois 60201

In "Reflections on Biochemistry" (1) written for a similar occasion as this one, Carl Cori notes the need to capture the flavor of a scientific period. Taking the cue from Carl Cori, I shall relate some of my early experiences with heme and the role of isotopes in the elaboration of its biosynthesis.

The use of stable isotopes for the study of cellular reaction was introduced by Rudolf Schoenheimer in 1935 (2). At that time very little was known about the biochemical reactions involved in the synthesis of amino acids and their utilization to other nitrogenous cellular constituents. The little that was known about the degradation or metabolism of some amino acids was usually deduced from the structural similarities of the amino acids to that of the possible product. The brilliant work of Garrod (3), who related biochemical reactions to genetics, was indeed helped by the obvious structural relationships of tyrosine to the aromatic compounds that accumulated in some individuals because of the lack of a particular functioning enzyme. The structural relationship of some amino acids to other compounds led to the suggestions, for example,

 ISBN 0-12-397580-8

that arginine is a direct precursor of guanidinoacetic acid; that proline, or pyrrolidine carboxylic acid or the indole ring of tryptophan are the likely precursors of the pyrrole structure of porphyrins; and that the imidazole group of histidine is a precursor of purines. The possible role of the simplest amino acid, glycine, in the elaboration of these more complicated molecules was not at all predicted or were concepts of biosynthesis sufficiently developed for one to suggest even on the theoretical grounds the possible role of glycine in the synthesis of these compounds. Furthermore, this lack of appreciation of the utilization of aliphatic amino acids for the synthesis of ring structure rose in part from the lack of consideration that, most likely, the more complicated structures in prebiotic times were made from small and readily available compounds.

Since Martin Kamen introduced ^{14}C, I would like to elaborate in some detail the role that isotopes played in the 1940's in the elucidation of the synthesis of heme and related compounds such as chlorophyll and later the corrin ring of vitamin B_{12}.

In the 1930's and early 1940's, the laboratory of Rudi Schoenheimer at Columbia University was especially concerned with the concept of the dynamic state of body constituents and especially on protein turnover (4). In 1943 I undertook, together with David Rittenberg, an elaborate study on "Some Interrelationships in General Nitrogen Metabolism in the Rat" (5). This involved the labeling of the protein of the rat with ^{15}N by feeding an amino acid which contained ^{15}N in its amino group. The study was then duely published in the *J. Biol. Chem.* and we then thought that it might be of some interest to study the turnover of plasma proteins in man. In order to do this experiment, I synthesized 66 g of glycine labeled with 35% excess ^{15}N. The cost of this amount of ^{15}N was about $1,000. Since one really did not know the effect of relatively large doses of the stable isotope of nitrogen and since we believed that the maximum incorporation into the protein could be achieved by the ingestion of the glycine in some continual manner, I ingested 1-g samples of glycine at hourly intervals for the next 66 hours. This experiment started in February, 1945.

At stated intervals, blood was withdrawn and after proper preparation the ^{15}N concentrations of different proteins were determined. Since the utilization of glycine for heme synthesis was not anticipated, there appeared little urgency to prepare hemin samples and at times the hemoglobin fractions were stored for work-up at some future time. The ^{15}N concentrations of the plasma proteins were determined and the data obtained were consistent with the concept of the dynamic state of the body constituents. The half-life of the total plasma proteins was estimated to be less than 5 days. The ^{15}N concentration of the heme appeared to reach a maximum value at about 20 days after the start of the experiment. Since the data were thus far not extraordinary and we expected a slow decline in the ^{15}N concentration of the heme, blood samples were taken after the 18th day at less frequent intervals. Samples were taken on the 28th, 38th, 60th, and 77th days and eventually hemin crystals were isolated from these samples. The hemin isolated from the blood sample taken on the 77th day was analyzed and, much to our surprise, the ^{15}N concentration was very similar to that taken on the 18th day; that is, it had much more ^{15}N than we anticipated. Since both Rittenberg and I were programmed to the concept that body constituents are in a continuous state of flux, we were indeed astonished to find that the ^{15}N analysis of the hemin sample taken on the 77th day was similar to the one taken on the 18th day. We then had a serious discussion in which he more than mildly suggested that I either mixed up the samples or somehow contaminated the sample and I, offended by such implication on my impeccable technique, suggested that his mass spectrometer was returning incorrect data. To settle the matter another hemin sample from the blood was isolated and the same result was found. We then realized that hemoglobin, in contrast to all other proteins previously examined by these techniques, was not in the dynamic state and that its ^{15}N concentration over a period of time would reflect the life span of the red blood cell.

A plot of the ^{15}N concentration of all the hemin samples against time in days gave us a curve shown in Fig. 1. An analysis of the curve demonstrated that the average life span of the human red blood cell is about 127 days, rather than 30 days that was quoted at that time (6). Furthermore, from the curve one can readily show that glycine is the nitrogenous precursor of heme or of porphyrins. For example, 10 days from the start of the experiment the ^{15}N concentration of the heme was 0.34 atom % excess. Since the average life time of the red blood cell is about 127 days, approximately only one-thirteenth of the cells were formed since the start of the experiment, and only these newly formed cells contained the isotopically labeled heme. Therefore, these newly formed red cells must have contained heme with an average ^{15}N concentration 13 times as high as the heme analyzed, or about 4.4 atom % ^{15}N excess. It was clear from these considerations that glycine, the amino acid fed, was the only compound that could have had as high an average ^{15}N concentration for the first 10 days. Other possible compounds (NH_3, glutamate, etc.) were eliminated by analysis of the ^{15}N concentration of the amino acids in the plasma proteins and that of the urea nitrogen. Further support for this conclusion was obtained subsequently by comparing the ^{15}N concentrations of the heme of rat hemo-

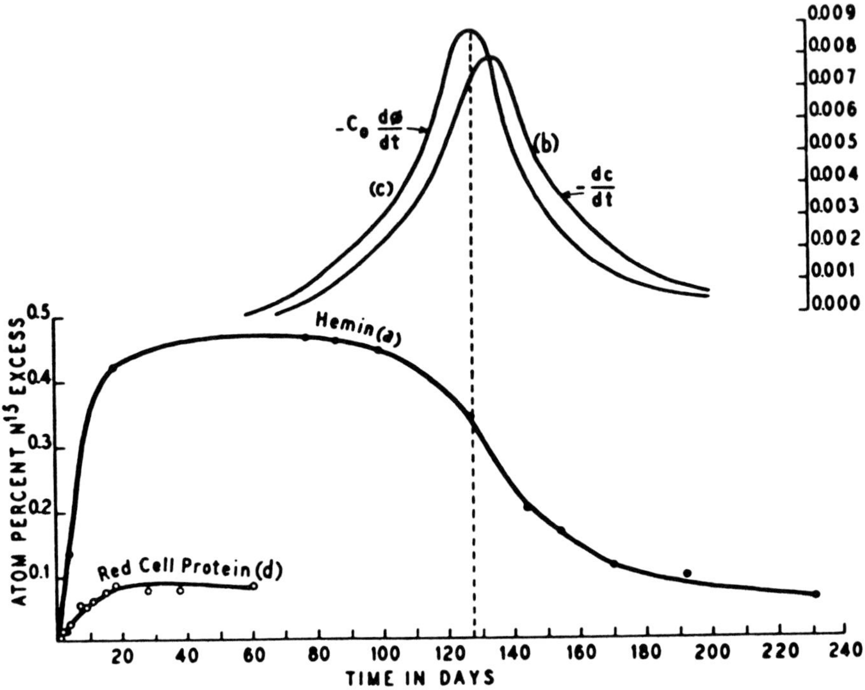

FIG. 1. *^{15}N Concentration in hemin after feeding ^{15}N-labeled glycine for 3 days.*

globin after the administration of a number of ^{15}N labeled amino acids. Unless a special transamination occurred, we assumed that the nitrogen atom of glycine was utilized along with its carbon atom. It is of interest to note that Hans Fischer in his last paper published (1944) found that formylacetone and glycine at room temperature at a pH 8 condensed to form a substance that gave an Ehrlich test for a pyrrole. In this paper, Fischer speculated about the possibility of glycine as the nitrogenous precursor of porphyrins (7).

In 1947 I was invited by Einar Hammersten, Professor of Biochemistry at the Karolinska Institute, to work in his laboratory since he was building a mass spectrometer for the analysis of stable isotopes. Among other things I did in Stockholm, we determined the life span of the red blood cell of the chicken, which was found to be about 28 days. On my return to Columbia University, I sought a simpler system, for obvious reasons, in order to continue our studies on heme synthesis. Probably my experience in Sweden led me to the suggestion that perhaps a nucleated red blood cell of a bird was capable of converting glycine to heme *in vitro*. Z. Dische was then working with pigeons. Upon consultation with him, it was suggested that I use a goose since he believed that goose red blood cells were more permeable to compounds than pigeon red cells. I ordered a goose and soon a beautiful white bird appeared from which I took wing blood and incubated it with ^{15}N-labeled glycine. Analysis of the heme demonstrated excellent incorporation of the ^{15}N from the glycine. In retrospect it was a long shot, for at that time few *in vitro* systems were called upon to carry out a long synthetic sequence of reactions. On repetition of the experiment, the same results were obtained. However, being rather cautious, I called upon the bird for a third time. As the bird was being delivered to my laboratory, my colleague and early mentor, G. L. Foster, an amateur ornithologist, said as he passed the bird in the hallway, "Dave, that is a nice looking duck." Regardless of the species, bird red cells can carry out the synthesis and thus we had a simpler system to explore for heme synthesis (8). I might add that we also demonstrated at that time globin synthesis with the duck erythrocyte, but did not pursue the mechanism of protein synthesis. Perhaps it was poor judgement, or it was a bit premature. Meanwhile, our new colleague, Irving London from the Department of Medicine, who was using our technique to measure the life span of the red cell of patients with blood dyscrasias, carried out a similar experiment with the blood of patients with sickle-cell anemia. Although the reticulocytosis was similar in all these disorders, only the blood of patients with sickle-cell anemia was capable of synthesizing heme from glycine. Apparently the sickle-cell blood contained more immature red cells than the blood of the patients with other anemias. This lack of correlation of heme synthesis with the reticulocyte count was borne out when London then introduced the use of immature reticulocytes of the rabbit as a system to study heme synthesis (9, 10).

We then turned our attention to the question of the utilization of the carbon atoms of glycine for porphyrin synthesis. It soon became apparent that not only must we determine which of the carbon atoms of glycine were utilized, but also the particular position in the porphyrin molecule in which the carbon atoms of glycine entered. Norman Radin, a graduate student, incubated the duck red cells with $^{15}NH_2$-$^{14}CH_2$-COOH and found that for each nitrogen atom, two α-carbon atoms of glycine were incorporated into the protoporphyrin; therefore, it appeared that eight carbon atoms of the porphyrin molecule arise from the α-carbon atom of glycine (11). In order to get some clue as to the mechanism of glycine utilization, Jonathan B. Wittenberg, also a graduate student, set out to degrade the heme in such a manner that one can isolate each carbon atom unequivocally from a particular position in the porphyrin.

On degradation of protoporphyrin synthesized from

$[2-^{14}C]$glycine, it was found that, indeed, eight carbon atoms are derived from the α-carbon atom of glycine and the positions in the molecule were located; four bridge carbon atoms and one carbon atom in each of the four rings. It was found that the carbon atom in the pyrrole rings, derived from the α-carbon atom of glycine were in the α position in the pyrrole ring underneath the vinyl and propionic acid side chains. We then suggested that a common precursor pyrrole is utilized for all four pyrrole rings and that the vinyl groups arose from the propionic acid side chains by dehydrogenation and decarboxylation (12).

While we were doing these experiments, a paper appeared from M. Kamen's laboratory that demonstrated to our, and I believe to their, surprise that the carboxyl group of glycine is not found in porphyrin synthesized from $[1-^{14}C]$glycine (13). This was confirmed by Radin in our laboratory (11). This negative experiment was of significant importance for it allowed me to formulate the important intermediate for porphyrin synthesis, for the question I raised concerned the mechanism by which the carboxyl carbon atom of glycine is separated from its α-carbon atom. This will be discussed a bit later.

As the origin of eight carbon atoms of the porphyrin were now known, the source of the remaining twenty-six carbon atoms remained to be determined. Just prior to our finding that glycine was the nitrogeneous source of porphyrins, Konrad Bloch working on the other side of the bench isolated hemin from a rat who had ingested deuterioacetic acid. He found deuterium in the hemin and although none of the carbon atoms of the pyrrole ring are bonded to hydrogen it appeared that acetic acid may be involved in porphyrin synthesis (14). Norman Radin incubated the duck red blood cells with $[1-^{14}C]$acetate and $[2-^{14}C]$acetate and found that probably both carbon atoms are indeed utilized for porphyrin formation and that the methyl-labeled acetate produced hemin six times more radioactive than an equivalent amount of carboxyl-labeled acetate of equal radioactivity (15). It seemed to us at that time that probably all the carbon atoms of the porphyrin are derived by the α-carbon atom of glycine and acetate. In order to determine the origin of each of the carbon atoms and to get some inkling of the mechanisms that are involved, Jonathan Wittenberg bled a number of ducks, pooled the blood, and divided it into two equal volumes. To one half, we added $[2-^{14}C]$acetate and ^{15}N-labeled glycine and to the other half $[1-^{14}C]$acetate and an equal amount of ^{15}N-labeled glycine. The labeled glycine was added in order to measure the amount of synthesis in each flask and, thus, we could relate the data obtained with methyl-labeled acetate to those obtained with carboxyl-labeled acetate. It was found on subsequent degradation that indeed all the remaining twenty-six carbon atoms are derived from acetate.

The distribution of the ^{14}C activities among the remaining twenty-six carbon atoms of the protoporphyrin allowed us to draw the following conclusions: (1) The acetic acid is utilized by its conversion to a four-carbon atom unsymmetrical compound; (2) that each side of the pyrrole is derived from the same four-carbon atom compound; (3) that the common precursor pyrrole contained acetic and propionic side chains in its β position; and (4) that the methyl groups in protoporphyrin arise by decarboxylation of the acetic acid side chains and that the vinyl groups arise by dehydrogenation and decarboxylation of the propionic acid side chains. These conclusions were arrived at mainly from the finding that comparable carbon atoms in each pyrrole ring had the same ^{14}C activity in each of the experiments (see Fig. 2) (16).

It appeared that some mechanism concerned with the metabolism of acetate should account for our findings which involved an asymmetric four-carbon atom compound arising from acetate and had the particular distribution of ^{14}C from the methyl and carboxyl carbon atoms. A likely mechanism seemed to be the reactions involved in the tricarboxylic acid cycle, although relative quantitative calculations concerning the distribution of the carbon atoms of acetate in intermediates of the tricarboxylic acid cycle had not previously been done.

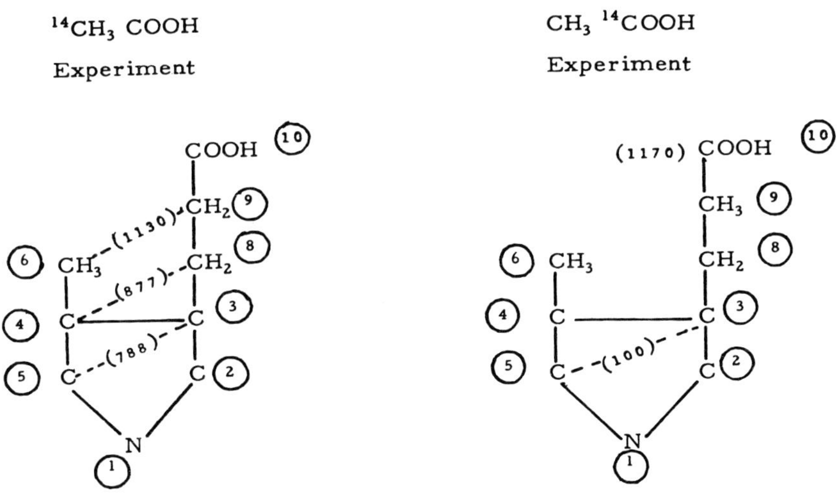

FIG. 2. Labeling patterns found in each of the experiments. The ^{14}C activities are given in parentheses. The pyrrole unit represented contains a carboxyl group which is found only in two rings of protoporphyrin.   are carbon atom numberings.

On the assumption that extraneous metabolic reactions could not obscure the theoretical distribution of the carbon atoms of acetate in members of the tricarboxylic acid cycle, we calculated the theoretical labeling pattern in the four-carbon atom unsymmetrical compound that would result from the introduction of acetate into the cycle after each complete turn of the cycle. If one ignores endogeneous dilution, one can calculate the relative distribution of ^{14}C activity in the four-carbon atom compound that would theoretically be found after each turn of the cycle after the entrance of the acetic acid. Let us first examine the result one would obtain from methyl-labeled acetate. Assuming a relative ^{14}C activity of 100 in the methyl group of the acetic acid, the α-ketoglutarate acid formed in the first turn of the cycle should have the relative ^{14}C distribution shown in Table 1. On formation of succinate, a symmetrical compound, the ^{14}C activity of the γ-carbon atom of α-ketoglutarate, 100 cpm, would be distributed between the two methylene carbon atoms and thus their relative activities would be 50 and 50 cpm. The newly formed oxaloacetate would, therefore, have the same distribution of radioactivity. The condensation of this newly formed oxaloacetate with more of the methyl-labeled acetate would give rise to α-ketoglutarate with the distribution of radioactivity shown after the second cycle. After each turn of the cycle, the radioactivities of the α and β carbon atoms would increase by the same increment and approach that of the methyl group of the acetate or that of the γ-carbon atom of the α-ketoglutarate. Therefore, a four-carbon unsymmetrical compound formed from α-ketoglutarate would contain three adjacent radioactive carbon atoms; the carbon atom adjacent to the carboxyl group should be most radioactive, whereas the other two carbon atoms should be somewhat less radioactive, but their ^{14}C activities should be equal to each other. Let us compare the ^{14}C distribution theoretically predicted after a finite number of cycles with that experimentally found. The four-carbon atom compound (Fig. 2) presumably utilized for pyrrole formation synthesized from methyl-labeled acetate does indeed have three adjacent carbon atoms arising from the methyl group of acetate; the one next to the carboxyl group has the highest activity (1130 cpm) and the other two carbon atoms, although labeled, are not quite equally labeled (877 and 788 cpm). This slight inequality of activities may suggest that the tricarboxylic acid cycle is not functioning as postulated theoretically. However, a valid correlation can be made, and if made these carbon atoms do indeed have the same radioactivity or arise equally from the methyl group of acetate. It can be seen from Fig. 2 that carbon atoms-5 and -3 are, in part, also derived from the carboxyl group of acetate. In the experiment in which carboxyl-labeled acetate was the substrate, the carboxyl carbon atom labeled these same carbon atoms to the

TABLE 1. Relative Distribution of ^{14}C Activity in Carbon Atoms of α-Ketoglutaric Acid Resulting from Utilization of ^{14}C-Labeled Acetate in Tricarboxylic Acid Cycle[a]

α-Ketoglutaric acid	*From ^{14}C-Methyl-labeled acetate (activity of methyl group = 100 c.p.m.)*				*From ^{14}C-Carboxyl-labeled acetate (activity of carboxyl group = 100 c.p.m.)*			
	No. of Cycles in Tricarboxylic Acid Cycle							
	1st	*2nd*	*3rd*	∞	*1st*	*2nd*	*3rd*	∞
COOH	0	0	0	0	100	100	100	100
CH_2	100	100	100	100	0	0	0	0
CH_2	0	50	75	100	0	0	0	0
C = O	0	50	75	100	0	0	0	0
COOH	0	0	25	50	0	50	50	50

[a] *The results are expressed in counts per minute.*

extent of 100 counts. Therefore, in the methyl-labeled acetate experiment carbon atoms 5 and 3 were diluted by the unlabeled carboxyl group by 100 counts because of the entrance of the unlabeled carboxyl carbon atom in these positions. The correction of 100 counts to carbon atoms 5 and 3 raises their value to 888, a figure close to 877. It appeared, therefore, that acetate is converted to the four-carbon atom compoundly utilized for pyrrole formation, via the tricarboxylic acid cycle since the relative ^{14}C activities experimentally found were in excellent agrement with those theoretically predicted.

The ^{14}C distribution pattern of the four-carbon atom compound that may arise from carboxyl-labeled acetate also can be postulated. It can be seen from Table 1 that the four-carbon atom compound formed from the α-ketogluterate in the carboxyl-labeled acetate experiment would contain radioactivity only in its carboxyl group after one or an infinite number of cycles. It can be seen in Fig. 2 that the labeling pattern of the four-carbon atom compound formed in the carboxyl-labeled acetate experiment is similar to that postulated in that most of the ^{14}C activity is in the carboxyl group. However, there was about 10% of the activity of the carboxyl group in the other terminal carbon atom. This is similar to the discrepancy mentioned above in the methyl-labeled acetate experiment. In order to explain these small nagging differences between the predicted ^{14}C distribution and found ^{14}C distribution, we postulated that perhaps the four-carbon atom compounds arose mostly from α-ketoglutarate and to a lesser extent from a symmetrical compound like succinic acid. In 1951 we suggested that the four-carbon asymmetric compound was most likely a succinyl co-enzyme compound, probably similar to acetyl-CoA. At that time, succinyl-CoA was not known to exist or was its formation from succinate suggested. However, if the four-carbon atom did arise from both α-ketoglutarate and succinate, the ^{14}C disetribution found would be readily explainable. This postulated explanation is illustrated in Fig. 3.

It appeared at this point that the postulated relationship of the tricarboxylic acid cycle and porphyrin formation could be experimentally tested with the use of labeled succinate. This relationship is shown in Fig. 4. This scheme suggests the existence of a reaction that had not been described, namely reaction C, and, furthermore, suggested that aspects of the tricarboxylic acid cycle could be studied by investigating porphyrin formation. The main point of this scheme is that the succinyl intermediate used for pyrrole synthesis can be derived from both α-ketoglutarate and succinate. This can be experimentally tested by two ways; (1) by determining the labeling pattern in protoporphyrin synthesized from carboxyl-labeled succinate and from methylene-labeled succinate; and (2) by studying the degree of utilization of the differently labeled

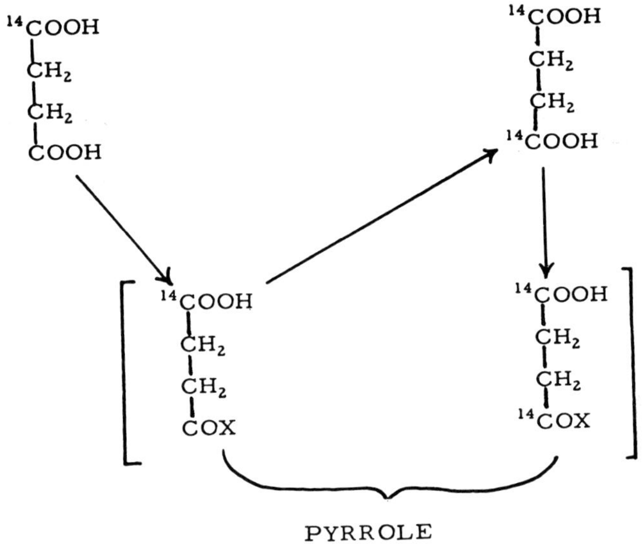

FIG. 3. The formation of the 4-carbon atom unsymmetrical compound from both α-ketoglutarate and from succinate. The ^{14}C distribution would suggest that in this biological system 90% of the succinyl derivative arises from α-ketoglutarate.

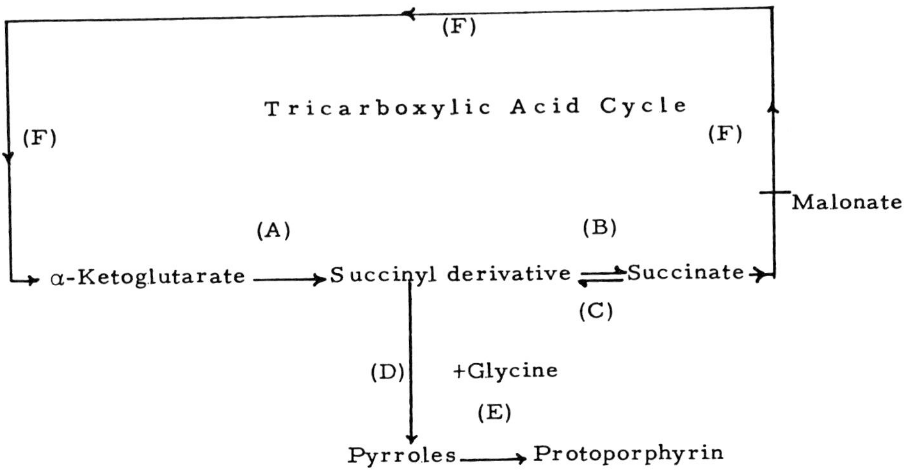

FIG. 4. The relationship of the citric acid cycle and protoporphyrin formation.

succinates for porphyrin synthesis in the presence or absence of malonate. Theoretically, ^{14}C-carboxyl-labeled succinate cannot give rise to ^{14}C-labeled protoporphyrin via reaction F, but only via reaction C. On entering the tricarboxylic acid cycle, i.e., in the oxidative direction of the cycle (reaction F), carboxyl-labeled succinate can give rise only to α-carboxyl labeled α-ketoglutarate. The resulting succinyl intermediate arising by oxidative decarboxylation and utilized for porphyrin formation would, therefore, contain no ^{14}C. However, if reaction C does occur, carboxyl-labeled succinate should produce labeled protoporphyrin. If this conclusion were correct, then carboxyl-labeled succinate should produce labeled protoporphyrin only by reaction C and, therefore, malonate, which blocks reaction F, should have no influence on the ^{14}C activity found in protoporphyrin synthesized from carboxyl-labeled succinate and labeling of porphyrin from carboxyl-labeled succinate in the presence of malonate would be evidence of the occurrence of reaction C.

On the other hand, methylene-labeled succinate, in contrast to carboxyl-labeled succinate, should produce labeled protoporphyrin via two pathways: (1) reaction C and (2) the oxidative direction of the tricarboxylic acid cycle (reaction F). Malonate should, in this case, have a marked inhibitory effect on the utilization of methylene-labeled succinate for porphyrin synthesis, and any ^{14}C activity in protoporphyrin made from methylene-labeled succinate in the presence of malonate should be the result via reaction C.

The ^{14}C activity of the hemin samples, obtained by incubating equal amounts of ^{14}C-carboxyl-labeled and ^{14}C-methylene-labeled succinate, with and without malonate, in duck blood preparations, were determined. It was found that the ^{14}C activities of the hemin samples made from methylene-labeled succinate were much higher than that made from carboxyl-labeled succinate and, whereas, malonate had a large inhibitory effect on the utilization of methylene-labeled succinate, it had essentially no inhibitory effect on the utilization of succinate labeled in its carboxyl groups. These results were in complete agreement with the postulation that reactions A and C exist and that methylene-labeled succinate forms labeled protoporphyrin by reactions C and F, whereas, carboxyl-labeled succinate via reaction C only (17).

Having established that two molecules of a succinyl derivative, two α-carbon atoms of glycine and one nitrogen atom of glycine are all involved in the formation of a pyrrole, we next considered possible mechanisms. As previously mentioned, the negative experiment by Grinstein, Kamen, and Moore established that the carboxyl group of glycine is not utilized for porphyrin synthesis. This was of great import. Since the carboxyl group of glycine must be separated from its α-carbon atom, some mechanism must exist for this separation and this mechanism would perhaps give us a clue as to the mechanism of pyrrole formation. Examining the possibilities one day with Charlotte Russell, it became apparent that if the succinyl derivative condensed onto the α-carbon atom of glycine to form a carbon-carbon bond, a β-keto acid would be formed and thus could readily decarboxylate and give a reasonable mechanism for the loss of the carboxyl group of glycine. Furthermore, the resulting product would be an amino ketone which were known to readily undergo a Knorr type of condensation to yield pyrroles. Formation of this pyrrole from two molecules of this amino ketone, δ-aminolevulinic acid, would give rise to the precursor pyrrole which we had previously postulated and it would contain the succinate and glycine carbon atoms in all the positions we previously had established. To this end, we immediately developed methods to synthesize this new compound, δ-aminolevulinic acid, with ^{15}N and ^{14}C and demonstrated it to be the source of all the atoms of protoporphyrin and later that of chlorophyll and the corrin ring of vitamin B_{12} (18, 19). (See Fig. 5.)

While our experiments with δ-aminolevulinic acid were in progress, Westfall in England isolated for the first time the compound called porphobilinogen by Waldenstrom, from the urine of patients with acute intermittent porphyria (20). The structure of porphobilinogen was then shortly established by Cookson and Rimington, and our postulated structure of the precursor pyrrole and porphobilinogen were one and the same (18, 21). It was later shown in several laboratories that 4 molecules of porphobilinogen cyclyze to form the tetracyclic structure of porphyrin and related compounds.

ACKNOWLEDGMENT

Over the years, this research was supported by grants from the American Cancer Society on the recommendation of the Committee on Growth of the National Research Council, from the Rockefeller Foundation, from the National Institutes of Health, United States Public Health Service, and from the National Science Foundation.

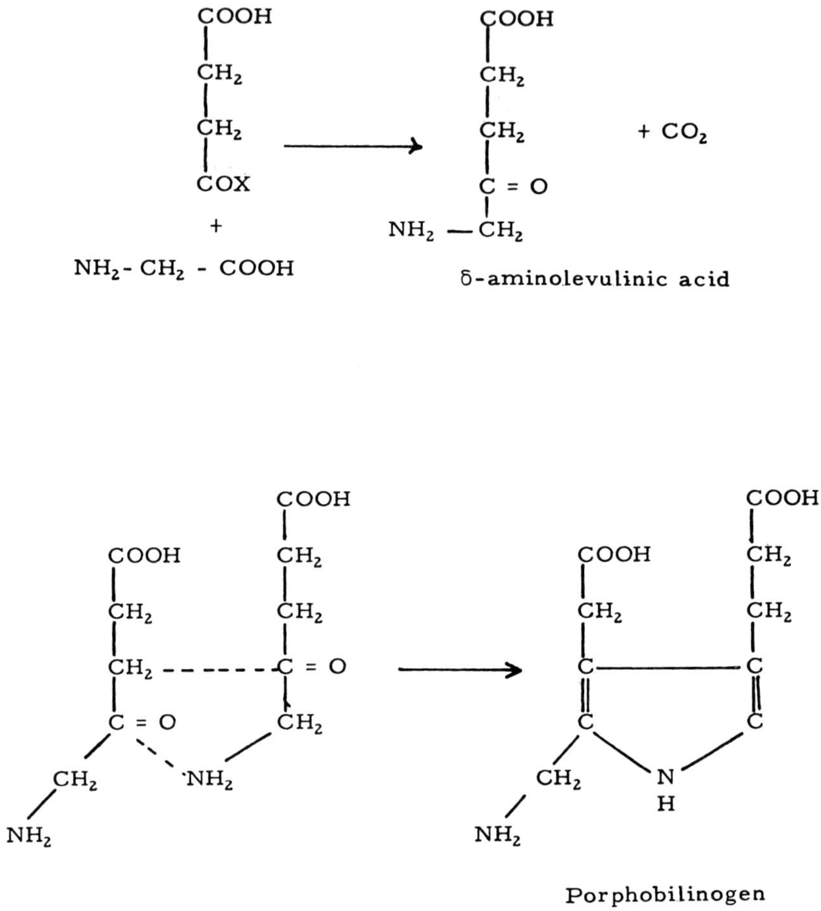

FIG. 5. The formation of δ-aminolevulinic acid and its conversion to porphobilinogen.

REFERENCES

1. Cori, Carl F. (1976). "Reflections on Biochemistry," In Honour of Severo Ochoa, Pergamon Press, New York.
2. Schoenheimer, R. and Rittenberg, D. (1935). *Science 82*, 156.
3. Garrod, A. E. (1923). "Inborn Errors of Metabolism," 2nd ed. Henry Frowde and Hodder and Stoughton Pub.,

London.
4. Schoenheimer, R. (1941). "The Dynamic State of Body Constituents." Harvard Press.
5. Shemin, D. and Rittenberg, D. (1944). *J. Biol. Chem.* *153*, 401.
6. Shemin, D. and Rittenberg, D. (1946). *J. Biol. Chem.* *166*, 627.
7. Fischer, H. and Fink, E. (1944). *Z. Physiol. Chem.* *280*, 123.
8. Shemin, D., London, I. M., and Rittenberg, D. (1948). *J. Biol. Chem.* *173*, 799; (1950). *J. Biol. Chem.* *183*, 757.
9. London, I. M., Shemin, D., and Rittenberg, D. (1948). *J. Biol. Chem.* *173*, 797.
10. London, I. M., Shemin, D., and Rittenberg, D. (1950). *J. Biol. Chem.* *183*, 749.
11. Radin, N. S., Rittenberg, D., and Shemin, D. (1950). *J. Biol. Chem.* *184*, 745.
12. Wittenberg, J. and Shemin, D. (1950). *J. Biol. Chem.* *185*, 103.
13. Grinstein, M., Kamen, M., and Moore, C. U. (1948). *J. Biol. Chem.* *174*, 767.
14. Bloch, K. and Rittenberg, D. (1945). *J. Biol. Chem.* *159*, 45.
15. Radin, N. S., Rittenberg, D., and Shemin, D. (1950). *J. Biol. Chem.* *184*, 755.
16. Shemin, D. and Wittenberg, J. (1951). *J. Biol. Chem.* *192*, 315.
17. Shemin, D. and Kumin, S. (1952). *J. Biol. Chem.* *198*, 827.
18. Shemin, D. and Russell, C. S. (1953). *J. Am. Chem. Soc.* *75*, 4873.
19. Shemin, D., Russell, C. S., and Abromski, T. (1955). *J. Biol. Chem.* *215*, 613.
20. Westfall, R. G. (1953). *Nature (London)* *170*, 614.
21. Cookson, G. H. and Rimington, C. (1953). *Nature (London)* *171*, 875.

Cytochromes: Chemistry and Structure

CHEMICAL EVOLUTION AND CHEMICAL INNOVATION

Fujio Egami

Mitsubishi-Kasei Institute of Life Sciences
Machida-shi, Tokyo 194, Japan

Admirable is the organism.
And man, a wonderful thing.
Nature created the organism
    with elements less than twenty.*
From nature came forth man
    by elements less than thirty.
What will the chemist create
    with about a hundred?
Shall something wonderful
    for the welfare of man.
Herein is human intelligence

**Elements essential for primitive bacteria: H, O, C, N, P, S, K, Mg, Mo, Zn, Fe, Mn, Se, Co - total 14. Even with Na, Cl, Ca, Cu - total 18.*

Essential elements for man: H, O, C, N, Ca, P, S, Na, K, Cl, Mg, I, Si, Mo, Zn, Fe, Cu, Cr, Mn, Se, Co - total 21. Even with F, V, Al, Ni, Sn - total 26.

 ISBN 0-12-397580-8

Protein is a wonderful substance
Enzymes, antibodies and many others,
All these are proteins.
From Nature came the protein
with the twenty amino acids.
What will the chemist create
with innumerable monomers?
They shall create great polymers
for the lives of human beings.
Herein is the intelligence
of the chemist of the future

Translated from Japanese (1)

Contrary to traditional chemical industry, a chemical industry is now required which will be in harmony with nature, life, and human welfare requiring less energy consumption. Thus, chemical processes so far studied by biochemists are attracting chemists and chemical engineers undertaking the innovation of chemistry and chemical industry.

Chemistry and chemical industry of the future should combine the advantages of both the traditional chemical processes and those of the chemical processes of the living systems.

What are the characteristics of traditional chemical processes and the chemical processes in living systems? In a word, the former is an energy chemistry and the latter an information chemistry or a system chemistry at the molecular level. Chemical reactions in the former are performed under conditions such as high temperature and high pressure intolerable to living organisms. In the latter the production of biocatalysts (enzymes) is prepared by genetic information, and regulated by the environmental information; enzymes thus produced are highly specific and conveniently localized depending upon the information proper to the molecules themselves. Good examples are the multienzyme systems such as the fatty acid synthesizing system or pyruvate oxidase system, and the multibiopolymer system such as the protein synthesizing system. Thus, chemical reactions proceed under mild conditions suitable to the organism. It is indeed a system chemistry at the molecular level (molecular system chemistry).

The differences between the two processes are shown in Table 1.

What brought about these differences? The organic chemistry in the Berzelius's epoch (1820's) was the chemistry of living systems or the present biochemistry. He defined organic chemistry as follows (2): organic chemistry is that branch of chemistry concerned with the study of the constituents of living organisms and the chemical changes taking place in them. Since then, organic chemistry has departed more and more from "chemis-

TABLE 1

	Traditional chem. reaction	*Chem. reaction of living system*
Condition	Severe	Mild
Energy consumption	Large	Small
Reaction system	Simple (Complex system is impossible because of the absence of information)	Very complex (ordered) (Possible because of the presence of information)
Product	Nonbiological - retained in the environments for a long time	Biological - easily biodegraded
Catalyst	Mostly toxic	Scarcely toxic
As a system	Primitive, gigantic system in chem. industry	Very advanced system at the molecular level
Species of materials and reactions	Unlimited	Very limited
Summary	Energy chemistry difficult to be in harmony with nature and life Unlimited reactions can be considered.	System chemistry (molecular level) in harmony with nature and life Small energy consumption depending upon clean solar energy Limited to reactions of living systems

try of living systems." In 1860 it was already defined as the "chemistry of carbon compounds" (3). Why this separation? Specialists in the history of chemistry or specialized organic chemists certainly have something to say about this. I, as a biochemist, believe that it came about in part from a general characteristic of most scientists. They don't like to be bogged down, and prefer to carry out research which will lead to clear-cut conclusions by easy methods. Thus, "organic chemists" have prefered to study easily crystallizable substances from among the constituents of organisms in the beginning and the middle of the 19th century. They extracted natural substances by organic solvents and obtained crystals after evaporation. These substances, generally insoluble in water, were dissolved in organic solvents and their chemical properties studied. Thus the main current of "organic chemistry" became something different from the chemistry of living systems containing about 70% water.

The cell contains hundreds of substances in a minute space. It certainly seems too complex to chemists. Chemists like simple systems, from which clear-cut conclusions may be deduced. Such a typical system is A + B → Products using a solvent and sometimes a catalyst. No chemist would mix even ten substances in a flask to make them react since this would certainly lead to an unexplainable disorder. Chemists have thus chosen easy ways to accelerate simple reactions under extreme conditions (high temperature, high pressure, etc.). Hence, traditional chemistry became "energy chemistry." The synthesis of complex organic compounds consists of a series of such elementary reactions. Thus, it is often difficult and expensive. In advanced chemical industry a plant consists of a series of reactors for each separate step and is regulated at the center of the plant. It is indeed a gigantic system with large energy consumption. Organic chemistry and organochemical industry have developed and prospered in this way. The present organochemical industry has thus become strong in the production of substances not occurring in nature, such as DDT, PCB, etc., but weak in the production of substances familiar to life such as D-glucose or sucrose.

Harmony with nature and life and the conserving of energy are urgently required for the chemical industry of the future. We should attempt a new start from Berzelius and renovate the chemical industry. This is the transformation from energy chemistry to information chemistry, and finally to molecular system chemistry.

As a simple example of this molecular system chemistry, I should like to cite the alcoholic fermentation by yeast extract. This system consists of substrates, organic, and inorganic cofactors and thirteen enzymes (multimolecular, multienzyme system) and the following overall reaction takes place under mild conditions:

$$C_6H_{12}O_6 + 2\ ADP + 2\ Pi \rightarrow 2\ C_2H_5OH + 2\ CO_2 + 2\ ATP + 2\ H_2O$$

Though the number of participating substances is considerable, there is no remarkable disorder. This is because a system chemistry at the molecular level is established depending upon the information characterized as the substrate specificity of enzymes. Furthermore, by the transfer of energy in this water soluble system, a thermodynamically unfavorable substance, ATP, is synthesized as a final product from the simple starting materials (ADP and P_i) in aqueous medium.

This example is one of the most primitive systems. In more advanced systems, depending upon the information carried by enzymes, the enzyme molecules self-assemble so as to carry out reactions smoothly. This is the case in the fatty acid synthesizing system and the pyruvate oxidation system. A much more advanced molecular system chemistry may be the protein synthesizing system in which a number of proteins and ribonucleic acids participate.

In conclusion, chemical processes in a living system are characterized by the ordered complexity of the system and the mildness of the conditions. It is an information chemistry or a molecular system chemistry with less energy consumption.

The coming innovation of the chemical industry will be achieved by the structural transformation from energy chemistry to information chemistry. It is not a less advanced system consisting of large reactors as found in the present chemical industry, but an advanced system chemistry. The innovation will be completed by the achievement of molecular system chemistry.

So far I have discussed the strong points of the chemistry of living systems. But it is neither one evolving supernaturally, or one created by human intelligence. It is nothing but a result of natural evolution. I hope that the coming chemistry and chemical industry to be brought about through human intelligence will prove far superior to the chemistry of living systems. Why do I hope so? That is because there is much more freedom than in the chemistry of living systems.

What is the weak point in the chemistry of living systems? It is the limitation of participating materials and environmental conditions.

As a biochemist, I have had no interest in the chemistry of Ge, Nb, Rh, etc., since a living system does not use these. But it has been proved by science and technology that these elements are very useful to human life. It is simply a predestination of living systems that these elements are not used by the living system, a predestination brought on by the virtual absence of these elements in the sea from where life originated and evolved (4).

Chemical evolution, the origin of life, and early biological evolution occurred in the primeval sea, and the common characteristics of organisms were such as to make adaptation to a sea environment. Thus life had to use materials relatively abundant in the primeval sea. As indicated in the beginning of this chapter, primitive organisms evolved from nature with less than twenty kinds of elements and even advanced organisms with less than thirty kinds of elements. This is to say that the starting materials were very limited and only those relatively abundant in the environment could be utilized (4).

Chemistry and chemical industry, however, can use even artificial elements besides all the natural elements, based upon their properties. The same is probably true for organic substances. Living systems apparently synthesize a wide variety of organic substances, but the starting materials are very limited. Twenty protein amino acids, several nucleic acid bases, sugars, etc., were just substances selected and more or less modified from those accumulated in the primeval sea in the course of chemical evolution. Chemistry and chemical industry without such limitations have greater possibilities.

Thus, the chemistry of the future and chemical industry should be characterized by harmony with nature and life and lower energy consumption just as in the chemistry of living systems; but contrary to the latter, it should be characterized by having freedom in the selection of materials. This is what I call "life chemistry," chemistry for the human life. How may it be realized?

It seems to me that there are several ways under consideration or being put into practice. Among them I should like to emphasize two with my special interest.

The first is the development of specific or highly selective catalysts. In this case, enzymes may be regarded as the models. But materials for the catalysts must be chosen without limitations, and catalysts exceeding enzymes in activity and selectivity and above all in broad applications will be obtained in future. Remarkable advance in this direction has already been achieved and there is no need to describe it. What I should like to emphasize is that specific catalysts once obtained can be used together in simultaneous and successive reactions performed in a reaction vessel. The multicatalytic system thus realized is nothing but the simplest molecular system chemistry.

The second way is to take the chemistry in the biosphere as a model.

The characteristics of the chemistry in the biosphere may be described as follows: chemical syntheses in mild conditions by ordered multimolecular systems depending upon the unlimited and clean energy of the sun. It is certainly fruitful to learn from it, and some success has been achieved. However, it seems sometimes so magnificent as if created by God and nearly impossible to be achieved by human technology. This is true, for example, in the case of photosynthesis by the chloroplast, protein synthesis on ribosomes, etc.

So I suggest learning from chemical evolution which has led to the establishment of magnificent living systems. It will suggest a most efficient way to be utilized by chemistry of the future and the chemical industry.

An experimental approach to chemical evolution has been pursued by M. Calvin, J. Oró, C. Ponnamperuma, L. E. Orgel, and many others (see 5, 6, 7) following the pioneering work by S. L. Miller (8). Generally speaking, it seems to elucidate chemical evolution itself, resulting in the origin of life.

However, what I should like to propose is different from this. My aim is to establish an ordered multimolecular system chemistry using the chemical evolution as a model. In this case, it is not necessary to feel hesitant in regard to the environmental conditions in chemical evolution. On the contrary, in order to accelerate chemical evolution, which has continued for several hundred million years, conditions such as temperature, pH, and species and concentration of catalysts may be changed considerably. The final aim is to establish an artificial chemical evolution superior for the human life to the natural chemical evolution.

What is important for this is to elucidate the essential principles in chemical evolution, resulting in molecular system chemistry, and based upon these principles, to establish the latter with much more freedom regarding the conditions and materials for experimentation.

From this viewpoint we have been studying chemical evolution. As a first step, based on the generally accepted concept that the life originated in the primeval sea, we have studied the synthesis of amino acids and peptides in a modified sea medium (9). As a standard modified sea medium, the following substances were used: HPO_4^{2-}, SO_4^{2-}, Mg^{2+}, Ca^{2+} each 0.010 *M*, Na^+, 0.015 *M*, K^+ 0.05 *M*, Cl^- 0.07 *M*, NO_3^-, 0.0005 *M*, MoO_4^{2-}, Fe^{3+}, Zn^{2+}, Cu^{2+}, Co^{2+}, and Mn^{2+}, each 0.0001 *M*. The pH was adjusted to 5.5. This does not mean that the pH of the primeval sea was 5.5. Though not certain, it is generally considered to have been about 7 to 8. However, at that pH, the yield of amino acids was very low, and so we have chosen pH 5.5. In a more acidic medium, the polymers produced might undergo decomposition. The concentration of Na^+ and Cl^- was decreased, because it is considered that they played no essential role in chemical and early biological evolution. The six transition elements (Mo, Fe, Zn, Cu, Mn, Co), essential not only for higher organisms but also for most bacteria, were added at a 10,000 - 100,000 times higher concentrations than that in sea water. It is believed that these elements acted as catalysts in chemical evolution (multicatalytic system) (4, 10).* Clays were added in the beginning, as it was generally accepted that they played important catalytic roles in chemical evolution since the time of J. D. Bernal (11). It was found, however, that clays were not necessary, at least for the formation of amino acids and amino acid polymers in the medium. So, in later experiments, clays were omitted. High temperature (105°C) was chosen in order to accelerate the reactions and above all to remove microbal contamination.

As starting materials we used formaldehyde, C_1-compound (3.0 or 0.5 *M*), and hydroxylamine (0.5 *M*), N_1-compound. There is no evidence that these compounds accumulated in the primeval sea. They are highly reactive - thus, J. Oró (12) used them as starting materials - and probably participated in chemical evolution without accumulation. Anyhow, under these experimental conditions, the reaction mixture was kept heated under a nitrogen atmosphere. After several days the reaction mixture turned yellow. When analyzed by an automatic amino acid analyzer, about 40 species of ninhydrin-positive compounds were detected. Based upon the retention time, taurine, Asp, Thr, Ser, homoserine, Glu, Pro, Gly, Ala, α-aminobutyric acid, Val, norvaline, Ile, Leu, norleucine, β-Ala, Lys, His, and Arg were tentatively assigned. Among these, Gly, Ala, Ser, Asp, Glu, Pro, β-Ala, and α-aminobutyric acid were further confirmed.

What is remarkable is that after about 10 days of incubation, amino acid oligomers or polymers were obtained which gave rise to amino acid mixture by HCl hydrolysis. As expected from the nature of the starting materials, this was not a simple straight chain of polypeptides consisting of glycine and L-amino acids, but rather oligomers and polymers generally of a complex nature containing glycine, natural, and unnatural amino acids. However, we could identify glycylglycine among these products (13). This finding is remarkable, since a thermodynamically unstable peptide bond was produced in the aqueous medium. This probably means that simultaneous and consecutive reactions mutually dependent on one another took place in the medium with an energy transfer thus bringing about the apparently improbable formation of a peptide bond.

The formation of a thermodynamically unstable glycylglycine in a modified sea medium is somewhat similar to the formation of ATP in an aqueous medium by a multienzyme system of alcoholic

**Since it seems that copper played no essential role in chemical evolution and in early biological evolution under an anoxygenic atmosphere (10), it has often been omitted in later experiments.*

fermentation. Of course, such reactions are quite common in living systems, but have been scarcely observed in traditional chemistry, particularly in chemical industry, in which a simple reaction such as A + B → Products without or with a single catalyst was always carried out in a reaction vessel.

Our studies on chemical evolution have only begun: yet it is noteworthy that peptide bonds were formed in the aqueous medium. Here we find a germ of molecular system chemistry and this suggests a promising future for these studies.

How many reaction steps will be required for the synthesis of glutamic acid, proline or glycylglycine from a C_1-compound and a N_1-compound in traditional organic chemistry?

It should be added that in our experiments, glycylglycine thus produced disappeared in the longer incubation as expected from thermodynamics. However, in natural chemical evolution which proceeded as an open system under the constant supply of light energy from the sun, such thermodynamically unstable substances could accumulate and lead finally to life through further complex structures. So it is expected that studies on chemical evolution using ultraviolet and visible light will yield many fruitful results which will give rise to the accumulation of thermodynamically unstable products.

Photosynthesis by green plants and other photosynthetic organisms have been extensively studied by plant physiologists, biochemists, biophysicists, and molecular biologists, including M. Kamen as one of the most distinguished pioneers. Taking this as a model for photochemical synthesis or transformation of light energy into chemical energy, considerable advance has been achieved using solar energy. However, because of the very marvelously ordered complexity of photosynthetic systems and because of the limitation of utilized wave lengths in the present photosynthesis, its value as a model is limited. So I propose to study the photochemical effects in chemical evolution and the gradual development of photochemical systems. On the primeval Earth, the atmosphere contained little dioxygen and no ozone layer existed. Thus very strong solar ultraviolet rays participated in chemical evolution. The photochemical study of chemical evolution will thus open a new way to the use of light energy in chemistry.

It has been often said that the first half of the 20th century was the generation of physical sciences and the latter half that of biological sciences or life sciences. I dare to add that in the 21st century there will come about the generation of chemistry and chemical industry.

By combining the advantages of the chemistry of living systems and that of traditional chemistry and chemical industry, the innovation of chemistry and chemical industry will be established. It is in harmony with nature, life, and human welfare and will certainly make way for an easier and more comfortable life in the 21st century.

The chemical synthesis of vitamin B_{12} was a brilliant success by R. B. Woodward. But in biosphere it is easily achieved by multienzyme systems of *Clostridia,* the most primitive bacteria closest to the origin of life (Klein and Cronquist, 14). The chemical faculty of *Clostridia* is nothing but a development of multicatalytic syntheses in chemical evolution. The studies on chemical evolution will show us how the marvelous chemical faculties of orderly multienzyme system were established, and the studies will further suggest a way to a novel chemistry and chemical industry, in which syntheses will be smoothly performed by ordered multicatalytic systems. These studies on chemical evolution will contribute to the innovation of chemistry.

REFERENCES

1. Egami, F. (1976). *Kagaku to Seibutsu 14,* 625 (in Japanese).
2. Berzelius, D. J. Jacob (1820). "Lehrbuch der Chemie," translated from Swedish by K. A. Blöde, Dresden, p. 6.
3. Kekulé, D. A. (1859). "Lehrbuch der organischen Chemie oder Chemie der Kohlenstoffverbindungen," Erlangen, p. 11.
4. Egami, F. (1974). *J. Mol. Evol. 4,* 113.
5. Calvin, M. (1969). "Chemical Evolution." Oxford Univ. Press, Oxford.
6. Gabel, N. M. and Ponnamperuma, C. (1972). "Exobiology," p. 95. North Holland, Amsterdam.
7. Miller, S. L. and Orgel, L. E. (1974). "The Origins of Life on the Earth." Prentice-Hall, London.
8. Miller, S. L. (1953). *Science 117,* 528.
9. Hatanaka, H. and Egami, F. (1977). *Bull. Chem. Soc. Japan 50,* 1147.
10. Egami, F. (1975). *J. Biochem. (Tokyo) 77,* 1165.
11. Bernal, J. D. (1949). *Proc. Phys. Soc. 62,* 527.
12. Oró, J., Kimball, A., Fritz, R., and Master, F. (1959). *Arch. Biochem. Biophys. 85,* 115.
13. Ochiai, T., Hatanaka, H., Ventilla, M., Yanagawa, H., Ogawa, Y., and Egami, F. (1978). "Origin of Life," Proc. 5th Intern. Conf. Origin of Life, p. 135. Center for Acad. Publ. Japan, Tokyo.
14. Klein, R. M. and Conquist, A. (1967). *Quart. Rev. Biol. 42,* 105.

CYTOCHROME c_1 - AN INERT CYTOCHROME*

Tsoo E. King
Young-Ling Chiang

Laboratory of Bioenergetics and Department of Chemistry
State University of New York
Albany, New York 12222

A wealth of information has accumulated about the structure and function relationship of cytochrome *c*, in contrast to our meager knowledge of cytochrome c_1 which is due partly to the unavailability, until just recently, of a pure preparation. The recent purification (1, 2) of cytochrome c_1 in gram quantities has thus allowed the investigation of some properties of this component. It is one, if not the, most inert cytochromes known. Inertness actually refers to the lack of the reaction characters of small molecule ligands. However, it forms well-defined complexes with other cytochromes and we interpret these complexes as representing the events possible *in situ*.

**Our studies of cytochrome c_1 were influenced by the beautiful work on a family of cytochromes, formerly known as RHP, ingeniously done by Professor Martin Kamen. The actual work began after a casual conversation with him many years ago by one of the authors who had some difficulty in c_1 purification at Professor Keilin's laboratory in the late 1950's at the Molteno Institute.*

 ISBN 0-12-397580-8

PURIFICATION

Since Yakushiji and Okunuki reported the isolation of a partially purified cytochrome c_1 (as well as cytochrome oxidase) (3, 4), numerous attempts of further purification have been undertaken and reported (5 - 11).

However, at least in our hands, these methods not only gave low purity and low yields but were also tedious and nonreproducible. Through years of trial and error by more than 10 senior people singly or collectively, eventually we have systematically used "sequential fragmentation" (12) and succeeded in gradually developing two simple methods (1, 2) for the isolation of practically pure cytochrome c_1 with reproducible results and relatively good yields. One of our methods uses β-mercaptoethanol extraction of the cytochrome b - c_1 complex or succinate-cytochrome c reductase. This method has been applied to yeast (13, 14) with equal success. The other method takes advantage of cleavage of the cytochrome b - c_1 complex by 2% cholate in the presence of 20% ammonium sulfate at room temperature. In both methods cytochrome c_1 remains in the solution and b cytochromes in the precipitate. Subsequently, conventional methods are then used to obtain the almost pure cytochrome c_1 (1, 2).

PROPERTIES

In a remarkable paper published in 1955, Keilin and Hartree (15) not only yielded their nomenclature of cytochrome *e* to c_1, because of the prior discovery by Yakushiji and Okunuki (1) who called the new cytochrome, c_1, but, from a hand spectroscope mounted on a microscope with limited methodology available to them, more importantly, they concluded: (1) cytochrome c_1 like cytochrome *c* is not autooxidizable; (2) it reacts with neither carbon monoxide nor cyanide; (3) unlike cytochrome *c*, but like other cytochromes, it is thermolabile and more intimately linked with "structural elements" of the cell. Not until about 20 years later have these properties been completely confirmed by using purified cytochrome c_1. Now it is universally agreed that cytochromes c_1 and *c* are distinct entities and essential electron (hydrogen)* carriers of the respiratory chain.

**The term, electron transfer, in this paper includes hydrogen transfer and transfer of other species, if they indeed exist.*

Prosthetic Group

The prosthetic group of cytochrome c_1 has been identified as an iron chelate of protoporphyrin IX, which cannot be dissociated from the protein by acid-acetone extraction. It can be dissociated by the silver sulfate method (16), which means the heme is covalently linked to the protein (1, 2, 16, 17).

Spectra

The absorption spectra of purified cytochrome c_1 prepared by our methods are well-defined and reproducible. The Soret band of ferricytochrome is located at 411 nm, indicating the low spin state of the heme iron. Upon reduction, the Soret band shifts to 417 nm and shows α and β bands at 552.5 and 525.5 nm, respectively (1). The shoulders that appear on both sides of β band are characteristic for this cytochrome. In addition, a weak, but distinct band is observed at 690 nm in the ferric state at room temperature (18). Upon reduction, the absorption of c_1 between 570 and 770 nm decreases greatly and the band at 690 nm is no longer discernible. The 695 nm absorption band with a low absorbance index has also been observed in cytochrome *c* and it is attributable to the methionine ligand (18, 19). At -160°C (1), reduced cytochrome c_1 exhibits a very colorful spectrum. The α band, which is relatively symmetrical at room temperature, splits into at least three absorption bands. The β band of c_1, which shows some fine structure even at room temperature, splits into a minimum of nine bands. The intensification of the α and β bands at low temperature is about 6- to 7-fold, whereas the Soret maximum is not so intensified or clearly split. No shift of these bands is found at low temperature, in great contrast to that of cytochrome *c* (20). These findings are in accord with the observation of Keilin and Hartree (15) on the heart muscle preparation, a kind of submitochondrial particle.

The CD spectra (21) of cytochrome c_1, both in oxidized and reduced forms, show a large positive Soret Cotton effect. They are completely different from those of the mammalian cytochrome *c* despite the resemblance of the absorption spectra and the identity of the prosthetic groups. Such a difference certainly reflects the difference of the protein structure surrounding the heme.

Subunits and Amino Acid Composition

Purified cytochrome c_1 is a soluble preparation with a minimal molecular weight of about 40,000, but hydrodynamic determinations* suggest the isolated cytochrome is perhaps in a pentameric form (22). Gel electrophoresis in dissociation media shows two subunits with molecular weights of 29,000 and 15,000 (2). The 29.000 subunit contains heme, while the other does not. Trumpower and Katki (23) have confirmed the above observation. The yeast cytochrome c_1 purified by Ross and Schatz (14) contains one major polypeptide chain with molecular weight of 31,000 and variable amounts of a minor band of molecular weight of 18,500. Whether the peptide without heme in beef heart cytochrome c_1 is an intrinsic component required for catalytic activity remains to be determined by experimentation.

The predominance of acidic (glutamic acid + aspartic acid) over basic (lysine + arginine) amino acids in cytochrome c_1 indicates that it is an acidic protein. Indeed, its isoelectric point has been found to be 5.35 (22). The amino acid composition has been determined (22) and a tryptic digested heme *a* core peptide containing 14 peptides isolated (23a). The sequence of the heme-containing subunit is nearing completion by collaboration with Professor Matsubara, a pupil of one original discoverer of this cytochrome, and his group in Osaka. One interesting feature of the sequence is the finding of existence of numerous "doublets" such as ...Pro-Pro..., ...Ala-Ala..., ...Gly-Gly....

Redox Properties and the Inertness

The kinetics of electron transfer between isolated cytochrome c_1 and cytochrome *c* have been reported (24). The second-order rate constant (k_1) for the electron transfer from ferrocytochrome c_1 to ferricytochrome *c* is 1.3×10^{-7} M^{-1} sec^{-1} at 10°C, pH 7.4, and the reverse reaction shows a rate constant (k_{-1}) of 4×10^6 M^{-1} sec^{-1}.+ Therefore, the equilibrium constant is about 3. The reaction rates toward ferricytochrome *c* are the same whether ferrocytochrome c_1 is in the soluble form or structured in a particulate succinate-cytochrome *c* reductase

**Hydrodynamic data are: sedimentation coefficient, 13.5 S, intrinsic viscosity, 0.0575 g/100 ml at 20°C; and partial specific volume, 0.732 cm³/g.*

+The date of second-order constants cited in Ref. 24 are in error by a factor of ×4. Some of the values described in this paper are newly estimated or the factor 4 multiplied.

(24). The incorporation of cytochrome c_1 into phospholipid vesicles does not affect the electron transport properties of cytochrome c_1 (25).

The reduction of ferricytochrome c_1 by ascorbate is slow with a pseudo-first-order rate constant of 2.5×10^{-2} sec^{-1}, which is about 20-fold slower than that of cytochrome *c* (26). However, tetramethyl-*p*-phenylenediamine reduces ferricytochrome c_1 and *c* at similar rates with second-order rate constants of 8×10^4 M^{-1} sec^{-1} and 4×10^4 M^{-1} sec^{-1}, respectively (24).

Ferrocytochrome c_1 is nonautooxidizable and becomes autooxidizable when exposed to pH above 11, 4 *M* urea, 0.05% SDS, or 1.5 *M* guanidine hydrochloride (18, 25). These reagents irreversibly denature cytochrome c_1 and the denatured c_1 becomes carbon monoxide reactive. Ferrocytochrome c_1 is more stable toward denaturants than the ferric form, similar to cytochrome *c*; on the other hand, the ferrous form of c_1 is more susceptible to trypsin digestion than the ferric form (23), in contrast to cytochrome *c*.

Cytochrome c_1 does not react with cyanide at neutral pH as evidenced by the absence of any spectral change upon addition of cyanide to ferricytochrome c_1 (18) in accordance with the results of Sekuzu *et al.* (6) using partially purified c_1. The observation reported by Schejter and Berke (27) concerning the reactivity with cyanide is very likely due to the use of a somewhat impure and denatured sample prepared by the procedure of Bomstein *et al.* (8, 9). On the other hand, cyanide can evidently reach the coordinated methionine-80 residue and displace this ligand in cytochrome *c* (28, 29). Azide, which rapidly reacts with cytochrome *c*, also does not react with cytochrome c_1 (22, 30). The autooxidation of reduced c_1 is much less than that of *c* (24, 30). The inertness of cytochrome c_1 toward ligands may be due to the inaccessibility of the heme group to these small compounds. The lack of a quenching effect of cytochrome c_1 on 8-anilino-1-naphthalene sulfonic acid fluorescence (31) also suggests the deeply buried nature of the heme group. The possibility that the inaccessibility of the heme group is the consequence of the self-aggregation of the isolated c_1 is not likely, since the similarity in properties of the isolated c_1 and the c_1 in membranes indicates that purified c_1 as an active preparation even kinetically (24). Self-aggregation may not have any effect on the function of purified cytochrome c_1. Moreover, it is not certain whether c_1 *in situ* is a monomer or oligomer, although *a priori* consideration would suggest the former.

Other Physical Properties

From spectral evidence (18, 26), cytochrome c_1 is in the low spin state at neutral pH and the heme iron is six-coordinated. One of the axial ligands is drawn from the lysine, histidine, or arginine residues of the protein (32) and the other most probably from methionine (31).

EPR studies (32) at as concentrated as 1 m*M* also show a low spin state for ferricytochrome c_1 heme iron. Three typical low spin *g* values (32) have been determined at 1.4°K; they are g_z = 3.34, g_y = 2.06, and g_x = 1.08. A small amount of extraneous copper, apparently introduced during the purification, was also detected. Although the copper signal interfered with the determination of the g_y signal, all three *g* values can nevertheless be clearly demonstrated. Orme-Johnson *et al.* (33) have detected only two signals from our purified cytochrome c_1 at 13°K; they are g_z = 3.33 and g_y = 2.1. At any rate, both laboratories have shown that these signals are quite different from cytochrome *c* at neutral pH. However, the g_z signal at 3.34 is very close to the g_z = 3.4 observed in ferricytochrome *c* at pH 11 (34, 35). It has been interpreted that the methionine ligand of cytochrome *c* at high pH was displaced by another strong field, a nitrogenous ligand which is probably from the lysine-79 residue. Therefore, it is very possible that ferricytochrome c_1 like cytochrome *c* at high pH also has a nitrogenous base (amine) ligand; the most probable one is also the lysine residue. Nevertheless, all evidence shows cytochrome c_1 may not have a histidine axial ligand (32). However, a ligand other than methionine cannot be decided upon even from the widely known but an unpublished most recent "truth table" of Blumberg and Peisach.

The midpoint redox potential of purified cytochrome c_1 in solution (23, 25), in artificial liposomes (25), or naturally existing in the mitochondrial membrane (36) is about 225 mV around neutrality. This value is about 35 mV lower than cytochrome *c* in solution (e.g. 36). In contrast, the E_m of cytochrome *c* is lowered by as much as 60 mV on binding the *c* inside mitochondrial particles or inside phospholipid vesicles (37).

CYTOCHROME c_1 COMPLEXES

Cytochrome c_1 - cytochrome c Complex

A stable cytochrome c_1-cytochrome *c* complex has been prepared (22). This complex is stable in low ionic strength and can be purified from Sephadex chromatography with a stoichiometry of c_1 to c of 1:1. This complex is dissociated in solutions of ionic strength higher than 0.05. Lysine residue of cytochrome c are evidenced to be involved in the complex formation, since no complex is formed when acetylated cytochrome c is used (26). The methionine sulfoxide derivative of cytochrome c, which has a substantially altered heme crevice, but with all lysine residues unmodified, did not form a complex with cytochrome c_1 at all (26). This indicates that complex formation is not due just to electrostatic attraction but also requires proper conformation. Although the association constant at neutral pH has not accurately been determined, it must be over 10^7 (cf. 26). From the results using a series of chemically modified cytochrome c (26), we believe optimum electron transfer can take place first in the formation of the complex; but the latter does not assure that electron transfer can, indeed, occur.

The c_1 - c complex differs in the summation of the properties of these two cytochromes from the individual components, such as CD (26) and fluorescence (30), indicating the binding sites are close to the heme crevices of both cytochromes and a drastic conformational change occurred in that region. The slight decrease of reaction of cyanide with c in the complex (26) suggest the binding site of cytochrome c_1 on c is on the left side (the methionine-80 side) of the exposed heme edge of c, which may not overlap with the binding domain proposed for cytochrome oxidase (37). Sedimentation velocity experiments of the complex reveal a single schlieren peak with a sedimentation coefficient of 16.2 S, which is equivalent to a molecular weight of 290,000. This figure further substantiates the unity ratio of cytochrome c_1 to c in terms of heme in the complex. It also indicates that the complex is a large aggregate, similar to the purified cytochrome c_1 (22, 26).

Cytochrome c_1 - Cytochrome Oxidase Complex

A rather unexpected complex with unity ratio in terms of heme has been prepared by ammonium sulfate fractionation of a mixture of cytochrome c_1 and cytochrome oxidase (22, 25). This complex is stable in media of high ionic strength (>1 *M* NaCl), but dissociated at pH deviated from neutral. An interesting observation found is that the complex is soluble and also stable to Sephadex chromatography in the absence of any detergent, in spite of the fact that cytochrome oxidase alone under these conditions is insoluble (25). Results from sedimentation studies show the combination of these two cytochromes mutually depolymerize each other. It is indeed intriguing to find that no reagent other than cytochrome oxidase can depolymerize isolated c_1 without denaturation (25). The interaction between oxidase and c_1 has been claimed by Yakushiji and Okunuki (4). Now the reduction of cytochrome oxidase by ferrocytochrome c_1 has been experimentally demonstrated by purified components. The oxidation of the reduced oxidase is also through its oxygen compound (38), although the rate of electron transfer from c_1 to the oxidase is many-fold slower than that from *c* to oxidase (38). The reasoning is detailed elsewhere (38). We consider the formation of the c_1-oxidase complex as a control or regulatory device.

The Ternary Complex

The ternary complex of c_1-*c*-oxidase was first postulated by Orii *et al.* (39). Recently, it has been prepared and purified to a heme ratio of unity from the purified cytochromes by any of the following methods (1) c_1 + *c* + oxidase, (2) c_1 + *c*-oxidase complex, (3) *c* + c_1-oxidase complex, and (4) c_1-*c* complex + oxidase (22, 25). All of these complexes are formed by pure protein - protein interactions, because lipid-free components are equally effective. The ternary complex prepared by any of the above methods shows the same properties indicating the binding sites are all different and not overlapping. All these complexes can be easily dissociated back to the original components by a slight increase of ionic strength or a small change of pH (25). The ternary complex shows higher activity in electron transfer media of low ionic strength (10 m*M*) than in high ionic strength (100 m*M*). This fact indicates the ternary complex may be more active than its components alone.

SUMMARY

This paper briefly reviews the methods of purification of cytochrome c_1. Reactions with small molecule ligands certainly show the extreme inertness of the cytochrome at neutral pH. However, it can form complexes with *c*, cytochrome oxidase, and *c* + oxidase. The cytochrome c_1 - *c* complex and the ternary complex are active, while cytochrome c_1 - oxidase may be used as a control device *in vivo* because the rate of electron transfer from c_1 to oxidase is many-fold lower than that of *c* to oxidase.

ACKNOWLEDGMENT

Support for experimental work was supplied by the National Science Foundation and the National Institutes of Health.

REFERENCES

1. Yu, C. A., Yu, L., and King, T. E. (1972). *J. Biol. Chem. 247,* 1012-1019.
2. Yu, C. A., Chiang, Y. L., Yu, L., and King, T. E. (1975). *J. Biol. Chem. 250,* 6218-6221.
3. Yakushiji, E. and Okunuki, K. (1940). *Proc. Imp. Acad. Tokyo 16,* 299-302.
4. Yakushiji, E. and Okunuki, K. (1941). *Proc. Imp. Acad. Tokyo 17,* 38-40.
5. Glaze, R. P. and Morrison, M. (1960). *Fed. Proc. 19,* 34.
6. Sekuzo, I., Orii, Y., and Okunuki, K. (1960). *J. Biochem. 48,* 214-225.
7. Green, D. E., Jarnefelt, J., and Tisdale, H. (1959). *Biochim. Biophys. Acta 31,* 34-46.
8. Bomstein, R., Goldberger, R., and Tisdale, H. (1960). *Biochem. Biophys. Res. Commun. 3,* 479-483.
9. Bomstein, R., Goldberger, R., and Tisdale, H. (1961). *Biochim. Biophys. Acta 50,* 527-543.
10. Rieske, J. S. (1967). *Methods Enzymol. 10,* 349-352.
11. Orii, Y. and Okunuki, K. (1969). *Annu. Rep. Biol. Works Fac. Sci. Osaka Univ. 17,* 1-14.
12. King, T. E. (1966). *Advan. Enzymol. 28,* 155-236.
13. Claisse, M. L. and Pajot, P. E. (1974). *Eur. J. Biochem. 49,* 49-59.
14. Ross, E. and Schatz, G. (1976). *J. Biol. Chem. 251,* 1991-1996.
15. Keilin, D. and Hartree, E. F. (1955). *Nature (London) 176,* 200-206.
16. Keilin, D. (1966). "The History of Cell Respiration and Cytochromes." Cambridge University Press, London and New York.
17. Lemberg, R. and Barrett, J. (1973). "Cytochromes," pp. 212-216. Academic Press, New York.
18. Kaminsky, L. S., Chiang, Y. L., and King, T. E. (1975). *J. Biol. Chem. 250,* 7280-7287.
19. Schechter, E. and Saludjian, P. (1967). *Biopolymers 5,* 788-790.
20. Estabrook, R. W. (1961). *In* "Haematin Enzymes" (J. E. Falk, R. Lemberg, and R. K. Morton, eds.), pp. 436-456.

Pergamon Press, London.
21. Yu, C. A., Yong, F. C., Yu, L., and King, T. E. (1971). *Biochem. Biophys. Res. Commun. 45,* 508-513.
22. Chiang, Y. L. (1976). Ph.D. Thesis, State University of New York at Albany, New York.
23. Trumpower, B. L. and Katki, A. (1975). *Biochemistry 14,* 3635-3641.
23a. Yu, L., Chiang, Y. L., Yu, C. A., and King, T. E. (1975). *Biochim. Biophys. Acta 379,* 33-42.
24. Yu, C. A., Yu, L., and King, T. E. (1973). *J. Biol. Chem. 248,* 528-533.
25. Chiang, Y. L. and King, T. E. (1979). *J. Biol. Chem. 254,* 1845-1853.
26. Chiang, Y. L., Kaminsky, L. S., and King, T. E. (1976). *J. Biol. Chem. 251,* 29-36.
27. Schejter, A. and Berke, G. (1968). *Biochim. Biophys. Acta 162,* 459-461.
28. Dickerson, R. E. and Timkovich, R. (1975). *In* "The Enzymes" (P. D. Boyer, ed.), 3rd ed., Vol. XI, pp. 489-492. Academic Press, New York.
29. George, P. and Tsoo, C. L. (1952). *Biochem. J. 50,* 440-448.
30. Kaminsky, L. S., Chiang, Y. L., Yu, C. A., and King, T. E. (1974). *Biochem. Biophys. Res. Commun. 59,* 688-692.
31. Brill, A. S. and William, R. J. P. (1961). *Biochem. J. 78,* 246-253.
32. Peisach, J., Blumberg, W. E., Chiang, Y. L., and King, T. E., unpublished results.
33. Orme-Johnson, N. R., Hansen, R. E., and Beinert, H. (1974). *J. Biol. Chem. 249,* 1928-1939.
34. Blumberg, W. E., Peisach, J., Hoffman, B., Stellwagen, E., Margoliash, E., Marchant, L., Tulloss, J., and Feinberg, B. (1973). *Fed. Proc. 32,* 469.
35. Lembeth, D. O., Campbell, K. L., Zand, R., and Palmer, G. (1973). *J. Biol. Chem. 248,* 8130-8136.
36. Dutton, P. L., Wilson, D. F., and Lee, C. P. (1970). *Biochemistry 9,* 5077-5082.
37. Ferguson-Miller, S., Brautigan, D. L., and Margoliash, E. (1978). *J. Biol. Chem. 253,* 149-159.
38. Orii, Y. and King, T. E. (1978). "Frontiers of Bioenergetics," pp. 833-842. Academic Press, New York.
39. Orii, Y., Sekuzo, I., and Okunuki, K. (1962). *J. Biochem. 51,* 204-215.

Note Added in Proof

This paper was originally written in 1978. Since then, some advances have been made, for example, the formation of cytochrome c_1-c complex requires a 15 K mitochondrial protein for the reaction of "one band" c_1 and c. Readers may be interested to consult a more complete review in Advances in Enzymology (in press).

CYTOCHROMES AND FERREDOXINS IN BACTERIAL ELECTRON TRANSPORT AND CLASSIFICATION

T. E. Meyer

Department of Chemistry
University of California at San Diego
La Jolla, California 92093

Postgate (1954) and Kamen and Vernon (1954a) were the first to isolate cytochromes from strictly anaerobic bacteria, in contrast to the viewpoint that cytochromes, as part of the electron transport pathway of mitochondria and aerobic bacteria, should not be found in anaerobes (Keilin, 1966). These studies laid the foundation for a comparative biochemistry of anaerobic bacterial electron transport, which has inspired contributions from many laboratories. A number of excellent reviews on the history, distribution, and characterization of cytochromes and ferredoxins have been written (Keilin, 1966; Kamen and Horio, 1970; Horio and Kamen, 1970; Kamen *et al.*, 1972; Kamen, 1973; Lemberg and Barrett, 1973; Lovenberg, 1973, 1977; Yamanaka and Okunuki, 1974; Palmer, 1975; Salemme, 1977; Bartsch, 1978a; Pettigrew, 1981). However, we are now entering a very productive period in the study of electron transport and it is appropriate to show the extent of progress and future developments to be expected. The following is a brief description of the properties, classification, and distribution of bacterial electron transport proteins. The functional role of cytochromes in

 ISBN 0-12-397580-8

photosynthetic electron transport pathways is suggested, and a molecular concept of the bacterial species as it relates to classification and identification is developed.

CYTOCHROMES

A relatively large number of distinct classes of cytochromes (or electron transport heme proteins) have been established primarily on the basis of physicochemical properties. Even within the more narrowly defined group of *c*-type cytochromes (which have protoheme covalently attached through condensation of protein cysteine residues with vinyl side chains) there is a considerable diversity of structure and function. The *c*-type cytochromes can be subdivided into three groups, which have differing extraplanar ligands to the heme iron. Furthermore, amino acid sequence studies have shown the peptide chains of these three cytochrome subgroups to be unrelated. Three-dimensional structures just now being completed will also show that typical representatives of these subgroups, namely "mitochondrial" cytochrome *c*, *Desulfovibrio* cytochrome c_3, and cytochrome c' are unrelated structurally.

The majority of known classes of *c*-type cytochromes are related to mitochondrial cytochrome *c* in having histidine (His) and methionine (Met) extraplanar ligands. The amino acid sequences of these proteins can be approximately aligned with heme attachment site and histidine ligand near the amino terminus and methionine ligand near the carboxyl terminus. In addition, the three-dimensional structures of the "His-Met" cytochromes are characterized by spatially adjacent amino and carboxyl terminal helices and alternating folds of peptide chain across the faces of the heme. The heme propionates are buried and hydrogen bonded to the protein rather than exposed to solvent as found, e.g., in myoglobin. Cytochrome c_2 is very similar to mitochondrial cytochrome *c* and can be easily aligned with it, but most of the bacterial his-met cytochromes are approximately sixteen amino acid residues smaller in the midsection and are so divergent that the sequences cannot be aligned accurately and the exact numbers of insertions and deletions remain unknown. The more important classes are: *Chlorobium* cytochromes *c*-555, *Pseudomonas* cytochromes *c*-551, algal cytochromes *c*-553, cytochromes c_4, and cytochromes c_5.

Cytochromes c_2

Rhodospirillum rubrum cytochrome c_2 was the first bacterial cytochrome to be isolated and characterized (Vernon, 1953). The amino acid sequence (Dus *et al.*, 1968) showed that it was closely related to mitochondrial cytochrome *c* and the three-dimensional structure (Salemme *et al.*, 1973) provided convincing proof of common origin or homology. Cytochromes c_2 have been isolated from eleven out of thirteen species of purple photosynthetic bacteria examined in the family *Rhodospirillaceae* (Bartsch, 1978a), and examples exist in non-photosynthetic bacteria (Vernon, 1956; Scholes *et al.*, 1971; Timkovich and Dickerson, 1976; Van Beeumen *et al.*, 1980) (see Table 1). The amino acid sequences of all known species of cytochrome c_2 have been determined and found to fall into two groups (Ambler *et al.*, 1976, 1979a). One group has a peptide chain virtually the same size as mitochondrial cytochrome *c*, whereas the other minimally has three- and eight-residue insertions plus a single residue deletion. Most of the known sequences appear to have diverged to the maximum extent allowed by the functional requirements of the protein. This limit to structural variation indicates that convergence is an important component of cytochrome c_2 evolution and that phylogenetic trees cannot be constructed. Obviously, if "trees" have no valid quantitative basis for the cytochromes c_2 alone, all-encompassing trees (Dayhoff, 1978; Dayhoff and Schwartz, 1978; Almassy and Dickerson, 1978) for the homologous bacterial His-Met cytochromes *c* are scientifically unsound. There is a continuum of structural changes between mitochondrial cytochromes *c* and the most divergent of the cytochromes c_2, yet the functional properties of these two classes appear to be quite different. For example, cytochromes c_2 react poorly with bovine cytochrome *c* oxidase but are quite active with bovine cytochrome *c* reductase (Kamen and Vernon, 1954b; Yamanaka and Okunuki, 1968; Errede and Kamen, 1978). The mitochondrial cytochromes *c* appear to have a relatively uniform redox potential near 250 mV, whereas the cytochromes c_2 are characterized by redox potentials which span a 250-400 mV range and which have complex pH dependences (Pettigrew *et al.*, 1975, 1978). There are probably as many as four separate functional roles for the cytochromes c_2, which will be discussed in a later section. As the first recognized class of bacterial cytochrome, c_2 is also the most thoroughly characterized and certainly one of the most interesting because of its very close and unique relationship to mitochondrial cytochrome *c*.

TABLE 1. Bacterial Electron Transport Protein Distributior

	Cytochromes									
Bacteria	L c_2	S c_2	Pseudomonas c-551	c'	membrane c-556-c-552	c_5	c_4	cd	Flavin c	c_3
Cyanobacteria										
Spirulina maxima										
Anacystis nidulans										X
Chlorobiaceae										
Chlorobium thiosulfatophilum									X	
Prothecochloris aestuarii										
Chromatiaceae										
Ectothiorhodospira halochloris						X				
Ectothiorhodospira mobilis		X		X						X
Ectothiorhodospira halophila				X		X				
Thiocapsa pfennigii				X	X	X				
Thiocapsa roseopersicina			?	X	X	?			X	
Chromatium vinosum			?	X	X	X			X	
Rhodospirillaceae										
Rhodopseudomonas gelatinosa			X	X	X	X				
Rhodospirillum tenue			X	X		X				X
Rhodopseudomonas globiformis		X								
Rhodomicrobium vannielli		X								

Cytochromes			Copper Proteins	Ferredoxins				
Chlor-obium c-555	algal c-553	c-556 and c-554		HiPIP	4-Fe-S "Bact."	2-Fe-S	1-Fe	H_2ase
	X		X			X		
	X		X			X		
X					X		X	X
X					X		X	
					X			
				X	X			
				X	X			
				X	X			X
				X	X			X
				X	X			
				X	X			
				X	X			
				X	X			X

Bacteria	L c_2	S c_2	Pseudomonas c-551	c'	membrane c-556-c-552	c_5	c_4	cd	Flavin c	c_3
					Cytochromes					
Rhodopseudomonas acidophila		X								
Rhodopseudomonas viridis		X			X					
Rhodospirillum molischianum		X		X						
Rhodospirillum fulvum		X		X						
Rhodospirillum rubrum	X			X						
Rhodospirillum photometricum	X			X						
Rhodopseudomonas spheroides	X			X						X
Rhodopseudomonas capsulata	X			X						X
Rhodopseudomonas sulfidophila	X			X						
Rhodopseudomonas palustris	X			X						X
Pseudomonadales										
Azotobacter vinelandii			X	X		X	X			
Pseudomonas stutzeri			X			X	X			
Pseudomonas aeruginosa			X			X	X	X		
Alcaligenes sp.				X	X					
Paracoccus denitrificans	X							X		
Paracoccus sp.			?	X			?	X		

Cytochromes			Copper Proteins	Ferredoxins				
Chlor-obium c-555	algal c-553	c-556 and c-554		HiPIP	4-Fe-S "Bact."	2-Fe-S	1-Fe	H_2ase
					X			
					X			
					X			
					X			
					~~X~~			X
					X			
		X			X			X
		X			X			X
		X			X			X
		~~X~~			X	X		X
					X	X		X
			X		X			
			~~X~~					
			?					X
				X				

	Cytochromes									
Bacteria	*L* c_2	*S* c_2	*Pseudomonas c-551*	*c'*	*membrane c-556-c-552*	c_5	c_4	*cd*	*Flavin c*	c_3
Desulfovibriones										✱
Desulfovibrio vulgaris										
Bacillaceae										
Clostridium pasteurianum										

[a]*See text for description of protein categories. X, Protein was determined to be present; ✱, at least two proteins or isozymes were isolated which fit the description; ?, protein which was isolated cannot be assigned with certainty. A blank space indicates absence of the protein or failure to detect it.*

Chlorobium Cytochromes *c*-555

Chlorobium cytochromes *c*-555 have been characterized in only two species in the green photosynthetic bacterial family *Chlorobiaceae* (Kamen and Vernon, 1954; Gibson, 1961; Meyer *et al.*, 1968; Olson and Shaw, 1969; Shioi *et al.*, 1972). The redox potentials are among the lowest found for His-Met cytochromes, being in the range 100-150 mV. Cytochrome *c*-555 may function in a manner analogous to that of the cytochromes c_2 as a direct electron donor to reaction center bacteriochlorophyll. However, there is doubt as to whether the light-reactive cytochrome *c*-553 is identical to the soluble cytochrome *c*-555 (Prince and Olson, 1976). The amino acid sequences (Van Beeumen *et al.*, 1976) are most similar to the cytochromes c_5 and the algal cytochromes *c*-553. The three-dimensional structure of *Chlorobium* cytochrome *c*-555 is clearly homologous to mitochondrial cytochrome *c*, cytochrome c_2, and *Pseudomonas* cytochrome *c*-551 (Korszun and Salemme, 1977).

Cytochromes			*Copper Proteins*	*Ferredoxins*				
Chlorobium c-555	*algal c-553*	*c-556 and c-554*		*HiPIP*	*4-Fe-S "Bact."*	*2-Fe-S*	*1-Fe*	*H_2ase*
					X		*X*	*X*
					X	*X*	*X*	*X*

Algal Cytochromes C-553

Algal cytochromes *c*-553 have been found in a large number of eukaryotic algae (Sugimura *et al.*, 1968), but have also been isolated from the blue-green bacteria (Holton and Myers, 1967; Aitken, 1976; Ho *et al.*, 1979), which have photosynthetic machinery so similar to algae that they were, until recently, treated as such. The amino acid sequences are very divergent and difficult to align with one another (Laycock, 1972; Pettigrew, 1974; Ambler and Bartsch, 1975; Aitken, 1976), but nevertheless show greater similarity to the *Chlorobium* cytochromes *c*-555 than to other bacterial cytochromes *c*. In contrast to the *Chlorobium* cytochromes *c*-555, with their very low redox potentials, the algal cytochromes *c*-553 have some of the highest known redox potentials, ranging from 300 to 400 mV (Sugimura *et al.*, 1968). Perini *et al.* (1964) showed that *Euglena gracilis* cytochrome *c*-552 was localized in the chloroplast, whereas an unusual cytochrome *c*-556 was a mitochondrial protein. Plant chloroplasts do not have a small, soluble *c*-type cytochrome similar to what is found in the algae, but both plants and algae have a membrane-bound protein (Wood, 1977; Gray, 1978) to which Hill and Scarisbrick (1951) first applied the designation cytochrome *f*.

Cytochromes c_5

Cytochrome c_5 was first characterized in *Azotobacter vinelandii* (Tissieres and Burris, 1956; Neumann and Burris, 1959; Swank and Burris, 1969), but has since been found in several *Pseudomonas* species (Horio, 1958; Ambler and Taylor, 1973). The only known amino acid sequence for cytochrome c_5 is that of *Pseudomonas mendocina* Ambler and Taylor, 1973). The sequence is unique in having an "extra" pair of cysteine residues near the C-terminus, in having only one aromatic residue, and in having a "ragged" N-terminus, which is probably due to proteolytic posttranslational modification. This cytochrome often appears to be a minor component of the cells, which may reflect its actual level of biosynthesis, or more likely may be due to the supposed proteolytic fragmentation or to possible limited solubilization from the membrane. Minor cytochrome components of a small number of purple bacterial species (Table 1) have similar physicochemical properties and may be closely related (Cusanovich and Bartsch, 1969; Meyer *et al.*, 1973; Bartsch, 1978a).

Pseudomonas Cytochromes c-551

Pseudomonas cytochrome *c*-551 is a major soluble component of many denitrifying *Pseudomonas* species (Horio, 1958; Horio *et al.*, 1960; Kodama and Shidara, 1969). Cytochrome *c*-551 is also a major component of buffer-extracted cells of the strictly aerobic, but nitrogen-fixing *Azotobacter vinelandii* (Meyer and Kamen, unpublished), but yields are drastically reduced with butanol extraction (Swank and Burris, 1969; Campbell *et al.*, 1973). A number of sequences have been determined (Ambler, 1963, 1974; Ambler and Wynn, 1973), and the three-dimensional structure is known (Almassy and Dickerson, 1978). Although the folding of the peptide chain is qualitatively similar to that of the larger mitochondrial cytochrome *c*, the hydrogen-bonding pattern is remarkably different, particularly in the greater amount of helical structure. *Pseudomonas* cytochrome *c*-551 and *Chlorobium* cytochrome *c*-555 appear to share the same large deletion at the "bottom" of the molecule when the structures are compared with mitochondrial cytochrome *c*. Cytochromes from a relatively small number of purple bacterial species (e.g., *Rhodospirillum tenue* and *Rhodopseudomonas gelatinosa*) have amino acid sequences similar to the *Pseudomonas* cytochromes *c*-551 (Ambler *et al.*, 1979b). On the other hand, the degree of sequence difference for the photosynthetic proteins is much greater than that within the order Pseudomonadales. It may not be coincidental that these are the same purple bacterial species which have cytochromes similar to *Pseudomonas* cytochromes c_5

(Table 1). The functional role of cytochrome *c*-551 in *Pseudomonas* and *R. gelatinosa* is obscure, but the cytochrome from *R. tenue* may function analogously to the cytochrome c_2 of other purple bacteria by virtue of its high redox potential.

Cytochromes c_4

Cytochrome c_4 is a protein first characterized in *Azotobacter vinelandii* (Tissieres and Burris, 1956; Neumann and Burris, 1959; Swank and Burris, 1969; Campbell *et al.*, 1973), but it is also found in the denitrifying *Pseudomonas* species (Ambler and Murray, 1973). The *Pseudomonas aeruginosa* cytochrome c_4 amino acid sequence is nearly complete (Ambler, 1977) and conclusively proves that the protein has two hemes attached to a peptide chain of about 180 residues as earlier reported (Swank and Burris, 1969; Campbell *et al.*, 1973). This diheme cytochrome appears to be the result of gene duplication and fusion of a smaller monoheme protein similar to *Pseudomonas* cytochrome *c*-551. A monoheme cytochrome from a halophilic *Paracoccus sp.* (Ambler, 1977), which is isolated in dimeric form, is closer to *Ps. aeruginosa* cytochrome c_4 in sequence than it is to *Pseudomonas* cytochrome *c*-551.

Desulfovibrio Cytochrome c-553

Desulfovibrio vulgaris and *D. desulfuricans* are the only *Desulfovibrio* species in which cytochrome *c*-553 has been found (LeGall and Bruschi-Heriaud, 1968; Yagi, 1969; LeGall and Forget, 1978; Fauque *et al.*, 1979). *Desulfovibrio vulgaris* cytochrome *c*-553 is related in sequence to the other His-Met cytochromes, but it is too divergent to confidently align with any of the other known classes (Bruschi and LeGall, 1972). The redox potential is probably as low as that of *Chlorobium* cytochrome *c*-555.

Flavocytochromes c

Flavocytochrome *c* was first discovered in *Chromatium vinosum* (Bartsch and Kamen, 1960) and, subsequently, in *Chlorobium thiosulfatophilum* (Meyer *et al.*, 1968), *Thiocapsa roseopersicina* (Bartsch, 1978a), and *Pseudomonas putida* (Hopper and Taylor, 1977). Intensive efforts to find it in other purple and green sulfur bacteria have been fruitless (Table 1). All three known types of flavocytochrome *c* are composed of dissimilar heme and flavin subunits (Meyer, 1970; Bartsch *et al.*, 1968; Yamanaka and Kusai, 1976; Hopper and Taylor, 1977). The flavin is covalently bound through an 8α-cysteinyl thioether (Kenney *et al.*, 1977) to a 46,000-56,000 molecular weight subunit, but the other subunit has a single heme and is 11,000 daltons in *Chlorobium* (Yamanaka and Kusai, 1976), it has two hemes and is 20,000 daltons in *Chromatium* (Kennel and Kamen, 1971; Yamanaka and Kusai, 1976), and it has a single heme and is 56,000 daltons in *Pseudomonas* (Hopper and Taylor, 1977). The flavins of *Chromatium* and *Chlorobium* flavocytochromes *c* react with small molecules such as sulfite, thiosulfate, cyanide, and mercaptans to form what appears to be covalent adducts, which in turn form charge transfer complexes with the protein (Meyer and Bartsch, 1976). It is surprising that the phototrophic bacterial flavocytochromes *c* act as sulfide dehydrogenases (Yamanaka and Kusai, 1976) only in those species which can utilize thiosulfate as electron source in addition to sulfide. The flavocytochromes *c* of *Pseudomonas putida* act as *p*-cresol dehydrogenases (Hopper and Taylor, 1977; Keat and Hopper, 1978).

Membrane Cytochrome c-556 c-552

Light-induced oxidation of cytochromes in *Chromatium vinosum* membranes was studied by Olson and Chance (1960) and Cusanovich *et al.* (1968). There are both high redox potential *c*-556 hemes and low redox potential *c*-552 hemes responsive to light. A single cytochrome band with molecular weight 45,000 was observed on SDS gel electrophoresis of a membrane fragment (Kennel and Kamen, 1971). This cytochrome was solubilized with detergent and had both high (325 mV) and low (8 mV) redox potential hemes similar to what was observed in membranes, but SDS gel electrophoresis of purified cytochrome showed a major and two minor heme-containing bands corresponding to 45,000, 29,000, and 23,000 daltons, respectively (Kennel and Kamen, 1971). There were approximately four hemes assuming a molecular weight of 45,000, but it is not known whether the minor bands observed on electrophoresis represent subunits or proteolytic fragments. Membrane-bound cytochromes *c*-556 *c*-552 have been observed in other purple bacteria (Olson and Nadler, 1965; Meyer *et al.*, 1973; Bartsch, 1978a) and a crystalline, soluble cytochrome with similar redox properties was found in a denitrifying *Alcaligenes* species (Iwasaki and Shidara, 1969a). Attempts to solubilize cytochrome *c*-556 *c*-552 from *Thiocapsa pfennigii* (Meyer *et al.*, 1973) and from *Chromatium vinosum* (Knaff *et al.*, 1979) resulted in release of protein with the low redox potential heme *c*-552 only.

Cytochrome cd

Cytochrome *cd* was discovered by Horio (1958) and purified by Yamanaka and Okunuki (1963). It has two hemes of different structure (Yamanaka, 1966) bound to a single peptide chain of about 63,000 daltons size (Kuronen and Ellfolk, 1972). It is an inducible protein found in most of the denitrifying bacterial species, where it acts as a nitrite reductase (Yamanaka and Okunuki, 1974). It is not the only dissimilatory nitrite reductase to be found in denitrifiers, e.g., an *Alcaligenes* species uses a copper-containing nitrite reductase (Miyata and Mori, 1969). Cytochrome *cd* is rapidly air-oxidized and serves as a useful model for study of the mechanism of electron transport as its reaction with a number of cytochromes as well as azurin has been studied (Yamanaka and Okunuki, 1968; Yamanaka, 1976; Wharton *et al.*, 1973). A crystal structure analysis has been initiated (Takano *et al.*, 1979).

Cytochrome c_3

The *Desulfovibrio* cytochromes c_3 are generally tetraheme proteins of 13,000 daltons size and have very low redox potentials (near minus 200 mV) (Postgate, 1956; Meyer *et al.*, 1971a; Yagi and Maruyama, 1971; DerVartanian and LeGall, 1974). The amino acid sequences are unrelated to any of the other cytochrome classes which have been studied (Ambler, 1968; Ambler *et al.*, 1969, 1971). The three-dimensional structure (Higuchi *et al.*, 1981) shows that both extraplanar ligands are histidines. These multi-heme proteins are probably the result of successive gene doubling events. A triheme cytochrome from *Desulfuromonas acetoxidans* appears to be related to the tetraheme cytochromes c_3 through deletion of a section of peptide chain which contained a heme binding site (Ambler, 1971). *Desulfovibrio* species have a second cytochrome c_3 of larger molecular weight, but these are less completely characterized than the 13,000 dalton proteins. The large cytochrome c_3 from *D. gigas* is 26,000 daltons (LeGall and Forget, 1978), but that from *D. vulgaris* is 85,000 daltons (Meyer and Kamen, unpublished). Cytochromes with electronic absorption spectra, redox potential, and presumably ligands identical to the *Desulfovibrio* cytochromes c_3 have been sporadically found in photosynthetic bacteria and algae (Holton and Meyers, 1967; Yamanaka *et al.*, 1967; Meyer *et al.*, 1971; Yoch *et al.*, 1978; Ho *et al.*, 1979), but these cytochromes appear to have a lower heme content and may not be homologous to the *desulfovibrio* cytochromes. On the other hand, structural relatedness would be an important result not only from the distributional and functional standpoints, but in charting mutational development of this multiheme protein. Three-dimensional structural analyses are forthcoming for additional tetraheme and triheme cytochromes (Frey *et al.*, 1976; Haser *et al.*, 1979). Cytochromes c_3 function as both effectors and electron acceptors for *Desulfovibrio* hydrogenase (Yagi, 1970). The cytochromes c_3 potentially can act as monomolecular electron transport chains (DerVartanian and LeGall, 1974) because the individual hemes differ by as much as 120 mV in redox potential (Niki *et al.*, 1979).

Cytochromes c'

Cytochrome *c'* is the most widely occurring *c*-type cytochrome known in bacteria and also one of the first to be discovered (Vernon and Kamen, 1954). As shown in Table 1, cytochromes *c'* are found in both purple photosynthetic bacterial families (Bartsch, 1978a) and in several denitrifying bacterial species in the genera *Pseudomonas, Paracoccus,* and *Alcaligenes* (Iwasaki, 1960; Kono and Taniguchi, 1960; Iwasaki and Shidara, 1969b) plus the strictly aerobic *Azotobacter vinelandii* (Yamanaka and Imai, 1972). Cytochrome *c'* is most often found in dimeric form, and the monomer peptide chain contains about 125 amino acid residues, with the single heme bound near the carboxyl terminus. The amino acid sequences (Ambler, 1973; Meyer *et al.*, 1975; Ambler *et al.*, 1979b; Ambler *et al.*, 1981) have accumulated a greater amount of amino acid substitution than have those of the cytochromes c_2 and are, therefore, more difficult to align and interpret. The helix content (Imai *et al.*, 1969) is much higher than for other cytochromes and the three-dimensional structure is unique (Weber *et al.*, 1980). Cytochromes *c'* are in an unusual high spin electronic configuration (Maltempo and Moss, 1976) and the heme has no sixth ligand at all. However, the cytochromes *c'* are unreactive with most potential ligands (Taniguchi and Kamen, 1965), in contrast to members of the globin family. In spite of this fact, there are low-spin forms of the cytochromes *c'* as revealed by amino acid sequence studies of *Rps. palustris* cytochrome *c*-556.

Palustris Cytochrome c-556

The monomeric cytochromes *c*-556 and the oligomeric cytochromes *c*-554 (Bartsch, 1978a) have electronic absorption spectra which are distinguished in red-shifted detail from the other low-spin cytochrome *c* spectra, but often have some superimposed high-spin character. The wavelength maxima and extinction coefficients are similar to those of the cyanide complexes of *c*-type cytochromes (Keilin, 1966). The amino acid sequence of *Rhodopseudomonas palustris* cytochrome *c*-556 (Ambler *et al.*, 1981) shows that it is in the same family as cytochrome *c*', with the heme bound near the C-terminus. The cytochromes *c*-554 are generally only trace components of bacteria and as a result are poorly characterized. The known distribution of both the cytochromes *c*-556 and *c*-554 is rather erratic (Table 1), but as their properties become more clearly defined, the relative number of species giving a positive test for occurrence should increase. There is one example of a cytochrome *c*-556 from a nonphotosynthetic source, the plant pathogen, *Agrobacterium tumefaciens* (Hirata and Fukui, 1968; Van den Branden *et al.*, 1975), based on amino acid sequence data as well as electronic absorption spectra (Van Beeumen *et al.*, 1980). The evidence for occurrence of both cytochrome *c*-556 and *c*-554 in the aerobic sulfur bacterium *Thiobacillus neapolitanus* is based only on electronic absorption spectra (Trudinger, Meyer, Bartsch, Kamen, in preparation).

FERREDOXINS

The ferredoxins (or electron transport non-heme iron proteins) are known to contain at least three types of chromophore, increasing in complexity from that provided by the single iron of rubredoxin to the 2-Fe-S cluster of "plant" ferredoxin, and culminating in the 4-Fe-S clusters of "bacterial" ferredoxin, and HiPIP. Each prosthetic group is covalently bound to the protein by cysteine residues and additionally stabilized by hydrogen bonding. The peptide chains of these four ferredoxin classes are apparently unrelated or very distantly related in amino acid sequence.

Bacterial Ferredoxins

Bacterial ferredoxin was discovered by Mortenson *et al.* (1962), and was first characterized from *clostridial* species, but has since been found to be one of the most commonly occurring of the ferredoxins (Lovenberg, 1973, 1977). *Clostridium* ferredoxin has two 4-Fe-S clusters and a sequence of 55 amino acid residues (Tanaka *et al.*, 1966), which is the result of a gene doubling event. The redox potentials of the 4-Fe-S clusters are nearly identical and near minus 400 mV (Sweeney and McIntosh, 1979). Some photosynthetic bacterial ferredoxins such as from *Chlorobium* and *Chromatium* also have two 4-Fe-S clusters with very low redox potentials, but the proteins contain a six-residue insertion plus an "extra" cysteine residue (Tanaka *et al.*, 1974, 1975; Hase *et al.*, 1978a), and *Chromatium vinosum* ferredoxin has a 21-residue C-terminal tail (Hase *et al.*, 1977a). The *Pseudomonas ovalis* bacterial ferredoxin sequence is related to those of *clostridia* but has a two-residue insertion, an extra cysteine, and a 51-residue C-terminal tail (Hase *et al.*, 1978b). The Fe-S clusters in *Azotobacter vinelandii* ferredoxin, which is almost identical to that of *Pseudomonas* (Hase *et al.*, 1978b), have remarkably different redox potentials (+350 and -420 mV) and the EPR spectra are atypical for both clusters (Sweeney *et al.*, 1975). The *Azotobacter* ferredoxin is unusual in that it has one 4-Fe-S cluster and one 3-Fe-S cluster (Stout, 1979; Ghosh *et al.*, 1981). The *Desulfovibrio* ferredoxins are even more divergent in having an apparently degenerate peptide chain with a single 4-Fe-S cluster (Travis *et al.*, 1971). *Bacillus stearothermophilus* ferredoxin has a single 4-Fe-S cluster in an 81-residue peptide chain, but there is little resemblance to other bacterial ferredoxins (Hase *et al.*, 1976). Although the above described bacterial ferredoxins all appear to be related to one another, they probably represent at least four separate subgroups of protein, which have different functional roles. In support of this idea, four bacterial ferredoxins have been described in *Rhodospirillum rubrum,* one of which has similar properties to the *Pseudomonas Azotobacter* proteins, another which may be similar to the *Chromatium* and *Chlorobium* ferredoxins, and one which has a single 4-Fe-S cluster (Yoch *et al.*, 1975, 1977).

HiPIP

HiPIP is a ferredoxin with a single 4-Fe-S cluster and a +350 mV redox potential, originally encountered in *Chromatium vinosum* (Bartsch, 1963). HiPIP has been found in eight species of purple bacteria representing two families and a halophilic, denitrifying *Paracoccus sp.* (Bartsch, 1978b). The amino acid sequences (Dus *et al.*, 1971; Tedro *et al.*, 1974, 1976, 1977, 1979) are unrelated to those of the bacterial ferredoxins and the three-dimensional structure is unique (Carter *et al.*, 1974). The sequences in fact are characterized by a tremendous amount of change which suggests that their functional role is likely to be rapidly evolving (Tedro *et al.*, 1979). The kinetics of oxidation-reduction are ionic strength dependent, but there is apparently nonconservation of charge at the as yet unknown site of electron transfer (Mizrahi *et al.*, 1976; Mizrahi, 1977), further suggesting nonconservation of function.

2-Fe-S Ferredoxins

"Plant" ferredoxin was characterized by Tagawa and Arnon (1962) and identified as the "photosynthetic pyridine nucleotide reductase" (PPNR) of San Pietro and Lang (1958). Plant ferredoxin has a single low redox potential 2-Fe-S cluster bound to four cysteines in a peptide chain of approximately 100 amino acid residues, and is found in plants, algae, and blue-green bacteria. The amino acid sequences of plant ferredoxin (Matsubara *et al.*, 1967), adrenal 2-Fe-S ferredoxin (Tanaka *et al.*, 1973), and *Pseudomonas putida* 2-Fe-S ferredoxin (Tanaka *et al.*, 1974), are related to one another, although each fills a different functional role. The three-dimensional structure of plant ferredoxin is unrelated to that of other ferredoxins (Fukuyama *et al.*, 1980). The strictly aerobic bacterium, *Halobacterium halobium,* has a 2-Fe-S ferredoxin very similar to plant ferredoxin in physicochemical properties (Kerscher *et al.*, 1976) and in amino acid sequence (Hase *et al.*, 1977b). A 2-Fe-S ferredoxin has been found in nitrogen-fixing cells of *Clostridium pasteurianum* (Hardy *et al.*, 1965; Cardenas *et al.*, 1976) and *Azotobacter vinelandii* (Shethna *et al.*, 1964; Der-Vartanian *et al.*, 1969) which has a single iron-sulfur cluster in a 24,000 dalton peptide chain and appears to be active in protecting nitrogenase against air inactivation (Scherings *et al.*, 1977). *Rhodopseudomonas palustris* has a 100,000 dalton 2-Fe-S ferredoxin (Yamanaka and Kamen, 1967; Meyer and Bartsch, unpublished), but it is still incompletely characterized and possible affinity to the other 2-Fe-S ferredoxins is unknown. This brief account does not include the complex ferredoxins which contain functional groups in addition to 2-Fe-S clusters, e.g., xanthine dehydrogenase, but it is likely that the 2-Fe-S ferredoxins are as varied a group as the bacterial ferredoxins and serve as many if not a greater number of functional roles.

Rubredoxins

Rubredoxin is a single-iron ferredoxin discovered by Lovenberg and Sobel (1965). Rubredoxin from anaerobic bacteria, such as *Clostridium* (Lovenberg and Sobel, 1965), *Desulfovibrio* (LeGall and Dragoni, 1966), *Peptococcus* (Bachmayer *et al.*, 1967a), *Megasphaera* (Atherton *et al.*, 1966), and *Chlorobium* (Meyer *et al.*, 1971b) has about 55 amino acid residues and a redox potential of about minus 50 mV. Iron-binding cysteine pairs at opposite ends of the peptide chain are notable characteristics of the sequences (Bachmayer *et al.*, 1967b, 1968; Bruschi *et al.*, 1976a,b). The crystal structure shows the iron to be tetrahedrally coordinated (Herriott *et al.*, 1970). The sole aerobic representative, from *Pseudomonas oleovorans* (Peterson *et al.*, 1966), appears to be the result of gene duplication and fusion (Benson *et al.*, 1971) and is the only rubredoxin for which a specific functional role has been identified.

Hydrogenase

Hydrogenase has been highly purified from a variety of bacterial sources as reviewed by Mortenson and Chen (1974). A recent symposium (Schlegel and Schneider, 1978) documents the latest advances in a rapidly developing field of study. Hydrogenase is generally believed to be a 4-Fe-S ferredoxin with minimal molecular weight approximately 50,000-60,000 daltons. However, there is still debate with respect to the number of iron sulfur clusters, subunit structure, and catalytic activity. The enzymes from typical anaerobes such as *Clostridium pasteurianum* (Nakos and Mortenson, 1971a,b; Chen and Mortenson, 1974; Erbes *et al.*, 1975) and *Desulfovibrio gigas* (Bell *et al.*, 1974; Hatchikian *et al.*, 1978) are soluble proteins which appear to have three 4-Fe-S clusters per 60,000 daltons, and are oxygen-labile. *Desulfovibrio vulgaris* also has a soluble hydrogenase with twelve irons and molecular weight 50,000 daltons, but it is oxygen-stable and is catalytically active with cytochrome c_3 as well as viologen dyes (LeGall *et al.*, 1971; Van Der Westen *et al.*, 1978). *Desulfovibrio vulgaris* strain Miyazaki has a membrane-bound hydrogenase, which was solubilized with pancreatin, resulting in an 89,000-dalton fragment composed of 28,000 and 59,000 dalton subunits (Yagi, 1970, Yagi *et al.*, 1976). Cytochrome c_3 acts both as an effector and as electron acceptor for this hydrogenase. Hydrogenases from purple photosynthetic bacteria are primarily membrane-bound proteins, which are reported to have a single 4-Fe-S cluster per 65,000-98,000 daltons, and to be relatively inactive with any electron carrier other than viologen dyes (Gitlitz and Krasna, 1975; Kakuno *et al.*, 1977; Adams and Hall, 1977; Gogotov *et al.*, 1978). Relative thermal stabilities of solubilized versus membrane-bound *Chromatium vinosum* hydrogenase was studied by Klibanov, *et al.*, 1980). The membrane-bound hydrogenases of the facultative anaerobe, *Proteus mirabilis* (Schoenmaker *et al.*, 1979) and the aerobic hydrogen bacterium *Alcaligenes eutrophus* (Schink and Schlegel, 1979) are reported to have molecular weight 100,000 - 200,000 and to have dissimilar subunits. A soluble hydrogenase from *Alcaligenes eutrophus* is different from the membrane-bound protein and is unique in its ability to interact with pyridine nucleotides (Schneider and Schlegel, 1976).

The functional role of hydrogenases is remarkably variable. The enzyme in *Clostridium* and the coliform bacteria evolves hydrogen as the end product of fermentation, whereas in the photosynthetic bacteria, in *Desulfovibrio* and in *Alcaligenes*, hydrogen is taken up as an oxidizable electron source. Hydrogenase may act in concert with nitrogenase in the nitrogen-fixing bacteria to recover lost reducing equivalents (Smith *et al.*, 1976). Hydrogenase figures heavily in solar hydrogen production schemes utilizing chloroplasts (Benemann *et al.*, 1973), but one of the major obstacles to implementation was in finding a hydrogenase which was both reactive with ferredoxins as cofactors and oxygen-stable. This problem was overcome when the enzymatically very active *Clostridium* hydrogenase was stabilized against oxygen inactivation through adsorption to polycationic exchange media (Klibanov *et al.*, 1978), and through addition of metal-chelating agents (Klibanov *et al.*, 1979).

COPPER PROTEINS

There are at least three classes of electron transport copper proteins, two of which are distantly related. Plastocyanin is a 10,000 dalton molecular weight protein discovered by Katoh (1960), which has a single, high redox potential copper atom bound tetrahedrally to two histidines, a cysteine, and a methionine residue (Colman *et al.*, 1978) and it appears to be widely distributed in plants, algae, and blue-green bacteria (Aitken, 1978). Azurin was first purified and characterized by Horio (1958a,b). Azurin commonly occurs in many denitrifying bacteria in the genera *Pseudomonas*, *Alcaligenes*, and *Bordetella* (Horio, 1958a,b; Suzuki and Iwasaki, 1962; Sutherland and Wilkinson, 1963). The amino acid sequences of the azurins are homologous to those of the plastocyanins, but azurin is one-third larger than the 100 amino acid residue peptide chain of plastocyanin (Ambler and Brown, 1967; Ambler, 1977; Aitken, 1978). The three-dimensional structures of both plastocyanin (Colman *et al.*, 1978) and azurin (Adman *et al.*, 1978) have been reported. The partial sequence of a copper protein from the denitrifier, *Achromobacter cycloclastes*, appears to be closer to plastocyanin than to azurin (Ambler, 1977). The 150,000 dalton and 69,000 dalton nitrite reductases from an *Alcaligenes sp.* and from *Achromobacter cycloclastes*, respectively, are copper proteins which seem to fill the same role as cytochrome *cd* in other denitrifying bacteria (Suzuki and Iwasaki, 1962; Iwasaki *et al.*, 1963; Miyata and Mori, 1969; Iwasaki and Matsubara, 1972; Iwasaki *et al.*, 1975). A 90,000 dalton copper protein from *Alcaligenes faecalis* does not reduce nitrite, but this enzyme activity is expressed by a cytochrome *cd* instead (Matsubara and Iwasaki, 1972).

FUNCTIONAL ROLE OF BACTERIAL ELECTRON TRANSPORT PROTEINS

The study of electron transport proteins is quite different from the study of enzymes in the sense that the physicochemical and molecular characteristics of the cytochromes and ferredoxins are often well known before a specific functional role can be determined. The comparative approach to study of electron transport followed in the Kamen laboratory has provided some insights into functional roles which would not have been possible by concentrating effort on one or a few species. Cytochromes are photooxidized as a secondary consequence of absorption of light by chlorophyll in photosynthetic bacteria (Duysens, 1954; Olson and Chance, 1960; Olson and Nadler, 1965; Dutton *et al.*, 1975), but identification of specific components of these photosynthetic electron transport pathways has been difficult. As an example, consider the purple bacteria. It appears that cytochrome c_2 is the direct electron donor to active center chlorophyll in *Rhodopseudomonas spheroides* *Rps. capsulata,* and *Rhodospirillum rubrum* (Kakuno *et al.*, 1971; Dutton *et al.*, 1975; Prince *et al.*, 1975; Hochman and Carmeli, 1977), and presumably in most other species which contain cytochrome c_2 (see Table 1). However, in *Chromatium vinosum* and *Rhodopseudomonas viridis,* a membrane-bound cytochrome *c*-556-*c*-552 serves in this capacity (Olson and Chance, 1960; Olson and Nadler, 1965; Cusanovich *et al.*, 1968; Tiede *et al.*, 1978). This places *Rps. viridis* in an anomalous position in that it has both cytochrome *c*-556 *c*-552 and cytochrome c_2, but only the membrane protein is rapidly oxidized by reaction center (Olson and Nadler, 1965). The cytochrome c_2, in this instance, may serve to poise the redox potential for optimal photophosphorylation. If the membrane-bound cytochrome *c*-556 *c*-552 takes precedence over cytochrome c_2 as donor to reaction center when both are present, then what of the bacteria which have neither? *Rhodospirillum tenue* appears to have neither cytochrome c_2 nor cytochrome *c*-556 *c*-552 (Meyer and Kamen, unpublished), but has a very high redox potential form of a protein related to *Pseudomonas* cytochrome *c*-551 (Ambler *et al.*, 1979b) which could fill that role. Therefore, there are at least three different classes of cytochrome, which appear to fill one particular functional role in purple bacterial photosynthesis, namely membrane-bound cytochrome *c*-556 *c*-552, soluble cytochrome c_2, and soluble cytochrome *c*-553. A characteristic of the proteins which appear to serve in this capacity is their very high redox potential, which is in the vicinity of 350-400 mV (Pettigrew *et al.*, 1975, 1978). The reaction center chlorophyll has a redox potential of about 450-500 mV in the purple bacteria (Cusanovich *et al.*, 1968; Case and Parson, 1973; Dutton *et al.*, 1975; Prince, 1978). *Rhodopseudomonas viridis* cytochrome c_2, which occurs with cytochrome *c*-556 *c*-552, but is inactive as donor, has a redox potential of 300 mV (Pettigrew, 1978). Cytochrome *c*-551, which occurs together with cytochrome *c*-556 *c*-552 in *Rhodopseudomonas gelatinosa*, has a redox potential estimated to be between 100 - 200 mV (Meyer, Cusanovich, and Kamen, unpublished). The apparent inactivity of *Rps. viridis* cytochrome c_2 and *Rps. gelatinosa* cytochrome *c*-551 as donors to active center chlorophyll may therefore be partially explained by their low redox potentials. From this argument, it might be supposed that the *Chlorobium* cytochromes *c*-555, with their relatively low redox potentials, might be inactive with green bacterial active center chlorophyll, but in this example, the reaction center has a redox potential of only 250 mV (Fowler *et al.*, 1971; Prince and Olson, 1976), which probably reflects the more anaerobic habitat of the green bacteria.

The purple photosynthetic bacterial electron transport pathway apparently can use any of three classes of cytochromes interchangeably, and it also appears that this pathway is not the only one in which cytochrome c_2, for example, may participate. Cytochrome c_2 has been given a dual role: (1) in photosynthetic electron transport, and (2) in aerobic respiration in those purple bacteria which are facultatively aerobic (Baccarini-Melandri *et al.*, 1978). In the nonphotosynthetic, denitrifying *Paracoccus denitrificans*, cytochrome c_2 is reported to function analogously to mitochondrial cytochrome *c* in a very similar pathway (Kamen and Vernon, 1955; Smith *et al.*, 1966; John and Whatley, 1975), but anaerobically grown cells have increased quantities of the cytochrome c_2, which presumably functions as electron donor to a cytochrome *cd* type of nitrite reductase (Newton, 1969; Timkovich and Robinson, 1979). Therefore, cytochrome c_2 may have four functions: (1) as immediate electron donor to reaction center chlorophyll in those species which do not have membrane-bound cytochrome *c*-556 *c*-552, (2) as electron donor to a bacterial cytochrome oxidase and ultimately oxygen in the facultative aerobes, (3) as electron donor to a cytochrome *cd* type of nitrite reductase in *Paracoccus denitrificans*, and (4) as a soluble redox potential poising pool in all purple bacteria in which it is found.

In plants, an analogy has been made between the electron transport pathway connecting photosystems I and II, and that section of the mitochondrial electron transport pathway known as complex III (Wood, 1977; Gray, 1978; Malkin and Posner, 1978). Both systems are composed of quinone, *b*-type cytochromes, a high redox potential iron-sulfur center, and a membrane-bound *c*-type cytochrome known as cytochrome *f* in plants and as cytochrome c_1 in mitochondria. Both plant cytochrome *f* and mitochondrial cytochrome c_1 are remarkably distinctive membranous, monoheme proteins with minimum molecular weight near 30,000 daltons (Nelson and Racker, 1972; Yu *et al.*, 1972). To extend the analogy between these sections of the respective pathways to its limit, these apparently similar electron transport chains may be the result of a complete duplication of a multigene unit. There is a strong tendency to look for unity in nature, in spite of the diversity emphasized in this paper, and at the risk of arousing latent feelings of mitochondrial chauvinism, it is tempting to extend the chloroplast/mitochondrial analogy to the purple bacteria. In support of this analogy, membrane-bound *b*-type cytochromes are recognized in such species as *Chromatium vinosum, Rhodopseudomonas spheroides, Rps. capsulata, Rhodospirillum rubrum,* and *Rhodomicrobium vannielii* (Morita and Conti, 1963; Kakuno *et al.*, 1971; Dutton and Wilson, 1974; Knaff *et al.*, 1979). There is also a membranous cytochrome *c*-553 in some, but apparently not all, of these same bacteria, which remains to be isolated and characterized (Morita and Conti, 1963; Wood, 1980; Meyer, Bartsch, and Kamen, unpublished). Quinones are identified as the primary electron acceptors, and there is an antimycin A binding site, plus a high redox potential iron-sulfur center (Dutton and Wilson, 1974). The aerobic electron transport pathway of some of these bacteria has been shown to be constitutive, whereas photosynthesis is induced by light and anaerobicity (Oelze and Drews, 1972). Unless portions of the photosynthetic and aerobic pathways are identical, observations of a constitutive aerobic pathway could lead to the wrong conclusions about photosynthesis. Such a difficulty might be avoided if strictly anaerobic bacteria are studied more intensively. As a working hypothesis, analogy between mitochondrial complex III and purple bacterial photosynthesis may be quite useful, provided proper distinction between fact and conjecture is maintained. It could explain why mitochondrial cytochrome *c* and cytochrome c_2 appear to have similar functional constraints imposed upon their structures (Ambler *et al.*., 1979a) and why cytochrome c_2 is as active as mitochondrial cytochrome *c* with bovine cytochrome *c* reductase (Errede and Kamen, 1977).

DEVELOPMENT OF A MOLECULAR CONCEPT OF THE BACTERIAL SPECIES BASED ON ELECTRON TRANSPORT PROTEIN STRUCTURE AND DISTRIBUTION

Bacterial taxonomy is based primarily on morphological examination and on simple physiological tests which can readily be scored as plus or minus, but in most instances, the detailed molecular basis for variable response to such tests is unknown and the reliability of these tests is sometimes questionable. Numerical taxonomy is an improvement over the old methods in attempting to statistically utilize as much data as can be gathered, but falls short of a natural classification. DNA cross-hybridization is an additional improvement, but is limited to similar species. Judging from the relative success in using protein sequences to construct evolutionary trees for animals, it has been assumed that sequences can provide long needed quantitative data for creation of a natural classification of bacteria (Dayhoff, 1978; Dayhoff and Schwartz, 1978; Almassy and Dickerson, 1978). Such a natural classification would obviously be extremely useful in medical, agricultural, and industrial identification if fully developed.

Ambler (1977) has determined the amino acid sequences of a large number of bacterial proteins, partly to examine feasibility of constructing phylogenetic trees patterned after those constructed from eukaryotic protein sequences. Because animals cannot transfer genetic information between species (by definition) and this process is well-documented for bacteria particularly in transfer of drug resistance factors, this potential difficulty has to be evaluated. While there is no hard evidence for transfer of electron transport protein genes between bacterial species, there are a number of anomalies which suggest just such occurrences (Ambler, 1977; Tedro *et al.*, 1977; Ambler *et al.*, 1979a,b). There are other possible explanations for these anomalies, but until each example is adequately dealt with, gene transfer remains the simplest of alternatives. Furthermore, it is apparent that the sequences of most species of cytochrome c_2 (Ambler *et al.*, 1979a) have reached a limit of change and are roughly equidistant from one another and from the mitochondrial cytochromes *c*. This result shows that evolutionary trees, in fact, cannot be constructed from bacterial protein sequences and a natural classification patterned after that of animals is virtually impossible even if gene transfer is relatively infrequent (Meyer and Kamen, in preparation).

This is not to say that protein structures are of no present use in bacterial classification. At the very least, both sequences and crystal structures will aid in classification of the proteins, allowing occurrences in the various species to be more accurately mapped so that protein distributions can be used in a manner similar to other bacteriological tests. More importantly, protein sequences can be used to formulate a molecular concept of the bacterial species, which at this writing is quite arbitrary. The question as to how much mutational variation is necessary to establish a species is the fundamental problem. If we assume for the moment that our observations with a limited number of species are generally applicable, then gene transfer exists, but is not of common occurrence for bacterial electron transport proteins. The sequences of cytochromes c_2 appear to be equidistant from one another and at a limit of change as inferred by the continuum of structural change with mitochondrial cytochrome *c*. Different strains of the same species differ from one another by a much smaller amount for those species which have been examined. A new purple bacterial species, therefore, might be constituted from a group of strains which have proteins differing from analogous proteins by amounts equal to those observed among previously recognized and distinctive species. Continuing to use the purple bacterial cytochromes c_2 as an example, *Rsp. molischianum* and *fulvum* appear to be incorrectly distinguished as two species because their cytochromes c_2 differ by a small amount, which is no more than recognized among strains of *Rps. palustris* (Ambler *et al.*, 1979a). *Rhodopseudomonas molischianum* and *fulvum* also have isocytochromes c_2 and cytochromes c', which differ to the same extent as found for their predominant cytochromes c_2. The three principle microbiological characters separating *molischianum* and *fulvum* are: *p*-aminobenzoate vitamin requirement, cellular size, and utilization of benzoate as carbon source. Each of these characters may be the result of a single mutation or gene transfer event, whereas the numbers of mutations required to change a single small protein sequence to the maximum allowed by its function may be in the hundreds. The numbers of mutations observed for cytochromes c_2 in strains of *Rps. palustris* for which there are no known microbiologically distinguishing characters may be extrapolated into hundreds of thousands for the whole bacterial genome.

The anomalous occurrence of cytochrome c_2 in *Paracoccus denitrificans* could be due either to gene transfer or to mutational deletion of photosynthetic ability. This might be determined by close comparison of *Paracoccus* and *Rps. capsulata*, that photosynthetic bacterium which has a cytochrome c_2 most similar to the *Paracoccus* cytochrome c_2. On the other hand, the numbers of mutations separating these species may be so large as to preclude formulation of definitive conclusions. This example serves to illustrate the difficulty in actually proving a gene transfer event which may have occurred in the distant past.

Another example may be cited - that of *Pseudomonas* cytochrome *c*-551. The cytochrome *c*-551 from strains of *Ps. fluorescens* shows a small amount of variability, whereas *Ps. aeruginosa* strains are essentially identical to one another (Ambler, 1974). The conservative *Ps. aeruginosa* situation provides an argument against neutral mutation (Ambler, 1974), but another explanation is that the microbiological tests for distinguishing *Ps. aeruginosa* may be among the most selective yet discovered. *Azotobacter vinelandii* cytochrome *c*-551, which is no more different than cytochromes *c*-551 of *Pseudomonas* species, is an anomaly, which might be ascribed to gene transfer, but could alternatively be explained if the cytochrome *c*-551 has reached a functionally imposed limit of change similar to that found in cytochromes c_2, which does not reflect the additional mutations which cause *Azotobacter vinelandii* to be placed in a genus separate from *Pseudomonas*.

Establishment of bacterial taxa above the species level may utilize both microbiological properties and distribution of well-defined protein classes such as described for the photosynthetic and the denitrifying bacteria and their allies. For example, both the distribution of electron transport proteins (Table 1) and such classic microbiological properties as utilization of sulfide and sulfur storage suggest that separation of the purple bacteria into two families has no valid basis. There is clearly an intermingling of characters between these artificially separate families. Similarity of electron transport proteins in *gelatinosa* and *tenue*, and among *globiformis*, *vannielli*, and *acidophila* suggest that present genera, which are decided primarily on gross morphology, are not natural, a conclusion already expressed by Pfennig (1977). This is to be contrasted with the relatively consistent species assignments. If morphology is not an absolutely reliable character for establishment of genera, what characters might be more selective? It has been found that the proteins of halophilic bacteria are very acidic (Larsen, 1967), and we have found that to be true of the HiPIP (Tedro *et al.*, 1977), cytochrome *c*', and cytochrome *c*-554-548 (Ambler, Meyer, and Kamen, unpublished), in a halophilic *Paracoccus* species. These protein sequences suggest that adaptation of *Paracoccus* to the halophilic environment was a slow process of accumulation of substitutions of acidic residues for basic ones. Therefore, halophilic, and to a lesser extent obligate marine bacteria should have diverged a great deal from fresh-water forms and halophily might be considered to be one of the more reliable bacteriological tests of species. There may be other ecological adaptations which require a very large accumulation of mutations, which may be useful in bacterial classification, and it is expected that the distribution and amino acid sequences of electron transport proteins will continue to contribute to a deeper understanding of bacteria and their almost limitless variety.

ACKNOWLEDGMENT

This work was supported by grant GM 18528 from the National Institutes of Health.

REFERENCES

Adams, M. W. W. and Hall, D. O. (1977). *Biochem. Biophys. Res. Commun. 77*, 730-737.

Adman, E. T., Stenkamp, R. E., Sieker, L. C., and Jensen, L. H. (1978). *J. Mol. Biol. 123*, 35-47.

Aitken, A. (1976). *Nature (London) 263*, 793-796.

Aitken, A. (1978). *In* "Evolution of Protein Molecules" (H. Matsubara and T. Yamanaka, eds.), pp. 251-263. Japan Scientific Soc. Press, Tokyo.

Almassy, R. J. and Dickerson, R. E. (1978). *Proc. Natl. Acad. Sci. U.S. 75*, 2674-2678.

Ambler, R. P. (1963). *Biochem. J. 89*, 349-378.

Ambler, R. P. (1968). *Biochem. J. 109*, 47P.

Ambler, R. P. (1971). *FEBS Lett. 18*, 351-353.

Ambler, R. P. (1973). *Biochem. J. 135*, 751-758.

Ambler, R. P. (1974). *Biochem. J. 137*, 3-14.

Ambler, R. P. (1977). *In* "The Evolution of Metalloenzymes, Metalloproteins and Related Compounds" (G. J. Leigh, ed.), pp. 100-118. Symposium Press, London.

Ambler, R. P. and Bartsch, R. G. (1975). *Nature (London) 253*, 285-288.

Ambler, R. P. and Brown, L. H. (1967). *Biochem. J. 104*, 784-825.

Ambler, R. P. and Murray, S. (1973). *Trans. Biochem. Soc. 1*, 162-164.

Ambler, R. P. and Taylor, E. (1973). *Trans. Biochem. Soc. 1*, 166-168.

Ambler, R. P. and Wynn, M. (1973). *Biochem. J. 131*, 485-498.

Ambler, R. P., Bruschi, M., and LeGall, J. (1969). *FEBS Lett. 5*, 115-117.

Ambler, R. P., Bruschi, M., and LeGall, J. (1971). *FEBS Lett. 18*, 347-350.

Ambler, R. R., Meyer, T. E., and Kamen, M. D. (1976). *Proc. Natl. Acad. Sci. U.S. 73*, 472-475.

Ambler, R. P., Daniel, M., Hermoso, J., Meyer, T. E., Bartsch, R. G., and Kamen, M. D. (1979a). *Nature (London) 278*, 659-660.

Ambler, R. P., Meyer, T. E., and Kamen, M. D. (1979b). *Nature (London) 278*, 661-662.

Ambler, R. P., Bartsch, R. G., Daniel, M., Kamen, M. D., McLellan, L., Meyer, T. E., and Van Beeumen, J. (1981). *Proc. Natl. Acad. Sci. U.S. 78*. In press.

Atherton, N. M., Garbett, K., Gillard, R. D., Mason, R., Mayhew, S. J., Peel, J. L., and Stangroom, J. E. (1966). *Nature (London) 212*, 590-593.

Baccarini-Melandri, A., Jones, O. T. G., and Hauska, G. (1978). *FEBS Lett. 86*, 151-154.

Bachmayer, H., Piette, L. H., and Yasunobu, K. T. (1967a). *Proc. Natl. Acad. Sci. U.S. 57*, 122-127.

Bachmayer, H., Yasunobu, K. T., and Whitely, H. R. (1967b). *Biochem. Biophys. Res. Commun. 26,* 435-440.

Bachmayer, H., Yasunobu, K. T., Peel, J. L., and Mayhew, S. (1968). *J. Biol. Chem. 243,* 1022-1030.

Bartsch, R. G. (1963). *In* "Bacterial Photosynthesis" (H. Gest, A. San Pietro, and L. P. Vernon, eds.), pp. 315-326. Antioch Press, Yellow Springs, Ohio.

Bartsch, R. G. (1978a). *In* "The Photosynthetic Bacteria" (R. K. Clayton and W. R. Sistrom, eds.), pp. 249-279. Plenum, New York.

Bartsch, R. G. (1978b). *Methods Enzymol. 53,* 329-340.

Bartsch, R. G. and Kamen, M. D. (1960). *J. Biol. Chem. 235,* 825-831.

Bartsch, R. G., Meyer, T. E., and Robinson, A. B. (1968). *In* "Structure and Function of Cytochromes" (K. Okunuki, M. D. Kamen, and I. Sekuzu, eds.), pp. 443-451. University Park Press, Baltimore, Maryland.

Bell, G. R., Le Gall, J., and Peck, H. D., Jr. (1974). *J. Bacteriol. 120,* 994-997.

Benemann, J. R., Berenson, J. A., Kaplan, N. O., and Kamen, M. D. (1973). *Proc. Natl. Acad. Sci. U.S. 70,* 2317-2320.

Benson, A., Tomoda, K., Chang, J., Matsueda, G., Lode, E. T., Coon, M. J., and Yasunobu, K. T. (1971). *Biochem. Biophys. Res. Commun. 42,* 640-646.

Boulter, D., Haslett, B. G., Peacock, D., Ramshaw, J. A. M., and Scawen, M. D. (1977). *Intern. Rev. Biochem. 13,* 1-40.

Bruschi, M. and LeGall, J. (1972). *Biochim. Biophys. Acta 271,* 48-60.

Bruschi, M., Bonicel, J., Bovier-Lapierre, G., and Couchoud, P. (1976a). *Biochim. Biophys. Acta 434,* 4-17.

Bruschi, M., Bonicel, J., Bovier-Lapierre, G., and Couchoud, P. (1976b). *Biochem. Biophys. Res. Commun. 70,* 615-621.

Campbell, W. H., Orme-Johnson, W. H., and Burris, R. H. (1973). *Biochem. J. 135,* 617-630.

Cardenas, J., Mortenson, L. E., and Yoch, D. C. (1976). *Biochim. Biophys. Acta 434,* 244-257.

Carter, C. W., Jr., Kraut, J., Freer, S. T., Xuong, N. H., Alden, R. A., and Bartsch, R. G. (1974). *J. Biol. Chem. 249,* 4212-4225.

Case, G. D. and Parson, W. W. (1973). *Biochim. Biophys. Acta 292,* 677-684.

Chen, J.-S. and Mortenson, L. E. (1974). *Biochim. Biophys. Acta 371,* 283-298.

Colman, P. M., Freeman, H. C., Guss, J. M., Murata, M., Norris, V. A., Ramshaw, J. A. M., and Venkatappa, M. P. (1978). *Nature (London) 272,* 319-324.

Cusanovich, M. A. and Bartsch, R. G. (1969). *Biochim. Biophys. Acta 189,* 245-255.

Cusanovich, M. A., Bartsch, R. G., and Kamen, M. D. (1968). *Biochim. Biophys. Acta 153,* 397-417.
Dayhoff, M. O. (1978). "Atlas of Protein Sequence and Structure," Vol. 5, Suppl. 3, National Biomedical Research Foundation, Silver Spring, Maryland.
Dayhoff, M. O. and Schwartz, R. M. (1978). *In* "Evolution of Protein Molecules" (H. Matsubara and T. Yamanaka, eds.), pp. 323-342. Japan Sci. Soc. Press, Tokyo.
DerVartanian, D. V. and LeGall, J. (1974). *Biochim. Biophys. Acta 346,* 79-99.
DerVartanian, D. V., Shethna, Y. I., and Beinert, H. (1969). *Biochim. Biophys. Acta 194,* 548-563.
Dus, K., Sletten, K., and Kamen, M. D. (1968). *J. Biol. Chem. 243,* 5507-5518.
Dus, K., Tedro, S. M., Bartsch, R. G., and Kamen, M. D. (1971). *Biochem. Biophys. Res. Commun. 43,* 1239-1245.
Dutton, P. L. and Wilson, D. F. (1974). *Biochim. Biophys. Acta 346,* 165-212.
Dutton, P. L., Petty, K. M., Bonner, H. S., and Morse, S. F. (1975). *Biochim. Biophys. Acta 387,* 536-556.
Duysens, L. N. M. (1954). *Nature (London) 173,* 692-693.
Erbes, D. L., Burris, R. H., and Orme-Johnson, W. H. (1975). *Proc. Natl. Acad. Sci. U.S. 72,* 4795-4799.
Errede, B. J. and Kamen, M. D. (1978). *Biochemistry 17,* 1015-1027.
Fauque, G., Bruschi, M., and LeGall, J. (1979). *Biochem. Biophys. Res. Commun. 86,* 1020-1029.
Fowler, C. F., Nugent, N. A., and Fuller, R. G. (1971). *Proc. Natl. Acad. Sci. U.S. 68,* 2278-2282.
Frey, M., Haser, R., Pierrot, M., Bruschi, M., and LeGall, J. (1976). *J. Mol. Biol. 104,* 741-743.
Fukuyama, K., Hase, T., Matsumoto, S., Tsukihara, T., Katsube, Y., Tanaka, N., Kakudo, M., Wada, K., and Matsubara, H. (1980). *Nature (London) 286,* 522-524.
Ghosh, D., Furey, W., Jr., O'Donnell, S., and Stout, C. D. (1981). *J. Biol. Chem. 256,* 4185-4192.
Gibson, J. (1961). *Biochem. J. 79,* 151-158.
Gitlitz, P. H. and Krasna, A. I. (1975). *Biochemistry 14,* 2561-2568.
Gogotov, I. N., Zorin, N. A., Zerebriakova, L. T., and Kondratieva, E. N. (1978). *Biochim. Biophys. Acta 523,* 335-343.
Gray, J. C. (1978). *Eur. J. Biochem. 82,* 133-141.
Hardy, R. W. F., Knight, E., Jr., McDonald, C. C., D'Eustachio, A. J. (1965). *In* "Non-Heme Iron Proteins: Role in Energy Conversion" (A. San Pietro, ed.), pp. 275-283. Antioch Press, Yellow Springs, Ohio.
Hase, T., Ohmiya, N., Matsubara, H., Mullinger, R. N., Rao, K. K., and Hall, D. O. (1976). *Biochem. J. 159,* 55-63.

Hase, T., Matsubara, H., and Evans, M. C. W. (1977a). *J. Biochem. 81,* 1745-1749.

Hase, T., Wakabayashi, S., Matsubara, H., Kerscher, L., Oester-Helt, D., Rao, K. K., and Hall, D. O. (1977b). *FEBS Lett. 77,* 308-310.

Hase, T., Wakabayashi, S., Matsubara, H., Evans, M. C. W., and Jennings, J. V. (1978a). *J. Biochem. 83,* 1321-1325.

Hase, T., Wakabayashi, S., Matsubara, H., Ohmori, D., and Suzuki, K. (1978b). *FEBS Lett. 91,* 315-319.

Haser, R., Payan, F., Bache, R., Bruschi, M., and LeGall, J. (1979). *J. Mol. Biol. 130,* 97-98.

Hatchikian, E. C., Bruschi, M., and LeGall, J. (1978). *Biochem. Biophys. Res. Commun. 82,* 451-461.

Herriott, J. R., Sieker, L. C., Jensen, L. H., and Lovenberg, W. (1970). *J. Mol. Biol. 50,* 391-406.

Higuchi, Y., Bando, S., Kusunoki, M., Matsuura, Y., Yasuoka, N., Kakudo, M., Yamanaka, T., Yagi, T., and Inokuchi, H. (1981). *J. Biochem. 89,* 1659-1662.

Hill, R. and Scarisbrick, R. (1951). *New Phytol. 50,* 98-111.

Hirata, H. and Fukui, S. (1968). *J. Biochem. 63,* 780-788.

Ho, K. K., Ulrich, E. L., Krogmann, D. W., and Gomez-Lojero, C. (1979). *Biochim. Biophys. Acta 545,* 236-248.

Hochman, A. and Carmeli, C. (1977). *Arch. Biochem. Biophys. 179,* 349-359.

Holton, R. W. and Myers, J. (1967). *Biochim. Biophys. Acta 131,* 362-374.

Hopper, D. J. and Taylor, D. G. (1977). *Biochem. J. 167,* 155-162.

Horio, T. (1958a). *J. Biochem. 45,* 195-205.

Horio, T. (1958b). *J. Biochem. 45,* 267-279.

Horio, T. and Kamen, M. D. (1970). *Annu. Rev. Microbiol. 24,* 399-428.

Horio, T., Higashi, T., Sasagawa, M., Kusai, K., Nakai, M., and Okunuki, K. (1960). *Biochem. J. 77,* 194-201.

Imai, Y., Imai, K., Ikeda, K., Hamaguchi, K., and Horio, T. (1969). *J. Biochem. 65,* 629-637.

Iwasaki, H. (1960). *J. Biochem. 47,* 174-184.

Iwasaki, H. and Matsubara, T. (1972). *J. Biochem. 71,* 645-652.

Iwasaki, H. and Shidara, S. (1969a). *J. Biochem. 66,* 775-781.

Iwasaki, H. and Shidara, S. (1969b). *Plant & Cell Physiol. 10,* 291-305.

Iwasaki, H., Shidara, S., Suzuki, H., and Mori, T. (1963). *J. Biochem. 53,* 299-303.

Iwasaki, H., Noji, S., and Shidara, S. (1975). *J. Biochem. 78,* 355-361.

John, P. and Whatley, F. R. (1975). *Nature (London) 254,* 495-498.

Kakuno, T., Bartsch, R. G., Nishikawa, K., and Horio, T.

(1971). *J. Biochem. 70,* 79-94.
Kakuno, T., Kaplan, N. O., and Kamen, M. D. (1977). *Proc. Natl. Acad. Sci. U.S. 74,* 861-863.
Kamen, M. D. (1973). *Proteins, Nucleic Acids, Enzymes (Tokyo) 18,* 753-773.
Kamen, M. D. and Horio, T. (1970). *Annu. Rev. Biochem. 39,* 673-700.
Kamen, M. D. and Vernon, L. P. (1954a). *J. Bacteriol. 67,* 617-618.
Kamen, M. D. and Vernon, L. P. (1954b). *J. Biol. Chem. 211,* 663-675.
Kamen, M. D. and Vernon, L. P. (1955). *Biochim. Biophys. Acta 17,* 10-22.
Kamen, M. D., Dus, K., Flatmark, T., and DeKlerk, H. (1972). *In* "Electron and Coupled Energy Transfer in Biological Systems" (T. E. King and M. Klingenberg, eds.), Vol. 1, pp. 243-324. Dekker, N.Y.
Katoh, S. (1960). *Nature (London) 186,* 533-534.
Keat, M. J. and Hopper, D. J. (1978). *Biochem. J. 175,* 649-658
Keilin, D. (1966). "The History of Cell Respiration and Cytochrome." Cambridge, University Press, London.
Kennel, S. J. (1971). Ph.D. Thesis, University of California at San Diego, La Jolla, California.
Kennel, S. J. and Kamen, M. D. (1971). *Biochim. Biophys. Acta 253,* 153-156.
Kenney, W. C., McIntire, W., and Yamanaka, T. (1977). *Biochim. Biophys. Acta 483,* 467-474.
Kerscher, L., Oesterhelt, D., Cammack, R., and Hall, D. O. (1976). *Eur. J. Biochem. 71,* 101-107.
Klibanov, A. M., Kaplan, N. O., and Kamen, M. D. (1978). *Proc. Natl. Acad. Sci. U.S. 75,* 3640-3643.
Klibanov, A. M., Kaplan, N. O., and Kamen, M. D. (1979). *Biochim. Biophys. Acta 547,* 411-416.
Klibanov, A. M., Kaplan, N. O., and Kamen, M. D. (1980). *Arch. Biochem. Biophys. 199,* 545-549.
Knaff, D. B., Worthington, T. M., White, C. C., and Malkin, R. (1979). *Arch. Biochem. Biophys. 192,* 158-163.
Komada, T. and Shidara, S. (1969). *J. Biochem. 65,* 351-360.
Kono, M. and Taniguchi, S. (1960). *Biochim. Biophys. Acta 43,* 419-430.
Korszun, Z. R. and Salemme, F. R. (1977). *Proc. Natl. Acad. Sci. U.S. 74,* 5244-5247.
Kuronen, T. and Ellfolk, N. (1972). *Biochim. Biophys. Acta 275,* 308-318.
Larsen, H. (1967). *In* "Advances in Microbial Physiology" (A. H. Rose and J. F. Wilkinson, eds.), Vol. 1, pp. 97-132. Academic Press, N.Y.
Laycock, M. V. (1972). *Can. J. Biochem. 50,* 1311-1325.
LeGall, J. and Bruschi-Heriaud, M. (1968). *In* "Structure and

Function of Cytochromes" (K. Okunuki, M. D. Kamen, and I. Sekuzu, eds.), pp. 467-470. University Park Press, Baltimore, Maryland.
LeGall, J. and Dragoni, N. (1966). *Biochem. Biophys. Res. Commun. 23,* 145-149.
LeGall, J. and Forget, N. (1978). *Methods Enzymol. 53,* 613-634.
LeGall, J., DerVartanian, D. V., Spilker, E., Lee, J.-P., and Peck, H. D., Jr. (1971). *Biochim. Biophys. Acta 234,* 525-530.
Lemberg, R. and Barrett, J. (1973). "Cytochromes." Academic Press, N.Y.
Lovenberg, W. (1973). "Iron Sulfur Proteins," Vol. 3. Academic Press, N.Y.
Lovenberg, W. and Sobel, B. E. (1965). *Proc. Natl. Acad. Sci. U.S. 54,* 193-199.
Malkin, R. and Posner, H. B. (1978). *Biochim. Biophys. Acta 501,* 552-554.
Maltempo, M. M. and Moss, T. H. (1976). *Quart. Rev. Biophys. 9,* 181-215.
Matsubara, T. and Iwasaki, H. (1972). *J. Biochem. 71,* 747-750.
Matsubara, H., Sasaki, R. M., and Chain, R. K. (1967). *Proc. Natl. Acad. Sci. U.S. 57,* 439-445.
Meyer, T. E. (1970). Ph.D. Thesis, University of California at San Diego, La Jolla, California.
Meyer, T. E. and Bartsch, R. G. (1976). *In* "Flavins and Flavoproteins" (T. P. Singer, ed.), pp. 312-317. Elsevier, N.Y.
Meyer, T. E., Bartsch, R. G., Cusanovich, M. A., and Mathewson, J. H. (1968). *Biochim. Biophys. Acta 153,* 854-861.
Meyer, T. E., Bartsch, R. G., and Kamen, M. D. (1971a). *Biochim. Biophys. Acta 245,* 453-464.
Meyer, T. E., Sharp, J. J., and Bartsch, R. G. (1971b). *Biochim. Biophys. Acta 234,* 266-269.
Meyer, T. E., Kennel, S. J., Tedro, S. M., and Kamen, M. D. (1973). *Biochim. Biophys. Acta 292,* 634-643.
Meyer, T. E., Ambler, R. P., Bartsch, R. G., and Kamen, M. D. (1975). *J. Biol. Chem. 250,* 8416-8421.
Miyata, M. and Mori, T. (1969). *J. Biochem. 66,* 463-471.
Mizrahi, I. A. (1977). Ph.D. Thesis, University of Arizona, Arizona.
Mizrahi, I. A., Wood, F. E., and Cusanovich, M. A. (1976). *Biochemistry 15,* 343-348.
Morita, S. and Conti, S. F. (1963). *Arch. Biochem. Biophys. 100,* 302-307.
Mortenson, L. E., and Chen, J.-S. (1974). *In* "Microbial Iron Metabolism" (M. Gibbs, A. Hollaender, B. Kok, L. O. Krampitz, and A. San Pietro, eds.), pp. 231-282. Academic Press, New York.

Mortenson, L. E., Valentine, R. C., and Carnahan, J. E. (1962). *Biochem. Biophys. Res. Commun. 7,* 448-452.
Nakos, G. and Mortenson, L. E. (1971a). *Biochim. Biophys. Acta 227,* 576-583.
Nakos, G. and Mortenson, L. E. (1971b). *Biochemistry 10,* 2442-2449.
Nelson, N. and Racker, E. (1972). *J. Biol. Chem. 247,* 3848-3853.
Newton, N. (1969). *Biochim. Biophys. Acta 185,* 316-331.
Neumann, N. P. and Burris, R. H. (1959). *J. Biol. Chem. 234,* 3286-3290.
Niki, K., Yagi, T., Inokuchi, H., and Kimura, K. (1979). *J. Am. Chem. Soc. 101,* 3335-3340.
Oelze, J. and Drews, G. (1972). *Biochim. Biophys. Acta 265,* 209-239.
Olson, J. M. and Chance, B. (1960). *Arch. Biochem. Biophys. 88,* 26-53.
Olson, J. M. and Nadler, K. D. (1965). *Photochem. Photobiol. 4,* 783-791.
Olson, J. M. and Shaw, E. K. (1969). *Photosynthetica 3,* 288-290.
Palmer, G. (1975). *In* "The Enzymes" (P. D. Boyer, ed.), Vol. XIIB, pp. 1-56. Academic Press, N.Y.
Perini, F., Schiff, J. A., and Kamen, M. D. (1964). *Biochim. Biophys. Acta 88,* 91-98.
Peterson, J. A., Basu, D., and Coon, M. J. (1966). *J. Biol. Chem. 241,* 5162-5164.
Pettigrew, G. W. (1974). *Biochem. J. 139,* 449-459.
Pettigrew, G. W. (1981). In press.
Pettigrew, G. W., Meyer, T. E., Partsch, R. G., and Kamen, M. D. (1975). *Biochim. Biophys. Acta 430,* 197-208.
Pettigrew, G. W., Bartsch, R. G., Meyer, T. E., and Kamen, M. D. (1978). *Biochim. Biophys. Acta 503,* 509-523.
Pfennig, N. (1977). *Annu. Rev. Microbiol. 31,* 275-290.
Postgate, J. R. (1954). *Biochem. J. 56,* Xl-Xll.
Postgate, J. R. (1956). *J. Gen. Microbiol. 14,* 545-572.
Prince, R. C. (1978). *Biochim. Biophys. Acta 501,* 195-207.
Prince, R. C. and Olson, J. M. (1976). *Biochim. Biophys. Acta 423,* 357-362.
Prince, R. C., Baccarini-Melandri, A., Hauska, G. A., Melandri, B. A., and Crofts, A. R. (1975). *Biochim. Biophys. Acta 387,* 212-227.
Salemme, F. R. (1977). *Annu. Rev. Biochem. 46,* 299-329.
Salemme, F. R., Freer, S. T., Xuong, N. H., Alden, R., and Kraut, J. (1973). *J. Biol. Chem. 248,* 3910-3921.
San Pietro, A. and Lang, H. M. (1958). *J. Biol. Chem. 231,* 211-229.
Scherings, G., Haaker, H., and Veeger, C. (1977). *Eur. J. Biochem. 77,* 621-630.

Schink, B. and Schlegel, H. G. (1979). *Biochim. Biophys. Acta 567,* 315-324.

Schlegel, H. G. and Schneider, K. (1978). "Hydrogenases: Their Catalytic Activity, Structure, and Function." E. Goltze KG, Gottingen.

Schneider, K. and Schlegel, H. G. (1976). *Biochim. Biophys. Acta 452,* 66-80.

Schoenmaker, G. S., Oltmann, L. F., and Stouthamer, A. H. (1979). *Biochim. Biophys. Acta 567,* 511-521.

Scholes, P. B., McLain, G., and Smith, L. (1971). *Biochemistry 10,* 2072-2076.

Shethna, Y. I., Wilson, P. W., Hansen, R. E., and Beinert, H. (1964). *Proc. Natl. Acad. Sci. U.S. 52,* 1263-1270.

Shioi, Y., Takamiya, K., and Nishimura, M. (1972). *J. Biochem. 71,* 285-294.

Smith, L. A., Hill, S., and Yates, M. G. (1976). *Nature (London) 262,* 209-210.

Smith, L., Newton, N., and Scholes, P. B. (1966). *In* "Hemes and Hemoproteins" (B. Chance, R. Estabrook, and T. Yonetani, eds.), pp. 395-403. Academic Press, N.Y.

Stout, C. D. (1979). *Nature (London) 279,* 83-84.

Sugimura, Y., Toda, F., Murata, T., and Yakushiji, E. (1968). *In* "Structure and Function of Cytochromes" (K. Okunuki, M. D. Kamen, and I. Sekuzu, eds.), pp. 452-458. University Park Press, Baltimore, Maryland.

Sutherland, I. W. and Wilkinson, J. F. (1963). *J. Gen. Microbiol. 30,* 105-112.

Suzuki, H. and Iwasaki, H. (1962). *J. Biochem. 52,* 193-199.

Swank, R. T. and Burris, R. H. (1969). *Biochim. Biophys. Acta 180,* 473-489.

Sweeney, W. V. and McIntosh, B. A. (1979). *J. Biol. Chem. 254,* 4499-4501.

Sweeney, W. V., Rabinowitz, J. C., and Yoch, D. C. (1975). *J. Biol. Chem. 250,* 7842-7847.

Tagawa, K. and Arnon, D. I. (1962). *Nature (London) 195,* 537-543.

Takano, T., Dickerson, R. E., Schichman, S. A., and Meyer, T. E. (1979). *J. Mol. Biol. 133,* 185-188.

Tanaka, M., Nakashima, T., Benson, A. M., Mower, H. F., and Yasunobu, K. T. (1966). *Biochemistry 5,* 1666-1681.

Tanaka, M., Haniu, M., Yasunobu, K. T., and Kimura, T. (1973). *J. Biol. Chem. 248,* 1141-1157.

Tanaka, M., Haniu, M., Yasunobu, K. T., Dus, K., and Gunsalus, I. C. (1974a). *J. Biol. Chem. 249,* 3689-3701.

Tanaka, M., Haniu, M., Yasunobu, K. T., Evans, M. C. W., and Rao, K. K. (1974b). *Biochemistry 13,* 2953-2959.

Tanaka, M., Haniu, M., Yasunobu, K. T., Evans, M. C. W., and Rao, K. K. (1975). *Biochemistry 14,* 1938-1943.

Taniguchi, S. and Kamen, M. D. (1965). *Biochim. Biophys.*

Acta 96, 395-428.
Tedro, S. M., Meyer, T. E., and Kamen, M. D. (1974). *J. Biol. Chem. 249,* 1182-1188.
Tedro, S. M., Meyer, T. E., and Kamen, M. D. (1976). *J. Biol. Chem. 251,* 129-136.
Tedro, S. M., Meyer, T. E., and Kamen, M. D. (1977). *J. Biol. Chem. 252,* 7826-7833.
Tedro, S. M., Meyer, T. E., and Kamen, M. D. (1979). *J. Biol. Chem. 254,* 1495-1500.
Tiede, D. M., Leigh, J. S., and Dutton, P. L. (1978). *Biochim. Biophys. Acta 503,* 524-544.
Timkovich, R. and Dickerson, R. E. (1976). *J. Biol. Chem. 251,* 4033-4046.
Timkovich, R. and Robinson, M. K. (1979). *Biochem. Biophys. Res. Commun. 88,* 649-655.
Tissieres, A. and Burris, R. H. (1956). *Biochim. Biophys. Acta 20,* 436-437.
Travis, J., Newman, D. J., LeGall, J., and Peck, H. D., Jr. (1971). *Biochem. Biophys. Res. Commun. 45,* 452-458.
Van Beeumen, J., Ambler, R. P., Meyer, T. E., Kamen, M. D., Olson, J. M., and Shaw, E. K. (1976). *Biochem. J. 159,* 757-774.
Van Beeumen, J., Tempst, P., Stevens, P., Bral, D., Van Damme, J., and De Ley, J. (1980). *Protides of the Biological Fluids 28,* 69-74.
Van Den Branden, C., Van Beeumen, J., De Ley, J., and Van De Mieroop, A. (1975). *Hoppe-Seyler's Z. Physiol. Chem. 356,* 1251-1258.
Van Der Westen, H. M., Mayhew, S. G., and Veeger, C. (1978). *FEBS Lett. 86,* 122-126.
Vernon, L. P. (1953). *Arch. Biochem. Biophys. 43,* 492-493.
Vernon, L. P. (1956). *J. Biol. Chem. 222,* 1035-1044.
Vernon, L. P. and Kamen, M. D. (1954). *J. Biol. Chem. 211,* 643-663.
Weber, P. C., Bartsch, R. G., Cusanovich, M. A., Hamlin, R. C., Howard, A., Jordan, S. R., Kamen, M. D., Meyer, T. E., Weatherford, D. W., Xuong, N. H., and Salemme, F. R. (1980). *Nature (London) 286,* 302-304.
Wharton, D. C., Gudat, J. C., and Gibson, Q. H. (1973). *Biochim. Biophys. Acta 292,* 611-620.
Wood, P. M. (1977). *Eur. J. Biochem. 72,* 605-612.
Wood, P. M. (1980). *Biochem. J. 189,* 385-391.
Yagi, T. (1969). *J. Biochem. 66,* 473-478.
Yagi, T. (1970). *J. Biochem. 68,* 649-657.
Yagi, T. and Maruyama, K. (1971). *Biochim. Biophys. Acta 243,* 214-224.
Yagi, T., Kimura, K., Daidoji, H., Sakai, F., Tamura, S., and Inokuchi, H. (1976). *J. Biochem. 79,* 661-671.

Yamanaka, T. (1966). *In* "Biochemistry of Copper" (J. Peisach, P. Aisen, and W. E. Blumberg, eds.), pp. 275-292. Academic Press, N.Y.

Yamanaka, T. (1976). *J. Biochem. 79,* 435-440.

Yamanaka, T. and Imai, S. (1972). *Biochem. Biophys. Res. Commun. 46,* 150-154.

Yamanaka, T. and Kamen, M. D. (1967). *Biochim. Biophys. Acta 131,* 317-329.

Yamanaka, T. and Kusai, A. (1976). *In* "Flavins and Flavoproteins" (T. P. Singer, ed.), pp. 292-301. Elsevier Press, Amsterdam.

Yamanaka, T. and Okunuki, K. (1963). *Biochim. Biophys. Acta 67,* 379-393.

Yamanaka, T. and Okunuki, K. (1968). *In* "Structure and Function of Cytochromes" (K. Okunuki, M. D. Kamen, and I. Sekuzu, eds.), pp. 390-403. University Park Press, Baltimore.

Yamanaka, T. and Okunuki, K. (1974). *In* "Microbial Iron Metabolism" (J. Nielands, ed.), pp. 349-400. Academic Press, N.Y.

Yamanaka, T., De Klerk, H., and Kamen, M. D. (1967). *Biochim. Biophys. Acta 143,* 416-424.

Yoch, D. C., Arnon, D. I., and Sweeney, W. V. (1975). *J. Biol. Chem. 250,* 8330-8336.

Yoch, D. C., Carithers, R. P., and Arnon, D. I. (1977). *J. Biol. Chem. 252,* 7453-7460.

Yoch, D. C., Carithers, R. P., and Arnon, D. I. (1978). *J. Bacteriol. 136,* 1018-1026.

Yu, C. A., Yu, L., and King, T. E. (1972). *J. Biol. Chem. 247,* 1012-1019.

THE STRUCTURE AND CATALYTIC MECHANISMS OF CYTOCHROME *c* PEROXIDASE

Thomas L. Poulos
Joseph Kraut

Department of Chemistry
University of California, San Diego
La Jolla, California 92093

INTRODUCTION

Yeast cytochrome *c* peroxidase (CCP) catalyzes the oxidation of ferrocytochrome *c* by hydroperoxides in the following reaction,

$$ROOH + 2\ cyt\ c\ (Fe^{2+}) + 2H^{+} \rightarrow ROH + H_2O + 2\ cyt\ c\ (Fe^{3+})$$

The precise biological role of CCP has yet to be established but the location of the enzyme between the inner and outer membrane of yeast mitochondria where the respiratory components are concentrated suggests that the function of CCP is closely linked to respiration. CCP is able to partially support oxidative phosphorylation (1) and can function as an alternate terminal electron sink in place of cytochrome oxidase in mitochondrial respiration (1). The electron transfer reaction catalyzed by CCP is usually found in enzymes, like cyto-

 ISBN 0-12-397580-8

chrome oxidase, that are much more complex and intimately associated with membranes. In contrast to cytochrome oxidase, however, CCP appears not to be bound to mitochondrial membranes and unlike most enzymes that interact with cytochrome *c*, CCP is a water-soluble monomer of 34,000 daltons containing a single, noncovalently bound heme. Finally, and perhaps the most significant property of CCP, at least from the crystallographer's point of view, is the ease with which high-quality crystals that diffract beyond 2.0 Å can be obtained.

The structure of CCP is of interest for two principal reasons. First, until now there has not been a detailed crystallographic analysis of any heme enzyme, which has left a considerable gap in our understanding of a uniquitous class of biochemical reactions. Second, knowledge of the three-dimensional structure of CCP should help to establish the stereochemistry of the complex formed between CCP and cytochrome *c*, thereby helping to elucidate the mechanism of inter-protein electron transfer reactions.

In addition, the reaction product between CCP and peroxides, compound I, is relatively stable, and consequently this important reaction intermediate has been the subject of many detailed solution studies (2, 3). Fortunately, it appears possible for CCP in the crystalline state to be converted to compound I with retention of crystallinity, and so we are presented with an opportunity to examine the structural consequences of this conversion. We hope to carry out such a study in the near future.

Here we report progress to date on the crystal structure of CCP and propose a possible mechanism for the CCP-catalyzed reduction of hydroperoxides.

We should also like to add a short personal note since Martin Kamen has had a direct influence on the development of our laboratory's interest in the structure and function of redox proteins. It was the early and close collaborations with Martin, Robert Bartch, and other members of Martin's laboratory that lead to determination in La Jolla of the structure of cytochrome c_2 (4) and the high potential iron protein from *Chromatium* (5), both of which had been purified and sequenced in Martin's laboratory. We are indeed fortunate that Martin and his group are so close by. He has, in no small way, continued to be a source of inspiration to our current investigations.

THE STRUCTURE OF CCP

Our current model of CCP is based on a 2.5 Å electron-density map phased on two isomorphous mercury derivatives (6).

In conjunction with the x-ray work, the sequence of CCP is being determined in the laboratory of Takashi Yonetani. He has kindly provided us with sequence data as it has become available. Thus far, seven continuous pieces of the sequence comprising 60 residues have been fitted to the electron-density map. Fortunately, these include several residues in the active site region.

In those portions of the molecule for which sequence information was not available, arbitrarily selected side chains were fitted to the electron-density map wherever possible. The present CCP model contains 266 of the expected 293 (7) residues, or about 80% of all non-hydrogen protein atoms. Fifteen residues at the N terminus have not been included since the electron-density map does not yet clearly indicate the exact geometry in this region. In addition, two surface loops will require modification, but any future changes are not likely to alter the essential features of our current model.

Figure 1 is a stereoscopic view of the CCP molecule based on the position of 266 alpha carbon atoms. Each half of the molecule is highlighted differently in Fig. 1 to emphasize the division of CCP into two clearly defined domains. Approximately one-half of the residues in our present model are distributed among ten helical segments, labeled A through J, in Fig. 1. The helices are, in turn, divided roughly equally between the two domains, with helices A through D lying in domain I and helices F through J in domain II. The E helix, located in the approximate center of the sequence, forms the principal connection between domains.

Fig. 1. Stereo model of the alpha carbon backbone of CCP. The ten helices are labeled A through J and the N region is circled. The C terminus is hidden from view. The proximal and distal histidines are shown in addition to the distal tryptopah and arginine.

There is relatively little β structure in the CCP molecule. We have located only three antiparallel β pairs, all of which lie on the surface, and the molecule contains no parallel β structure. The two longest β pairs are found in domain II, one of them connecting the F and G helices. One strand of this β pair lies just below the heme propionates. The second β pair in domain II begins at the C terminus of helix G. The remaining β pair lies in domain I and begins at the abrupt turn in the polypeptide chain at the C terminus of helix B.

STRUCTURE AT THE HEME CREVICE

The heme sits in a crevice between the two domains, with the B helix in domain I and the C-terminal region of helix F in domain II forming the walls of the crevice. The heme is deeply embedded within this crevice and is surrounded by hydrophobic residues. Only one pyrrole ring edge is exposed to the surrounding medium.

The propionates lie just above one strand of a β pair that begins at the C-terminal end of helix F. One of the propionates appears to hydrogen bond to a side chain, probably histidine, which extends from the upper strand of the β pair.

As we observed earlier (6), the fifth axial heme ligand is NE2 of a proximal histidine side chain extending from the C-terminal end of helix F (Fig. 1). This identification has now been confirmed by Yonetani's partial sequence data. The sixth axial ligand appears to be a water molecule, as judged by a finger of electron density protruding from the heme iron atom.

The sequence data have also enabled us to confirm the identification of three residues, a histidine, an arginine, and a tryptophan on the distal side of the heme plane. We believe the side chains of these residues are important to the catalytic activity of CCP. All three side chains extend from helix B (see Fig. 1). Figure 2 is a close-up stereoscopic view showing the stereochemistry in this region of the heme crevice. Both the distal tryptophan and arginine side chains are within van der Waal's contact of the heme, with the tryptophan above the most buried surface of the heme and the arginine above the most exposed surface of the heme. The distal histidine is analogous to the distal histidine (His-E7) in the globins. In fact, as shown in Fig. 2, NE 2 of the distal histidine can hydrogen bond with the axial water ligand as in metmyoglobin (10).

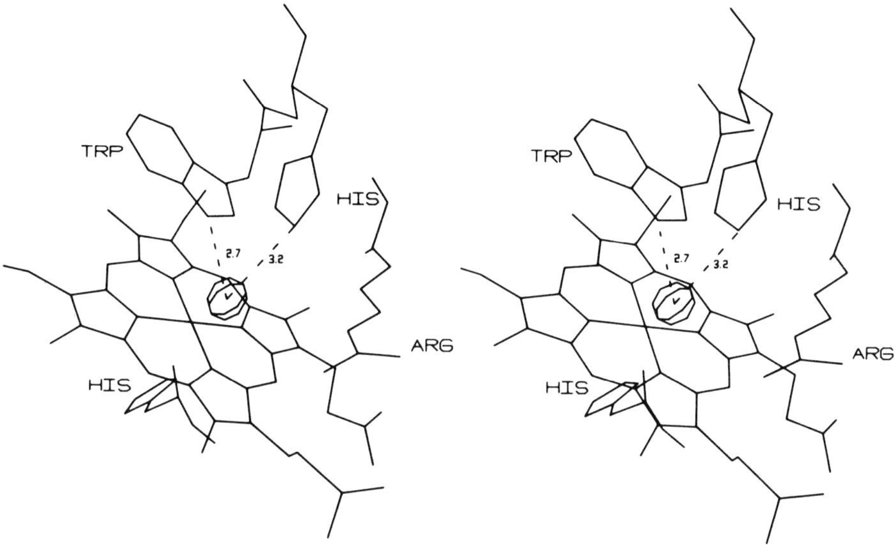

Fig. 2. Stereo view of the heme crevice showing the interactions between the heme, its axial ligands and the three distal residues, Arg, Trp, and His.

Though the pocket at the sixth axial coordination position is clearly large enough to accept a water molecule, some repositioning of adjacent structure is probably necessary to accommodate a larger moiety, like hydroperoxide. We have observed by difference Fourier techniques that a cyanide ligand indeed does induce a general expansion around the ligation pocket by causing both the tryptophan and arginine side chains to move away (6).

The most likely route into the ligation pocket is through the opening between the arginine quandinium group and the heme-pyrrole ring just below the arginine side chain (Figs. 1 and 2).

PEROXIDASE CATALYTIC MECHANISM

With the elucidation of the structure of the heme crevice in CCP, it now becomes possible to discuss the stereochemistry

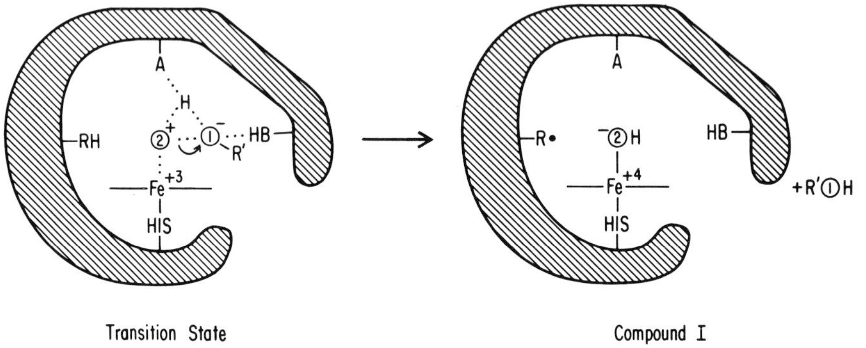

Fig. 3. Schematic representation of the transition state and compound I in the reaction between CCP and hydroperoxides. The transition state shows stabilization of charges on 01 and 02 of the substrate by side chains A (histidine) and B (arginine). R• in compound I is the amino acid side chain free radical which we believe to be a tryptophan side chain in CCP.

of peroxidase catalysis. A wide body of evidence indicates that in peroxidase catalysis, reduction of hydroperoxides involves heterolytic cleavage of the RO-OH as well as acid/base catalysis (11 - 16). Heterolytic fission of a peroxide necessitates charge separation in the transition state, RO^- ... $\overset{+}{O}H$. Thus, the peroxidase active site must be designed to stabilize these charges and possibly to facilitate proton transfer from $\overset{+}{O}H$ to the leaving group, RO^-. A general mechanism for charge stabilization and acid/base catalysis is depicted in Fig. 3. We propose that A = histidine, B = arginine, and R = tryptophan in CCP. This proposal is based on simple model building experiments in which a hypothetical peroxide molecule was inserted into the CCP active site. We have assumed that the OH of an RO-OH substrate will occupy the same position as the axial water ligand. Consequently, the position of RO is sterically limited by the presence of the R group which must extend out of the heme crevice. The resulting stereochemistry shows that the distal histidine and distal arginine are ideally situated for stabilization of the transition state. NE2 of the distal histidine hydrogen bonds to both 01 and 02 (see Fig. 3), thus stabilizing a positive charge on 02 and serving as an acid/base catalyst by facilitating proton transfer from 02 to 01, while the guanidinium group can stabilize the negative charge on 01 by donating a hydrogen bond from NE. The result is transient production of a potent oxidizing center at 02. Meanwhile, the edge of the distal indole ring is only 2.7 Å from 02, and is a likely candidate for oxidation. Thus, the oxidizing center at 02 can abstract both one electron from the heme iron and a hydrogen atom from the indole ring, leaving Fe^{4+}-OH^{-} and R•, an indole-free radical. The presence of free radical delocalized over the indole aromatic system offers a reasonable explanation for the intense free radical signal observed in CCP compound I (2).

We believe this reaction mechanism is applicable to other peroxidases as well. Welinder and Mazza (17) have reported the sequences around the presumed distal histidine in horseradish isoperoxidase C and four isoenzymes of turnip peroxidases. All these peroxidases possess a histidine and an arginine at the same relative position as in CCP. Such invariances strongly imply a functional role for these residues and indicate that the same stereochemistry, involving charge stabilization and acid/base catalysis, is a feature of peroxidase mechanisms in general.

Nevertheless, two important questions remain: (1) Why does CCP compound I exhibit an intense free radical ESR signal while HRP does not; and (2) why is CCP compound I red while HRP compound I is green? There is obviously a significant difference in the electronic structure of the heme. We believe these differences are related to the initial site of free radical formation in both HRP and CCP. If we further examine the sequences of HRP and turnip peroxidases, we find that phenylalanine occupies the same position as the distal tryptophan in CCP. Therefore, phenylalanine should be close to the hydroperoxide molecule in HRP and could serve as a source of one oxidizing equivalent, yielding a phenyl radical. Yet HRP compound I generates only a very weak free radical signal (18) while the absorption spectrum of HRP compound I is very similar to model heme compounds that have been oxidized to cationic heme radicals (19). In addition, such heme radicals are not expected to generate intense free radical ESR signals because of strong interactions between the radical and the iron (19). To resolve these difficulties, let us assume the following equilibrium results after the formation of compound I.

$$\underset{\langle\text{red}\rangle}{\text{R}\cdot + \text{heme}(\text{Fe}^{4+})} \rightleftarrows \underset{\langle\text{green}\rangle}{\text{R:} + \text{heme}^{\dot{+}}(\text{Fe}^{4+})}$$

where R• is either the phenyl radical in HRP or indole radical in CCP, and $\text{heme}^{\dot{+}}(Fe^{4+})$ is the green, cationic heme radical. Such an intramolecular electron transfer equilibrium is possible since the indole ring of the tryptophan makes contact with the heme, and we expect the phenyl ring in HRP to be similarly situated. Because the indole radical is significantly more stable than the phenyl radical relative to the heme cationic radical, in CCP the above equilibrium lies to the left while in HRP it lies to the right.

ACKNOWLEDGMENT

This work was supported by Grant PCM 77-08554 from the National Science Foundation and by Grants GM10928 and RR00757 from the National Institutes of Health.

REFERENCES

1. Erecinska, M., Oshino, N., Loh, P., and Brocklehurst, E. (1973). *Biochim. Biophys. Acta 292,* 1.
2. Yonetani, T., Schleyer, H., and Ehrenberg, A. (1966). *J. Biol. Chem. 241,* 3240.
3. Yonetani, T. (1976). *In* "The Enzymes," (P. Boyer, ed.), Vol. 13, p. 345. Academic Press, N.Y.
4. Salemme, F. R., Freer, S. T., Xuong, Ng. H., Alden, R. A., and Kraut, J. (1973). *J. Biol. Chem. 248,* 3910.
5. Carter, C. W., Kraut, J., Freer, S. T., Xuong, Ng. N., Alden, R., and Bartsch, R. G. (1974). *J. Biol. Chem. 249,* 4212.
6. Poulos, T. L., Freer, S. T., Alden, R. A., Xuong, Ng. H., Edwards, S. L., Hamlin, R. C., and Kraut, J. (1978). *J. Biol. Chem. 253,* 3730.
7. Takio, K. and Yonetani, J., unpublished results.
8. Hoard, J. T. (1971). *Science 174,* 1295.
9. Hoard, J. T. (1975). "Porphyrins and Metalloporphyrins," p. 317. Biomedical Press.
10. Nobbs, C. L., Watson, H. C., and Kendrew, J. C. (1966). *Nature (London) 209,* 339.
11. Schonbaum, G. R. and Chance, B. C. (1976). *In* "The Enzymes," (P. Boyer, ed.), Vol. 13, p. 363. Academic Press, N.Y.
12. Schonbaum, G. R. (1973). *Abst. 2b 9th Intern. Congr. Biochem.,* p. 48.
13. Jones, P. and Sugget, A. (1968). *Biochem. J. 110,* 621.
14. Coulson, A. F. W. and Yonetani, T. (1975). *Biochemistry 14,* 2389.
15. Schonbaum, G. R. and Lo, S. (1972). *J. Biol. Chem. 247,* 3353,
16. Morrison, M. and Schonbaum, G. R. (1976). *Annu. Rev. Biochem. 45,* 861.

17. Welinder, K. G. and Mazza, G. (1977). *Eur. J. Biochem. 73,* 353.
18. Aasa, R., Vanngard, T., and Dunford, H. B. (1975). *Biochim. Biophys. Acta 391,* 259.
19. Dolphin, D., Forman, A., Borg, D. C., Fajer, J., and Felton, R. H. (1971). *Proc. Natl. Acad. Sci. U.S. 68,* 614.

INVESTIGATION OF CYTOCHROMES AT SUBZERO TEMPERATURES

Pierre Douzou

INSERM and Institut de Biologie Physico-Chimique
Paris, France

INTRODUCTION

It was my good fortune to be among those honoring Martin D. Kamen and to remember the good old days when Martin was in Paris and occasionally worked in this laboratory. Highlights of these days are unforgettable, and are part of our personal treasures.

Martin was one of the first people to watch and encourage our efforts to carry out enzymological procedures in fluid-mixed solvents at subzero temperatures and communicated our first significant work to the National Academy of Sciences. This work achieved the temporal resolution, step by step, of the peroxidatic reaction catalyzed by horseradish peroxidase (1),

$$H_2O_2 + 2\ AH \xrightarrow{E} 2\ A_{ox} + 2\ H_2O_1$$

 ISBN 0-12-397580-8

a reaction worked out, in particular, by B. Chance using fast kinetic techniques which allowed the detection of two consecutive intermediates, namely compounds I and II.

Cooling ⟷ heating cycles allowed us to obtain the thermal resolution of this reaction and to obtain "pure" compounds I and II then analyzed by several spectroscopic techniques. Thus at -40°C,

$$\text{HRP }(Fe_p^{3+} - H_2O) + H_2O_2 \rightleftharpoons \text{compound I (stable 99\%)}$$

then,

$$\text{compound I} + AH \longrightarrow \text{compound II (stable 100\%)} + A_{ox}$$

at 0°C,

$$\text{compound I} + AH \longrightarrow \text{HRP }(Fe_p^{3+} - H_2O) + A_{ox} + 2\ H_2O$$

Such was the first result indicating that reactions involving long-lived intermediates could be investigated at subzero temperatures to make resolution easier and to improve their step-by-step analysis.

It is largely under the convincing influence of Martin that we turned our attention toward the study of cytochromes at subzero temperatures. Once again, Martin was at the origin of an adventure carried out in Paris, the first recordings of absorption spectra of cytochrome *c* in mixed solvents at temperatures under 0°C.

Since then, intensive work has been carried out on various systems at subzero temperatures which I will briefly summarize.

CRYOENZYMOLOGY IN AQUEOUS ORGANIC ENVIRONMENTS

To overcome the problem of the equilibrium freezing point of aqueous solutions, fluid aqueous organic mixtures have been used in low-temperature studies. We have mentioned above the potential of such procedures on studies of peroxidase, and we will recall the successful attempts made on hydrolytic enzymes, luciferase, and enzymes in the crystalline state, reviewed elsewhere (2, 3).

Low-temperature procedures allow the accumulation of kinetic and thermodynamic information and permit stabilized intermediates to be analyzed by a variety of spectroscopic and chemical techniques.

ISOLATION OF TRANSIENT INTERMEDIATES: OXYFERROUS COMPOUNDS OF CYTOCHROME P-450

Many enzyme - substrate intermediates, too short-lived to be obtained at sufficient concentrations, can be stabilized in fluid-mixed solvents at subzero temperatures and then accumulated and analyzed by various physical techniques. However, such intermediates may be contaminated by an excess of substrates or reaction products, interferring with physicochemical investigation and forbidding their use as "pure" reactants in subsequent processes. Purification can be achieved by column chromatography at subzero temperatures.

This procedure has been successfully used to purify the unstable oxygenated compound of bacterial cytochrome P-450 from the reducing system used for its preparation. While low temperature inhibits the spontaneous decay of the protein component, the LH-20 Sephadex gel shows an abnormal affinity toward the reducing agent (proflavin) which is thus retained on the column.

We have shown that it is possible to stabilize the oxygenated complexes of cytochrome P-450 in supercooled fluid media. However, in order to form these oxygenated compounds the cytochrome P-450 must be reduced either enzymatically or chemically. In both cases, these results are in the presence of an excess of reducing agents in the sample. The chemical reduction can be obtained by either the addition of an excess of dithionite or photochemically in the presence of acridine orange and methyl viologen (4) or of proflavin sulfate.

If the enzymatic reactions which follow the formation of these oxygenated compounds are to be studied, the reducing agents or degradation products, including thermodecomposition products (5, 6), must be removed from the system.

In order to isolate the oxygenated compound, this one (Fe_S^{2+} - O_2) was stabilized at -40°C and was applied on the top of a precooled (-20°C) 10 cm × 2.5 cm Sephadex LH-20 column, which had been degased under vacuum before packing in order to avoid irregularities in flow rate. The flow rate was 0.4 ml min^{-1} and the time required to obtain the pure complex was 30 min. The concentration in the peak fraction was 10 μ*M* in 1.3 ml, which was only 3 times less than the starting solution. The recovery was nearly 100%.

Figure 1 shows the uv-visible absorption spectrum of the complex in the peak sample (diluted 10 times in aqueous buffer and therefore containing 5% ethylene glycol), measured at 2°C immediately after chromatography and corrected for camphor absorption. The absence of contaminating proflavin is obvious. Furthermore, there is no noticeable change of the 280 nm absorption after decomposition of the complex into Fe_S^{3+}.

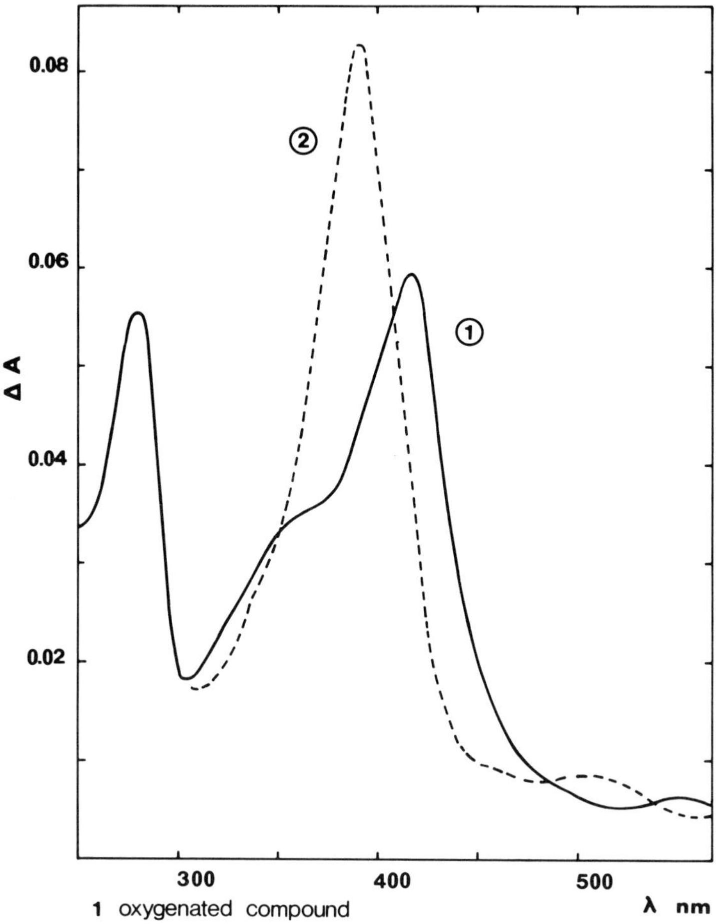

1 oxygenated compound

2 ferric form after thermal decomposition of the compound

Fig. 1

From this spectrum, the ratio of the absorbancies at 392 and 280 nm may be measured, $A_{392}/A_{280} = 1.48$, which is in good agreement with the previous value of 1.47 obtained by Yu and Gunsalus (7).

The solution of pure compound may be stored frozen at 77°K, or in fluid state at -35°C, without any appreciable reoxidation. The spontaneous decay of purified $m^{rs}_{O_2}$ at higher temperatures is identical to the autodecomposition of the non-chromatographed compound.

An alternative procedure was also used to prepare the less stable substrate-free cytochrome P-450 ($Fe^{2+}-O_2$). This oxyferrous cytochrome was prepared in the presence of camphor which ensured a higher stability for the complex and the conversion $Fe_S^{2+} - O_2 \rightarrow Fe_{O_2}^{2+}$ occurred during elution.

When prepared from the substrate-free ferric cytochrome, the oxyferrous compound $Fe_{O_2}^{2+}$ decays much faster than the camphor-bound compound.

The uv-visible spectrum at -45°C of the complex is obviously not contaminated by proflavin. Since the binding of camphor has no measurable effect on the Soret and visible bands of the oxygenated compound, the recording of the optical spectrum alone does not allow one to ascertain the absence of bound camphor. Furthermore, free camphor may only be detected by its absorption at 290 nm (ε = 0.03 mM^{-1}) and is thus obscured by the higher absorption of the protein at the same wavelength. The absence of free and bound camphor was, therefore, tested by the autoxidation rate of the compound at different temperatures and the values obtained for the purified compound are in excellent agreement with those obtained when $Fe_{O_2}^{2+}$ is prepared from the camphor-free ferric cytochrome. When camphor is added to the $Fe_S^{2+} - O_2$ solution before decomposition, the autoxidation rate constant falls on the Arrhenius plot of $Fe_S^{2+} - O_2$ decay.

Stock solutions of concentrated camphor-bound and camphor-free oxyferrous compounds may be stored at low temperature (77°K) without appreciable decay (8, 9). This allows one to make comparative studies of their respective decomposition mechanisms and products.

The complexes obtained are pure and it is very important to note that they are also completely free of hydrogen peroxide which was present at rather high concentration in the initial solution. The presence of peroxide is due to the rapid reoxidation of proflavin by the excess oxygen added in order to form the oxyferro cytochrome.

When studying the reactions involved in oxygen metabolism it is very important that excess oxidoreduction agents and decomposition products are completely absent as they might interact with the short-lived oxidizing species or protein intermediates that may be produced during the oxygen activation process. These purified complexes can be used to study the last and essential steps of the hydroxylation cycle.

DYNAMICS OF ELECTRON TRANSPORT IN CYTOCHROME P-450 SYSTEMS

This cycle has been resolved to show four main steps in the sequence of cytochrome P-450 reactions (4, 10). These steps are shown in the scheme in Fig. 2 and involves the binding of the substrate camphor (RH) by the free ferric cytochrome

$$(Fe^{3+} + RH \rightleftharpoons Fe^{3+} - RH$$

the reduction of the enzyme-substrate complex

$$(Fe^{3+} \xrightarrow{+e^-} Fe^{2+} - RH)$$

The binding of molecule oxygen to give the ternary oxyferrous compound

$$\left(Fe^{2+} - RH + O_2 \rightleftharpoons Fe^{2+} \begin{matrix} \diagup RH \\ \diagdown O_2 \end{matrix}\right)$$

and the uptake of a second electron leading to the hydroxylated substrate (R - OH) and the restoration of the free enzyme (Fe^{3+}).

Elementary mechanisms at the molecular level are not yet established with sufficient detail; their study requires a tight control of each step and such control can be obtained by low temperature studies, whereas kinetic studies can be carried out by fast techniques. The reaction pathway is not modified in mixed solvents such as ethylene glycol - water mixtures (1:1, v/v) and we have seen that subzero temperatures permit slowing down of the reactions and, in the present case, uncoupling of the cycle and the stabilization of the oxyferrous compounds

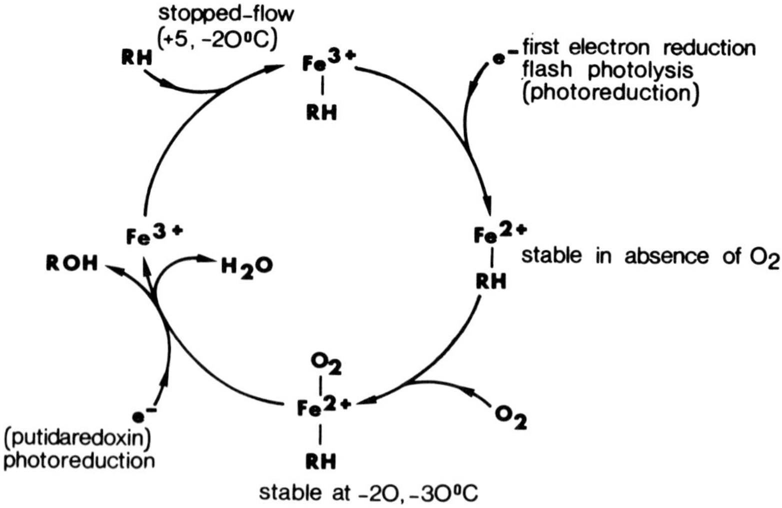

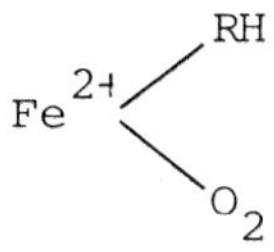

A combination of subzero temperatures and fast kinetic techniques is very useful in gaining kinetic and thermodynamic information on the elementary steps of the cycle. In addition, should be mentioned the study of camphor (RH) binding with the ferric cytochrome by stopped flow between +5° and -20°C to obtain kinetic rate constants of formation and dissociation of the Fe^{3+}·RH complex (11) and also the activation of the system by light representing a fast, clean, and versatile means of introducing electrons at various levels of the cycle (see Fig. 2): first by an electron-reducing interaction of the ternary oxygenated (oxyferrous) compound with reducing gradients, and reduction of the putidaredoxin - cytochrome complex (12).

These investigations of the bacterial methylene hydroxylase allowed a partial temporal resolution of the two electron transfers from putidaredoxin to cytochrome. This was made possible essentially through the differential effect of solvent and temperature on rate constants. The main interest of such studies is to give access not only to the enzyme - substrate interaction, but also to the dynamics of protein - protein interaction, which, through the sequential formation and dissociation of functional multienzyme complexes, is the basis of reactivity and control of multicomponent functions. This procedure could be applied to many other soluble or membraneous multicomponent systems.

AQUEOUS ENVIRONMENTS

Attractive though such procedures may be, they are not free from problems, mainly associated with the presence of high concentrations of organic cosolvents, even if it can be demonstrated that the cosolvent does not affect the specific activity of the enzyme or the pathway of the reaction. Thus, it is well documented that the organic cosolvent used, e.g., methanol, dimethyl sulfoxide, *N*,*N*-dimethylformamide, or ethylene glycol, causes drastic changes in the peculiar long-range intermolecular order which is the unique feature of liquid water. As a result such organic cosolvents markedly influence the solvation and transport behavior of ions and the dissociation equilibria of acids and bases. Since the conformational stability of proteins in solution is governed by the balance of these interactions (and others, such as the hydrophobic interaction), it is to be expected that major changes in the medium composition will affect any process which depends on the solvation interactions of amino acid residues.

To avoid cosolvent effects and yet to investigate at subzero temperatures, we have tried to use supercooled water. Water in oil emulsions supersaturated with water and oil in soluble surfactants involve water droplet diameters from 1 to 5 μm and these droplets supercooled down to -35 to -40°C. We have applied this emulsion procedure on a number of enzyme systems (13 - 15). While such attempts were successful, the procedure is not free from problems; thus, the turbidity of the emulsions precludes analytical techniques which rely on optical transparency, whereas their very high viscosity jeopardizes kinetic studies. These limitations prompted us more recently to use water-soluble surfactants and, in fact, micellar solutions.

Many surfactants form micelles in organic solvents as well as in aqueous solutions. The former are, however "inverted" (reversed) compared with the aqueous phase, since the polar head groups build the polar core of the micelle shielded by the apolar hydrocarbon tails. Such inverted micelles are stable for hours and days.

A large number of experimental data support the view of the lipophilic micelles as consisting of a spherical or near-spherical water core surrounded by a monolayer of amphiphilic molecules. Solutions with a low ratio of water to amphiphilic compounds appear to contain micromicelles forming transparent and fluid solutions. Figure 3 represents a schematic description of such micelles.

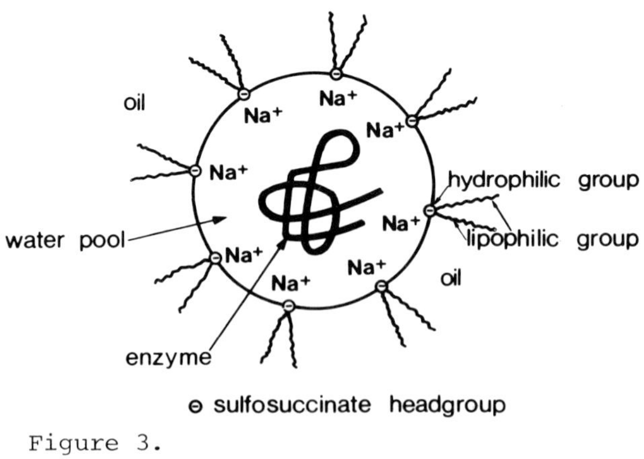

Figure 3.

Amphiphilic compounds used in this procedure are characterized by possessing in the same molecule two groups which differ greatly in their solubility relationship. These are (1) a hydrophilic group which tends to be water soluble and hydrocarbon insoluble, and (2) a lipophilic group which tends to be hydrocarbon soluble and water insoluble. One of the most useful amphiphilic compounds is the diisooctylsodium sulfosuccinate in which both the hydrophilic and lipophilic functions are strong and not too unequally matched. This anionic surfactant, termed aerosol OT (AOT), with a mol. wt. 444,57, is characterized by its pronounced solubilizing capacity in nonpolar organic solvents. The dry solutions of such surfactant are able to solubilize relatively large amounts of water (1 - 6%, w/v). The "solubilizate" is added with a microsyringe and the ternary solution is shaken and treated in an ultrasonic bath at 4°C to ensure that the "solubilizate" is taken up by the micelles.

Water-soluble molecules are also soluble in water solubilized by nonaqueous solutions of the oil-soluble surfactant. This phenomenon is referred to as secondary solubilization of the water-soluble molecules, and we have been able to dissolve several enzymes representative of hydrolytic enzymes, flavoproteins, hemoproteins, etc. The solubility of such enzymes largely depends on the ratio of concentration in water and surfactant, as well as on the viscosity of micellar solutions. Solutions involving nonpolar solvents such as heptane, silicone oils of a viscosity lower than 1 cps, and containing 1 to 3% of water readily dissolve the proteins. The stability of these proteins can be established through absorption spectra. Absorption spectra of cytochrome *c* and cytochrome P-450, respectively, have been recorded down to -50°C and show their usual characteristics.

These recordings clearly show the possibility of carrying out investigations at relatively broad range of subzero temperatures. Solutions remain perfectly clear ("colorless" according to the classical terminology) until they become turbid at lower temperatures. Water solubility diagrams have been established for several solutions and it has been found that experiments of enzyme-water clusters were possible, in some cases, down to -60°C. It may be suggested that water clusters supercool, but no direct evidence of such a process is at present available, except the fact that solutions remain colorless and that enzyme activity still occurs. It is well known that the nature of water phase in reversed micelles is quite different from that of ordinary bulk water and that there are strongly restricted motions of water dipoles. This is certainly not the only specific property of water clusters. We have found that the charged heads of amphiphilic compounds developed strong electrostatic potentials and induced special physicochemical conditions influencing markedly enzyme activity.

Thus, although the physical nature of inverted micelles has been extensively investigated in recent years, relatively little information is presently available about the static and dynamic properties of such micelles. Further experiments are designed to investigate these systems in more detail and to improve the possibility of investigating enzyme systems, including cytochromes, in pure aqueous environments at subzero temperatures.

ACKNOWLEDGMENT

Work on Cryobiochemistry is supported by the Institut National de la Santé et de la Recherche Médicale and mainly carried out in the Unité 128 in Montpellier, France.

REFERENCES

1. Douzou, P., Sirex, R., and Travers, F. (1970). *Proc. Natl. Acad. Sci. U.S. 66,* 787.
2. Douzou, P. (1977). "Cryobiochemistry: An Introduction." Academic Press, N.Y.
3. Douzou, P. (1977). *Advan. Enzymol. 45,* 157-268.
4. Gunsalus, I. C., Lipscomb, J. D., Marshall, V., Frauenfelder, H., Greenbaum, E., and Münck, E. (1972). *In* "Biological Hydroxylation Mechanisms" (G. S. Boyd and R. M. S. Smellie, eds.), p. 135. Academic Press, N.Y.
5. Gunsalus, I. C., Meeks, J. R., Lipscomb, J. D., Debrunner, P., and Münck, E. (1974). *In* "Molecular Mechanisms of Oxygen Activation" (O. Hayaishi, ed.), p. 559. Academic Press, N.Y.
6. Lipscomb, J. D., Sligar, S. G., Namtved, H. J., and Gunsalus, I. C. (1976). *J. Biol. Chem. 251,* 1116-1124.
7. Yu, C. A. and Gunsalus, I. C. (1970). *Biochem. Biophys. Res. Commun. 40,* 1431.
8. Balny, C., Debey, P., and Douzou, P. (1976). *FEBS Lett. 69,* 231.
9. Debey, P., Balny, C., and Douzou, P. (1976). *FEBS Lett. 69,* 236.
10. Sligar, S. G., Lipscomb, J. D., Debrunner, P. D., and Gunsalus, I. C. (1974). *Biochem. Biophys. Res. Commun. 61,* 290-296.
11. Lange, R., Hui Bon Hoa, G., Debey, P., and Gunsalus, I. C.

(1977). *Eur. J. Biochem.* 77, 479.
12. Debey, P. and Balny, C., personal communication.
13. Douzou, P., Debey, P., and Franks, F. (1977). *Nature (London) 268*, 466.
14. Douzou, P., Debey, P., and Franks, F. (1978). *Biochim. Biophys. Acta 523*, 1.
15. Douzou, P., Balny, C., and Franks, F. (1978). *Biochimie 60*, 151.

REDUCTION KINETICS OF *Chromatium vinosum* CYTOCHROME *c'*

*W. Greg Miller**
Michael A. Cusanovich

Department of Biochemistry
University of Arizona
Tucson, Arizona 85721

INTRODUCTION

The cytochromes *c'* have long been a major interest of Martin Kamen dating back to their discovery by Vernon and Kamen in 1954 (1). This atypical class of cytochromes which qualifies as a *c*-type cytochrome on the basis of heme attachment has spectral properties closely resembling those of hemoglobin and myoglobin, that is, high-spin heme proteins. The literature on cytochrome *c'* has been reviewed a number of times by the Kamen laboratory (2 - 4) and should be consulted for references to the early work. Studies on the oxidation-reduction properties (5 - 7), spectral properties (5, 7 - 11), ligand binding (12 - 14), solvent perturbation (8), and magnetic properties (15 - 17) have been reported. Although con-

**Present address: Department of Pathology, Medical College of Virginia, MCU Station, Richmond, VA 23298.*

 ISBN 0-12-397580-8

sistent with the anomalous behavior of this class of cytochromes, the studies mentioned above have been difficult to interpret due to a lack of structural information. The amino acid sequences of cytochrome *c*' from nine different sources are known (18 - 20) but only recently has progress been made in the elucidation of the three-dimensional structure of a representative of the cytochrome *c*' class (reference 21 and Weber and Salemme, personal communication).

In light of Martin Kamen's interest in the cytochrome *c*', we are presenting here a study of the oxidation-reduction kinetics of *Chromatium vinosum* cytochrome *c*'. The *Chromatium* protein is typical of the cytochrome *c*' class with a molecular weight of 29,000 at a neutral pH, an isoelectric point of 5.0, and an oxidation-reduction potential of -5 mV at pH 7.0 (22, 23). The effects of pH and organic solvents on the absorption spectra of several cytochromes *c*' have been reported. It has been found that the cytochromes *c*' undergo a two-step spectral transformation in the pH range 5 - 14 (5, 8, 24). The neutral pH form, termed "type I," is clearly high-spin; the intermediate pH form (pK = 9.1 for *Chromatium* cytochrome *c*'), termed "type II," is predominantly high-spin with some low-spin character; and the alkaline pH form (pK = 10.8 for *Chromatium* cytochrome *c*'), "type III," is low-spin (8). Alcohols convert type I to type III without the appearance of the type II spectrum (8). Analysis of the EPR spectrum of *Chromatium* ferricytochrome *c*' at low temperature indicates a single species in the pH region 3 to 7.4, a transition with pK = 9.8 to another high-spin species, and multiple paramagnetic species above pH 11.6 (16). The pK 9.8 transition is in reasonable agreement with the type I to type II conversion observed in the absorption spectrum. However, the type II to type III conversion monitored by absorption spectra occurs at a much lower pH.

The kinetics of ligand binding by the cytochromes *c*' have been studied in some detail (13, 14) and are consistent with a five-step mechanism for carbon monoxide binding to ferrocytochrome *c*' which involves an access controlled step and subsequent binding of two molecules of CO per heme iron. A plausible explanation is that the heme moiety is buried in the interior of the molecule and not directly accessible to solvent (14). Gibson and Kamen (13) noted that the reduction of *Chromatium* cytochrome *c*' by sodium ascorbate was kinetically complex. However, no detailed studies of the oxidation-reduction kinetics of the cytochromes *c*' have been reported.

It is our objective to determine if the kinetic characteristics for the reduction of cytochrome *c* are exhibited by the cytochromes *c*'. Moreover, these atypical cytochromes *c* offer an opportunity to investigate the role of heme iron spin state in electron transfer reactions.

MATERIALS AND METHODS

Chromatium vinosum was grown heterotrophically and cytochrome *c*′ isolated and purified as described previously (25). *Chromatium* cytochrome *c*′ was rechromatographed until a spectral ratio A_{280}/A_{398} of 0.32 to 0.35 was obtained. Heme concentration was determined by absorbance of the pyridine hemochrome (0.1 *M* NaOH plus 25% pyridine) based on reduced minus oxidized millimolar extinction coefficient at 550 nm of 19.1.

Cytochromes *c*′ from *Rhodopseudomonas sphaeroides, Rhodopseudomonas gelatinosa,* and *Rhodopseudomonas palustris* were isolated as described (26). Sperm whale myoglobin was purchased from Worthington and purified by chromatography on DEAE-cellulose (27).

FMN and riboflavin were obtained from Sigma Chemical and FAD from Calbiochem. Lumiflavin was kindly supplied by Dr. Gordon Tollin of this department. The flavins were purified by chromatography on DEAE-cellulose prior to use. Thionine and indigodisulfonic acid were obtained from K & K Laboratories, gallocyanin from Allied Chemical, and prune from Esbe Laboratory Supplies; all were used without further purification. Sodium dithionite was obtained from Hardman and Holden (Manchester, England) and the SO_2^- anion radical concentration was determined in different solvents as described previously (28). All kinetic studies were conducted in a Durrum-Gibson stopped-flow spectrophotometer. Anaerobic reservoir vessels for the stopped-flow spectrophotometer (29) were constructed for reactions using reductants other than sodium dithionite. These anaerobic reservoir vessels were attached to the filling ports of the driving syringes to allow deoxygenated reactants to be transferred to the driving syringes of the stopped-flow spectrophotometer without oxygen leakage. The reduced forms of the various organic redox compounds were prepared in the reservoir vessel with hydrogen gas using platinum on asbestos as a catalyst.

Solvent perturbation studies were conducted in 0.1 *M* potassium phosphate, pH 7.0, supplemented with various amounts of organic solvent. Phosphate ions were sparingly soluble at high concentrations of the organic solvents used (approximately 40%) so 0.01 M phosphate solutions were used under these conditions. Changes in the junction potential of the pH electrode in the various mixed solvents necessitated correction of the observed pH to obtain an actual pH of 7.0 in the solutions.

RESULTS

Reduction by Sodium Dithionite

The reaction of *Chromatium* ferricytochrome *c*' with sodium dithionite is kinetically complex as has been reported for horse heart cytochrome *c* (28, 30). All data obtained can be accurately described by equation 1 which represents the simultaneous reduction by SO_2^- and $S_2O_4^{2-}$.

$$k_{obs} = k_1(S_2O_4^{2-}) + k_2K_{eq}^{\frac{1}{2}}(S_2O_4^{2-})^{\frac{1}{2}} \quad (1)$$

Where k_1 and k_2 are the rate constants for $S_2O_4^{2-}$ and SO_2^- reduction. K_{eq} was determined in different solvents by EPR as described previously (28).

The rate constants for reduction of *Chromatium* cytochrome *c*' by $S_2O_4^{2-}$ and SO_2^- were found to be 7.3×10^5 $M^{-1}sec^{-1}$ and 1.9×10^9 $M^{-1}sec^{-1}$, respectively, at pH 7.0 in 0.1 *M* potassium phosphate. As observed for the reduction of cytochrome *c* (28, 30), the rate of reduction by SO_2^- is much larger than that for reduction by $S_2O_4^{2-}$. However, in the presence of 0.1 *M* potassium phosphate pH 7.0, *Chromatium* cytochrome *c*' is reduced 50 times faster by SO_2^- than is cytochrome *c* (Table I). Similarly, $S_2O_4^{2-}$ reduction is 10 times faster. As observed with cytochrome *c* (28), increasing ionic strength decreases the rate of reduction of cytochrome *c*' by both SO_2^- and $S_2O_4^{2-}$. Debye-Huckel plots (31) cannot be made for ionic strengths below 0.1 as the reduction process becomes too rapid to accurately measure, thus we have extrapolated available data (μ = 0.1 to 1.2) to infinite dilution. Although highly questionable in terms of the absolute values, these estimates do provide the sign of the charge at the site of electron transfer and a lower limit for the rate constants. The rate constant thus obtained for $S_2O_4^{2-}$ reduction at infinite dilution is 2×10^6 $M^{-1}sec$ and that for SO_2^- reduction at infinite dilution is 3×10^9 $M^{-1}sec^{-1}$, with an estimated net charge of +0.5 on the protein at the site of reduction for both SO_2^- and $S_2O_4^{2-}$.

Reduction of Other Cytochromes c'

The rate of reduction of cytochromes *c*' from several sources was investigated to ascertain possible correlations between the common features of these cytochromes and their reduction kinetics. Results of this survey, together with midpoint potentials at pH 7 and isoelectric points of the various cytochromes, are given in Table I. The rate constants for reduction

Table I. Reduction of Various High-Spin Heme Proteins by Dithionite[a]

Heme Protein	pI	$E_{m,7}$ (mV)	k_1[b] $(M^{-1} sec^{-1}) \times 10^{-5}$	k_2[b] $(M^{-1} sec^{-1}) \times 10^{-7}$
Horse heart cytochrome *c*	10.1	+260	0.5	6
Myoglobin	7.2	+146	negligible	0.44
Chromatium vinosum cytochrome *c'*	5.0	-5	7.3	190
Rhodopseudomonas palustris cytochrome *c'*	9.4	+110	28.0	4.3
Rhodopseudomonas gelatinosa cytochrome *c'*	9.6	?	7.0	35
Rhodopseudomonas sphaeroides cytochrome *c'*	4.6	+30	4.0	311

[a]Reaction in 0.1 *M* potassium phosphate, pH 7.0; heme concentration approximately 2 μ*M*; 20°C; monitored in Soret region.

[b]See text.

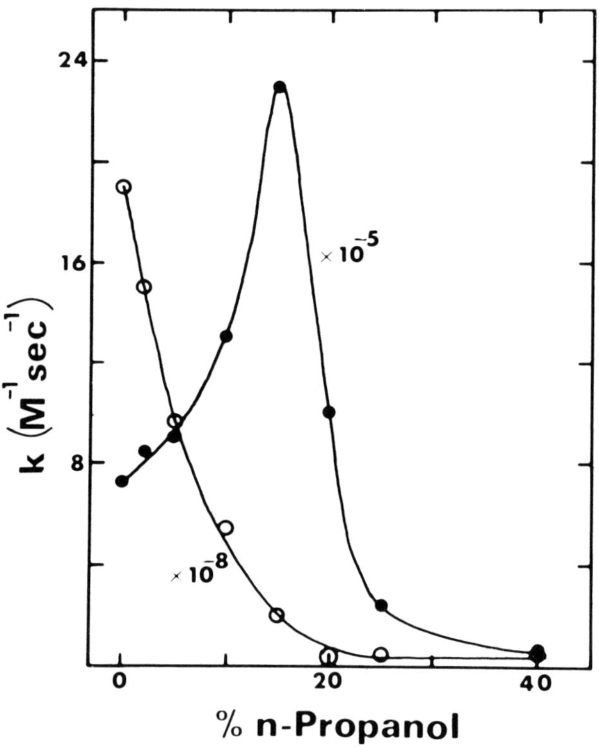

Fig. 1. Effect of n-propanol on the second-order rate constant for the reduction of Chromatium cytochrome c′ by SO_2^- (open circles) and $S_2O_4^{2-}$ (filled circles). Buffer: 0.1 M potassium phosphate, pH 7.0; heme concentration, 3 μM; temperature, 20°C; n-propanol was added to give the indicated concentrations.

by $S_2O_4^{2-}$ varies by an order of magnitude between the various cytochromes c′ while the rate constant for SO_2^- reduction varies over a 50-fold range. In all cases, reduction by SO_2^- is substantially faster than by $S_2O_4^{2-}$ and reduction by either species does not appear to correlate with the oxidation-reduction potentials or isoelectric points of the cytochrome *c*'.

Also included in Table I is the rate of reduction of myoglobin by $S_2O_4^{2-}$ and SO_2^-. In agreement with Lambeth and Palmer (30), we found that the rate of reduction of myoglobin by $S_2O_4^{2-}$ is negligible with a plot of k_{obs} versus $(S_2O_4^{2-})^{1/2}$ linear over the entire $(S_2O_4^{2-})$ range investigated (10 - 500 μ*M*).

Effect of pH

The effect of pH on the rate of reduction of *Chromatium* ferricytochrome *c*' by dithionite is summarized in Table II. Increasingly, alkaline pH has a similar effect on the rate of reduction by both $S_2O_4^{2-}$ and SO_2^-. Reduction by either species has a maximum at pH 7 followed by sharp changes in the rate constant centered around pH 9.5 and 11.

Effect of Organic Solvents

Figure 1 presents the variations in the second-order rate of reduction of *Chromatium* ferricytochrome *c*' with dithionite as a function of the concentration of *n*-propanol in the reaction mixture. The rate of reduction by $S_2O_4^{2-}$ increases by a factor of 3 between 0 and 15% *n*-propanol after which it decreases to a value an order of magnitude less than that in the absence of *n*-propanol. The rate of reduction by SO_2^- decreases

Table II. Effect of pH on the Reduction of Chromatium Cytochrome c' by Dithionite

pH[a]	$S_2O_4^{2-}$ Rate Constant $(M^{-1}sec^{-1}) \times 10^5$	SO_2^- Rate Constant $(M^{-1}sec^{-1}) \times 10^{-7}$
5.5	3.9	32
6.2	5.4	56
7.0	7.4	180
8.0	2.6	46
9.0	1.9	45
10.0	0.3	12
11.0	0.003	4.9
12.0	0.007	1.6
13.0	0.030	1.3

[a] *Buffer: 25 mM Tris, 25 mM potassium phosphate, 25 mM sodium acetate and 25 mM glycine adjusted to the indicated pH. Heme concentration 3 μM, at 20°C.*

TABLE III. Reduction of Chromatium Cytochrome c' by Various Electron Donors

Reductant[a]	$E_{m,7}$ [b]	Net Charge at pH 7.0	k_f [d] ($M^{-1}sec^{-1}$)	k_r [d] ($M^{-1}sec^{-1}$)
Lumiflavin	-207	0	TFTM[c]	?
Flavin mononucleotide	-219	-1	5.2×10^7	1.2×10^4 [e]
Riboflavin	-208	0	3.9×10^7	1.4×10^4 [e]
Flavin adenine dinucleotide	-219	-2	1.4×10^7	1.2×10^4 [e]
Indigo disulfonate	-125	-2	3.8×10^7	3.4×10^5 [e]
Thionine	+ 63	0	6.5×10^4	9.7×10^4
Gallocyanin	+ 21	-1	1.5×10^5	9.6×10^4
Prune	+ 58	0	1.6×10^4	3.9×10^4

[a]Reaction in 0.1 M potassium phosphate, pH 7.0; monitored at a suitable isosbestic point of the reductant; isoalloxazines measured at 2°C, others at 20°C. Heme concentration 2 μM.

[b]See reference 36 for potentials.

[c]TFTM, too fast to measure with stopped-flow.

[d]Defined in the text.

[e]Calculated from the equilibrium constant assuming a simple equilibrium and the measured value of k_f.

almost two orders of magnitude with a midpoint for the transition at approximately 14% *n*-propanol. Ethanol yields results identical to that found for *n*-propanol except that greater alcohol concentrations are required to achieve the same effect.

Reduction by Other Electron Donors

Assuming that reductants interact with ferricytochrome *c*' in a specific manner, a homologous series of reductants, which differ in side chain substituents, should reflect the nature of the electron transfer interaction site. The results of an

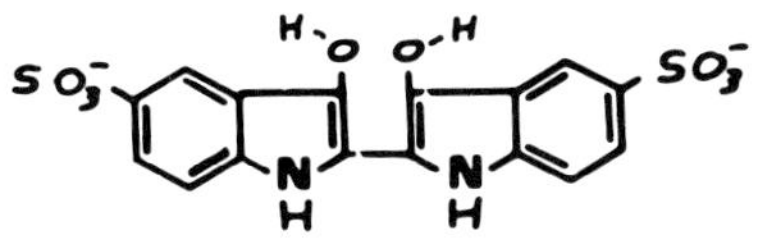

INDIGODISULFONATE

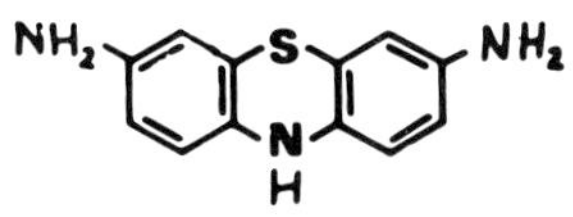

THIONINE

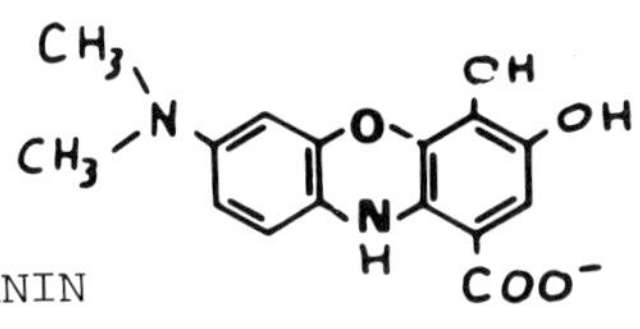

GALLOCYANIN

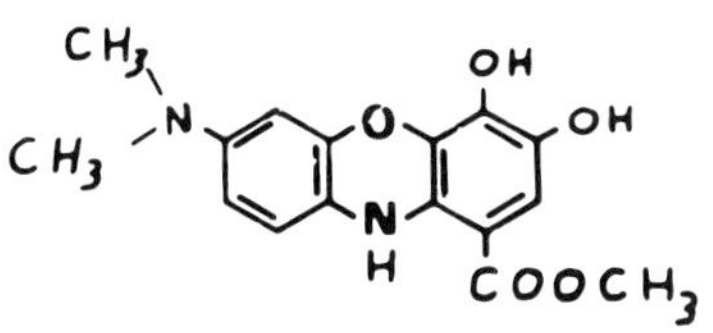

PRUNE

Fig. 2. Structures of some organic redox compounds given in Table III.

investigation of the reaction of *Chromatium* cytochrome *c*' with several different reductants is summarized in Table III, while Fig. 2 gives the structure of the less well known reductants used.

Reduction of cytochrome *c*' by isalloxazines and indigo disulfonate is rapid, precluding the use of pseudo first-order conditions. Thus, reactions were conducted with equal quantities of reductant and cytochrome (in terms of equivalents of electrons). The oxidation-reduction potential difference between the isoalloxazines and cytochrome *c*' were such that complete reduction occurred under equimolar conditions. The fact that second-order plots were linear for reduction by the isoalloxazines and indigo disulfonate suggested that a simple second-order mechanism applied.

The reduction of *Chromatium* cytochrome *c*' by thionine, gallocyanin, and prune was performed under pseudo first-order

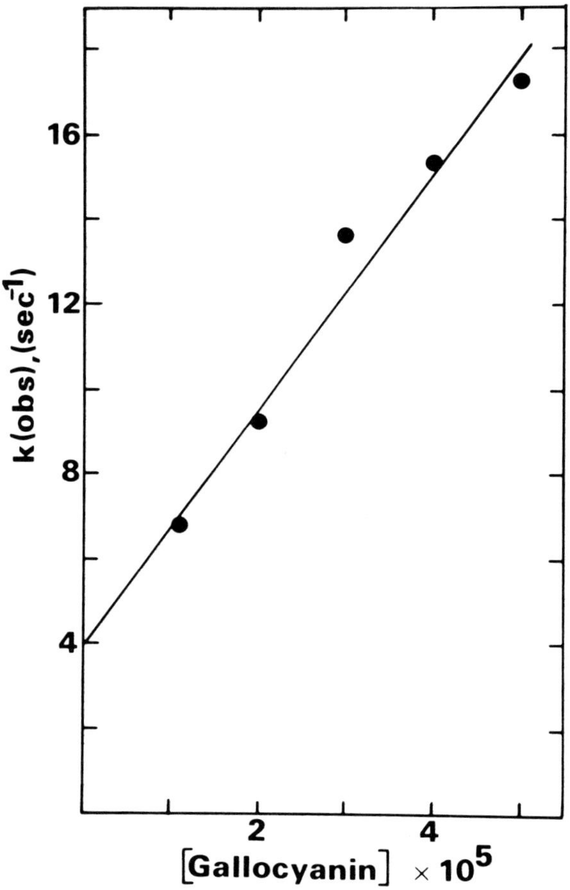

Fig. 3. Second-order plot for the reduction of Chromatium cytochrome c' by gallocyanin. Buffer: 0.1 M potassium phosphate, pH 7.0; heme concentration, 3 μM; temperature, 20°C; wavelength monitored, 429 nm.

conditions. However, the oxidation-reduction potentials of these reductants were quite similar to the cytochrome *c*' potential so that even under pseudo first-order conditions there was an appreciable contribution of the reverse reaction to the k_{obs} measured. Figure 3 presents typical data for the effect of reductant concentration on k_{obs}. As expected, the cytochrome *c*' was not fully reduced by these electron donors unless a large excess was used. The rate constants reported in Table III for the forward, k_f (reduction), and the reverse, k_r (oxidation), reactions of these compounds with cytochrome *c*' were obtained by digital computer fitting of the observed absorbance change as a function of time to the integrated rate expression for a reversible second-order reaction. The concentration of each reductant was taken to be the number of equivalents of electrons per liter.

DISCUSSION

The reduction of cytochrome *c*' by sodium dithionite resembles that observed for horse heart cytochrome *c* in that both SO_2^- and $S_2O_4^{2-}$ are active reducing species. This observation is in sharp contrast to the case of myoglobin where only SO_2^- is active (30). Cytochrome *c*' does not bind anionic ligands in the ferriform (5, 24) implying that the heme iron is buried in the interior of the molecule and hence not available for direct electron transfer iron by anionic reductants. Thus it would appear that, like cytochrome *c* and c_2 (32, 33), only a portion of the heme (heme edge) is available for interaction with reductants in the cytochrome *c*'.

The effect of ionic strength on the reduction by *Chromatium* cytochrome *c*', a protein negatively charged at pH 7, clearly points to a positive charge(s) at the site of electron transfer. This observation suggests a specific site of electron transfer on the protein surface with a charge different from the net protein charge. This latter result is similar to that obtained with *Rhodospirillum rubrum* and *Rhodopseudomonas sphaeroides* cytochrome c_2 (34). Based on the effect of ionic strength and pH on the reduction of cytochrome *c*' by both SO_2^- and $S_2O_4^{2-}$, we conclude that both reducing species act at the same site.

It is not possible to quantitatively analyze the effect of pH on the reduction kinetics in terms of specific amino acid side chains without some structural information. However, the changes in rate constants observed in pH regions 5.5–7 are not coupled to any detectable spectral (32) or EPR (16) changes. Thus, the ionizations affecting the kinetics in this pH region are not heme-linked, indicating that acid side chains removed from the immediate heme vicinity are involved. In the alkaline pH region, large changes in the rate constant for reduction by both SO_2^- and $S_2O_4^{2-}$ occur between pH 9 and 10 and between pH 10 and 12. These pH regions are suspiciously close to the pK values for the spectral transitions (type I to type II, pK = 9.1, type II to type III, pK = 10.8, ref. 35) suggesting that changes in spin state are coupled to changes in the kinetics. However, a definitive analysis cannot be made as the structural changes associated with the spectral changes are unknown. The differences between the kinetic/spectral changes and the EPR results (16) cannot be explained at this time. However, the different temperatures used for the EPR and kinetic/spectral studies may be important.

Studies were made on the effects of organic solvent on the kinetics of reduction of *Chromatium* cytochrome *c*' for two reasons: (1) to ascertain the requirement for cytochrome integrity for efficient electron transfer, and (2) to deduce if any correlation existed between solvent-induced spectral changes (5, 8, 32) and changes in the rate of reduction. We observed very substantial effects of *n*-propanol and ethanol on the kinetics of reduction at low alcohol concentrations, leadin to the conclusion that protein integrity was required. Moreover, the effects of organic solvent on the reduction kinetics were seen at much lower solvent concentrations than those required for spectral perturbations (35). Thus, there was no direct correlation between solvent-induced spectral changes and the rate constants for reduction. These observations suggest that it is the structure in the vicinity of the site of interaction of the cytochrome and reactant that are of primary importance.

It can be proposed that the site on the cytochrome *c*' molecule which interacts with an electron donor has a specific conformation which imposes restrictions on the size or shape of molecules which can effectively interact to transfer electrons to the cytochrome. Thus, differences in rate of reduction of cytochrome *c*' by reductants with the same molecular nucleus, but with different side chain substituents, may reflect the steric requirements and chemical composition of the electron-accepting site. A survey of several types of reductants includ ing isoalloxazines, phenoxazines, phenothiazines and indigo dye have widely varying rate constants for reduction of cytochrome *c*' which do not directly correlate with the oxidation-reduction potentials of the reductants. This lack of correlation suggest that steric and chemical interactions between reductant and cytochrome plays an important role in controlling rates of electron transfer and that thermodynamic considerations are of lesser importance. For example, the rate constant for reduction of cytochrome *c*' by the isoalloxazines investigated is measurably increased when the size of the side chain is decreased, suggesting that the bulky side chain hinders interaction between the flavins and the cytochrome. Admittedly, the kinetic changes are small; however, this is what would be expected for small alterations in interaction.

The reaction of gallocyanin with cytochrome *c'* is significantly faster than that of prune, yet both molecules are phenoxazines with the difference that in prune the carboxyl side chain is methyl esterified. Thus, the negative charge on gallocyanin apparently allows more efficient interaction with the electron-accepting site of cytochrome *c'* than is possible with prune, a result consistent with the identification of a positively charged region of cytochrome *c'* as the electron-accepting site.

In the absence of detailed structural information, no specific statements about the pathways of electrons to and from the heme group of *Chromatium* cytochrome *c'* can be made. However, we can conclude that the site(s) of electron transfer is not close to the heme iron, is in the vicinity of positively charged amino acid side chains, and is sterically hindered. Further, nonpolar interactions may play a role although no direct data bearing on this point are available.

The similarity in responses of cytochrome *c* and cytochrome *c'* to dithionite, as opposed to that of myoglobin, is remarkable. As the cytochromes *c'* closely resemble myoglobin in spectral properties, the kinetic mechanisms could have been expected to be similar. However, this is clearly not the case. Thus, it appears from these studies with nonphysiological reductants that the cytochromes *c'* are properly classified as *c*-type cytochromes.

CONCLUSION

The reduction of *Chromatium vinosum* cytochrome *c'* by sodium dithionite and organic redox compounds (including isoalloazines, indigo dyes, phenothiazines, and phenoxazines) has been investigated. Cytochrome *c'* undergoes reduction by sodium dithionite in a manner similar to that reported for horse heart cytochrome *c*, that is, reduction by both SO_2^- and $S_2O_4^{2-}$. The major difference noted was that *Chromatium* cytochrome *c'* is reduced by both SO_2^- and $S_2O_4^{2-}$ much more readily than cytochrome *c*. From results obtained with SO_2^-, $S_2O_4^{2-}$ and organic dyes, and studies on the effect of pH, ionic strength, and solvent dyes, and studies on the effect of pH, ionic strength, and solvent perturbation, the following conclusions are appropriate: (1) The reduction kinetics of cytochrome *c'* are typical of those observed with other *c*-type cytochromes; (2) the site of electron transport (with inorganic or small organic reductants) is cationic and easily perturbed by organic solvents; and (3) the rate constant for reduction of *Chromatium* cytochrome *c'* is influenced by charge, steric constraints, and possibly the nonpolar character of the reductant.

ACKNOWLEDGMENTS

This work was supported by National Science Foundation Research Grant PCM 75-21009 and a NIH Career Development Award (K04EY00013) to M.A.C.

REFERENCES

1. Vernon, L. P. and Kamen, M. D. (1954). *J. Biol. Chem. 211*, 643.
2. Kamen, M. D. (1955). *Bacteriol. Rev. 19*, 250.
3. Bartsch, R. G. (1968). *Annu. Rev. Microbiol. 22*, 181.
4. Bartsch, R. G. (1981). *In* "The Photosynthetic Bacteria" (R. K. Clayton, ed.), in press. Plenum, N.Y.
5. Cusanovich, M. A., Tedro, S. M., and Kamen, M. D. (1970). *Arch. Biochem. Biophys. 141*, 557.
6. Kakuno, T., Hosoi, K., Higuti, T., and Horio, T. (1973). *J. Biochem. (Japan) 74*, 1193.
7. Horio, T., Kamen, M. D., and Deklerk, H. (1961). *J. Biol. Chem. 236*, 2783.
8. Imai, Y., Imai, K., Sato, R., and Horio, T. (1969). *J. Biochem. (Japan) 65*, 225.
9. Kamen, M. D., Horio, T., Bartsch, R. G., and Hannon, S. (1973). *Proc. Natl. Acad. Sci. U.S. 70*, 1851.
10. Imai, Y., Imai, K., Ikeda, K., Hanaguchi, K., and Horio, T. (1969). *J. Biochem. (Japan) 65*, 629.
11. Yong, F. C. and King, T. E. (1970). *J. Biol. Chem. 245*, 2457.
12. Taniguchi, S. and Kamen, M. D. (1963). *Biochim. Biophys. Acta 74*, 438.
13. Gibson, Q. H. and Kamen, M. D. (1966). *J. Biol. Chem. 241*, 1969.
14. Cusanovich, M. A. and Gibson, Q. H. (1973). *J. Biol. Chem. 248*, 822.
15. Ehrenberg, A. and Kamen, M. D. (1965). *Biochim. Biophys. Acta 102*, 333.

16. Maltempo, M. M., Moss, T. H., and Cusanovich, M. A. (1974). *Biochim. Biophys. Acta 342,* 290.
17. Maltempo, M. M. (1975). *Biochim. Biophys. Acta 379,* 95.
18. Ambler, R. P. (1973). *Biochem. J. 135,* 751.
19. Meyer, T. E., Ambler, R. P., Bartsch, R. G., and Kamen, M. D. (1975). *J. Biol. Chem. 250,* 8416.
20. Ambler, R. P., unpublished.
21. Weber, P. and Salemme, F. R. (1977). *J. Mol. Biol. 117,* 815.
22. Cusanovich, M. A. (1971). *Biochim. Biophys. Acta 236,* 238.
23. Kamen, M. D., Dus, K. M., Flatmark, T., and Deklerk, H. *In* "Electron and Coupled Energy Transfer in Biological Systems" (T. E. King and M. Kleingenberg, eds.), Vol. I, p. 243. Dekker, N.Y.
24. Horio, T. and Kamen, M. D. (1961). *Biochim. Biophys. Acta 48,* 266.
25. Cusanovich, M. A. (1967). Ph.D. Thesis, University of California at San Diego, San Diego, California.
26. Meyer, T. E. (1970). Ph.D. Thesis, University of California at San Diego, San Diego, California.
27. Brown, W. D. (1961). *J. Biol. Chem. 236,* 2238.
28. Miller, W. G. and Cusanovich, M. A. (1975). *Biophys. Struct. Mech. 1,* 97.
29. Fox, J. L. and Schenkken, D. M. (1970). *Rev. Sci. Inst. 41,* 1637.
30. Lambeth, D. O. and Palmer, G. (1973). *J. Biol. Chem. 248,* 6095.
31. Frost, A. A. and Pearson, R. G. (1971). "Kinetics and Mechanism," p. 150. Wiley, N.Y.
32. Salemme, F. R., Kraut, J., and Kamen, M. D. (1973). *J. Biol. Chem. 248,* 7701.
33. Swanson, R., Trus, B. C., Mandel, N., Mandel, G., Kallai, O. B., and Dickerson, R. E. (1977). *J. Biol. Chem. 252,* 759.
34. Wood, F. E., Post, C. B., and Cusanovich, M. A. (1977). *Arch. Biochem. Biophys. 184,* 586.
35. Miller, W. G. (1973). Ph.D. Thesis, University of Arizona, Arizona.
36. Clark, W. M. (1960). "Oxidation-Reduction Potentials of Organic Systems." Williams & Wilkins, Baltimore, Maryland.

TOWARD A COMMON STRUCTURAL CONCEPT FOR P-450 HEME PROTEINS

Karl M. Dus

Edward A. Doisy Department of Biochemistry
St. Louis University Medical Center
St. Louis, Missouri 63104

It gives me special pleasure to contribute to a volume dedicated to Martin D. Kamen, this account of our recent work on P-450 heme proteins and heme peptides derived from these proteins by BrCN degradation. It was my association with Martin Kamen during the formative years of my research career that further deepened my already keen interest in cytochromes and focused my attention on those oxygen and carbon monoxide reactive heme proteins which, due to their atypical spectral properties and, poorly understood, but potentially most important biological function, present a formidable challenge to the chemist, biochemist, and biophysicist alike.

While I was working at the Massachusetts General Hospital I had become aware of Martin's presence at Brandeis. I was at once strongly attracted by his towering intellect, his fascination with cytochromes, and his warm, outgoing personality. I did not stop until I had a chance to work with him, first for a short time period at Brandeis and, subsequently for many years at UCSD in La Jolla. I shall never forget this wonderful time of scientific excitement, in a very scenic and casual

 ISBN 0-12-397580-8

environment, and the continuous support of Martin's insight, understanding, and guidance. This was a time of growth and development for me thanks to the unusual opportunities afforded by Martin's thoughtful provision of daily interaction with excellent visiting scholars from around the globe. Most of all, I cherished Martin's strong personal concern for every member of his research team and the almost unlimited extent of freedom in scientific inquiry and approach which was so conducive to novel, imaginative ideas and experimentation. In short, I felt so secure, comfortable, and, at the same time, intellectually stimulated that I kept forever postponing the inevitable decision to start on an independent scientific venture. However, it then developed quite unexpectedly and required only a relatively minor adjustment for me to switch from cytochromes *c* and *c*', and their function in bacterial photosynthesis to cytochromes P-450 and biological O_2 fixation.

To set the stage for my special interest in comparative structural characterization of these enzymes it seems useful to give a brief overview of their properties and organization in living systems (1 - 3). Mixed-function oxidations mediated by the group of heme proteins commonly referred to as cytochromes P-450 involve binding, activation, and cleavage of the dioxygen molecule, one atom of which is incorporated into the substrate, which is also tightly bound to the active site of the enzyme. These mixed-function oxidases or monoxygenases play a key role in biosynthesis and biodegradation of many natural compounds especially fatty acids, steroids, and aliphatic and aromatic hydrocarbons. In addition, the monoxygenases associated with the endoplasmic reticulum of the liver, kidney, lung, etc., catalyze the biotransformation of drugs, carcinogens, pesticides, and other xenobiotics. In this activity their principal biological function appears to be the conversion of lipophilic compounds which otherwise would accumulate in membranes and disrupt the characteristic organization of proteins and phospholipids to much more hydrophilic compounds which can be metabolized further by more conventional enzymes or conjugated for transport and excretion. In general, this process is designed to metabolize and remove toxic substances but it does not always lead to detoxification and may, in some instances, generate more toxic intermediates from initially rather innocuous compounds.

Due to the wide distribution and general importance of monoxygenases much attention has been focused recently on their isolation and characterization, and the detailed study of their mechanism of action. But despite much effort most of the membrane-associated enzymes have resisted final purification and even for those which have yielded to the efforts the need for solubilization with the aid of detergents seems to preclude crystallization which could lead to refined structural analysis by X-ray diffraction. In contrast, bacterial monoxygenases are usually soluble in aqueous buffers and crystallizable, and thus hold the promise of providing crucial data which seem inaccessible with detergent-solubilized preparations.

Monoxygenases consist of a truncated electron transfer chain which always includes the O_2 and substrate reactive cytochrome P-450 and an NADH- or NADPH-specific flavoprotein reductase. In addition, there can be an iron-sulfur redox protein as a third component to carry electrons from the reductase to P-450, and also to bind to the heme protein as a specific effector in the product-forming step of the oxygenation cycle. Although the redoxin is found invariably in monoxygenases from prokaryotes and mitochondria, the corresponding enzymes of the endoplasmic reticulum lack this component; but phosphatidylcholine was shown to be essential for reconstitution of activity in microsomal monoxygenases. The designation P-450 relates to the shift of the maximum of the reduced Soret band to ∿450 nm which accompanies formation of the photodissociable heme protein-CO complex. Similarly, the maximum of the oxidized Soret band shifts concomitantly with the binding of substrates and other small organic molecules. Depending on the chemical nature of the ligand these shifts may be to lower (type I) or higher wavelength (type II). The cyclic process of oxygen activation and cleavage by P-450 heme proteins requires two one-electron reduction steps at different energy levels which generate short-lived intermediates with characteristic Soret maxima (1). From these spectral perturbations and other physicochemical evidence it was concluded that O_2 and CO bind to the heme iron while substrates, inhibitors, and other small organic compounds bind to a site in the proximity of the heme. Since the binding of these compounds occurs at or close to the surface of the heme proteins and since the above mentioned spectral characteristics are essential for their classification as P-450 heme proteins, it may be assumed that the heme environment of P-450s, including the specific binding sites of substrates and inhibitors, share a unique design which is expressed by similar structural and mechanistic features.

This was the concept I shaped in my mind when I was first introduced to the P-450 field by Irwin C. Gunsalus. From his unique attitude toward science, and through his constant encouragement and support of my activities and ideas I learned a great deal beyond the important details of this unfolding field of biochemistry. Stimulated through our frequent informal discussions I focused my imagination on the enormous yet still mostly undeveloped potential of monoxygenase systems, and soon I was fascinated by the variety of P-450 catalyzed reactions and the special beauty of the bacterial model system which had been developed in *Pseudomonas putida* (2). In addition, I sensed a commitment on my part to contribute to the final purification of the camphor hydroxylase components.

ISOLATION OF HEME PEPTIDES OF CYTOCHROME P-450$_{CAM}$

The contrast between the ready supply of the soluble and highly purified P-450$_{CAM}$ of the camphor hydroxylase of *P. putida* and the poor availability of purified heme protein preparations of membrane-associated monoxygenases at that time quickly convinced me that it was essential to develop simple but generally applicable procedures with the pure bacterial enzyme which hopefully could be extended later to include many of the less accessible but biologically very important members of the P-450 heme protein family (3). In view of the size of P-450s, ranging from 42,000 to 60,000 daltons, I spent much effort and time in searching for approaches which would enable me to focus directly on the structure of the heme environment without losing much time and material in the study of less characteristic portions of these enzymes. Obviously, information on the structure of the heme environment would be relevant to the molecular basis of the spectral shifts and could have profound implications for the mechanism of action as well. Furthermore, such an approach would permit the use of partially purified enzyme preparations as long as the absence of other heme proteins was assured. While this aspect in itself is of great practical significance in that it circumvents the need for costly purification procedures, it would also generate comparable portions for the study of common structural features of these enzymes which so far had resisted substantial purification, and provide a glimpse at how these structures can be modulated to accommodate different substrates.

Due to my extensive previous exposure to heme peptides of *c*-type cytochromes, their purification and characteristic properties, I had high hopes that similar techniques could also be applied to P-450 heme proteins. Of course, it would be necessary to use gentler procedures because the heme group of *b*-type cytochromes, like P-450, are not covalently attached to the apoprotein. And so I embarked on a long and arduous survey of conditions under which P-450$_{CAM}$ could be fragmented without significant loss of heme. These early studies proved that in contrast to cytochrome *c* all commonly used proteolytic enzymes yielded very small fragments of the heme environment of P-450$_{CAM}$ from which the heme was readily dissociated during subsequent manipulations. Any attempts to purify a heme peptide, large or small, from these digests failed. In contrast, limited degradation of the camphor-saturated cytochrome P-450_{CAM} with BrCN seemed to maintain the original heme environment as signified by retention of the Soret absorption of the substrate complex (4). Unfortunately, separation of the BrCN digest on Sephadex G-75 revealed the presence of numerous heme-containing fragments and the quantity of these components in the digest decreased with their size. Continued digestion caused increasing decoloration. Surprisingly though, even the smallest of the BrCN-derived heme-containing fragments still showed the same Soret maximum as the enzyme-complex (∿390 nm). This finding was encouraging at the time but the contrast to the enzymatic digestions was puzzling, and the spectral properties of the heme extracted from P-450_{CAM} proved to be strikingly similar to those of the heme-associated peptide fraction. Eventually we isolated enough of the small heme-containing band which emerged from the Sephadex column as the last of all peptide bands to show by amino acid composition that no undegraded methionines remained and that the amino acids occurred in roughly integral quantities. The sum of these residues (47) was in reasonable agreement with the size of the peptide estimated from its elution volume (Table I) (4). It was also of great importance that we recognized that heme peptide material from the ascending slope of the heme peptide fraction exhibited a Soret maximum at 356 instead of the 390 nm maximum seen at the descending slope of this peak (Fig. 1, insert on upper left). We assumed that the different maxima refer to different heme orientations, and possible different chelating groups. Although the yield of this relatively small heme peptide accounted for only a few percent of the heme protein digested, several additional experiments encouraged me to pursue this line of investigation further. First, I noticed that the BrCN-derived fragments were inherently more stable than the heme-containing fragments generated by enzymatic digestion of P-450_{CAM}, and did not seem to lose the heme group as readily upon further manipulations whether the maximum was at 356 or at 390 nm. The heme could be, however, extracted from the heme peptide by the acid-acetone-ether procedure used to extract the heme from P-450_{CAM} indicating that no unexpected covalent link had been formed during the conditions used in the BrCN degradation.

We then prepared antibodies against P-450_{CAM} in rabbits and successfully developed a radioimmunoassay (RIA) after incorporating ^{125}I exclusively into tyrosines of P-450_{CAM}. Taking advantage of the high sensitivity of this technique we tested all our digests and isolated heme-containing BrCN-derived fragments for their cross reactivity (Table I). To our great satisfaction we were able to prove that the peptides generated enzymatically from P-450_{CAM} had very little or no cross reactivity while the BrCN-derived heme peptides showed significant cross

TABLE I

BrCN-Derived Hemepeptides of P-450 Hemeproteins

Amino Acids	P-450$_{CAM}$	P-450$_{LM-2}$	P-450$_{LM-4}$	P-450$_{SCC}$	P-450$_{11\beta}$
$CySO_3H$	1	1	1	1	1
Asx	3	4	4	3	3
$MetSO_2$	-	-	2	-	-
Thr	2	2	2	2	2
Ser	3	3	4	2	4
Glx	3	4	6	5	6
Pro	3	2	2	2	2
Gly	8	3	6	3	4
Ala	6	3	6	3	6
Val	4	2	2	3	4
Ile	2	2	2	2	2
Leu	3	5	4	4	6
Tyr	1	1	1	2	1
Phe	2	2	2	2	1
His	1	1	2	1	2
Lys	2	2	2	3	3
HSerL.	1	1	1	1	1
Arg	2	3	1	2	3
Total	47	41	50	41	51

reactivity (Fig. 2) (4). Thus the implication was made that the heme binding site, and probably also the substrate binding site, are indeed located on the surface of the protein, that the cleavage with BrCN and the subsequent gel filtration step did not substantially alter the conformation of this portion of the polypeptide chain, and that the heme Soret maximum at either 356 or 390 nm assured the continued presence of the heme, although the smallest fragment was only 47 residues long.

To account for the whereabouts of the substrate on the heme peptide, one would have to postulate that after cleavage at the methionines the heme group was trapped in a pouch-like pocket the exit of which was blocked by the substrate. Subsequent optimization of the reaction conditions for BrCN cleavage improved the yields of the small heme peptide so that it was recovered in 10 - 15%. The presence of excess camphor, the use

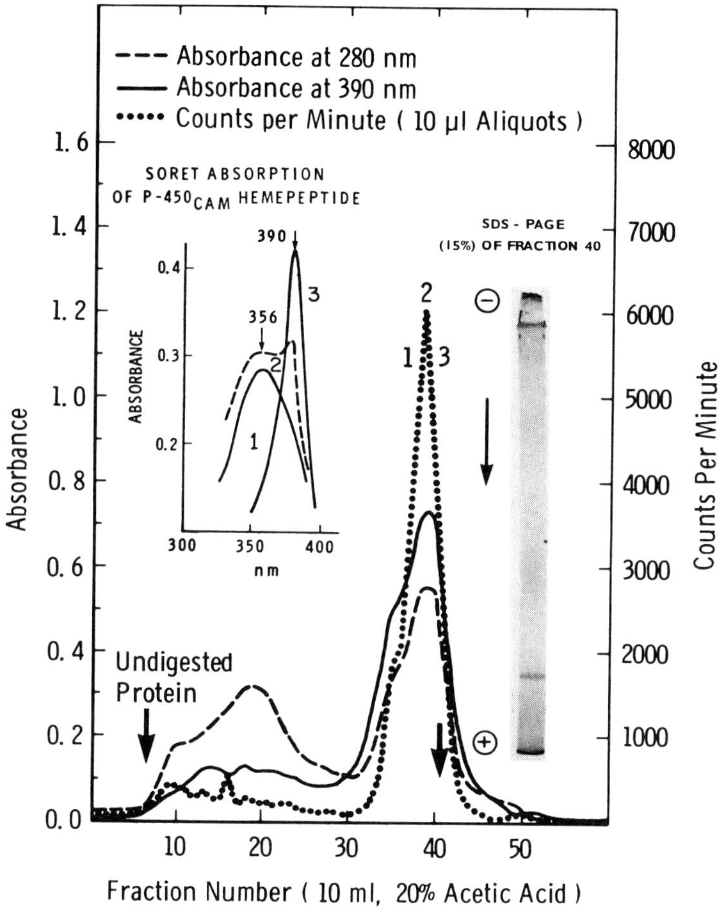

Fig. 1. Cytochrome P-450$_{CAM}$*, monolabeled with 1-(4-azido-[*^3H*]phenyl)imidazole (API): Resolution of BrCN-derived peptides on Sephadex G-75 and properties of heme peptide. The Sephadex G-75 column (4.2 × 48.0 cm, 665 ml, 1.67 ml/min) was equilibrated with 20% acetic acid, R. T., loaded with 250 nmoles radiolabeled API-P-450*$_{CAM}$ *BrCN digest, wrapped in foil, and the effluent was monitored for O.D. at 280 and 390 nm, and for tritium. The peptide concentration was also determined by automatic amino acid analysis on acid-hydrolyzed aliquots. Inset on the upper left: The spectra of the material eluting with the ascending slope (1), the center of the peak (2), and the descending slope (3) of the heme peptide fraction were recorded in 20% acetic acid. Inset on the upper right: The polyacrylamide gel electrophoresis of the heme peptide fraction (No. 40) was carried out in 0.1% SDS using a 15% cross-linked running gel in the system of Laemmli (25).*

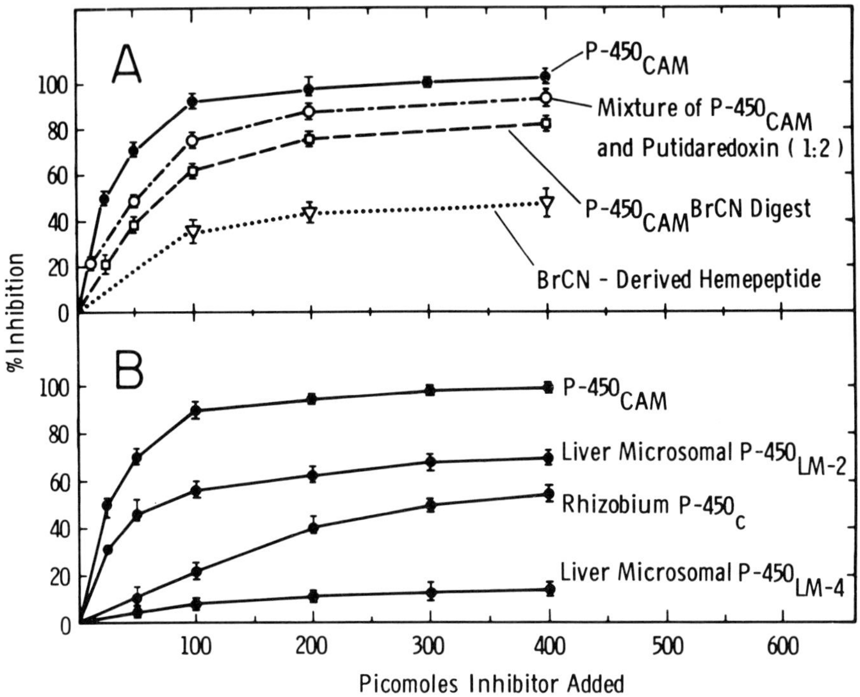

Fig. 2. Inhibition of binding between ^{125}I-labeled cytochrome P-450_{CAM} and its antibody by cross-reacting heme proteins and heme peptides. To unlabeled antigen in 150 mM NaCl - 50 mM sodium phosphate buffer, pH 7.5, a constant volume of antibody was added to give a total volume of 0.30 ml. After 2 hr incubation (R.T.) one equivalent of ^{125}I-labeled P-450_{CAM} was added; after an additional 48 hr incubation at 10°C the precipitate was collected by centrifugation, washed once with cold buffer, and dissolved in 0.2 ml of 0.2 N NaOH for counting with the ^{14}C window in a Beckman liquid scintillation counter, Model LS-30. The inhibition of binding by cross-reacting material is based on maximal P-450_{CAM}-anti P-450_{CAM} antibody binding normalized to 100%.

of 70% formic acid as the solvent, and especially a high protein concentration (7 - 10 mg protein/ml) seemed to be important factors contributing to these improvements.

COMPARATIVE STUDIES ON P-450 HEME PROTEINS AND THEIR HEME PEPTIDES

It was most important next to determine whether this technique of limited BrCN digestion could be applied to other P-450 heme proteins as well. Since no protein preparations were available at this time, we isolated microsomes from phenobarbital and from methylcholanthrene-induced rats and digested them directly with BrCN in the presence of excess inducer. From the supernatant of the digests we eventually isolated heme peptide fractions with spectral characteristics and sizes resembling our small heme peptide from P-450_{CAM}. Radioimmunoassays with anti P-450_{CAM} antibodies indicated the possible presence of cross-reacting material. The success of this simple and far-reaching experiment was very exciting and prompted me to talk to Minor J. Coon who had pioneered the isolation and purification of rabbit liver microsomal monoxygenases (5); he was, at that time, probably the only person who had highly purified P-450 heme proteins of the drug hydroxylase P-450_{LM-2}, and the arylhydrocarbon hydroxylase, P-450_{LM-4}. Stimulated by his strong interest in our comparative approach, heme peptide work, and immunochemical experiments, and with his support and generous supply of purified P-450s, we made rapid progress in applying our techniques to the microsomal enzymes. Soon we were able to show unequivocally that P-450_{LM-2} of rabbit liver microsomes and P-450_{CAM} of *P. putida* have comparable amino acid compositions, heme peptides (Table I) (6), and immunochemical characteristics (Fig. 2) (6, 7). Not only did we find mutual cross reactivity between these two heme proteins and their respective antibodies but we established that anti P-450_{CAM} antibodies inhibit benzphetamine demethylation by the reconstituted drug hydroxylase (7). However, a surprising observation by Lu and his colleagues put these immunochemical experiments into a different light. They found that by Ouchterlony double diffusion using anti-P-450_{LM-2} sera, they could not see any precipitin lines or arcs indicative of cross reaction with P-450_{LM-4} (P-448) of rat liver microsomes (8). Similar observations with rabbit liver microsomal proteins were then also reported from Coon's laboratory (9). From these findings it seemed for a while that there was no structural relationship between the drug hydroxylase and the aromatic hydrocarbon hydroxylase. We repeated these experiments with anti-P-450_{CAM} antibodies using our RIA and found, indeed, negligible cross reactivity with P-450_{LM-4} (Fig. 2) (10). After BrCN digestion of this protein, however, there was substantial cross reactivity with some fragments of the digest, presumably the heme peptides. This seemed to confirm our earlier findings with heme peptide fractions obtained by BrCN degradation of methylcholanthrene induced rat liver microsomes. Thus we proceeded with the assumption that there probably existed within these two heme proteins portions with structural similarity but that they were hidden from the surface of $P\text{-}450_{LM\text{-}4}$ which, after all, was sufficiently larger than $P\text{-}450_{LM\text{-}2}$ or $P\text{-}450_{CAM}$ to permit construction of an all new surface over the skeleton of $P\text{-}450_{LM\text{-}2}$-like material.

Another P-450 type heme protein, which was discovered rather early (11), was brought to my attention by Cyril Appleby from the CSIRO in Canberra who had found this heme protein in *Rhizobium japonicum* grown under N_2-fixing conditions. He had developed one of the first successful purification procedures in the P-450 field and was able to demonstrate the existence of a family of at least three different heme proteins (called a, b, and c) of the $P\text{-}450_{Rh}$ type in *Rhizobia* (12). Unfortunately, the biological significance of these heme proteins in N_2 fixation is still poorly understood, although a function in facilitated O_2 diffusion has been proposed (13). In view of the tight binding of typical P-450 substrates and inhibitors, and the strong possibility of the existence of a proper iron-sulfur redox protein, however, it seems difficult to accept the diffusion hypothesis unmodified.

As soon as we had received small samples of these heme proteins, we tested them for cross reactivity with our RIA using anti $P\text{-}450_{CAM}$ antibodies and found substantial cross reactivity, although somewhat less than with $P\text{-}450_{LM\text{-}2}$ (Fig. 2) (14). To prove our point about similarity of antigenic determinants even further we immobilized a particularly suitable fraction of medium affinity anti-$P\text{-}450_{CAM}$ antibodies on Sepharose 4B and substantially purified *Rhizobium* P-450 *c* by immunoadsorption chromatography (14). Subsequent isoelectric focusing resolved this purified fraction further into two individual heme proteins (14).

From these experiments I became convinced that the strong resemblance between $P\text{-}450_{CAM}$ and $P\text{-}450_{LM\text{-}2}$ was not incidental and that we would find many more structurally related P-450 heme proteins. This was, in fact, confirmed when we started to investigate the steroid-metabolizing P-450 heme proteins of adrenocortical mitochondria, i.e., $P\text{-}450_{SCC}$ of the cholesterol side chain cleavage enzyme and $P\text{-}450_{11\beta}$ of the steroid 11β-hydroxylase, both of which yielded comparable small heme peptides (Table I). For these studies, highly purified preparations of mitochondrial heme proteins were provided by Colin R. Jefcoate (15). John Bumpus, a graduate student in my laboratory, has extended this work to adrenocortical microsomes from which he isolated several P-450 heme proteins in respectable purity. Thus, we have at this time about a dozen P-450 heme proteins a various stages of investigation, and it is gratifying to see that the comparative approach to this group of heme proteins has revealed unique aspects that point to definite similarity among them.

PHOTOAFFINITY LABELING STUDIES OF P-450 HEME PROTEINS

During the last few years we started a new line of enquiry into the detailed structural features, which may be common to many P-450 heme proteins, based on photoaffinity labeling of their substrate and inhibitor binding sites with suitable arylazido derivatives, especially 1-(4-azidophenyl)imidazole (API) (16). This line of research is made possible by the ready availability of heme peptides via BrCN degradation and was initially designed to prove the presence of the substrate or inhibitor on heme peptides derived from the corresponding enzyme-substrate or inhibitor complex. In turn, this approach now adds a new dimension to our investigation of P-450-derived heme peptides because it aims at the exact localization of these covalently linked labels, down to the C atoms to which they are attached, and at determining their distance and orientation relative to the heme group. Through multifunctional photoaffinity labels (PALs) one might even get access to important features of the folding of the polypeptide chain in the heme environment which perhaps could give us the answer to the unexpected rigidity of conformation in this area of the protein. Furthermore, the photocovalently attached labels could be expected to enhance significantly the conformational stability of the heme peptide and provide added insurance against inadvertant loss of heme (Fig. 1). The opportunity to incorporate radioactivity via the PALS either in form of ^3H or ^{14}C opened the way for the development of a micro procedure.

The first objective of this research project was the synthesis of suitable PALs (Fig. 3) (16), which would essentially retain the high binding affinity and specificity of the unmodified substrate or inhibitor molecules and, once bound, could be photolyzed readily to highly reactive intermediates capable of binding to the nearest C atom without discrimination. As a further precondition the specific photocovalent attachment had to reach completion under conditions that altered the native conformation of the heme protein only slightly, if at all. After a short survey of the literature we followed essentially the pioneering researches of S. J. Singer (17) and R. W. Porter (18) and settled for the use of arylazides which can be readily synthesized and photolysed to the highly reactive nitrenes under sufficiently mild conditions. Fortunately, the arylazido derivatives of most inhibitors active on P-450 heme

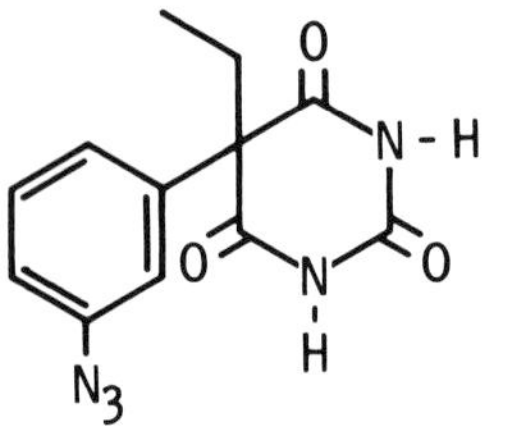

3' - Azidophenobarbital
3' - APB

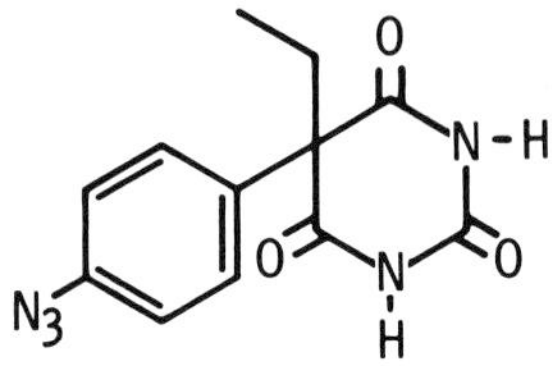

4' - Azidophenobarbital
4' - APB

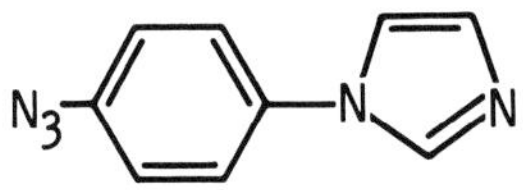

1 - (4 - Azidophenyl) - Imidazole
API

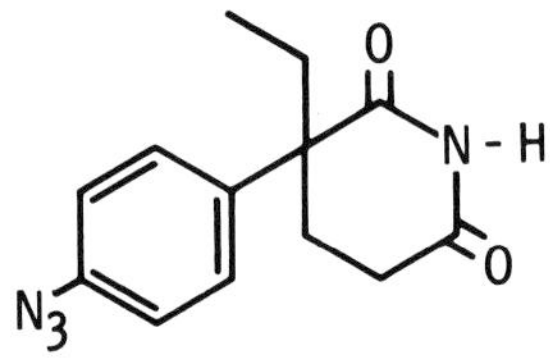

4' - Azidoglutethimide
4'-AGL

Fig. 3. Photoaffinity labels (PALs). All PALs were used in ninefold molar excess over protein. Photolysis of the P-450 heme protein-PAL complex was carried out (450 W Hanovia medium pressure quartz-mercury lamp fitted with filter solutions) at 313 nm, 5°C, for 4 half-lives of the respective PAL.

proteins seem to be binding rather tightly and specifically; but this is not the case with the majority of substrate derivatives. While the azido group was readily incorporated into phenobarbital, glutethimide, and *N*-phenylimidazole (Fig. 3), its incorporation into the norbornane structure of camphor proved to be a real challenge which we have not yet successfully met despite the synthesis of at least a dozen compounds. Traditionally the most powerful inhibitors for P-450 heme proteins are molecules which contain a group that can serve as a ligand to the heme iron such as the imidazole in *N*-phenylimidazole or the pyridine ring in metyrapone. These compounds also have the greatest potential for our comparative photoaffinity labeling studies because we may assume that they will label the same characteristic position relative to the heme in all P-450 heme proteins to which they bind tightly. As shown in Table I

TABLE II

Photoaffinity Labeling with 1 - (4-Azidophenyl) Imidazole (API)

ISOLATED P-450 HEMEPROTEIN	SOURCE	K_D^1	% PHOTOCOVALENT ATTACHMENT[2]
$P\text{-}450_{CAM}$	P. putida	1.07 μM	114
$P\text{-}450_{LM\text{-}4}$	Rabbit Liver Microsomes	1.18 μM	74
$P\text{-}450_{LM\text{-}2}$		2.25 μM	88
$P\text{-}450_{11\beta}$	Bovine Adrenocortical Mitochondria	26.9 μM	80
$P\text{-}450_{SCC}$		33.9 μM	111

[1]Apparent Spectral Dissociation Constant

[2]Applying a 9-Fold Molar Excess of API over Hemeprotein, at 313 nm, 5h.

we have used API successfully to label a group of five P-450 heme proteins with high affinity of binding as indicated by their typical type II spectral transition and corresponding apparent spectral dissociation constant (K_D). The PAL also afforded the highest photocovalent incorporation into the specific binding sites of $P\text{-}450_{CAM}$, $P\text{-}450_{LM\text{-}2}$, $P\text{-}450_{LM\text{-}4}$, $P\text{-}450_{SCC}$, $P\text{-}450_{11\beta}$ (Table II), and even *Rhizobium* P-450*c* which is not shown in Table II because its K_D was only 55 μ*M*. Reconstitution experiments with extensively labeled $P\text{-}450_{CAM}$ (Fig. 4) and $P\text{-}450_{LM\text{-}2}$ showed that the labeled enzymes can no longer catalyze the respective reactions of camphor hydroxylation and benzphetamine demethylation. Pancreatic ribonuclease, which only labels to 3% with API because it has no specific binding sites, was used as a macromolecular scavenger of nitrene radicals to avoid adventitious labeling of P-450s and thus to assure specificity.

Although we had anticipated that a covalently attached label would stabilize the BrCN-derived heme peptide (Fig. 1) the actual results of our experiments in which photocovalently

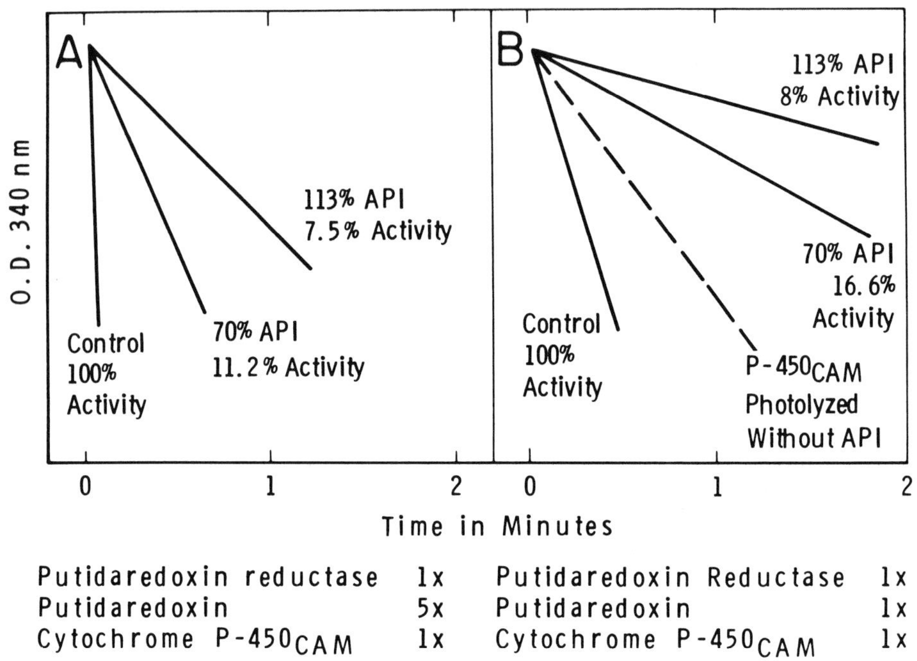

Fig. 4. Reconstitution of camphor hydroxylation with API-labeled cytochrome P-450$_{CAM}$. The residual activity of mono-labeled preparations in the reconstituted camphor hydroxylase was monitored for NADH oxidation at 340 nm.

API-labeled P-450$_{CAM}$ was degraded with BrCN substantially exceeded our expectations. It was found that the small, 47 residue-containing heme peptide when separated by gel filtratio on Sephadex G-75 now accounted for almost 80% of the original amount of heme in the protein and about 75 - 80% of the radioactivity of the incorporated PAL (Fig. 1). This small heme peptide contained almost stoichiometric amounts of heme and label. Thus we speculate that the fortuitous increase in yield might b due to greater exposure of the two methionines at the hinge regions after labeling so that the cleavage reaction goes farther toward completion after affinity labeling. Also, a smaller los of heme during the reaction may be occasioned by the increased stability of the active site after covalent attachment of API. This finding proves unequivocally that the substrate and inhibitor binding sites are located in the immediate heme environment as previously implied (4). Similar results were obtained with APB-labeled P-450$_{LM-2}$ except that the recovery of radiolabeled heme peptide was not in excess of 65%, in this case due to a less extensive degradation by BrCN (10, 19).

When the photoaffinity labeling procedure with APB was applied under identical conditions to P-450_{LM-2} and P-450_{LM-4}, both induced by phenobarbital in rabbit liver, the resulting BrCN-derived heme peptides showed significant cross reactivity with anti-P-450_{CAM} antibodies (6, 10, 20). Hence it is my belief that the heme peptides contain one of the most important antigenic determinants of these P-450 heme proteins, and that this determinant may be closely similar in structure for a large number of proteins of the P-450 heme protein family. This concept is further supported by the unusual conformational stability of this portion of the molecule which brings to mind the situation found in antibody molecules where functional portions of the molecule exist as separate structural domains interconnected by thin, exposed hinge regions. Analogously, it may be postulated that the heme and substrate binding sites of P-450 represent a tightly folded, functionally and structurally preserved domain to which the selective BrCN cleavage has provided an unexpected, convenient access due to the fact that two methionines seem to be fortuitously located, one each, in the two surrounding hinge regions. It is of interest to note that the heme peptide of P-450_{LM-4} still contains two internal methionines which apparently were not accessible for cleavage under the conditions applied (Table I). In a comparison of the amino acid compositions of five heme peptides it can be seen that all of them contain at least one residue each of cysteine, histidine, and tyrosine. These residues are currently under consideration as ligands to the heme iron in the heme peptide. Whether these residues are also the residues of concern for ligation to the heme iron in the protein is more difficult to answer at this point, although it is most likely, based on several pieces of evidence that accrued recently (21, 22). ESR evidence on the NEM_5-P-450_{CAM} derivative (23) would imply that the cysteine residue found in the heme peptide is indeed the one implicated as ligand to the heme iron in the parent protein. The opposite ligand is most likely histidine (22) or tyrosine (26). The ready availability of the heme peptide has now opened the way for isolation of site-specific antibodies for the antigenic determinant associated with the heme peptide. The specific antibodies will, in turn be used to further improve the study of cross-reacting heme peptides.

Various affinity chromatography schemes have evolved from the use of derivatives of *N*-phenylimidazole. Aminophenylimidazole immobilized on Sepharose is used in our laboratory for several purification procedures. For instance, it is employed to remove unlabeled P-450_{CAM} from incompletely API-labeled preparations to produce 100% labeled preparations. It is also very useful in the purification of P-450 heme proteins which bind *N*-phenylimidazole less tightly than does P-450_{CAM} so that they may be released under conditions which will not destroy their active conformation.

Now that essentially complete BrCN degradation can be employed for the generation of heme peptides due to the protective effect of the covalently linked inhibitor derivative, API, we are in the process of improving our procedure for release and purification of heme peptides further to avoid, if possible any dislocation and realignment of the heme within the binding region. For this purpose P-450 heme proteins are linked to API which is immobilized on Sepharose via a carboxyl and with the help of a cleavable spacer group first by affinity, and after photolysis covalently. The covalent adduct is then digested with BrCN, and the released fragments are removed by washing with buffer. Finally, the purified heme peptide is released from the resin by cleavage at the sensitive bond in the spacer. During the entire procedure the heme and the covalently linked inhibitor are expected to remain immobile until the heme peptide is released by the final cleavage, and hopefully even after this event. The feasibility of all steps of the procedure has been established, and we are now proceeding with the preparation of purer heme peptides in higher yields than ever before.

From many of the results discussed above, it has probably become apparent that the point of attachment of the heme to the apoprotein via ligands to the heme iron depends critically on the spacing and availability of a unique sulfhydryl of a cysteine residue. This special SH group may be common to P-450s as a group since all members so far investigated have shown very similar spectral characteristics. A second cysteine sulfhydryl may also be shared by many P-450 heme proteins as a group essential for substrate binding and orientation. At least in P-450_{CAM} a cysteine SH group has been definitely implicated to serve this function (Fig. 5) (23, 24). The occurrence of these two cysteine residues together with their functional involvement in the binding of heme and substrate, respectively, will be exploited to establish important and hopefully immutable reference points for the comparison of amino acid sequences derived from different P-450 heme proteins. We have recently developed a scheme that permits us to break the linear polypeptide chain of P-450_{CAM} at the heme-liganding cysteine selectively and we have synthesized isobornyl mercaptan which binds to P-450_{CAM}, covalently. For this purpose the reagent is introduced as mixed pyridyl disulfide and linked up to the protein by disulfide exchange to avoid interaction of the sulfhydryl with the heme iron. By this procedure it forms a disulfide bridge with the substrate-protected cysteine sulfhydryl and provides a highly selective marker for this cysteine residue (24, 27). It is hoped that similar SH-bearing derivatives of respective substrates can be developed for other P-450 heme proteins.

In summary, the different approaches we have used are in-

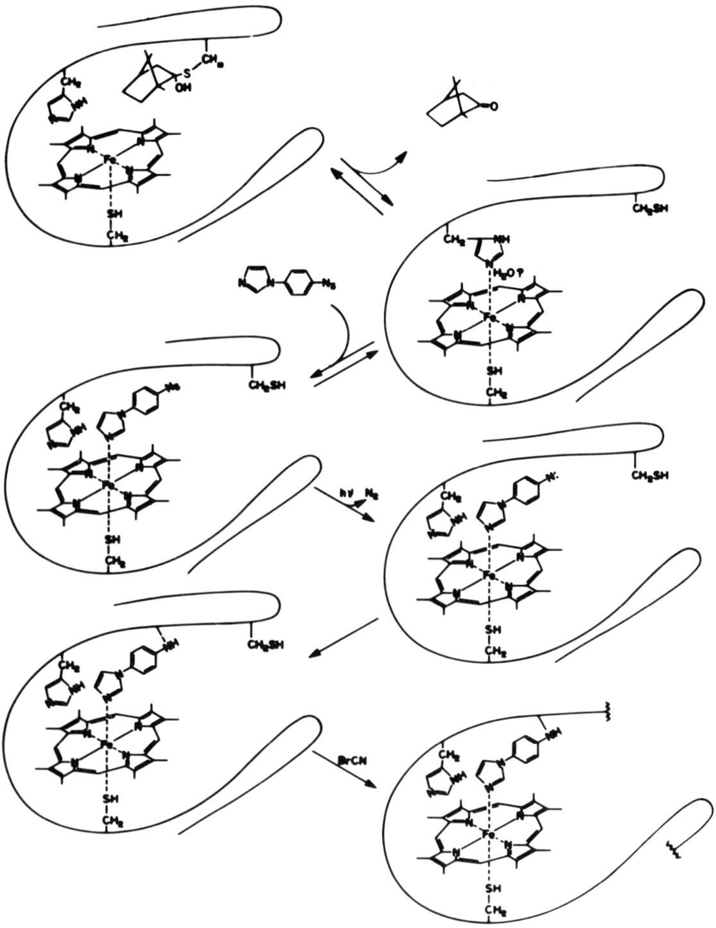

Fig. 5. Preparation of photocovalently labeled P-450 heme peptides - a conceptual outline.

tended to be mutually supportive and have the common goal to search out and establish the salient structural features of P-450 heme proteins that make them uniquely suited for their biological function in O_2 fixation. The data available to date are only indicative of what can be expected in the pursuit of these studies, but especially the existence of a tightly folded structural domain functional in heme and substrate (or inhibitor) binding clearly points to a structural design common to many, if not all, P-450 heme proteins. This common molecular design must nevertheless provide enough freedom for the individual P-450 heme protein to modulate preexisting structural elements according to the requirements of specific substrates; and it must permit adjustment for highly selective interaction with an appropriate redoxin-type electron carrier as well.

REFERENCES

1. Gunsalus, I. C., Meeks, J. R., Lipscomb, J. D., Debrunner, P., and Muenck, E. (1974). Bacterial monoxygenases - the P-450 cytochrome system. *In* "Molecular Mechanisms of Oxygen Activation" (O. Hayaishi, ed.), pp. 559-613. Academic Press, New York.
2. Hedegaard, J. and Gunsalus, I. C. (1965). Mixed function oxidation. IV. An induced methylene hydroxylase in camphor oxidation. *J. Biol. Chem. 240,* 4038-4043.
3. Dus, K. (1976). On the structure and function of cytochromes P-450. *In* "The Enzymes of Biological Membranes" (A. Martonosi, ed.), Vol. 4, pp. 199-238. Plenum, New York.
4. Dus, K. (1975). On the structure of putidaredoxin and cytochrome $P\text{-}450_{CAM}$ and their mode of interaction. *Advan. Exp. Med. Biol. 58,* 287-309.
5. Coon, M. J. and Lu, A. Y. H. (1969). Fatty acid ω-oxidation in a soluble microsomal enzyme system containing P-450. *In* "Microsomes and Drug Oxidations" (J. R. Gillette, A. H. Conney, G. J. Cosmides, R. W. Estabrook, J. R. Fouts, and G. I. Mannering, eds.), pp. 151-160. Academic Press, New York; Haugen, D. A., Van der Hoeven, T. A., and Coon, M. J. (1976). Purified liver microsomal cytochrome P-450-separation and characterization of multiple forms. *J. Biol. Chem. 250,* 3567-3570.
6. Dus, K., Litchfield, W. J., Miguel, A. G., Van der Hoeven, T. A., Haugen, D. A., Dean, W. L., and Coon, M. J. (1975). Immunochemical and compositional comparison of cytochromes $P\text{-}450_{CAM}$ of *P. putida* and $P\text{-}450_{LM}$ of phenobarbital-induced rabbit liver microsomes. *Advan. Exp. Med. Biol. 58,* 47-53
7. Dus, K., Litchfield, W. J., Miguel, A. G., Van der Hoeven, T. A., Haugen, D. A., Dean, W. L., and Coon, M. J. (1974). Structural resemblance of cytochrome P-450 isolated from

Pseudomonas putida and from rabbit liver microsomes. *Biochem. Biophys. Res. Commun. 60,* 15-21.

8. Thomas, P. E., Lu, A. Y. H., Ryan, D., West, S. B., Kawalek, J., and Levin, W. (1976). Multiple forms of rat liver cytochrome P-450: immunochemical evidence with antibody against cytochrome P-448. *J. Biol. Chem. 251,* 1385-1391.
9. Dean, W. L. and Coon, M. J. (1977). Immunochemical studies on two electrophoretically homogeneous forms of rabbit liver microsomal cytochrome P-450: P-450_{LM-2} and P-450_{LM-4}. *J. Biol. Chem. 252,* 3255-3261.
10. Dus, K., Carey, D., Goewert, R., and Swanson, R. A. (1977). Microsomal heme proteins P-450_{LM-2} and P-450_{LM-4}: Comparative structural studies. *In* "Microsomes and Drug Oxidations" (V. Ullrich, ed.), pp. 95-102. Pergamon Press, Oxford.
11. Appleby, C. A. (1968). Properties of soluble heme proteins P-450 purified from *Rhizobium japonicum* bacteroids. *In* "Structure and Function of Cytochromes" (K. Okunuki, M. D. Kamen, and I. Sekuzu, eds.), pp. 666-679. University of Tokyo Press, Tokyo) and University Park Press, Baltimore.
12. Appleby, C. A. and Daniel, R. M. (1973). *Rhizobium* cytochrome P-450: a family of soluble separable heme proteins. *In* "Oxidases and Related Systems" (T. E. King, H. S. Mason, and M. Morrison, eds.), pp. 515-528. University Park Press, Baltimore, Maryland.
13. Wittenberg, J. B., Bergersen, F. J., Appleby, C. A., and Turner, G. L. (1974). Facilitated oxygen diffusion. The role of leghemoglobin in N_2-fixation by bacteroids isolated from root nodules. *J. Biol. Chem. 249,* 4057-4066.
14. Dus, K., Goewert, R., Weaver, C. C., Carey, D., and Appleby, C. A. (1976). P-450 hemeproteins of *Rhizobium japonicum*: purification by affinity chromatography and relationship to P-450_{CAM} and P-450_{LM-2}. *Biochem. Biophys. Res. Commun. 69,* 437-445.
15. Jefcoate, C. R. (1977). Cytochrome P-450 of Adrenal Mitochondria. *J. Biol. Chem. 252,* 8788-8796; Hanukoglu, I., Spitsberg, V., Bumpus, J. A., Dus, K. M., and Jefcoate, C. R. (1981). Adrenal mitochondrial cytochrome P-450_{SCC}. *J. Biol. Chem. 256,* 4321-4328.
16. Swanson, R. A. and Dus, K. M. (1978). Specific covalent labeling of cytochrome P-450_{CAM} with 1-(4-azidophenyl)imidazole an inhibitor-derived photoaffinity probe for P-450 heme proteins. *J. Biol. Chem. 254,* 7238-7246.
17. Ruoho, A. E., Kiefer, H., Roeder, P. E., and Singer, S. J. (1973). The mechanism of photoaffinity labeling. *Proc. Natl. Acad. Sci. U.S. 70,* 2567-2571.
18. Fleet, G. W. J., Porter, R. R., and Knowles, J. R. (1969). Affinity labelling of antibodies with aryl nitrene as

reactive group. *Nature (London) 224*, 511-512; Fleet, G. W. J., Knowles, J. R., and Porter, R. R. (1972). Labelling of a specific antibody against the photoprecursor of an arylnitrene. *Biochem. J. 128*, 499-508.

19. Swanson, R. A. and Dus, K. (1977). Localization and nature of substrate binding sites on cytochrome P-450 by photoaffinity labeling. *Fed. Proc. 36*, 664.
20. Dus, K. M., Litchfield, W. J., Hippenmeyer, P. J., Bumpus, J. A., Obidoa, O., Spitsberg, V., and Jefcoate, C. R. (1980). Comparative immunochemical studies of cytochrome $P\text{-}450_{CAM}$ of *Pseudomonas putida* and of cytochrome $P\text{-}450_{SCC}$ of bovine adrenocortical mitochondria. *Eur. J. Biochem. 111*, 307-314.
21. Stern, J. O. and Peisach, J. (1974). A model compound study of the CO-adduct of cytochrome P-450. *J. Biol. Chem. 249*, 7495-7498.
22. Chevion, M., Peisach, J., and Blumberg, W. E. (1977). Imidazole, the ligand *trans* to mercaptide in ferric cytochrome P-450. An EPR study of proteins and model compounds. *J. Biol. Chem. 252*, 3637-3645.
23. Lipscomb, J. D., Harrison, J. E., Dus, K. M., and Gunsalus, I. C. (1978). Cytochrome $P\text{-}450_{CAM}$: SS-dimer and -SH derivative reactivities. *Biochem. Biophys. Res. Commun. 83*, 771-778.
24. Dus, K. M., Bumpus, J. A., and Murray, R. I. (1981). Significance of SH groups for the structure and function of P-450 hemeproteins, ISOX-5. Tokyo, July 26-29, 1981, Abstracts p. 4 (lecture S-1-4).
25. Laemmli, U. K. (1970). *Nature (London) 227*, 680-685.
26. Ruckpaul, K., Rein, H., Ballou, D. P., and Coon, M. J. (1980). Second derivative spectroscopic studies of the essential components of the liver microsomal monooxygenation system. *In* "Microsomes, drug oxidations, and chemical carcinogenesis" (M. J. Coon *et al.*, eds.), Vol. *1*, pp. 37-45.
27. Murray, R. I and Dus, K. M. In preparation.

Thiobacillus novellus CYTOCHROME OXIDASE

Tateo Yamanaka
Keiko Fujii
Yoshihiro Fukumori

Department of Biology
Faculty of Science
Osaka University
Toyonaka, Osaka 560, Japan

The cytochrome *a*-type oxidase is widely distributed not only in eukaryotes but also in prokaryotes, although other types of the terminal oxidases are known to function in the bacterial respiratory systems (1, 2). As most studies on the cytochrome *a*-type enzyme of bacteria have been done only spectrophotometrically or spectroscopically, little is known of its molecular features, although a few of them have been purified (3, 4). However, it is expected that the oxidase differs in some of its properties from organism to organism as judged from its absorption spectrum; a minute difference in the α peak position attributable to the enzyme is observed among bacteria (e.g., refs. 1, 2). Further, as we have already found, many kinds of cytochromes *c* with different enzymatic properties occur in bacteria (5, 6). This will mean that cytochrome oxidases with different properties may also occur, as cytochrome *c* functions as the direct electron donor for cytochrome oxidase.

We have recently succeeded in purifying cytochrome *a*-type oxidase derived from *Thiobacillus novellus* (7). The oxidase

 ISBN 0-12-397580-8

is similar to the mitochondrial enzyme in spectral properties but its specificity for cytochrome *c* greatly differs from that of the latter enzyme. In the present investigation, some properties of the purified *T. novellus* oxidase have been studied, and its specificity for cytochrome *c* has been compared from an evolutionary point of view with those of *Pseudomonas aeruginosa* nitrite reductase (=*Pseudomonas* cytochrome oxidase) and cow cytochrome oxidase.

PURIFICATION

Cells grown autotrophically were treated with a French pressure cell and the resulting suspension centrifugally fractionated. The particulate fractions spun down between 10,000 and 100,000 *g* were suspended in 0.1 *M* sodium phosphate buffer, pH 7.5, containing 0.2 *M* KCl and 0.5% Triton X-100. The suspension was centrifuged at 100,000 *g* after it had been allowed to stand overnight. The supernatant thus obtained was fractionated with $(NH_4)_2SO_4$ and the precipitate formed between 25 and 50% saturation was collected by centrifugation. The precipitate obtained was dissolved in a minimum volume of 10 m*M* Tris-HCl buffer, pH 8.0, containing 0.5% Triton X-100. The resulting solution was charged on a DEAE-cellulose column after it had been dialyzed against the same buffer as was used for the dissolution. The enzyme was adsorbed and a green band appeared on the column. It was eluted with 10 m*M* Tris-HCl buffer, pH 8.0, containing 0.5% Triton X-100 and 100 m*M* NaCl. A green enzyme solution was obtained. Finally, the enzyme solution was subjected to chromatography on a Sephadex G-100 column The green solution thus obtained was used for the preparation of *T. novellus* cytochrome oxidase.

SPECTRAL PROPERTIES

The oxidase has absorption peaks at 428 and around 600 nm in the oxidized form, and the peaks at 442 and 602 nm in the $Na_2S_2O_4$ reduced form. The pyridine ferrohemochrome of the enzyme shows peaks at 430 and 588 nm. These facts indicate that the enzyme has heme *a* as the prosthetic group. When the solution of the reduced enzyme is bubbled with CO, the absorption peaks appear at 431 and 599 nm. The spectral characteristics of the enzyme mentioned above appear very similar to those of the mitochondrial oxidase.

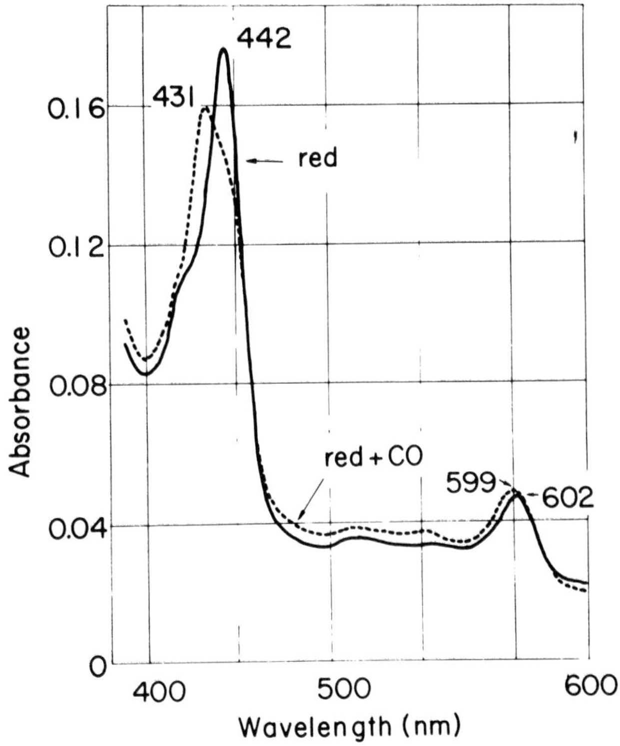

FIG. 1. Absorption spectra of T. novellus cytochrome oxidase. The enzyme was dissolved in 10 mM Tris-HCl buffer, pH 8.0, containing 0.5% Triton X-100. ——, Reduced with $Na_2S_2O_4$*; ----, the solution of the reduced enzyme was bubbled with CO.*

REACTION WITH CYTOCHROME *c*

Thiobacillus novellus cytochrome *c* resembles eukaryotic cytochromes *c* in its reactivities with certain redox enzymes (8, 9). It reacts with cow cytochrome oxidase (10, 11) more rapidly than with *Pseudomonas aeruginosa* nitrite reductase (= *Pseudomonas* cytochrome oxidase) (12), and it reacts with yeast cytochrome *c* peroxidase (13), *T. novellus* sulfite-cytochrome *c* reductase (14), and *Nitrosomonas europea* hydroxylamine oxidoreductase (15) as rapidly as eukaryotic cytochromes *c* (Table I). Since cytochrome *c* functions as the direct elec-

TABLE I. *Reactivities of Various Cytochromes c with Several Redox Enzymes*[a]

Cytochrome c		Relative reactivity				
Organism	α Peak (nm)	Nitrite reductase	Cytochrome oxidase	Cytochrome c peroxidase	Sulfite-cytochrome c reductase	Hydroxylamine-cytochrome c oxidoreductase
Pseudomonas aeruginosa	551	100	0	0	0	0
Nitrosomonas europaea	552	56	0	0	0	0[b]
Azotobacter vinelandii	551	25	0	0	0	ND[c]
Thermus thermophilus[d]	552	140	3.7	7.0	ND	2.7
Chlorobium limicola F. thiosulfatophilum	555	55	19	0	0	30
Chromatium vinosum[e]	553 (550)	16	0	17	ND	ND
Thiobacillus novellus	550	6.0	23	106	100	91
Saccharomyces oviformis	550	4.9	100	100	104	40
Tuna	550	8.7	109	120	117	128
Horse	550	2.5	124	88	97	100
Cow	550	0.53	73	ND	106	70

[a]Reaction conditions are given in Ref. 9. The reactivities of cytochromes c were expressed as relative values; the molecular activity (moles of cytochrome c reduced or oxidized per mole of enzyme) or specific activity was taken as 100%, which was observed when each enzyme reacted with its native cytochrome c, with the exception of cow oxidase and the N. europaea oxidoreductase where yeast and horse cytochromes c were used, respectively, in place of their native cytochromes c.

[b]Native electron acceptor for the enzyme is cytochrome c-554, but not cytochrome c-552.

[c]Not determined. [d]Cited from Ref. 16. [e]T. Yamanaka and Y. Fukumori, unpubl.

tron donor for cytochrome oxidase, the resemblance to the eukaryotic cytochromes *c* of the *T. novellus* cytochrome suggests that cytochrome oxidase of this organism may have similar enzymatic properties to the mitochondrial enzyme.

As Table II shows, the purified oxidase of *T. novellus* does not react or reacts very poorly with the "bacterial-type cytochromes *c* such as cytochrome *c*-551 (*P. aeruginosa*) and cytochrome *c*-552 (*N. europaea*). However, it reacts rapidly with cytochromes *c* of *T. novellus, Saccharomyces oviformis, Candida krusei,* tuna, and bonito. It reacts very poorly with horse and cow cytochromes *c*. Human, housefly, and wheat cytochromes *c* react slowly with the enzyme, but their reactivity is still higher than seen with horse and cow cytochromes *c*.

TABLE II. Reactivities of Various Cytochromes c with Thiobacillus novellus Cytochrome Oxidase[a]

Cytochrome c		Relative reactivity
Organism	α Peak (nm)	
Pseudomonas aeruginosa	551	0.78
Nitrosomonas europaea	552	2.5
Chlorobium limicola F. thiosulfatophilum	555	18
Chromatium vinosum	553 (550)	0
Rhodospirillum rubrum	550	1.5
Thiobacillus novellus	550	100
Spirulina platensis	554	3.8
Saccharomyces oviformis	550	77
Candida krusei	549	290
Wheat	550	26
Tuna	550	93
Bonito	550	91
Housefly (larva)	550	12
Cow	550	4.7
Horse	550	4.7
Man	550	14

[a]Reactions were performed in 50 mM phosphate buffer at pH 5.5 and at 18°–23°C. The reactivity with the oxidase of cytochromes c was expressed as relative value; the molecular activity (moles of cytochrome c oxidized per mole of enzyme) per min was taken as 100% which was observed when the enzyme reacted with cytochrome c-550 (T. novellus).

The reactions catalyzed by the oxidase are affected by salt (e.g., phosphate) concentration. The oxidation rates of many cytochromes *c* rapidly decrease as phosphate concentration increases, while those of *S. oviformis* and *C. krusei* cytochromes *c* increase first with phosphate concentration and then decrease as the salt concentration increases (Fig. 2). The inhibitory effect of phosphate seems to be competitive with cytochrome *c* (Figs. 3A, B).

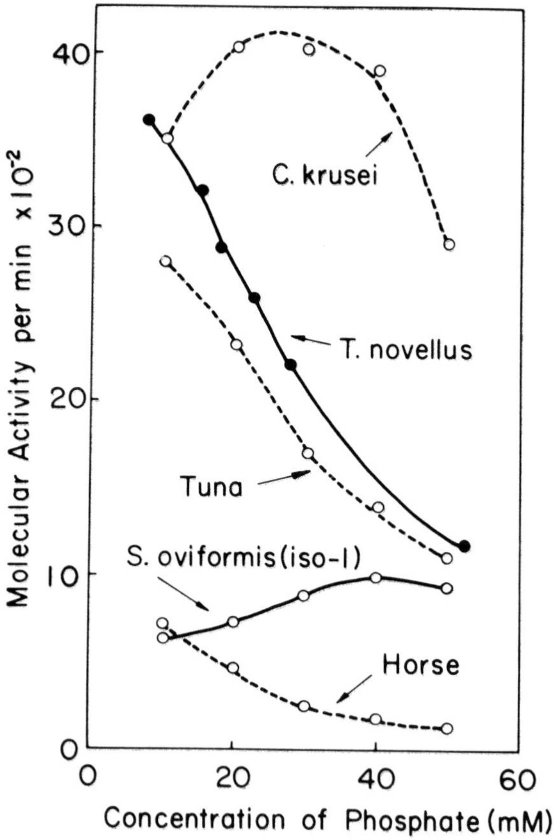

Fig. 2. Effect of phosphate concentration on oxidation rates of cytochrome c catalyzed by T. novellus cytochrome oxidase. The reactions were performed at 18°C. The concentrations of cytochromes c were 12-17 μM, and that of the enzyme 21 nM. A total volume of the reaction mixture was 1.05 ml.

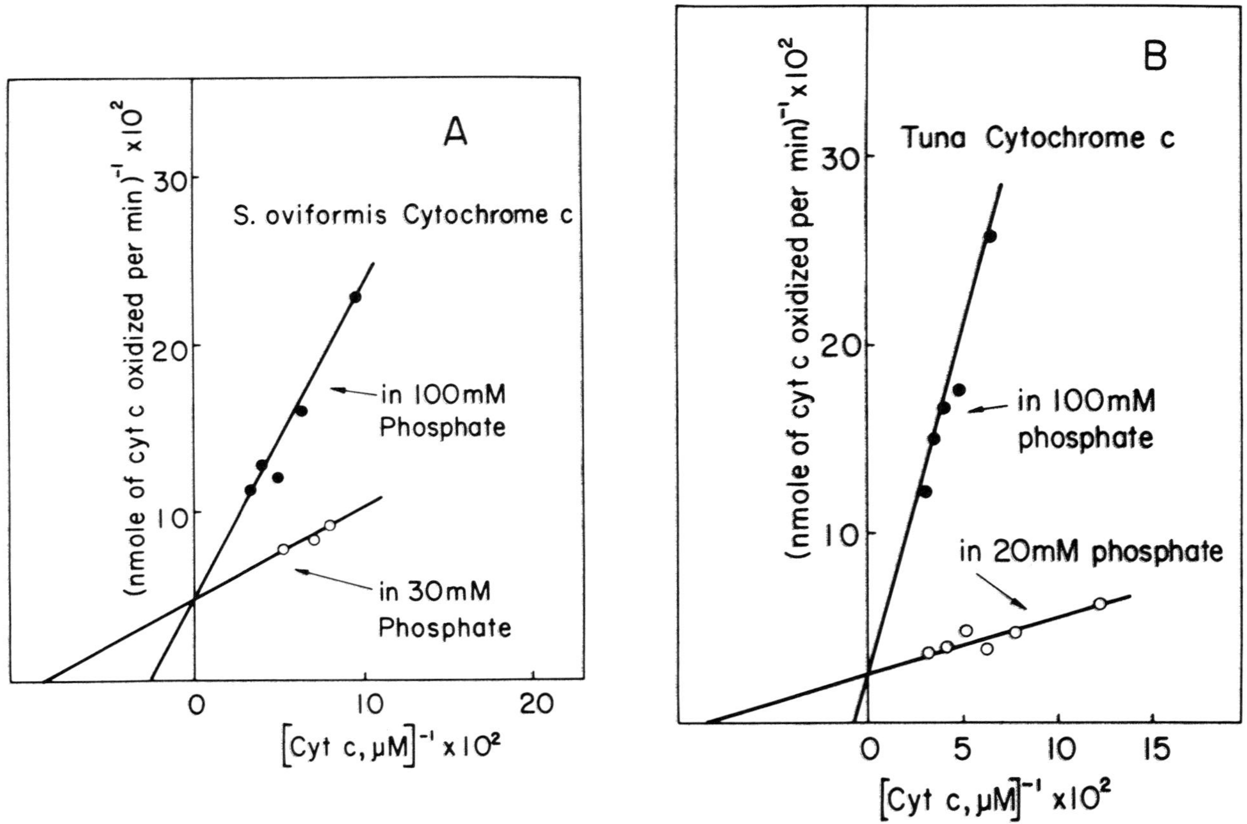

Fig. 3. Lineweaver-Burk plots for oxidation of cytochrome c catalyzed by T. novellus cytochrome oxidase. (A) S. oviformis; (B) tuna.

When the primary structures between the cytochromes *c* which react rapidly with the enzyme and the proteins which react poorly are compared, it is observed that the former cytochromes *c* have Tyr at the 46th position; the latter proteins have Phe at the same position (e.g., ref. 17). The fact that human cytochrome *c*, which has also Tyr at the 46th position, reacts with the enzyme more rapidly than the horse and cow cytochromes may support the idea that Tyr in this position is necessary for reaction with the enzyme. Tyr in this position seems to also affect the reactivities of eukaryotic cytochromes *c* with the *P. aeruginosa* enzyme; this enzyme reacts comparatively rapidly with tuna, bonito, and yeast cytochromes *c* as compared with cow and horse cytochromes *c*, while its reactivity with carp cytochrome *c* is as low as those with horse and cow cytochromes *c* (5, 6). It may be said that the *T. novellus* enzyme can detect the difference between the marine bony fish and yeast cytochromes *c* and other eukaryotic cytochromes *c* more distinctly than the *P. aeruginosa* enzyme. In any case, the *T. novellus* oxidase can distinguish between some of eukaryotic cytochromes *c*.

EVOLUTIONARY ASPECTS

Although the reactions catalyzed by the *T. novellus* oxidase are greatly affected by phosphate concentration and the mode of this salt effect is not the same for all cytochromes *c*, the reactivity of cytochrome *c* with the enzyme at a certain concentration of phosphate buffer seems to reflect the evolutionary relationship of the host organism of the cytochrome. In Fig. 4, the reactivities of various cytochromes *c* with the oxidase are compared to those with *P. aeruginosa* nitrite reductase and cow cytochrome oxidase. The host organisms of cytochrome *c* which react rapidly with each oxidase vary from the primitive to higher forms as the host organisms of the oxidase vary from *P. aeruginosa* to cow via *T. novellus*. We have already predicted that cytochrome oxidase and cytochrome *c* might have coevolved during the evolution of the organisms (5, 6). The experimental results mentioned above seem to verify the prediction. Further, it seems likely that the specificity for cytochrome *c* of cytochrome oxidase has changed continuously during the evolution of the organisms; a serious gap between prokaryotic and eukaryotic enzymes is not seen as far as specificity is concerned. However, a great difference in molecular features may be seen between both enzymes, especially in subunit structure. The molecule of the mitochondrial oxidase is composed of 5 to 7 subunits (18 - 21). In the *S.*

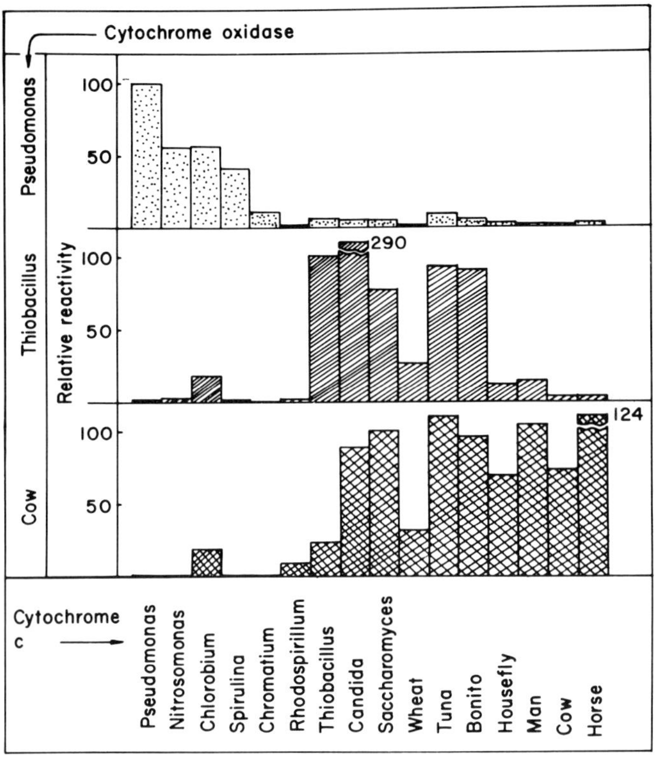

Fig. 4. Comparison of the specificities for cytochrome c of cytochrome oxidases derived from P. aeruginosa, T. novellus and cow (see note added in proof).

cerevisiae oxidase three of the seven subunits are biosynthesized on mitochondrial ribosomes while the remaining four are biosynthesized on cytoplasmic ribosomes (22). The endosymbiotic hypothesis as to the evolution from prokaryote to eukaryote presumes that the mitochondrion has been derived from an aerobic bacterium (23). If the hypothesis is correct, it is expected that the molecule of the bacterial oxidase may be composed of subunits similar to those of the mitochondrial oxidase biosynthesized on mitochondrial ribosomes. Our preliminary experiments have determined that the molecule of the *T. novellus* oxidase seems to be composed of two kinds of subunits with molecular weights of 23,000 and 32,000, respectively. The determination of the subunit structure of the bacterial oxidase is now in progress in our laboratory. (See refs. 25 and 26.)

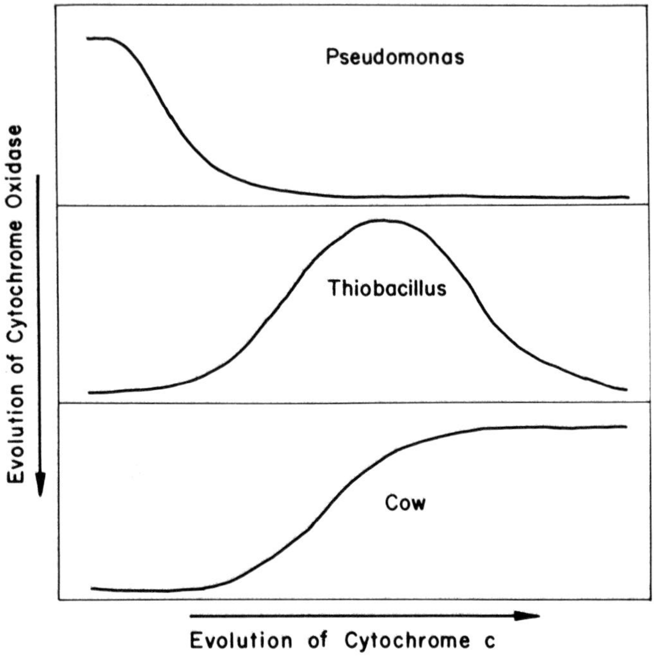

Fig. 5. A diagrammatical presentation of coevolution of cytochrome oxidase with cytochrome c.

SUMMARY

Cytochrome oxidase of *Thiobacillus novellus* has been purified and its properties determined. The oxidase has heme *a* as the prosthetic group, judging from its absorption spectrum and from that of its pyridine ferrohemochrome. The enzyme reacts to carbon monoxide in the similar way to that of the mitochondrial cytochrome oxidase. The reactions catalyzed by the enzyme are greatly affected by phosphate concentration. It reacts rapidly with cytochromes *c* of *Saccharomyces oviformis*, *Candida krusei*, tuna, and bonito, in addition to *T. novellus* cytochrome *c*. Eukaryotic cytochromes *c* which react rapidly with the enzyme have Tyr at the 46th position of their primary structures. Some evolutionary aspects of cytochrome oxidase are discussed on the basis of its reactivity with cytochrome *c*, and the experimental results obtained in the present investigation seem to verify the idea that cytochrome *c* has coevolved with cytochrome oxidase.

ACKNOWLEDGMENTS

The authors wish to thank Professor H. Matsubara for his interest and encouragement throughout this work. We are also grateful to Drs. N. Tanaka and M. Kakudo (Institute for Protein Research, Osaka University, Japan) for their generosity for supplying bonito cytochrome *c*.

REFERENCES

1. Keilin, D. (1966). "The History of Cell Respiration and Cytochrome." Cambridge University Press, Cambridge.
2. Kamen, M. D. and Horio, T. (1970). *Annu. Rev. Biochem. 39*, 673-700.
3. Erickson, R. H., Hooper, A. B., and Terry, K. (1972). *Biochim. Biophys. Acta 283*, 115-166.
4. Sugiyama, Y., Kitano, K., and Kanzaki, J. (1973). *Agric. Biol. Chem. 37*, 1607-1612.
5. Yamanaka, T. (1972). *Advan. Biophys. 3*, 227-276.
6. Yamanaka, T. (1973). *Space Life Sci. 4*, 490-504.
7. Yamanaka, T. and Fukumori, Y. (1977). *FEBS Lett. 77*, 155-158.
8. Yamanaka, T. and Kimura, K. (1974). *FEBS Lett. 48*, 253-255.
9. Yamanaka, T. (1975). *J. Biochem. 77*, 493-499.
10. Okunuki, K. (1966). *In* "Comprehensive Biochemistry" (M. Florkin and E. Stotz, eds.), Vol. 14, pp. 232-308. Elsevier, Amsterdam.
11. Orii, Y. (1978). *Advan. Biophys. 11*, 285-308.
12. Yamanaka, T. (1964). *Nature (London) 204*, 253-255.
13. Yonetani, T. and Ray, G. S. (1965). *J. Biol. Chem. 240*, 2503-2508.
14. Charles, A. M. and Suzuki, I. (1966). *Biochim. Biophys. Acta 128*, 522-534.
15. Hooper, A. B. and Nason, A. (1965). *J. Biol. Chem. 240*, 4044-4057.
16. Hon-nami, Y. and Oshima, T. (1977). *J. Biochem. 82*, 769-776.
17. Dickerson, R. E. and Timkovich, R. (1975). *In* "The Enzymes" (P. D. Boyer, ed.), Vol. XI, pp. 397-547. Academic Press, N.Y.
18. Mason, T. L., Poyton, R. O., Wharton, D. C., and Schatz, G. (1973). *J. Biol. Chem. 248*, 1346-1354.
19. Weiss, H., Sebald, W., and Bücher, T. (1971). *Eur. J. Biochem. 22*, 19-26.

20. Briggs, M., Kamp, P. F., Robinson, N. C., and Capaldi, R. A. (1975). *Biochemistry 15,* 5123-5128.
21. Maeshima, M. and Asahi, T. (1978). *Arch. Biochem. Biophys. 187,* 423-430.
22. Poyton, R. O. and Schatz, G. (1975). *J. Biol. Chem. 250,* 752-761.
23. Margulis, L. (1970). "Origin of Eukaryotic Cells." Yale University Press, New Haven, Connecticut.
24. Yamanaka, T. and Fukumori, Y. (1981). *Plant and Cell Physiol. 22,* 1223-1230.
25. Yamanaka, T., Fujii, K., and Kamita, Y. (1979). *J. Biochem. 86,* 821-824.
26. Yamanaka, T. and Fujii, K. (1980). *Biochim. Biophys. Acta 591,* 53-62.

Note Added in Proof

Recently, it has been found that the reactivity with cytochrome *c* of the cow enzyme varies a little with the conditions where the enzyme preparation has been stored before use. However, the profile of specificity for cytochrome *c* of the enzyme does not differ much, as a whole, between the different preparations (24).

THE STRUCTURE AND CLASSIFICATION OF CYTOCHROMES *c*

R. P. Ambler

Department of Molecular Biology
University of Edinburgh
Edinburgh EH9 3JR, Scotland

INTRODUCTION

Before 1953 there was only cytochrome *c*. In that year, 25 years before this Symposium, two short papers were published (Elsden *et al.*, 1953; Lenhoff and Kaplan, 1953). One was from Martin Kamen and his associates, the other from the organizer of this meeting. Each paper reported the occurrence of a bacterial cytochrome *c* which had spectral properties similar to the Keilin cytochrome *c* of higher organisms, but which had quite a different protein moiety.

Since 1953, a large number of different cytochromes *c* have been recognized (Lemberg and Barrett, 1973), and attempts have been made to classify them on the basis of such physical properties as spectrum, redox potential, and isoelectric point. The fundamental property of any protein is its amino acid sequence, and now that sufficient sequence information is becoming available a logical classification for cytochromes *c* has become possible.

 ISBN 0-12-397580-8

The sequences of mitochondrial cytochrome *c* from many different organisms has been determined (Dickerson and Timkovich, 1975; Dayhoff, 1972; 1979), and it has been demonstrated that a sequence classification of the proteins closely resembles the phylogeny of the organisms derived from fossil and morphological evidence (Fitch and Margoliash, 1970). This finding has prompted the hope that sequence methods can be used to derive phylogenies for organisms like bacteria which leave an inadequate fossil record and which may have diverged in the remote pre-Cambrian past. Cytochromes *c* have a very wide distribution in bacteria (Keilin, 1925; Yaoi and Tamiya, 1928), and are particularly abundant in some organisms with specialized systems of energy metabolism (Table I), including some types of metabolism which have been postulated to have evolved very early in the history of life on earth. A comparison of bacterial cytochrome *c* structures and an analysis of their distribution provide an ideal system with which to assess the validity of macromolecular methods of establishing phylogenies, resulting in a useful and enjoyable controversy between optimists and pessimists (Almassy and Dickerson, 1978; Ambler *et al*., 1979a; Dayhoff, 1979).

I began my own involvement with cytochromes *c* as a graduate student with Fred Sanger. We wanted a small protein that would be amenable to both structural and genetic analysis with which to investigate gene-protein colinearity, and Sydney Brenner suggested that *Pseudomonas* cytochrome *c*-551 (Lenhoff and Kaplan, 1956; Horio, 1958) might be suitable. While purification of the protein (I must have been one of the last students

TABLE I. Bacterial Taxa Containing Members That Are Known to Produce Large Amounts of Soluble Cytochromes c[a]

1.	*Photosynthetic bacteria*
	Rhodospirillum, Rhodopseudomonas, Rhodomicrobium, Chlorobium, Chromatium, blue-green bacteria
2.	*Sulfate and sulfur reducing bacteria*
	Desulfovibrio, Desulfuromonas
3.	*Nitrate reducing bacteria*
	Pseudomonas, Alcaligines/Achromobacter, Thiobacillus, Paracoccus, Spirillum, Bacillus
4.	*Methane using bacteria*
	Methylococcus, Methylosinus
5.	*Nitrogen fixing bacteria*
	Azotobacter

[a]*Apart from Bacillus and some species of Desulfovibrio and blue-green bacteria, all these organisms have G+C contents in the range 55 - 70%.*

TABLE II. Proteins in Pseudomonas aeruginosa That Contain Heme-c [a]

1.	*Cytochrome c-551 class ID one heme-c/82 res. (Horio, 1958)*[b]
2.	*Cytochrome c_4 dimeric class IC two heme-c/∿185 res. (Fig. 3 and c.f. Ambler and Murray, 1973)*[c]
3.	*Cytochrome c_5 class IE one heme-c/>85 res. (Ambler, 1973; Ambler and Taylor, 1973)*[d]
4.	*Cytochrome c peroxidase class ? two heme-c/∿500 res. (Lenhoff and Kaplan, 1956; Soininen et al., 1973)*
5.	*Cytochrome cd (= nitrite reductase) class IV one heme-c, one heme-d/∿600 res. (Horio, 1958; Nagata et al., 1970)*

[a]*When grown under high aeration, Ps. aeruginosa produces a cytochrome oxidase of heme-a type (Azoulay and Couchoud-Beaumont, 1965).*

Cytochrome c' has not yet been reported in Ps. aeruginosa, but is found in A. vinelandii (Yamanaka and Imai, 1972), an organism that contains at least the first three of the c-cytochromes shown in the Table.

Ps. aeruginosa produces large amounts of azurin (Horio, 1958), a protein that appears to have a similar function to cytochrome c-551.

Ps. aeruginosa is a common organism, and strains can be isolated from a wide range of habitats. Nevertheless there seems to be little variation between strains (Ambler, 1974).

[b]*A similar protein is present in Azotobacter vinelandii (Ambler, 1973) called "cytochrome c_4-minor" by Swank and Burris (1969) and in Pseudomonas stutzeri the cytochrome c-552(I) of Kodama and Shidara (1969).*

[c]*A similar protein is present in A. vinelandii (Swank and Burris, 1969; Ambler and Murray, 1973) and Ps. stutzeri isolated from a particulate fraction as cytochrome c-552(II) by Kodama and Shidara (1969).*

[d]*A similar protein is present in A. vinelandii (Swank and Burris, 1969; Ambler, 1973). Cytochromes c_5 are particularly difficult to purify, and yields are very erratic. Forms with similar absorption spectrum but higher molecular weight are often observed, and it is possible that the cytochrome is a fragment derived by proteolysis from a larger molecule such as 4 or 5.*

who was helped with spectroscopy by Professor Keilin) and the structural analysis proved feasible (Ambler, 1963), the system proved intractable to genetic analysis, probably because of the existence of alternative routes for electron transport of this organism (Table 2). The sequence showed only a remote resemblance to that of mitochondrial cytochrome *c* (Margoliash *et al.*, 1961), and I was fascinated by the problems of interspecies variation in proteins and by the esthetic and practical advantages of working with colored proteins. Through the guidance of Robin Hill I was very lucky to have the opportunity to come to LaJolla for a year (1962 - 1963) to learn about bacterial cytochromes from Martin Kamen. This visit has led on into a very happy and productive symbiosis, helped by long visits to Edinburgh by Terry Meyer and shorter visits (normally coinciding with the Music Festival) by Martin Kamen himself.

THE DEFINITION OF CYTOCHROME *c*

Cytochromes *c* are defined as cytochromes in which there is a covalent linkage between the prosthetic heme group side chains and the residues of the protein. With the exceptions of the proteins from the mitochondria of some protozoa, all cytochromes *c* so far characterized have the heme group attached by two thioether links to two cysteine residues separated by two (or in rare cases, four) amino acid residues. In the exceptional proteins, the heme group is bound by only a single thioether cysteine link. The proof of this unusual structure was one of the first fruits of the La Jolla - Edinburgh symbiosis (Pettigrew and Meyer, 1971).

Mitochondrial cytochrome *c* and many bacterial cytochromes *c* are small stable molecules that contain only one heme group per polypeptide chain, and are readily solubilized and purified. However the definition also includes some larger and more complex molecules. Some, such as the cytochrome *f* of photosynthetic eukaryotes (Davenport and Hill, 1951; Wood, 1977), are tightly bound to a membrane. Additional prosthetic groups may also be attached to the protein, such as flavin in the flavocytochromes *c* of photosynthetic bacteria (Bartsch *et al.*, 1968; Bartsch, 1978) or heme *d* in *Pseudomonas* cytochrome oxidase (Horio, 1958; Nagata *et al.*, 1970). Cytochromes *c* are also known that contain more than one heme group per polypeptide chain. The cytochrome c_4 of *Azotobacter vinelandii* and *Pseudomonas* contains two hemes (Fig. 1; Ambler and Murray, 1973), the cytochrome c_7 (Ambler, 1971) of "*Chloropseudomonas ethylica*" (now known to be the *Desulfuromonas* component of the consortium; Pfennig and Biebl, 1976) three hemes, and the cytochromes c_3 of *Desulfovibrio* species four hemes (Postgate, 1956; Ambler *et al.*, 1969).

Another class of proteins are the spectrally distinct high-spin cytochromes *c*' (Kamen and Vernon, 1955), which are widely distributed in both photosynthetic and nonphotosynthetic

```
                10         20         30         40         50         60         70         80         90
(A)  AGDAAAGQAKAAVCGACHGABbbgsapp-fpKLAGQGERYLLKQMHDIKDGKRTVLE//EMTGLL-TBLSbzdIaaladYASQKMSVGMAlbbpvaqgEA

(B)  AGDAAAGEDKIGTCVACHGTDGQGLAPI-YPNLTGQSATYLESSIKAYRDGQRKGGNAALMTPMA-QGLSDEDIADIAAYYSSQE

     LFRGGKIAEGMPACTGCHGsspvgiatagfphLGGQHATYVAKQLTDFREGTRNDDGTKIMQSIAAIKLSNKDIAAISSYIQGLH
```

Fig. 1. (A) Tentative and incomplete amino acid sequence of Pseudomonas aeruginosa cytochrome c_4, a dimeric class IC protein, aligned with (B), the amino acid sequence of "halotolerant micrococcus" (ATCC 12084) cytochrome c-554(548) (Hori, 1963). The latter protein is isolated as a very stable dimer. In sequence (A), the residues shown in lower case letters have not yet been ordered. There may be unidenitrified residues in the gap marked || between residues 56 and 59. The evidence for the compositions, sequence, and overlapping of some of the other segments is poor. Residues identical in sequences (A) and (B) are joined with vertical lines. The one-letter notation is that recommended by the IUPAC-IUB Commission on Biochemical Nomenclature. A, alanine; B = D or N; C, cysteine; D, aspartic acid; E, glutamic acid; F, phenylalanine; G, glycine; H, histidine; I, isoleucine; K, lysine, L, leucine; M, methionine; N, asparagine; P, proline; Q, glutamine; R, arginine; S, serine; T, threonine; V, valine; W, tryptophan; Y, tyrosine; Z = E or Q.

bacteria. As the spectra of the reduced native proteins lack the distinctive α and β peaks of the low-spin cytochromes *c* they cannot be detected in impure preparations. The heme is covalently bound to the protein through thioether links to cysteine residues as in other cytochromes *c* (Barrett and Kamen, 1961; Kennel *et al.*, 1972).

The amino acid sequence of the mitochondrial cytochrome *c* from horse heart was one of the earliest protein structures to be determined (Margoliash *et al.*, 1961). When the first bacterial protein sequence, *Pseudomonas* cytochrome *c*-551 (Ambler, 1963), was determined, very little structural similarity could be seen except in the immediate neighborhood of the heme attachment site. In contrast, the cytochrome c_2 from *Rhodospirillum rubrum* has a sequence (Dus *et al.*, 1968) that clearly resembles mitochondrial cytochrome *c*, although it is longer and deletions are necessary before optimal matching is obtained (Fig. 2). Three dimensional structures of each of these molecules have now been determined (Takano *et al.*, 1973; Almassy and Dickerson, 1978; Salemme *et al.*, 1973) which show that all three proteins do indeed fold in the same general way, and enable the sequence of the small *Pseudomonas* cytochrome *c*-551 to be aligned with the longer sequences in a structurally meaningful way.

STRUCTURAL VARIETY IN CYTOCHROMES *c*

The functional versatility of cytochromes *c* is shown by their wide range of redox potentials, from +350 to -200 mV, and by their occurrence in bacteria with specialized systems of energy metabolism (Table I). Some bacteria produce several different cytochromes *c*, such as the versatile and ubiquitous *Pseudomonas aeruginosa* (Table II), which produces at least five distinct proteins, an abundance that can be more than matched in photosynthetic bacteria such as *Rhodopseudomonas palustris* (Bartsch, 1978). There is now enough amino acid sequence information available to form the basis of a structural classification of cytochromes *c*, depending on the location of the heme binding sequence in the polypeptide chain (Table III). Some well-known cytochromes, such as mitochondrial cytochrome c_1 and the true cytochrome *f* of chloroplasts, have not yet been characterized in sequence, and further classes may have to be created to accommodate them.

In the class I cytochrome *c*, which include all the most studied proteins, the heme is attached near the *N*-terminus of a single polypeptide chain of 80-120 residues (but see Fig. 1). In most (if not all) of these proteins, a methionine residue

Mitochondrial cytochrome *c*

```
                                  10        20                  30        40
(A) Equus caballus          @GDVEKGKKIFVQKCAQCHTVEKGG-------KHKTGPNLHGLFGRKTGQAPGFT-
(B) Euglena gracilis        @GDAERGKKLFESRAAQCHSAQKG--------VNSTGPSLWGVYGRTSGSVPGYA-
```

Cytochrome c_2

```
(C) Rps. globiformis        GSAPPGDPVEGKHLFHTICILCHTDIKG--------RNKVGPSLYGVVGRHSGIEPGYN-
(D) Rm. vannielii              AGDPVKGEQVF-KQCKICHQVGPTA-------KNGVGPEQNDVFGQKAGARPGFN-
(E) Rps. viridis                QDAASGEQVF-KQCLVCHSIGPGA-------KNKVGPVLNGLFGRHSGTIEGFS-
(F) R. acidophila              AGDPDAGQKVF-LKCAACHKIGPGA-------KNGVGPSLNGVANRKAGQAWGFA-
(G) R. molischianum iso-1       ADAPP--PAF-NQCKACHSIDAG--------KNGVGPSLSGAYGRKVGLAPNYK-
(H) R. fulvum iso-1             ADAP---TAF-NQCKACHSIEAG--------KNGVGPSLSGAYGRKVGLAPNYK-
(I) R. molischianum iso-2       ADAP---AGF-TLCKACHSVEAG--------KNGVGPSLAGVYGRKAGTISGFK-
(J) R. fulvum iso-2             ADAP---PAF-GMCKACHSVEAG--------KNGVGPSLAGVYGRKAGTLAGFK-
(K) R. rubrum                  EGDAAAGEKV-SKKCLACHTFDQGG-------ANKVGPNLFGVFENTAAHKDNYA-
(L) R. photometricum           AGDAAVGEKIAKAKCTACHDLNKGG-------PIKVGPPLFGVFGRTTGTFAGYS-
(M) Rps. palustris 2.1.6        QDAAKGEAVF-KQCMTCHRAD----------KNMVGPALGGVVGRKAGTAAGFT-
(N) Rps. palustris 2.1.37       QDAKAGEAVF-KQCMTCHRAD----------KNMVGPALGGVVGRKAGTAAGFT-
(O) Rps. sphaeroides          QEGDPEAGAKAF-NQCQTCHVIVDDSGTTIAGRNAKTGPNLYGVVGRTAGTQADFKG
(P) Rps. capsulata 2.3.1        GDAAKGEKEF-NKCKTCHSIIAPDGTEIV-KGAKTGPNLYGVVGRTAGTYPEFK-
(Q) Rhodopseudomonas TJ12      EGDAAVGEKEF-TKCKTCHSITAPDGTDIV-KGGKVGPNLYGVIGRZAGSYPBFR-
```

Figure 2

```
            50         60        70                   80        90      100
(A)   YTDANKN---KGITWKEETLMEYLENPKKYI------PG--------TKM--IFAGIKKKTEREDLIAYLKKATNE
(B)   YSNANKN---AAIVWEEETLHKFLENPKKYV------PG--------TKM--AFAGIKAKKDRQDIIAYMKTLKD


(C)   YSEANIK---SGIVWTPDVLFKYIEHPQKIV------PG--------TKM--GYPGQPDPQKRADIIAYLETLK
(D)   YSDAMKN---SGLTWDEATLDKYLENPKAVV------PG--------TKM--VFVGLKNPQDRADVIAYLKQLSGK
(E)   YSDANKN---SGITWTEEVFREYIRDPKAKI------PG--------TKM--IFAGIKDEQKVSDLIAYLKQFNADGSKK
(F)   YSDANKN---SGLTWDEATFKEYITAPQKKV------PG--------TKM--TFPGLPNEADRDNIWAYLSQFKADGSK
(G)   YSPAHLA---SGMTIDDAMLTKYLANPKETI------PG--------NKMGAAFGGLKNPADVAAVIAYLKTVK
(H)   YSAAHLA---SGMTIDEAMLTNYLANPKATI------PG--------NKMGASFGGLKKPEDVKAVIEYLKTVK
(I)   FSDPHIK---SGLTWDEPTLTKYLADPKTVI------PG--------NKM--VFAGLKNPDDVKAVIEYLKTLK
(J)   FSDPHAK---SGLTWDEPTLTKYLADPKGVI------PG--------NKM--VFAGLKNPADVAAVIAYLKSL
(K)   YSESYTEMKAKGLTWTEANLAAYVKDPKAFVLEKSGDPK------AKSKM--TFK-LTKDDEIENVIAYLKTLK
(L)   YSPGYVTMGQKGHTWDDNALKAYLLDPKGYVQAKSGDPK------ANSKM--IFR-LEKDDDVANVIAYLHTMK
(M)   YSPLNHNSGEAGLVWTQENIIAYLPDPNAYLKKFLTDKGQADKATGSTKM--TFK-LANDQQRKDVAAYLATLK
(N)   YSPLNHNSGEAGLVWTADNIINYLNDPNAFLKKFLTDKGKADQAVGVTKM--TFK-LANEQQRKDVVAYLATLK
(O)   YGEGMKEAGAKGLAWDEEHFVQYVQDPTKFLKEYTGDAK------AKGKM--TFK-LKKEADAHNIWAYLQQVAVRP
(P)   YKDSIVALGASGFAWTEEDIATYVKDPGAFLKEKTDDKK------AKSGM--AFK-LAK--GGEDVAAYLASVVK
(Q)   YKDSIVELGESGFVWTAEEIAVYMQDPTAFLKEKTGNKR------ARSGM--THK-QNK--NQEDVIAFLASVVTE
```

Fig. 2

about three-fourths of the way along the chain acts as the sixth iron ligand. An attempt has been made to subdivide the class (Table IV), but some proteins have sequence characteristics of more than one of the subdivisions. Both the sequence similarities and features in the three-dimensional structures of some of these proteins can be plausibly explained by evolutionary divergence from a common ancestor (Almassy and Dickerson, 1978), but the most aberrant of the class I cytochromes (e.g., *Desulfovibrio* cytochrome *c*-553; Bruschi and Le Gall, 1972) are not linked to the rest of the class through known intermediate forms, so an independent origin and convergence cannot be excluded as an explanation for some of the structural

Fig. 2. Amino acid sequences of Rhodospirillaceae cytochromes c_2 aligned with selected mitochondrial cytochrome c sequences. Residues are numbered from the horse (A) sequence. Underlined residues are referred to in this caption.

Sequences (A) - (J) are of class IB proteins, and (K) - (Q) of class IA. Sequences (C) and (Q) are newly determined and are reported here for the first time. The other sequences are taken from Ambler et al. (1979a) where the original references are given. Rhodopseudomonas strain TJ12 is a marine capsulata (P. Weaver, personal communication).

In sequence (B) residue 86 is ε-N-trimethyl lysine. In sequence (F) the evidence for the amide ascriptions among residues 87 - 90 is weak. In sequences (I) and (J) position 67 contains phenylalanine (about 30%) as well as the predominant tyrosine. In sequence (J) only the compositions of regions 32-36 and 93 - 97 are known. In sequence (L) there is conflicting evidence, some of which suggests that there is an additional glycine residue between residues 56 and 57. Sequence (P) is for Rps. capsulata strain 2.3.1; in strain St. Louis, residue 75+4 is leucine (not threonine) and residue 78 is threonine (not serine).

The one-letter notations used are defined in the caption to Fig. 1. @, α-N-acetyl-.

TABLE III. Cytochrome *c* Sequence Classes[a]

Class I

Heme attachment site near N-terminus. Single polypeptide chain of 80-120 residues (but see Fig. 1). Sixth iron ligand in methionine residue at about position 60.
Example: Mitochondrial cytochrome *c* (and see Table IV).

Class II

Heme attachment site near C-terminus. Single polypeptide chain of about 120 residues. Can have either high- or low-spin spectrum.
Examples: high spin, *Rhodospirillum rubrum* cytochrome *c'*; low spin, *Rhodopseudomonas palustris* cytochrome *c*-556 (Bartsch, 1978).

Class III

Multiple hemes, more than one attached per 30 residues of polypeptide chain. Low redox potential.
Examples: three heme, *Desulfuromonas* ("*Chloropseudomonas*") cytochrome c_7 (Ambler, 1971); four heme, *Desulfovibrio* cytochrome c_3 (Postgate, 1956).

Class IV

Complex proteins which have other types of prosthetic group in addition to heme *c*.
Examples: *Chromatium vinosum* flavocytochrome *c* (Bartsch, 1978); *Ps. aeruginosa* cytochrome *cd* (nitrite reductase) (Horio, 1958).

[a]For lack of sequence information, chloroplast true cytochrome *f*, mitochondrial cytochrome c_1 and *Ps. aeruginosa* cytochrome *c* peroxidase cannot yet be included in this scheme.

TABLE IV. Subdivision of Sequence Class I Cytochrome c[a]

Class IA long cytochrome c_2

Contain several extra loops when compared with class IB, although three-dimensional structure is generally similar. The insertions are in different places in the different members of this class.
Examples: Rhodospirillum rubrum cytochrome c_2 (Dus et al., 1968); Paracoccus denitrificans cytochrome c-550 (Timkovich et al., 1976).

Class IB mitochondrial cytochrome c

Examples: Euglena gracilis cytochrome c-558 (Pettigrew and Meyer, 1971); Rhodopseudomonas viridis cytochrome c_2.

Class IC split-α-band cytochromes c

Possess a widened or split α band of lowered absorptivity. The splitting does not seem to correlate with dimerization. Sequence characteristics include -M------LS---I----- Y- (sixth ligand region) and aromatic and proline residues after the heme-binding site.
Examples: monomeric, algal "soluble cytochrome f"; pseudodimeric, "halotolerant micrococcus" cytochrome c-554 (548); dimeric, cytochrome c_4 (Fig. 1).

Class ID Pseudomonas cytochrome c-551

Sequence characteristics include several proline residues around the sixth ligand methionine and a tryptophan residue near the C-terminus.
Examples: Azotobacter vinelandii cytochrome c_4-minor; R. tenue cytochrome c553 (Ambler et al., 1979b).

Class IE cytochrome c_5

Sequence characteristics include extra cysteine residues (probably as a short disulfide loop) and an absence of aromatic residues near the C-terminus. N-terminal heterogeneity and purification difficulties suggest that they may be derived from larger molecules by proteolysis.
Example: Pseudomonas mendocina cytochrome c_5 (Ambler and Taylor, 1973).

---

[a]Further classes will be necessary to cover all known class I cytochromes c (e.g., **Desulfovibrio** vulgaris cytochrome c-553 (Bruschi and LeGall, 1972).

similarities. The cytochromes c_4 of *Azotobacter* and *Pseudomonas* are covalent dimers (Fig. 1) of two different class I polypeptide chains joined end-to-end. Other proteins that qualify for inclusion in class I, such as *Azotobacter/Pseudomonas* cytochrome c_5, may be derived by proteolysis from a larger functional molecule.

In the class II cytochromes *c*, the single heme group is attached near to the *C*-terminus of a polypeptide chain of about 120 residues. The earliest members of the class to be characterized were the high-spin cytochromes *c*', which have a wide distribution in bacteria. Low-spin examples of the class have now been recognized in *Agrobacterium* (Francois, 1976) and in *Rps. palustris* (Bartsch, 1978). In the latter organism there are two class II cytochromes with different but related sequences, a cytochrome *c*' and a low-spin cytochrome *c*-556.

The proteins of class III have been characterized so far only in the sulfate-reducing bacteria. They possess several heme groups attached to a single, short polypeptide chain, and have very low redox potentials. Multi-heme cytochromes with very low redox potentials have also been found in *Rhodospirillum tenue* and in cyanobacteria (Bartsch, 1978), but have not yet been characterized in sequence. There is little structural information yet available about the large complex cytochromes *c* of class IV.

INTERSPECIES DIFFERENCES IN CYTOCHROMES *c*

Mitochondrial cytochromes *c* have been the classical system for correlating the phylogeny of the whole organism with that of a single gene (Fitch and Margoliash, 1970). The results show good topological agreement between the phylogenies, and some of the few anomalies may probably be explained as due to errors in published sequences. Opinions have altered as to whether interspecies differences have arisen through neutral mutation and "genetic drift" (Margoliash, 1972) or selection (Margoliash *et al.*, 1976). The balance of evidence from this and other systems currently favors the belief that only a small proportion of the observed interspecies variation is nonselective.

Mitochondrial isocytochromes *c* are known in yeast (Sherman *et al.*, 1965), and from different tissues in mammals (Hennig, 1975). Isocytochromes *c* with clearly related sequences have been characterized in two bacterial species (Fig. 2).

Bacterial classification is taxonomic not phylogenetic, and the extent of subdivision of a group of microorganisms is often related to the medical or commercial importance of the organisms

The sequences of the cytochromes *c*-551 from several species of *Pseudomonas* have been examined (Ambler and Wynn, 1973; Ambler, 1974), and show that interspecies differences are large. The amount of difference corresponds to many hundreds of millions of years of evolutionary divergence in the functionally similar mitochondrial cytochrome *c* even for species pairs such as *Ps. aeruginosa* and *Ps. fluorescens* which have until recently been difficult to distinguish by bacteriological methods.

Results from both *Pseudomonas* (Ambler, 1974) and *Rhodopseudomonas* suggest that there are major discontinuities between species in their cytochrome sequences, and that in at least some species selection is acting to restrict severely sequence polymorphism.

The only group of bacteria that have yet been studied in sufficient detail at the sequence level for enough information to be available to be of taxonomic use are the Rhodospirillaceae, the nonsulfur purple photosynthetic bacteria (Ambler *et al.*, 1979a, 1979b). The cytochromes c_2 from 12 species have now been sequenced (Fig. 2). Despite the functional differences, the sequences of some of these proteins (class IB, Table IV) are close in sequence to mitochondrial cytochromes *c* (Ambler *et al.*, 1976). In contrast, some species of Rhodospirillaceae altogether lack cytochrome c_2, and in them it is apparently replaced by a cytochrome of class ID that resembles *Pseudomonas* cytochrome *c*-551 in sequence (Ambler *et al.*, 1979b).

Most species of the Rhodospirillaceae also produce cytochrome *c*' (Bartsch, 1978), and the sequence of this protein has been determined for all the organisms shown in Fig. 2 in which it occurs (R. P. Ambler and T. E. Meyer, unpublished results). While the proteins all show some sequence similarity, particularly around the heme attachment site, the interspecies differences are very large. We have not yet worked out a suitable way of displaying the sequences that will accommodate this degree of difference but yet show the patches of good alignment that exist (but see Ambler *et al.*, 1979b). The sequences of the cytochromes *c*' from *Rhodospirillum fulvum* and *Rhodospirillum molischianum* are shown in Fig. 3 as the three dimensional structure of the *R. molischianum* protein has now been determined (P. Weber and F. R. Salemme, personal communication). The amount of difference between the cytochromes *c*' of these two organisms (Fig. 3) is very similar to that between their iso-1 and iso-2 cytochromes c_2 (Fig. 2), and is so small that on these grounds it would be justified to lump them together in the same species.

QQSKPE$\genfrac{}{}{0pt}{}{D}{E}$LLKLRQGLMQTLKSQW$\genfrac{}{}{0pt}{}{V}{A}$PIAGFAAGKADLPADAAQRAENM$\genfrac{}{}{0pt}{}{AM}{VL}$VAKLAPIGWAKGTEA-

-LPN$\genfrac{}{}{0pt}{}{G}{S}ETK\genfrac{}{}{0pt}{}{P}{A}$EAFG$\genfrac{}{}{0pt}{}{S}{A}KSA\genfrac{}{}{0pt}{}{E}{Q}F\genfrac{}{}{0pt}{}{L}{M}$EGWKA$\genfrac{}{}{0pt}{}{L}{M}A\genfrac{}{}{0pt}{}{T}{A}$ESTKLAAAAKAGPDALKAQAAATGKVCKACHEEFKQD

Fig. 3. Amino acid sequences of Rhodospirillum molischianum and Rhodospirillum fulvum cytochromes c'. The sequences are identical except at 11 places, where the R. molischianum residue is shown above the R. fulvum residue. The one-letter notations used are defined in the caption to Fig. 1.

THE INTERPRETATION OF BACTERIAL CYTOCHROME *c* SEQUENCE INFORMATION

Amino acid or nucleotide sequence information, in sufficient quantity, may provide the means by which "the long-sought phylogenetic tree of the prokaryotes seems now within our grasp" (C. R. Woese, personal communication). An alternative possibility is that bacterial evolution proceeds both by the gradual processes of mutation and selection and by the assimilation of genes for single functions or for whole metabolic pathways from other organisms. Such horizontal gene transfer could introduce such confusion and inconsistencies into whole organism phylogenies that the mere acquisition of more data for more independent genes will not clarify it. We have no information about the frequency of successful gene assimilation or about the time scale for protein evolution or speciation in bacteria. We have no evidence at all to suggest that either a gene product (such as cytochrome *c*) or an assemblage of genes (such as *Pseudomonas aeruginosa* or *Rhodospirillum rubrum*) should be stable through geological time.

A problem that gravely complicates attempts to build phylogenies of bacterial groups from protein sequence data is the sporadic distribution of proteins within a set of organisms. This problem is often ignored (Almassy and Dickerson, 1978). As an example, cytochrome *c*-551 has been isolated from only four of the well-characterized species of *Pseudomonas*, *Ps. aeruginosa*, *Ps. fluorescens* biotype C, *Ps. stutzeri* and *Ps. mendocina*. Despite careful attempts to isolate it from *Ps. putida* and from the other biotypes of *Ps. fluorescens* the protein has not been found, although these organisms produce at least some of the cytochromes *c* shown in Table II (Ambler, 1977). Comparable cases of sporadic distribution occur in the Rhodospirillaceae, such as the absence of cytochrome *c*' from just one of several isolates of *Rhodopseudomonas palustris* (Bartsch, 1978). It is difficult to prove the total absence of a gene, and in some cases mentioned above the gene may be present but is expressed at only a low level. The superproduction of cytochrome *c*-551 by denitrifying cultures of *Ps. aeruginosa* may be a pathological reaction to unnatural conditions, and expression at a low and difficult-to-detect level "normal." *Bacillus licheniformis* contains quite high levels of soluble cytochromes *c* (they co-purify with the β-lactamase), but as it is one of the very few members of the genus that can denitrify vigorously, the cytochrome genes may well have been transferred in by horizontal gene transfer from another denitrifying organism, and so could not yield information about *Bacillus* phylogeny.

At present we do not have sufficient information to be able to speculate about either the outlines or the rules of bacterial evolution with any reasonable chance of success. This can be demonstrated by the low predictive success of speculations that have been published so far (e.g., Almassy and Dickerson, 1978). When we look at the sequence of a cytochrome *c* from a new source we have little idea as to which if any known sequence it will resemble, and completely unexpected close similarities are found. An example of this is the close similarity in sequence between *Rhodopseudomonas capsulata* cytochrome c_2 (Fig. 2) and the cytochrome *c*-550 of *Paracoccus denitrificans* (Timkovich *et al.*, 1976). While it is easy to speculate that aerobic respiration may have evolved from photosynthetic electron transport, it is difficult to understand why the *Rps. capsulata* sequence is so much closer to that of the respiratory protein than it is to any of the other Rhodospirillaceae proteins (Fig. 2), even though the origin of aerobically respiring microorganisms is set in the remote past. Another example is the cytochrome c_2 of *Rhodopseudomonas globiformis* (C, Fig. 2), which is a class IB protein with an N-terminal segment that closely resembles in sequence some plant mitochondrial cytochromes *c*. A genetic connection would seem very unlikely, and the most likely explanation must be molecular convergence to a favored three-dimensional structure.

We have recently looked at the sequences of the predominant low-spin cytochromes *c* in representatives of three different species of the genus *Ectothiorhodospira* (R. P. Ambler, T. E. Meyer and J. F. Imhoff, unpublished results), a group chosen for study because it is in some ways intermediate between the Rhodospirillaceae and the Chromatiacae (Pfennig, 1977). From two of the species the cytochrome *c* is a small class I sequence with slight affinities to *Chlorobium* cytochrome *c*-555 (Van Beeumen *et al.*, 1976), while in the third species the cytochrome *c* is clearly of the mitochondrial sequence type (class IB), but contains several unprecedented substitutions and alterations, such as a tryptophan residue immediately after the sixth ligand methionine residue, and a high arginine content.

CONCLUSION

Our present duty is open-mindedly and without preconceptions to gather more information about the stability and variability of all sorts of genes in natural populations of prokaryotes. While all types of genes - coding and noncoding, for essential and peripheral enzymes, for RNA and for structural proteins - should be studied, the advantages of continuing and extending the study of cytochromes *c* remain, even in the present era of rapid and accurate DNA sequencing of accessible genes.

ACKNOWLEDGMENT

These studies have been supported in part by grants to R.P.A. from the Medical Research Council and to M. D. Kamen from the NIH (GM-18528) and the NSF (BMS-75-13608). Most of the experimental work described has been done in close collaboration with M. D. Kamen, R. G. Bartsch, and T. E. Meyer, and with the enthusiastic assistance of Margaret Daniel and Lesley McLellan.

REFERENCES

Almassy, R. J. and Dickerson, R. E. (1978). *Proc. Natl. Acad. Sci. U.S.A. 75,* 2674-2678.
Ambler, R. P. (1963). *Biochem. J. 89,* 349-378.
Ambler, R. P. (1971). *FEBS Lett. 18,* 351-353.
Ambler, R. P. (1973). *Syst. Zool. 22,* 554-565.
Ambler, R. P. (1974). *Biochem. J. 137,* 3-14.
Ambler, R. P. (1977). *In* "Evolution of Protein Molecules" (H. Matsubara and T. Yamanaka, eds.), pp. 311-322. University of Tokyo Press, Tokyo.
Ambler, R. P. and Murray, S. (1973). *Biochem. Soc. Trans. 1,* 162-164.
Ambler, R. P. and Taylor, E. (1973). *Biochem. Soc. Trans. 1,* 166-168.

Ambler, R. P. and Wynn, M. (1973). *Biochem. J. 131*, 485-498.
Ambler, R. P., Bruschi, M., and Le Gall, J. (1969). *FEBS Lett. 5*, 115-117.
Ambler, R. P., Meyer, T. E., and Kamen, M. D. (1976). *Proc. Natl. Acad. Sci. U.S.A. 73*, 472-475.
Ambler, R. P., Daniel, M., Hermoso, J., Meyer, T. E., Bartsch, R. G., and Kamen, M. D. (1979a). *Nature (London) 278*, 659-660.
Ambler, R. P., Meyer, T. E., and Kamen, M. D. (1979b). *Nature (London) 278*, 661-662.
Azoulay, E. and Couchoud-Beaumont, P. (1965). *Biochim. Biophys. Acta 110*, 301-311.
Barrett, J. and Kamen, M. D. (1961). *Biochim. Biophys. Acta 50*, 573-575.
Bartsch, R. G. (1978). *In* "The Photosynthetic Bacteria" (R. K. Clayton and W. R. Sistrom, eds.), pp. 249-279. Plenum, New York.
Bartsch, R. G., Meyer, T. E., and Robinson, A. B. (1968). *In* "Structure and Function of Cytochromes" (K. Okunuki, M. D. Kamen, and I. Sekuzu, eds.), pp. 443-451. University of Tokyo Press, Tokyo.
Bruschi, M. and Le Gall, J. (1972). *Biochim. Biophys. Acta 271*, 48-60.
Davenport, H. E. and Hill, R. (1952). *Proc. Roy. Soc. London B139*, 327-345.
Dayhoff, M. O. (1972). "Atlas of Protein Structure and Sequence," Vol. 5. National Biomedical Research Foundation, Washington, D.C.
Dayhoff, M. O. (1979). "Atlas of Protein Sequence and Structure," Vol. 5, Supplement 3. National Biomedical Research Foundation, Washington, D.C.
Dickerson, R. E. and Timkovich, R. (1975). *In* "The Enzymes," 3rd ed., Vol. 11, (P. D. Boyer, ed.), pp. 397-547. Academic Press, New York.
Dus, K., Sletten, K., and Kamen, M. D. (1968). *J. Biol. Chem. 243*, 5507-5518.
Elsden, S. R., Kamen, M. D., and Vernon, L. P. (1953). *J. Amer. Chem. Soc. 75*, 6347-6348.
Fitch, W. M. and Margoliash, E. (1970). *Evolu. Biol. 4*, 67-109.
Francois, G. (1976). Thesis, University of Ghent, Belgium.
Hennig, B. (1975). *Eur. J. Biochem. 55*, 167-183.
Hori, K. (1963). *J. Biochem. (Tokyo) 53*, 354-363.
Horio, T. (1958). *J. Biochem. (Tokyo) 45*, 195-205.
Kamen, M. D. and Vernon, L. P. (1955). *Biochim. Biophys. Acta 17*, 10-22.
Keilin, D. (1925). *Proc. Roy. Soc. London B98*, 313-339.
Kennel, S. J., Meyer, T. E., Kamen, M. D., and Bartsch, R. G. (1972). *Proc. Natl. Acad. Sci. U.S.A. 69*, 3432-3435.

Kodama, T. and Shidara, S. (1969). *J. Biochem. (Tokyo) 65,* 351-360.
Lemberg, R. and Barrett, J. (1973). "Cytochromes." Academic Press, London.
Lenhoff, H. M. and Kaplan, N. O. (1953). *Nature (London) 172,* 730-731.
Lenhoff, H. M. and Kaplan, N. O. (1956). *J. Biol. Chem. 220,* 967-982.
Margoliash, E. (1972). *Harvey Lectures 66,* 177-247.
Margoliash, E., Ferguson-Miller, S., Brautigan, D. L., and Chaviano, A. H. (1976). *In* "Structure-Function Relationships of Proteins" (R. Markham and R. W. Horne, eds.), pp. 145-165. Elsevier/North-Holland, Amsterdam.
Margoliash, E., Smith, E. L., Kreil, G., and Tuppy, H. (1961). *Nature (London) 192,* 1121-1127.
Nagata, J., Yamanaka, T., and Okunuki, K. (1970). *Biochim. Biophys. Acta 221,* 668-671.
Pettigrew, G. W. and Meyer, T. E. (1971). *Biochem. J. 125,* 46P.
Pfennig, N. (1977). *Ann. Rev. Microbiol. 31,* 275-290.
Pfennig, N. and Biebl, H. (1976). *Arch. Microbiol. 110,* 3-12.
Postgate, J. R. (1956). *J. Gen. Microbiol. 14,* 545-572.
Salemme, F. R., Freer, S. T., Xuong, Ng. H., Alden, R. A., and Kraut, J. (1973). *J. Biol. Chem. 248,* 3910-3921.
Sherman, F., Taber, H., and Campbell, W. (1965). *J. Mol. Biol. 13,* 21-35.
Soininen, R., Ellfolk, N., and Kalkkinen, W. (1973). *Acta Chem. Scand. 27,* 1106.
Swank, R. T. and Burris, R. H. (1969). *Biochim. Biophys. Acta 180,* 473-489.
Takano, T., Kallai, O. B., Swanson, R., and Dickerson, R. E. (1973). *J. Biol. Chem. 248,* 5234-5255.
Timkovich, R., Dickerson, R. E., and Margoliash, E. (1976). *J. Biol. Chem. 251,* 2197-2206.
Van Beeumen, J., Ambler, R. P., Meyer, T. E., Kamen, M. D., Olson, J. M., and Shaw, E. K. (1976). *Biochem. J. 159,* 757-774.
Wood, P. (1977). *Eur. J. Biochem. 72,* 605-612.
Yamanaka, T. and Imai, S. (1972). *Biochem. Biophys. Res. Commun. 46,* 150-154.
Yaoi, H. and Tamiya, H. (1928). *Proc. Imp. Acad. Japan 4,* 436-439.

Phosphorylation, Oxidation, and Photosynthesis

FIFTY YEARS OF ATP

Fritz Lipmann

The Rockefeller University
New York, New York

I first met Martin Kamen in St. Louis; it might have been in 1946. At that time, he was involved in nitrogen fixation in bacteria. When I came into his office, I was very impressed by the enormous recording machine standing behind his desk which he apparently used quite frequently to tide him over dull minutes. I was very happy to hear and see him play for the first time; I have learned a lot about him, which has increased my knowledge of him as a unique human being. We had some conversations about music, and I realize what a great difference there is between a musician who can expertly play an instrument, in this case the viola he loves, and a music listener, such as myself, who cannot play anything. He told me that unlike most other composers, Mozart used two violas in the great quintets instead of an extra cello. He also told me that whenever Mozart played in a quartet he played the viola. I would like to note the German word for viola, the "Brātsche," which is what Mozart called it and which I think is very expressive.

Now to ATP. When I entered the Meyerhof laboratory in 1927, creatine phosphate (CrP) had just been discovered; it had

 ISBN 0-12-397580-8

been isolated from muscle and was the first phosphorylated compound the hydrolysis of which seemed, in some manner. to be involved in muscle contraction. That worried Meyerhof a good deal because he could not see any place there for it. Nevertheless, he put the compound, which had a very acid-labile N-P bond, into the calorimeter and measured the heat liberated by its hydrolysis. He found that it was 10-12 kilocalories, and this was about three to four times as high as it was for ordinary phosphate esters such as hexose 6-phosphate or glycerophosphate. As my first task in the laboratory, he asked me to see if there might be a link between breakdown and the tensionless contraction of frog rectus caused by acetylcholine. I could not find any breakdown of CrP. But Lundsgaard studied the contraction of a muscle that occurred without formation of lactic acid because the muscle had been poisoned with iodoacetate to inhibit glycolysis. He found that CrP was hydrolyzed in proportion to the tension developed, and contractibility lasted as long as CrP was available. This came as a great shock to Meyerhof because he had rather strongly believed that there was a causative relationship between lactic acid formation and contraction.

I will now briefly recapitulate the mechanism of glycolysis. The first step is normally an input of two phosphates from two ATPs into the phosphorylation of glucose which yields fructose 1', 6'-diphosphate. The next reaction is the splitting of fructose diphosphate into two triose phosphates, which was first imaginatively postulated by Embden and later confirmed as yielding: dihydroxyacetone-P $\longleftrightarrow$ P-glyceraldehyde. The latter constantly disappears through oxidation, and dihydroxyacetone-P converts to it during glycolysis, the course of which is shown for one-half glucose:

Hexose/2 $\longleftarrow$ ∿P

P-glyceraldehyde +P

P-glyceryl∿P -2H $\longrightarrow$ ∿P

P-glycerate $-H_2O$

P∿enolpyruvate $\longrightarrow$ ∿P

pyruvate + 2H

lactate

Although Harden and Young had shown as far back as 1909 that alcoholic fermentation, later found to be mechanistically equivalent to muscle glycolysis, is dependent on the presence of inorganic phosphate (P_i), it took about 30 years to find the mechanism of phosphate participation. It was clarified in 1939 by Warburg through the observation that P_i is essential for the oxidation of phosphoglyceraldehyde + P_i to yield a ∿P by forming 3'-phosphoglyceryl∿phosphate as oxidation product. Thereby, the energy of oxidation is saved and transferred to ADP to "fill up" ATP. The remaining 3'-phosphate moves into the 2'-position and the 2'-phosphoglycerate is dehydrated to phospho∿enolpyruvate, a second ∿P per half glucose which is also transferred to ADP. The glucose is metabolized by the use of two energy-rich phosphates and yields four energy-rich phosphates. Thus, two extra ATPs are yielded as the net energetic yield derived from glycolysis.

I have gone into this because it essentially vindicates Meyerhof's proposition that glycolysis yields energy for contraction, but by ATP and not by lactic acid formation. It is the inner mechanism of the glycolytic reaction that generates the ATP which actually drives the muscle. The energy turnover in glycolysis is

glucose + 2 ATP ⟶ fructose diphosphate + 2 ADP (-2 ATP)

fructose-1',6'-diphosphate + 2 P_i + 4 ADP + 2 NAD ⟶

2 lactic acid + 4 ATP + 2 NADH

⟨balance: + 2 ATP (net)⟩

But, we have anticipated the discovery of ATP. This was from 1928 - 1929 and occurred in two stages. The first preliminary stage was published by Lohmann and Meyerhof on the isolation of pyrophosphate from muscle extract. The second preliminary stage was that Embden and Schmidt isolated 5'-adenylic acid from muscle. Then, Fiske and Subbarow became interested in this Embden-Schmidt product. Their experiments were done with very fresh muscle extract whereby degradation was apparently prevented since they found only traces of adenosine 5'-monophosphate (AMP), but isolated instead adenosine 5'-triphosphate (ATP). They thus considered the adenylic acid of Embden and Schmidt was likely to be a degradation product of their ATP. Meanwhile, Lohmann suspected his pyrophosphate to be a breakdown product and indeed found that $Ba(OH)_2$, which he used in the isolation procedure, split off pyrophosphate from ATP. He corrected himself and published later in the same year (1929), having also isolated ATP, which he called adenyl pyrophosphate. It actually is a good name, since it singles out the last two energy-rich phosphates.

Here was ATP, then, and it was pretty well established that it was really connected with muscle contraction, although as Herman Kalckar has mentioned, it is taking a long time to settle the question and the mechanism by which energy is transferred from ATP to the muscle is still not completely understood in detail. Herman Kalckar and I, in two articles that appeared in the same year (1941), enlarged on the role of ATP in cellular energy transmission. I took up the problem after I had started working on the oxidation of pyruvate which, instead of being reduced as in glycolysis, was oxidized in respiration to acetate and CO_2. Thus, pyruvate was, so to say, at the crossroads between glycolysis and respiration. It was not known at that time that mitochondria catalyze this reaction, but I wanted very much to bring it into homogeneous solution. I then discovered that a suspension of acetone powder of *Lactobacillus delbruekii* (*longissimus*) was active in the oxidation of pyruvate to acetate and CO_2. I became interested in this procedure, using a bacterial extract dried in the desiccator over sulfuric acid. It was quite stable and gave a good extract for the oxidation of pyruvate. Then, accidentally, I found that phosphate was essential for this oxidation. I wanted to find out whether methylene blue could be reduced by this enzyme + pyruvate, which I wanted to measure manometrically. Therefore, I shifted from phosphate to bicarbonate buffer, and to my great surprise I could not get any oxidation except on addition of phosphate. I thought immediately that probably here I might have found an indication of acetyl phosphate as intermediary. So I proceeded to devise methods for its isolation and succeeded in obtaining a rather pure silver salt of acetyl phosphate from the products of this oxidation.

That acetyl phosphate was a metabolic intermediary was a revelation to me because pyruvate oxidation thus yielded an "active" acetate as well as an "active" phosphate. I had already measured the transfer of phosphate activated during pyruvate oxidation to ADP which indicated a phosphokinase was present in the crude extracts. To prove an acetyl transfer from acetyl phosphate was more difficult. Yet, an activated acetate had been indicated by Schoenheimer's work on the use of deuterated acetate. When deuteroacetate was injected into rats it was traced to having been incorporated into many compounds including some amino acids, but steroids and fatty acids, in particular. It became fairly clear, therefore, that an active acetate was indicated here as a precursor. And here I had acetyl phosphate as a fine candidate for it. I was greatly disappointed when I went to considerable pains to find a suitable acetyl transfer reaction in animal tissue. I chose the acetylation of sulfonamide by ATP + acetate, which was abundant in pigeon liver extracts. When I tested acetyl phosphate, however, it was very rapidly split but I could still see that it was not active as acetyl donor. On autolysis of the extract, transfer activity disappeared and was replaced by a boiled extract. This led me to finding coenzyme A, a coenzyme essential for acetylation (1). It turned out to be a secondary acetyl carrier and is necessary for the activation of acetate by ATP in this extract, which we found to go by displacement of pyrophosphate. Paul Berg then identified acetyl adenylate as the intermediary.

Feldberg and Mann and Nachmansohn and John had seen that there is also a coenzyme involved in the acetylation of choline by ATP + acetate in brain extracts. We tested for and found that purified CoA likewise supplements this acetylation. We went on to isolate the coenzyme further; it was discovered to contain pantothenic acid by tests in the laboratory of Roger Williams, who had discovered this vitamin. Snell *et al.* then isolated a new form of pantothenic acid which was called the *B. bulgaricus* factor. There, pantothenic acid was present peptidically linked to thioethanolamine. This turned out to be the form in which pantothenic acid is bound in CoA, and Lynen showed that acetate is linked to it by way of a thioester link. Eventually, acetyl phosphate in bacterial systems was found by Stadtman and Barker to react with an enzyme they called transacetylase which was shown by Stadtman *et al.* to be CoA-dependent; it catalyzed the reaction:

$$\text{acetyl}\sim\text{phosphate} + \text{CoA} \longrightarrow \text{acetyl-CoA} + \text{phosphate}$$

The acetyl∼phosphate thus was a two-headed molecule since it could transfer acetyl groups to acetyl∼acceptors, but also acted as phosphoryl∼acetate in the transfer of ∼P to ADP.

In the meantime, with Chou *et al.*, we had separated from pigeon liver the ATP-acetate reaction from the acetyl-sulfonamide acceptor enzyme. By connecting the latter with the bacterial transacetylase and acetyl∼phosphate + CoA, we finally could show that acetyl phosphate was indeed an active acetate. It could transfer here its acetyl to sulfonamide in the mixture of bacterial activation enzyme and liver acceptor enzyme.

There is one curious reaction described by Harting and Chance (2). They found that acetaldehyde can be oxidized with the phosphoglyceraldehyde oxidation enzyme from animals, and showed that acetyl phosphate is formed. With this reaction one can actually transfer the acetyl in an animal enzyme system to CoA and to an acetyl∼acceptor enzyme donated by acetyl phosphate.

This has actually been only an introduction to make clear what I had in mind when I wrote the 1941 article, which was to propose that the phosphoryl transfer from ATP was a reaction that most likely was involved in a large number of group activations, and that group activation was an essential part of the biosynthesis of large molecules. As an example, I postulated then, without knowing it in any detail, that ATP would deliver energy for protein synthesis by activating the carboxyl group of the amino acid in a similar manner to what we have seen to be the activation of the carboxyl group of acetate. This later proved to be true and many group activations also proved to be dependent on ATP.

I had many bad hours over being told that this was all wrong and that there was nothing like energy-rich phosphates or energy-transferring molecules. Once I got a letter from a chemistry professor telling me that I should rapidly retract all that nonsense; it only confused his students. I realized the confusion of the students was somewhat justified because Linus Pauling in his book on the chemical bond (3) distinguishes between a strong link and a weak link. The strong link, for example, is a carbon-to-carbon linkage which is produced with liberation of a lot of free energy and therefore is a "strong" link difficult to pull apart. It was a picture opposite to what I had talked about when using the term energy-rich for an easy to hydrolyze bond. I therefore introduced the term high group potential. Some people prefer instead to call it group transfer potential in energy-rich links, which is expressed by the release of free energy by hydrolysis. I had not made this quite evident in the 1941 paper.

I would like to emphasize once more that in biosynthetic reactions the substrate has to be preactivated before it goes into biosynthesis since the synthetic reaction requires an excess of energy which is transferred from the energy-rich compounds characterized above. This, I think, is quite analogous to what the organic chemist does in organic synthesis. To use a very simple comparison: in the Schotten-Baumann reaction, he uses acetic anhydride to introduce an acetyl group to a molecule just as the living organism uses a mixed anhydride, acetyl phosphate.

REFERENCES

1. Lipmann, F. (1954). The metabolic function of pantothenic acid. *In* "The Vitamins" (P. György and W. N. Pearson, eds.), Vol. II, p. 598. Academic Press, New York.
2. Harting, J. and Chance, B. Discussion remark. *Fed. Proc. 12,* 714 (1953).
3. Pauling, L. (1960). "The Nature of the Chemical Bond," 3rd ed., Chapters 3 and 4 and Appendix XII, Cornell University Press, Ithaca, N.Y., 1960.

All other references can be found in two papers published in *Trends Biochem. Sci.*: Discovery of creatine phosphate in muscle. *2*, 21 (1977), and Discovery of the adenylic acid system in animal tissues. *4*, 22 (1979), or in Herman Kalckar (1969) "Biological Phosphorylations," Prentice Hall, Englewood Cliffs, New Jersey.

BIOLOGICAL PHOSPHORYLATIONS, ^{32}P AND FIFTY YEARS OF ATP

*Herman M. Kalckar**

Huntington Laboratories
Massachusetts General Hospital
and
Department of Biological Chemistry
Harvard Medical School
Boston, Massachusetts 02114

Rereading Martin Kamen's dry rendition of the "radio-phosphorus" correspondence between the Ernest Lawrence group in Berkeley and George Hevesy in Copenhagen in 1939 (1), brings to life important events from the 1930's, which shaped the future of the next 40 or 50 years. It was Hevesy and Bohr who organized the first primitive production of significant amounts of ^{32}P in 1935. It was injected into mice and rats and traced into bones and phospholipids.

At that time, my late brother Fritz learned from Fritz Lipmann (who was one of our mutual friends in Denmark in 1935) that I was wrestling with some problems generated by my disclosure of a new type of tissue particle phosphorylation, stringent aerobic phosphorylation (or, as Ochoa named it in 1940, "oxidative phosphorylation").

When Fritz Kalckar, being close to Bohr and Hevesy, challenged me by asking in his low-keyed way, "How come you are not joining the Hevesy adventure on radio-phosphorus?", I could only reply that unfortunately the Hevesy ^{32}P mix was too toxic for my particle system. The ^{32}P, which Hevesy was to receive 3

**Present address: Chemistry Department, Boston University, Boston, MA 02215.*

 ISBN 0-12-397580-8

to 4 years later from Kamen, was much purer and would probably have launched my project into the orbit of radioisotopes. However, by early 1939 I had left Copenhagen.

I should like to relate my own impelling reasons for my going to California. Fritz spent a fabulous spring in Berkeley where he met J. R. Oppenheimer, Ernest Lawrence and his family, and other brilliant physicists.

I was greatly influenced by Fritz's enthusiasm about the great California schools. My American adventure thus began in February, 1939, at Cal Tech. On July 1, I enrolled in Van Niel's laboratory course in microbiology at Hopkins Marine Station of Stanford University, Pacific Grove. In Van Niel's workshop I first learned about the electrifying work done by Sam Ruben and Martin Kamen on "dark" incorporation of radioactive CO_2 into bacterial cells.

At Cal Tech, I joined the Kerckhoff Laboratories of Biological Sciences. At the time, the chairman was Thomas Hunt Morgan, who founded this biology division in 1928. Linus Pauling and his pupils were next door at Crellin and Gates Laboratories of Chemistry. One of his associates, came to play a great role in shaping my ideas about "energy-rich bonds," - the late Charles Coryell. The other Cal Tech scholars who were willing to listen to my preachings about phosphorylations were Henry Borsook, Professor of Biochemistry, and Norman Horowitz, one of the advanced graduate students in biology who received his Ph.D. in June, 1939.

PHOSPHORYLATION KINASES AND ENERGY CHANGE

The combination of scientific problems which were gathered in Pasadena and at Pacific Grove, somehow got me started on an extensive bioenergetics review which, thanks to Pauling, appeared in Chemical Reviews (2). My 1940 stay with Carl and Gerty Cori, Washington University, St. Louis, Missouri also developed harmoniously. They had become convinced that my puzzling observations of 1937 on the existence of the compulsory stringent aerobic phosphorylation ("oxidative phosphorylation") should be pursued. I was also lucky to have had the opportunity to work with Sidney Colowick, an advanced graduate student in the department. I felt also that I should learn methods of enzymology having been exposed by the Coris to this highly advanced technology. Sidney and I were trying to purify yeast hexokinase and soon ran into a puzzle. We found that hexokinase catalyzes the transfer of only one of the phosphoryl groups in ATP, forming ADP and one molecule of glucose 6-phosphate. We found that both of the groups could be transferred upon addition of a heat-stable muscle protein which we called "myokinase" (3). By late 1941, I tried to obtain documentation for the mode of action of myokinase. By means of a highly specific muscle enzyme, 5'-adenylate deaminase, I soon managed to demonstrate that myokinase catalyzes a reversible "phosphoryl dismutation" (4). In this reaction, ADP can function as a phosphoryl acceptor (for ATP) as well as donor (to 5'-AMP). The enzyme, later given the more operational name "adenylate kinase," has evoked interest in several fields including Dan Atkinson's formulation of the adenylate energy charge (5).

My active interest in the use of radioisotopes was a different story. It started in a somehow left-handed way. In 1941, my Rockefeller grant, which had been extended a full year (partly, I suppose, to help me over the trauma of the Nazi invasion of Denmark on April 9, 1940), finally expired. Carl Cori managed, however, to get me a Commonwealth Fund which supported me for a full year as a research associate at Washington University Medical School. I found myself "resonating" between two departments, the new one being Radiology (Dr. Sherwood Moore) and my old beloved Pharmacology Department (Drs. Gerty and Carl Cori). In this new job, biomedical studies with radioisotopes were specifically stipulated by the Fund as well as by the new department. This small biochemistry section was later to be expanded to a highly important center for molecular biology when Martin Kamen arrived in 1945, although I did not meet him there until 1949.

Setting up a small biochemistry section in the Department of Radiology was easy enough; the chairman was always cooperative. However, formulating a biochemistry problem tailored to the use of ^{32}P was not too easy in 1941. I was most fortunate to have with me two able graduate students, Jean Dehlinger and Alan Mehler.

We administered ^{32}P inorganic phosphate by intravenous or intraperitoneal injection into rabbits or frogs. We then removed the extracellular fluid with the highly radioactive phosphorus by perfusion of one of the hind legs with chilled phosphate-free Ringer solution (the other leg served as control). This perfusion created conditions which lent themselves to ^{32}P analyses of the muscle, albeit only resting ones.

We found that even in resting muscles one of the near rate-limiting steps was the rate of transport of inorganic phosphate into the muscle. Hence, the increase of labeling from rest to contraction was not dramatic, since in the resting muscle the degree of labeling of the pyrophosphate had already reached close to or exceeded 50% of that of the intracellular inorganic phosphate (6, 7).

In a further study of distribution of ^{32}P in ATP in frog muscle, we encountered a perturbing phenomenon. In resting frog muscle, the middle phosphoryl repeatedly showed a specific

TABLE I

P^{32} Concentrations in Phosphocreatine and Adenosine Triphosphate from Frog Muscle (cf. protocol 3)[a,b]

	P^{32}, *per cent of inorganic P activity*			
	Exp. 1	*Exp. 2*	*Exp. 3*	*Exp. 4*
Phosphocreatine P	10			20
Adenosine triphosphate, groups 2 and 3	25	28	22	20
Adenosine triphosphate, group 2	38	41	27	19
			26.5[c]	
Adenosine triphosphate, group 3 (terminal)	9	11	9	20
Adenosine triphosphate, group 1	0.5–1		1	

[a] *From Kalckar et al. (7).*

[b] *The frogs were held at a temperature of about 20° in Experiments 1, 2, and 4; in Experiment 3, at 5°. Prior to analysis, the extracellular* ^{32}P *was removed by perfusion.*

[c] *Phosphate group 2 liberated as orthophosphate by hydrolysis of the adenosine diphosphate with potato adenylpyrophosphatase instead of hexokinase plus myokinase.*

radioactivity 3 to 4 times as high as the terminal phosphoryl (7). This could be explained if myokinase activity were relatively very low in resting frog muscle at temperatures below 25°C. Perhaps AMP might be phosphorylated directly from acyl P, whereas ADP might be phosphorylated primarily from the storage of phosphocreatine (see Table I).

Another possibility would be that inorganic pyrophosphate, which has been restored to respectability by recent observations (8), could act as a phosphoryl donor of 5'-AMP, forming ADP. The inorganic pyrophosphate might in turn be formed by phosphorylation of inorganic phosphate by acyl phosphate.

It seems that the glycolytic and oxidative phosphorylation of ADP and 5'-AMP in frog muscle may occur at one place and the relatively slower myokinase activity be confined to another compartment. There are recent reports (9) that ATP is not necessarily equilibrated with ADP and AMP in mammalian liver. Labeled adenosine was found to preferentially incorporated into ATP, whereas adenine or hypoxanthine were incorporated into AMP and ADP as well (9). One option would be the existence of a pathway from adenosine which leads to a compartmentalized ATP pool. The existence of "comparticles" (10) in rat diaphragm has been suggested recently in connection with some aberrant ^{32}P labeling patterns of ATP from mitochondrial activity, which takes into account myokinase activity (10).

I realize that our old observations on the slow response of myokinase in resting frog muscle may be in conflict with Atkinson's formulation of the universal physiological role of myokinase, alias adenylate kinase, in living organisms (5). I would welcome a discussion of this aberration in the broad context in which Atkinson presents it.

FIFTY YEARS AGO

At the time this volume appears, its publication will probably coincide rather closely with the fifty years' anniversary of the discovery of ATP by Fiske and Subbarow (11, 11a) and by Lohmann (12). Two years earlier, in 1927, Fiske and Subbarow had discovered phosphocreatine (13; see also 14). Late, in 1929, my mentor at the University of Copenhagen, Einar Lundsgaard, then formulated the new physiological design for muscle contraction, phosphorylation energetics. Lundsgaard showed that phosphocreatine and its fission are more fundamental to the thermodynamics of muscle contraction than is lactic acid formation (15). His documentation was supported vigorously by Fritz Lipmann and Gustav Embden from the beginning, and later by Lohmann, who also initiated the enzymology of ATP (16). The multitude of enzymatic reactions in which ATP is involved has been reviewed this year by Lipmann (17), and even this paper covers only a fraction of its many reactions.

EPILOGUE

In his "Valediction for a Symposium on Bioenergetics" (18), Martin Kamen proposed that "triggerings of special syntheses or degradations" are dependent not only on the adenylate energy charge, but also on time-dependent fluctuations in the adenine nucleotide concentrations. These may be signals for transitions from one type of metabolism to another (18).

REFERENCES

1. Kamen, M. D. (1973). *Proc. Conf. Historical Development Bioenergetics,* p. 261. American Academy of Arts and Sciences.
2. Kalckar, H. M. (1941). *Chem. Rev. 28,* 71-178.
3. Colowick, S. P. and Kalckar, H. M. (1941). *J. Biol. Chem. 137,* 789.
4. Kalckar, H. M. (1942). *J. Biol. Chem. 143,* 299.
5. Atkinson, D. E., and Walton, G. M. (1967). *J. Biol. Chem. 242,* 3239.
6. Kalckar, H. M. (1944). *J. Biol. Chem. 242,* 3239; also *154,* 267.
7. Kalckar, H. M., Dehlinger, J., and Mehler, A. (1944). *J. Biol. Chem. 154,* 275.
8. Wood, H. G. (1977). *Fed. Proc. 36,* 2197.
9. Rapaport, E., and Zamecnik, P. C. (1976). *Proc. Natl. Acad. Sci. U.S. 73,* 3122.
10. Bessman, S., Borreback, B., Geiger, P. J., and Ben-Or, S. (1978). *In* "Metabolic Creatine Kinase and Hexokinase Role of Compartmentalization in Metabolic Regulation" (P. A. Screr and R. W. Estabrook, eds.). Academic Press, N.Y.
11. Fiske, C. H. and Subbarow, Y. (1929). *Science 70,* 381-383.

11a. Kalckar, H. (1969). "Biological Phosphorylations," pp. 34-37. Prentice-Hall, Englewood Cliffs, N.J.

12. Lohmann, K. (1929). *Amer. J. Physiol. 90,* 166 (Abstr. 13).
13. Fiske, C. H. and Subbarow, Y. (1927). *Science 65,* 401. (See also ref. 11a, pp. 54-56.)
14. Fiske, C. H. and Subbarow, Y. (1929). *J. Biol. Chem. 81,*

629-679.
15. Lundsgaard, E. (1929-1930). *Biochem. Z. 217,* 162-177; ibid. (1930). *227,* 51-83. (Engl. transl., pp. 344-353.)
16. Lohmann, K. (1934). *Naturwissenschaften 22,* 409. (Engl. transl. 11a, pp. 57-61.)
17. Lipmann, F. (1979). "Trends in Biochemical Science 50 Years Ago," Vol. 4, pp. 22-24.
18. Kamen, M. D. (1972). *In* "Horizons of Bioenergetics" (A. San Pietro and H. Gest, eds.), pp. 259-268. Academic Press, N.Y.

CYTOCHROMES AND REDOX SYSTEMS IN PHOTOSYNTHESIS

Robin Hill

University of Cambridge
Department of Biochemistry
Cambridge, England CB2 1QW

That the vegetable creation should restore the air which is spoiled by the animal part of it looks like a rational system and seems to be of a piece with the rest.
Benjamin Franklin

--and the search for a rational system goes on.

The quotation is from a letter to Joseph Priestley in 1773 about his discovery of the amelioration of the air by a growing plant. This system has now to be interpreted in terms of the activities of the mitochondria and the chloroplasts; both these subcellular structures contain cytochromes and their functions concern oxidation and reduction processes. Looking back again to the 18th century, Mirabeau had described Franklin as the man "able to restrain both thunderbolts and tyrants." May we honor Martin Kamen with this epidiectic phrase? Restraining the thunderbolts might represent the production of radioisotopes ^{11}C and ^{14}C. The restraint of tyrants could represent the successful law suit which liberated so many of his fellow scientists from arbitrary bonds.

 ISBN 0-12-397580-8

My own meeting with Martin Kamen in 1952 was at the bench. Dr. Robert Emerson, who organized a working party in his laboratory in Urbana, Illinois, introduced us. Martin was very keen to find some extracellular reaction with the photosynthetic bacteria in line with what had been found with chloroplasts isolated from a green plant. Leo Vernon was also with us. All three of us were amazed at the quantity of *c*-type cytochrome coming from *Rhodospirillum rubrum*. This was the beginning of the association with Martin which, happily, has continued until the present time.

My first postgraduate research was a biochemical study of the respiratory pigment, hemoglobin, suggested and inspired by F. G. Hopkins. The separation of the molecule into globin and heme and the heme into porphyrin and iron led to my joining (as a crude youngster) David Keilin in his work, following his discovery of the true respiratory function of the intracellular heme pigments. He had observed them in all types of the aerob organisms including animals, plants, and bacteria. He gave the name cytochrome to the system composed of these pigments. At this time his first paper (Keilin, 1925) had just been published. The problem now remained to examine the system more thoroughly *in vitro*. The starting point was the isolation of cytochrome *c*, which was the soluble component. We obtained successful results from a strain of yeast. However, our results with mammalian muscle were not quite successful, that is, until eventually Albert Szent-Györgyi, then in Hopkins' laboratory, showed us how to proceed using trichloracetic acid. This seemed to be a violent reagent, but the heme protein stood up to it. The use of trichloracetic acid with *R. rubrum* seemed to have led Kamen and Vernon (1955) to the discovery of a new class of heme proteins. At first this was called *Rhodospirill* heme protein (RHP). The type of heme protein that it should b classed with was left until later when Martin came over to England. These heme proteins had to be distinguished from hemoglobin. They resemble hemoglobin in the absorption spectr of their ferrous states, but differ in that they are oxidized directly to the ferric state with molecular oxygen and do not combine reversibly with it. These heme proteins were placed i the class cytochromoids.

In 1933, I had developed a rapid spectroscopic method for measuring the oxygen dissociation curve of hemoglobin. I was fortunate to be in contact with Joseph Barcroft in connection with determining the oxygen affinity of different hemoglobins, particularly, as it related to his work on fetal respiration. The spectroscopic method required only a small amount of either a dilute solution of the hemoglobin or a dilute suspension of red blood cells. This suggested the examination of a model for oxygen requirement of red muscle by relating extracellular respiration to intercellular respiration. The transport of oxygen from the blood to respiring cells could be related to the oxygen affinities of two hemoglobins together with the oxygen affinity for the intracellular cytochrome system of Keilin. The use of basic lead acetate in animal biochemistry was brilliantly introduced by Hugo Theorell (1932). It gave a relatively stable solution of myoglobin from a small quantity of perfused red muscle. The high affinity for oxygen of myoglobin suggested its use in measuring oxygen production by isolated chloroplasts. Having long desired to study plant biochemistry, in 1937, I thus approached the study of photosynthesis. The production of oxygen then became a problem separate from the reduction of the carbon dioxide.

Just about this same time, Rubin, Hassid, and Kamen actually followed the photochemically induced uptake of the carbon from carbon dioxide by a green plant. Their highly critical work with the short-lived isotope ^{11}C confirmed the dual nature of the process of photosynthesis. This seemed to correspond exactly with the original physiological studies of F. F. Blackman (1905) many years before. The process was to be interpreted as two sharply distinguished parts, the "light" and "dark" reactions, whose individual activities combined to determine the rate.

The production of oxygen by isolated chloroplasts seemed to suggest a system similar to that of intracellular respiration, but working in reverse. The presence of cytochromes in chloroplasts suggested a stepwise build up of oxidation-reduction potential during the process of photosynthesis. The discovery by Robert Emerson (1957, 1958) of the abolition of "red drop" in quantum efficiency for photosynthesis by simultaneous illumination with light of two frequencies led to the interpretation of the process in terms of two light reactions. These two aspects were combined in the hypothesis which Fay Bendall and I (1960) proposed. This came to be called "the Z scheme" and it survived because the chain of steps was represented by components in addition to the cytochromes. The so-called steps were imagined to be redox spans whose magnitude was in some way related to the quantum energy of the light being absorbed.

In 1963 the book "Primary Processes in Photosynthesis" by Martin Kamen was published. It was a joy to very many of us involved in the study of the subject. He noted (p. 85) "...that absorbancy is independent of light intensity...," as long as no change is occurring in the light absorbing system. It follows that, in steady state illumination, when many of the reaction centers are open, the rate is limited only by the intensity of the light. The quantum requirement then is found to be constant over a range of light intensities. This forms an experimental basis for the determination of the efficiency for conversion of the energy by the reaction centers into a form of chemical-free energy. I feel constrained to offer my subjective view of this problem in what follows even though it is so very near to the darkest eras in Martin's picture of the spectrum of ingorance.

For the biochemical interpretation of a photochemical process, it is not primarily of interest to compare the action of a green leaf to that of a heat engine. The efficiency of energy conversion can be defined as the rate of appearance of a chemical product divided by the rate of absorption of the radiant energy. When, in Blackman's sense, the light intensity is limiting the rate, there is found to be a stoichiometric relationship between the number of quanta absorbed and the number of molecules changed by the photochemical process. The idea of a heat engine giving the same efficiency of energy con version independently of the difference between two temperatures brings a new dimension for any interpretation of the pro cess. If the rate of the photochemical process is to be deter mined by the difference between two temperatures, the efficien cy within the Blackman limits is surely to be determined by th energy of the photons, within a narrow region of the spectrum which is absorbed by chlorophyll.

The simplest procedure is to represent this photochemical process in terms of three temperatures. The temperature of th radiation which determines the rate is given by the formula of Planck and can be represented by T_1 in the classical thermodynamic sense. T_2 can represent the temperature of the photochemical system. The quantum energy can be made to depend on a third temperature θ. This can be defined as the temperature at which the quanta being absorbed by a reaction center corresponds to the mean quantum energy for complete equilibrium black-body radiation. The efficiency for activation of the photochemical system is then to be represented approximately by $1 - T_2/\theta$. The process of conversion of the activation ener gy is to be regarded as irreversible. The matter of biochemical interest is the efficiency of energy conversion as defined here and determined experimentally for a partial process observed with the biological system.

The proposal for considering the three temperatures may be derived from a detailed consideration of a reversible Carno cycle with complete equilibrium radiation. This was indicated in a former publication (Hill, 1977). The third temperature will be given by

$$\theta = hc/2.701\ k\lambda$$

where c/λ is the mean frequency of the light being sbsorbed, referring to a reaction center and k is Boltzmann's constant. The constant 2.701 is the ratio of the two integrals representing total energy and number of photons for the complete radiation as given by the formula of Planck. It follows from a thermodynamic argument that any direct conversion of the energy of a quantum of radiation to any form of external work must involve degradation of quanta to lower frequencies. Thus, while the photochemical system remains at a constant temperature part of the energy of the photons is degraded to heat. It can be assumed that the more the energy being degraded the less the tendency is for a reverse process involving the external work. Hence it would seem that the most favorable limit for conversion would be approximately $h\nu/2$. In the generally accepted diagram showing the relation between the two light reactions it seems that the two redox spans can be represented as being within the range of half the energy of the quanta absorbed by the chlorophyll *a*--and the search for a rational system goes on.

REFERENCES

Blackman, F. F. (1905). *Ann. Bot. 19,* 281.

Emerson, R. (1957). *Science 125,* 746.

Emerson, R. (1958). *Science 127,* 1059.

Hill, R. (1937). *Nature (London) 139,* 881.

Hill, R. (1977). "Photosynthetic Organelles," Special Issue Plant Cell Physiol., p. 47.

Hill, R. and Bendall, F. (1960). *Nature (London) 186,* 136.

Kamen, M. D. (1963). "Primary Processes in Photosynthesis," pp. 4 and 85. Academic Press, N.Y.

Kamen, M. D. and Vernon, L. P. (1955). *Biochim. Biophys. Acta 17,* 10.

Keilin, D. (1925). *Proc. Roy. Soc. (London) B98,* 312.

Ruben, S., Hassid, W. Z., and Kamen, M. D. (1939). *J. Amer. Chem. Soc. 61,* 661.

Theorell, A. H. T. (1932). *Biochem. Z. 1,* 252.

THE COMPARATIVE BIOCHEMISTRY OF PHOTOSYNTHESIS: MILESTONES IN A CONCEPTUAL ZIGZAG

Howard Gest

Photosynthetic Bacteria Group
Department of Biology
Indiana University
Bloomington, Indiana 47405

Research on photosynthetic bacteria during the late 19th century indicated that such organisms do not produce molecular oxygen. The anaerobic nature of bacterial photosynthesis was eventually conclusively established, and the search for an answer to the question of why photosynthetic bacteria do not evolve O_2 became a central issue for the "comparative biochemistry of photosynthetic processes." This chapter outlines the history of hypotheses and research advances made in clarifying the basic differences between oxygenic and anaerobic photosyntheses.

THE PANDEMIC OF OVERLY COMPARATIVE BIOCHEMISTRY

The notion that the chemical activities of cells of different kinds show both common as well as grossly divergent features is old, and was emphasized by Pasteur in his pioneer-

 ISBN 0-12-397580-8

ing research on the causes of fermentations. Ruminations on this theme permeated early investigations on the enzymatic steps of glycolysis and the alcoholic fermentation, and from Meyerhof's monograph of 1924 (32), one can recapture the mood of comparative biochemistry on the eve of the realization that glucose catabolism by muscle and yeast cells occurs through an almost identical series of reactions. Attempts to formalize comparative biochemical principles during the late 1920's led to the doctrine of a movement which, in my opinion, became a kind of psychological pandemic of "overly comparative biochemistry." This affected biologists and biochemists in several ways. Some victims became biochemically blurred and unable to clearly distinguish organisms of very different types (an extreme symptom - the view that all metabolic activities are intrinsically similar, consisting of series of reactions which could be generally represented by the equation: $AH_2 + B \rightarrow A + BH_2$). Others saw metabolic differences in somewhat sharper focus and some, with mild infections, vacillated somewhere in-between.

Bioenergetics, in one way or another, seemed to exacerbate the syndrome, and a diagnosis in terms of philosophical attitudes plus a hopeful note of amelioration were later offered by H. A. Krebs and H. L. Kornberg in their comprehensive essay on "Energy Transformations in Living Matter" (28): "Discussions have taken place from time to time about the 'unity' or 'disunity' of the biochemical organization of different types of organism. Some authors consider it as more profitable to emphasise different rather than common features. Thus Cohen (1954) suggests that comparative biochemistry should concern itself with 'the study of the origin, nature and control of biochemical variability.' On the other hand, Stanier (1954), following Kluyver (1931) and van Niel (1949) defines the task of the comparative biochemist as 'to seek for the common biochemical principles which are expressed in all forms of life.' These differences merely reflect different attitudes towards the same facts. Some authors prefer to lay stress on endless variety and complexity, and others to trace the common ancestry from which the variety has sprung, but most will agree on the need for both approaches" (28a).

COMPARATIVE BIOCHEMISTRY OF PHOTOSYNTHESIS; ETIOLOGICAL AGENT OF THE DILEMMA

In the area of photosynthetic processes, the "culprit" was identified quite early as molecular oxygen. Failure to observe O_2 production during light-dependent growth of photosynthetic bacteria was reported as early as 1883 (10). Seventy years later, experiments were still being conducted to verify the anaerobic nature of bacterial photosynthesis. This remarkable fact underlines the depth of the "attitude problem" of comparative biochemistry. As matters developed, two questions came into particular focus: (1) Why do the photosynthetic bacteria not produce O_2? (2) Why do the bacteria require an accessory H (e) donor - other than water - for photosynthetic CO_2 reduction?

The search for answers to these questions form the fabric of development of the gross comparative biochemistry of photosynthetic processes, which eventuated as a conceptual zigzag that began in earnest about 1930. The following is an account, as I see it, of the main course of the journey. To help convey the flavor of the chronicle, I make liberal use of direct quotations from various authors.

1883: Attempts to demonstrate oxygen production by photosynthetic bacteria give negative results (10); see ref. 40 for description and evaluation of researches during the 1930's claiming demonstration of O_2 evolution.

1930: Photosynthetic bacteria cannot use water as a hydrogen donor

In the mid-1920's, van Niel began systematic quantitative studies on the physiology of photosynthetic sulfur bacteria; these soon led him to the conclusion that such organisms cannot use water as a hydrogen donor, but instead use H_2S for CO_2 reduction. In his words (38):

"For the process of photosynthetic carbon dioxide assimilation (or reduction) we may then say that the reaction

$$CO_2 + 2H_2A = CH_2O + H_2O + 2A$$

requires various and special compounds of hydrogen for various and special organisms. From which it follows that the photosynthetic activity of the chlorophyll-bearing organisms, in which H_2A represents H_2O, represents only one very special instance of a group of possible photosynthetic activities. This would mean that the purple sulphur bacteria can use H_2S as a hydrogen donor but cannot use H_2O."

Accordingly, the photosynthesis of the purple sulfur bacteria was seen as a process fundamentally different from that of the green plant. At this stage of research, CO_2 reduction was firmly entrenched as a central feature of photosynthetic processes. Matters soon became complicated by observations on the physiological patterns of various purple bacteria in respect to organic substances. With certain organic compounds, photometabolism (in growing cultures or resting cell suspensions) was characterized by a net production of CO_2, whereas with others

CO_2 was consumed. Net formation of CO_2 no doubt was a disconcerting phenomenon, and research efforts tended to be mainly directed at explaining the situation with organic substrates whose metabolism was coupled with CO_2 utilization.

1935 - 1941: Development of the comparative-biochemical water-cleavage hypothesis of photosynthesis; photosynthetic bacteria can use water as a hydrogen donor, but only in a special way

By 1935, van Niel (39) had elaborated the argument that in photosynthetic bacteria, simple organic compounds can function as H_2A (i.e., as H donors for CO_2 reduction) in the generalized equation of photosynthesis. More significantly, the generalized equation was amplified with a hypothetical mechanism that provided the possibility of a "unified" conception of photosynthesis. The essence of the scheme, aimed at reconciling all the seemingly discordant facts, was summarized in the equations:

a. $4[\text{chlorophyll}\cdot H_2O + h\nu \rightarrow \text{chlorophyll}\cdot OH + H]$

b. $CO_2 + 4H \rightarrow CH_2O + H_2O$

c. $2[2\ \text{chlorophyll}\cdot OH + H_2A \rightarrow 2\ \text{chlorophyll}\cdot H_2O + 2A]$

$$CO_2 + 2H_2A + 4h\nu \rightarrow CH_2O + H_2O + 2A$$

Thus, water was invoked as the "direct" H donor for CO_2 reduction in all photosyntheses [i.e., reactions (a) and (b) represented a common denominator in all photosynthetic processes]. Now, the different forms of photosynthesis could be explained by idiosyncrasies in respect to reaction (c): in green plants, H_2A is water and O_2 is evolved; alternatively, in photosynthetic bacteria, the "dehydro-chlorophyll (hydrate)" is reduced to its original state with hydrogen derived from H_2S, H_2, organic compounds, etc. In effect, this scheme proposed what later came to be known as "noncyclic" electron flow, with chlorophyll acting as a redox catalyst. In fact, the printed discussion of van Niel's paper (39) suggests that it was a hairs-breadth from conceptions advanced almost 25 years later -

"Dr. Burk: Why put water in the first equation of your scheme, and not just chlorophyll and dehydro-chlorophyll? The equation balances just as well without the H_2O."

"Dr. van Niel: Because of the evidence presented by Stoll for the existence of a chlorophyll hydrate, and on account of certain suggestive energetic considerations."

The comparative water cleavage scheme was developed in detail in van Niel's extensive review of 1941 (40), which ends with the statement: "As a simple working hypothesis it seems to present some advantages; its value can only be judged by future developments." Despite a gradual accumulation of facts difficult to rationalize by the scheme, it enjoyed great popularity until the early 1960's as the basis for design of experiments and interpretation of results.

1945 - 1949: Can light energy be converted to chemical bond energy in the form of ATP?; photoproduction of H_2

The possibility that light energy might be convertible to chemical energy in the form of "energy-rich" phosphate esters was suggested in 1943 by Ruben (36), but experimental studies during the next several years failed to generate unambiguous supporting evidence. In the meantime, investigators with strong backgrounds in photochemistry and physics argued strenuously, mainly on theoretical grounds, against the possibility that such mechanisms could play a significant role in photosynthesis [e.g., see Rabinowitch (35)]. Moreover, an experimental study (3) with spinach grana, tobacco leaves, and *Chlorella vulgaris* using a radioactive tracer gave negative results: "Using radioactive phosphorus, no direct connection between gross formation of organic phosphorus compounds and photosynthesis or photochemical reductions has been found to occur."

The investigation of this problem using $^{32}P_i$ as a tracer formed the major part of my Ph.D. thesis research with Martin Kamen, and the results (18) - with both green algae and *Rhodospirillum rubrum* - showed that illumination caused a marked increase in P_i uptake by intact cells and also a great stimulation in turnover of ^{32}P between soluble and insoluble cell fractions. We concluded that "The results obtained in this investigation indicate either that phosphorylation is an integral feature of the photosynthetic process proper or that nonrelated 'dark' phosphorylation reactions are greatly stimulated in some way as a consequence of illumination." Even though the experiments with *R. rubrum* were conducted using anaerobic conditions, eliminating the possibility of oxidative phosphorylation, negative feedback inhibition from the Photochemical Establishment apparently led us to add the vague qualifier relating to possible stimulation of a "dark" process.

It is of interest that in 1947, Wassink (43) discussed bacterial photosynthesis in terms of hydrogen transfer between the accessory H donor and CO_2 being "driven" in some fashion by light energy - but the time was not yet ripe for the right connections to be made.

On the comparative biochemical front, new fuel was added by the discovery (19) that purple bacteria produce H_2 in large quantity by a light-dependent process during photoheterotrophic growth with certain amino acids as nitrogen sources. This seemed, at first, to be analogous to the photoproduction of H_2

by anaerobically adapted green algae first observed by Gaffron and Rubin (13). For some time to come, the mechanism of light-dependent H_2 formation by photosynthetic bacteria was interpreted on the basis of the water cleavage hypothesis; schemes were devised (e.g., see 20) in which an "imbalance" would occur resulting in an excess of the "photoreductant" moiety HX, the presumed precursor of H_2 ($4\ HX \rightarrow 2H_2 + 4X$). Among other difficulties, there was no obvious physiological rationale for explaining why a seemingly large fraction of the "photoreductant" should be oxidized by evolution of H_2. Eventually, it became evident that photoproduction of H_2 by green algae and photosynthetic bacteria occur through different mechanisms, with different physiological meanings.

1953 - 1954:

Photosynthesis as a photoelectric phenomenon

Levitt (29, 30) proposes that green plant photosynthesis is basically a photoelectric phenomenon in that absorption of light by a chlorophyll molecule results in expulsion of an electron, and he draws schemes in which chlorophyll acts as a redox catalyst for transfer of electrons from H_2O to an organic disulfide acceptor. "The entire process may thus be visualized as a flow of electrons actuated by light; or, essentially, as a photoelectric current flowing from water through the chlorophyl to the disulfide" (29). This conception, the "electron flow mechanism of photosynthesis," was generally accepted some years later as a basic feature of all photosynthetic processes.

Cytochrome *c* discovered in *R. rubrum*

Vernon (41) discovers the presence of large quantities of cytochrome *c* in *R. rubrum* cells grown photosynthetically under anaerobic conditions, and soon therafter cytochrome is also observed in the strict anaerobe *Desulfovibrio desulfuricans* - "thus the absence of cytochromes from obligate anaerobes can no longer be taken as axiomatic" (34). These findings were the harbingers of an avalanche of research of great import for understanding the light-dependent energy metabolism of photosynthetic bacteria and of energy conversion processes of organisms that use inorganic oxidants such as sulfate and nitrate as substitutes for O_2.

Photosynthetic bacteria surely do not produce O_2

The experiments of Johnston and Brown (25) employing ^{18}O as a tracer confirm that O_2 is not produced by *R. rubrum*. Engelmann's original observations (10) are vindicated once again, and we know of no exceptions to the rule that algae and higher plants are the only organisms known to produce O_2.

Light-induced phosphorylation demonstrated with "chromatophores" of *R. rubrum*

Frenkel (11) discovers that pigmented particles obtained by disruption of *R. rubrum* cells can rapidly phosphorylate ADP when illuminated anaerobically in the absence of appreciable quantities of electron donors or acceptors; and at about the same time, a similar observation is made by Arnon *et al.* (2) with green plant chloroplast preparations. Thus, an important common denominator of photosynthetic processes is more clearly revealed, setting the stage for the concept of cyclic electron flow. Evidently, the interaction of light with chlorophyll generates both the "reductant" and "oxidant" necessary for an anaerobic electron transfer sequence that can drive the phosphorylation of ADP; accordingly, "external" sources of electrons and terminal oxidant are not required.

1959 - 1963: A fertile period of "synthesis," marked by considerable zigzagging; the pandemic begins to subside

We now come to a point in the chronology recent enough so that a detailed accounting of events seems unnecessary. Indeed, it would be difficult to accurately follow the flow of ideas and research accomplishments between 1959 and 1963 because these years of intense activity were characterized by significant shifts in positions taken by individual investigators on various issues.

By and large, those who had a strongly ingrained commitment to the comparative water cleavage theory ignored, for several years, the discovery of *in vitro* anaerobic photophosphorylation in 1954. As of 1959, it seemed that a concensus "suddenly" began to develop for a revised view of the comparative biochemistry of photosynthesis, embracing the following tenets (for detailed discussions, see ref. 1, 33, 37):

a. In all photosynthetic systems, the interaction of light with chlorophyll initiates an electron flow (as in Levitt's model) that drives "photophosphorylation."

b. In oxygenic photosynthesis, the ultimate electron donor for the net reduction of pyridine nucleotide is water ("noncyclic" electron flow). Photosynthetic bacteria, however, cannot use water as a H(e) donor and must, therefore, generate reducing power from an "accessory" H(e) donor.

c. CO_2 reduction, per se, is not really a fundamental general feature of photosynthesis; the essence of photosynthetic processes is the use of light energy to produce ATP (and reducing power in oxygenic photosynthesis).

In respect to point (c), it should be noted that earlier preoccupation with CO_2 reduction as a central issue simply ignored well known observations that clearly indicated the photometabolism of organic compounds such as malate, pyruvate, etc. by purple bacteria was photoheterotrophic rather than photoautotrophic (see ref. 14 and 33). The influence of overly comparative biochemistry in interpretation of the photometabolism of organic compounds by purple bacteria was aptly summarized by S. S. Cohen (9) as follows:

"About 15 years ago I participated in the rediscovery of the long neglected hexose monophosphate shunt and helped to relate it to the origin of ribose. This work was done in the face of a world of biochemists committed, perhaps for the sake of simplicity, to the Embden-Meyerhof pathway. The plethora of alternative pathways of carbohydrate metabolism revealed in the recent past has developed our considerable respect for the evolutionary ingenuity of organisms, at least in these metabolic areas. There may be no possible analogy to protein synthesis at this point because there might be a selective advantage in minimizing the number of critical pathways for the biosynthesis of protein, just as there would be such an advantage in maximizing the ability to metabolize exogenous carbon sources to utilizable intermediates. However, insofar as broadening the latter capability tends to increase enzyme, that is proteins, the degree of synthesis and the presence of optimal equipment for survival must be balanced and compromised.

The role of organic substrates in bacterial photosynthesis has recently been reexamined (R. Y. Stanier, *Bacteriol. Rev. 25*, 1, 1961). The efforts to generalize the mechanisms of bacterial and green plant photosynthesis had led to a relatively simple scheme in which the organic substrates essential for purple sulfur bacteria and the purple nonsulfur bacteria were supposed to serve primarily as a reducing agent (even as does H_2S) to handle the oxidizing fragment generated in the photochemical cleavage of water. An early study of the stoichiometry of acetate utilization in certain bacterial photosyntheses was not accepted as a demonstration of a reductive photoassimilation of this carbon source because it did not fit the generalized theory. These recent studies of Stanier and his associates on the role of organic substrates in the photometabolism of bacteria (R. Y. Stanier, *Bacteriol Rev. 25*, 1, 1961) show that these compounds do serve as a carbon source, as had been shown earlier by Gaffron and by Gest. Furthermore, the routes of the utilization of acetate in such photoassimilations are different in the purple sulfur or nonsulfur bacteria from those in the green sulfur bacteria. Thus development of our knowledge on the mechanisms of photosynthesis provides another interesting instance in which the predilection for simplicity has impeded the development of understanding."

<u>Net reducing power in bacterial photosynthesis; energy-linked electron transfer</u>

The problem of how photosynthetic bacteria generate net reducing power from accessory H(e) donors proved to be particularly vexatious. With donors of sufficiently low redox potential, such as H_2, reduced pyridine nucleotide could evidently be generated by dark reactions. On the other hand, with donors of relatively high potential it could be shown with chromatophore preparations that light is required to effect net reduction of pyridine nucleotide (12). To explain this, most investigators invoked a "noncyclic" electron flow mechanism (as in oxygenic photosynthesis); accordingly, the action of light presumably caused a transfer of electrons from bacteriochlorophyll to pyridine nucleotide, and these were "replaced" by electrons derived from the accessory donor via a dark process. In other words, bacteriochlorophyll acts as a redox carrier in the transfer of electrons from donor to acceptor. In some versions of this hypothesis, net formation of reduced pyridine nucleotide was assumed to occur by such "noncyclic" electron flow with all accessory donors, regardless of their redox potentials.

A more plausible alternative to the noncyclic electron flow explanation was suggested by the studies of Chance and colleagues on the succinate-dependent reduction of NAD^+ by mammalian mitochondria. Their investigations established that such electron transfer against the thermodynamic gradient could be driven by ATP and "energy-rich intermediates" associated with oxidative phosphorylation (7, 8). Various kinds of experiments with *R. rubrum* (intact cells and chromatophores) indeed provided evidence consistent with the occurrence of energy-linked ("reverse") electron flow in the net generation of reducing power in bacterial photosynthesis (4, 5, 15). In addition, other studies (6) showed that claims for the existence of an antimycin-resistant "noncyclic photophosphorylation" (i.e., a phosphorylation driven by light-stimulated "noncyclic" electron flow) in *R. rubrum* chromatophores were based on an experimental artifact.

These developments led to proposal at the first international conference on bacterial photosynthesis (Yellow Springs, 1963) of the scheme shown in Fig. 1.

At the time, biochemists generally showed a curious reluctance to accept Chance's thermodynamically reasonable concept of energy-linked electron flow and the experimental evidence in its support. Similarly, the proposal for energy-linked electron flow in bacterial photosynthesis was received with skepticism.

In the meantime, a radically new view of light-induced electron transport in oxygenic photosynthesis was advanced by Hill and Bendall (23) - the "Z" scheme. According to this conception (see Fig. 2) there is a linear "noncyclic" electron transport in which two light-driven "segments" are connected by a "dark," energy-yielding sequence of electron flow.

1963 - 1969: Noncyclic electron flow mechanisms for bacterial photosynthesis gradually disappear; reinterpretation of the mechanism of photoproduction of H_2 by the bacteria

One might say that during this period, the "dust was settling." Two distinct metabolic patterns could be clearly discerned in photosynthetic bacteria (16):

Photoheterotrophic pattern. An organic compound serves as

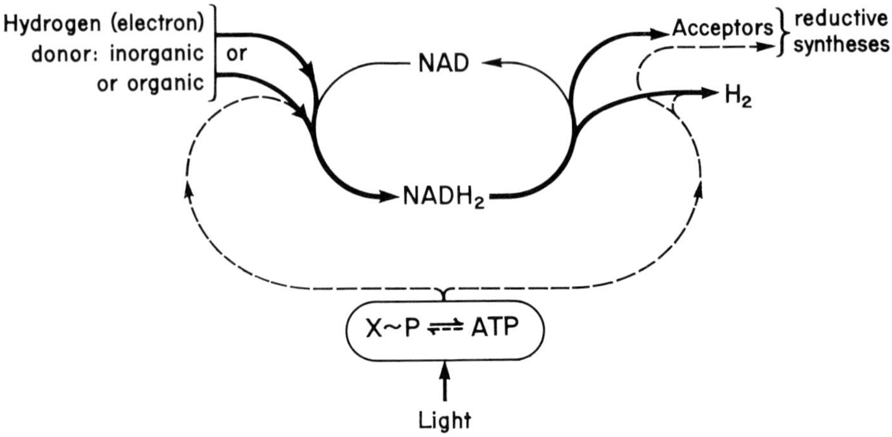

Fig. 1. "Scheme for hydrogen (electron) flow from donors to acceptors and photoproduction of H_2 in bacterial photosynthesis. The pyridine nucleotide may be either NAD or NADP; for convenience, only the former is shown"; after Gest (ref. 15).

the source of carbon and through typical heterotrophic mechanisms directly provides the intermediates necessary for biosynthesis of cellular constituents; quantitatively, reduction of carbon dioxide is not necessarily a major process. ATP and its energy-rich precursors are formed by an "autotrophic" mechanism, that is, by light-induced cyclic phosphorylation. Reducing power is obtained by dark oxidation of organic substrates (or intermediates), and, in some instances, this requires an energy contribution from the phosphorylation system.

Photoautotrophic pattern. The reductive pentose cycle operates as a major mechanism and together with other pathways enables the conversion of carbon dioxide to all cellular constituents. As in the above, the ATP required for these conversions, and energy for other purposes, is furnished by the cyclic phosphorylation system. The inorganic "accessory" electron donor provides net reducing power through dark oxidation or by energy-linked reverse electron transfer.

This picture was reinforced by studies of Keister and Minton (27) with chromatophores, from which they concluded that "all of the NAD^+ reduction observed in *R. rubrum* is by an energy-linked process" (they refer to light-dependent NAD^+ reductio[n] processes).

Thus, in respect to light-dependent electron transport and ATP synthesis, the comparative biochemistry of photosynthesis could be summarized by the diagrams of Fig. 2 (16).

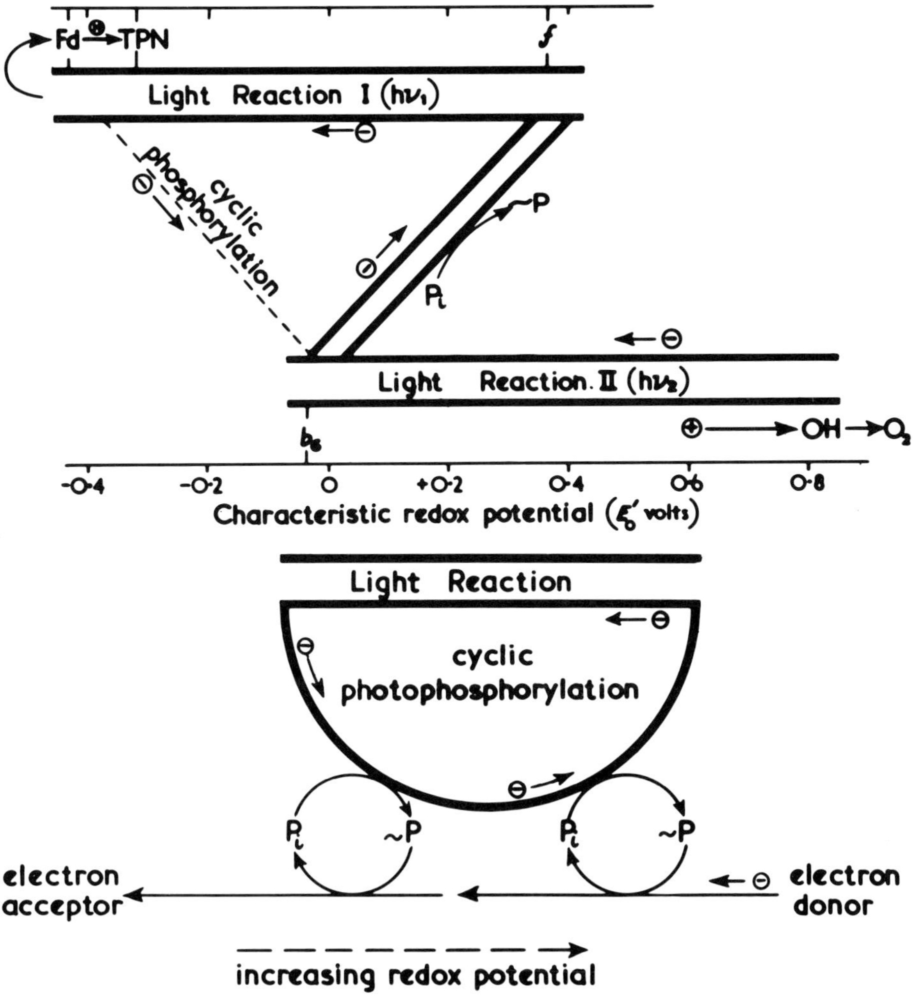

Fig. 2. Schemes for light-driven electron transport and phosphorylation in oxygenic and anaerobic photosyntheses (after Gest, ref. 16). (Top) Oxygenic photosynthesis, according to Hill (22): "Electrons are transported against the thermodynamic gradient in the light reactions. Light reaction II results in formation of an electron and a positive hole; discharge of an hydroxyl ion by the positive hole forms OH, the precursor of oxygen." (Bottom) "Reverse" electron transfer driven by the cyclic phosphorylation system of photosynthetic bacteria. "The heavy curved line represents the pathway of electron transfer in cyclic photophosphorylation; a number of catalysts, including cytochromes, have been implicated. It is suggested that reverse

Physiological significance of light-dependent H_2 production by photosynthetic bacteria

Superficially, photoproduction of H_2 by photosynthetic bacteria appears to represent a wasteful dissipation of energy and reducing power, and this should have been an early clue to the physiological significance of the process. The latter, however, became apparent only after conditions were established for determining the maximum yield of H_2 obtainable from organic substrates. Surprisingly, it was found (21) that certain tricarboxylic acid cycle intermediates could be completely catabolized to H_2 and CO_2 by illuminated resting cells of *R. rubrum* (e.g., succinate + $4H_2O \rightarrow 4CO_2 + 7H_2$). Moreover, the photoproduction of H_2 was inhibited by inhibitors of cyclic photophosphorylation. These and other experiments (4, 5) supported the scheme given in Fig. 1, and suggested that light-dependent H_2 formation signified a regulatory device for "energy idling." Thus, the energy-dependent production of H_2 could be construed as a control mechanism that aids the cell in maintaining the "pools" of reduced pyridine nucleotide and ATP at levels consistent with the temporal demands of the biosynthetic machinery. This mechanism is operative only in cells in which the nitrogenase system is present and uninhibited (by NH_4^+, etc.), and recent experiments (42) have conclusively confirmed that *in vivo* photoproduction of H_2 is catalyzed by the nitrogenase enzyme complex (note: since green algae are incapable of synthesizing nitrogenase and, consequently, cannot fix N_2, it is evident that photoproduction of H_2 by such organisms occurs through a different kind of mechanism).

A recent study (24) of photoproduction of H_2 by growing cultures of *Rhodopseudomonas capsulata* revealed that H_2 formation also occurs under conditions in which growth rate is limited by light intensity, i.e., when the ATP supply is limiting. In other words, it appears that during growth in circumstances of energy stress with nitrogenase present but not functioning to provide cellular nitrogen, some unregulated energy-dependent electron flow to nitrogenase occurs, resulting in H_2 production. This must represent a bioenergetic burden to the cell, of magnitude and significance yet to be determined.

(Fig. 2 cont'd.) electron transfer (horizontal long, solid arrows) may also be promoted by energy-rich non-phosphorylated precursors of ∿P; note, however, that the energy coupling could equally well be expressed in terms of Mitchell's chemiosmotic hypothesis."   , electron;   , positive hole; P_i, inorganic phosphate; ∿P, ATP or related energy-rich phosphorylated precursors; Fd, ferredoxin; TPN, nicotinamide adenine dinucleotide phosphate; f, cytochrome f; b_6, cytochrome b_6.

The 1970's; epilogue and prologue of the future

This account of the transit to our present position attempts to identify major "milestones" in a tortuous route, and necessarily omitted are a number of second-order perturbations (e.g., the movement of ferredoxin into and out of schemes for cyclic electron transport in photosynthetic bacteria; at this writing, if there are functions of ferredoxin aside from participation in H_2 reduction and H_2 evolution, they are still unknown). The possibility that NAD^+ can be reduced in bacterial photosynthesis by a "direct photoreduction" (i.e., by noncyclic electron flow with bacteriochlorophyll acting as a redox catalyst) continues to be occasionally suggested (e.g., see ref. 26) merely because the estimated midpoint redox potential of the "primary acceptor" in *Chlorobium* (but not in other types of photosynthetic bacteria) seems to be more negative than that of the pyridine nucleotides. Perhaps we have arrived at the last, tottering bastion of the notion that there is a noncyclic electron flow in bacterial photosynthesis.

In reviewing the literature on generation of net reducing power in bacterial photosynthesis some years ago (17), it was inevitable that comments on comparative biochemistry would be forthcoming, and some of them are appropriate here:

> "Without doubt, the search for a 'unitary' mechanism has contributed to the fitful history of research on the comparative biochemistry of photosynthetic processes, which has witnessed protagonists shifting their positions up and back and frequently being right for the wrong reasons. Perhaps this was due in large measure to the fact that there are numerous possible levels of correlation in comparative biochemical reasoning. It is not always evident or true that the proper feature has been seized upon for comparison. Historically, close examination of apparent disunities has led to many of the great advances in biochemistry. Some remarks of Lipmann (17a) are pertinent: 'It is astonishing to realize that the more one proceeds with the understanding of the working of the organism, the more one becomes concerned with methodological problems. Strangely, the prying into the mystery of life reduces more and more to an unravelling of a sometimes rather unusual and unexpected methodology of the cell.' "

As for the future, we are on the threshold of an exciting era in which the biochemical genetics and molecular biology of photosynthetic bacteria (31) will undoubtedly furnish new and incisive ways of elucidating the evolutionary relationships between bacterial and oxygenic photosyntheses.

ACKNOWLEDGMENT

Research of the author is supported by a grant from the U.S. National Science Foundation.

REFERENCES

1. Arnon, D. I. (1959). Conversion of light into chemical energy in photosynthesis. *Nature (London) 184,* 10-21.
2. Arnon, D. I., Allen, M. B., and Whatley, F. R. (1954). Photosynthesis by isolated chloroplasts. *Nature (London) 174,* 394-396.
3. Aronoff, S. and Calvin, M. (1948). Phosphorus turnover and photosynthesis. *Plant Physiol. 23,* 351-358.
4. Bose, S. K. and Gest, H. (1962). Hydrogenase and light-stimulated electron transfer reactions in photosynthetic bacteria. *Nature (London) 195,* 1168-1171.
5. Bose, S. K. and Gest, H. (1963). Relationships between energy-generation and net electron transfer in bacterial photosynthesis. *In* "Energy-Linked Functions of Mitochondria" (B. Chance, ed.), pp. 207-218. Academic Press, New York.
6. Bose, S. K. and Gest, H. (1963). Bacterial photophosphorylation: regulation by redox balance. *Proc. Natl. Acad. Sci. U.S. 49,* 337-345.
7. Chance, B. (1961). The interaction of energy and electron transfer reactions in mitochondria. V. The energy transfer pathway. *J. Biol. Chem. 236,* 1569-1576.
8. Chance, B. and Hollunger, G. (1961). The interaction of energy and electron transfer reactions in mitochondria. IV. The pathway of electron transfer. *J. Biol. Chem. 236,* 1562-1568.
9. Cohen, S. S. (1963). On biochemical variability and innovation. *Science 139,* 1017-1026.
10. Englemann, T. W. (1883). *Bacterium photometricum.* Ein Beitrag zur vergleichenden Physiologie des Licht - und Farbensinnes. *Pflügers Arch. Physiol. 30,* 95-124.
11. Frenkel, A. W. (1954). Light-induced phosphorylation by cell-free preparations of photosynthetic bacteria. *J. Amer. Chem. Soc. 76,* 5568-5569.
12. Frenkel, A. W. (1958). Simultaneous reduction of diphosphopyridine nucleotide and oxidation of reduced flavin mononucleotide by illuminated bacterial chromatophores. *J. Amer. Chem. Soc. 80,* 3479-3480.

13. Gaffron, H. and Rubin, J. (1942). Fermentative and photochemical production of hydrogen in algae. *J. Gen. Physiol. 26,* 219-240.
14. Gest, H. (1951). Metabolic patterns in photosynthetic bacteria. *Bacteriol. Rev. 15,* 183-210.
15. Gest, H. (1963). Metabolic aspects of bacterial photosynthesis. *In* "Bacterial Photosynthesis" (H. Gest, A. San Pietro, and L. P. Vernon, eds.), pp. 129-150. Antioch Press, Yellow Springs, Ohio.
16. Gest, H. (1966). Comparative biochemistry of photosynthetic processes. *Nature (London) 209,* 879-882.
17. Gest, H. (1972). Energy conversion and generation of reducing power in bacterial photosynthesis. *In* "Advances in Microbial Physiology" (A. H. Rose and D. W. Tempest, eds.), Vol. 7, pp. 243-282. Academic Press, London and New York.

17a. Lipmann, F. (1956). *In* "Currents in Biochemical Research 1956" (D. E. Green, ed.), p. 241. Interscience, New York.

18. Gest, H. and Kamen, M. D. (1948). Studies on the phosphorus metabolism of green algae and purple bacteria in relation to photosynthesis. *J. Biol. Chem. 176,* 299-318.
19. Gest, H. and Kamen, M. D. (1949). Photoproduction of molecular hydrogen by *Rhodospirillum rubrum. Science 109,* 558-559.
20. Gest, H., Kamen, M. D., and Bregoff, H. M. (1950). Studies on the metabolism of photosynthetic bacteria. V. Photoproduction of hydrogen and nitrogen fixation by *Rhodospirillum rubrum. J. Biol. Chem. 182,* 153-170.
21. Gest, H., Ormerod, J. G., and Ormerod, K. S. (1962). Photometabolism of *Rhodospirillum rubrum*: Light-dependent dissimilation of organic compounds to carbon dioxide and molecular hydrogen by an anaerobic citric acid cycle. *Arch. Biochem. Biophys. 97,* 21-33.
22. Hill, R. (1965). The biochemists' green mansions: the photosynthetic electron-transport chain in plants. *In* "Essays in Biochemistry" (P. N. Campbell and G. D. Greville, eds.), Vol. 1, pp. 121-151. Academic Press, London and New York.
23. Hill, R. and Bendall, F. (1960). Function of the two cytochrome components in chloroplasts: a working hypothesis. *Nature 186,* 136-137.
24. Hillmer, P. and Gest, H. (1977). H_2 metabolism in the photosynthetic bacterium *Rhodopseudomonas capsulata*: H_2 production by growing cultures. *J. Bacteriol. 129,* 724-731.
25. Johnston, J. A. and Brown, A. H. (1954). The effect of light on the oxygen metabolism of the photosynthetic bacterium, *Rhodospirillum rubrum. Plant Physiol. 29,* 177-182.

26. Jones, O. T. G. (1977). Electron transport and ATP synthesis in the photosynthetic bacteria. *In* "Microbial Energetics" (B. A. Haddock and W. A. Hamilton, eds.), (27th Symposium of the Society for General Microbiology), pp. 151-183. Cambridge University Press, Cambridge.
27. Keister, D. L. and Minton, N. J. (1969). Energy-linked reactions in photosynthetic bacteria. III. Further studies on energy-linked nicotinamide-adenine dinucleotide reduction by *Rhodospirillum rubrum* chromatophores. *Biochemistry 8*, 167-173.
28. Krebs, H. A. and Kornberg, H. L. (1957). Energy transformations in living matter. *Ergeb. Physiol. Biol. Chem. Exp. Pharmakol. 49*, 212-298.
28a. Cohen (1954) and Stanier (1954): *In* "Cellular Metabolism and Infections" (E. Racker, ed.), p. 84 and p. 3, respectively). Academic Press, New York; Kluyver (1931): "The Chemical Activities of Microorganisms," University of London Press; van Niel (1949): *In* "Photosynthesis in Plants" (J. Franck and W. E. Loomis, eds.), p. 437. Iowa State College Press, Ames, Iowa.
29. Levitt, L. S. (1953). Photosynthesis as a photoelectric phenomenon. *Science 118*, 696-697.
30. Levitt, L. S. (1954). The role of magnesium in photosynthesis. *Science 120*, 33-35.
31. Marrs, B., Wall, J. D., and Gest, H. (1977). Emergence of the biochemical genetics and molecular biology of photosynthetic bacteria. *Trends Biochem. Sci. 2*, 105-108.
32. Meyerhof, O. (1924). "Chemical Dynamics of Life Phaenomena." Lippincott, Philadelphia and London.
33. Ormerod, J. G. and Gest, H. (1962). Hydrogen photosynthesis and alternative metabolic pathways in photosynthetic bacteria. *Bacteriol. Rev. 26*, 51-66.
34. Postgate, J. R. (1954). Presence of cytochrome in an obligate anaerobe. *Biochem. J. 56*, xi.
35. Rabinowitch, E. I. (1945). "Photosynthesis and Related Processes," Vol. I. Wiley (Interscience), New York.
36. Ruben, S. (1943). Photosynthesis and phosphorylation. *J. Amer. Chem. Soc. 65*, 279-282.
37. Stanier, R. Y. (1961). Photosynthetic mechanisms in bacteria and plants: development of a unitary concept. *Bacteriol. Rev. 25*, 1-17.
38. van Niel, C. B. (1930). Photosynthesis of bacteria. *In* "Contributions to Marine Biology," pp. 161-169. Stanford Univ. Press, Stanford.
39. van Niel, C. B. (1935). Photosynthesis of bacteria. *Cold Spring Harbor Symp. Quant. Biol. 3*, 138-150.
40. van Niel, C. B. (1941). The bacterial photosyntheses and their importance for the general problem of photosynthesis *Advan. Enzymol. 1*, 263-328.

41. Vernon, L. P. (1953). Cytochrome *c* content of *Rhodospirillum rubrum*. *Arch. Biochem. Biophys. 43*, 492-493.
42. Wall, J. D., Weaver, P. F., and Gest, H. (1975). Genetic transfer of nitrogenase-hydrogenase activity in *Rhodopseudomonas capsulata*. *Nature (London) 258*, 630-631.
43. Wassink, E. C. (1947). Photosynthesis as a light-sensitized transfer of hydrogen. *Antonie van Leeuwenhoek J. Microbiol. Serol. 12*, 281-293.

STUDIES ON THE OXYGEN EVOLVING SYSTEM OF PHOTOSYNTHESIS: ARE THERE NONLINEAR MISSES?

J. Lavorel

Laboratoire de Photosynthese du C.N.R.S.
91190 Gif-sur-Yvette, France

INTRODUCTION

A remarkable property of the oxygen evolving system (OES) first observed by Joliot *et al.* (1969) is the order four periodicity of the emission when induced by short, saturating flashes, spaced a few tenths of seconds apart. According to the linear four-step model of Kok *et al.* (1970), this periodicity is simply related to a cycle of four oxidation states (so-called S states) in the OES corresponding to the sequential removal of four electrons to two H_2O molecules in order to evolve one O_2 molecule. Similar order four periodicities have been observed in several phenomena associated to photosystem II and their occurrence is now recognized as the standard signature of events thought to take place on the donor side of System II with the OES intact (for a recent review, see Radmer and Cheniae, 1977). Not less remarkable is the damping of the sequence as flashing proceeds, the amplitude of the oscillations gets smaller and smaller until a steady, constant O_2 yield is

 ISBN 0-12-397580-8

reached. In the above model, damping is ascribed mainly to a "miss" factor - and secondarily to a "double hit" factor - affecting the photochemical functioning of centers. These add a small, random contribution to the rotation of the S cycle, resulting ultimately - and irrespective of the initial situation in an invariant population of states. As proposed, the miss hypothesis is purely formal. To quote Radmer and Cheniae (1977), "one would like to establish whether these parameters (misses and double hits) are merely fudge factors to get computed values to match experimental values, or whether they have real biological and technical significance, and can be varied by appropriate means." Surprisingly, however, little attention has been paid among workers in the field to the miss hypothesis and to the nature and signficance of its possible mechanism. The common attitude has been to accept a photochemical imperfection as the only possible source of disorder, to ignore the difficulties which this hypothesis raises, and to casually propose mechanisms without detailed experimental proof. The mechanisms most commonly proposed to specify the miss hypothesis have been either of the *a priori* type (a center may be momentarily photoinactive because the primary acceptor Q is reduced or because the primary donor P_{680} is oxidized) or of the *a posteriori* type (a pair of charges may back-react on the center, instead of being further separated and stabilized by transfer to secondary carriers).

There are two serious difficulties in the photochemical miss hypothesis (Lavorel, 1976b). First, a substantial quantum deficit must be the consequence of misses. In *Chlorella*, the deficit would be from 20 to 40% as deduced from damping figures found in the literature, the exact but unknown value depending on whether the transition parameters (misses, single hits, double hits) are identical for all states or not, a point not yet settled. Oddly enough, the deficit would be less in isolated chloroplasts than in *Chlorella*, since the miss coefficient is less for the former ($\alpha \cong 0.1$) than for the latter ($\alpha \cong 0.17$). These consequences seem very unlikely in view of our general belief of a very high quantum yield of the primary photoreactions and in view of the uncertainty concerning the integrity of isolated chloroplasts. Second, damping and yield must be inversely correlated under conditions where the two quantities can be varied on the same material. Such is indeed the case with chloroplasts in the absence of electron acceptor when Q becomes reduced in the course of a flash sequence (Radmer and Kok, 1973). But there are numerous exceptions where, on the contrary, both yield and damping are decreased in parallel (see e.g., Maison and Lavorel, 1977; Lavorel, 1978)

Thus, the photochemical hypothesis is either wrong or incomplete. We have, therefore, explored the possibilities afforded by an alternative principle, that of "conservative" misses, meaning by this a type of disorder not necessarily associated to any photochemical defect. In particular, the E - P (or mobility) model, which is of the conservative type, predicts that, under some circumstances, damping and yield will be directly correlated (Lavorel, 1978). However, we are still far from a solution. For one thing, the E - P model cannot be tested unequivocally in the present state of our knowledge and thus remains speculative. Also the correlated variations of damping and yield are observed in such diverse circumstances as to suggest no specific site or mechanism for the postulated disorder. This led us to attempt an altogether different approach. The question after all may be asked whether misses - in this case truly photochemical - could not be considered as an artifact produced by the very high photon density and intensity inherent in sequence experiments when using electronic flashes. If the answer were positive, it would solve the first problem, because the quantum yield at low light intensity would not be subject to this artifact. However, the second problem of correlation would still remain. The present paper describes an attempt to answer this question. The result is partly negative, but it leads to the investigation of another problem, which is that of relaxation of misses and finally suggests another scheme where both the difficulties with quantum yield and damping-yield correlation might be rationalized.

MATERIAL AND METHODS

Chlorella pyrenoidosa was grown and harvested daily as previously described (Lavorel and Lemasson, 1976).

Oxygen emission was monitored polarographically at room temperature using a "rate" electrode similar to that described by Joliot and Joliot (1968). The electrolyte composition was: growth medium diluted twice and KCl 0.1 *M*. The concentration of the algal sample (chlorophyll : 350 μg ml^{-1}) was calculated to yield a single cell layer after sedimentation onto the platinum electrode chamber (0.15 mm thickness). Conventional O_2 yield sequences were run using a Xenon flash lamp (General Radio, Stroboslave) at a flashing period of 0.3 sec. For double-flashes experiments, another identical flash lamp was positioned above the first one; both flash lamps were separately saturating. In such experiments, the delay between the two flashes was set by a pulse generator (Tekelec Airtronic, TE 10). Sequences were always recorded after a dark adaptation of 200 sec following a 20-flash sequence. Turnover experiments with flash pairs were performed at steady-state after at least 20 flashes.

As explained below, the main part of this study involved the comparison of O_2 yields induced by "rectangular" flashes of variable intensity and duration. The latter were obtained using a pulsed argon laser (Spectra Physics, laser model 164, exciter model 265, acousto-optic coupler and driver models 365 and 465). The emission consists of some 15 lines in the violet-green part of the spectrum, 60% of the intensity being concentrated within two lines (514.5 and 488.0 nm). Due to the small size of the laser beam (1.5 mm diameter) and its limited power output (e.g., about 4 10^{-5} J in a 10 μs pulse), the amperometric method had to be specially modified. The platinum chamber was reduced to the laser beam section to exclude any oxygen electrochemical current from the nonilluminated portion of the electrode that would only contribute noise to the signal. The electrode response was thus reduced 8- to 10-fold as compared to that in the conventional set-up with electronic flashes. In addition, this laser system produces a low continuous background between pulses. Although this is much lower in intensity than the pulse itself, it can contribute a large fraction of the O_2 signal at the low pulse frequency (40 Hz), which had to be chosen in this study. The background light was minimized by passing the laser beam through a rotating sector (at the above defined frequency) and synchronizing the laser pulse within a narrow window (about 40 μs) during the sector rotation.

The overall response was improved by the following procedure (see Fig. 1a). Trains of n successive laser pulses separated by dark periods of equal total duration were selected by intercepting the beam with an electromechanical shutter. With such a procedure, the electrode cannot distinguish the individual O_2 pulses within a train - it actually integrates them and thus responds at the lower shutter frequency. The amplitude of the signal as a function of n, the number of pulses in a train, is as shown in Fig. 1b; it "saturates" at about n = 20. This "saturation" is due to the kinetics of the electrode response and not to any kinetic limitation within the photosynthetic chain. This is of some importance for later discussion; it is simply verified by noting that the shape of the curve does not change after decreasing the light intensity 10 times (see Fig. 1b). In order to compare the light power at different settings of the pulse duration, part of the laser beam (∿10%) was directed to a light power meter (Spectra Physics, Model 404). No attempt was made to determine the actual light intensity at the electrode surface because of the uncertainties in measuring the transmission of the optical system. The energy density at saturation of O_2 emission could, however, be estimated as 100 μJ cm^{-2}, in reasonable agreement with other published figures (see, e.g., Den Haan, 1976). The O_2 response induced by the background laser light (even limited) was not negligible as compared to the sum of background and pulse. But

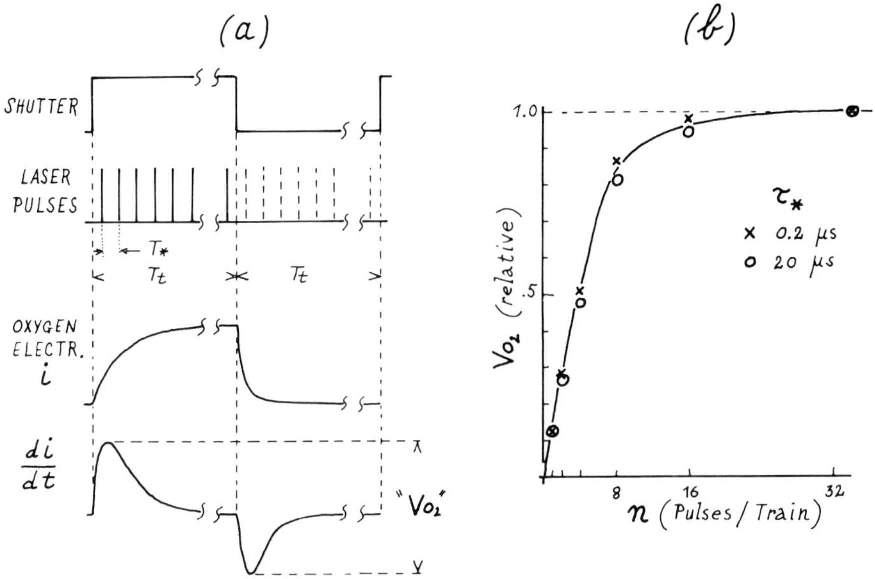

Fig. 1. (a) Time diagram of the O_2 measuring protocol. Laser pulses are synchronized to a rotating sector at a frequency T^{*-1} ($T^* = 25$ ms). Train of n pulses are selected at a lower frequency $(2\ T_t)^{-1}$ ($T_t = 0.64$ s usually) with the help of electro-mechanical shutter ($n = T_t/T^*$). The O_2 electrode integrates the individual O_2 pulses during the train. This signal is electrically differentiated and the peak amplitude is taken as V_{O_2} according to the protocol in (a) versus the number of pulses in a train; the latter is varied by changing T_t. Notice that the relative response is the same for weak pulses ($\tau^* = 0.2$ μs) and for saturating pulses ($\tau^* = 20$ s), both types of the same intensity.

as the resolution of the part in the signal corresponding to the pulse required uncertain assumptions, it was decided to consider only the total O_2 response as a function of the total light density per pulse.

The transition parameters in the Kok standard model were computed by applying σ analysis (Lavorel, 1976a) to consecutive Y_n (O_2 yield) taken from experimental sequences. This procedure yields symmetrical (or quasisymmetrical) functions of the transition parameters, free of any assumption concerning the distribution of misses, etc., among states. The main damping coefficient, σ_1, in particular, is the sum of the miss coefficients of all states ($\sigma_1 = \sum_i \alpha_i$).

RESULTS AND DISCUSSION

Observation of a Nonlinear Quantum Deficit

According to the working hypothesis, isoactinic flashes (i.e. of constant light density) of different intensities might not always yield the same O_2 emission. In other words, the reciprocity law might not be obeyed at very high intensities, thus producing a nonlinear quantum deficit in O_2 yield. The occurrence of nonlinear quenching for *in vivo* chlorophyll fluorescence has already been described (Mauzerall, 1976) and thus could support speculation about similar effects for *in vivo* chlorophyll photochemistry.

A 100-fold intensity range was chosen by using flashes of 0.1 and 10 μs. The lower limit resulted from practical considerations: the strongest laser pulses available could not be much shorter than this; the higher limit was *a priori* fixed to be well below the turnover time of System II (∿600 μs; see Vater *et al.*, 1968) in order to avoid too many double hits. A deficit in O_2 yield was consistently found with the high intensity 0.1-μs flash as compared to the low intensity 10-μs one. The results of a number of such comparison, independently run over a period of several months, is shown in Table I. It should be noted that during this period, the experimental set up was frequently modified (alignment of the beam, change in beam diameter, in electrode area, in measuring protocol). Figure 2a shows the O_2 responses induced by 0.1- and 10-μs pulses as a function of their light density D. In line with the working hypothesis, the deficit tends to zero with D decreasing. The maximum possible deficit could not be observed directly because the upper limit of the 0.1-μs pulse light density (about 20 μJ cm^{-2}) did not permit us to reach more than about 50-60% of the light-saturated O_2 yield. While attempting to overcome this limitation by using double (or multiple) pulses a few 10 μs apart instead of single pulses in the pulse train protocol, it was observed that the difference in O_2 yield for the two types of isoactinic pulses was now almost completely eliminated (Fig. 2b).

One might argue that, when making such comparisons, one is actually detecting an excess of O_2 emission induced by long flashes over that induced by short ones, instead of a deficit the other way around. This could obviously result from a sizeable contribution of double hits in long flashes. Two lines of evidences are against this reasoning. In the first place, if this were true, the increase in O_2 yield induced by isoactinic flashes as a function of their duration τ^* should match the turnover kinetics of System II relaxation. As shown in Fig. 3a, the value thus computed (∿50 μs) is much too short in com-

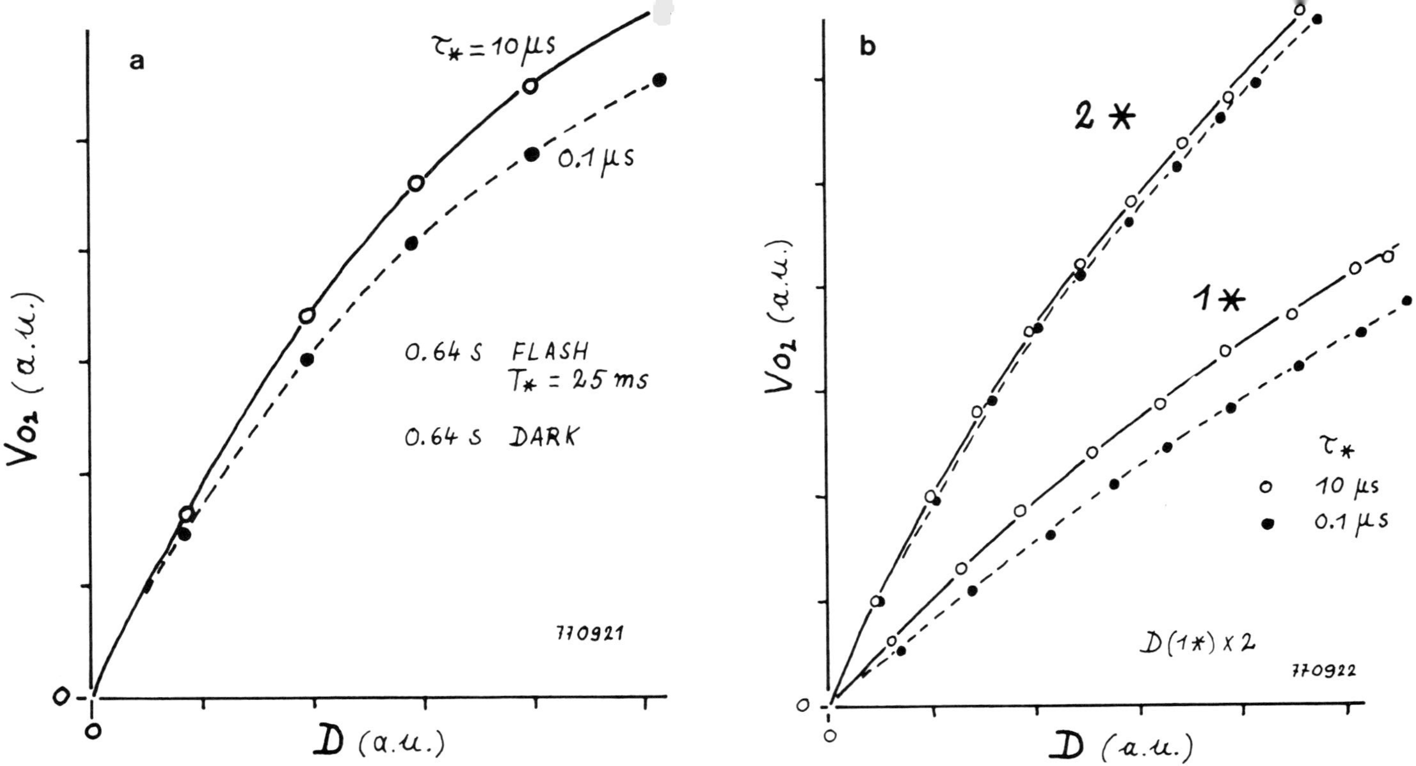

Fig. 2. Steady-state O_2 emission (V_{O_2}) as a function of energy of light density (D) of laser pulses following the protocol in Fig. 1a. Pulse separation $T^* = 25$ ms; light or dark period $T_t = 0.64$ S. Two pulse durations are shown: $\tau^* = 10$ μs or 0.1 μs. (a) Single pulse experiment (b) double pulse experiment (τ^*) two pulses at 20-μs interval are given during each T^* period; the D scale has been expanded twice for comparison with a single pulse experiment (1).*

TABLE I. Comparison of Steady State O_2 Emission Induced by Isoactinic Laser Pulses[a]

Flash duration (μs) (1) Short	(2) Long	Relative O_2 deficit for short flash $[V(2) - V(1)]/V(2)$[b]
0.1	10	0.13
0.1	10	0.15
0.1	10	0.17
0.1	10	0.17
0.05	20	0.19
0.1	10	0.13
0.4	10	0.14
0.4	10	0.12
0.1	5	0.12
0.1	10	0.20
0.1	10	0.06
1	10	0.09
0.5	10	0.30
0.1	10	0.10
0.1	10	0.07
0.1	10	0.10

[a]Isoactinic pulses were used of maximum possible intensity, corresponding to ∿50% saturation of O_2 emission.
[b]$V(2)$, $V(1)$: steady-state O_2 emission rate induced by long, short laser pulses respectively.

parison with the known turnover time of System II. In the second place, the actual System II turnover time in the present experimental conditions may be directly measured using saturating (see, however, Fig. 4 below) long flashes in pairs with variable separation Δt (Fig. 3b). The initial slope of the curve indicates a time constant of 1.3 ms, now a value more than twice the known value. Notice also that the zero intercept of the curve with the ordinate axis is positive, as if the first flash of the pair alone were not really saturating.

Fig. 3. (a) O_2 emission induced by isoactinic laser pulses of variable duration τ^*; the constant light density of the pulses corresponds to the maximum D value on the abscissa of Fig. 2. The turnover first-order kinetics (dashed line) which would fit the data has a time constant $\tau \cong 50$ μs. (b) Determination of System II turnover rates with laser pulses of $\tau^* = 10$ or 20 μs. $V_1{*}$ = steady-state O_2 emission using single flashes; $V_2{*}$ = same using pair of pulses with spacing Δt. Notice the positive (0.06) initial offset.

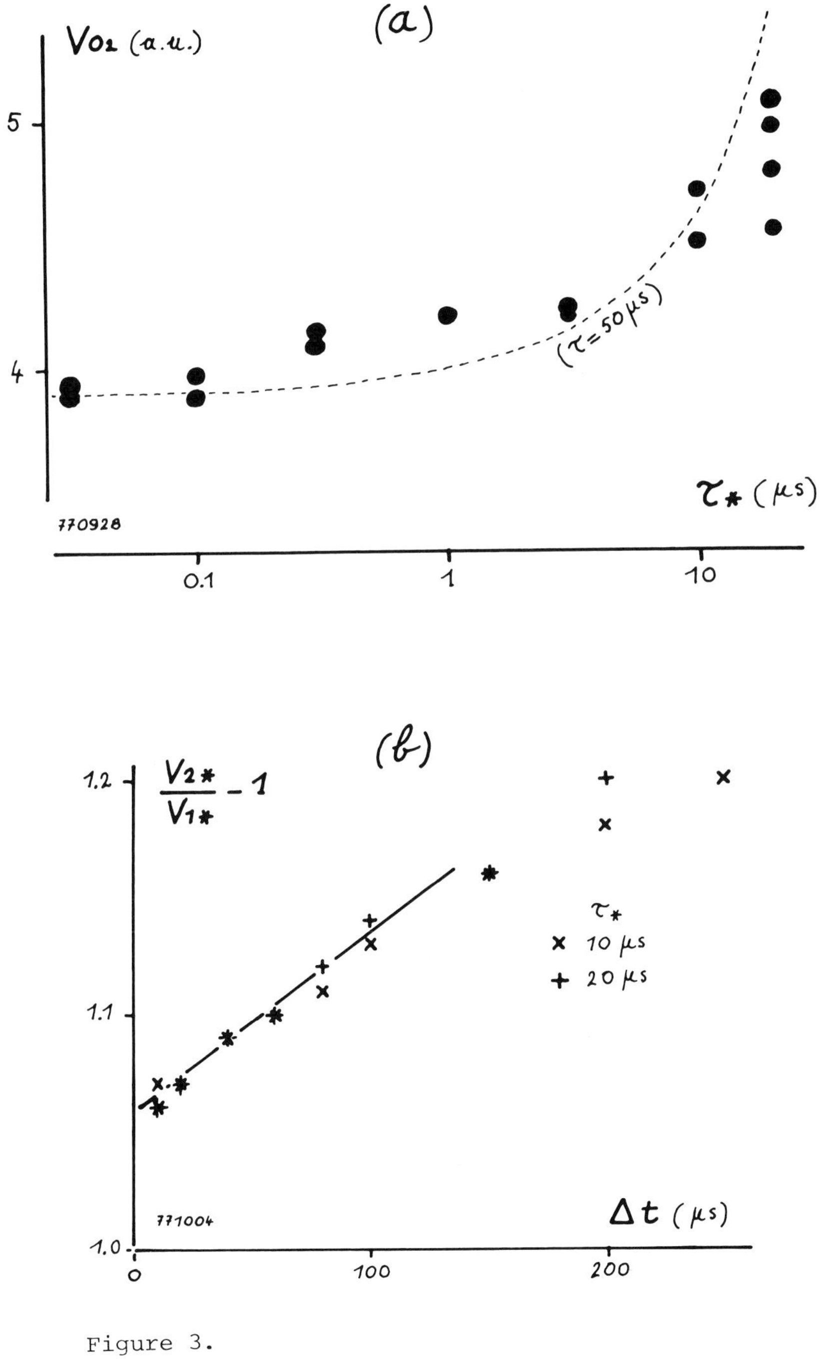
(a)
Vo2 (a.u.)
5
4
(τ=50μs)
τ* (μs)
770928
0.1
1
10
(b)
1.2
V2*/V1* - 1
1.1
1.0
τ*
× 10 μs
+ 20 μs
771004
Δt (μs)
0
100
200

Figure 3.

However, doubling the flash energy (τ^* from 10 to 20 μs in Fig. 3b) did not modify this initial offset. Probably another manifestation of this effect is seen in Fig. 4 showing the light saturation curve of O_2 yield for a 10 μs pulse and also the light curve for the excess O_2 yield resulting from another identical pulse fired 10 μs following the first one. The latter reaches a steady non-zero level much higher than expected from the System II turnover time. The problems raised by the initial offset as well as by the too long turnover time deduced from the initial slope in Fig. 3b will be discussed later. Anyhow, it is clear that the effect of flash intensity in the 0.1 - 10 μs range cannot be ascribed to double hitting.*

So far, the conjecture of nonlinear misses seems agreeably verified. No artifactual explanation can be found for the above results. In particular, the comparison of the single and double pulse experiments (Fig. 2b) appears quite convincing. What seems plausible is that the postulated nonlinear effect thought to produce the relative O_2 yield deficit is able to relax in a time of the order of 10 μs. In effect, assuming that during a short, high-intensity pulse a fraction α of centers are missed (nonlinear miss coefficient = α), then a pair of such pulses spaced by Δt will display a compound miss coefficient equal to α^2, provided the condition prevailing before the first pulse has had the opportunity to be fully reestablished during Δt, i.e., before firing the second pulse. As Fig. 2 shows, α is $\sim$ 0.01 - 0.04, which is indeed negligible. If true, the above interpretation introduces the important concept of relaxation of the miss condition; it is seen that this relaxation is very fast in the case of nonlinear misses.

Relaxation of Misses Responsible for Damping in O_2 Yield Sequences

The next question then is whether the above demonstrated nonlinear miss explains the damping of O_2 yield induced by a conventional sequence of electronic flashes. As noted above, the shortest laser pulses best displaying the nonlinear property were not saturating; therefore an attempt to solve the problem directly - comparing sequences with single and paired

**More exactly, if one wants to ascribe the nonlinear effect to double hits, this must be a special kind of fast System II turnover which has no correspondence with the known limiting step of System II relaxation (see also the latter part of this section).*

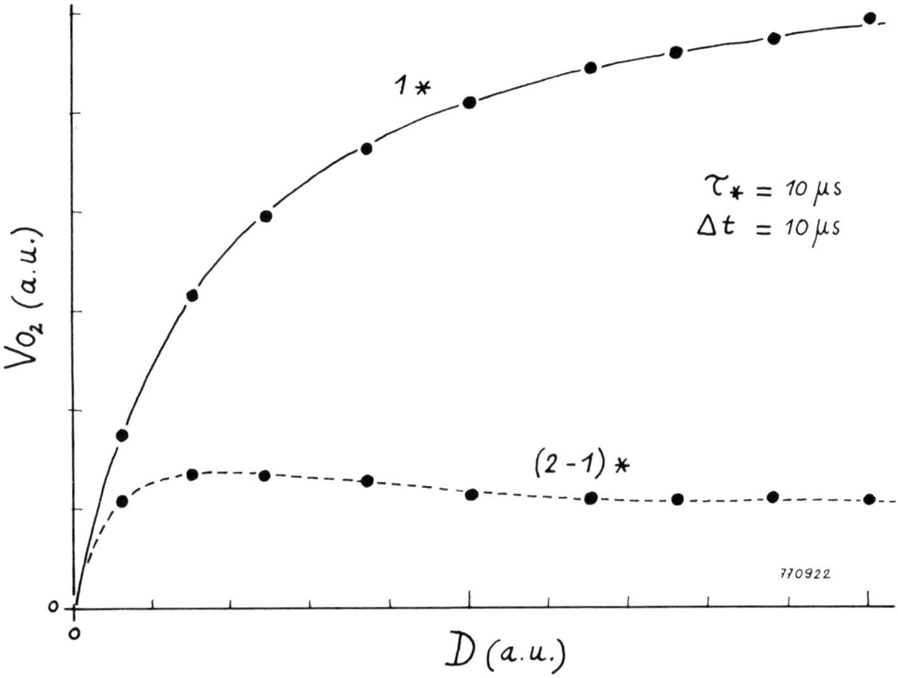

Fig. 4. Light saturation curve of O_2 emission induced by laser pulses of τ = 10 μs (1*, continuous line). The dashed line ((2 - 1)*) is the curve for the additional O_2 emission induced by a second identical pulse following the first one after Δt = 10 μs.*

pulses, similarly to Fig. 3b - was out of question. Besides, the likelihood of the explanation was *a priori* uncertain, for several reasons. Although the α value estimated from Fig. 2 was of the right order of magnitude, extrapolating short laser pulses to electronic flashes might not be warranted. The latter, due to their long "tail," could possibly be rather similar to the long laser pulses which, as we have seen, exhibit lesser nonlinear effect than the short ones. Weiss *et al.* (1971) compared O_2 sequences using electronic flashes or 10 ns nitrogen laser pulses; the damping characteristics was apparently not dependent upon flash duration. With all the above reservations, it was still of interest to investigate on the relaxation time of the actual misses operating in a conventional electronic flash sequence. The result might shed light indirectly on the miss mechanism.

Figure 5 shows Y_n (O_2 yield) sequences performed with

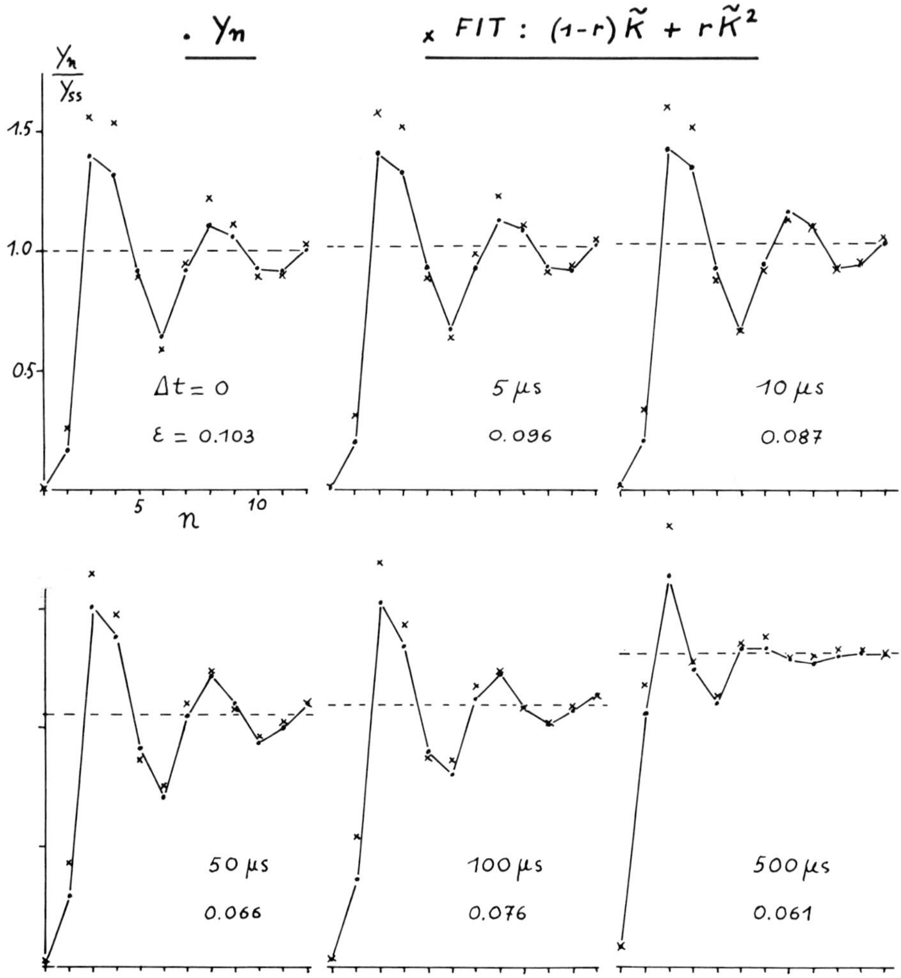

Fig. 5. O_2 yield (Y_n) in sequences of saturating electronic flashes in pairs separated by Δt; spacing of pairs is 0.3 s. Y_n values are normalized to the steady-state emission Y_{ss} of the first sequence ($\Delta t = 0$). (X) computed Y_n values using the recurrence law shown (see text); ε is the quadratic error of the fit defined as $[(\sum_i (Y_i - Y_i')^2 / \sum_i Y_i^2]^{1/2}$; Y_i, Y_i' = measured, computed values.

pairs of electronic flashes separated by several Δt values, each pair being 0.3 sec apart from the next as usual. It is seen that, up to $\Delta t \sim 50$ μs, the second flash in the pair does not affect significantly the pattern of oscillations, and in particular its damping. If the property displayed in Fig. 2b had been carried over in the present experiment, one would have expected a large decrease of the damping for Δt larger than a few 10 μs, because of the second flash largely compensating for the nonlinear misses incurred during the first flash. Evidently, this is not so. What the second flash does is merely to permit more double hits as Δt increases. This is clearly seen by the change of Y_{ss}, the steady-state O_2 yield, versus Δt and by a parallel progressive modification of overall pattern of oscillations.*

First, it may be concluded that the nonlinear miss mechanism does not appear to take any important part in the damping of O_2 yield in sequences of saturating electronic flashes. This does not mean necessarily that it is not operating at all during such flashes: possibly, the effective duration of a saturating electronic flash (i.e. taking into account its tail) is large enough for the nonlinear effect to be rapidly compensated because of its fast relaxation and of the occurrence of "physical" double hits.+ Thus damping of Y_n sequence must have a different origin. Apparently, no fast relaxation is involved. Instead, it seems that the "intrinsic" miss condition relaxes at the same rate as that of reopening of centers. In other words, it seems that a refractory period immediately follows each flash; during this period, compensation of miss by another trial (as is the case for short laser pulses) is not possible and the end of this period seems to coincide approximately with the relaxation of centers to the photoactive state. This is borne out by comparing the experimental double flashes sequences with computed ones (crosses in Fig. 5) where the usual transition matrix $\tilde{K}$ is replaced by

$$(1 - r)\tilde{K} + r\,\tilde{K}^2 \qquad (1)$$

Thus, $\tilde{K}$ is the matrix which describes the recurrence of states in Kok's model. Its elements were computed using σ analysis,

**See the change in relative value of Y_2 as Δt increases. Also, σ analysis points to an increase of γ, the double-hit coefficient; for $\Delta t > 100$ μs, the analysis breaks down due to a prominent contribution to O_2 yield of the second flash of the pair.*

+What is meant by "physical" double hits is a succession of several exciton arrivals to the same center, even though only one of them will give rise to the quantum conversion. This should be distinguished from "photochemical" double hits which imply that a center has had time to relax to the photoactive state between two exciton trappings.

assuming that the transition parameters were independent of states. *r* is the fraction of centers which has reopened at Δt; it is computed from the variation of Y_{SS} versus Δt. Although the fit in this particular instance is not very good, at least it does not get worse as r increases (see quadratic error ε in Fig. 5). Then, assuming Eq. (1) is a valid description of the data, one notes that the second term ($r\,\tilde{K}^2$) implies that the transition probabilities (including misses) are set to their equilibrium values as soon as the center reopens. This is equivalent to the above statement of a refractory period. Thus another tentative conclusion is that, since the limiting step in the relaxation of System II centers are thought to be on their acceptor side, misses also must be somehow associated with the acceptor side. This conclusion would support an earlier hypothesis (e.g. Radmer and Kok, 1973) according to which misses arise from the fraction of centers with the acceptor Q being in the reduced state at the moment of the flash; this point will be discussed in greater detail later.

Anomalous Turnover Results

Turnover experiments with electronic flashes were performed to further investigate the problems raised by the data in Fig. 3b and 4. The results show overall biphasic relaxation kinetics (Fig. 6 insert), as already reported by Vater *et al.* (1968); the first phase has a time constant of ∿600 μs, as shown by a more resolved series of measurements, if one excepts an initial, sigmoidal part (Fig. 6, curve 1). The latter is reminiscent of the turnover experiment of Bouges-Bocquet (1973) relative to the stabilization step $S_2^* \rightarrow S_3$; in fact, in a steady-state relaxation experiment, one is also dealing with this very step. Below saturation (Fig. 6, curve 2), the sigmoidal shape is no longer seen, the initial slope is smaller, and a definite positive initial offset appears. These features seem to correspond exactly to those found in the laser pulse turnover experiment of Fig. 3b.* This would suggest that the pulses in the experiment of Fig. 3b were not saturating, which is inconsistent with the fact that doubling the energy (τ* from 10 to 20 μs) did not change the results. To solve this contradiction one could think of the cross-section inhomogeneity of

**Concerning a light saturation effect for the initial offset the case of Fig. 6, curve 2 is even more striking than that of Fig. 3b: in the former, two electronic flashes are always fired even at $\Delta t = 0$, whereas, in the latter, the point $\Delta t = 0$ really refers to a one flash experiment and therefore the light energy is double as soon as Δt is > 0.*

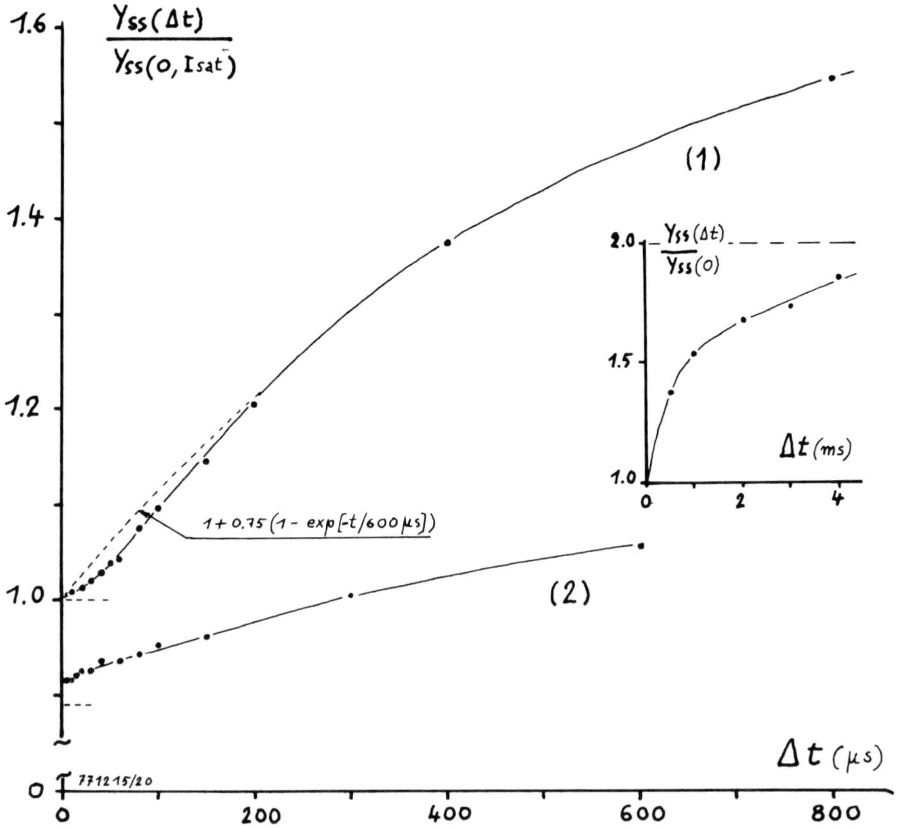

Fig. 6. Determination of System II turnover with electronic flashes in pairs separated by Δt. $Y_{SS}(\Delta t)$*: steady-state* O_2 *yield induced by the flash pair;* Y_{SS} *(0, Isat): same for saturating flash pair at* $\Delta t = 0$. *(1) saturating flashes; dashed line: turnover first-order kinetics assuming time-constant* τ = *600 μs and amplitude of first phase is 0.75 (compared to 1 for complete restoration) - insert = same on extended time scale. (2) subsaturating flashes; notice the positive initial offset (dashed horizontal line indicates the level corresponding to* $\Delta t = 0$*).*

the laser beam: the peripheral part of the illuminated sample receives less light than does the central one and it would require a higher total light power in order to reach saturation. Although this effect must occur to some extent in the present set up, it is unlikely to be the complete explanation. Probably a relevant factor is the duration of the laser pulse (10 or 20 μs; see Figs. 3b and 4). In this connection, it is interesting to refer to the experiment of Glaser *et al.* (1976) showing a biphasic light saturation curve of the P_{680} absorption change when using 20-μs excitation flashes, whereas the corresponding curve for P_{700} was monophasic. These authors considere the difference as evidence of an anomalous, fast turnover of P_{680}. It is clear, however, as discussed above, that the "normal" System II relaxation and therefore the "normal" double-hitting cannot explain the effect. One would have to think in terms of a fast turnover occurring only during light. This is difficult to understand at present.

CONCLUSIONS

Possible Significance of Nonlinear Misses

We discussed above kinetic reasons why apparently nonlinea misses did not contribute to damping in electronic flashes sequences. Another reason should be considered; it arises in dealing with the possible significance of the nonlinear effect. *A priori*, two distinct types of nonlinear loss mechanism may be conceived: antenna type or center type. In the antenna type, two excitons can annihilate each other because their lifetimes and spatial reaches mutually overlap or because one decays by giving rise to a parasitic trap of limited lifetime that will catch the other. There is ample evidence for this kind of mechanism in the field of nano- or picosecond *in vivo* chlorophyll fluorimetry (see Geacintov *et al.*, 1977). Also the carotenoid triplet quencher could be of this type, although Den Haan (1976) considers it as being associated to the center. The center-type mechanism is more difficult to imagine: for instance, one needs a short-lived state of the center after charge separation which upon absorption of another quantum would revert to the ground state, thus negating the previous photochemical conversion (Lavorel, 1978). In any case, the two types have distinct dependences upon the flash light density. Clearly, in the antenna type, it should always be possible to make up for the loss by oversaturation. That is, a large enough light input, in spite of exciton losses in the antenna, will always saturate all the centers, because their concentration is finite and small relative to that of the antenna chlorophyll. In the center type, no such compensation is possible. As already discussed, the only possible compensation stems from the fast relaxation of the miss condition. Thus the absence of nonlinear miss in electronic flash experiments could also indicate that we are dealing with an antenna-type mechanism.

For the various reasons given above, one should expect that subsaturating electronic flashes exhibit some nonlinear effect which should contribute to damping together with the intrinsic misses and the nonsaturation misses. However, verification of this point by σ analysis is not possible due to complicating side effects (Lavorel, unpublished). Compensation of nonlinear miss by fast relaxation should be possible, presumably not during the short effective duration of a single flash, but during the time elapsed between two flashes in a pair. This would then explain the positive initial offset in the case of curve 2 of Fig. 6.

Properties of Intrinsic Misses

Evidently, we are far from a complete understanding of the intrinsic miss mechanism. The present results, however, contribute new information. It is clear from the sequence experiment with paired electronic flashes (Fig. 5) that intrinsic misses cannot be compensated for or suppressed by repeated excitation. This excludes a once favored hypothesis: the *a posteriori* mechanism by recombination. Previous results (Maison and Lavorel, 1977) were also against this hypothesis. Another (tentative) conclusion of the above experiment is that misses and Q^- relax at the same rate. This together with other published evidence gives weight to the idea of an *a priori* mechanism where Q^- would be responsible for the inactive state, as formerly proposed by several authors (Lemasson and Barbieri, 1971; Radmer and Kok, 1973; Delrieu, 1974; Diner, 1977). The objection that it was difficult to imagine a randomization of the inactive state among centers in this hypothesis (Lavorel, 1976b, see also Joliot *et al.*, 1971) can be answered if one takes into account the equilibrium $Q^-B \rightleftharpoons QB^-$, where B is the secondary acceptor proposed by Bouges-Bocquet (1974) and Velthuys and Amesz (1974). The attribution of intrinsic misses to Q^- is also interesting with respect to the quantum yield problem which any type of photochemical hypothesis raises (see Introduction). The fact that centers in state Q^- are inactive does not mean that they will contribute nothing to the quantum yield under low-intensity illumination. In fact, those centers belong to closed (fluorescent) photosynthetic units which may transfer their excitation to neighboring open units (see Lavorel and Joliot, 1972). This process is probably a very efficient way for reclaiming excitation that would otherwise be lost. To summarize the argument, whereas Q^- centers are "missed" using flash excitation because they are photochemically inactive, they nevertheless contribute to the overall yield by increasing the photon cross-section of units with active centers when using low-level, continuous excitation.

A Possible Solution of the Damping-Yield Paradox

The idea of miss relaxation introduced in this study might provide a solution of the damping-yield paradox. As was pointed out in the Introduction, the well-documented fact that variations of yield and of damping are often in direct correlation seems to be an absolute contradiction in the context of the photochemical miss hypothesis. Of course, one can resolve it with additional, *ad hoc* assumptions, but there is actually a simpler solution to the problem. Let us assume a photochemical mechanism of the *a priori* type and, for generality, let us write the equilibrium:

$$\text{active} \underset{k_a}{\overset{k_i}{\rightleftharpoons}} \text{inactive} \qquad (2)$$

For simplicity we shall assume that the equilibrium rate constants are independent of the states. As already noted (e.g., Delrieu, 1974) this equilibrium specifies a miss mechanism in the sense of Kok's model provided it is fast as compared to the flashing period, T*. Only under this condition can the equilibrium fraction of inactive centers $K/(1+K)$ - with $K = k_i/k_a$ - be identified with the miss probability, α. What was called miss relaxation above, is simply the tendency to the steady-state condition of the active/inactive equilibrium after the perturbation brought in by the flash. In this connection, it is interesting to consider what would occur if the relaxation were not complete between successive flashes. Let us take the extreme case where the equilibrium is very slow as compared to T*. Then, evidently the single hit transition probability is unity for those centers which happen to be active during the whole sequence and zero for the other centers; in other words, the latter are not seen at all. In this extreme case, the sequence keeps oscillating indefinitely; damping is zero. Yet in both cases, fast or slow equilibrium, the yield is the same, e.g., referring to the steady state.*

**In the hypothetical case of zero damping, it would seem that Y_{ss} is not defined. Actually Y_{ss} is always equal to $\overline{Y}$, a computed quantity (see Lavorel and Lemasson, 1976) which, in this case, is the average of four consecutive Y_n's.*

TABLE II. Effect of the Equilibrium Active/Inactive on the Damping of Simulated Sequences[a]

Flashing period, T^* (arbitrary unit)	Extent of equilibration[b] (%)	σ_1	σ_2	σ_3
1	∼100	0.524	0.106	0.011
0.5	99.7	0.521	0.103	0.001
0.2	90.0	0.420	0.023	-0.003
0.1	68.3	0.238	-0.056	0.014
0.05	43.7	0.106	-0.056	0.031

[a]*Sequences have been computed according to the generalized model explained in the Appendix. They were analyzed as previously described (Lavorel, 1976a) to yield the damping coefficients σ_1, σ_2, and σ_3; σ_1 is the sum of miss coefficients; σ_2 and σ_3 are more complicated functions of the transition coefficients. Normally with negligible double hits, $\sigma_1 > \sigma_2 > \sigma_3$; if double hits are not negligible σ_2 may be smaller than σ_3. Initial vector state: $S^{(o)} = \{0.275, 0.675, 0.05, 0\}$ Rate constants (arbitrary units): $k_i = 1.5$, $k_a = 10$ throughout.*

[b]*$1 - \exp[-(k_i+k_a)T^*]$. This function describes the time course of the reaction toward the equilibrium.*

$$Y_{ss} = 0.25/(1+K) \tag{3}$$

In intermediate situations, where the equilibrium turnover rate is commensurate with $1/T^*$, it follows that the damping should correspond to a miss coefficient comprised between the maximum $\sigma = K/(1+K)$ and zero. This point has been verified by σ analysis on model sequences generated with variable turnover rates of the equilibrium active/inactive (see Appendix for the recurrence law). The result of an analysis is shown in Table II for a constant value of the equilibrium constant (hence a constant yield) and variable values of the individual rate constants k_i and k_a. Several interesting properties appear in this simulation: (1) except at the beginning, the sequences are almost regular,[+] i.e., consecutive trains of Y's give consistently the same values of σ_i's (not shown); (2) when the miss relaxation

[+]*It is interesting that in spite of its different structure the recurrence law of this model (see appendix) obeys the σ relation which was derived from Kok's recurrence law. Isomorphism is a vexatious characteristic of many models.*

tends to become incomplete during T* the computed double-hit coefficient γ is significantly different from zero. The first property has been often observed in experimental sequences (Lavorel and Lemasson, 1976). The second property may be qualitatively understood by noticing that, due to the cyclic structure of the S states model, there is a kinetic equilvalence between forward steps and lack of step advance during several flashing periods, which must be the case when the miss relaxation is slow. This raises the question of the real significance of computed double hits in O_2 sequences.

The above treatment generalizes the *a priori* mechanism with active/inactive equilibrium by making allowance of the rate of miss relaxation. It shows that, while the yield is determined only by the equilibrium constant, the damping is determined in a complex manner by both rate constants of the equilibrium and by the flashing period. Therefore, contrary to the asymptotic situation considered up to now, damping and yield cannot be correlated in a unique way.

In conclusion, while the hypothesis attributing misses to the equilibrium $Q^-B \rightleftharpoons QB^-$ yields satisfactory solutions to the quantum yield problem and to the damping - yield correlation problem, it is not yet established as valid. Another important difficulty remains: Diner (1977) noticed that the equilibrium constant $K = (QB^-)/(Q^-B)$ obtained from independent experiment is 15 - 20, which could account for part of the misses. Actually, in *Chlorella* if misses were totally ascribed to this equilibrium, K should be ~ 2.* Thus, either one of the above estimations of K is wrong or additional (or different) mechanisms of miss must be found.

SUMMARY

The possibility was investigated that misses, which according to the model of Kok *et al.* (1970) are responsible for the damping of O_2 emission in sequences of electronic flashes, could result from a nonlinear photochemical effect. The latter would only be associated with the high light intensity of electronic flashes, while being negligible under low intensity continuous excitation. Accordingly the O_2 emission induced in *Chlorella* by isoactinic argon laser pulses has been studied over a 100-fold intensity ranges. A deficit of 10 - 20% was consistently observed for 0.1-μs pulses of relative intensity 1

**Computed on the basis of: 50% centers in the state $Q^-B + QB^-$ and 0.17 average miss coefficient in Chlorella.*

as compared to 10-μs pulses of relative intensity 0.01. It is shown that this nonlinear effect cannot be ascribed to double hits with 10 μs pulses.

The deficit is not observed when using pairs of pulses spaced at about 10 μs instead of single pulses. This can be understood if the system, after incurring a nonlinear miss, relaxes very rapidly to the condition preexisting to the laser pulse. On the contrary, sequences of saturating electronic flashes, single or in pairs spaced a few 10 μs, exhibit no difference in damping. It is thought that the nonlinear deficit is compensated when using electronic flashes due to their effective duration and to the light density dependence of the nonlinear effect.

The above experiment also indicates that the intrinsic misses relax much slower than the nonlinear ones. In fact, it seems that the intrinsic misses relax in parallel with the reopening of System II centers. This led to a reexamination of the hypothesis that intrinsic misses are associated with the reduced state of Q, the primary acceptor of System II. Because of the possibility of exciton migration between units, this hypothesis can be reconciled with the finding of a maximum quantum yield under limiting light intensity. The anomalous direct correlation between yield and damping in O_2 sequences can also be explained if the miss relaxation rate is taken into account.

ACKNOWLEDGMENT

This work has been supported in part by a contract with the Communauté Economique Européene (Contract No. 013-76-ESF).

APPENDIX

The generalized Kok's model with active-inactive equilibrium corresponds to the scheme:

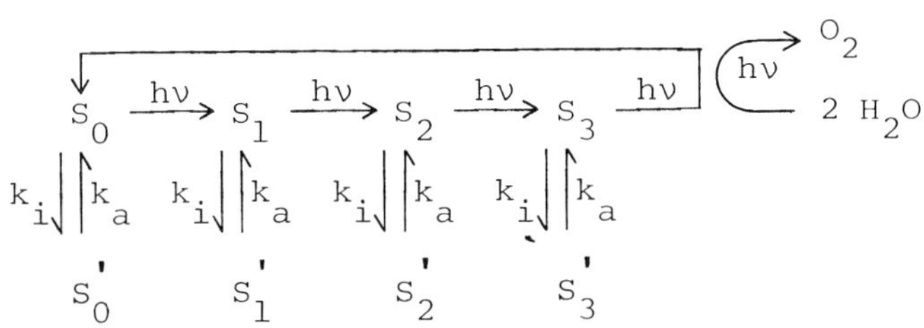

(A)

S_i and S_i' are the active and inactive states, respectively (i is modulo 4). Two recurrence relations between concentrations of states after flash n and after flash n + 1 are now needed to describe the system:

$$S_i^{(n+1)} = a\, S_{i-1}^{(n)} + b\, S_{i-1}^{'(n)}$$
$$S_i^{'(n+1)} = a' S_i^{'(n)} + b' S_i^{(n)} \qquad \text{(B)}$$

where

$$a = \frac{k_a + k_i f}{k_a + k_i} \qquad b = \frac{k_a g}{k_a + k_i}$$
$$a' = \frac{k_i + k_a f}{k_a + k_i} \qquad b' = \frac{k_i g}{k_a + k_i} \qquad \text{(C)}$$

and

$$f = \exp[-(k_a + k_i)T^*] \qquad g = 1 - f$$

In the asymptotic equilibrium limit, f = 0 and g = 1; the coefficients in Eq. (C) take on the meaning of the transition coefficients in the usual Kok's model:

$$\text{single hit:} \quad \beta = a = b = \frac{k_a}{k_a + k_i}$$
$$\text{miss:} \quad \alpha = a' = b' = \frac{k_i}{k_a + k_i} \qquad \text{(D)}$$

In this case, the two recurrence relations (Eq. B) can be lumped together into the usual relation of Kok's model (without double hits):

$$S_i^{(n+1)} = \alpha\, S_i^{(n)} + \beta\, S_{i-1}^{(n)}$$

where S_i now represents a state irrespective of its active or inactive condition.

REFERENCES

Bouges-Bocquet, B. (1973). *Biochim. Biophys. Acta 292,* 772-785.
Bouges-Bocquet, B. (1974). *Biochim. Biophys. Acta 314,* 250-256.
Delrieu, M. J. (1974). Thesis, University of Paris Sud.
Den Haan, G. A. (1976). Thesis, University of Leiden.
Diner, B. (1977). *Biochim. Biophys. Acta 460,* 247-258.
Geacintov, N. E., Breton, J., Swenberg, C. E., and Paillotin, G. (1977). *Photochem. Photobiol. 26,* 629-638.
Glaser, M., Wolff, C., and Renger, G. (1976). *Z. Naturforsch. 31c,* 712-721.
Joliot, P. and Joliot, A. (1968). *Biochim. Biophys. Acta 153,* 625-634.
Joliot, P., Barbieri, G., and Chabaud, R. (1969). *Photochem. Photobiol. 10,* 309-329.
Joliot, P., Joliot, A., Bouges, B., and Barbieri, G. (1971). *Photochem. Photobiol. 14,* 287-305.
Kok, B., Forbush, B., and McGloin, M. (1970). *Photochem. Photobiol. 11,* 457-475.
Lavorel, J. (1976a). *J. Theoret. Biol. 57,* 171-185.
Lavorel, J. (1976b). *FEBS Lett. 66,* 164-167.
Lavorel, J. (1978). *In* "Photosynthetic Oxygen Evolution" (H. Metzner, ed.), pp. 249-268. Academic Press, London.
Lavorel, J. and Joliot, P. (1972). *Biophys. J. 12,* 815-831.
Lavorel, J. and Lemasson, C. (1976). *Biochim. Biophys. Acta 430,* 501-516.
Lemasson, C. and Barbieri, G. (1971). *In* "Second International Congress on Photosynthesis" (G. Forti, M. Avron, and A. Melandri, eds.), pp. 107-114. W. Junk N.V. The Hague.
Maison, B. and Lavorel, J. (1977). *In* "Photosynthetic Organelles, Structure and Function" (S. Miyachi, S. Katoh, Y. Fujita, and K. Shibata, eds.), pp. 55-65. Japanese Society of Plant Physiologists, Tokyo.
Mauzerall, D. (1976). *Biophys. J. 16,* 87-91.
Radmer, R. and Kok, B. (1973). *Biochim. Biophys. Acta 314,* 28-41.
Radmer, R. and Cheniae, G. (1977). *In* "Topics in Photosynthesis," Vol. 2: Primary Processes of Photosynthesis" (J. Barber, ed.), Chapt. 8, pp. 303-348. Elsevier, Amsterdam.
Vater, J., Renger, G., Stiehl, H. H., and Witt, H. T. (1968). *Naturwissenschaften 55,* 220-221.
Velthuys, B. R. and Amesz, J. (1974). *Biochim. Biophys. Acta 253,* 298-301.
Weiss, C., Solnit, K. T., and Von Gutfeld, R. J. (1971). *Biochim. Biophys. Acta 253,* 298-301.

PHOTOPHOSPHORYLATION AFTER SINGLE TURNOVER LIGHT FLASHES IN CHROMATOPHORES FROM *Rhodospirillum rubrum*

Margareta Baltscheffsky
Arne Lundin

Department of Biochemistry
Arrhenius Laboratory
University of Stockholm
S-106 91 Stockholm, Sweden

INTRODUCTION

Electron transport-coupled phosphorylation in biological membranes has been an intensely studied field for many decades. An increasing number of these studies have been performed on chromatophores from photosynthetic bacteria, no doubt challenged by the statement by Martin Kamen 15 years ago that the photosynthetic bacterium *Rhodospirillum rubrum* "is the nearest thing to a schmoo we have in biology. It has an infinite capacity for surprises: and it can outsmart any investigator" (1).

The chemiosmotic hypothesis, proposed by Mitchell (2), is now widely recognized as a plausible explanation of the events leading to the synthesis of ATP in membranes. The protonmotive force generated in the vectorial electron transport system gives rise to chemical energy in the form of ATP via proton translocation through the coupling factor ATP synthetase complex. The protonmotive force consists of both an electrical potential and a proton gradient component, which are inter-

 ISBN 0-12-397580-8

changeable and used in different proportions in different systems. The same mechanism appears to operate both in photosynthetic systems, where light energy is captured and converted to chemical energy in ATP, and in respiratory systems, where energy is conserved when substrates are oxidized by molecular oxygen. One of the several remaining questions is whether the charge separation and the proton redistribution are transmembraneous or more localized to their nature (3).

Why are photosynthetic membranes advantageous objects for the study of electron transport-coupled energy conversion reactions? One obvious reason is that the energy source - light - may be delivered in precise and predetermined packages, i.e., light flashes, which, if the duration is short enough, may elicit only one turnover of each electron transport chain. This has been widely applied to the study of electron transport reactions in photosynthesis, and in some instances also to the study of phosphorylation reactions (4 - 8). In the latter case, however, the detection of low concentrations of ATP has caused some technical limitations of the flash technique.

In chromatophores from photosynthetic bacteria, studies of ATP formation with flashing light were pioneered by Nishimura and Geller in the early 1960's (4, 5) and, recently, a detailed study was performed by Jackson, Saphon, and Witt (6 - 8). In both these studies the accurate measurement of ATP after very short, single flashes was not possible, and the possible contribution from other reactions, obscuring the final results, could not be excluded.

In this paper, we describe some of the recent results we have obtained with an improved method, which makes it possible to measure accurately the ATP production after each separarate single turnover flash.

RESULTS AND DISCUSSION

A well known, extremely sensitive technique, for measuring low concentrations of ATP is the firefly luciferin-luciferase system. A newly developed version of this system (9) has greatly facilitated its use. Extensive purification of the luciferase together with carefully worked out conditions to minimize the decay of luminescence due to product inhibition has resulted in a preparation which has a comparatively stable luminescence (typically 5% decay in 15 min) and a linear dependence of luminescence on the ATP concentration from 10 p*M* up to 1 μ*M* ATP. It turned out that one could simply mix this luciferase system with the chromatophore assay for photophosphorylation, and directly measure the ATP production after even short light flashes (10).

In our first series of experiments we used a flash unit giving flashes of 1-ms duration, resulting in approximately two turnovers of the electron transport chain in *R. rubrum* chromatophores. Figure 1 (10) shows the ATP synthesis following a 1-ms flash. The amount of ATP formed in the flash in this case was equivalent to approximately 1 ATP/200 Bchl. The rise time of the luminescence in Fig. 1, $t_{1/2}$ about 200 ms, is the rise of the luciferase luminescence, as reported in the literature (11). This is a limiting condition with this meas-

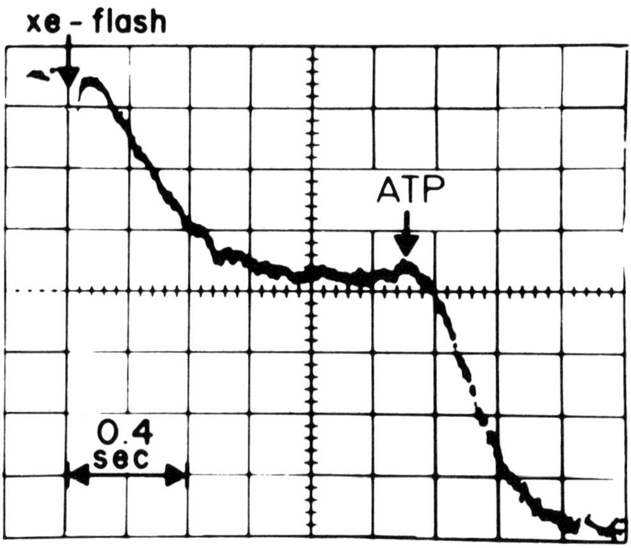

Fig. 1. Photophosphorylation after a single 1-ms light flash. The first increase (downward) of luminescence represents the ATP formed after the flash and the second increase the addition of 2.4 × 10^{-7}M ATP. The reaction mixture contained in a final volume of 1.5 ml: 0.1 M glycylglycine buffer, pH 7.75, 10 mM $Mg(Ac)_2$, 2 mM P_i, 0.1 mM Na-succinate, 0.02 mM ADP, 0.14 mM luciferin, 0.1% bovine serum albumin, about 6 µg luciferase, and chromatophores equivalent to 39 µM Bchl.

uring technique, and telling us only that the ATP formation proceeds faster than what is recorded. Based on our assumption (12) that all ATPase molecules contain at least 1 mole tightly bound ATP per mole ATPase, the value of approximately 1 ATP/200 Bchl after the 1-ms flash indicates that almost all ATPase have turned over once (12), since the bound, acid-extractable ATP also corresponds approximately to 1 ATP/200 Bchl (10). The ATP released into the medium after the flash is newly formed ATP, and the amount of ATP released in the absence of added ADP is much lower, only 2% of that in the presence of ADP.

A shorter flash with a duration of only 5 μs gives rise to only one turnover of the cyclic electron transport system (Fig. 2). The ATP concentration after the first flash is 2.75 n*M*, corresponding to 1 ATP/3650 Bchl. After two turnovers, i.e., two flashes, the concentration is 7.5 n*M*. After three flashes it is 13 n*M*. In other words, the first flash yields about 50% as compared with the consecutive flashes. Table I shows that this proportion between the flash yields is very consistent. In different preparations with different

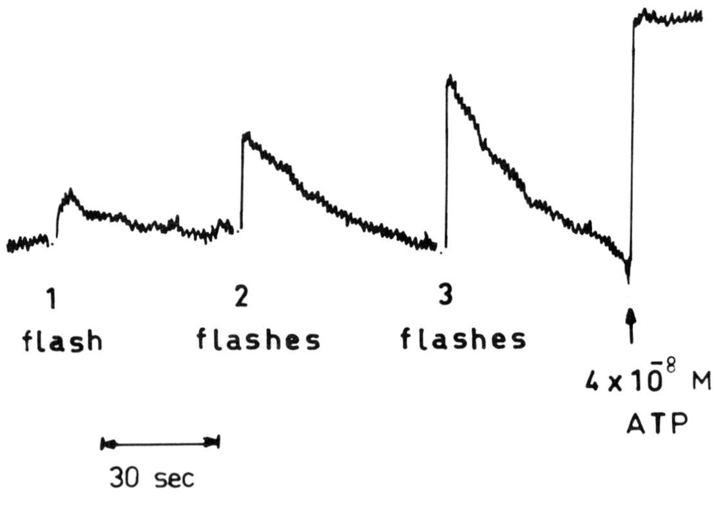

Fig. 2. Photophosphorylation after single and multiple 5 μsec flashes. Luminescence increase is upward in the figure. The reaction mixture in a final volume of 2 ml was the same as that in Fig. 1 with the exception of Bchl which was 10 μM.

Table I. Yield of ATP after each single turnover flash. Reaction conditions as in Fig. 2.

Flash no.	*ATP [nM]*		
	Expt I	*Expt II*	*Expt III*
1	2.6	2.5	5.6
2	5.0	4.8	10.3
3	5.1	5.3	10.3
4	5.1	5.4	12.7

phosphorylating capability the first flash always appears to yield about 50% of the ATP obtained after the second and following flashes. One of the possible explanations for this may be sought in the alternating site hypothesis for the reaction mechanism of the ATPase enzyme proposed by Boyer (13).

Uncouplers, such as FCCP (fluorocarbonyl cyanide phenylhydrazone), totally inhibit the flash-induced photophosphorylation in the concentration range which inhibits phosphorylation under continuous light (Fig. 3).

It has earlier been reported that valinomycin, which has only a limited effect on photophosphorylation under steady state conditions, completely inhibits the flash-induced phosphorylation (7). We have confirmed this. Nigericin, on the other hand, has no effect on flash-induced phosphorylation, which is not surprising since it is very unlikely that a single turnover would yield any sizeable proton gradient. The internal buffering capacity of chromatophores is considerable, and Melandri and Baccarini-Melandri have recently shown that as much as 10 - 15 turnovers are required in order to build up a proton gradient competent to drive ATP synthesis (15).

The above data strongly support the concept that a membrane potential is the main or even the only driving force for this flash-induced ATP synthesis. It has been shown by Leiser and Gromet-Elhanan (14) that the membrane potential gives an important contribution to phosphorylation in *R. rubrum* chromatophores. They subjected chromatophores to an acid-base jump, which established a ΔpH as large as 3 pH units, and could show that in order to obtain any appreciable ATP synthesis in the dark, they had to complement the ΔpH with a diffusion potential.

The earlier studies on flash-induced phosphorylation (6 - 8) were done with chromatophores from *Rhodopseudomonas sphaeroides*. In these chromatophores it was reported that antimycin causes only a marginal inhibition of this phosphorylation. We reinvestigated the question in the *R. rubrum* system and found that in our hands the inhibition by antimycin even in

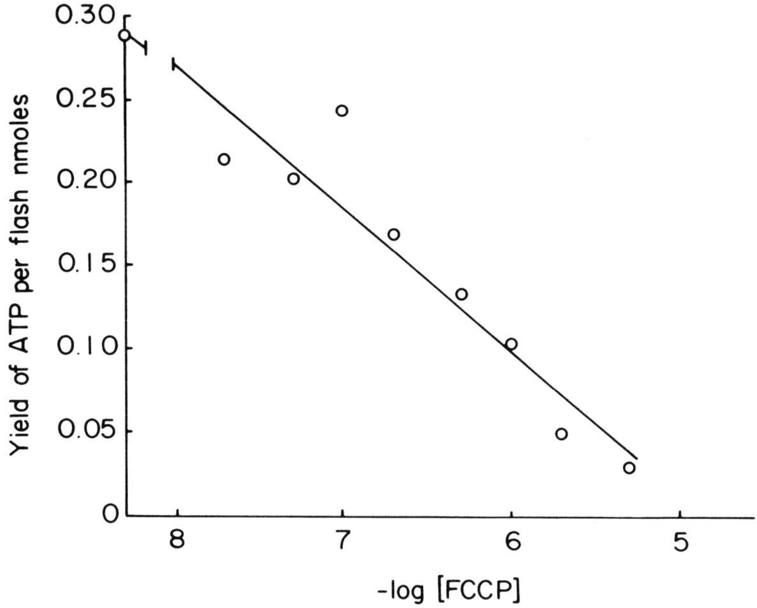

Fig. 3. Inhibition of flash-induced photophosphorylation by FCCP. Reaction conditions as in Fig. 1.

the first turnover is sizeable. Table II shows that the inhibition is 60% after the first flash and, as could be expected increases as the number of flashes in each flash group increases. These data seem to confirm earlier proposals (16) that there exist two coupling sites for ATP formation in the photosynthetic electron transport. One of these, between the cytochromes *b* and c_2, has been identified for quite some time (17); the other one, could be situated between the Fe-quinone acceptor and cytochrome *b*. Both sites would be involved in the flash-driven photophosphorylation in *R. rubrum*, and give rise to a membrane potential without the contribution of a transmembrane proton gradient since the inhibition in the presence of valinomycin and K^+ ion is total. This, in turn, points to the existence of a transmembrane charge separation resulting from the transport of one electron also between cytochromes *b* and c_2.

TABLE II. Inhibition by antimycin on phosphorylation after first and consecutive flashes.[a]

No. of flashes in flash group	Yield of ATP [nM] -antimycin	Yield of ATP [nM] +2 μM antimycin	Inhibition (%)
1	2.5	1.0	60
2	7.3	1.6	78
3	12.6	2.0	84
4	18.0	2.3	87

[a]*Reaction conditions as in Fig. 2. 200 ms between 5μs flashes in flash groups.*

The energetics of the flash-induced phosphorylation in *R. rubrum* still remains to be elucidated. The magnitude of the membrane potential in *R. rubrum* chromatophores is difficult to estimate, although values in the range 89 - 250 mV under continuous illumination have been reported (18 - 20). The measuring technique, therefore, appears to be crucial. The magnitude when measured with a potential sensitive probe after a single turnover light flash is about 20% (21) of that under continuous illumination or maximally 50 mV. Energetically this seems insufficient for ATP formation unless we assume that the chromatophore population is not homogeneously activated by the flash, and that certain chromatophores have a sufficient, higher potential, whereas others have practically none.

In summary, it is encouraging that through the combination of the stable luciferase technique with the single turnover light flash technique new information along the parameters reported has been possible to obtain. Some essential questions are still unresolved, as those discussed above, but future work will hopefully cope with these problems. Thus one may look forward to an increasingly clear elucidation of the true nature of the mechanism for light-induced phosphorylation.

REFERENCES

1. Kamen, M. D. (1963). *In* "Bacterial Photosynthesis" (R. K. Clayton, ed.), p. 445. Academic Press, New York.
2. Mitchell, P. (1968). "Chemiosmotic Coupling and Energy Transduction." Glynn Res. Ltd., Bodmin.

3. Williams, R. J. P. (1978). *Biochim. Biophys. Acta 505,* 1-44.
4. Nishimura, M. (1962). *Biochim. Biophys. Acta 57,* 88-95.
5. Geller, D. M. (1967). *J. Biol. Chem. 242,* 40-55.
6. Saphon, S., Jackson, J. B., Lerbs, V., and Witt, H. T. (1975). *Biochim. Biophys. Acta 408,* 58-66.
7. Saphon, S., Jackson, J. B., and Witt, H. T. (1975). *Biochim. Biophys. Acta 408,* 67-82.
8. Jackson, J. B., Saphon, S., and Witt, H. T. (1975). *Biochim. Biophys. Acta 408,* 83-92.
9. Lundin, A., Richardsson, A., and Thore, A. (1976). *Anal. Biochem. 75,* 611-620.
10. Lundin, A., Thore, A., and Baltscheffsky, M. (1977). *FEBS Lett. 79,* 73-76.
11. De Luca, M. and McElroy, W. D. (1974). *Biochemistry 13,* 921-925.
12. Harris, D. A. and Baltscheffsky, M. (1979). *Biochem. Biophys. Res. Commun. 86,* 1248-1255.
13. Boyer, P. (1977). *Trends Biochem. Sci. 2,* 38-41.
14. Leiser, M. and Gromet-Elhanan, Z. (1974). *FEBS Lett. 43,* 267-270.
15. Melandri, B. A. and Baccarini-Melandri, A. (1979). *In* "Cation flux across biomembranes" (L. Packer and Y. Mukohata, eds.), pp. 209-218. Academic Press, New York.
16. Baltscheffsky, H. and Arvidsson, B. (1962). *Biochim. Biophys. Acta 65,* 425-428.
17. Baltscheffsky, M. (1967). *Biochem. Biophys. Res. Commun. 28,* 270-276.
18. Schuldiner, S., Padan, E., Rottenberg, H., Gromet-Elhanan, Z., and Avron, M. (1974). *FEBS Lett. 49,* 174-177.
19. Kell, D. B., Ferguson, S. J., and John, P. (1978). *Biochim. Biophys. Acta 502,* 111-126.
20. Bashford, C. L., Baltscheffsky, M., and Prince, R. C. (1979). *FEBS Lett. 97,* 55-60.
21. Chance, B. and Baltscheffsky, M. (1975). *In* "Biomembranes" (H. Eisenberg, E. Katchalski-Kazir, and L. A. Manson, eds.), Vol. 7, pp. 33/55. Plenum, New York.

COMPOSITION AND DEVELOPMENT OF THE BACTERIAL PHOTOSYNTHETIC APPARATUS

Gerhart Drews

Institut für Biologie II
Albert-Ludwigs-Universität
78 Freiburg i. Br., Federal Republic of Germany

Photosynthetic bacteria are unique objects to study photosynthesis, cell differentiation, metabolic reactions in the broad field between photolithotrophy and chemoorganotrophy, and ecology. The exciting results from many of these investigations have stimulated interest in these organisms and created interdisciplinary studies. The attractiveness of this group of microorganisms is manifested by special meetings on photosynthetic prokaryotes held regularly in Europe and the United States.

Martin Kamen, in his inconspicuous but efficient manner, has influenced this process of mutual stimulation and cooperation.

I became acquainted with the photosynthetic prokaryotes in the laboratory of Buder, where I learned to enrich for *Chromatium* in a Winogradsky column and to study phototaxis of cyanobacteria. Later, in Jena, I discovered, in cooperation with Werner Niklowitz, by electron microscopy the lamellar structures of chromatophores in cells of *Rhodospirillum molischianum* (16) and in cells of cyanobacteria (17). I became

 ISBN 0-12-397580-8

aroused by these and other observations to study the problem of cell and membrane differentiation in photosynthetic bacteria.

By this time Martin Kamen had published a series of important papers on the hydrogen, nitrogen, and carbon metabolism of photosynthetic bacteria. Although I followed these studies with great interest, I did not meet Martin personally until 1967, when we had a long talk in a restaurant close to Place de Luxembourg in Paris with regard to many problems of electron transport, bacteriochlorophyll mutants, and membrane development. Since this time our paths have intersected pleasantly. Numerous discussions by word or letter and experimental cooperation in Freiburg have stimulated my scientifi work. Our common love for music and interest in the many problems of humanism have broadened the basis of our mutual interest and friendship.

It is my purpose here to discuss our recent results on composition and biosynthesis of the intracytoplasmic membranes of *Rhodopseudomonas capsulata* and to propose a model of the photosynthetic apparatus of this organism.

The bacterial photosynthetic apparatus is an integral membrane constituent. All light-dependent processes, beginning with the absorption of light quanta, the generation of an electric field and proton gradient across the membrane, and the resulting production of ATP, occur within this compartment. The prerequisite of photosynthetic energy conservation is the presence of bacteriochlorophyll (Bchl) *a* or *b* in a suitable molecular arrangement in the membrane. Photosynthetic apparatuses contain at least two spectral forms of Bchl: the antenna and the phototrap (reaction center) molecules. Most of the photons are absorbed by antenna Bchl which is the dominating spectral form (20 - 2000 molecules per reaction center). The transfer of excitation energy from antenna Bchl to reaction centers (RC) is a very efficient process. Trapping of energy occurs in the reaction centers when the electronic excitation has been converted to separate oxidized and reduced species. When the RC are "closed" most of the energy is dissipated as fluorescence (25). The vectorial processes, resulting in membrane energetization, presuppose an asymmetric arrangement of light-harvesting antenna pigment complexes, reaction centers, electron transport particles, and coupling factors within or on the membrane and a spatial cooperation between these subparticles.

Linear and circular dichroism and polarization fluorescence measurements suggest a regular orientation and close interactions of antenna Bchl molecules in the membrane (2, 3, 12, 14, 26, 27, 31). The orientation could be due to an association of the esterified phytol tail of Bchl with the fatty acid chains of phospholipids in the lipid bilayer of membranes. But the isolation of pigment - protein complexes from membranes which retain the *in vivo* spectrum and reactivity, indicates that pigment - protein interactions are important for the orientation and functional speciality of Bchl molecules in the membrane (8, 27). In reaction centers, Bchl - Bchl interactions also contribute to the shape of the *in vivo* spectrum and reactivity of Bchl molecules (8, 18, 23, 27, 31).

Rhodopseudomonas (R.) *capsulata, R. palustris,* and *R. sphaeroides* have photosynthetic units of similar composition and supramolecular structure of varying size (6, 9, 10, 29). The photosynthetic apparatus of these bacteria is composed of multiple Bchl - protein complexes.

THE PIGMENT - PROTEIN COMPLEXES OF *Rhodopseudomonas capsulata*

Reaction Center

The literature on bacterial reaction centers has been recently reviewed (4, 6, 7, 25, 30). The reaction center (RC) of *R. capsulata* is very similar to that of *R. sphaeroides*. The RC preparation isolated from the carotenoidless strain Ala^+ (15) contains polypeptides with apparent mol. wt. 28,000, 24,000, and 21,000. Each RC contains 4 molecules Bchl and 2 molecules bacteriopheophytin. Bchl is bound to the M(24,000) and L(21,000) polypeptides.

Antenna Bacteriochlorophyll - Protein Complexes

Rhodopseudomonas capsulata contains two light-harvesting (LH)-Bchl complexes (9). Both Bchl-protein complexes LH I and II have been isolated and purified. LH I has been isolated from the mutant Ala^+ lacking carotenoids and LH II from the mutant Y5 lacking RC and LH I(9). LH I of *R. capsulata* has a single major IR absorption peak at approximately 875 nm and consists of two polypeptides with a mol. wt. of about 12,000 and 6000. The molar ratio Bchl:protein in the LH I-Bchl-protein complex was found to be approximately one. LH I-Bchl is synthesized in a constant ratio relative to RC (28).

LH-Bchl II has two absorption maxima at 802 and 855 nm. The conclusion that the two Bchl species are associated with the same protein is based on the observation that the relative absorptivities are fairly constant. Membranes and the LH-Bchl II complex isolated from the Y5 mutant have the typical

IR spectrum (9). However, it will be shown later that B 802 and B 855 seem to be associated with different polypeptides. The isolated LH II Bchl-protein complex contains three polypeptides (mol. wt. approximately 14,000, 10,000, and 8,000). The molar ratio Bchl per carotenoids (3:) is not changed significantly during purification. The phospholipid content of the purified complex is low (approx. 1 µg lipid P/1 mg protein). The smallest units of the isolated complex, retaining the native absorption spectrum, separated by SDS-polyacrylamide gel electrophoresis, has a mol. wt. of about 95,000. This complex contains only the polypeptide 10 and 8k dalton polypeptide. The molar ratio Bchl:protein (8 + 10 k dalton proteins) was found to be approximately four (9). Quadruplets of the two lower molecular weight polypeptides, associated with 16 mol Bchl, aggregate *in vitro* to form a 95,000 complex (9). The MW of the whole complex was found to be approximately 170,000 (unpublished).

Tryptic digestion of the 14,000 k dalton polypeptide in membrane preparations does not modify the spectral features of the complex (Fig. 1). The degradation of the two smaller polypeptides in the isolated complex LH II results in a shift of the absorption maxima from 802 and 855 nm to 770 and 680 nm. Treatment of membranes with high trypsin concentrations (1 mg/mg protein) first splits the 14 k dalton protein, later the 8 k dalton protein, and, finally, the 10 k dalton protein. The loss of the 802 nm absorption peak occurs concomitantly with that of the 8 k dalton polypeptide, while the loss of the 855 nm peak parallels that of the 10 k dalton polypeptide (Feick and Drews, unpublished data) (Fig. 2). From these data we suggest that the LH-Bchl II complex of *R. capsulata* contains two Bchl species: Bchl B 802, which is associated with the 8 k dalton polypeptide, and Bchl B 855, which is associated with the 10,000 dalton polypeptide. This hypothesis is supported by data from CD and absorption spectra of Bchl LH II of *R. capsulata* (33 - 35). The appearance of B 802 + B 855 is synchronous with the biosynthesis of the 8 and 10 k dalton polypeptides. The third polypeptide of the LH II complex (14 k dalton) is not associated with Bchl and in photosynthetically induced cells is synthesized after the two smaller polypeptides (28). The function of this polypeptide is unknown. However, it is of interest to note that the 14,000 dalton polypeptide is not present in cells of *R. palustris*.

Electron Microscopic Studies

Cells of *R. capsulata,* cultivated under different oxygen partial pressures or light intensities, were studied by electro

(A)

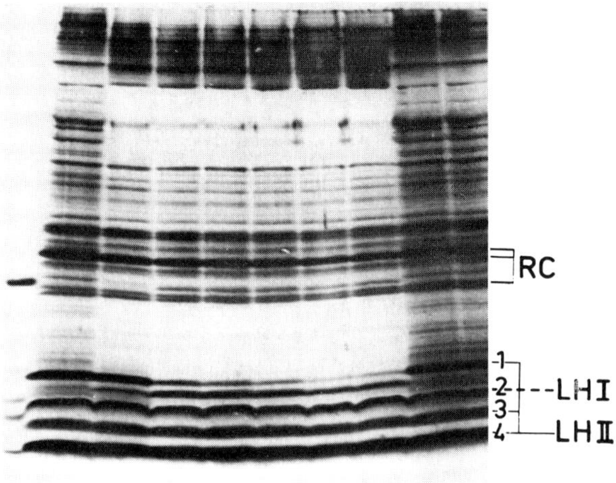

(B)

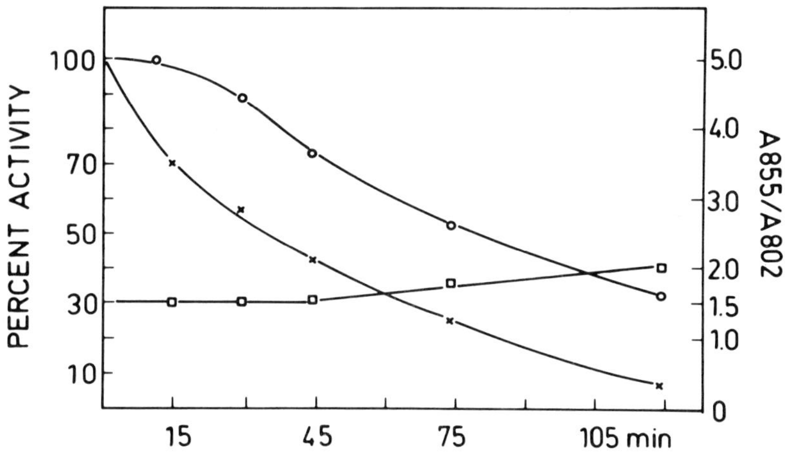

Fig. 1. (A) Polyacrylamide gel electrophoresis (12a) of membranes of R. capsulata 37b4 (wild type) after treatment with trypsin (60 µg trypsin per mg protein). 12 to 16% acrylamide gradient. RC, reaction center polypeptides. LH 1, 2, 3, 4 = polypeptides of light-harvesting (LH) Bchl protein complexes (LH I = band 2, LH II = bands 1, 3, 4). Apparent mol. wt. 1 = 14,000, 2 = 12,000, 3 = 10,000, 4 = 8,000. The samples (after stopping reaction with trypsin inhibitor, washing of membranes and denaturation by heating in SDS buffer) run from the top to the bottom. From left to right: trypsin; untreated control at 0°C; 0, 10, 20, 40, 90, 150 min of trypsin treatment at 26°C; untreated control (26°C). The 14,000 polypeptide is splitted. (B) Enzyme activity and absorption spectrum of mem-

microscopy. By freeze-fracturing, membranes were split within the hydrophobic inner regions. The exposed plasma face (PF) of the fracture faces of cytoplasmic and intracytoplasmic membranes exhibited particles which have a diameter of about 9.5 nm in membranes of *R. capsulata*. The density of particles in the cytoplasmic membrane was found to be increased when the cells were subjected to a lowering of the oxygen partial pressure in dark cultures from 400 to 0.5 mm Hg (pO_2). The number of particles per μm^2 in the P face of the cytoplasmic membrane was found to be 2688 at 9 hr, 3116 at 2 hr, 3688 at 3.5 hr, and 4448 at 5 hr incubation at 0.5 mm Hg (pO_2). By comparison, in cells grown anaerobically under high light intensity, the P face of the cytoplasmic membrane contains 3567 particles per μm^2. After 3.5 hr of incubation at 0.5 mm Hg (pO_2) the total Bchl concentration in the membrane fraction increased from approximately 0.1 to 8.7 nmol and the concentration of RC Bchl from approximately 9 to 260 pmol (28, 36, 37).

The density of particles in the P face of intracytoplasmic membranes was generally higher than in the cytoplasmic membrane of the same culture. In comparison, 5170 (intracytoplasmic membrane) and 3728 (cytoplasmic membrane) particles per μm^2 have been counted in cells grown anaerobically under high light intensities. Membranes of aerobically (400 mm Hg, pO_2) grown cells contained 3785 (intracytoplasmic membrane) and 2810 (cytoplasmic membrane) particles per μm^2 (unpublished results). Although no distinct classes of particles were discernible and the particles with a diameter of 9.5 nm dominate under all culture conditions, it is obvious that particles with a diameter less than 9.5 nm slightly increase relative to particles with a diameter higher than 9.5 nm in cells actively synthesizing photosynthetic apparatus (unpublished).

The clear correlation between increase of particle density and formation of the photosynthetic apparatus tempted us to assume that a group of these particles represents subunits of the

(Fig. 1 cont'd.) branes of R. capsulata Y 5 (pho^-, LH-Bchl II^+) after treatment with trypsin (8 μg trypsin per mg protein). □, Ratio absorbance 855/802 nm; ×, succinate oxidase; ○, cytochrome c oxidase (EC 1.9.3.1). Ordinate (left): percentage of enzymatic activity of zero-time control. Ordinate (right): absorbance ratio 855/802 nm. Abscisse: time of trypsin incubation at 26°C. During short periods of incubation with low concentrations of trypsin the in vivo absorption spectrum and the ratio 855/802 nm are not changed, but succinate oxidase activity decreased quickly, followed by a decrease of activity of cytochrome c oxidase. The 14 k dalton protein is digested.

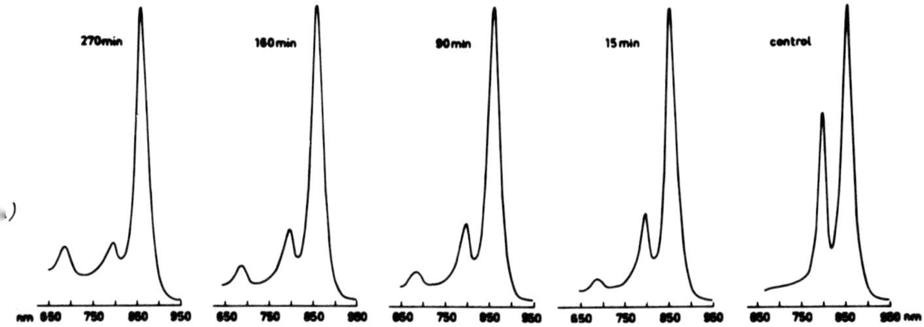

(B)

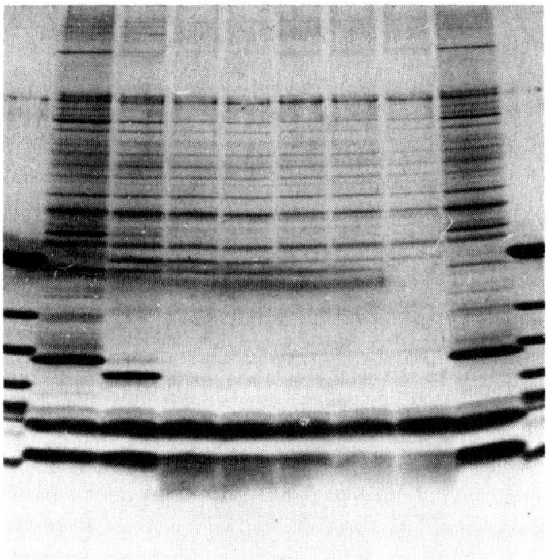

Fig. 2 . (A) Influence of trypsin (1 mg per mg protein) on the IR absorption spectrum of membranes of R. capsulata Y 5. The disappearance of the 802 nm absorption peak seems to occur concomitantly with the lost of the 8 k dalton polypeptide, while the 855 nm absorption maximum is unchanged. The disappearance of the 855 nm absorption maximum (not shown here) occurs during digestion of the 10 k dalton polypeptide with chymotrypsin (1 mg/mg protein). (B) Influence of trypsin (1 mg per mg protein) on polypeptide composition of R. capsulata Y 5 membrane. Polyacrylamide gel electrophoresis (12 - 16% acrylamide). From left to right: trypsin; untreated control 0°C, 0, 10, 20, 40, 90, 150 min of trypsin treatment at 26°C; untreated control (26°C). The 14,000 polypeptide is quickly digested, while the 8,000 polypeptide is more slowly digested.

photosynthetic apparatus. To test this theory a model of the photosynthetic unit was developed based on the analytical data of the pigment - protein complexes of *R. capsulata* and compared with the electron microscopic data.

The Organization of the Photosynthetic Apparatus

The RC particles of *R. rubrum* and *R. sphaeroides* consist of hydrophobic proteins and belong to the class of intrinsic membrane proteins (7). The heavy (H) polypeptide, however, is accessible on the cytoplasmic side of the membrane for iodination and for reaction with antibodies against RC (7, 19, 24, 32, and unpublished results). It was reported that cytochrome c_2 binds to RC (22) and antibodies against the M and L (medium and light) polypeptides react with RC on the external membrane surface (7). From these observations it was concluded that RC particles span across the membrane. It is unknown whether a single polypeptide of RC penetrates the membrane or the whole complex (total mol. wt. approx. 150,000) spans across the membrane.

The proteins which are associated with Bchl to form the LH I complex (B 875) seem to be localized in the hydrophobic interior of the membrane. They are not accessible to proteases and are solubilized by detergent treatment only. The light energy, absorbed by B 802 + 855 (LH II) Bchl, seems to be transferred by excitation migration via LH I (B 875) Bchl to th phototrap (13, 25, 34). The LH I Bchl is synthesized concomitantly with and in a constant ratio to RC - Bchl (28). From these data it is reasonable to assume that approx. 20 - 25 LH I Bchl - protein complexes are associated with RC to form a particle of approximately 500,000 mol. wt. The structural and functional mutual dependence of RC and LH I complexes is furthe substantiated by the observation that mutants having RC but no LH I have never been isolated and that in mutants lacking RC the LH complex is very labile (38).

The LH II (B 802 + 855) - Bchl complex is the variable com ponent of the photosynthetic unit (1, 6, 28, 39). The ratio of LH II Bchl per RC Bchl determines the size of the photosynthetic unit. The unit size is dependent on light intensity, oxygen partial pressure, the phase of differentiation or the concentration of substrates or metabolites (6). The low-molecular weight (8 + 10 k dalton) polypeptides of the LH II complex have a high content of apolar amino acids (65%), are insoluble in aqueous buffer solutions, chaotropic agents, or mild detergents (e.g., cholate), but are soluble in chloroform-methanol (40). In membrane preparations these proteins are digested only by high concentrations of trypsin and much later than external proteins as succinate dehydrogenase (unpublished). It is concluded that the 8 + 10 k dalton polypeptides, associated with Bchl, are localized in the hydrophobic interior of the membrane in contact to the RC-LH I particles.

The third polypeptide of the LH II complex (14 k dalton) has a higher polarity than the smaller polypeptides of the complex and is rapidly digested comparable to external proteins. It is assumed that this polypeptide reaches the plasma surface of the membrane (Fig. 3). It is synthesized with a clear lag phase to the 8 + 10 k dalton polypeptides (28). The orientation and interaction of antenna Bchl molecules with each other and with proteins in the membrane tempts one to assume that Bchl molecules are associated at specific sites with the proteins of the complex. The proteins aggregate possibly with specific attachment sites to each other. The lipid bilayer might influence the orientation of these integral proteins.

The present results on composition and localization of Bchl - protein complexes in the membrane of *R. capsulata* are summarized in a model (41, 42).

Intracytoplasmic membranes of photosynthetic bacteria grown under low light conditions have a high protein content. Approximately 70% of these proteins participate in formation of the Bchl - carotenoid - protein complexes of the photosynthetic apparatus. The LH II-Bchl complexes presumably fill the gaps between the particles consisting of reaction centers and LH I-Bchl complexes, allowing exciton transfer to several reaction centers.

The high number of integral pigment protein complexes per area of photosynthetic membranes reduce the rotational mobility of fatty acid chains in the lipid bilayer (11). The aggregation of these complexes may slow down long-distance lateral diffusion of proteins in the membrane and, consequently, decelerate a lateral equilibration of membrane components (39).

Numerous studies have shown that cytoplasmic and intracytoplasmic membranes form a continuous membrane system (6, 20), but both membranes are clearly different in function and composition (21, 39). This seems to be due to a differential incorporation of functional units into the membrane (43). The amount and composition of new synthesized units and the site and kinetics of incorporation are variable and under strict control. The mechanisms of this regulation is as yet unknown. Clearly, the differentiation by selective synthetic processes must be faster than equilization by diffusion and membrane flow.

ACKNOWLEDGMENT

This work was supported by grants from the Deutsche Forschungsgemeinschaft. I wish to express my thanks to R. Feick, J. Golecki, and J. Shiozawa for unpublished results and helpful discussions.

REFERENCES

1. Aagaard, J. and Sistrom, W. R. (1972). *Photochem. Photobiol. 15,* 209-225.
2. Breton, J., Becker, J. F., and Geacintov, N. E. (1973). *Biochem. Biophys. Res. Commun. 54,* 1403-1409.
3. Breton, J. and Roux, E. (1971). *Biochem. Biophys. Res. Commun. 45,* 557-563.
4. Clayton, R. K., Rafferty, C. N., Clayton, B., and Barouch, Y. (1977). *Proc. 4th Intern. Congr. Photosynthesis,* Reading, pp. 45-54.
5. Cogdell, R. J. and Crofts, A. R. (1978). *Biochim. Biophys. Acta 502,* 409-416.
6. Drews, G. (1978). *Current Top. Bioenerg. 8b,* 161-207.
7. Feher, G. and Okamura, M. Y. (1976). *Brookhaven Symp. Biol. 28,* 183-194.
8. Feher, G., Hoff, A. J., Isaacson, R. A., and Ackerson, L. C. (1975). *Ann. N.Y. Acad. Sci. 244,* 239.
9. Feick, R. and Drews, G. (1978). *Biochim. Biophys. Acta 501,* 499-513.
10. Firsow, N. N. and Drews, G. (1977). *Arch. Microbiol. 115,* 299-306.
11. Fraley, R. T., Jameson, D. M., and Kaplan, S. (1978). *Biochim. Biophys. Acta 511,* 52-69.
12. Geacintov, N. E., Van Nostrand, F., Pope, M., and Finkel, J. B. (1971). *Biochim. Biophys. Acta 226,* 486-491.

12a. Laemmli, U. K. (1970). *Nature (London) 227,* 680-685.

13. Monger, T. G. and Parson, W. W. (1977). *Biochim. Biophys. Acta 460,* 393-407.
14. Morita, S. and Miyazaki, T. (1971). *Biochim. Biophys. Acta 245,* 151-159.
15. Nieth, K. F., Drews, G., and Feick, R. (1975). *Arch. Microbiol. 105,* 43-45.
16. Niklowitz, W. and Drews, G. (1955). *Arch. Mikrobiol. 23,* 23, 123-129.
17. Niklowitz, W. and Drews, G. (1956). *Arch. Mikrobiol. 24,* 134-146.
18. Norris, J. R., Druysan, N. E., and Katz, J. J. (1973). *J. Amer. Chem. Soc. 95,* 1680.
19. Oelze, J. (1978). *Biochim. Biophys. Acta 509,* 450-461.
20. Oelze, J. and Drews, G. (1972). *Biochim. Biophys. Acta 265,* 209-239.
21. Parks, L. C. and Niederman, R. A. (1978). *Biochim. Biophys. Acta 511,* 70-82.
22. Prince, R. C., Baccarini-Melandri, A., Hauska, G. A., Melandri, B. A., and Crofts, A. R. (1975). *Biochim. Biophys. Acta 387,* 212-227.
23. Reed, D. W. and Ke, B. (1973). *J. Biol. Chem. 248,*

3041-3045.
24. Reed, D. W., Raveed, D., and Reporter, M. (1975). *Biochim. Biophys. Acta 387,* 368-378.
25. Sauer, K. (1975). *In* "Bioenergetics of Photosynthesis" (Govindjee, ed.), pp. 115-181. Academic Press, N.Y.
26. Sauer, K. and Austin, L. A. (1978). *Biochemistry 7,* 2011-2019.
27. Sauer, K., Dratz, E. A., and Coyne, L. (1968). *Proc. Natl. Acad. Sci. U.S. 61,* 17-24.
28. Schumacher, A. and Drews, G. (1978). *Biochim. Biophys. Acta 501,* 183-194.
29. Takemoto, J. (1974). *Arch. Biochem. Biophys. 163,* 515-520.
30. Thornber, J. P., Dutton, P. L., Fajer, J., Forman, A., Holten, D., Olson, J. M., Parson, W. W., Prince, R. C., Tiede, D. M., and Windsor, M. W. (1977). *Proc. 4th Intern. Congr. Photosynthesis,* Reading, pp. 55-70.
31. Vermeglio, A. and Clayton, R. K. (1965). *Biochim. Biophys. Acta 449,* 500-515.
32. Zürrer, H., Snozzi, M., Hanselmann, K., and Bachofen, R. (1977). *Biochim. Biophys. Acta 460,* 273-279.
33. Talsky, G., Rijgersberg, C. P., van Grondelle, R., Feick, R., and Drews, G. (1980). *Arch. Microbiol. 127,* 217-222.
34. Feick, R., van Grondelle, R., Rijgersberg, C. P., and Drews, G. (1980). *Biochim. Biophys. Acta 593,* 241-253.
35. Bolt, J. D., Sauer, K., Shiozawa, J. A., and Drews, G. (1981). *Biochim. Biophys. Acta 635,* 535-541.
36. Golecki, J., Drews, G., and Bühler, R. (1979). *Cytobiologie 18,* 381-389.
37. Golecki, J. R., Schumacher, A., and Drews, G. (1980). *Europ. J. Cell Biol. 23,* 149-156.
38. Drews, G., Wevers, P., and Dierstein, R. (1979). *FEMS Microbiology Letters 5,* 139-142.
39. Garcia, A. F., Drews, G., and Reidl, H.-H. (1981). *J. Bacteriol. 145,* 1121-1128.
40. Shiozawa, J., Cuendet, P., Zuber, H., and Drews, G. (1980). *Europ. J. Biochem. 111,* 455-460.
41. Drews, G. and Oelze, J. (1981). *Adv. Microb. Physiol. 22,* 1-92.
42. Oelze, J. and Drews, G. (1981). *In* "Organization of prokaryotic cell membranes," Vol. II (B. K. Ghosh, ed.), pp. 131-195. CRC Press, Boca Raton, Florida.
43. Dierstein, R., Schumacher, A., and Drews, G. (1981). *Arch. Microbiol. 128,* 376-383.

ACTIVE COMPONENTS BOUND TO THE CHROMATOPHORE MEMBRANE OF *Rhodospirillum rubrum*

Nozomu Nishi
Gilbu Soe
Tomisaburo Kakuno
Jinpei Yamashita
Takekazu Horio

Division of Enzymology
Institute for Protein Research
Osaka University, Suita-shi, Osaka 565

Since 1954 when Vernon and Kamen (1) reported the purification and characterization of cytochromes c_2 and c' from light-grown cells of the photosynthetic bacterium, *Rhodospirillum rubrum,* Kamen and his associates including three of us (T.K. for 1974 - 1976, J.Y. for 1966 - 1968 and T.H. for 1959 - 1961) have mainly studied cytochromes and non-heme iron proteins from several kinds of photosynthetic bacteria. These results are partly described in a recent review by Bartsch (2), who has been indispensable for the Kamen's history as a biochemist.

Our work in the field of photosynthesis at Osaka since 1961 has been mainly focused on biochemical studies of the components active in the electron transport system and the ATP-forming system of the photoheterotroph, *R. rubrum.* This occasion allows us to summarize our studies to the present, and for which are greatly indebted to Prof. Martin D. Kamen for his encouragement. One of the authors (T.H.) wishes to also take this opportunity to express his thanks to Dr. Robert G. Bartsch, Dr. Jack Newton and Dr. Joseph A. Orlando, who

 ISBN 0-12-397580-8

were working together with him in the group of Prof. Kamen at Brandeis University and taught him modern techniques and ways of thinking.

INTRACELLULAR LOCATIONS OF CYTOCHROMES

In *R. rubrum*, the contents of cytochromes c_2 and c' in cells grown anaerobically in the light are three and six times as high as those in cells grown aerobically in the dark, respectively; the content of *b*-type cytochrome is nearly the same for light- and dark-grown cells (3, 4). These cytochromes are bound partly to the cytoplasmic membrane and partly to the chromatophore membrane. The amounts of membrane-bound cytochromes c_2, c' and *b* are nearly the same for light- and dark-grown cells, and dark-grown cells incubated in the light, which produce chromatophores without proliferation (3). This indicates that when cells are grown in the light, cytoplasmic cytochromes c_2 and c' increase in amount, while membrane-bound cytochromes c_2 and c' do not increase, and that the cytoplasmic membrane has a limited number of sites capable of binding cytochromes c_2 and c', which is only slightly changed by formation of chromatophores (see below). It is known that light-grown cells each have hundreds of intracellular tubular organelles (5, 6), which are formed by invagination of the cytoplasmic membrane under anaerobic conditions in the light (or under oxygen tension as low as 4 mm Hg in the dark) (7, 11). Conceivably, when light-grown cells are disrupted, the tubular organelles are converted into chromatophores, which are membraneous, closed vesicles having an average diameter of 600 Å (membrane width = 55 Å) (volume = 1.13×10^8 Å^3) (6). The average content per cell of bacteriochlorophyll, ubiquinone-10, and carotenoids (in the wild-type strain) significantly increases with the progressively increasing number of chromatophores, while the total amount of protein hardly or only slightly increases; the extent of increase of bacteriochlorophyll and ubiquinone-10 is greater than 200 times and 2 times, respectively (3). This suggests that the NADH and succinate oxidation systems (see below) which exist on the cytoplasmic membrane in dark-grown cells are mostly diverted to connect with the photosynthetic, cyclic electron transport system (abbreviated as photosynthetic electron transport system) on the chromatophore membrane in light-grown cells.

The functions of cytochromes c_2 and c', abundant in the cytosol of light-grown cells, are not certain, except for the possibility that the cytochrome c' acts as an oxidase (see below) if the cell happens to be exposed to an atmosphere containing molecular oxygen, and that the redox states of the cytochrome c_2 ($E_{m,7}$ = +0.308 v) and the cytochrome c' ($E_{m,7}$ = +0.009 v) (12) easily render conditions suitable for a maximum coupling between the photosynthetic electron transport system and the ATP-forming system of chromatophores; photosynthetic ATP formation is most active when the E_h of the cytosol is +0.12 V (13, 14). Chromatophore-bound cytochromes c_2 and c' can be solubilized by a mixture of deoxycholate and cholate (12). These cytochromes thus solubilized are the same as those most abundant among the several forms (isocytochromes) present in the cytosol (15). Cytochrome *b*-558 ($E_{m,7}$ = -0.125 v) has been purified from light-grown cells without the use of detergents (15). Apparently, this cytochrome is different from chromatophore-bound cytochrome *b* ($E_{m,7}$ = -0.16 V in the bound state) (12).

FUNCTIONS OF CHROMATOPHORE-BOUND CYTOCHROMES

On the average, each chromatophore contains about 5 molecules each of cytochromes c_2, c', and *b* (12). This value is about one-fifth the number of the photoreaction units present in each chromatophore (see below). With respect to absorbance spectra, chromatophore-bound cytochrome c' resembles a typical *b*-type cytochrome, although solubilized cytochrome c' (high spin) differs significantly from typical *b*- and *c*-type cytochromes (low spin). However, the bound cytochromes c' and *b* are easily distinguishable from each other, because of the significant difference between the $E_{m,7}$ values; in the bound state, $E_{m,7}$'s are -0.16 V for cytochrome *b* and +0.017 V for cytochrome c'. With chromatophores under anaerobic conditions in the dark, cytochrome *b* is reduced by NADH, cytochrome c' and POC_{+60} (see below) are reduced by succinate, and ubiquinone-10, POC_{+275} (see below), and cytochrome c_2 are reduced by both substrates (12, 15 - 20). Antimycin A inhibits the reduction of cytochrome c_2 and POC_{+275}, but not of the others. This suggests that cytochrome *b*, ubiquinone-10, POC_{+275}, and cytochrome c_2 are involved in the NADH oxidation system, whereas cytochrome c', POC_{+60}, ubiquinone-10, POC_{+275}, and cytochrome c_2 are involved in the succinate oxidation system. Thus, the last three components are common to both systems. These three components are almost completely reduced by either NADH or succinate; that means that the NADH and succinate oxidation systems are interconnected. Taken together with the observation that the photosynthetic electron transport system is inhibited by antimycin A, this suggests that ubiquinone-10, POC_{+275}, and cytochrome c_2 are essential redox components for the photosynthetic electron transport system (16).

Anaerobically in the light, chromatophores catalyze the reduction of NAD^+ by succinate (21, 23). This reduction is inhibited by antimycin A. When the reaction mixture previously illuminated is allowed to stand anaerobically in the dark, the photosynthetically formed NADH is hardly reoxidized by the photosynthetically formed fumarate. However, the oxidation of the NADH occurs aerobically in the dark. The oxidation of added $NADH/NAD^+$ (1:1) ($E_{m,7}$ = -0.32 V) by added succinate/fumarate (various ratios) occurs anaerobically in the dark, provided that the E_h of the succinate/fumarate redox couple is made more positive than -0.04 V; consequently, the NADH formed by the oxidation of succinate anaerobically in the light is only slightly oxidized by the oxidation product, fumarate (24). This oxidation of NADH by fumarate anaerobically in the dark is not inhibited by antimycin A.

Both the NADH-cytochrome *b* system and succinate-cytochrome *c*' system are connected to the photosynthetic electron transport system, possibly, at the level of ubiquinone-10. Nevertheless, the cytochromes *b* and *c*' are functionally isolated fro each other. At the present time, there is no decisive evidence showing that cytochrome *b* or/and cytochrome *c*' are directly involved in the photosynthetic electron transport system. Howeve it may be rational to speculate that the NADH-cytochrome *b* system is more "tightly" connected to the photosynthetic electron transport system than the succinate-cytochrome *c*' system in order to provide a condition which could be advantageous for the supply of NADH to the general assimilative reactions.

Oxygen uptake by dark-grown cells is inhibited by CO. The action spectrum of the effect of light in diminishing the inhibitory effect of CO on the O_2 uptake resembles the absorbance spectrum of cytoplasmic cytochrome *c*' in the reduced form in the presence of CO (25 - 27). When the membrane is "intact" the membrane-bound cytochromes *b* and *c*' do not combine with CO at all, although both cytochromes react with CO when the membrane is less intact (16). Possibly, cytoplasmic cytochrome *c*' acts as an oxidase in the whole dark- and light-grown cells. "Intact" chromatophores also show O_2 uptake. This O_2 uptake appears to be carried out at the levels of flavin (16) and modified bacteriochlorophyll (26).

LOCATIONS OF UBIQUINONE-10, NON-HEME IRON, REACTION CENTER, NADH DEHYDROGENASE, AND ADP IN CHROMATOPHORE MEMBRANE

Chromatophores catalyze the ATP-P_i exchange and ATP hydrolysis as partial reactions of ATP formation (28). When chromatophores are depleted of 85-95% of ubiquinone-10, photosyntheti

ATP formation is lost, and the rates of ATP-P_i exchange, of ATP hydrolysis, of reduction of cytochrome c_2 and POC_{+275} by NADH and succinate, and of oxidation of NADH and succinate by O_2 are significantly lowered (16, 19, 20, 29 - 31). These activities are restored by readdition of the quinone. To the contrary, the reduction of cytochrome *b* by NADH and the reduction of cytochrome *c*' and POC_{+60} by succinate are hardly influenced by the removal of the quinone (19, 20).

In polarography, ubiquinone-10 bound to chromatophores exhibits polarographic oxidation and reduction waves having an $E_{1/2}$ of +0.050 V (32). This suggests that the quinone molecules exist, protruding their quinone moieties beyond the outer surface of the vesicles in such a way that the moieties can make electrochemical contact with the electrode.

In chromatophores, the reaction center bacteriochlorophyll shows two different kinds of light-induced absorbance changes (Liac), one at 860 nm (Liac-860) ($E_{m,7}$ = +0.45 V) and the other at 890 nm (Liac-890) ($E_{m,7}$ < -0.1 V) (33). Liac-890 also appears to react with the polarographic electrode ($E_{1/2}$ = -0.17 V), suggesting that it exists in the outside of the vesicles.

When Triton X-100 is added to chromatophores, two other polarographic redox waves become observable; their $E_{1/2}$'s are +0.060 V (POC_{+60}) and +0.275 V (POC_{+275}) (19). Possibly, POC_{+60} and POC_{+275} correspond to the non-heme iron proteins detected by ESR, which have g = 1.93 and g = 4.3, respectively (34). POC_{+275} may be the component with $E_{m,7}$ = +0.2 V, which has been proposed as an antimycin A-sensitive site (35). Both POC's can be solubilized from chromatophores by treatment with *n*-butanol (36). Even in the solubilized state, Triton X-100 is required for exhibition of the polarographic redox waves, indicating that non-heme irons exist at loci buried in the protein moieties so that the irons can not make electrochemical contact with the electrode. POC_{+60} has been purified; its absorbance spectrum has a sharp peak at 408 nm in the oxidized form, but no corresponding sharp maximum in the reduced form.

The NADH dehydrogenase which is functional in the NADH-cytochrome *b* system is loosely bound to chromatophores and it has been purified without the use of detergents (37). The purified enzyme does not possess any flavin, although flavins are required for exhibition of the activity. When the flavin-free enzyme is added to depleted chromatophores without addition of flavin compounds, NADH can reduce chromatophore-bound cytochrome *b*, perhaps, through the bound FAD. When antiserum toward the purified enzyme is added to "intact" chromatophores, the reduction of cytochrome *b* by NADH is inhibited, indicating that the enzyme is bound to the outer surface of the vesicles (38, 39).

PHOTOREACTION UNIT, REACTION CENTER, AND QUINONE-PROTEIN

When chromatophores are sonicated with a mixture of 2% deoxycholate and 1% cholate in 0.1 *M* Tris buffer (pH 8) and then subjected to repeated molecular-sieve chromatography in the presence of a mixture of 0.3% deoxycholate and 0.1% cholate chromatophores are divided into a bacteriochlorophyll-rich fraction (Bchl-F), a reaction center fraction (RC-F), and a quinone protein fraction (QP-F) (40-45). Almost all the bacteriochlorophyll present in chromatophores is collected in Bchl-F and RC-F QP-F has a molecular weight of about 1.1×10^4 daltons, an contains 1 ubiquinone-10 molecule in each molecule (ubiquinone-10 protein).

RC-F has a molecular weight of about 1.6×10^5 daltons is composed of five protein species of about 3.1×10^4, 2.5×10^4, 2.2×10^4, 1.2×10^4, and 1.1×10^4 daltons, and contains 3.8 bacteriochlorophyll, 2.0 bacteriopheophytin, and 1.9 acid-labil iron in each protein complex, but no quinone ($A_{280\ nm}/A_{802\ nm} = 2.31$). Possibly, the protein complex of RC-F is composed of each one of the three protein species of 3.1×10^4, 2.5×10^4, and 2.2×10^4 daltons, and about 4 each of the two protein species of 1.2×10^4, and 1.1×10^4 daltons. The values have beer revised by Tanaka *et al.* from those in (45). The absorbance spectrum of RC-F due to bacteriochlorophyll is similar to that of the reaction center purified with the use of *N,N*-dimethyllaurylamine oxide (LDAO) (46). It was reported that the LDAO reaction center is composed of three protein species of about 2.8×10^4, 2.4×10^4, and 2.1×10^4 ($A_{280\ nm}/A_{803\ nm} = 1.22$) (46). Added ferricyanide bleaches the absorbance peak at 865 r of RC-F to the same extent as that of the LDAO reaction center, indicating that RC-F contains a component for Liac-860. With RC-F, only 10% of the absorbance change brought about by ferricyanide is induced by the light. When RC-F is preincubated with QP-F, the light-induced absorbance change increases up to 70% of the extent caused by adding ferricyanide. It should be noted that Liac-890 has not been observed in either RC-F or the LDAO reaction center. Liac-890 becomes observable when RC-F is supplemented with other components solubilized with a mixture c cholate and deoxycholate from chromatophores, which have not been characterized. RC-F shows the activity for photooxidase; a substrate amount of cytochrome c_2 is oxidized aerobically in the light.

Bchl-F has a molecular weight of about 7.5×10^5, is composed of 9 protein species of about 4.0×10^4, 3.8×10^4, 3.1×10^4, 2.5×10^4, 2.2×10^4, 1.2×10^4, 1.1×10^4, 1.0×10^4, and 0.9×10^4 daltons, and contains about 30 bacteriochlorophylls, 3-4 ubiquinone-10, 90 phosphates, and 4 nonheme irons in each, but no phospholipid. The values have been revised by Tanaka *et al.* from those in (45). The protein species of 1.2×10^4 daltons contains 2 bacteriochlorophylls in each. Of the 9 protein species described above, the five protein species of 3.1×10^4, 2.5×10^4, 2.2×10^4, 1.2×10^4, and 1.1×10^4 daltons are due to the reaction center complex (RC-F); RC-F can be dissociated from Bchl-F by treatment with LDAO. T. Ueki and M. Kataoka (Prof. T. Mitsui, Department of Biophysical Engineering, Faculty of Engineering Science, Osaka University, Toyonaka-shi, Osaka 560) (70, 71) found that Bchl-F shows the same definite X-ray diffraction pattern as chromatophores do. They suppose that pattern implies the presence of a sixfold rotational symmetry. The most outside diameter of the hexagonal cluster of the protein complex of Bchl-F is estimated to be about 120 Å. Doubtless, the protein complex of Bchl-F maintains the same highly ordered assembly as in chromatophores. Bchl-F shows the light-induced absorbance change for reaction center (Liac-860). The ratio of the extent of the change to the content of bacteriochlorophyll is the same as that with chromatophores, indicating that the protein complex of Bchl-F is the unit for photochemical apparatus (photoreaction unit). Bchl-F, if prepared from the wild-type strain, contains carotenoids. This suggests that carotenoids are also bound to one of the 9 protein species described above. Each chromatophore of average size is estimated to contain 20 - 30 photoreaction units.

Tanaka *et al.* have obtained evidences showing that Bchl-F (photoreaction unit) is composed of 1 each of the five-protein species of 4.0×10^4, 3.8×10^4, 3.1×10^4, 2.5×10^4, and 2.2×10^4 daltons, 12 each of the two-protein species of 1.2×10^4 and 1.1×10^4 daltons, and 6 each of the two-protein species of 1.0×10^4 and 0.9×10^4 daltons. Matsuda *et al.* have succeeded in the reconstitution of the photosynthetic electron transport system by mixing and preincubating cytochrome c_2 in the reduced form and Bchl-F in the presence of a mixture of cholate and deoxycholate at concentrations as low as 0.001%. The system thus reconstituted does not oxidize the cytochrome in the dark, but oxidizes only a small part of the cytochrome in the light, reaching a steady state. The level of the steady state is significantly increased in the presence of antimycin A; the increased level is raised with increasing concentrations of the cytochrome, reaching to the sum of two times the molar amount of the ubiquinone-10 and the molar amount of the non-heme iron. Km for cytochrome c_2 in the reduced form is as low as 65 - 80 nM, indicating that Bchl-F has a specific site capable of tightly binding the cytochrome.

CYTOCHROME-BINDING PROTEINS

Limited amounts of cytochromes c_2 and c' are tightly bound to the membranes, although the cytochromes are extremely water-soluble. When chromatophores are sonicated with 1% cholate in 0.1 *M* Tris buffer (pH 8) and then subjected to molecular-sieve chromatography, fractions are obtained which contain cytochromes c_2 and c', but which have molecular weights significantly higher than those of the free cytochromes. There is a possibility that the membranes possess cytochrome c_2- and cytochrome *c*-binding proteins.

NTP-NDP KINASE, ATPase, AND BOUND ADP

Chromatophores catalyze ATP hydrolysis, ATP-P_i exchange and ATP-ADP exchange in the dark (28, 47). When chromatophores are repeatedly washed with 0.1 *M* Tris buffer (pH 8), almost all the ATP-ADP exchange activity is removed, although the other activities as well as photosynthetic ATP formation are hardly influenced (48). However, when the well-washed chromatophores are sonicated, an appreciable amount of ATP-ADP exchange enzyme (NTP-NDP kinase) is solubilized. The enzyme thus solubilized is the same as that abundant in the cytosol. NTP-NDP kinase has been purified and characterized. The antiserum for the enzyme fails to inhibit photosynthetic ATP formation.

The ATPase purified from acetone-dried powder of chromatophores is active in the presence of Ca^{2+}, but not in the presence of Mg^{2+} or Mn^{2+} (Ca-ATPase) (49, 50). Ca-ATPase is converted into Mg- and Mn-ATPase when proper pH indicators and/or detergents are present (51). The following conversion mechanism has been proposed by us. In the presence of Ca^{2+}, the enzyme hydrolyzes ATP into ADP and P_i with repeated turnover because of the reaction ($Ca^{2+} + P_i + H^+ \rightarrow CaHPO_4$). In the presence of Mg^{2+} or Mn^{2+}, the ATP-hydrolyzing reaction does not proceed with repeated turnover, because the proton produced by a single turnover of ATP hydrolysis is tightly held by the catalytic site buried in a hydrophobic region of the enzyme. If appropriate pH indicators and/or detergents are added, the tertiary structure of the hydrophobic region is slightly modified so as to form a channel, through which the proton can leak out of the catalytic site; thus, the ATP-hydrolyzing reaction proceeds with repeated turnover. If the channel is partially opened so that other appropriate pH indiators such as 2,4-dinitrophenol have access to the catalytic site, the leakage of the proton is stimulated by the function of the dyes as proton carriers; thus, the Mg- and Mn-ATPase activity is stimulated.

Recently, a factor capable of converting Ca-ATPase into Mg- and Mn-ATPase (conversion factor) has been purified from chromatophores (52, 53). The conversion factor is palmitoleic acid ($C_{18:1}\Delta 9$). In addition to the ability for the conversion, the fatty acid appears to act as a factor capable of binding the ATPase on the membrane.

On the average, 16 molecules of ADP are tightly bound to each chromatophore (54). This bound ADP is not exchanged with either added ADP or added ATP in the light or in the dark, and can be effectively extracted when 3% sodium dodecyl sulfate and then 0.5 *M* perchloric acid are added to chromatophores. When P_i is added to chromatophores in the light, the bound ADP is phosphorylated to form bound ATP. This phosphorylation is not inhibited by oligomycin. There is a possibility that this phosphorylation of bound ADP is the primary step of light-induced phosphorylation. Subsequent phosphoryl transfer from the resulting bound ATP to added ADP is inhibited by the antibiotic so bound on the membrane as to prevent access of added ADP to the bound ATP.

COUPLING MECHANISM BETWEEN ELECTRON TRANSPORT SYSTEM AND ATP-FORMING SYSTEM

The volume of one chromatophore is $1/10^4 - 1/10^6$, the volume of one mitochondrion or chloroplast (6). Nevertheless, chromatophore suspensions can exhibit a light-induced pH change in a manner similar to chloroplast suspensions (proton uptake in the light) (55, 56). With chromatophore suspensions, the light-induced pH change is optimum around pH 5, where both activities for photosynthetic and oxidative ATP formation are negligibly low. For the most part, the change which occurs at pH 5 requires the presence of high concentrations of inorganic cation salts; Rb^+, Na^+, Cs^+, Li^+, K^+, Mn^{2+}, Mg^{2+}, and Ca^{2+} are equally effective (57). These cations are abbreviated as M^+ (see below). On the average, each chromatophore possesses 1.2×10^4 ionizable groups having an apparent pK_a of 4.1 to 4.3, which are supposed to be mostly the carboxyl groups of chromatophore-bound proteins. In the photosynthetic electron transport system of chromatophores, ubiquinone-10 becomes more reduced, while cytochrome c_2 and non-heme iron (perhaps, POC_{+275}) become more oxidized, in the light than in the dark. A membrane potential difference may be generated. The reduction of the quinone requires protons from water in addition to electrons from the electron transport system with the consequence that photosynthetic electron transport is inhibited by deuterium water (58). Possibly, the cation-dependent pH change is brought about by acidifying particular loci (X^-) (perhaps carboxyl groups of bound proteins) on the chromatophore membrane by the protons liberated from the reduced quinone upon its oxidation, in a manner similar to the process which occurs in cation exchangers (X^-) ($X^-M^+ + H^+ \rightarrow XH + M^+$) (57). At pH 8, where photosynthetic ATP formation is optimum, the light-induced pH change is significantly lower than that at pH 5, and hardly influenced by high concentrations of cations, because the cation exchange reaction described above does not take place. The light-induced pH change at pH 8 is explainable by the uptake of protons from the medium into the bound ubiquinone-10 upon its reduction in the light (57). This implies that the light-induced pH change at pH 8 is not brought about by the transfer of protons from the medium to the inner phase of chromatophore vesicles.

The minimum volume of one membraneous, closed vesicle (not specified), in which one H^+ and one OH^- are allowed to stand "all the time," (the volume of a solution/the number of H^+ or OH^- at 10^{-7} M) is calculated to be about 1.66×10^{10} Å^3. The average volume of each chromatophore is about 1.13×10^8 Å^3 i.e., 1/147 the minimum volume mentioned above (57). According to the chemiosmotic coupling hypothesis for oxidative and photosynthetic phosphorylations (59 - 62), the total potential difference of the protons across the membrane of closed vesicles (the proton-motive force, Δp) consists of the sum of the electronic potential difference ($\Delta\psi$) and the pH difference across the membrane (ΔpH). The average volume of one chromatophore indicates that ΔpH is not obtainable, unless the total volume of the chromatophores present in the reaction mixture is used in the calculation.

Earlier, Hosoi *et al.* (63, 64) and Oku *et al.* (65) suggested that the association of chromatophore-bound ATPase (E^-) with one of the two protons (originally suggested as it would be the latent NTP-NDP kinase), which are liberated from reduced ubiquinone-10 upon its oxidation, ($E^- + H^+ \rightarrow EH$) provides the motive force for the formation of ATP from ADP and P_i. The protonated state of the enzyme (energized enzyme) (EH) may be stabilized in the presence of added ADP (66).

In chromatophores, about 85% of the ATPase activity in the dark is influenced by the redox potential of the medium (14, 67). The optimum E_h's are +0.12 V for ATP formation in the light, $\leq$ +0.07 V for ATP-P_i exchange in the dark, +0.15 V for ATPase activity in the dark, and $\geq$ +0.18 V for pyrophosphatase activity in the dark (68). The E_h-dependent ATPase and pyrophosphatase activities, but not the independent ones, are depressed when ubiquinone-10 is extracted from the membrane, and restored by the readdition of the quinone (30, 31, 69). At the present time, we believe that in chromatophores, the driving force for ATP formation is the protonation of the ATPase by the protons liberated from the reduced ubiquinone-10 upon its oxidation.

ACKNOWLEDGMENT

This series of studies was supported by a grant for the exchange visitor fellowship to T. H. from the Rockefeller Foundation, New York, by U.S. Public Health Service Grant GM-09246 from the National Institutes of Health, Bethesda, Md. and by grants including No. 354250 (1978) and No. 311909 (1978-1981) from the Ministry of Education, Science and Culture of Japan, Tokyo.

REFERENCES

1. Vernon, L. P. and Kamen, M. D. (1954). *Arch. Biochem. Biophys. 51,* 122-138.
2. Bartsch, R. G. (1978). *In* "Photosynthetic Bacteria" (R. K. Clayton and W. R. Sistrom, eds.), pp. 249-279. Plenum, New York.
3. Ishimoto, M. and Yamashita, J. (1971). "Protein, Nucleic Acid and Enzyme," Vol. 16, pp. 822-830. Kyoritsu Shuppan Co., Tokyo (in Japanese).
4. Geller, D. M. (1962). *J. Biol. Chem. 237,* 2947-2954.
5. Schachman, H. K., Pardee, A. B., and Stanier, R. Y. (1952). *Arch. Biochem. Biophys. 38,* 245-260.
6. Oda, T. and Horio, T. (1964). *Exptl. Cell Res. 34,* 414-417.
7. Tuttle, A. L. and Gest, H. (1959). *Proc. Natl. Acad. Sci. (U.S. 45,* 1262-1269.
8. Cohen-Bazire, G. and Kunisawa, R. (1963). *J. Cell Biol. 16,* 401-419.
9. Holt, S. C. and Marr, A. G. (1965). *J. Bacteriol. 89,* 1402-1412.
10. Newton, J. (1960). *Biochim. Biophys. Acta 42,* 34-43.
11. Lascelles, J. (1959). *Biochem. J. 72,* 508-518.
12. Kakuno, T., Bartsch, R. G., Nishikawa, K., and Horio, T. (1971). *J. Biochem. 70,* 79-94.

13. Horio, T. and Kamen, M. D. (1962). *Biochemistry 1,* 144-153.
14. Horiuti, Y., Nishikawa, K., and Horio, T. (1968). *J. Biochem. 64,* 577-587.
15. Bartsch, R. G., Kakuno, T., Horio, T., and Kamen, M. D. (1971). *J. Biol. Chem. 246,* 4489-4496.
16. Nisimoto, Y., Kakuno, T., Yamashita, J., and Horio, T. (1973). *J. Biochem. 74,* 1205-1216.
17. Horio, T. and Yamashita, J. (1964). *Biochim. Biophys. Acta 88,* 237-250.
18. Yamashita, J., Kamen, M. D., and Horio, T. (1969). *Arch. Mikrobiol. 66,* 304-314.
19. Erabi, T., Higuti, T., Sakata, K., Kakuno, T., Yamashita, J., Tanaka, M., and Horio, T. (1976). *J. Biochem. 79,* 497-503.
20. Higuti, T., Erabi, T., Kakuno, T., and Horio, T. (1975). *J. Biochem. 78*, 51-56.
21. Frenkel, A. W. (1958). *J. Amer. Chem. Soc. 80*, 3479-3480.
22. Vernon, L. P. and Ash, O. K. (1959). *J. Biol. Chem. 234,* 1878-1882.
23. Vernon, L. P. (1959). *J. Biol. Chem. 234,* 1883-1888.
24. Horio, T., Yamashita, J., and Nishikawa, K. (1963). *Biochim. Biophys. Acta 66,* 37-49.
25. Horio, T. and Yamashita, J. (1963). *In* "Bacterial Photosynthesis" (H. Gest, A. San Pietro, and L. P. Vernon, eds.), pp. 275-306. Antioch Press, Yellow Springs, Ohio.
26. Horio, T. and Taylor, C. P. S. (1965). *J. Biol. Chem. 240,* 1772-1775.
27. Taniguchi, S. and Kamen, M. D. (1965). *Biochim. Biophys. Acta 96*, 395-428.
28. Horio, T., Nishikawa, K., Katsumata, M., and Yamashita, J. (1965). *Biochim. Biophys. Acta 94,* 371-382.
29. Okayama, S., Yamamoto, N., Nishikawa, K., and Horio, T. (1968). *J. Biol. Chem. 243,* 2995-2-99.
30. Horio, T., Nishikawa, K., Okayama, S., Horiuti, Y., Yamamoto, N., and Kakutani, Y. (1968). *Biochim. Biophys. Acta 153,* 913-916.
31. Yamamoto, N., Hatakeyama, H., Nishikawa, K., and Horio, T. (1970). *J. Biochem. 67,* 587-598.
32. Erabi, T., Higuti, T., Kakuno, T., Yamashita, J., Tanaka, M., and Horio, T. (1975). *J. Biochem. 78,* 795-801.
33. Okayama, S., Kakuno, T., and Horio, T. (1970). *J. Biochem. 68,* 19-29.
34. Higuti, T., Shiga, T., Kakuno, T., and Horio, T. (1974). *J. Biochem. 75,* 1363-1371.
35. Yamashita, J., Yoshimura, S., Matuo, Y., and Horio, T. (1967). *Biochim. Biophys. Acta 143,* 154-172.
36. Morton, R. K. (1955). *Methods Enzymol. 1,* 25-51.
37. Horio, T., Bartsch, R. G., Kakuno, T., and Kamen, M. D.

(1969). *J. Biol. Chem. 244,* 5899-5909.
38. Nisimoto, Y., Yamashita, J., and Horio, T. (1973). *J. Biochem. 73,* 515-521.
39. Nisimoto, Y., Yamashita, J., and Horio, T. (1973). *J. Biochem. 73,* 523-528.
40. Nishi, N., Soe, G., and Yamashita, J. (1977). *Proc. 50th Congr. Jap. Biochem. Soc. 49,* 976 (*Seikagaku,* in Japanese).
41. Nishi, N., Soe, G., and Yamashita, J. (1978). *Proc. 19th Congr. Jap. Soc. Plant Physiol.,* p. 81 (in Japanese).
42. Nishi, N., Kakuno, T., and Yamashita, J. (1978). *Proc. 51th Congr. Jap. Biochem. Soc. 50,* 677 (*Seikagaku,* in Japanese).
43. Nishi, N., Soe, G., Kakuno, T., and Yamashita, J. (1978). *Proc. 4th Congr. Jap. Bioenerg. Group,* pp. 52-54 (in Japanese).
44. Nishi, N., Soe, G., and Yamashita, J. (1978). *Proc. 6th Intern. Biophys. Congr.,* p. 179.
45. Nishi, N., Kataoka, M., Soe, G., Kakuno, T., Ueki, T., Yamashita, J., and Horio, T. (1979). *J. Biochem. 86,* 1211-1224.
46. Feher, G. and Okamura, M. Y. (1976). *In* "Photosynthetic Bacteria" (R. K. Clayton and W. R. Sistrom, eds.), pp. 1-43. Plenum, New York.
47. Horio, T., Nishikawa, K., and Yamashita, J. (1974). *J. Biochem. 55,* 327-332.
48. Yamamoto, N., Horiuti, Y., Nishikawa, K., and Horio, T. (1972). *J. Biochem. 72,* 599-613.
49. Baccarini-Melandri, A. and Melandri, B. A. (1971). *Methods Enzymol. 23,* 556-561.
50. Johansson, B. C., Baltscheffsky, M., Baltscheffsky, H., Baccarini-Melandri, A., and Melandri, B. A. (1973). *Eur. J. Biochem. 40,* 109-117.
51. Soe, G., Nishi, N., Kakuno, T., and Yamashita, J. (1978). *J. Biochem. 84,* 805-814.
52. Soe, G., Nishi, N., Kakuno, T., Yamashita, J., and Horio, T. (1979). *In* "Cation Flux Across Biomembranes" (Y. Mukohata and L. Packer, eds.), in press. Academic Press, N.Y.
53. Soe, G., Nishi, N., Kakuno, T., Yamashita, J., and Horio, T. (1979). *J. Biochem.,* in press.
54. Yamamoto, N., Yoshimura, S., Higuti, T., Nishikawa, K., and Horio, T. (1972). *J. Biochem. 72,* 1397-1406.
55. von Stedingk, L.-V. and Baltscheffsky, H. (1966). *Arch. Biochem. Biophys. 117,* 400-404.
56. Jackson, J. B., Crofts, A. R., and von Stedingk, L.-V. (1968). *Eur. J. Biochem. 6,* 41-54.
57. Nishi, N., Sakata-Sogawa, K., Soe, G., and Yamashita, J. (1977). *J. Biochem. 82,* 1267-1279.

58. Kakuno, T., Hosoi, K., Higuti, T., and Horio, T. (1973). *J. Biochem. 74,* 1193-1203.
59. Mitchell, P. (1961). *Nature (London) 191,* 144-148.
60. Mitchell, P. (1966). "Chemiosmotic Coupling in Oxidative and Photosynthetic Phosphorylation." Glynn Research Ltd., Bodmin, Cornwell.
61. Mitchell, P. (1966). *Biol. Rev. 41,* 445.
62. Mitchell, P. (1972). *Mitochondria/Biomembranes, FEBS Symp. 28,* pp. 353-370.
63. Hosoi, K., Yoshimura, S., Soe, G., Kakuno, T., and Horio, T. (1973). *J. Biochem. 74,* 1275-1278.
64. Hosoi, K., Soe, G., Kakuno, T., and Horio, T. (1975). *J. Biochem. 78,* 1331-1346.
65. Oku, T., Hosoi, K., Soe, G., Kakuno, T., and Horio, T. (1974). *J. Biochem. 76,* 233-235.
66. Horio, T., Nishikawa, K., and Yamashita, J. (1966). *Biochem. J. 98,* 321-329.
67. Horio, T., Nishikawa, K., Horiuti, Y., and Kakuno, T. (1968). *In* "Comparative Biochemistry and Biophysics of Photosynthesis" (K. Shibata, A. Takamiya, A. T. Jagendorf, and R. C. Fuller, eds.), pp. 408-424. University of Tokyo Press, Tokyo.
68. Horio, T., Yamashita, J., Nishikawa, K., Kakuno, T., Hosoi, K., Suzuki, J., and Yoshimura, S. (1973). *In* "Organization of Energy-Transducing Membrane" (M. Nakao and L. Packer, eds.), pp. 239-249. University of Tokyo Press, Tokyo.
69. Nishikawa, K., Hosoi, K., Suzuki, J., Yoshimura, S., and Horio, T. (1973). *J. Biochem. 73,* 537-553.
70. Ueki, T., Kataoka, M., and Mitsui, T. (1976). *Nature 262,* 809-810.
71. Kataoka, M. and Ueki, T. (1981). *J. Biochem. 89,* 71-78

STUDIES ON CAROTENES IN BACTERIAL REACTION CENTERS

Françoise Reiss-Husson
Ileana Agalidis

Laboratoire de Photosynthèse
C.N.R.S., 91190 Gif-sur-Yvette, France

The characterization of bacterial reaction centers has advanced rapidly since the initial isolation of these complexes from *Rhodopseudomonas spheroides* (1) and *Rhodospirillum rubrum* (2). In these two cases, blue-green mutants were originally used; most of the investigations thereafter were devoted to these carotenoid-less preparations, particularly to reaction centers derived from *Rps. spheroides* R26. Although it now seems likely that the findings obtained with *Rps. spheroides* R26 reaction centers have a general application to other photosynthetic bacteria (3), exceptions are possible (4). Studies of other strains and species thus still deserve attention.

In carotenoid-less strains, the only reaction center pigments are 4 bacteriochlorophylls and 2 bacteriopheophytins. Yet in wild-type strains a third component, carotene, is present. In our first attempt to isolate *Rps. spheroides* reaction centers from a wild-type strain, using a cationic detergent, we noticed that carotene was indeed part of the pigment complement of the complex (5). Further progress in the purifi-

 ISBN 0-12-397580-8

cation (6) as well as the choice of another detergent (lauryl-dimethylamine oxide) for isolation (7) did not remove this carotenoid, which was identified as spheroidene and shown to be tightly bound to the reaction center protein (8). Another carotene, spirilloxanthin, was found to be present in wild-type *R. rubrum* (9). These findings were extended by Cogdell *et al.* (10) to several carotenoid-containing *Rps. spheroides* strains. In each of these, 1 mol of a specific carotenoid per mole of reaction center was present; the pigment was bound to the same polypeptide chains as were the bacteriochlorophylls and bacteriopheophytins. Energy transfer from the carotenoid to bacteriochlorophyll was shown to be an efficient process; the reverse pathway, that is, removal of energy from the reaction center chlorophyll by transfer to the carotene and formation of a carotenoid triplet, was demonstrated as a deexcitation mechanism (11).

Thus carotene appeared as an integral and functional component of *Rps. spheroides* and *R. rubrum* reaction centers. Puzzling properties of this bound carotene were observed at the same time. These include circular dichroism (10) and Raman resonance spectrum (12), both of which were strikingly different from those observed for bulk carotene in the chromatophores. These differences might arise either from a particular local environment of the carotene as compared to the bulk pigment, or from a different conformation: the bulk carotene being obviously in all-trans form, the reaction center bound being a stereoisomer. Here we will review the experimental evidence in favor of the second hypothesis, which appears to apply to a number of strains and species.

As resonance Raman spectroscopy is a powerful tool for distinguishing between carotene stereoisomers (13), the resonance Raman spectra of spheroidene in reaction centers isolated from 2 wild-type strains (Y and 2.4.1) were carefully recorded, and compared to those of all-trans and central mono-*cis*-β-carotene (14). All - trans and 15,15'-*cis*-β-carotene spectra differed in intensities and (or) positions of a number of bands. The salient features of the *cis* compound included: (a) an additional pair of strong bands at 1237, 1245 cm^{-1}, in the so-called fingerprint region of vibrational spectra of polyenes, known to be sensitive to the conformation of the chain and the nature of the end groups (13); (b) a shift in frequency and a modification in structure of the ν_1 band (located at 1540 cm^{-1} for the *cis* isomer); (c) the intensification of bands situated at 955 cm^{-1} and 487 cm^{-1}, and several other spectral changes in the 750 - 900 cm^{-1} range. When compared to bulk spheroidene of the chromatophores, spheroidene of the reaction center presented the same differences (Fig. 1): (a) an additional strong band at 1241 cm^{-1}; (b) a shift in frequency and alteration in structure of the ν_1 band at 1540 cm^{-1}; (c) an increase in relative intensities of bands at 956 cm^{-1} and 490 cm^{-1}, and several spectral changes in the 750 - 900 cm^{-1} range (not shown in Fig. 1). It was immediately apparent from these data that spheroidene bound to wild-type *Rps. spheroides* reaction centers adopted a *cis* conformation.

This *cis* conformation was insensitive to the oxidoreduction state of the reaction center. It was not temperature dependent: recording the spectra at room temperature or at temperatures down to 25°K did not alter the spectral characteristics. It was strongly dependent on the pigment - protein interactions: extraction of spheroidene by conventional methods yielded the all-trans form only. However, a rapid extraction of all the reaction center pigments (30 s) followed by immediate freezing at 30°K allowed us to record a Raman spectrum still indicative of a *-cis* conformation (Fig. 1); yet differences in relative intensities were indicative either of an intermediate *cis* form between the native *cis* and the all-trans conformation, or to further strain exerted on the *cis* form by binding to the protein.

Electronic absorption spectra of wild-type *Rps. spheroides* reaction centers, although not demonstrative of, were also consistent with spheroidene assuming a *cis* conformation in reaction centers. Difference absorption spectra between reaction centers of wild and carotenoid-less strains revealed the visible $^1B \leftarrow {}^1A$ transition bands of spheroidene (Fig. 2), which had lower extinction coefficients and wavelengths with respect to those of bulk carotene, as expected for a *cis* form. The difference spectra persistently exhibited a series of positive bands in the 300 - 420 nm region. Two of these, situated at 403 and 335 nm, had variable extinction coefficients and most likely arose from impurities. As shown by the 370 - 360 nm peak and trough, the Soret band in the wild-type reaction center was approximately 15% higher than that of R26 reaction center, and slightly red shifted. We attribute these effects to spheroidene - bacteriochlorophyll interactions. A persistent positive band was observed at 348 nm, sometimes together with a weaker band at 328 nm. These bands might represent the so-called "cis" band ($^1C \leftarrow {}^1A$) transition of spheroidene.

The occurrence of a carotene *cis* isomer in the wild-type *Rps. spheroides* center could have been a singularity, but Raman spectroscopy of various reaction centers demonstrated that it was quite general. Similar conclusions were drawn from carotenes in reaction centers isolated from mutants of *Rps. spheroides*: Ga strain (containing chloroxanthin) and Gl C (containing neurosporene) and from wild-type *R. rubrum* (containing spirilloxanthin) and *Rps. viridis* (probably containing 1,2-dihydrolycopene). The close similarity between the resonance Raman spectra of these preparations suggested that the carotenes assumed a nearly identical *cis* conformation. As in

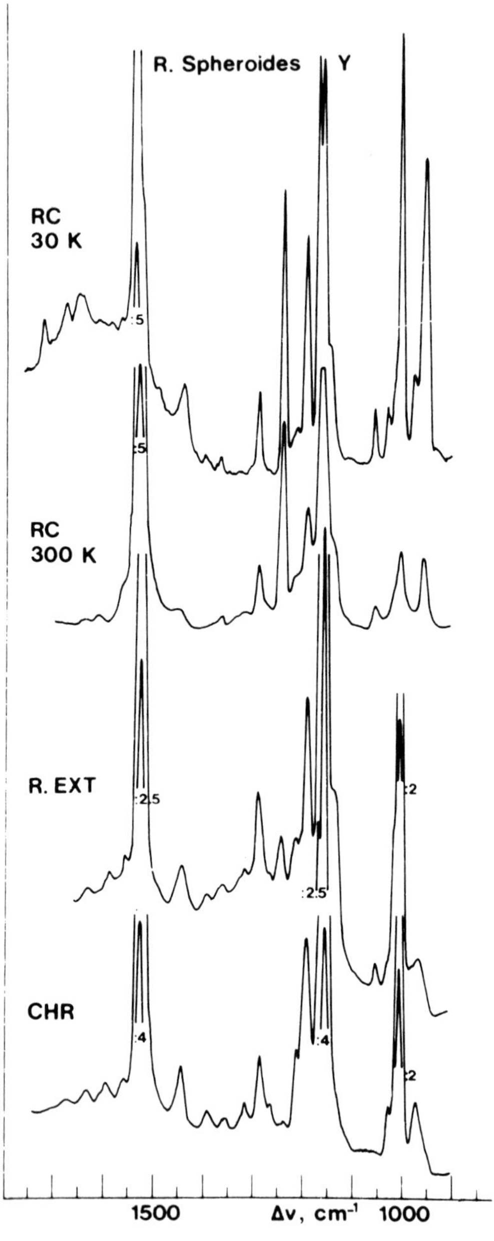

Fig. 1. Resonance Raman spectra of spheroidene from Rps. spheroides Y (wild type), 900 - 1700 cm^{-1} region, excitation 496.5 nm; resolution at 1000 cm^{-1}: 6 cm^{-1}. From top to bottom RC 30 K and 300 K: reaction centers at 30 and 300 K. R. EXT, spheroidene in petroleum ether solution, rapidly extracted from reaction centers and frozen at 30 K. CHR, in chromatophores at 30 K (14).

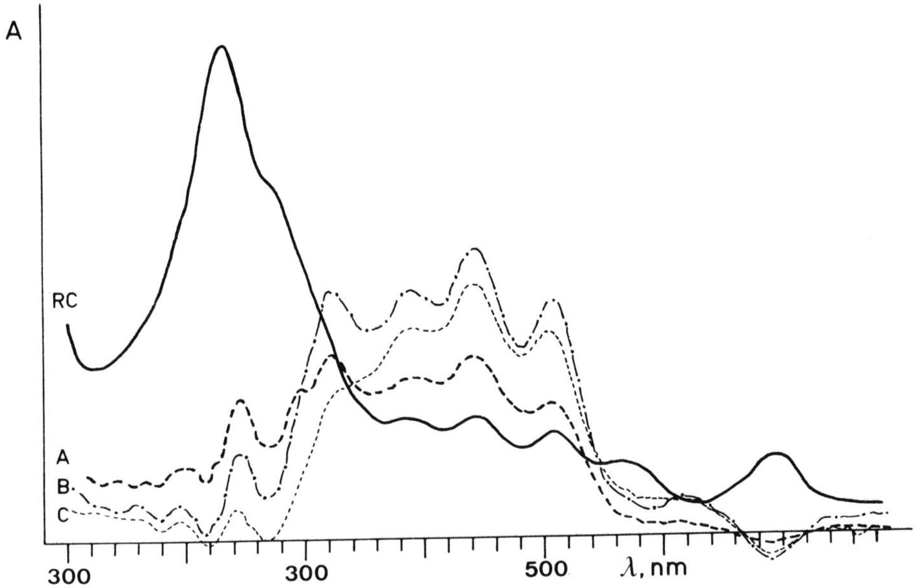

Fig. 2. (A - C) Differential absorption spectra of wild-type Rps. spheroides reaction centers minus R26 Rps. spheroides reaction centers (20°C), 300 - 650 nm region. (A - C) The results from experiments on three different cultures of each type. The Q_X BChl band in the wild-type reaction centers was overcompensated by 10% in (A), and by 20% in (B) and (C). RC, absolute absorption spectrum of wild-type Rps. spheroides (14).

the wild-type *Rps. spheroides* reaction center, this native form was unstable and quickly isomerized upon extraction of the carotene from the protein. However, precise identification of the conformation should await studies of a series of isolated C_{40} carotenoid isomers. The discussion of the available data (14) which we will not develop here, suggested that in these reaction centers the carotene might assume an hindered, di-*cis* form.

In a recent investigation, Boucher *et al.* (15) demonstrated that in the *R. rubrum* reaction center, carotenes confer protection against photooxidative destruction of bacteriochlorophyll. They showed, moreover, that *R. rubrum* reaction centers isolated from the carotene-less mutant G9 could be recombined with any of a number of carotenes: spirilloxanthin, spheroidene and spheroidenone, chloroxanthin, in a nearly 1:1 molar ratio. Difference spectra of the recombined preparations, namely, the presence of two very intense bands in the 350 - 400 nm range (i.e., 387 and 393 nm for spheroidene) led them to conclude that in the native *R. rubrum* reaction center, as well as in the recombined one, carotene assumed a mono-*cis* conformation. However, the intensity of these uv absorption bands are much too intense, and their maxima too red-shifted, to be identified as a "cis" band. Indeed, the "cis" band of isolated carotene "cis" isomer is known to be of low absorbance (16). It seems more plausible that they originate from impurities as well as bacteriochlorophyll contributions (similar to those we have detected in *Rps. spheroides,* see above and Fig. 2). This led us to investigate the structure of spheroidene in recombined spheroidene - R26 *Rps. spheroides* reaction centers (17).

By following an experimental procedure similar to that described by Boucher *et al.* (15), we observed that all-trans spheroidene could be bound to the carotenoid-less reaction center isolated from R26 *Rps. spheroides,* up to a nearly 1:1 molar ratio. Spheroidene bound to R26 reaction centers exhibited resonance Raman spectra quite similar to those of native spheroidene in wild-type reaction centers: indeed, no differences in band frequencies could be observed, but only some small differences in relative intensities. Thus binding to R26 reaction center protein converted all-trans spheroidene into a *cis* isomer. A differential absorption spectrum between spheroidene - R26 reaction center and R26 reaction center was computed (Fig. 3). It displayed absorption bands of the bound carotene at 503, 471, and 444 nm ($^1B \leftarrow {}^1A$ transition), and a weak 344 nm band that we ascribe to the "cis" band. A positive band at 368 nm indicated hyperchromism (∿15%) and a slight red shift of the bacteriochlorophyll Soret band, presumably induced by the presence of spheroidene. All these characteristics were strongly similar to those observed in wild-type reaction centers (Fig. 2). However, more precise measurements disclosed slight shifts of the visible bands of spheroidene bound to the R26 complex, as compared to native spheroidene in wild-type reaction centers.

These preliminary results indicate that spheroidene might be bound to the R26 reaction center as a "cis" isomer, very close or identical to the one present in the wild-type reaction center. The slight differences observed in Raman intensities and in the position of the electronic absorption bands might reflect a different local environment of spheroidene bound to the R26 reaction center. The site of binding is probably the same as in the wild-type reaction center, or at least in similar proximity to P 870. Indeed, spheroidene when bound to the R26 reaction center exhibited light-induced absorbance changes identical to those observed in the wild-type reaction center (17).

In summary, these observations implicate the existence of a high affinity binding site for carotene in reaction center proteins of *Rps. spheroides, Rps. viridis,* and *R. rubrum.* This

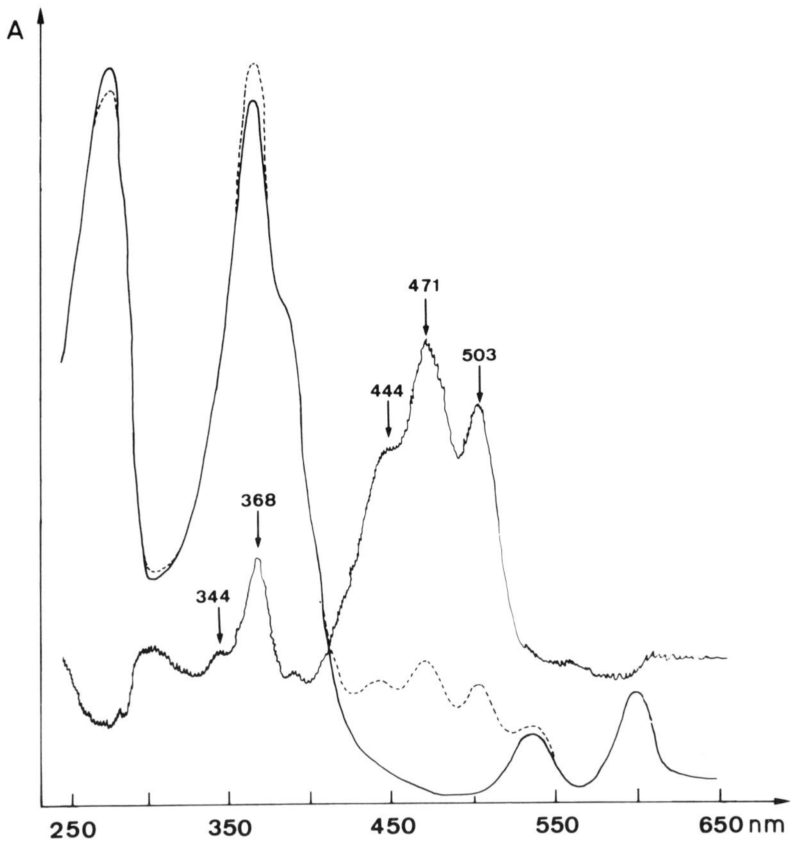

Fig. 3. A (——) and B (- - -): absolute absorption spectra of R26 Rps. spheroides reaction center and of spheroidene - R26 reaction center. C, Differential absorption spectra (B) minus (A).

binding site has probably the same structure in the three species, as it constrains the carotenes to a similar *cis* conformation. As the carotene functions seem to be in protection of bacteriochlorophyll against photodestruction (15) and participation to deexcitation processes (11), the binding site should be quite close to those of the other chlorophyll pigments. Yet in reaction centers of these species, polypeptide chains are known to be not strictly identical, and the chemical nature of the chlorophyll pigments varies (4). Despite these gross differences, it seems possible that the pigment binding site is located in a core region of invariant structure. Confirmation of this hypothesis should await the elucidation of the reaction center structure which still remains to be done.

ACKNOWLEDGMENTS

The work reviewed from our laboratory could not have been done without the original participation of M. D. Kamen, who introduced one of the authors (F.R.H.) to the field of bacterial photosynthesis; M. D. Kamen was head of our research team in Gif-sur-Yvette from 1965 to 1969, and his constant interest and advice were of invaluable help. We are greatly indebted to all the other members of our team, particularly to G. Jolchine, and to M. Lutz for his preeminent contribution to the Raman experiments.

REFERENCES

1. Reed, D. W. and Clayton, R. K. (1968). *Biochem. Biophys. Res. Commun.* *30*, 471-475.
2. Gingras, G. and Jolchine, G. (1968). *In* "Progress in Photosynthesis Research" (H. Metzner, ed.), Vol. I, pp. 209-216. Tübingen.
3. Ke, B. (1978). *Current Topics in Bioenerget.* *7*, 76-139.
4. Gingras, G. (1978). *In* "The Photosynthetic Bacteria" (R. K. Clayton and W. R. Sistrom, eds.), Plenum, New York.
5. Jolchine, G., Reiss-Husson, F., and Kamen, M. D. (1969). *Proc. Natl. Acad. Sci. U.S.* *64*, 650-653.
6. Reiss-Husson, F. and Jolchine, G. (1972). *Biochim. Biophys. Acta* *256*, 440-451.
7. Jolchine, G. and Reiss-Husson, F. (1974). *FEBS Lett.* *40*, 5-8.
8. Jolchine, G. and Reiss-Husson, F. (1975). *FEBS Lett.* *52*, 33-36.
9. Van der Rest, M. and Gingras, G. (1974). *J. Biol. Chem.* *249*, 6446-6453.
10. Cogdell, R., Parson, W. W., and Kerr, M. A. (1976). *Biochim. Biophys. Acta* *430*, 83-93.
11. Cogdell, R., Monger, T. G., and Parson, W. W. (1975). *Biochim. Biophys. Acta* *408*, 189-199.
12. Lutz, M., Kleo, J., and Reiss-Husson, F. (1976). *Biochem. Biophys. Res. Commun.* *69*, 711-717.
13. Callender, R. and Honig, B. (1977). *Annu. Rev. Biophys.*

Bioeng. 6, 33-57.
14. Lutz, M., Agalidis, I., Hervo, G., Cogdell, R. J., and Reiss-Husson, F. (1978). *Biochim. Biophys. Acta 503,* 287-303.
15. Boucher, F., Van der Rest, M., and Gingras, G. (1977). *Biochim. Biophys. Acta 461,* 339-357.
16. Zeichmeister, L. (1962). "Cis-Trans Isomeric Carotenoids, Vitamins A and Arylpolyenes." Springer Verlag, Wien.
17. Agalidis, I., Lutz, M., and Reiss-Husson, F. (1980). *Biochim. Biophys. Acta 589,* 264-274.

THE RELATIONSHIP BETWEEN PHOTOSYNTHESIS AND NITROGEN FIXATION REVISITED

J. W. Newton

Northern Regional Research Center
Agricultural Research Service
U.S. Department of Agriculture
Peoria, Illinois 61604

I do not think it has been fully appreciated what an important impact Professor Kamen's fundamental discoveries have had upon the present and probable future course of agricultural research. Two of his favorite problem areas, photosynthesis and nitrogen fixation, are currently receiving renewed attention by agricultural scientists. It is not generally recognized that it was Professor Kamen who, more than anyone else, called our attention to the curious relationship between photosynthesis and nitrogen fixation, and in particular, showed that these processes could exist simultaneously in the same cell in the anaerobic photosynthetic bacteria.

I first met Martin while I was engaged in graduate studies on nitrogen fixation in the laboratories of P. W. Wilson and R. H. Burris at the University of Wisconsin. The landmark papers of Gest and Kamen had recently appeared, and a whole new area of biological nitrogen fixation research had been opened up with the discovery of nitrogen fixation and photohydrogen production in photosynthetic bacteria. I was aware of the discovery by Ruben and Kamen of radioactive carbon and of their

 ISBN 0-12-397580-8

extensive use of radioactive tracers to solve a host of fundamental questions in biology, including the origin of oxygen in photosynthesis. When the opportunity arose to pursue postdoctoral work with Martin, I eagerly took it hoping that some of this astute serendipity might rub off on me. I joined his laboratory at Washington University Medical School in 1956. It was an exciting time. The groups at Washington University in microbiology and biochemistry, with which we were closely associated, consisted of Kornberg's group, beginning their work on DNA; Strominger and Glaser studying cell walls; Stritmatter and Velick, studying enzymology of cytochrome *b*; Drysdale and Mildred Cohn applying tracers to study enzyme mechanisms; and a host of scientists drawn to Washington University by the Cori-Lowry tradition of excellence in biochemistry.

Martin, as usual, was involved in new discoveries, the latest being the existence of cytochromes in obligately anaerobic microorganisms. I was fortunate to be able to explore with him the significance of this finding. Leo Vernon, who preceded me, had isolated a variety of new heme proteins from photosynthetic bacteria, and Robert Bartsch had begun purification of these materials, opening up new areas for biochemical investigation. The general significance of cytochromes in a variety of photo- and chemoautotrophs had become apparent. Arnon, working with chloroplasts, and Frenkel, studying *Rhodospirillum* in Professor Lipmann's laboratory, had recently discovered photosynthetic phosphorylation. Since both of these systems were derived from aerobic organisms, it was of fundamental interest to determine whether the phosphorylation process itself was oxygen dependent. Consequently, we initiated a series of studies on the obligately anaerobic photosynthetic bacterium, *Chromatium,* a purple sulfur bacterium. Our studies, as well as extensive work on the chloroplast system in Professor Arnon's laboratory, left little doubt that photophosphorylation was an essentially anaerobic process. What we never were convinced of was whether this phosphorylation was dependent upon electron flow through a carrier system or upon raising an electron to a higher energy level and allowing it to "relax" in some unrecognized manner. Peter Mitchel's theoretical analyses of the problem had appeared at about that time and provided an intellectually satisfying, if preliminary, analysis of this question, which is still being actively pursued in many laboratories.

The Kamen laboratory moved to Brandeis in 1957, where N. O. Kaplan was assembling a group of biochemists emphasizing a multidisciplinary department engaged in study of a broad array of biological problems. As always, Martin was both a source of encouragement and a catalyst to many of us to pursue any problem of fundamental interest. I left this laboratory in 1960 armed with enough ideas and problems to last anyone through several careers. Two of these areas, photosynthesis and nitrogen fixation, and, in particular, the relationship between the two processes, have continued to hold my attention, as they have held Professor Kamen's for many years.

Recent general concern about the energy requirements for crop production has led agricultural researchers to contemplate a variety of ways to lower these energy costs. It has been estimated that about one-third of the energy input into corn production in the United States is in the form of nitrogen fertilizer. Thus, nitrogen input into crops is an area where large potential benefits for agriculture can be anticipated and upon which attention is being focused by many researchers.

Several years ago, our laboratory initiated a formal program of research on biological nitrogen fixation. Our initial approach to this problem was to contemplate what an ideal nitrogen-fixing system in plants would be like, to see if such a system might already occur in nature, and then to attempt a systematic examination of the properties of such a system with the ultimate goal of applying the information to an agronomically important crop.

I have listed in Table I the properties that we considered most desirable in an "ideal" system. Our considerations were to try to overcome the vexing problem of establishing a nitrogen-fixing symbiont in soil and to overcome the barriers in the plant that impede intake of the symbiont. Furthermore, because nitrogen fixation normally consumes so much energy, we would prefer to have a system either driven photosynthetically or which puts a modest strain upon the plant as source of energy for fixation.

It turns out that such a system not only exists in nature, but is already commonly used in agriculture in the Orient. The plant is the tiny floating water fern *Azolla*, which contains, in a cavity in its leaf, a nitrogen-fixing blue-green algal symbiont that supplies all of the nitrogen needs of the fern. The plant has been widely used in southeast Asia for centuries as a green manure in rice paddies. *Azolla* is capable of rapid growth in paddies and can provide, through growth and decay, as much as 100 kg/ha nitrogen for rice crops grown without any fertilization. Despite extensive use of this plant in such forms of agriculture, relatively little research had been done with it, and many aspects of the symbiosis are unclear. A major obstacle has been the inability of most investigators to isolate and grow the

Table I. Properties of an Ideal Symbiont

1. *Easily introduced into plant, no soil or root barrier*
2. *High nitrogenase; supply plant needs*
3. *Energy cost to plant low; photosynthetic symbiont*

blue-green algal (cyanobacterial) symbiont or to reestablish the symbiosis in algal-free plants. Recently, we have developed a procedure with which we were able to obtain pure cultures of cyanobacteria from *Azolla*. The method consists of isolating algal material from enzymatic digests of leaves of the fern, using these preparations as massive inocula into nitrogen-free media, and selecting out pure strains of cyanobacteria capable of growth *in vitro*. Although we have not yet reestablished the isolated cyanobacteria in *Azolla*, we have obtained strains suitable for physiological studies that are derived from plant material.

Figure 1 shows some of the preparations. Digestion of fern leaves with a pectinase-cellulase mixture releases algal "bundles" from the leaf. These bundles conform to the shape of the leaf cavity from which they are derived. The bundle contains a variety of cyanobacterial cell types, which are released upon rupture of the bundle. Figure 1 also shows a pure culture of a heterocyst-containing cyanobacterium isolated from our preparation when grown on nitrogen-free media in the light.

We are conducting investigations of the general physiology of the symbiosis and its isolated partners. In general, the isolated blue-greens resemble *Anabaena sp.* in their morphology. However, unlike most cyanobacteria, these strains from *Azolla* are more heterotrophic in character and can be readily grown on fructose in the dark. Furthermore, data in Table II show that such strains have their highest nitrogenase activity when grown under these conditions.

Elaboration of high levels of nitrogenase activity when grown heterotrophically is the most unique property of the cyanobacteria isolated from *Azolla*. What are the other characteristics of this symbiosis? Studies with intact ferns and extracts indicate that very high levels of free ammonia are formed by these algae when in the plant. Our studies have shown that as much as 50% of the nitrogenous pool in the fern can be ac-

TABLE II. Nitrogenase Activity of Isolated Cyanobacteria

	Specific Activity[a]	
	Autotrophic, light	*Heterotrophic, dark*
Anabaena flos-aquae control	*10*	*10*
Azolla cyanobacterium	*10*	*28*

[a]*n moles acetylene reduced/mg protein/min. Assays in Argon-acetylene; light.*

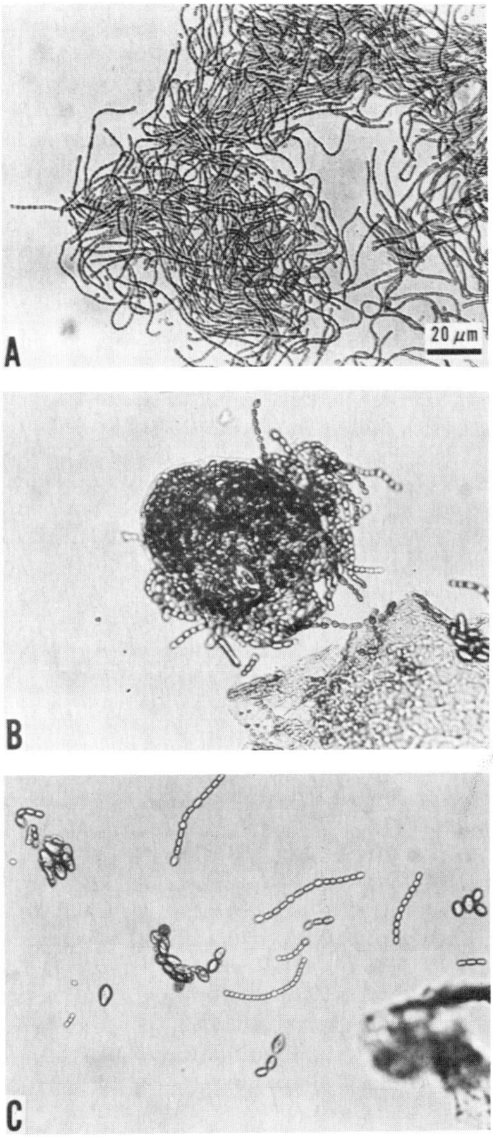

Fig. 1. Types of algal (cyanobacterial) preparations. (A) Pure culture of cyanobacteria. (B) Bundle released from plant digest. (C) Cells released from bundle showing swollen, morphologically altered cells.

Table III. Characteristics of Azolla Symbiont

1. *Ammonia excretion when in plant*
2. *Permeability and morphology change*
3. *Heterotrophic nitrogen fixation high*

counted for as free ammonia in nitrogen-fixing plants. Furthermore, several observations we have made with isolated bundle preparations indicate that the cyanobacteria in leaves have lost their permeability barrier to large molecules, as well as to ammonium ions. Thus, most of the symbiont cells appear to be in a senescent state, which explains why it is difficult to isolate viable algae from the symbiosis. Our observations are summarized in Table III.

An examination of Tables I and III reveals a striking property of this "ideal symbiont." Although the cells are photosynthetic, they appear to prefer a heterotrophic route for nitrogen fixation, as indicated in studies with isolated cyanobacteria. Choice of a photosynthetic system to study as an ideal symbiosis has now led us to the notion that, in fact, all symbionts may be essentially heterotrophic nitrogen fixers - even cells that are potentially capable of supplying photosynthetically fixed nitrogen to the plant. The high rate of nitrogen fixation and growth of the fern appears to dictate that large amounts of plant carbon be diverted to the fixation process. It is also of interest that the preferred heterotrophic substrate for our cyanobacterial isolates, fructose, is found as a major pool component in the fern. In many ways, therefore, the algal symbiont of *Azolla* resembles the root nodule bacteria, or *Rhizobia*, more than that of a photosynthetic system.

We have thus again revisited the relationship between nitrogen fixation and photosynthesis. The two processes do not occur in nature within the same plant cell. We now have indications that even when both partners of a nitrogen-fixing symbiosis are photosynthetic, the system appears to prefer a heterotrophic mode of fixation. I can only hope that we do not have to wait another 20 years for new insights into this problem. However, when the new leads come, I shall not be at all surprised if they are again provided by Professor Kamen.

REFERENCE

Lindstrom, E. S., Newton, J. W., and Wilson, P. W. (1952). *Proc. Natl. Acad. Sci. U.S. 38*, 392-396.

NITROGENASE: PREPARATIVE METHODS AS A BASIS FOR NEW INVESTIGATIVE APPROACHES OR WHEN LESS CYTOCHROME IS BETTER

C. E. McKenna
H. T. Nguyen
C. W. Huang
M. C. McKenna
J. B. Jones
P. J. Stephens

Department of Chemistry
University of Southern California
Los Angeles, California 90007

Nitrogen tantalizes mankind with the paradox of poverty in the midst of plenty (M. D. Kamen in *Sci. Amer.* Mar. 1953, p. 38).

Purification "of ... cell-free nitrogen-fixing" preparations has proved a trying task (Mahler and Cordes (1971) *In* "Biological Chemistry," 2nd ed., p. 759. Harper and Row, N.Y.).

MARTIN KAMEN AND NITROGEN FIXATION

On his journey from cyclotrons to cytochromes, Martin Kamen developed a comprehensive interest in two fundamental processes, photosynthesis and nitrogen fixation. The contributions made by him in the latter area are typically diverse. In 1949 he made the important discovery (with H. Gest) that the non-sulfur purple bacterium *Rhodospirillum rubrum* can fix nitrogen (1), thus revealing a significant new class of N_2-fixing microorganisms, the photosynthetic bacteria (2). In

 ISBN 0-12-397580-8

collaboration with Ruben and Hassid a decade earlier, Martin Kamen suggested what remains to this day the most sensitive direct assay for N_2 fixation - detection of radioactive ammonia from $^{13}N_2$ (3). Appropriately enough, employment of this method is restricted to laboratories having direct access to a cyclotron due to the short (10 min) half-life of ^{13}N. Despite this rather specialized requirement, the practicality of the technique has been confirmed by other workers (4). In addition to these and other scientific contributions, Martin Kamen has promoted research activity in the area of N_2 fixation at several levels. To cite a few relatively recent examples, he stimulated work on the *nif* gene and on biochemical aspects of nitrogen fixation during the late 1960's at the University of California at San Diego. From 1972 to the present time he has served on the Science Advisory Committee of the C. F. Kettering Foundation, directly influencing the mission-oriented research programs in N_2 fixation (and photosynthesis) at the Kettering Laboratories in Yellow Springs, Ohio. Early in 1975 Martin moved to the University of Southern California (receiving joint appointments in three different departments), where he assumed a favorite role - that of institutional demiurge. Of decisive importance to N_2 fixation work at USC was his creation of an *ad hoc* interdepartmental entity, the Chemical-Biological Development Laboratory, which greatly strengthened the biochemical infrastructure available there to young researchers in a number of fields.

INTRODUCTION

The use of crop rotation to enhance soil nitrogen content was known to the ancient Chinese (5) and Romans (6), but recognition that atmospheric N_2 is fixed biologically did not come until the early 19th century (7). The ability of free-living aerobes, such as *A. vinelandii*, and anaerobes, such as *C. pasteurianum,* to grow on N-free medium was established around 1900 (7). The postulate that a specific enzyme [assigned the name nitrogenase by Burk (8)] was responsible for this extraordinary biochemical capacity first received clear expression in the 1930's, but demonstration of cell-free nitrogen fixation in a preparation from a N_2-fixing microorganism was not accomplished until 1960 (9). Six years later, Bulen and co-workers succeeded in demonstrating that cell-free N_2 fixation in *Azotobacter* extracts was catalyzed by two essential protein components and proceeded in the presence of an exogenous electron donor (e.g., dithionite) only with concomitant hydrolysis of ATP to ADP and inorganic phosphate (10).

One nitrogenase component (Fe-Mo protein, Mo-Fe protein, protein I, azofermo or molybdoferrodoxin) has been shown to have a mol. wt. of 2.2 to 2.8 × 10^5 daltons and contains 1 - 2 Mo with 22 - 36 Fe atoms, as well as a considerable amount of "acid-labile" S^{2-}; the other component (Fe protein, protein II, azofer or azoferredoxin) contains 4 Fe and 4 S^{2-} per mol. wt. of 5.5 to 6.5 × 10^4 daltons (11). As suggested by the uncertain limits given above, molecular weight and especially bound metal content data have fluctuated disconcertingly over the past 10 years, even for preparations obtained from one particular microorganism by the same laboratory. This is partly due to purification problems caused by the extreme O_2 lability of the two components, which necessitates rigorous anaerobic separation, handling, and storage procedures.

N_2 fixation is known to be confined (unfortunately) to the procaryotes, including heterotrophic bacteria such as *A. vinelandii, C. pasteurianum,* and *K. pneumoniae*; the agriculturally important symbiotic bacteria (e.g., *R. japonicum*), which fix N_2 in association with certain plants; blue-green algae; and others (7, 12). Some types, such as *R. rubrum* which is a photosynthetic N_2-fixing bacterium, have been recognized comparatively recently (1, 7).

Nitrogenase-catalyzed N_2 fixation accounts for hundreds of millions of tons of NH_3 added to the soil annually and thus is of obvious agrobiological importance. Chemical N_2 fixation by the industrial Haber-Bosch process (13) produçes only a fraction of this total. The Haber-Bosch process has had major economic impact chiefly in the more developed countries and suffers from inflating energy and capital investment costs. For this reason the practical incentives to understanding nitrogenase in structural and functional terms are great. Chemists and biochemists are also intrigued by the now virtually certain occurrence of hitherto unprecedented types of natural metal clusters in nitrogenase Fe-Mo proteins (14, 15) and hope to derive fundamental, possibly far-reaching insights into reductive activation and energy transduction by elucidating the mechanisms involved in the unique catalytic properties of the enzyme.

Approaches* to investigating nitrogenase as a catalyst may be conveniently subdivided into: (1) the chemical reactivity

**A complementary genetic approach (16) is based on the concept of introducing the nif gene from an existing N_2-fixing microorganism into nonsymbiotic plant genomes. Apart from the as yet unresolved problem of effecting this genetic transfer, expression of the transferred gene in an agriculturally significant way still faces serious obstacles to success in that nitrogenase biosynthesized in plant cells would require systems to prevent O_2 inactivation of the enzyme, metabolic adjustment to the larger ATP demand (at least 8 ATP/N_2 molecule fixed), and integration of subcellular NH_3 output with overall nitrogen metabolism in the plant.*

(substrate-inhibitor) approach (including kinetic analyses and equilibrium binding studies), largely focused on the catalytic mechanism; (2) spectroscopic methods, mostly directed to definition of metal ion structures, states, and electron-transfer behavior; and (3) the model system approach which in default of detailed structural information about the enzyme (other than the presence of Mo and Fe) has necessarily been limited to mimicking nitrogenase chemical activities or spectroscopic properties. We shall concern ourselves here primarily with studies carried out directly on nitrogenase using approaches (1) and (2), with emphasis on the latter.

Substrate Studies

During 1965 - 1968 a burst of discoveries by a number of laboratories established that nitrogenase is a versatile catalyst capable of reducing several classes of small unsaturated molecules: N_2O, N_3^-, HCN and certain of its organic derivatives and $C_2H_2^2$. Reductions of these adventitious (or alternate) substrates resemble N_2 fixation in that they are CO-inhibited and are coupled obligatorily to MgATP hydrolysis. This dramatic efflorescence of chemical studies was brief and ended rather abruptly: virtually no new substrates were found for almost a decade afterward (2). However we have discovered that cyclopropene is a very effective nitrogenase substrate and can be used as a chemical probe of the active site for reduction (17a - f).

Spectroscopic Studies

During the past 10 years spectroscopic approaches have increasingly played a major role in biochemical nitrogenase research. Epr and Mössbauer spectra were first reported in 1971 (18) and 1970 (19), respectively, and the first EXAFS spectra of Mo in the Fe-Mo protein were described in 1976 (20). Although, in the latter two cases the initially reported spectra were subsequently found to be artifactual in the sense that the spectroscopic protein samples had been inadvertently air-oxidized (21 - 22), extensive work on active native components continuing to the present has gradually led to improved understanding of nitrogenase metal clusters (23, 24). The picture that has emerged suggests that during turnover electrons are transferred from a reduced state of the Fe protein to the Fe-Mo protein, which thereupon becomes "super-reduced;" in this state it (presumably) effects reduction of bound substrate (25).

Cluster extrusion experiments (26) [also referred to as cluster displacement experiments (27)] indicate that virtually all of the Fe in Fe proteins and about half the Fe in Fe-Mo proteins can be recovered in "classical" cubane-type 4Fe-4S clusters such as occur in bacterial ferredoxins or the HIPIP protein from *C. vinosum*; however, the Fe protein epr is anomalous and the epr-Mössbauer properties of the native Fe-Mo pro-

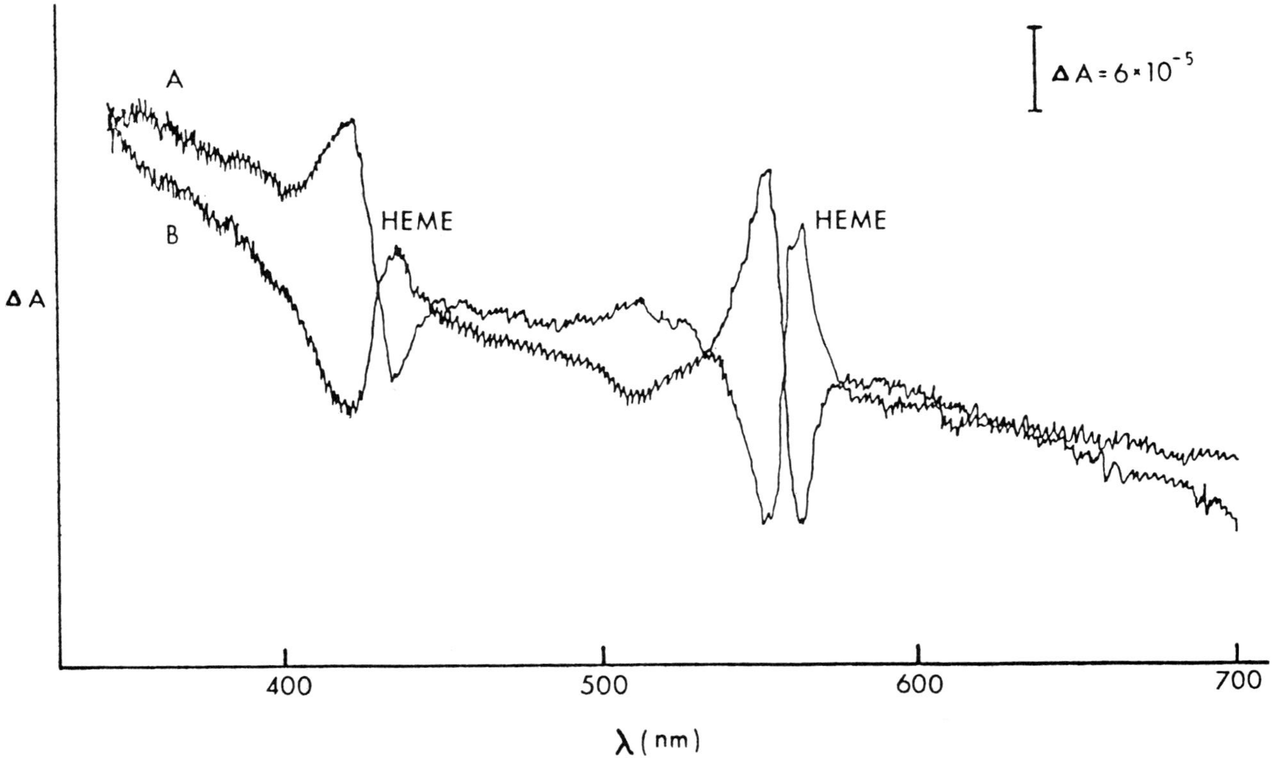

Fig. 1. Visible MCD spectrum (Cary[61]) of Av1, SA ≅ 1600 (17a). The protein sample was 5.0 mg/ml in 0.025 M Tris, pH 7.4 and 0.25 M NaCl, with 1.6 mM dithionite. The cell path length was 1 mm. A = (+) field; B = (−) field; applied field ±4 Tesla.

tein are not fully consistent with the presence of such clusters (11, 14, 28). Several important states of the component Fe-Mo centers Fe protein and Fe-Mo protein are diamagnetic, and thus inaccessible by epr (29). Both epr and Mössbauer require cryogenically cooled samples, and the freezing process introduces potential complications (30). EXAFS can provide nearest neighbor information on Mo in Fe-Mo protein but is incapable of specifying ligand geometry and has failed to provide evidence of substrate-Mo interactions thus far.

From the foregoing discussion, it should be clear that application of new spectroscopic techniques, preferably ones complementing existing methods and also permitting use of solution samples at room temperature (nitrogenase is generally assayed at 30°) would be beneficial. Careful analysis of the literature suggests that routine provision of highly active, homogeneous N_2ase components in large amounts is a *sine qua non* for many reasonable new approaches, especially spectroscopic ones. Several interesting spectroscopic approaches to N_2ase study, e.g., CD, MCD, and NMR are quite insensitive compared to, for example, epr. It is instructive to consider some of the *prima facie* evidence in the literature that purification difficulties have resulted in important work being done on nitrogenase protein samples having suboptimal activity relative to the then existing state-of-the art value (one assumes that other things being equal, it is generally safest to utilize the most active proteins available in adequate quantities). Thus, in 1973 Shah and Brill published a procedure which yielded *A. vinelandii* Fe-Mo protein of specific activity (SA) 1638 and Fe protein of specific activity (SA) 1815 (31). Quoting this procedure, in 1974 Kleiner and Chen published analytical data on MW and metal content in Av1* and Av2 with reported activities of only 900 - 1100 (32). A 1975 paper by Brill and Shah (33) presented results of kinetic studies performed using Av2 of SA 800. More recently, an important EXAFS study was done on Av1 having an initial activity of 1100 (23). We ourselves first reported in 1976 on a new substrate discovery using Av2 of about 1200 SA (Av1-SA was about 1600), relying at that time on a different purification method (17a). Obtaining reproducibly high activity in serial "routine" purifications has evidently been a frustrating activity for workers in the past.

We were forced to confront the protein quantity and quality issue by preliminary attempts to observe nitrogenase

The "Sussex nomenclature" - Av1: Fe-Mo protein from A. vinelandii; Av2 Fe protein from A. vinelandii. Thus Kp1/Kp2 (K. pneumoniae), Cp1/Cp2 (C. pasteurianum); etc.

component CD and MCD spectra in the near-ir (34), visible, and near-uv regions. Other workers (35, 36) had reported visible CD to be absent from Fe proteins, and very weak or absent in Fe-Mo proteins; MCD spectra were not previously recorded. We have recently been successful in measuring CD and MCD in simple Fe-S proteins at room temperature (2Fe-2S and 4Fe-4S proteins). This work showed that CD and MCD together are diagnostic of cluster type and oxidation state (37, 38). Our first experiment told us that fairly elevated concentrations of Av1 or Av2 would yield both CD and MCD over the near iv-visible-near uv range and thus established that previous reports on the non-existence of visible CD would be attributed to experimental insensitivity rather than intrinsic absence of the phenomenon. The preliminary MCD spectra of the Fe-Mo protein in particular also made it clear to us that a trace heme contaminant present in our then current protein preparation would spectrally dominate the visible MCD (Fig. 1).

The problem of heme-like features in absorbance spectra of Av1 preparations was not new. The visible spectrum of repeatedly recrystallized Av1, SA = 1400 published in 1971 - 1972 displayed marked bands near 557, 525, and 420 nm (39). (In the then-standard purification method, the Av1 protein was eluted from a DEAE-cellulose column just after a cytochrome-rich fraction.) While working as a guest of the Kamen laboratory at UCSD, the author obtained a small Av1 sample of nearly equivalent activity, but with an absorbance spectrum almost completely lacking the features mentioned above, which were ascribed to a heme-containing impurity (40). Av1 absorbance spectra with little or no discernible heme bands were later published by Shah and Brill and by Bulen (41). Unfortunately, levels of "cytochrome" impurity giving at worst an unobtrusive bump in the Soret region of Fe-Mo absorbance spectra proved to interfere seriously in the visible MCD spectra (Fig. 1), due to the higher anisotropy value associated with the heme chromophore. This forced us to search for a preparative methodology that would yield very large amounts of "cytochrome"-free (by MCD) Av1 and Av2.

If the time needed to culture the source bacteria is included, the logistics of nitrogenase purification are best worked out in terms of activity-units (produced) per person-months (invested). This ratio ideally is much larger than the corresponding ratio for activity-units consumed in productive experiments. To achieve a favorable balance for enzyme-intensive experiments such as CD/MCD spectroscopy, the economy of effort realizable by working on a multigram preparative scales is essential. We took as a starting point the published procedure giving the best SA values for both components (31). A drawback of this method was the small scale described - 48 mg for Av1 and 10 - 31 mg Av2. The final step (preparative electrophoresis) in the Av2 procedure is difficult and not readily amenable to scale-up. We, therefore, set out to design a purification method suitable for large scale, while yielding highly active, heme-free Av1 and Av2.

Given our desired scale, multikilogram amounts of N_2-fixing bacteria were required, corresponding to batch cultures on the order of 10^3 liters. Again a favorable output/consumption ratio is important. We sought to solve this problem while circumventing the unavailability of correspondingly large fermentation equipment in our laboratory by the simple expedient of using a chemostat. We noticed that activity levels in crude extracts harvested continuously at optimal log phase cell density were significantly higher than in extracts obtained from cells harvested from cultures grown in simple batch fermentors. It was also observed that, in general, better final activities in isolated components were associated with higher initial crude extract activity. Although batch culture seems to have been used exclusively in earlier work, continuous culture is attractively suited for preparative growth of aerobic N_2 fixers. The selectivity of the medium allows unsophisticated equipment to be used: in the past, horse troughs have been utilized as batch fermentors for *Azotobacter* (42)!

We have been successful in evolving methods to prepare reproducibly several grams of both Av1 and Av2 with specific activities of 2000 or better. No detectable "cytochrome" bands can be observed in the Av1 (or Av2) visible MCD. The working time required beginning with initial cell culture is about 3 - 5 weeks, including time spent on essential analytical work. The availability of such amounts of pure nitrogenase components has made it possible to obtain excellent CD and MCD spectra, establishing the existence of these chiroptical phenomena in nitrogenase (43).

These spectra demonstrate directly that the reduced Fe protein in solution at 25°C contains a (C^{3-}) 4Fe-4S cluster; provide the first direct evidence for MgATP binding to the oxidized Fe-protein; and show that novel metal cluster chromophores are present in the Fe-Mo protein (43).

The macroscale preparative method underlying our CD/MCD spectroscopic investigations of nitrogenase will be presented below.

RESULTS AND DISCUSSION

Preparative-Scale Continuous Culture of A. vinelandii

A simple 100 liter chemostat designed in our laboratory for preparative growth of N_2-fixing *Azotobacter* is shown schematically in Fig. 2.

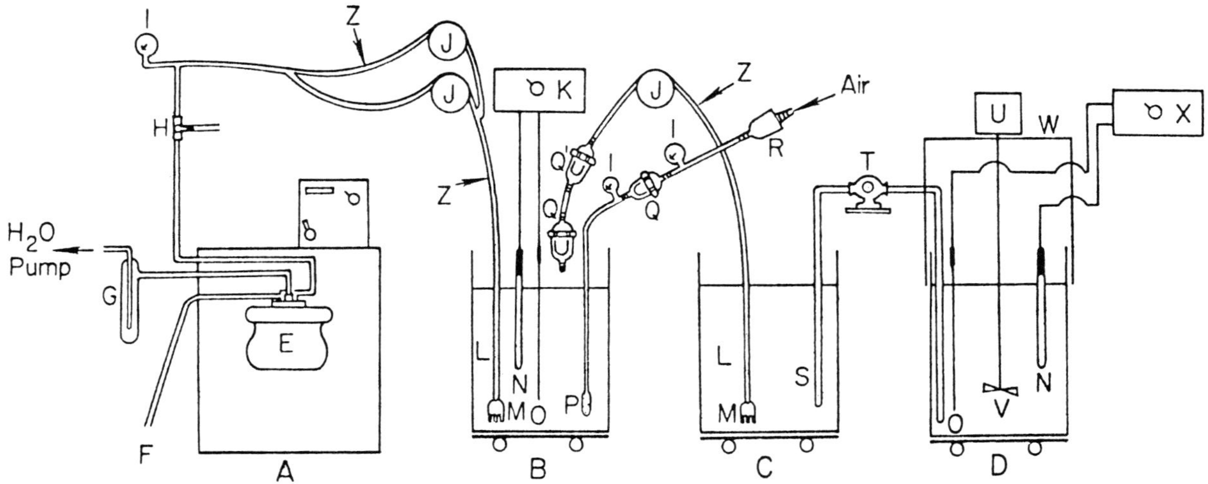

Fig. 2. Chemostat used for preparative growth of Azotobacter vinelandii: A, centrifuge; B, fermentor drum; C, feeder drum; D, mixer drum; E, continuous flow rotor[†]*; F, effluent tube; G, trap for scavenger pump; H, three-way valve; I, pressure gauge; J, peristaltic pump; K, temperature controller (1 KW); L, feeding tube for fermentor drum; M, flange; N, heater rod; O, temperature probe; P, air sparger; Q, fine water filter; Q', coarse water filter; R, air filter; S, feeding tube for feeder drum; T, Flotec pump; U, stirring motor, V, mixing paddle; W, supporting frame for motor; X, temperature controller (2 KW).*

[†]*Beckman JCF-Z.*

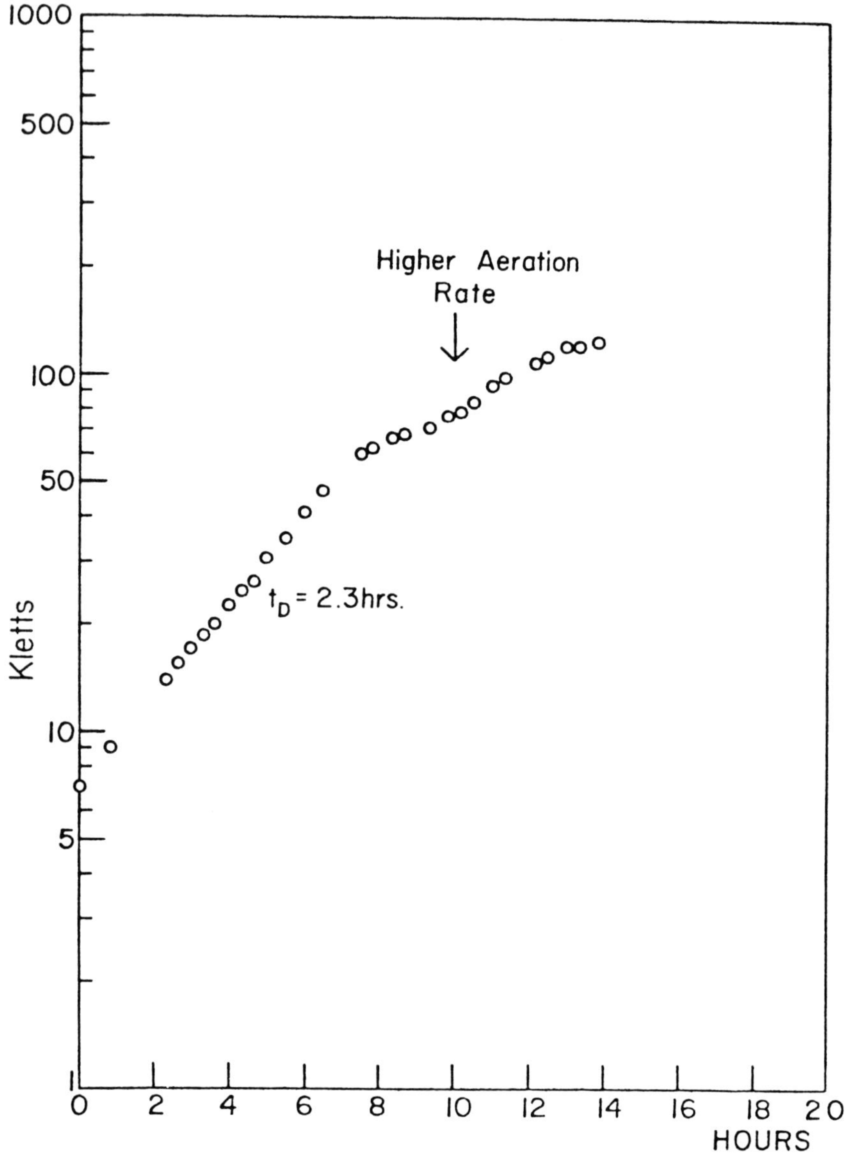

Fig. 3. Fermentor growth curve of Azotobacter vinelandii. t_D = *doubling time.*

In a chemostat operated at constant volume V with continuous influx of fresh medium at rate (dV_m/dt) and continuous efflux of culture at rate (dV_c/dt), it is self-evident that

$$dV_m/dt = -dV_c/dt = F$$

where F = rate of flow through the chemostat. The dilution rate for any species in solution D = F/V. If the bacteria are growing exponentially, then the net time change in their concentration

$$dC_b/dt = \mu \cdot C_b - D \cdot C_b$$

$$= (\mu - D) \cdot C_b$$

since $\mu = \ln 2/t_D$ where t_D = culture doubling time. Culture cell density can be maintained at any value within log phase by setting $\mu = D$, requiring that

$$F = (V/t_D) \cdot \ln 2$$

In practice, harvest yield is naturally maximized at the highest cell density consistent with logarithmic growth. This point can be estimated from the fermentor growth curve (Fig. 3) under postinoculation batch culture conditions, which also give a value for t_D (1.7 - 2.3 h). Provided that the culture is not allowed to pass too far out of log phase, setting F as described will cause the cell density to relax back to a level close to optimum. Depending on the efficiency of the air spargers and the age of the culture, harvest cell densities corresponding to 110 - 160 Klett units (1.6 - 2.2 g/liter) are easily achieved. At an average flow of 35 liters/hr, a full JCF-Z rotor (500 - 600 g cell paste) is obtained every 8 hr, giving a daily output of 1.7 kg. Runs producing over 5 kg cell paste, or the equivalent of a 2500-liter culture in a simple batch fermentor, are routinely accomplished without noticeable microscopic contamination. Use of a pure inoculum, microbiological filtration of the air and medium feeds (Fig. 2), uninterrupted air flow, and maintenance of a constant high dilution rate (avoidance of prolonged post steady-state growth) are factors contributing to the absence of significant contaminants.

PREPARATION OF NITROGENASE COMPONENTS

The purification is outlined in Scheme I; yield and activity data are summarized in Table I. The overall yields are equivalent to, or better than, those cited for a recently published

Scheme I

Purification of Nitrogenase
from *Azotobacter vinelandii**

Part A

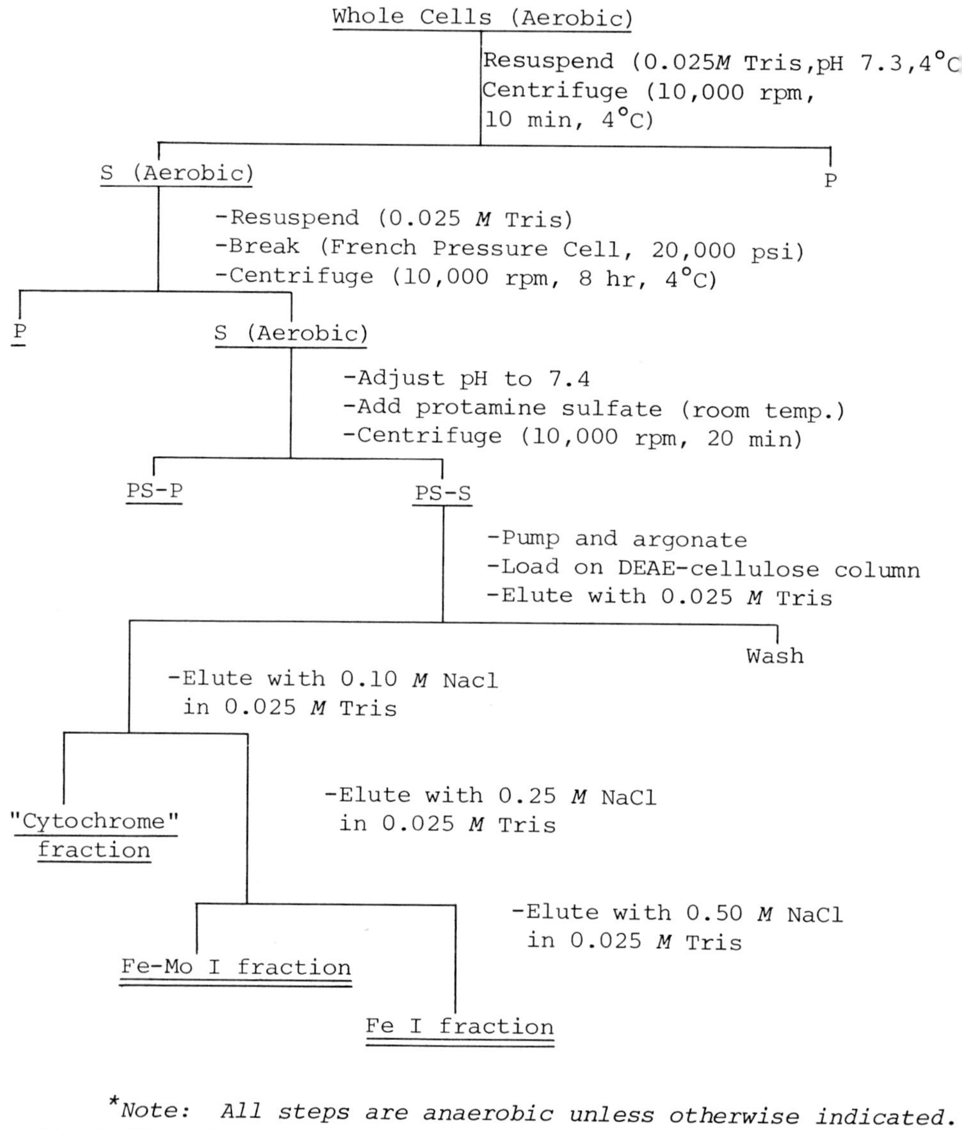

**Note: All steps are anaerobic unless otherwise indicated. All buffers have a pH of 7.4, S, supernatant; P, pellet.*

Part B

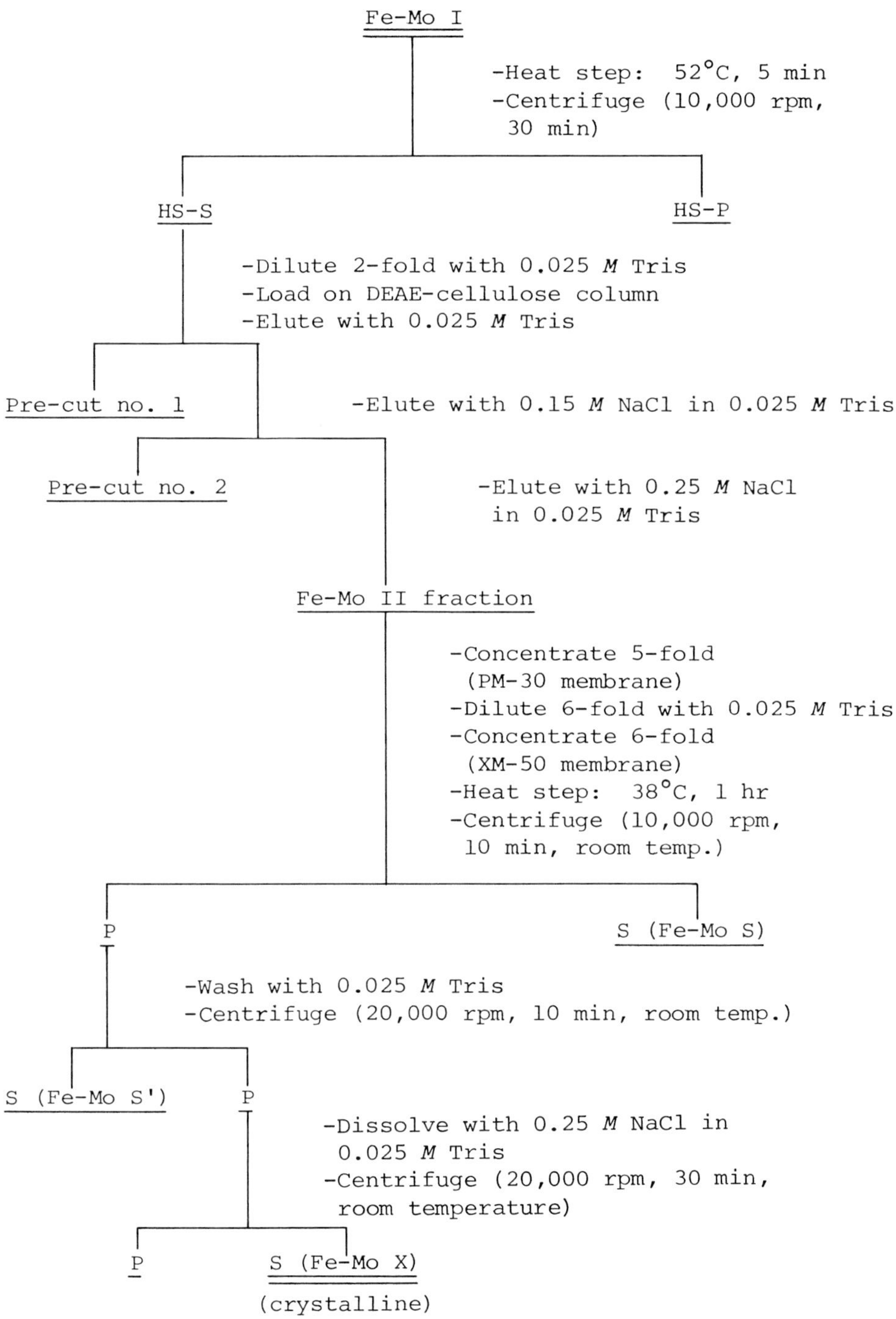

Fe-Mo I
-Heat step: 52°C, 5 min
-Centrifuge (10,000 rpm, 30 min)
HS-S
HS-P
-Dilute 2-fold with 0.025 *M* Tris
-Load on DEAE-cellulose column
-Elute with 0.025 *M* Tris
Pre-cut no. 1
-Elute with 0.15 *M* NaCl in 0.025 *M* Tris
Pre-cut no. 2
-Elute with 0.25 *M* NaCl in 0.025 *M* Tris
Fe-Mo II fraction
-Concentrate 5-fold (PM-30 membrane)
-Dilute 6-fold with 0.025 *M* Tris
-Concentrate 6-fold (XM-50 membrane)
-Heat step: 38°C, 1 hr
-Centrifuge (10,000 rpm, 10 min, room temp.)
P
S (Fe-Mo S)
-Wash with 0.025 *M* Tris
-Centrifuge (20,000 rpm, 10 min, room temp.)
S (Fe-Mo S')
P
-Dissolve with 0.25 *M* NaCl in 0.025 *M* Tris
-Centrifuge (20,000 rpm, 30 min, room temperature)
P
S (Fe-Mo X)
(crystalline)

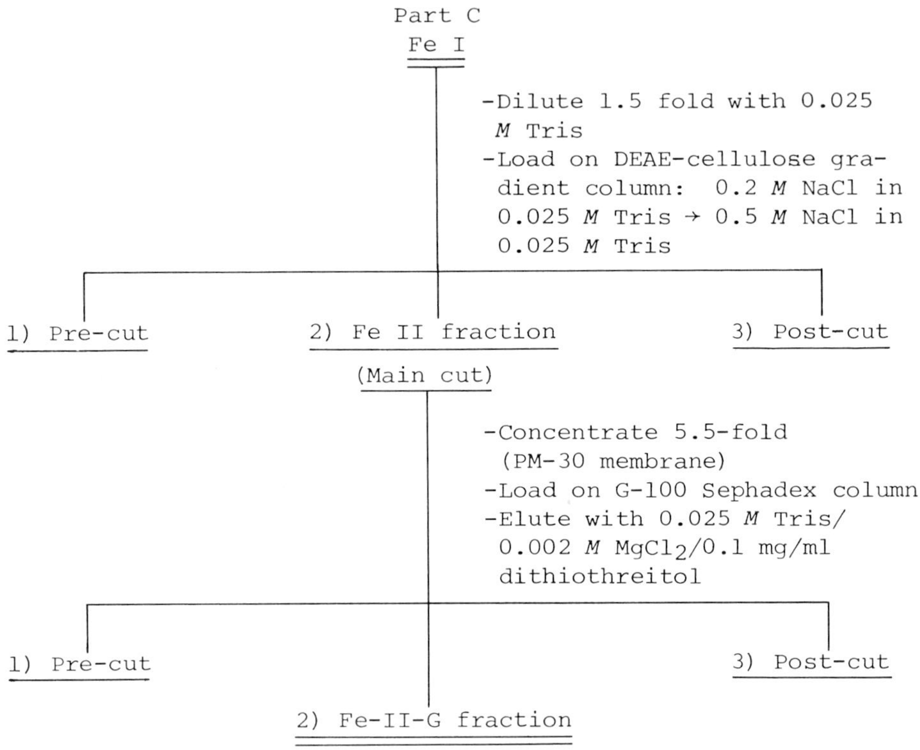

version (44) of the Shah and Brill procedure (31); the scale is 10^2 larger; and the SA values equal or exceed those previously published. The purified enzyme components are essentially homogeneous by anaerobic gel electrophoresis. Most importantly, the visible absorbance and MCD spectra of both proteins show no trace of cytochrome-like features. We have presented baseline-corrected and field/concentration-normalized MCD spectra of these Av1 and Av2 samples elsewhere (43); here, the complete absence of heme contributions to the uncorrected Av1 MCD spectrum is clearly demonstrated in Fig. 4, which may be compared directly to that in Fig. 1.

CONCLUSION

The development of a reliable, convenient macroscale procedure for isolation of heme-free *A. vinelandii* nitrogenase components has enabled us to demonstrate that contrary to

TABLE I. *Purification of Fe-Mo and Fe Proteins from 1 kg of Azotobacter vinelandii*

Fraction	Vol. (ml)	Total activity (units)[b]	Total protein (mg)	SA[a] (units/mg)	Recovery (%)
Fe-Mo protein					
1. Crude Extract	1815	4.54×10^6	1.03×10^5	44	100
2. Protamine sulfate fractionation (PS-S)	2495	4.31×10^6	8.76×10^4	49	95
3. DEAE-Cellulose-I (Fe-Mo I)	701	9.33×10^6	2.10×10^4	444	100[c]
4. 52 C Supernatant (HS-S)	715	8.77×10^6	1.20×10^4	730	94
5. DEAE-Cellulose-II (Fe-Mo II)	330	6.31×10^6	6.01×10^3	1050	72
6. Crystallized (Fe-Mo X)	158	4.10×10^6	2.05×10^3	2000	65
Fe protein					
1. Crude extract	1815	4.54×10^6	1.03×10^5	44	100
2. Protamine sulfate fractionation (PS-S)	2495	4.31×10^6	8.76×10^4	49	95
3. DEAE-Cellulose-I (Fe-I)	400	4.26×10^6	6.00×10^3	710	100
4. DEAE-Cellulose-II (Fe-II)	115	3.63×10^6	2.57×10^3	1410	85
5. G-100 Sephadex (Fe-II-G)	190	3.38×10^6	1.61×10^3	2100	79

[a]SA, specific activity (nmoles of ethylene formed/min per mg of protein).
[b]Units, nmoles of ethylene formed/min.
[c]Av2 added to assay to saturate Av1.

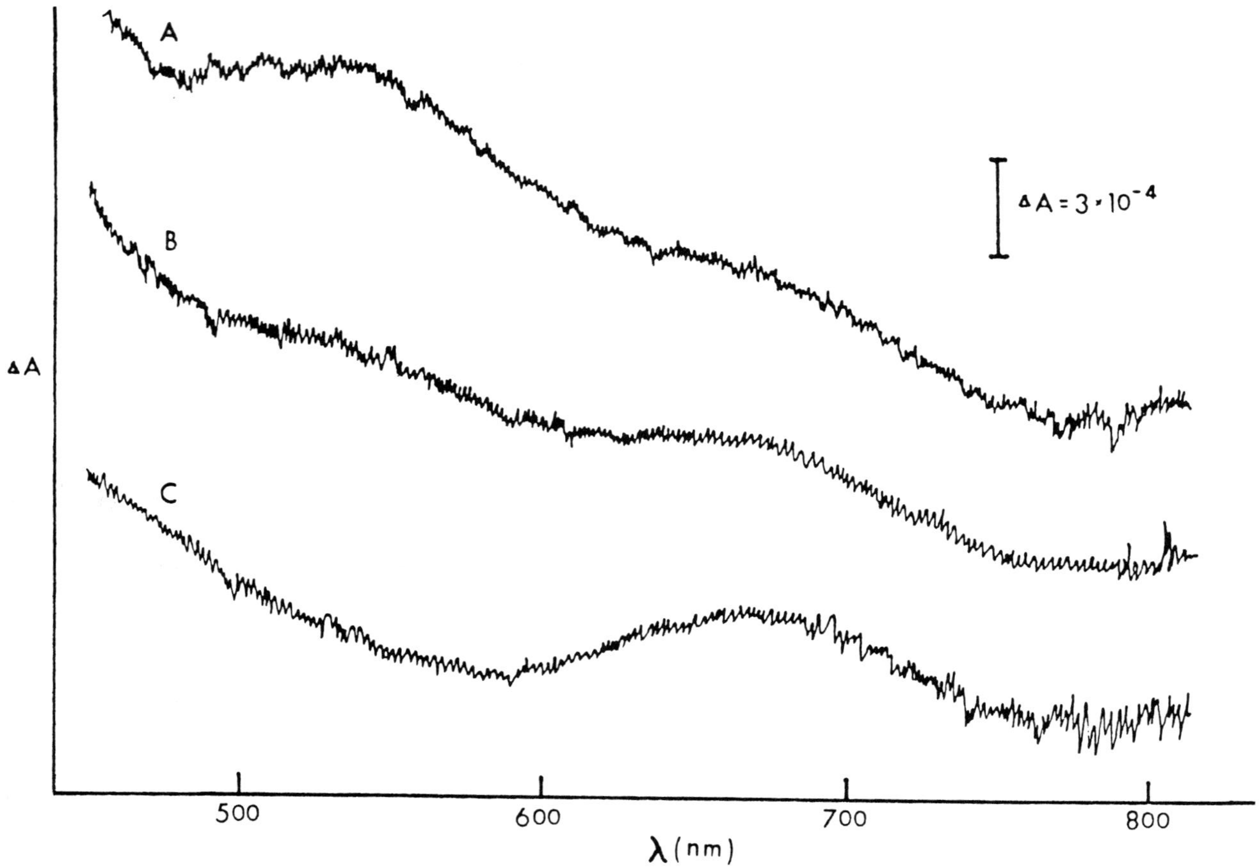

Fig. 4. Visible MCD spectrum (Cary 6l) of Av1, SA = 2000 (this work). The protein sample was 22 mg/ml in 0.025 M Tris, pH 7.4, and 0.25 M NaCl, with 1.6 mM dithionite. A = (+) field, B = zero field (CD), C = (-) field, applied field ±4 Tesla.

previous reports, CD and MCD spectra are obtainable for the *Azotobacter* Fe-Mo and Fe proteins. As we describe elsewhere (43), the spectra are interpretively of great value and create the basis for further development of a new, potentially powerful physical probe of nitrogenase - chiroptical spectroscopy. The large amounts of enzyme afforded have also greatly facilitated development in our laboratory of new chemical approaches to probe nitrogenase active site function (17 a - d, f). These were among the first kinetic studies to utilize consistently well-defined mixtures of highly purified, optimally active *A. vinelandii* components. We expect that the routine availability of 5 g or more of nitrogenase will lead to introduction of yet additional new physical and other methods previously made impractical by sample size limitations.

ACKNOWLEDGMENT

We thank USDA (CEM[c]K), the Frasch Foundation (CEM[c]K) and NSF (CEM[c]K/PJS) for their support.

REFERENCES

1. Kamen, M. D. and Gest, H. (1949). *Science 109,* p. 560.
2. Burns, R. C. and Hardy, R. W. F. (1975). *In* "Molecular Biology, Biochemistry, and Biophysics," Vol. 21: Nitrogen Fixation in Bacteria and Higher Plants (A. Kleinzeller, G. F. Springer, and H. G. Wittman, eds.). Springer-Verlag, New York.
3. Ruben, S., Hassid, W. Z., and Kamen, M. D. (1940). *Science 91,* p. 578.
4. Nicholas, D. J. D., Silvester, D. J., and Fowler, J. F. (1961). *Nature 189,* 634.
5. Jia Si Xie (1978). *In* "Chi Ming Yao Su" (544) cited in "Zong Guo Gu Dai Ke Ji Cheng Jiu" (Scientific and Technological Achievements in Ancient China) Research Institute for Natural Sciences History ed., pp. 356-358. Chinese Youth Publisher, Beijing, China.
6. Wilson, P. W. (1940). "The Biochemistry of Symbiotic Nitrogen Fixation." University of Wisconsin Press, Madison, Wisconsin.
7. Stewart, W. D. P. (1966). "Nitrogen Fixation in Plants." The Athlone Press, London.
8. Burk, D., Lineweaver, H., and Horner, C. K. (1934). *J.*

Bacteriol. *27,* 325.
9. Carnahan, J. E., Mortenson, L. E., Mower, H. F., and Castle, J. E. (1960). *Biochim. Biophys. Acta 38,* 188.
10. Bulen, W. A. and LeComte, J. R. (1966). *Proc. Natl. Acad. Sci. U.S. 56,* 979.
11. Eady, R. R. and Smith, B. E. (1979). *In* "Treatise on Nitrogen Fixation" (R. C. Burns, ed.) Section II, ch. 2. Wiley-Interscience, New York.
12. Postgate, J. R. (ed.) (1971). "The Chemistry and Biochemistry of Nitrogen Fixation." Plenum, New York.
13. Haber, L. F. (1968). *Endeavour 27,* 150.
14. Orme-Johnson, W. H., Davis, L. C., Henzl, M. T., Averill, B. A., Orme-Johnson, N. R., Munck, E., and Zimmerman, R. (1977). *In* "Recent Developments in Nitrogen Fixation" (W. Newton, J. R. Postgate, and C. Rodriguez-Barrueco, eds.), pp. 131-178. Academic Press, New York.
15. Rawlings, J., Shah, V. K., Chisnell, J. R., Brill, W. J., Zimmerman, R., Munck, E., and Orme-Johnson, W. H. (1978). *J. Biol. Chem. 253,* 1001.
16. Malavich, T., Bachhuber, M., Howe, M., Bishop, E., Maier, J., and Brill, W. J. (1977). *In* "Recent Developments in Nitrogen Fixation" (W. Newton, J. R. Postgate, and C. Rodriguez-Barrueco, eds.), pp. 331-336. Academic Press, New York.
17a. McKenna, C. E., McKenna, M-C., and Higa, M. T. (1976). *J. Amer. Chem. Soc. 98,* 4657.
17b. McKenna, C. E., Huang, C. W., Jones, J. B., McKenna, M-C., Nakajima, T., and Nguyen, H. T. (1980). *In* "Nitrogen Fixation" Vol. I, (W. E. Newton and W. Orme-Johnson, eds.), pp. 223-235. University Press, Baltimore, Maryland.
17c. McKenna, C. E., McKenna, M-C., and Huang, C. (1979). *Proc. Natl. Acad. Sci. U.S. 76,* 4773.
17d. McKenna, C. E. and Huang, C. (1979). *Nature (London) 280,* 609.
17e. McKenna, C. E., Nakajima, T., McKenna, M-C., and Newton, W. E. (1982). *J. Am. Chem. Soc. 104.*
17f. McKenna, C. E., Jones, J. B., Eran, H., and Huang, C. (1979). *Nature (London) 280,* 611.
18. Dalton, H., Morris, J. A., Ward, M. A., and Mortenson, L. E. (1971). *Biochemistry 10,* 2066.
19. Kelly, M. and Lang, G. (1970). *Biochim. Biophys. Acta 223,* 86.
20. Cramer, S. P., Eccles, T. K., Kutzler, F. W., Hodgson, K. O., and Mortenson, L. E. (1976). *J. Amer. Chem. Soc. 98,* 1287.
21. Smith, B. E. and Lang, G. (1974). *Biochem. J. 137,* 169.
22. Cramer, S. P.,Hodgson, K. O., Gillum, W. O., and Mortenson, L. E. (1978). *J. Amer. Chem. Soc. 100,* 3398.

23. Orme-Johnson, W. H., Hamilton, W. D., Jones, T. L., Tso, M-Y. W., Burris, R. H., Shah, V. K., and Brill, W. J. (1972). *Proc. Natl. Acad. Sci. U.S. 69,* 3142.
24. Orme-Johnson, W. H. and Davis, L. C. (1977). *In* "Iron-Sulfur Proteins" (W. Lovenberg, ed.), Vol. 3, p. 15. Academic Press, New York.
25. Burris, R. H. and Orme-Johnson, W. H. (1976). *In* "Proceedings of the First International Symposium on N_2 Fixation" (W. E. Newton and C. J. Nyman, eds.), pp. 208-233. WSU Press, Pullman, Washington.
26. Gillum, W. O., Mortenson, L. E., Chen, J. S., and Holm, R. H. (1977). *J. Amer. Chem. Soc. 99,* 584.
27. Averill, B. A., Bale, J. R., and Orme-Johnson, W. H. (1978). *J. Amer. Chem. Soc. 100,* 3034.
28. Zimmermann, R., Munck, E., Brill, W. J., Shah, V. K., Henzl, M. T., Rawlings, J. and Orme-Johnson, W. H. (1978). *Biochim. Biophys. Acta 537,* 185.
29. Watt, G. (1980). *In* "Molybdenum Chemistry of Biological Significance" (W. E. Newton and S. Otsuka, eds.), p. 3. Plenum Press, New York.
30. Davis, L. C. and Orme-Johnson, W. H. (1976). *Biochim. Biophys. Acta 452,* 42.
31. Shah, V. K. and Brill, W. J. (1973). *Biochim. Biophys. Acta 305,* 445.
32. Kleiner, D. and Chen, C. H. (1974). *Arch. Mikrobiol. 98,* 93.
33. Davis, L. C., Shah, V. K., and Brill, W. J. (1975). *Biochim. Biophys. Acta 403,* 67.
34. Osborne, G. A., Cheng, J. C., and Stephens, P. J. (1973). *Rev. Sci. Inst. 44,* 10.
35. Yates, M. G. and Planque, K. (1975). *Eur. J. Biochem. 60,* 467.
36. Eady, R. R., Smith, B. E., Cook, K. A., and Postgate, J. R. (1972). *Biochem. J. 128,* 655.
37. Stephens, P. J., Thomson, A. J., Dunn, J. B. R., Keiderling, T. A., Rawlings, J., Rao, K. K., and Hall, D. O. (1978). *Biochemistry 17,* 4770.
38. Stephens, P. J., Thomson, A. J., Keiderling, T. A., Rawlings, J., Rao, K. K., and Hall, D. O. (1978). *Proc. Natl. Acad. Sci. U.S. 75,* 5273.
39. Burns, R. C. and Hardy, R. W. F. (1972). *Methods Enzymol. 24B,* 480-496.
40. McKenna, C. E. (1971). Ph.D. Thesis, Univ. of Calif., San Diego.
41. Bulen, W. A. and LeComte, J. R. (1972). *Methods Enzymol. 24B,* 456.

42. Bulen, W. A., personal communication.
43. Stephens, P. J., McKenna, C. E., Smith, B. E., Nguyen, H. T., McKenna, M-C., Thomson, A. J., Devlin, F., and Jones, J. B. (1979). *Proc. Natl. Acad. Sci. U.S. 76,* 2585.
44. Emerich, D. W. and Burris, R. H. (1978). *Methods Enzymol. 53D,* 314-329.

STRUCTURE AND FUNCTION OF MEMBRANE-BOUND PROTEINS FROM *M. phlei*

A. F. Brodie[*]
V. K. Kalra[†]
G. Kumar
A. J. Jacobs
S. H. Lee

Department of Biochemistry
University of Southern California
School of Medicine
Los Angeles, California 90033

A number of bacterial systems have been used to study membrane related phenomena, such as oxidative phosphorylation and mechanism of active transport of solutes, which exhibit a strict requirement for structural organization and spatial orientation. Studies of the mechanisms of the bioenergetic processes, such as, oxidative phosphorylation and active transport of solutes requires a basic understanding of the nature, composition, and vectorial orientation of the component parts necessary to carry out these functions.

Dr. Kamen had an early influence in providing an understanding in the bioenergetic mechanisms of photophosphorylation and the role played by certain respiratory carriers, in particular, the cytochromes. Methods which are commonly used to study biochemical processes depend on the resolution of the system into component parts, however, the gentlest methods used to resolve these processes into component parts results in either a loss of function or in a drastic alteration in the mechanisms of energy transduction and active transport. To some degree, the bacterial membrane systems can be resolved into component parts without

[*]*Deceased January 24, 1981.*
[†]*Address correspondence to this author.*

 ISBN 0-12-397580-8

loss of activity and provide an ideal tool for gaining insight concerning the relationship of the structure to the function of these membrane-associated processes.

Different types of membrane structures, which differ in size and vectorial orientation have been obtained from *Mycobacterium phlei* cells (1). By studying the bioenergetic processes in membrane structures which have different orientation of respiratory components and transport carrier proteins, information can be obtained which should be useful in discerning the mechanism of oxidative phosphorylation and active transport of solutes.

Various types of membrane structures have been resolved from *M. phlei* cells (2). Protoplast ghosts prepared from *M. phlei* by osmotic shock treatment have right side out oriente[d] structure (3). Sonication of these ghosts or whole cells results in the formation of membrane vesicles (ETP) which are mostly (85 - 92%) oriented inside out (IO) (3). Negative staining of these vesicles revealed that the membrane structure contain spherical bodies or repeating units (90 - 120 Å) attached to the membrane by a stalk (23 Å).

NATURE OF RESPIRATORY CHAIN

The respiratory carriers of *M. phlei* closely resemble thos[e] of the mammalian system in chemical composition (4, 5). The only major difference of particular interest was the observatio[n] that a naturally occurring menaquinone is utilized in the electron transport sequence of carriers instead of the mammalian benzoquinone. The menaquinone has been identified as MK_9 (II-H) (6, 7). This menaquinone plays an important role in the electron transport chain since it serves as a cofactor shuttling electrons between FAD and cytochrome *b* on both NAD^+ and the malate-vitamin K reductase pathways. Irradiation of the membrane vesicles (electron transport particles, ETP) with light at 360 nm results in the destruction of the natural menaquinone (8) with a concommitant loss of both oxidation and phosphorylation with NAD^+ or succinate as electron donors. Restoration of these activities occurs on addition of the natural quinone, vitamin K_1, or closely related homologues (8). Succinoxidase activity was also destroyed by irradiation but was not restored by the addition of quinones (9). The light-sensitive factor on the succinoxidase pathway involves a protein and a low molecula[r] weight component which has not been fully characterized. Both components, the protein and low molecular weight component, which may be a new cofactor, are required for light sensitivity at this site in the succinoxidase pathway (9). The sequence of

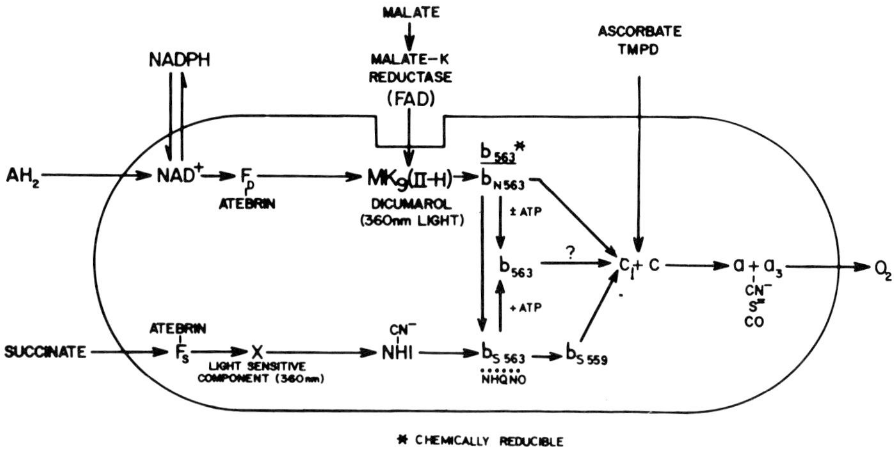

Fig. 1

respiratory carriers in *M. phlei* are shown in Fig. 1. The NAD^+- and succinate-linked chains converage at the level of cytochrome *c* utilizing different types of cytochromes *b* reducible either with succinate or NAD^+-linked substrates (10). There are three active forms of cytochromes *b* present in these membranes, cytochrome b_N563 reducible by NAD^+-linked substrate and the other two with λ_{max} at 559 and 563 (10, 11) when reduced by either succinate or NADH. Lanyi (12) has described two *b*-type cytochromes in a halophilic bacterium whereas many investigators have established multiple nature of *b*-type cytochromes in mitochondria (13).

The topological localization of *b*-type cytochromes has been difficult to ascertain because of the hydrophobic nature of these proteins. It has been demonstrated using inhibitor and kinetic studies with vitamin K9 (II-H) of *M. phlei* that one or more forms of cytochrome *b* are used as the electron acceptor from the napthoquinone. Fragments of the respiratory chain have been extracted by detergent treatment of membrane vesicles. It is becoming clear that the different types of cytochrome *b* are contained in different complexes in the respiratory chain. Preliminary ^{125}I labeling studies indicate that cytochrome *b* is on the inside of the cell membrane. It is known that cytochromes *b* interact with cytochromes c_1 and *c* on the terminal side of the electron transport chain.

MALATE-VITAMIN K REDUCTASE

The oxidation of malate in *M. phlei* membrane system has been shown to occur by two distinct pathways (4, 5, 14). One pathway involving malic dehydrogenase was found to be dependent on the addition of NAD^+ and utilized the respiratory chain common to all NAD^+-linked substrates (4). The other pathway involved oxidation of malate via a NAD^+-independent pathway (4).

The oxidation by this NAD^+-independent pathway involves an enzyme malate-vitamin K reductase and FAD, and bypasses the initial segment of the chain, but converges with the respiratory chain at the level of particulate bound MK_9 (II-H), thus providing a means of obtaining energy by bypassing a site of phosphorylation in the respiratory chain. The malate-vitamin K oxidoreductase (MKR) is unique with respect to other malic dehydrogenases since it requires vitamin K and a terminal electron acceptor dye such as thiazolyl blue tetrazolium (14, 15).

Malate-vitamin K reductase enzyme can be obtained in solubilized form upon sonication of cells followed by differential centrifugation (14, 15). The pellet (ETP) still retains 8 - 10% of total MKR activity. Restoration of activity in the pellet (particulate fraction) occurs upon the addition of solubilized enzyme and FAD. The addition of either component alone does not stimulate malate oxidation (16). The MKR of the particulate fraction (ETP: IO vesicles) which contains endogenous vitamin MK_9 (II-H) exhibited enzymatic activity in the absence of exogenous vitamin K and phospholipid. However, for maximal enzymatic activity it required the addition of both these two components. On the other hand, protoplast ghosts (oriented RO) exhibited 85% of the total malate-vitamin K reductase activity (3) which was dissociated from ghost upon sonication, indicating that the enzyme is loosely bound to the cytoplasmic side of membrane. This is further substantiated by the observation that residual activity in the particulate fraction (IO) was completely released upon washing with either 10% glycerol or 0.25 *M* sucrose (17).

CYTOCHROME *c*

Cytochrome *c*-552 from *M. phlei* has been isolated and purified to homogeneity (18). It has a molecular weight (12,600 ± 400 daltons) and amino acid composition that is quite similar to *Baccillus subtilis*. *Mycobacterium phlei* cytochrome *c*, however, contains a reduced number of lysine residues and an increased content of valine and arginine residues when compared to *B. subtilis* (18). These differences in amino acid content become evident when the isoelectric points of the two cytochromes *c* are compared. *Mycobacterium phlei* cytochrome *c*-552 (pI = 4.70 ± 0.05) is more acidic than *B. subtilis* cytochrome *c*-550 (pI = 8.65 ± 0.05) (19).

The acidic *M. phlei* cytochrome *c* binds preferentially to phosphatidylethanolamine (18). Phosphatidylcholine and phosphatidylserine bind 4 and 10% as much *M. phlei* cytochrome *c*, respectively, as phosphatidylethanolamine will bind. Electrostatic interaction of cytochrome *c* to phospholipids was strongly suggested when it was observed that the cytochromes *c* could be released from the phosphatidylethanolamine vesicles by incubating them in 0.15 *M* KCl.

To determine if cytochrome *c* was asymmetrically located in the membrane, ^{125}I-lactoperoxidase (20) labeling studies on two membrane preparations from *M. phlei*, membrane vesicles (ETP, electron transport particles) (21), which are oriented inside out (85 - 92%), and protoplast ghosts (3), which have membranes oriented the same as the normal whole cells, have conclusively demonstrated that cytochrome *c* is located on the outer surface of protoplast ghosts. This was further demonstrated by leaching cytochrome *c* from the protoplast ghost membranes by washing them in 0.15 *M* KCl (22, 23). These results are in agreement with studies deriving the localization of cytochrome *c* in mammalian mitochondrial preparations (24, 25), which also release cytochrome *c* on washing with 0.15 *M* KCl.

The thermodynamic electrochemical potential of *M. phlei* cytochrome *c* was +259 mV. This number is in the mid-range for all cytochromes *c*. Different cytochromes *c* have thermodynamic electrochemical potentials ranging from approximately 103 - 405 mV (26). Kamen has observed that isoelectric points and thermodynamic electrochemical potentials do not have any correlation to cytochrome oxidase or cytochrome reductase activities and consequently to cytochrome *c* function (26) (Table I) (26a).

Studies on the kinetics of NADH-cytochrome *c* reductase catalyzed reduction of cytochrome *c* from *M. phlei*, horse heart, and yeast show that the relative rate constants for the reduction of these three cytochromes *c* are independent of the oxidation-reduction potential and apparently the isoelectric point. The relative rate constants for the reduction of yeast cytochrome *c* and horse heart cytochrome *c* were slower by 20% and 46% than the reduction of *M. phlei* cytochrome *c* when *M. phlei* NADH cytochrome *c* reductase was used as the reductant. When yeast NADH cytochrome reductase was used as reductant, *M. phlei* and horse heart cytochrome *c* were less effective than the yeast cytochrome by 17 and 38%, respectively, when the relative rate constants were compared to yeast cytochromes *c*'s relative rate constant. As Kamen has stated (26), it appears that the three-

TABLE I. Comparison of Properties of Cytochrome *c*

	M. phlei	*S. cerevisiae*	Horse hear
Oxidation-reduction potential (mV)	259[a]	282[b]	260[c]
Isoelectric point	4.70 ± 0.05[a]	9.85[b]	10.04[c]
Percentage relative rate constant with *M. phlei* NADH cytochrome *c* reductase	100[a,d]	80[a]	54[a]
Percentage relative rate constant with yeast NADH cytochrome *c* reductase	83[a]	100[a,d]	62[a]

[a]See ref. (18).
[b]See ref. (26a).
[c]See ref. (26).
[d]100%, maximum relative rate constant obtained for cytochrome *c* and NADH cytochrome *c* reductase from same species.

dimensional structure of the cytochrome *c* molecule is the most important factor for determining the reductase and oxidase activities of cytochrome *c*.

The cytochrome *c* of the respiratory chain donates electrons to the cytochrome $a + a_3$ complex. Evidence has been obtained which shows that this complex spans across the membrane in *M. phlei* as has been shown in the mammalian system. The *M. phlei* system also contains cytochrome *O* common to certain bacteria. The role and function of cytochrome *O* has not been determined.

COUPLING FACTOR - LATENT ATPase

Coupling factor-ATPase is another example of a membrane-bound protein associated with the respiratory chain and localized on the inner membrane. The electron transport particles (ETP) of *M. phlei* are capable of exhibiting the phosphorylation coupled to the oxidation of the substrate. Washing of these particles with low ionic buffer or 0.25 *M* sucrose results in the formation of depleted electron transport particles (DETP) which are capable of oxidation but does not exhibit coupled phosphorylation. The sucrose supernatent protein which shows MG^{2+}-ATPase activity upon treatment with trypsin is referred to as coupling factor latent ATPase because the addition of it to DETP membrane in the presence of 4 m*M* $MgCl_2$ results in the restoration of phosphorylation coupled to the oxidation of the substrate. However, in protoplast ghosts the latent ATPase can not be removed by washing with 0.25 *M* sucrose since it exists on the cytoplasmic side of the membrane (3). Trypsin treatment of the ghosts results in the unmasking of the ATPase activity, which suggest that ATPase has become accessible to the outer surface of the membrane (27). Use of iodination technique, which specifically labels proteins of the outer surface of the membranes, as well as studies with the antibody to latent ATPase further indicates that the ATPase moiety becomes accessible to the outer surface of the membrane upon trypsin treatment of intact ghosts (27).

The coupling factor latent ATPase has been purified 200-fold to homogeneity in a single step using Sepharose-ADP conjugate affinity chromotography (2, 28). The purified latent ATPase also exhibits the coupling factor activity suggesting that both the coupling factor and the ATPase activity are associated with the same protein moiety. The molecular weight of coupling factor latent-ATPase has been determined to be 404,000 ± 6% by pore gradient polyacrylamide gel electrophoresis (29). The latent ATPase from *M. phlei* exhibits five subunits on SDS gel electrophoresis i.e., α (64,000), β (53,000), γ (33,000), δ (14,000), and ε (8,000). The subunit stoichiometry has been arrived at by relative areas of subunit peaks and appears to be $\alpha_3\beta_3\gamma\delta\varepsilon$. A comparison of molecular weight and subunit composition of other energy-transducing ATPase has been shown in Table II.

The latent ATPase from *M. phlei* exhibits three binding sites for ADP and one for AMP-PNP per molecule of enzyme (30). It has also been found that Mg^{2+}, which is very essential for the ATPase activity, has no effect on the binding of the nucleotides to the enzyme. Trypsin treatment of the enzyme results in the severalfold enhancement of ATPase activity and a loss of the ability to rebind to membrane vesicles depleted of coupling fac-

TABLE II. Molecular Properties of Energy Transducing ATPase (F_1)[a]

Source	Mol. Wt. (×10^{-3})	Subunit Structure	Subunit mol. wt. (×10^{-3}) α	β	ν	δ	ε	Reference
Beef Heart	360	$\alpha_3\beta_3\gamma\delta\varepsilon$	53	50	33	17.3	17.5	Lambeth, D. O. et al., FEBS Lett. 17, 330 (1971)
Rat Liver	384	$\alpha_2\beta_2\gamma_2\delta_2\varepsilon$	62.5	57	36	12	7.5	Bragg, P. D., Hou, C., Arch. Biochem. Biophys. 167, 311 (1975)
Yeast	340	--	60.5	58	32.5	13.5	8	Tzagoloff, A. V., Meagher, P., J. Biol. Chem. 246, 7328 (1971)
Spinach Chloroplast	342	$\alpha_3\beta_3\gamma\delta\varepsilon$	59	56	37	17.5	13	Nelson, N. et al., J. Biol. Chem. 248, 2049 (1973)
E. coli	340	$\alpha_3\beta_3\gamma\delta\varepsilon$	56	52	32	21	11.5	Vogel, G. and Steinhart, R., Biochemistry 15, 208-216 (1976)
M. lysodeikticus	368	$\alpha_3\beta_3\delta(\gamma)\varepsilon$	60	50	(18)	30	25	Carreira, J. et al., Biochim. Biophys. Acta 436, 183 (1976) Carreira, J. et al., Mol. Cell. Biochemistry 10, 67-76 (1976)
M. phlei	404	$\alpha_3\beta_3\gamma\delta\varepsilon$	64	53.5	33	14	8	Lee, S. H. et al., J. Biol. Chem. 252, 2084 (1977) Ritz, C. J. & Brodie, A. F., Biochem. Biophys. Res. Commun. 75, 933 (1977)
S. faecalis	385	$\alpha_3\beta_3\gamma\delta\varepsilon$	60	55	37	20	12	Schnebli, H. P. & Abrams, J., J. Biol. Chem. 245, 1115 (1970) Abrams, A. et al., Biochemistry 15, 5560 (1976)
Thermophilic bacterium	380	$\alpha_3\beta_3\gamma\delta\varepsilon$	56	53	52	15	11	Yoshida, M. et al., J. Biol. Chem. 250, 7910 (1975)

[a]Taken in part from thesis work done by Carolyn J. Ritz-Gold entitled "Trypsin Treatment of Solubilized BCF_1 From Mycobacterium Phlei: Structure-Function Studies and Theoretical Modeling."

tor latent ATPase (29). The binding of ADP to the enzyme decreased by 33% upon trypsin treatment while that of AMP-PNP remains unchanged. Maximal stimulation of ATPase activity by trypsin treatment is accompanied by a decrease of about 20,000 daltons in molecular weight of the enzyme. This is associated with the conversion of the α subunit (64,000) of latent enzyme to an α" subunit (58,000) via α' (61,000) intermediate (29). The computer modeling studies on the changes in the primary structure of α subunits of latent ATPase and the expression of ATPase activity during tryptic proteolysis (31) seems to implicate the alpha subunit structure as a major factor for maintaining the latency of the enzyme *in vivo*. On the other hand, a role of epsilon (8000) subunit as a major regulator of ATPase activity does not seem likely as only very small change in epsilon subunit is observed during trypsin activation of ATPase activity. The active role of α subunit in ATP hydrolysis in ATPase has been further substantiated by the affinity labeling studies (32) using the 2',3'-dialdehyde derivative of ATP (dial ATP). The covalent modification of enzyme with dial ATP, results in the inactivation of ATPase activity, but does not effect coupling factor activity of the enzyme. This has been ascertained by the reconstitution of the dial ATP-treated DETP membrane. These reconstituted membranes exhibit P/O ratio similar to that observed in untreated membrane vesicles, demonstrating that the sites involved in the coupling phosphorylation of ADP to ATP are not effected by dial ATP treatment and tend to corroborate the earlier studies on rat liver and bovine heart ATPase (33 - 35) in which it has been shown that an ADP site distinct from ATP hydrolysis site in the ATPase moiety may be involved in oxidative phosphorylation process.

One molecule of dial ATP covalently binds to 1 mol of enzyme with the complete inactivation of ATPase activity. The binding of dial ATP to trypsin-treated enzyme remained unaltered indicating that trypsin does not affect the ATP binding site of the enzyme. The dial ATP binds to the α subunit of ATPase which shows that the α subunit contains the ATP binding site. These studies reveal that α subunit of latent ATPase is involved in ATP hydrolysis. It is pertinent to mention here that although the ATPase has three α subunits only 1 mol of dial ATP per mole of enzyme is sufficient to block the ATPase activity (32). This suggests that the interactions between the ATP binding sites show strong negative cooperativity such that only one of the three subunits can bind ATP at a time. Similar type of negative cooperativity in ATP binding has also been shown in *E. coli* F_1, where only 1 mol of 8-azido ATP binds to the α subunit with complete loss of ATPase activity (36). These observations lead to the concept that the subunits are functionally asymmetric by nature. The computer modelling studies by Ritz (31) have also substantiated the functional asymmetry of the α subunit in the latent ATPase from *M. phlei*. The α subunit of *E. coli* ATPase has also been implicated for the ATPase activity in F_1 from *unc A* mutant of *E. coli* which has a defective α subunit (37) and is incapable of ATPase activity and oxidative phosphorylation. The addition of α subunit of F_1 from wild-type *E. coli* to the F_1 from *unc A* mutant of *E. coli* restored the ATPase activity indicating that α subunit of *E. coli* ATPase is responsible for ATPase activity. In chloroplast ATPase (38 - 40), while 1 mol of 7-chloro-4-nitrobenzo-2-oxa-1,3-diazole (NBD-Cl) binds to β subunit with the inhibition of ATPase activity, the antibody against β subunit failed to inhibit ATPase activity. Whereas, the antibody against α subunit was capable of inhibiting ATPase activity. In beef heart mitochondrial ATPase the β unit binds with the photoaffinity analog of ATP (41) indicating the presence of the catalytic site on β subunit. Both α and β subunits of various energy-transducing ATPases appear to be involved in the oxidative phosphorylation (38, 42 - 44). In *M. phlei* ATPase (32), the sites involved in the coupled phosphorylation have been studied using the 2',3'-dialdehyde derivative of ADP (dial ADP). The differential labeling of ATPase with ^{14}C dial ADP in the presence and absence of ADP reveal that ADP has the binding site on both α and β subunit of the ATPase molecule. The covalent binding of dial ADP to the ATPase results in the loss of coupling factor activity. The P/O ratio of the membranes reconstituted with DETP and dial ADP-treated ATPase show a reduction of 60 to 80% in phosphorylation and has no effect on oxidation. The dial ADP, on the other hand, has no effect on the ATPase activity as well as on the ability of the modified ATPase to rebind to the DETP membrane, as can be shown by the DCCD sensitivity of the reconstituted membranes. These experiments further substantiate the presence of different sites for ATP synthesis and ATP hydrolysis.

MEMBRANE PROTEINS INVOLVED IN AMINO ACID TRANSPORT

Whole cells or protoplast ghosts prepared from *M. phlei* were shown to actively accumulate various amino acids against a concentration gradient (45, 46). Uptake of these amino acids have demonstrated that not all of the amino acids utilize the same carrier or binding proteins, as well as efflux of amino acids were mediated by different membrane sites (47). On the basis of mutual inhibition of uptake with different amino acids it was shown that the binding site of proline was different from those of all other amino acids studied (47). Other groups of amino acids share a common binding site: (i) lysine, histidine or arginine; (ii) valine, leucine, and isoleucine; (iii) tryptophan, tyrosine, and phenylalanine; (iv) glutamic acid and aspartic acid.

Sonication of the protoplast ghost or whole cells results in the formation of membrane vesicles and are mostly (85 to 92%) oriented inside out. Those membrane vesicles contain carrier and/or binding proteins only for proline, glutamine, and glutamic acid (48). Active transport of proline in membrane vesicles exhibited a strict requirement for substrate oxidation and was inhibited by anaerobiosis and respiratory inhibitors. The uptake of proline proceeded against concentration gradient with succinate, generated NADH, exogenous NADH, or ascorbate-TPD as substrate. The latter two electrons donors were the most effective for the transport of proline and for oxidation but were the least efficient for oxidative phosphorylation. In contrast, succinate and ascorbate-TPD were equally effective for active transport with protoplast ghosts. The succinate-dependent transport of amino acids thus differs in different types of membrane preparation and appears to be dependent upon the orientation of the membranes, respiratory carriers, as well as membrane-bound carrier and/or binding proteins.

A proline carrier protein was isolated from the membrane vesicles by treatment with cholate and Triton X-100. The isolated crude proteins were purified by sucrose density gradient centrifugation, gel filtration, and isoelectric focusing techniques (1, 49). Molecular weight of the purified protein appeared to be 20,000 ± 1,000, respectively. This does not exclude the possibility that the active protein may be oligomeric. The reconstitution of proline transport was demonstrated when the purified proline carrier protein was incubated with the detergent-extracted vesicles. This reconstituted system was specific for proline transport and required substrate oxidation and the presence of Na^+ ions, as in the intact membrane vesicle (6.25 μM).

Purified carrier protein was also incorporated into phospholipid liposomes which contained an artificial electron acceptor, potassium ferricyanide. A rapid uptake of proline into these proteoliposomes was demonstrated by the addition of external ascorbic acid and benzoquinone serving as a lipid-soluble hydrogen carrier (50). A proton gradient was generated by the reduction of the entrapped ferricyanide by ascorbate oxidation with benzoquinone, and it was collapsed by proton conducting uncoupler, CCCP. The uptake of proline into this model liposomal system was dependent upon a proton gradient across the lipid bilayer and an obligatory requirement for Na^+.

Recently, a glutamine carrier and/or binding proteins was also solubilized from the membrane vesicles of *M. phlei* by treatment with cholate and cationic detergent (*N*-benzyl-*N*,*N*-dimethyl-β-ethanol ammonium ion). Glutamine transport in detergent-treated membrane vesicles was drastically reduced as compared to untreated membrane vesicles. The reconstitution of glutamine transport was also demonstrated when the solubilized proteins was incubated with the protein-depleted membrane vesicles, as well as a model proteoliposomal systems, which were described above.

These studies on the nature and topology of respiratory carriers and transport proteins should be useful in delineating the mechanism of energy transduction for the synthesis of ATP and transport of solutes.

REFERENCES

L. Brodie, A. F., Kalra, V. K., Lee, S. H., and Cohen, N. S. (1979). *Methods Enzymol. 55,* 175-200.
2. Brodie, A. F., Lee, S. H., Prasad, R., Kalra, V. K., and Kosmakos, F. (1974). *In* "Lipmann Symposium" (D. Richter, ed.), pp. 115-130. Walter de Gruyter, Berlin.
3. Asano, A., Cohen, N. S., Baker, R. F., and Brodie, A. F. (1973). *J. Biol. Chem. 248,* 3386-3397.
4. Asano, A. and Brodie, A. F. (1964). *J. Biol. Chem. 239,* 4280-4291.
5. Brodie, A. F. and Adelson, J. (1965). *Science 149,* 265-269.
6. Brodie, A. F. (1961). *Fed. Proc. 20,* 995.
7. Gale, P. H., Arison, C. H., Trenner, N. R., Page, A. C., Jr., Folkers, K., and Brodie, A. F. (1963). *Biochemistry 2,* 200-203.
8. Brodie, A. F. and Ballantine, J. (1960). *J. Biol. Chem. 235,* 226-231.
9. Krishnamurti, C. R. and Brodie, A. F. (1969). *Science 164,* 302-304.
10. Cohen, N. S., Bogin, E., Higashi, T., and Brodie, A. F. (1973). *Biochem. Biophys. Res. Commun. 54,* 800-807.
11. Cohen, N. S. and Brodie, A. F. (1975). *J. Bacteriol. 123,* 162-173.
12. Lanyi, J. K. (1969). *J. Biol. Chem. 244,* 2864-2869.
13. Wikstrom, M. K. F. (1973). *Biochim. Biophys. Acta 301,* 155-193.
14. Asano, A. and Brodie, A. F. (1963). *Biochem. Biophys. Res. Commun. 13,* 423-427.
15. Asano, A., Kaneshiro, T., and Brodie, A. F. (1965). *J. Biol. Chem. 240,* 895-935.
16. Murthy, P. S., Bogin, E., Higashi, T., and Brodie, A. F. (1969). *J. Biol. Chem. 244,* 3117-3124.
17. Imai, K. and Brodie, A. F. (1973). *J. Biol. Chem. 248,*

7487-7494.
18. Jacobs, A. J., Kalra, V. K., Cavari, B., and Brodie, A. F. (1981). *Arch. Biochem. Biophys. 194,* 531-541.
19. Miki, K. and Okunuki, K. (1969). *J. Biochem. 66,* 845-854.
20. Phillips, D. R. and Morrison, M. (1971). *Biochem. J. 91,* 222-233.
21. Brodie, A. F. (1959). *J. Biol. Chem. 234,* 398-404.
22. Jacobs, E. and Sanadi, D. R. (1960). *J. Biol. Chem. 235,* 531-534.
23. Lenaz, G. and MacLennan, D. H. (1966). *J. Biol. Chem. 241,* 5260-5262.
24. Kroger, A. and Klingenberg, M. (1970). *Vitamins Hormones 28,* 533-574.
25. Nicholls, P. and Chance, B. (1974). *In* "Molecular Mechanisms of Oxygen Activation" (O. Hayaishi, ed.), pp. 479-534. Academic Press, New York.
26. Errede, B. and Kamen, M. D. (1978). *Biochemistry 17,* 1015-1027.
26a. Margoliash, E. and Schejter, A. (1966). *Advan. Protein Chem. 21,* 113-286.
27. Nakagawa, H., Lee, S. H., Kalra, V. K., and Brodie, A. F. (1977). *J. Biol. Chem. 252,* 8229-8234.
28. Higashi, T., Kalra, V. K., Lee, S. H., and Brodie, A. F. (1975). *J. Biol. Chem. 250,* 6541-6548.
29. Ritz, C. J. and Brodie, A. F. (1977). *Biochem. Biophys. Res. Commun. 75,* 933-939.
30. Lee, S. H., Kalra, V. K., Ritz, C. J., and Brodie, A. F. (1977). *J. Biol. Chem. 252,* 1084-1091.
31. Ritz, C. J. (1978). Thesis, University of Southern California, School of Medicine, Department of Biochemistry, Los Angeles.
32. Kumar, G., Kalra, V. K., and Brodie, A. F. (1979). *J. Biol. Chem. 254,* 1964-1971.
33. Catterall, W. A. and Pedersen, P. L. (1972). *J. Biol. Chem. 247,* 7969-7976.
34. Pedersen, P. L. (1975). *Biochem. Biophys. Res. Commun. 64,* 610-616.
35. Penefsky, H. S. (1974). *J. Biol. Chem. 249,* 3579-3585.
36. Verheijen, J. H., Postma, R. W., and Vandam, K. (1978). *Biochim. Biophys. Acta 502,* 345-353.
37. Dunn, S. D. (1978). *Biochem. Biophys. Res. Commun. 82,* 596-602.
38. Nelson, N., Deters, D. W., Nelson, H., and Racker, E. (1973). *J. Biol. Chem. 248,* 2049-2055.
39. Nelson, N. (1976). *Biochim. Biophys. Acta 456,* 314-338.
40. Deters, D. W., Racker, E., Nelson, N., and Nelson, H. (1975). *J. Biol. Chem. 250,* 1041-1047.
41. Wagenvoord, R. J., Vanderkraan, I., and Keimp, A. (1977). *Biochim. Biophys. Acta 460,* 17-24.

42. Philasoph, S., Binder, A., and Gromet-Elhanan, Z. (1977). *J. Biol. Chem. 252,* 8747-8752.
43. Yoshida, M., Sone, N., Hirata, H., and Kagawa, Y. (1977). *J. Biol. Chem. 252,* 3480-3485.
44. Yoshida, M., Okamoto, H., Sone, N., Hirata, H., and Kagawa, Y. (1977). *Proc. Natl. Acad. Sci. U.S. 74,* 936-940.
45. Hirata, H., Kosmakos, F. C., and Brodie, A. F. (1974). *J. Biol. Chem. 249,* 6965-6970.
46. Prasad, R., Kalra, V. K., and Brodie, A. F. (1976). *J. Biol. Chem. 251,* 2493-2498.
47. Israeli, E., Kalra, V. K., and Brodie, A. F. (1977). *J. Bacteriol. 130,* 729-735.
48. Prasad, R., Kalra, V. K., and Brodie, A. F. (1975). *Biochem. Biophys. Res. Commun. 63,* 50-56.
49. Lee, S. H., Cohen, N. S., Jacobs, A. J., and Brodie, A. F. (1977). *J. Supramol. Struct. Suppl. 1,* 136.
50. Lee, S. H. and Brodie, A. F. (1978). *Biochem. Biophys. Res. Commun. 85,* 788-794.

THE DOPAMINE HYDROXYLATING SYSTEM OF BOVINE ADRENAL CHROMAFFIN GRANULES

Torgeir Flatmark

Department of Biochemistry
University of Bergen
Bergen, Norway

INTRODUCTION

The hormones adrenaline and noradrenaline are stored in the adrenal medulla within specific storage organelles, the chromaffin granules (for review, see ref. 1). When examined by electron microscopy, these organelles appear as spherical membrane-limited particles containing an osmiophilic dense core and with diameters ranging from 150 to 350 nm (bovine adrenal) (2). Our recent morphometric measurements have revealed that the chromaffin granules occupy approx. 13.5% of the cytoplasmic volume (3). The osmiophilic dense core, which is more generally termed the matrix, consists of stores of catecholamines, adenine nucleotides, and a group of proteins, including chromogranin A and the enzyme dopamine β-monooxygenase, and some phospholipids (1). Biochemical and morphological evidence suggests that both hormones are released from the chromaffin cells by the process of exocytosis (4). The fate of the empty chromaffin granules remaining in the cell is not yet known.

 ISBN 0-12-397580-8

The chromaffin granules are unique in their capacity to store high concentrations of catecholamines (on an average approx. 600 m*M* for the total population of chromaffin granules) and adenine nucleotides, notably ATP (on an average approx. 150 m*M*), giving a molar ratio of catecholamines to ATP of about 4 to 1 (1). We have also found (5, 6) that the granules contain high concentrations of ascorbate in the fully reduced form (on an average approx. 20 m*M* for the total population of chromaffin granules). In addition, the granules contain Ca^{2+} and Mg^{2+} at about the same concentration as ascorbate (7), as well as some proteins (chromogranins) and phospholipids (1). Based on [^{1}H]NMR spectroscopy (8 - 11) evidence has been presented for a loose network of interactions between the various components. The storage complex is a dynamic viscous solution in which the chromogranin A protein assists in the stabilization of the internal phase which through ternary complex formation, protein:ATP:catecholamines, is more viscous than a conventional solution of any of the components alone (11). The solution then has a lower osmotic pressure than expected from the sum of the analytical concentration of its components.

The membrane, which is an energy-transducing membrane, is composed of 35 - 40% of protein on a dry weight basis containing relatively few polypeptides (1). We have found, in addition to a [H^{+}]ATPase (12, 13) and dopamine β-monooxygenase (14, 15), a unique high-potential *b*-type cytochrome (16, 17) and a NADH:acceptor oxidoreductase (18). All experimental evidence (3) indicates that the membrane proteins, with the exception of dopamine β-monooxygenase, are located on the outer phase of the granule membrane.

THE DOPAMINE HYDROXYLATING SYSTEM OF CHROMAFFIN GRANULES

Catecholamines are formed in brain, chromaffin cells, sympathetic nerves, and sympathetic ganglia from their amino acid precursor tyrosine by a sequence of enzymatic steps. The classical pathway of noradrenaline biosynthesis (Fig. 1) involves two steps which are typical monooxygenase reactions, i.e., the aromatic hydroxylation of tyrosine, catalyzed by an iron enzyme (tyrosine 3-monooxygenase) and the side-chain hydroxylation of dopamine (in β-position), catalyzed by a copper enzyme (dopamine β-monoxygenase). Hydroxylation of dopamine to noradrenaline and storage of catecholamines in their biologically active reduced form are the most characteristic primary properties of the chromaffin granules of the adrenal medulla. The β-hydroxylation of dopamine is catalyzed by the copper enzyme dopamine β-monooxygenase (dopamine β-hydroxylase EC 1.14.17.1), a mixed-function oxidase (for review, see

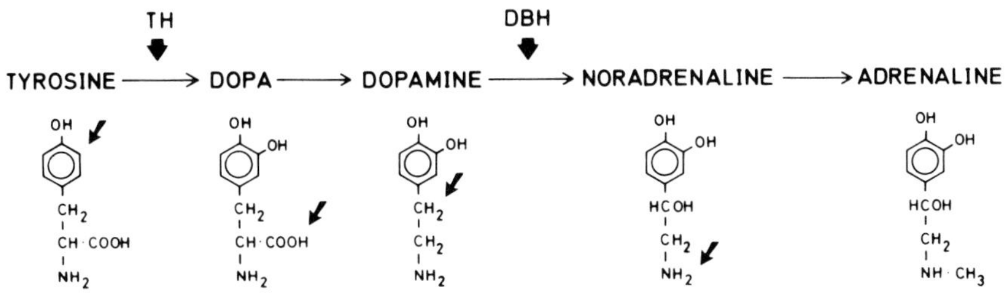

Fig. 1. Pathway of catecholamine biosynthesis.

ref. 19):

$$\text{Dopamine} + \text{H}\cdot\text{ascorbate}^- + O_2 \longrightarrow \text{noradrenaline} + \text{dehydroascorbate} + H_2O \qquad (1)$$

The enzyme is located within the limiting membrane of the chromaffin granules (20). When the organelles are subjected to hypoosmotic lysis (Fig. 2), approx. 50% of the enzyme is released together with the low molecular weight components (1).

Recent studies in this laboratory have revealed that the dopamine β-monooxygenase of the granule ghosts is largely ac-

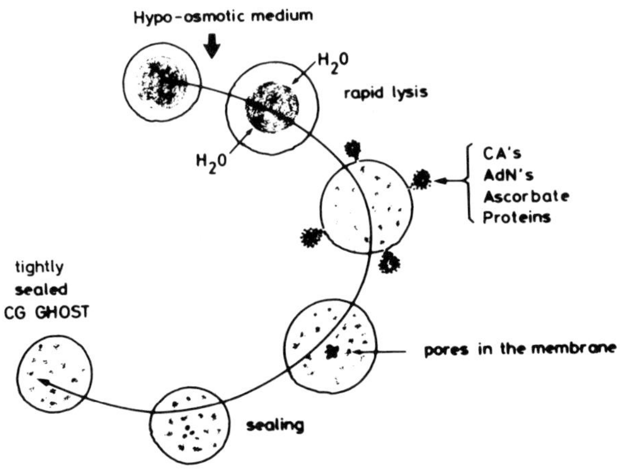

Fig. 2. The scheme indicates the overall process of colloid-osmotic lysis of the chromaffin granules (CG) and resealing of the granule ghosts.

counted for by the amphiphilic form (probably membrane-bound) of the enzyme (14), which also exists in a latent or masked form (21). Thus, the substrate to be hydroxylated (dopamine) as well as the immediate electron donor (H·ascorbate$^-$) have to be transported from the cytosol through the membrane into the matrix compartment. Thus, the hydroxylation of dopamine in the intact organelles or their ghost preparations involves a reaction sequence of at least four steps (see Fig. 3), i.e., (i) a vectorial transport of H^+, generating a pH gradient which is required for the energized uptake of dopamine, (ii) dopamine uptake against a concentration gradient, (iii) ascorbate uptake, (iv) hydroxylation of dopamine, and possibly also (v) regeneration of ascorbate in its fully reduced form.

It has been found difficult to study the details of any of the partial reactions (i) to (v) in the intact chromaffin granules due to the fact that the organelle contains large amounts of stored compounds, i.e., catecholamines, adenine nucleotides, and Ca^{2+}/Mg^{2+} in complexes of high molecular weight (8 - 11) and a matrix of high internal buffering capacity. We have found chromaffin granule ghosts to be much more convenient particularly for studies on reactions (i) to (iii). When subjected to hypoosmotic lysis (Fig. 2), the chromaffin granules are depleted, within seconds of >95% of their low molecular weight components, and a large percentage of the high molecular weight components, and then very efficiently reseal retaining a number of properties characteristic of the intact organelle,

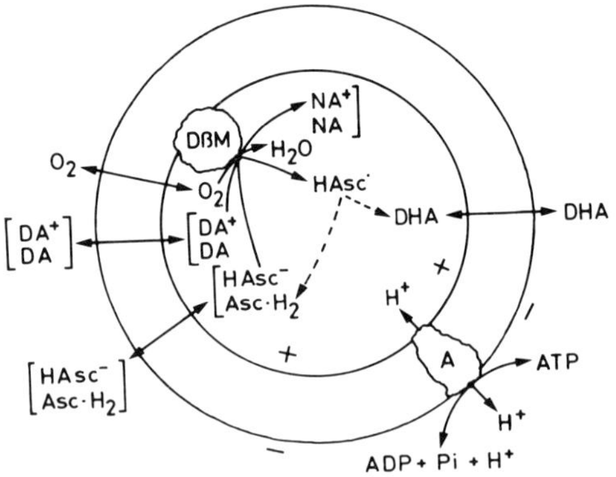

Fig. 3. Model of the dopamine hydroxylating system in chromaffin granule ghosts. DA, dopamine; $AscH_2$, ascorbic acid; $HAsc^-$, monovalent ascorbate anion; HAsc•, ascorbate free radical; DHA, dehydroascorbate; $\rightarrow H^+$, vectorial transport of proton; DβM, dopamine β-monooxygenase; A, (H^+)ATPase.

e.g., a low passive permeability to protons (22), which is important from a functional point of view.

Vectorial H^+ Transport and Dopamine Uptake

Using a spectrophotometric method with bromothymol blue as an indicator of intravesicular pH, we have found (22, 23) that the ATP-dependent proton translocation is well preserved in the ghosts and that the direction of this proton translocation is inward as proposed for intact chromaffin granules (Fig. 4A).

Our results indicate that resealed chromaffin granule ghosts take up dopamine by two independent mechanisms, i.e., by an energy-independent mechanism (when $\Delta pH \simeq 0$; lower curve in Fig. 4B) and an energy-dependent uptake (when $\Delta pH > 0$, acid inside), which revealed all criteria of an active transport mechanism. First, dopamine was taken up by a process driven by ATP hydrolysis (Fig. 4B); the rate of transport correlates to the rate of generation of ΔpH by the vectorial transport of H^+ ($\longrightarrow H^+$) mediated by the (H^+)ATPase of the membrane (Fig. 4A). Second, preincubation of the granule ghosts with a proton ionophore (FCCP) abolishes the time-dependent uptake of dopamine (Fig. 4B) as well as the internal acidification (Fig. 4A). Third, the dopamine taken up in an energy-dependent manner is rapidly released by NH_4Cl which abolishes the $\Delta\mu_{H^+}$ (23). Thus, $\Delta\mu_{H^+}$ is required not only to take up, but also to retain, dopamine in this system. Finally, the energy-dependent transport system is capable of dopamine uptake against a 5-fold concentration gradient in the ghosts. The rate of dopamine uptake revealed a hyperbolic function of the external concentration over the range studied (23). We have also found that the total amount of dopamine taken up can be increased if divalent cations and adenine nucleotides (ATP and ADP) are trapped inside during the lysis procedure (Fig. 2), indicating that complex formation is important in order to generate the high concentration gradients characteristic of the intact chromaffin granules (O. C. Ingebretsen and T. Flatmark, unpublished data).

Uptake and Storage of Ascorbate

Based on the finding by Friedman and Kaufman (24) that ascorbate functions as an efficient electron donor in the dopamine β-monooxygenase reaction, as well as our discovery of a high content of ascorbate stored within the chromaffin granules (5), it is likely that this vitamin functions as the immediate electron donor in this hydroxylation reaction. Thus, we have so far been unable to find an alternative candidate, and the

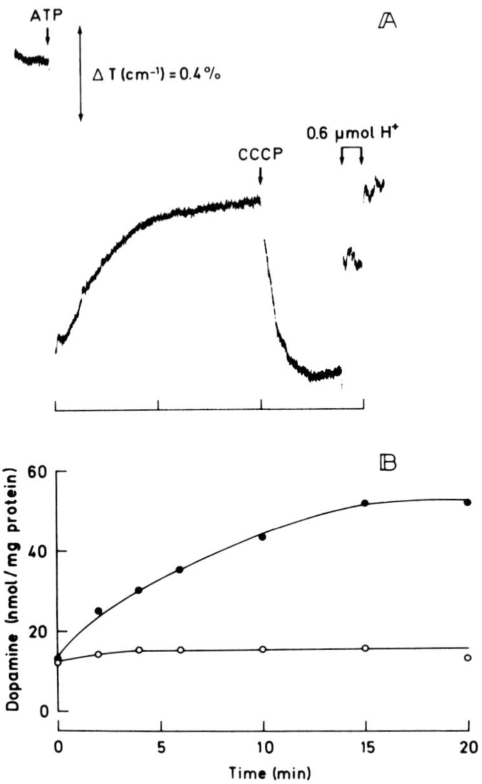

Fig. 4. (A) Time course of changes in transmittance at 618 minus 648 nm following the addition of 10 mM MgATP, 16 μM CCCP, and pulses of HCl to a suspension of chromaffin granule ghosts loaded with bromothymol blue (13 μM). The ghosts, 0.82 mg of protein/ml, were equilibrated at 35°C for 5 min with oligomycin (5 μg/mg of protein) prior to the addition of ATP. (B) Time course of dopamine uptake in chromaffin granule ghosts driven by ATP hydrolysis assayed in the absence (●) and presence (○) of the proton ionophore FCCP (50 μM). The ghosts, 1.66 mg of protein/ml, were equilibrated for 5 min with oligomycin (5 μg/mg of protein) at 35°C prior to the simultaneous addition of 1.8 mM dopamine and 4.8 mM MgATP.

involvement of ascorbate in noradrenaline biosynthesis seems to represent one of the most clearly defined biochemical roles of the vitamin.

There are a number of potential reactions in which ascorbate may take place within the chromaffin granules. In addition to the dopamine β-monooxygenase reaction (Eq. 1), ascor-

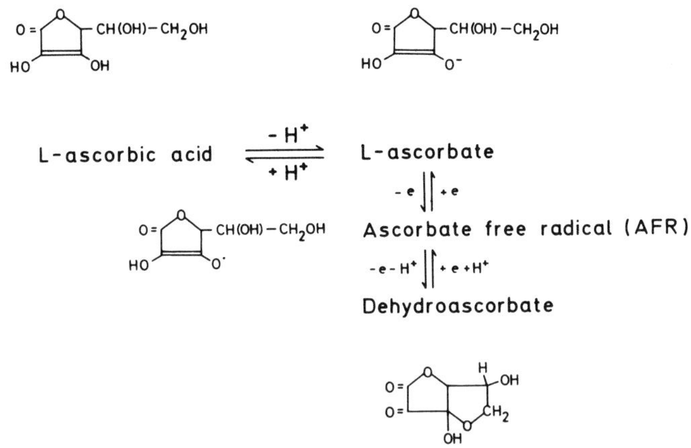

Fig. 5. Ionization of ascorbic acid and redox properties of ascorbate anion (based on refs. 26 - 29). The pK values for the two ionization stages are 4.18 (at O-3) and 11.6 (at O-2) in aqueous solution at 37°C (30).

bate may react with dioxygen or other oxidants present in these organelles, e.g., the quinone form of catecholamines. At the intravesicular pH of about 5.5 in the intact organelle (25), where the vitamin exists mainly as the H-ascorbate$^-$ ion (Fig. 5), the oxidation of ascorbate by O_2 is negligible. Thus, Lagercrantz (31) was able to detect the aerobic formation of ascorbate free radical (Fig. 5) only above pH 6.6. Additional protection is likely to be afforded by complex formation with Ca^{2+} and Mg^{2+} (32, 33) since the intravesicular content of Ca^{2+}, Mg^{2+} (7) is almost stoichiometric to that of ascorbate (5, 6). A molecular interaction of ascorbate with catecholamines by hydrogen bonding (33) and protein (34) is also possible, and a low activity of bulk water may also have a stabilizing effect on the intragranular ascorbate (33). The efficiency of these stabilizing factors is most clearly demonstrated by the rapid oxidation of ascorbate to dehydroascorbate when the chromaffin granules are lysed (Fig. 2), e.g., in the absence of added EDTA (5, 6).

Since the chromaffin granules contain catecholamines at a concentration of approx. 0.5 *M* and 1 mol of ascorbate is required for each mole of dopamine hydroxylated (Eq. 1), it is evident that the chromaffin granules must have an appropriate mechanism for transporting also this substrate into the matrix. The high K_m' value of ascorbate, i.e., 0.9 - 1.25 m*M* (35, 36), in the hydroxylation reaction indicates that the maintenance of the ascorbate concentration at a high level in the matrix (5, 6) is essential for the hydroxylation of dopamine. In contrast to dopamine (23), however, it has not yet been possible to demonstrate an ATP-driven uptake of ascorbate in chromaffin granule ghosts, and an energy-independent mechanism of ascorbate uptake, therefore, seems most likely (O. C. Ingebretsen and T. Flatmark, unpublished data).

Hydroxylation of Dopamine

When both dopamine and ascorbate are present inside the chromaffin granules or granule ghosts, noradrenaline is formed according to Eq. (1) catalyzed by the enzyme dopamine β-monooxygenase. The water-soluble form of the enzyme, isolated from the bovine adrenal medulla, has been extensively studied (for review, see ref. 19), and some of its physicochemical properties are given in Table I.

This protein contains four copper atoms per tetramer of 290,000 dalton, and a variable fraction of this copper exists as Cu(II) in the isolated enzyme (41, 42). The half-reduction potential of the enzyme-bound copper is about +370 mV (41), and all the copper can be oxidized by $K_3Fe(CN)_6$ to give paramagnetic (41, 42) Cu(II). EPR has provided no evidence for more than one type of copper in dopamine β-monooxygenase and of the three types of copper found in the blue oxidases (41, 42).

TABLE I. Physicochemical Parameters of the Water-Soluble Form of Dopamine β-Monooxygenase

Parameter	*Estimated value*	*Ref.*
$A^{1\%}_{280\ nm}$	12.4	(37)
$\Delta\ \underline{\varepsilon}(M^{-1}cm^{-1})$ at 680 nm[a]	40	(19)
Mol. wt.	*290,000 (tetramer)*	(38)
N-terminal	*2 Ala and 2 Ser/tetramer of 290,000*	(39)
Cu content	*4 Cu/tetramer of 290,000 (specifically bound)*	(40, 42)
EPR spectroscopy	*All Cu is paramagnetic as Cu(II)*	(41, 42)
$E_{m,7}$ (mV)	+370	(41)

[a]*Oxidized minus reduced.*

Kinetic studies in this laboratory (41, 43, 44) have indicated a sequential mechanism for the enzyme reaction, and evidence has been presented for only one copper atom per catalytic center (for review, see ref. 19). The mechanism of O_2 activation is, however, not yet known. Rapid-freeze technique and EPR spectroscopy (42) have revealed that ascorbate reduces the enzyme-bound copper at a rate (t/2 is 30 - 45 ms at pH 7.0 and 0.9 m*M* ascorbate) which is faster than the overall catalytic rate; about 10% of the enzyme was oxidized to Cu(II) during steady-state catalysis in the presence of ascorbate. These results support an electron transfer function of the enzyme-bound copper during catalysis, and indicate that the reduction by ascorbate is not the rate-limiting step. The reduced enzyme, however, needs additional reductant for completion of the catalytic cycle as it, in the presence of the substrate to be hydroxylated (e.g., tyramine) and O_2, will neither hydroxylate tyramine to octopamine nor oxidize the Cu(I) at a rate consistent with catalysis (44).

Although experimental evidence has been presented for the ascorbate^{2-} ion as the reductant for redox proteins acting as single electron acceptors, e.g., single-site heme and copper redox proteins (45, 46), this ascorbate species comprise only approx. 10^{-4}% of the total ascorbic acid present at the intravesicular pH of about 5.5 in the intact organelle (25). Thus, approx. 95% of the total ascorbic acid is present as H·ascorbate$^-$ which is, therefore, the most likely reducing agent in the hydroxylation reaction. The demonstration by EPR spectroscopy of a small signal of a free radical during catalysis with dopamine as the substrate (47) supports the conclusion that the hydroxylation is likely to proceed via an ascorbate-free radical (see Fig. 6). It is not yet known, however, if this radical is eliminated by its possible function as an electron donor in the hydroxylation reaction or by dismutation into ascorbate and dehydroascorbate (48). Although it has been proposed that ascorbate-free radical is a more potent reducing agent than H·ascorbate$^-$ (45), the dismutation reaction may effectively lower its steady-state concentration to insignificantly low values (48). Considering also the high K_m' value of ascorbate in the hydroxylation reaction (35, 36), the high intragranular concentration of ascorbate (5, 6), and the rapid reduction of enzyme-bound Cu(II) by ascorbate (42), it is likely that H·ascorbate$^-$ represents the reducing agent in the overall hydroxylation reaction.

OTHER REDOX ACTIVE COMPONENTS IN CHROMAFFIN GRANULE GHOSTS AND POSSIBLE SIGNIFICANCE IN THE HYDROXYLATION REACTION

The monooxygenases commonly appear in multienzyme systems where they are coupled to short electron transfer chains. Although preliminary studies (49) suggested the possibility that the hydroxylation of dopamine in chromaffin granule ghosts is stimulated by NADH, and possibly mediated by the NADH:acceptor oxidoreductase activity present in granule ghost preparations, more detailed studies do not support this conclusion (O. Terland and T. Flatmark, unpublished data). However, experimental evidence has been obtained that preparations of chromaffin granule ghosts contain an NADH:ascorbate-free radical oxidoreductase activity (T. Flatmark, unpublished data). Further studies are, however, required to prove the physiological significance of this reaction, notably in the regeneration of ascorbate in the adenal medulla.

REGULATORY MECHANISMS IN NORADRENALINE BIOSYNTHESIS

The complex problem of the control of noradrenaline biosynthesis has been a matter of much discussion. The pathway consists of a series of enzyme-catalyzed reactions (Fig. 1), and based on *in vitro* studies of the individual enzyme in the pathway it is generally believed that the first step, i.e., hydroxylation of tyrosine, is the rate-limiting step in the peripheral nervous system as well as in the brain (50). However, other possibilities should also be considered, notably that the hydroxylation of dopamine may be a regulatory step. Of the three enzymes in the pathway, only dopamine β-monooxygenase is localized within the limiting membrane of the chromaffin granules (51) and bovine splenic nerve granules (52). Therefore, all the three substrates (dopamine, ascorbate and O_2) have to cross the granule membrane.

There is a substantial body of evidence which suggests that the granule membrane has a specific mechanism of catecholamine uptake (for review, see ref. 53) including dopamine (23). Thus, the intragranular concentration of dopamine is low (54, 55), and the availability of dopamine to the enzyme may be limited by the energy state of the chromaffin granule membrane (23, 53). The transport process is competitively inhibited by other amines in the cytosol, e.g., adrenaline and noradrenaline (53).

A regulatory mechanism based on the availability of ascorbate is less likely. Thus, the estimated concentration of ascorbate in the matrix of bovine adrenal chromaffin granules (approx. 20 m*M*) (6) is approx. 20 times higher than the K_m'-value (35, 36). Furthermore, no inhibition of dopamine hydroxylation was observed in scorbutic guinea pigs in which the adrenal content of ascorbate was lowered to only about 5% of the normal value (56), and there is so far no evidence of an alternative electron donor in the hydroxylation of dopamine. These findings, therefore, support the presence of an efficient mechanism of storage and regeneration of ascorbate in the adrenal medulla.

Dioxygen has also to penetrate the granule membrane in order to reach its reaction site at the dopamine β-monooxygenase. The increased solubility of dioxygen in the lipids may favor diffusion across the membrane and thus reach the enzyme bound to the inner phase of the granule membrane (Fig. 3).

Regulation at the level of the concentration of enzyme (dopamine β-monooxygenase) is unlikely due to the very high content of the enzyme in chromaffin granules. However, experimental evidence has been presented that the matrix enzyme, which accounts for approx. 50% of the total amount of enzyme in the organelle (1), may be packed in the viscous ternary complex of protein, catecholamines, and adenine nucleotides (11), which results in a permeability barrier to the substrates. Furthermore, a potent inhibitor of the dopamine β-monooxygenase reaction has been demonstrated in the matrix of the bovine organelle (57), and recent studies in this laboratory indicate that the activity of the matrix enzyme may contribute very little to the overall noradrenaline synthesis in intact, mature chromaffin granules (O. Terland and T. Flatmark, unpublished data).

The possibility should also be considered that several mechanisms may, in fact, play some role in the regulation of the noradrenaline biosynthetic pathway. It should be particularly mentioned that the regulation may be different in adrenal medulla and in the noradrenergic neurons for several reasons, including differences in the matrix storage complex and the release mechanism (58).

CONCLUDING REMARKS

This report describes some of the characteristic features of the dopamine hydroxylating system of the chromaffin granules of the bovine adrenal medulla. Information obtained through the studies on chromaffin granule ghosts and the matrix fraction has provided some insight into a complexity that can be superimposed on the picture provided by studies on the isolated water-soluble form of the enzyme dopamine β-monooxygenase. In particular, the significance of the granule membrane and the matrix storage complex in the regulation of noradrenaline biosynthesis is discussed, and it is evident that more experiments are required to provide a better understanding of the molecular events involved in the regulation of this hydroxylating system.

ACKNOWLEDGMENTS

The studies reported here were carried out in collaboratio with Dr. Ole C. Ingebretsen, Dr. Torbjørn Ljones, Dr. Tore Skotland, and Dr. Ole Terland. I also wish to acknowledge the assistance of Mr. Erik Grung, Ing. Sissel Wahlstrøm Jacobsen and Ing. Sissel Vik. Figure 4 is reproduced with the permissio of *J. Biol. Chem.*

Research on which this report is based was made possible by grants from the Norwegian Research Council on Cardiovascular Diseases and Nordisk Insulin Foundation.

REFERENCES

1. Winkler, H. (1976). *Neuroscience 1,* 65.
2. Coupland, R. E. (1968). *Nature (London) 217,* 284.
3. Kryvi, H., Flatmark, T., and Terland, O. (1979). *Eur. J. Cell. Biol. 20,* 76.
4. De Robertis, E. D. P. and Ferreira, A. V. (1957). *Exp. Cell Res. 12,* 568.
5. Terland, O. and Flatmark, T. (1975). *FEBS Lett. 59,* 52.
6. Ingebretsen, O. C., Terland, O., and Flatmark, T. (1980). *Biochim. Biophys. Acta 628,* 182.
7. Phillips, J. H., Allison, Y. P., and Morris, S. J. (1977). *Neuroscience 2,* 147.
8. Daniels, A. J., Korda, A., Tanswell, P., Williams, A., and Williams, R. J. P. (1974). *Proc. Roy. Soc. B187,* 353.
9. Daniels, A. J., Williams, R. J. P., and Wright, P. E. (1976). *Nature (London) 261,* 321.
10. Sharp, R. R. and Richards, E. P. (1977). *Biochim. Biophys Acta 497,* 260.
11. Daniels, A. J., Williams, R. J. P., and Wright, P. E. (1978). *Neuroscience 3,* 573.
12. Ingebretsen, O. C. and Flatmark, T. (1977). *Abstr. Commun. Meet. Fed. Eur. Biochem. Soc. 11,* No. A4-15 601.

13. Apps, D. K. and Glover, L. A. (1978). *FEBS Lett. 85,* 254.
14. Skotland, T. and Flatmark, T. (1979). *J. Neurochem. 32,* 1861.
15. Bjerrum, O. J., Helle, K. B., and Bock, E. (1979). *Biochem. J. 181,* 231.
16. Flatmark, T. and Terland, O. (1971). *Biochim. Biophys. Acta 253,* 487.
17. Silsand, T. and Flatmark, T. (1974). *Biochim. Biophys. Acta 395,* 257.
18. Terland, O. and Flatmark, T. (1973). *Biochim. Biophys. Acta 305,* 206.
19. Skotland, T. and Ljones, T. (1979). *Inorg. Perspect. Biol. Med. 2,* 151.
20. Laduron, P. M. (1975). *FEBS Lett. 52,* 132.
21. Terland, O. and Flatmark, T. (1975). *Abstr. Commun. Meet. Fed. Eur. Biochem. Soc. 10,* No. 1098.
22. Flatmark, T. and Ingebretsen, O. C. (1977). *FEBS Lett. 78,* 53.
23. Ingebretsen, O. C. and Flatmark, T. (1979). *J. Biol. Chem. 254,* 3833.
24. Friedman, S. and Kaufman, S. (1965). *J. Biol. Chem. 240,* 4763.
25. Johnson, R. G. and Scarpa, A. (1976). *J. Biol. Chem. 251,* 2189.
26. Hvoslef, J. (1968). *Acta Crystallogr. B24,* 23.
27. Hvoslef, J. (1969). *Acta Crystallogr. B25,* 2214.
28. Hvoslef, J. (1971). *Acta Crystallogr. B28,* 913.
29. Hvoslef, J. (1972). Ph.D. Thesis, University of Oslo, Oslo.
30. Karrer, P., Schwarzenbach, K., and Schopp, G. (1933). *Helv. Chim. Acta 16,* 302.
31. Lagercrantz, C. (1964). *Acta Chem. Scand. 18,* 562.
32. Hvoslef, J. and Kjellevold, K. E. (1974). *Acta Crystallogr. B30,* 2711.
33. Lewin, S. (1974). *In* "Vitamin C. Recent Aspects of Its Physiological and Technological Importance" (G. G. Birch and K. J. Parker, eds.), p. 221. Applied Science, London.
34. Fiddick, R. and Heath, H. (1967). *Biochim. Biophys. Acta 136,* 206.
35. Foldes, A., Jeffrey, P. L., Preston, B. N., and Austin, L. (1973). *J. Neurochem. 20,* 1431.
36. Aunis, D., Miras-Portugal, M.-T., and Mandel, P. (1973). *Biochim. Biophys. Acta 327,* 313.
37. Skotland, T. and Ljones, T. (1977). *Intern. J. Peptide Protein Res. 10,* 311.
38. Ljones, T., Skotland, T., and Flatmark, T. (1976). *Eur. J. Biochem. 61,* 525.
39. Skotland, T., Ljones, T., Flatmark, T., and Sletten, K.

(1977). *Biochem. Biophys. Res. Commun. 74,* 1483.
40. Skotland, T. and Ljones, T. (1979). *Eur. J. Biochem. 94,* 145.
41. Ljones, T., Flatmark, T., Skotland, T., Petterson, L., Bäckstrøm, D., and Ehrenberg, A. (1978). *FEBS Lett. 92,* 81.
42. Skotland, T., Petterson, L., Bäckstrøm, D., Ljones, T., Flatmark, T., and Ehrenberg, A. (1980). *Eur. J. Biochem., 103,* 5.
43. Ljones, T. and Flatmark, T. (1974). *FEBS Lett. 49,* 49.
44. Skotland, T., Ljones, T., and Flatmark, T. (1978). *Biochem. Biophys. Res. Commun. 84,* 83.
45. Yamazaki, I. (1962). *J. Biol. Chem. 237,* 224.
46. Al-Ayash, B. I. and Wilson, M. T. (1979). *Biochem. J. 177,* 641.
47. Blumberg, W. E., Goldstein, M., Lauber, E., and Peisach, J. (1965). *Biochim. Biophys. Acta 99,* 187.
48. Foerster, G. V., Weiss, W., and Staudinger, H. (1965). *Ann. Chem. 690,* 166.
49. Flatmark, T. and Terland, O. (1971). *Abstr. Commun. Meet. Fed. Eur. Biochem. Soc. 7,* 292.
50. Udenfriend, S. (1968). *In* "Adrenergic Neuro-transmission" (G. E. W. Wolstenholme and M. O'Connor, eds.), p. 3. J. & A. Churchill Ltd., London.
51. L. M. Laduron (1975). *FEBS Lett. 52,* 132.
52. Stjärne, L. and Lishajko, F. (1967). *Biochem. Pharmacol. 16,* 1719.
53. Njus, D. and Radda, G. K. (1978). *Biochim. Biophys. Acta 463,* 219.
54. Eade, N. R. (1958). *J. Physiol. (London) 141,* 183.
55. Lishajko, F. (1968). *Acta Physiol. Scand. 72,* 255.
56. Nagatsu, T., van der Schoot, J. B., Levitt, M., and Udenfriend, S. (1968). *J. Biochem. 64,* 39.
57. Foldes, A., Jeffrey, P. L., Preston, B. N., and Austin, L. (1972). *Biochem. J. 126,* 1209.
58. Lagercrantz, H. (1971). *Acta Physiol. Scand. Suppl. 366,* 1.

Protein Structure

CONCERTED HYDROGEN BOND CHANGES IN PROTEIN CONFORMATIONAL TRANSITIONS

Herrick Baltscheffsky

The Salk Institute for Biological Studies
San Diego, California 92112
and
Department of Biochemistry
Arrhenius Laboratory
University of Stockholm, S-10691
Stockholm, Sweden

INTRODUCTION

Hydrogen bonds, specific as they are both in a chemical and in a directional sense, and the secondary α and β structures to which they contribute in various polypeptides, may be considered as potential "growing points" in molecular evolution (1 - 3). Such "growing points" would tend, as the expression implies, to be not only conserved but also elaborated upon in the evolutionary process to give rise to more useful and, step-by-step, more intricate three-dimensional arrangements. Our picture of functionally important temporary β-structural patterns in bacterial ferredoxin apoproteins (4) has led me to regard polypeptide structures, which allow concerted hydrogen bond changes to occur, as being even more pronounced evolutionary "growing points" (3). Some examples of such apparently functional changes of secondary structure have been given and a protein secondary structural change hypothesis for the mechanism of electron transport coupled phosphorylation

 ISBN 0-12-397580-8

has, very tentatively, been proposed (3). A mechanism of this sort would establish a similarity, at the fundamental levels of molecular structure and function, between biological transfer of energy and of information: a reversible, concerted hydrogen bond pattern formation (3).

This chapter will focus on specific concerted changes in the hydrogen bond patterns of α-helices or β structures which appear to be directly involved in polypeptide function. Although it is currently difficult to ascertain what general significance changes of this kind may have in protein function, mention should be made of a related new qualitative concept, proposed by Blumenfeld, of the physical mechanism of elementary acts in enzyme catalysis (5).

It would be too far-reaching to try to discuss, in this context, possible evolutionary or functional aspects of concerted hydrogen bond changes in oligopeptides, oligo- or polynucleotides, or oligonucleotide - oligopeptide complexes (6, 7). Although it is realized that various combinations of hydrogen bond change, electrostatic, hydrophobic, and other effects may together contribute to functional protein conformational changes, they will not be treated here. Electrostatic effects in proteins have recently been discussed by Perutz (8) and hydrophobic effects by Tanford (9). Only concerted hydrogen bond changes involving the α and β structures of the protein backbone will be discussed here. Hydrogen bonds linking side chains of amino acid residues or side chain with backbone, for example side-chain COO^- with backbone NH, will thus not be considered.

ON CONCERTED HYDROGEN BOND CHANGES IN GLUCAGON, INSULIN, AND PROTAMINE

The smallest polypeptides to be discussed are the hormones glucagon (29 amino acids) and insulin (22 in the A chain, 30 in the B chain). Chou and Fasman (10) predicted that residues 19 - 27 of glucagon, in apparent agreement with various experimental observations, could undergo "an $\alpha \rightarrow \beta$ transition" (a transition from α-helix to extended structure). X-ray data obtained by Saraki *et al.* (11) suggested that the polypeptide chain of residues 1 - 5 is flexible in the crystalline glucagon trimer and that the rest of the molecule essentially is an α-helix. The helical form was considered to be in equilibrium with a random coil form in solution and to be stabilized either as trimers or by association with a receptor by hydrophobic interactions. Earlier observations by Bornet and Edelhoch (12) had already indicated that a hydrophobic environment caused glucagon to attain secondary structure.

Whereas the predicted extended structure of glucagon does not appear to have been shown to exist experimentally, the more recent work of Bentley *et al.* (13) on the structure of insulin in 4-zinc crystals, as compared to 2-zinc crystals, is of interest in the present context. It was found that the first eight B-chain residues in one of the two insulin molecules in each asymmetric unit undergo a large change in conformation, from extended in 2-zinc crystals to α-helical in 4-zinc crystals. The transformation between 2- and 4-zinc insulin crystals is reversible.

Protamine has recently been reported by Warrant and Kim (14) to change its conformation, on binding to yeast phenylalanine tRNA, from a random coil to a structure containing α-helices. They proposed that all protamines have the potential to form a structure composed of three or four α-helical domains connected by two or three flexible joints, and that each α-helical domain can lie approximately along either groove of double-helical DNA.

The secondary structural changes in glucagon, insulin, and protamine, whether predicted or actually observed, were all assumed to be closely related to the biological functions of these compounds.

PROTEINS INVOLVED IN OXIDATION - REDUCTION

As was mentioned above, there are indications (4) that concerted hydrogen bond changes in ferredoxin apoproteins may occur and play a functional role. Having found that the predicted anti-parallel β structures in bacterial ferredoxins would orient cysteine residues to give an almost ideal spacing for initial binding of Fe_4S_4-clusters (4) and would change conformation in connection with the making of the holoproteins, we were alerted to look for other apparently functional changes of protein secondary structures.

Another instance of such a change in an oxidation - reduction protein has recently been reported. Brändén and Eklund (15) claim that binding of reduced coenzyme to liver alcohol dehydrogenase causes the last five of residues 46 - 55, which are arranged in an α-helix in the apoenzyme, to unwrap and become part of a randomly coiled loop region. Loop movement, in combination with domain twist and coenzyme binding is combined to completely shield off one access direction to the active site (15).

The concerted hydrogen bond change in liver alcohol dehydrogenase, which appears to be caused by the addition of nucleotide to the apoprotein, is of particular interest in connection with the great similarities observed between the nucleotide-binding domains in many different dehydrogenases and kinases. These domains in many of the proteins contain a β-pleated sheet structure consisting of six parallel strands, with α-helices situated between all or most of the β strands (the arrangements of α and β structures in the nucleotide-binding domains of some kinases differ more or less extensively from this pattern). It should be pointed out that the observed helix → random coil transformation region is not in the nucleotide-binding domain but in the catalytic part of the alcohol dehydrogenase (15).

MUSCULAR PROTEINS AND CALCIUM BINDING

As early as 1971, Harrington (16) proposed a mechanism for contraction in the skeletal muscle involving a fast, cooperative α helix → coil transition in the α helix-rich "hinge" region of myosin. This region is adjacent to the globular ATPase heads. Although no such well-defined conformational change in myosin has been observed, the proposal may be related to a number of findings with proteins involved in contractile processes.

In the bacteriophage T4 sheath protein, which, with a molecular weight between 50,000 and 60,000 appears to be the simplest structure that could be described as a contractile organelle. Venyaminov *et al.* (17) have observed a decrease in the α-helical content and an increase in the β-form content during sheath contraction. The changes in circular dichroism indicated a drop from 20 to 10% α-helix and a similar 10% increase in β structure upon contraction. Very recently Nagy *et al.* (18) reported a calcium-induced increase in α-helix from 20% to 38% in a fragment of troponin C that contains one intact Ca^{2+}-Mg^{2+} binding site (site III). This fragment contained residues 90 - 120 and the helix formed upon binding of Ca^{2+} to the loop region 103 - 112 consisted of the residues 94 - 102 or 103. A preexisting helix involving residues 112 - 120 was independent of Ca^{2+}. Thus the unoccupied Ca^{2+} binding site has an α-helix on one side and a loop on the other, whereas there are α-helices on each side of this site when occupied by Ca^{2+}.

Various fluorescence, reactivity, spectral, and NMR data were in agreement with the proposed specific Ca^{2+}-induced coil → α-helix change described above and with the assumption that the conformational change in the troponin C fragment reflects an identical change in the intact troponin C molecule

(18). These results also bring into focus earlier findings that troponin C and myosin light chains show enough sequence homology to indicate a common ancestor (19) and that in the more primitive molluskan muscle Ca^{2+} triggers contraction by interaction with myosin (20) and not, as in vertebrates, with troponin C.

When mixed with pure rabbit actin, the molluskan myosin exerts Ca^{2+}-dependent ATPase activity, and one of its three light chains, which itself lacks ATPase activity and does not bind Ca^{2+}, is required for the Ca^{2+} dependence of the ATPase (20). These findings may be related to the dependence on Ca^{2+} of the solubilized multi-subunit coupling factor ATPases from spinach chloroplasts (21, 22) or chromatophores from the photosynthetic bacterium *Rhodospirillum rubrum* (23). The Ca^{2+} dependence of these ATPase activities emerges only when the coupling factor protein complexes are solubilized from their respective membranes. However, it is an open question whether these Ca^{2+}-dependent activities, or any of them, are linked to conformational changes similar to those occurring upon the binding of Ca^{2+} to the troponin C segment. In all these instances, important but largely uncharted quaternary interactions, where concerted hydrogen bond changes could well have essential functions, would add to the complexity of the picture.

DISCUSSION

This attempt to give an overview of predicted and experimentally supported functional concerted hydrogen bond changes in α and β structures of proteins leads to the following questions: if concerted hydrogen bond changes in polypeptides occur more abundantly and are of greater functional significance than what is currently known, what experiments can be done to show it? The fact that about one-half of the examples given above have been reported only recently would seem to allow the assumption that more cases will be found in the near future. However, I would like to suggest a direct experimental approach to test, specifically, our ferredoxin apoprotein β structure prediction, and generally, the concept that soluble proteins may be found under appropriate conditions to reversibly go from one secondary structural conformation to another. It involves closer study of the electron carriers ferredoxin and cytochrome *c*.

As has been pointed out recently (4), methods exist for the reversible quantitative removal of iron and "labile sulfur" from ferredoxins (24, 25). Cytochrome *c* apoprotein has been prepared and found to exist in water or salt solution as a random coil with no significant helical structure remaining from the native protein (26, 27). On the other hand, the helix probability profile for cytochrome *c* apoprotein predicts α-helix configuration potential for more than 70% of the α-helical regions actually found in the heme-containing protein (28).

Both with native ferredoxin and cytochrome *c* and with their apoproteins, careful and reversible variation of parameters such as pH, temperature, and, a perhaps more physiological variable, hydrophobicity of the medium seem to be one way of looking into the question of whether concerted hydrogen bond pattern changes can be reversibly obtained. We are just beginning experiments along these lines.

ACKNOWLEDGMENTS

This work was supported by The Swedish Natural Science Research Council Grant No. K2292-100 and by NASA Grant No. NGR-05-067-001. My thanks are due to Drs. L. E. Orgel, M. Levitt and E. Stellwagen for valuable discussions.

REFERENCES

1. Baltscheffsky, H. (1977). *In* "Living Systems as Energy Converters" (R. Buvet *et al.*, eds.), pp. 81-88. Elsevier, North-Holland, Amsterdam.
2. Baltscheffsky, H., von Heijne, G., and Blomberg, C. (1978). *In* "Evolution of Protein Molecules" (H. Matsubara and T. Yamanaka, eds.), pp. 141-147. Japan Scientific Societies, Tokyo.
3. Baltscheffsky, H. (1978). *In* "Energy Conversion in Biological Membranes" (G. Schäfer and M. Klingenberg, eds.), 29 Mosbacher Colloquium, pp. 3 - 18.
4. von Heijne, G., Blomberg, C., and Baltscheffsky, H. (1978). *Origins of Life 9,* 27-37.
5. Blumenfeld, L. A. (1976). *J. Theoret. Biol. 58,* 269-284.
6. Orgel, L. E. (1968). *J. Mol. Biol. 38,* 381-393.
7. Carter, F., Jr. and Kraut, C. S. (1974). *Proc. Natl. Acad. Sci. U.S. 71,* 283-287.
8. Perutz, M. F. (1978). *Science 201,* 1187-1191.
9. Tanford, C. (1978). *Science 200,* 1012-1018.
10. Chou, P. Y. and Fasman, G. D. (1975). *Biochemistry 14,*

2536-2541.
11. Saraki, K., Dockerill, S., Adamiak, D. A., Tickle, T. J., and Blundell, T. (1975). *Nature (London) 257*, 751-757.
12. Bornet, H. and Edelhoch, H. (1971). *J. Biol. Chem. 246*, 1785-1792.
13. Bentley, G., Dodson, E., Dodson, G., Hodgkin, D., and Mercola, D. (1976). *Nature (London) 261*, 166-168.
14. Warrant, R. W. and Kim, S. H. (1978). *Nature (London) 271*, 130-135.
15. Brändén, C.-I. and Eklund, H. (1978). *CIBA Symp. No. 60*, pp. 63-76.
16. Harrington, W. F. (1971). *Proc. Natl. Acad. Sci. U.S. 68*, 685-689.
17. Venyaminov, S. Y., Rodikova, L. P., Metlina, A. L., and Poglazov, B. F. (1975). *J. Mol. Biol. 98*, 657-664.
18. Nagy, B., Potter, J. D., and Gergely, J. (1978). *J. Biol. Chem. 253*, 5971-5974.
19. Collins, J. H. (1974). *Biochem. Biophys. Res. Commun. 58*, 301-308.
20. Szent-Györgyi, A. G., Szentkiralyi, E. M., and Kendrick-Jones, J. (1973). *J. Mol. Biol. 74*, 179-203.
21. Vambutas, V. K. and Racker, E. (1964). *J. Biol. Chem. 240*, 2660-2667.
22. McCarty, R. E. and Racker, E. (1968). *J. Biol. Chem. 243*, 129-237.
23. Johansson, B. C., Baltscheffsky, M., Baltscheffsky, H., Baccarini-Melandri, A., and Melandri, B. A. (1973). *Eur. J. Biochem. 40*, 109-117.
24. Malkin, R. and Rabinowitz, J. C. (1966). *Biochem. Biophys. Res. Commun. 23*, 822-827.
25. Thompson, C. L., Johnson, C. E., Dickson, D. P. E., Cammack, R., Hall, D. O., Weser, U., and Rao, K. K. (1974). *Biochem. J. 139*, 97-103.
26. Stellwagen, E., Rysavy, R., and Babul, G. (1972). *J. Biol. Chem. 247*, 8074-8077.
27. Fisher, W. R., Taniuchi, H., and Anfinsen, C. B. (1973). *J. Biol. Chem. 248*, 3188-3195.
28. Lewis, P. N. and Scheraga, H. A. (1971). *Arch. Biochem. Biophys. 144*, 576-583.

PREDICTION OF THE SECONDARY STRUCTURE OF PROTEINS

Gerald D. Fasman

Graduate Department of Biochemistry
Brandeis University
Waltham, Massachusetts 02254

The conformation of biologically active proteins and polypeptides is one of the main determinants of the high degree of specificity of their reactivity *in vivo*. In turn, the native conformation of a protein is determined by its amino acid sequence (1). X-ray diffraction techniques have successfully elucidated the three-dimensional structure (conformation) of over 60 proteins (2, 3). However, this number is small compared to the number of protein sequences (>700) which have been completed (4). Many of these proteins, such as, histones, membranes, and ribosomal proteins have not yet been crystallized, so other techniques must be utilized to yield structural information. Because of this enormous backlog of proteins whose structure is unknown, and the conviction that the amino acid sequence determines conformation, many efforts have been made to predict protein secondary structure from sequence data. Many predictive methods are available and these have been reviewed in articles on protein folding and sequence analysis; two recent reviews summarize the various predictive algorithms and evaluate their efficacy (5, 6). Since these reviews were written two

 ISBN 0-12-397580-8

other methods have been published (7, 8).

No attempt will be made herein to discuss the relative merits of these predictive algorithms, but rather the Chou-Fasman method (9, 10) will be used as an example to demonstrate the general utility of this approach to assist in understanding the general phenomena of structure and function of polypeptides and proteins.

These predictive methods can locate α-helical, β-sheet, and β-turn regions in proteins with an 80% degree of accuracy. The predictive projections of protein conformation have been applied in the following ways:

1. Initial delineation of secondary structure in X-ray crystallographic studies.
2. Starting conformations for energy minimization procedures.
3. Detecting regions with potential for conformational changes.
4. Recognition of structural domains in homologous sequences.
5. Suggesting the rational design of synthetic analogs for experimental testing to see whether conservation or change in conformation will produce alteration or retention in biological activity, e.g., structure and biological function relationships.

A brief description of the Chou-Fasman (9, 10) method will be given and its utility demonstrated in terms of the above outlined potentials. A statistical survey of 15 proteins, whose X-ray structure had been determined, was made and the helix and β-sheet conformation potentials of all 20 amino acids were established in their hierarchial order. Tables on the frequency of helical and β-sheet boundary residues in these 15 proteins were also published (9). The helix and β-sheet conformational parameters P_α and P_β (where $P_\alpha = f_\alpha/\langle f_\alpha \rangle$ and $P_\beta = f_\beta/\langle f_\beta \rangle$; f_α and f_β are the frequency of residues in the helix and β regions; $\langle f_\alpha \rangle$ and $\langle f_\beta \rangle$ are the average frequency of residues in the helix and β regions) were utilized with a set of empirical rules to predict the α and β regions in proteins. The method's simplicity and relative degree of accuracy (≃75%) are the main reasons for its wide use (see approximately 90 examples in Appendix in ref. 5).

PREDICTION RULES

A more recent analysis of 29 proteins containing 4741 amino acids whose X-ray structure has been published has produced a slightly revised set of conformational parameters.

These are shown in Table I (11), in their hierarchial order, and classified as: H_α (strong helix former), h_α (helix former), I_α (weak helix former), i_α (helix indifferent), b_α (helix breaker), B_α (strong helix breaker), H_β (strong β former), etc., as for the helix. Using the following empirical rules (abbreviated), one locates the secondary structures of proteins:

1. A cluster of four helical residues (h_α or H_α) out of six residues along the protein sequence will initiate a helix. The helical segment is extended in both directions until sets of tetrapeptide breakers $\langle P_\alpha \rangle < 1.00$ are reached (e.g., b_4, b_3i, b_2i_2, etc.). Certain amino acids have positional preferences, for example, proline, aspartic acid, and glutamic acid are found with a high frequency at the N-terminal helical end, and histidine, lysine, and arginine at the C-terminal helical end. Furthermore, proline cannot occur in the inner helix or at the C-terminal helical end. Any segment with the $P_\alpha \geq 1.03$ as well as $\langle P_\alpha \rangle > \langle P_\beta \rangle$ is predicted as helical.

2. A sequence of three β formers or a cluster of three β formers out of four or five residues along the protein sequence will initiate a β-sheet. The β-sheet is propagated in both directions as long as the sheet contains less than one-third β breakers. It is terminated by the same set of tetrapeptide breakers as found for the helical sections (above). Any segment with $\langle P_\beta \rangle \geq 1.05$, as well as $\langle P_\beta \rangle > \langle P_\alpha \rangle$ is predicted as β-sheet.

3. Any segment containing overlapping α and β residues is resolved through conformational boundary analysis. These three basic rules determine the prediction of protein secondary structures (helix, β-sheet, and random coil).

β TURNS

There remains one additional important secondary structure, the β turn which involves four consecutive residues in a protein where the polypeptide chain folds back on itself by nearly 180°. It is these chain reversals which give a protein its globularity. Venkatachalam (12) first characterized three types of turns in a tetrapeptide where there is a hydrogen bond between the CO group of residue i and the NH group of residue i + 3. Analysis of proteins revealed many β turns present in their secondary structures (for review see Ref. 6). Utilizing the X-ray atomic coordinates from 29 proteins, Chou and Fasman (13) computed the $C_i^\alpha - C_{i+3}^\alpha$ distances of all 4651 tetrapeptides. Those whose distances were below 7 Å and not in a helical region were considered as β turns. A total of 457 β bends were located, and 32% of all the amino acid residues in the 29 proteins

TABLE 1. Conformational Parameters for α-Helical, β-Sheet and β-Turn Residues in 29 Proteins[a]

	P_α			P_β			P_t		f_i		f_{i+1}		f_{i+2}		f_{i+3}
Glu	1.51	H_α	Val	1.70	H_β	Asn	1.56	Asn	0.161	Pro	0.301	Asn	0.191	Trp	0.167
Met	1.45		Ile	1.60		Gly	1.56	Cys	0.149	Ser	0.139	Gly	0.190	Gly	0.152
Ala	1.42		Tyr	1.47		Pro	1.52	Asp	0.147	Lys	0.115	Asp	0.179	Cys	0.128
Leu	1.21		Phe	1.38	h_β	Asp	1.46	His	0.140	Asp	0.110	Ser	0.125	Tyr	0.125
Lys	1.16	h_α	Trp	1.37		Ser	1.43	Ser	0.120	Thr	0.108	Cys	0.117	Ser	0.106
Phe	1.13		Leu	1.30		Cys	1.19	Pro	0.102	Arg	0.106	Tyr	0.114	Gln	0.098
Gln	1.11		Cys	1.19		Tyr	1.14	Gly	0.102	Gln	0.098	Arg	0.099	Lys	0.095
Trp	1.08		Thr	1.19		Lys	1.01	Thr	0.086	Gly	0.085	His	0.093	Asn	0.091
Ile	1.08		Gln	1.10		Gln	0.98	Tyr	0.082	Asn	0.083	Glu	0.077	Arg	0.085
Val	1.06		Met	1.05		Thr	0.96	Trp	0.077	Met	0.082	Lys	0.072	Asp	0.081
Asp	1.01	I_α	Arg	0.93	i_β	Trp	0.96	Gln	0.074	Ala	0.076	Thr	0.065	Thr	0.079
His	1.00		Asn	0.89		Arg	0.95	Arg	0.070	Tyr	0.065	Phe	0.065	Leu	0.070
Arg	0.98	i_α	His	0.87		His	0.95	Met	0.068	Glu	0.060	Trp	0.064	Pro	0.068
Thr	0.83		Ala	0.83		Glu	0.74	Val	0.062	Cys	0.053	Gln	0.037	Phe	0.065
Ser	0.77		Ser	0.75	b_β	Ala	0.66	Leu	0.061	Val	0.048	Leu	0.036	Glu	0.064
Cys	0.70		Gly	0.75		Met	0.60	Ala	0.060	His	0.047	Ala	0.035	Ala	0.058
Tyr	0.69	b_α	Lys	0.74		Phe	0.60	Phe	0.059	Phe	0.041	Pro	0.034	Ile	0.056
Asn	0.67		Pro	0.55	B_β	Leu	0.59	Glu	0.056	Ile	0.034	Val	0.028	Met	0.055
Pro	0.57	B_α	Asp	0.54		Val	0.50	Lys	0.055	Leu	0.025	Met	0.014	His	0.054
Gly	0.57		Glu	0.37		Ile	0.47	Ile	0.043	Trp	0.013	Ile	0.013	Val	0.053

[a] *P_α, P_β, P_t are conformational parameters of helical, β-sheet and β-turns. f_i, f_{i+1}, f_{i+2}, f_{i+3} are bend frequencies in the four positions of the β-turn. H_α, H_β, etc., as defined previously (10). From Ref. 5, 11).*

occurred in these bends. The α-helix occupies 38% of the residues and 20% of the residues are found in the β sheet. These frequencies for all the amino acids, in the helical, β sheet and β turns are obtained when their occurrence in each conformational state is divided by their total occurrences in the 29 proteins. The frequencies of residues in the 1st, 2nd, 3rd, and 4th positions of β turns for all the residues were obtained. These positional preferences of amino acids in β turns can be seen in Table 1, which are based on 408 turns (5, 11).

Some residues are found to have a dramatic positional preference in the β turn: for example, proline in the 2nd (30%) but not in the 3rd position (3%); residues with the highest β-turn potential in all 4 positions (P_t values) are asparagine, glycine, proline, aspartic acid, and serine, with the most hydrophobic residues showing the lowest bend potential. On the other hand, in regions adjacent to β turns many hydrophobic residues were found, in either α - α, α - β, or β - β interactions. The probability of β turn occurrence at position i is calculated from $p_t = (f_i \times f_{i+1} \times f_{i+2} \times f_{i+3})$, and those tetrapeptides with $p_t > 0.75 \times 10^{-4}$, as well as $\langle P_t \rangle > 1.00$ and $\langle P_\alpha \rangle < \langle P_t \rangle > \langle P_\beta \rangle$ are predicted as β turns (14). The percentage of bend and non-bend residues predicted correctly for 29 proteins by this algorithm is $\%_{t+nt} = 70\%$, while 78% of the β turns were localized correctly within ±2 residues (14). Applying the above conditions of prediction, the secondary structure, the α, β, random coil, and β turns can be delineated. Since β turns are often located at the boundaries of both α and β regions, they may play an important role initiating the tertiary folding of proteins.

APPLICATIONS OF THE PREDICTIVE METHOD

The predictive method may serve as a *guideline for the direction of folding* in low-resolution electron density maps of proteins, as in the case of TMV protein (15). More recently, crystallographers have found the empirical rules of Chou and Fasman to be consistent with their X-ray results (15, 16 - 19).

The conformation of proteins can be estimated by several physicochemical techniques, e.g., circular dichroism, infrared spectroscopy, and NMR. The predictive method can supplement these experimental methods since it is able to locate where the secondary structural regions are in the amino acid sequence, as well as regions in proteins having both helical and β-forming potentials and, therefore, the potential for *conformational changes*. For example, the X-ray crystallographic studies of concanavalin A (20, 21) showed only 2% helix in the native structure. However, 55% helicity can be induced in concanavalin A with 70% chloroethanol (22). The predictive scheme (10) correctly located all 12 β-sheet regions in concanavalin A with only one overpredicted β region. In addition, it showed that a total of 47% of its residues in 13 regions also have α potential, although many of these had still higher β potentials. Similarly, elastase has 7% helicity as shown from X-ray diffraction (23), but circular dichroism studies showed that it assumed 35% helicity in sodium dodecyl sulfate (24). The predictive method (10) showed that there are 79 residues in 15 regions with helical potential accounting for 33% helicity. Hence, the easily computed $\langle P_\alpha \rangle$ and $\langle P_\beta \rangle$ values for the α and β segments in proteins may assist in elucidating the regions potentially capable of undergoing conformational change. It is interesting to note that the B1-7 region of insulin was predicted as β sheet (10) with $\langle P_\beta \rangle = 1.15 > \langle P_\alpha \rangle = 1.07$ in agreement with the X-ray data. Since B1-7 also has α potential, it is not surprising that this region was found to be helical in 4-Zn insulin in 6% NaCl (25). Thus by computing $\langle P_\alpha \rangle$ and $\langle P_\beta \rangle$ values for the α and β segments in proteins, one may elucidate the regions potentially capable of undergoing conformational change.

The prediction of the conformation of glucagon, a hormone containing 29 amino acid residues, offers an excellent example of the potential of the method. Utilizing the conformational parameters for helix, β sheet, β turns and random coil, Chou and Fasman (26) predicted two conformational states for glucagon. They showed that the conformational sensitivity of glucagon may be due to residues 19 - 27 which have both α-helical potential ($\langle P_\alpha \rangle = 1.19$) as well as β-sheet potential ($\langle P_\beta \rangle = 1.25$). Thus in predicted form (a), residues 5 - 10 with $\langle P_\beta \rangle = 1.08 > P_\alpha = 0.86$ adopt a β conformation, while residues 19 - 27 form a helical region (31% α, 21% β). In predicted form (b), both regions, residues 5 - 10 and residues 19 - 27, are β-sheets (0% α, 52% β). Circular dichroism studies (27) of glucagon solutions (12.6 mg/ml) yield 33% α and 20% β supportive of form (a). Infrared studies of glucagon gels and fibrils (28, 29) have a predominant β conformation, consistent with form (b). In addition, three reverse β turns were predicted at 2 - 5, 10 - 13, and 15 - 18, suggesting this small polypeptide has the potential to fold into a relatively compact structure. Thus it appears that glucagon has different α and β conformations under different concentration conditions. Hence, residues 19 - 27 may be involved in an $\alpha \rightarrow \beta$ transition.

The *in vivo* concentration is probably too small to elicit the β conformation, but this conformational state may be induced upon binding of glucagon to its receptor site. As the conformational state of region 19 - 27 is sensitively balanced between α and β states, it is predicted that replacement of one or more residues of high β potential in this region with strong α formers would lock the conformation in the helical state. If this hypothesis for receptor binding is correct, namely, the necessity of the β structure, then this homolog of glucagon would be biologically inactive as a result of its inability to bind. It is also feasible to lock the β conformation by suitable substitutions. Thus, the predictive scheme offers a working hypothesis whereby the structure of the biologically active hormone may be arrived at. The α → β transition of glucagon in solution has recently been followed by means of circular dichroism (30), providing evidence for the potential conformational change predicted.

The conformational parameters P_α, P_β, and P_t (Table 1) are expedient for detecting regions in proteins with potential for *conformational changes due to mutations* or changes in solvent conditions. The *lac* repressor - *lac* operator interaction of *Escherichia coli* provides an excellent example of the specificity of protein binding to DNA (31). The amino acid sequence of the *lac* repressor, a polypeptide subunit containing 347 amino acid residues, has been determined. Its secondary structure was predicted to contain 37% α-helix and 35% β-sheet while the trypsin-resistant core (residues 60 to 327) has 29% helix and 41% β-sheet (32). The extensive β-sheets predicted in the 215 - 324 region may be responsible for tetramer stabilization found in both the *lac* repressor and the core. These β sheets are almost devoid of charge and would have an extremely hydrophobic nucleus. There are 23 predicted β turns in the *lac* repressor, made up of 50% charged and polar residues (serine and threonine), which would be found on the surface conferring solubility.

Examination of five *lac* repressor mutants yields significant information regarding conformational requirements for repressor function. Mutant AP46 has a $Ala^{53} \rightarrow Val^{53}$ replacement and a loss of repressor activity, causing a predicted α → β transition at residues 52 - 57. Several *amber* mutants at Gln^{26} (Leu, Ser, Tyr) still cause repression and no conformational change is predicted. In mutant AP309 a $Ser^{16} \rightarrow Pro^{16}$ change is incurred with loss of biological activity and a predicted β turn at 14 - 17 is lost by this mutation. Thus, in these examples, it is possible to correlate biological activity with definite secondary structures, and the loss of activity upon the induction of a conformational change brought about by a mutation.

Sickle cell hemoglobin (HbS) differs from normal adult hemoglobin (HbA) in that there is a $Glu^6 \rightarrow Val^6$ mutation in both β chains (33). This mutation causes aggregation of HbS inside the erythrocyte upon deoxygenation resulting in the sickling phenomenon. The possible conformational consequence of this mutation has been predicted (34). The region at the N-terminus of the β chains of HbA, residues 1 - 6, has $\langle P_\alpha \rangle = 1.03 > \langle P_\beta \rangle = 0.99$ indicating α-helical potential and unfavorable β sheet formation. However, the β chain $Glu^6 \rightarrow Val$ mutation involves not only the replacement of a charged polar residue by an uncharged hydrophobic residue, but more importantly causes a drastic increase in the β-sheet potential. This region now has $\langle P_\beta \rangle = 1.21 > \langle P_\alpha \rangle = 0.96$. Hence, a single amino acid replacement could disrupt the first turn (residues 4 - 6) of the A-helix and convert the 1 - 6 region into a β sheet. These β regions may be responsible for the aggregation phenomena found in HbS.

The enthusiastic response to the predictive model of Chou and Fasman (9, 10) is evident by the numerous citations in the literature showing increasing usage of the method (35). By means of this formulation, the coenzyme-binding domains of glutamate dehydrogenases were located with the aid of homologous sequence comparisons (36), a model of the fd gene 5 DNA-binding protein was proposed as two long β sheets (37), and a three-dimensional structure of proinsulin was constructed (38). Durham and Butler (39) were gratified that the predicted secondary structure of tobacco mosaic virus (TMV) fitted well with other data based on low resolution X-ray results as well as immunological and chemical evidence. Using the P_α, P_β, and P_c conformational parameters, Parry (40) showed that α-tropomyosin is almost 100% α-helical with possible nonhelical regions confined to the N- and/or C-terminals. He also noted that short portions of the amino acid sequence with low helical potential show a well defined 40-residue period. The predictive method was also extensively applied to indicate the predominant α and β regions in many of the ribosomal proteins sequenced in Wittmann's laboratory (41, 42 and references herein). Experimental studies showed a correlation of the helical-forming potential of synthetic ribonuclease S (1 - 20) peptides with binding to the S protein (43). The helical residue Glu^9 of ribonuclease S peptide was replaced by Leu and Gly. While the binding affinity of the Leu S peptide to S protein is only a factor of three less than that for RNase S (1 - 20), that for the Gly peptide is about 20-fold less. The stronger binding of the Leu peptide than the Gly peptide corresponds to the higher helical potential of Leu ($P_\alpha = 1.34$) than Gly ($P_\alpha = 0.53$), so that the Leu peptide has the greater tendency to adopt the helical conformation which is a prerequisite for complex formation.

Other applications of the Chou-Fasman method include the elucidation of β-sheet and helical regions in the biotin subunit of transcarboxylase (44), the proposal of an elongated antiparallel β-pleated sheet in the hypothalamic (cyclic) tetradecapeptide, somatostatin (45), the identification of α, β, and β-turn regions in the epidermal growth factor (46), and the correlation of the predicted conformation of hemerythrin with circular dichroism and X-ray data (47). More extensive analysis was applied to predicting the conformation of the α_1-acid glyco protein (48), where it was found that the carbohydrate moiety of 9 glycoproteins are situated predominately in β-turn regions (49). Upon analysis of 30 phosphorylated sites in 14 different proteins, it was found that 80% were found to exist in regions predicted as β turns, usually on Ser or Thr (50). By a combination of four predictive methods (51), a tertiary model of plastocyanin was constructed. More recently, the Chou-Fasman model was successfully applied to elucidating the structure of staphylococcal enterotoxin B, showing excellent agreement between the predicted conformation (11% α, 34% β, 55% coil) and circular dichroism studies (9% α, 38% β, 53% coil) (52). In addition, the method was used to examine the effect of mutational amino acid replacements on the secondary structure predicted for glutamate dehydrogenase (53), as well as to propose the NAD domain of aldolase based on a complete secondary structural prediction of the enzyme (54). Furthermore, the prior conformational prediction of triose phosphate isomerase (Phillips, personal communication) and fl bacteriophage coat protein (Tanford, personal communication) correlated well with later experimental data from X-ray crystallographic (16) and physicochemical measurements (55). Using the Chou-Fasman method, Low *et al.* (17) predicted three β regions (2 - 6), 12 - 16, and 33 - 37) and two of the β turns, 18 - 21 and 47 - 50, in erabutoxin b, in agreement with their X-ray analysis (observed β: 4 - 9, 12 - 14, and 34 - 40). Hence, the relative accuracy and simplicity of the Chou-Fasman method have resulted in its extensive use in predicting the conformation of biological macromolecules.

The β-turn conformation can now be predicted with the same degree of accuracy as the α-helical and β-sheet regions in proteins (14). As mentioned above, the three-dimensional structure of proinsulin has been predicted (38). Examining the C peptide sequence of 10 mammalian species showed a remarkable conservation of predicted conformation with a β turn at residues 15 - 18 flanked by two helices. Utilizing data from 29 proteins (5), it has been shown that a high β-turn potential exists in the 12 - 17 region for this series and, more importantly, none outside of it (14). Although no biological role has been assigned to the C peptide, the present prediction shows that the β-turn conservation in proinsulin is probably necessary for directing the proper folding of C peptide helices, which possibly masks the receptor binding region of the hormone thus making the precursor, proinsulin, inactive.

The importance of the β turn is also seen in the predicted structure of seven homologous proteinase inhibitors. There are only 5 other invariant residues, as well as 6 half-cystine residues (of approximately 60 residues), however, the β-turns are conserved with onset at residues 12, 25, 37, and 41, as shown by by X-ray studies for the bovine pancreatic trypsin inhibitor (14).

Conformational homologies were elucidated in growth hormones (56), immunoglobulins (57), and the neurotoxins. Fifty-seven snake venom toxins were predicted and a common distribution of secondary structure was detected throughout these toxins. The results also highlight the contrasts between short and long neurotoxins and neurotoxins and cytotoxins (58 - 60). The conformational parameters have also been used to deduce the probable polypeptide conformation on prebiotic earth (61), to test structural convergence during protein evolution (62), and to construct genealogical trees (63). Recent phytogenetic studies include rubredoxins (64) and muscular parvalbumins (65).

Finally, the predictive method may suggest the rational design of synthetic analogs for experimental testing to see whether conservation or changes in conformation will produce alteration or retention of hormonal or enzymatic activity [for example, ribonuclease S peptide (43), proinsulin C peptide (66) secretin (67, 68), and the region 75 - 120 of human growth hormone (56)].

The structure of the pentapeptide enkephalins was predicted as a β bend (69) and this has recently been shown by X-ray crystallography to be true in the solid state (70). The synthesis of a model β turn was based on the predictive rules and the circular dichroism spectra of this conformation was obtained from this model compound (71). It has been suggested, based on the predictive scheme, that the polypeptide secondary structure may direct the specificity of prohormone conversion (72).

Comparisons of the various predictive methods have been published for adenylate kinase (73) and T4 phage lysozyme (74). The prediction of lysozyme was severely criticized (74), however, in a more recent publication (75), at a 2.4 Å resolution, it has been found that the predictive method (9, 10) yielded results of 70% accuracy.

Although unquestionably the function of a protein depends on its unique three-dimensional structure, one may still learn much from the prediction of the secondary structure of proteins Although it might be considered judicious to wait for a perfect predictive algorithm to be developed, the wealth of knowledge accumulated during the last decade regarding protein conformational prediction should be continually tested, refined, and applied. Although great caution should be exercised in the application of predictive models, it would be ultraconservative not to utilize this knowledge of secondary structure to aid our further understanding of protein conformation and biological activity. The prediction of secondary structures should not be a end in itself but a means for furthering our understanding of protein conformation through synthetic analogs, corroborative experimental studies, sequence and conformational homology comparisons, and tertiary structure model building. Such an approach will hopefully provide great insight into the principles of protein folding.

ACKNOWLEDGMENTS

Preparation of this manuscript was generously supported in part by Grants from the U.S. Public Health Service (GM 17533), and the National Science Foundation (PCM-76-21856). This is Publication No. 1386 from the Graduate Department of Biochemistry, Brandeis University, Waltham, MA 02254. The author expresses his appreciation to Dr. Peter Chou, a former collaborator, who contributed royally to the development of the method.

REFERENCES

1. Anfinson, C. B., Haber, E., Sela, M., and White, F. H., Jr. (1961). *Proc. Natl. Acad. Sci. U.S. 47,* 1309-1314.
2. Liljas, A. and Rossmann, M. G. (1974). *Annu. Rev. Biochem. 43,* 475-507.
3. Matthews, B. (1976). *Annu. Rev. Phys. Chem. 27,* 493-523.
4. Dayhoff, M. O. (1972). "Atlas of Protein Sequence and Structure," Vol. V, and Suppl. 1 and 2 (1973, 1976). Nat. Biomed. Res. Found., Silver Spring, Maryland.
5. Chou, P. Y. and Fasman, G. D. (1978). *Advan. Enzymol. 47,* 45-148.
6. Chou, P. Y. and Fasman, G. D. (1978). *Annu. Rev. Biochem. 47,* 251-276.
7. Levitt, M. and Greer, J. (1977). *J. Mol. Biol. 114,* 181-293.
8. Garnier, J., Osguthorpe, D. J., and Robson, B. (1978). *J. Mol. Biol. 120,* 97-120.
9. Chou, P. Y. and Fasman, G. D. (1974). *Biochemistry 13,* 211-222.
10. Chou, P. Y. and Fasman, G. D. (1974). *Biochemistry 13,* 222-245.
11. Chou, P. Y. and Fasman, G. D. (1977). *In* "Peptides," (M. Goodman and J. Meienhofer, eds.), Proc. 5th Amer. Peptide Symp., pp. 284-287. Wiley, New York.
12. Venkatachalam, C. M. (1968). *Biopolymers 6,* 1425-1436.
13. Chou, P. Y. and Fasman, G. D. (1977). *J. Mol. Biol. 115,* 135-175.
14. Chou, P. Y. and Fasman, G. D. (1979). *Biophys. J. 26,*

367-384.
15. Durham, A. C. H. and Butler, P. J. G. (1975). *Eur. J. Biochem. 53,* 397-404.
16. Banner, D. W., Bloomer, A. C., Petsko, G. A., Phillips, D. C., Pogson, C. I., Wilson, I. A., Corran, P. H., Furth, A. J., Milman, J. D., Offord, R. E., Priddle, J. D., and Waley, S. G. (1975). *Nature (London) 255,* 609-614.
17. Low, B. W., Preston, H. S., Sato, A., Rosen, L. S., Searl, J. E., Rudko, A. D., and Richardson, J. S. (1976). *Proc. Natl. Acad. Sci. U.S. 73,* 2991-2994.
18. Olsen, K. W., Moras, D., Rossmann, M. G., and Harris, J. I. (1975). *J. Biol. Chem. 250,* 9313-9321.
19. Mavridis, I. M. and Tulinsky, A. (1976). *Biochemistry 15,* 4410-4417.
20. Edelman, G. M., Cunningham, B. A., Reeke, G. N., Jr., Becker, J. W., Waxdal, M. J., and Wang, J. L. (1972). *Proc. Natl. Acad. Sci. U.S. 69,* 2580-2584.
21. Hardman, K. D. and Ainsworth, C. F. (1972). *Biochemistry 11,* 4910-4919.
22. McCubbin, W. D., Oikawa, K., and Kay, C. M. (1971). *Biochem. Biophys. Res. Commun. 43,* 666-674.
23. Shotton, D. M. and Watson, H. C. (1970). *Nature (London) 225,* 811-816.
24. Visser, L. and Blout, E. R. (1971). *Biochemistry 10,* 743-752.
25. Bentley, G., Dodson, E., Dodson, G., Hodgkin, D., and Mercola, D. (1976). *Nature (London) 261,* 166-168.
26. Chou, P. Y. and Fasman, G. D. (1975). *Biochemistry 14,* 2536-2541.
27. Srere, P. A. and Brooks, G. C. (1969). *Arch. Biochem. Biophys. 129,* 708-710.
28. Gratzer, W. B., Bailey, E., and Beaven, G. H. (1967). *Biochem. Biophys. Res. Commun. 28,* 914-919.
29. Epand, R. M. (1971). *Can. J. Biochem. 49,* 166-169.
30. Moran, E. C., Chou, P. Y., and Fasman, G. D. (1977). *Biochem. Biophys. Res. Commun. 77,* 1300-1306.
31. Bourgeois, S. and Pfahl, M. (1976). *Advan. Protein Chem. 30,* 1-99.
32. Chou, P. Y., Adler, A. J., and Fasman, G. D. (1975). *J. Mol. Biol. 96,* 29-45.
33. Ingram, V. M. (1959). *Biochim. Biophys. Acta 36,* 402-411.
34. Chou, P. Y. (1974). *Biochem. Biophys. Res. Commun. 61,* 87-94.
35. Current Contents *19,* No. 8, 5 (1976).
36. Wootton, J. C. (1974). *Nature (London) 252,* 542-546.
37. Anderson, R. A., Nakashima, Y., and Coleman, J. E. (1975). *Biochemistry 14,* 907-917.
38. Snell, C. R. and Smyth, D. G. (1975). *J. Biol. Chem. 250,*

6291-6295.
39. Durham, A. C. H. and Butler, P. J. G. (1975). *Eur. J. Biochem. 53,* 397-404.
40. Parry, D. A. D. (1975). *J. Mol. Biol. 98,* 519-535.
41. Stadler, H. (1974). *FEBS Lett. 48,* 114-116.
42. Yaguchi, M. and Wittmann, H. G. (1978). *FEBS Lett. 88,* 227-230.
43. Dunn, B. M. and Chaiken, I. M. (1975). *J. Mol. Biol. 95,* 497-511.
44. Wood, H. G. and Zwolinski, G. K. (1976). *Crit. Rev. Biochem. 4,* 47-122.
45. Holladay, L. A. and Puett, D. (1976). *Proc. Natl. Acad. Sci. U.S. 73,* 1199-1202.
46. Holladay, L. A., Savage, C. R., Jr., Cohen, S., and Puett, D. (1976). *Biochemistry 15,* 2624-2633.
47. Klotz, I. M., Klippenstein, G. L., and Hendrickson, W. A. (1976). *Science 192,* 335-344.
48. Aubert, J. P. and Loucheux-Lefebvre, M. H. (1976). *Arch. Biochem. Biophys. 175,* 400-409.
49. Aubert, J. P., Biserte, G., and Loucheux-Lefebvre, M. H. (1976). *Arch. Biochem. Biophys. 175,* 410-418.
50. Small, D., Chou, P. Y., and Fasman, G. D. (1977). *Biochem. Biophys. Res. Commun. 79,* 341-346.
51. Wallace, D. G. (1976). *Biophys. Chem. 4,* 123-130.
52. Munoz, P. A., Warren, J. R., and Noelken, M. E. (1976). *Biochemistry 15,* 4666-4671.
53. Brett, M., Chambers, G. K., Holder, A. A., Fincham, J. R. S., and Wootton, J. C. (1976). *J. Mol. Biol. 106,* 1-22.
54. Stellwagen, E. (1976). *J. Mol. Biol. 106,* 903-911.
55. Nozaki, Y., Chamberlain, B. K., Webster, R. E., and Tanford, C. (1976). *Nature (London) 259,* 335-337.
56. Pena, C., Stewart, J. M., Paladini, A. C., Dellacha, J. M., and Santome, J. A. (1975). *In* "Peptides: Chemistry, Structure, and Biology," (R. Walter and J. Meienhofer, eds.), pp. 523-528. Ann Arbor Sci. Publ., Ann Arbor, Michigan.
57. Low, T. L. K., Liu, Y. S. V., and Putnam, F. W. (1976). *Science 191,* 390-392.
58. Eterovic, V. A. and Ferchmin, P. A. (1977). *Intern. J. Peptide Protein Res. 10,* 245-251.
59. Hseu, T-H., Liu, Y-C., Wang, C., Chang, H., Hwang, D-M., and Yang, C-C. (1977). *Biochemistry 16,* 2999-3006.
60. Dufton, M. J. and Hider, R. C. (1977). *J. Mol. Biol. 115,* 177-193.
61. Brack, A. and Orgel, L. E. (1975). *Nature (London) 256,* 383-387.
62. Salemme, F. R., Miller, M. D., and Jordan, S. R. (1977). *Proc. Natl. Acad. Sci. U.S. 74,* 2820-2824.
63. Goodman, M. and Moore, G. W. (1977). *J. Mol. Evol. 9,*

121-130.
64. Vogel, H., Brushi, M., and LeGall, J. (1977). *J. Mol. Evol. 9,* 111-119.
65. Goodman, M. and Pechere, J. F. (1977). *J. Mol. Evol. 9,* 131-158.
66. Vogt, H. P., Wollmer, A., Naithani, V. K., and Zahn, H. (1976). *Hoppe-Seyler's Z. Physiol. Chem. 357,* 107-115.
67. Fink, M. L. and Bodanszky, M. J. (1976). *J. Amer. Chem. Soc. 98,* 974-977.
68. Bodanszky, M. and Fink, M. L. (1976). *Bioorg. Chem. 5,* 275-282.
69. Bradbury, A. F., Smyth, D. G., and Snell, C. R. (1976). *Nature (London) 260,* 165-166.
70. Smith, G. D. and Griffin, J. F. (1978). *Science 199,* 1214-1216.
71. Kawai, M. and Fasman, G. D. (1978). *J. Amer. Chem. Soc. 100,* 3630-3632.
72. Geisow, M. J. (1978). *FEBS Lett. 87,* 111-114.
73. Schulz, G. E., Barry, C. D., Friedman, J., Chou, P. Y., Fasman, G. D., Finkelstein, A. V., Lim, V. I., Ptitsyn, O. B., Kabat, E. A., Wu, T. T., Levitt, M., Robson, B., and Nagano, K. (1974). *Nature (London) 250,* 140-142.
74. Matthews, B. W. (1975). *Biochim. Biophys. Acta 405,* 442-451.
75. Remington, S. J., Anderson, W. F., Owen, J., Eyck, L. F. T., Grainger, C. T., and Matthews, B. W. (1978). *J. Mol. Biol. 118,* 81-98.

SUBSTRATE SPECIFICITY AND CHARACTERISTICS OF π ALCOHOL DEHYDROGENASE AND OTHER HUMAN LIVER ADH ISOENZYMES

Bert L. Vallee
William A. Frey
Werner P. Dafeldecker

Center for Biochemical
and Biophysical Sciences and Medicine
Harvard Medical School
Boston, Massachusetts 02115

William F. Bosron
Ting-Kai Li

Department of Medicine and Biochemistry
Indiana University School of Medicine
Indianapolis, Indiana

INTRODUCTION

The addictive, pathologic, and pharmacologic consequences of alcohol intake are related directly to the chemical properties of ethanol and/or its metabolic products. Detailed examination of the enzyme involved in its elimination is critical to the delineation of the etiology and pathological mechanisms underlying alcoholism. The human liver alcohol dehydrogenases have been available in suitable purity and quantity only recently. Hence, biochemical knowledge regarding their potential relevance to these processes had to be based on studies of such enzymes isolated from other species, primarily yeast, rat, and horse. Moreover, it has been difficult to unravel the complex interrelationships of psychosocial and biologic factors known to accompany alcoholism. Because of these and other handicaps, possible biochemical bases for the disorders accompanying alcohol abuse have remained obscure. In recent years, however, increasingly convincing evidence has appeared suggesting the ex-

 ISBN 0-12-397580-8

istence of genetic links between alcohol metabolizing capacity and alcoholism in certain individuals (1, 2). Because liver alcohol dehydrogenase (EC 1.1.1.1), ADH, is the principal enzyme involved in ethanol elimination (3), a study of the genetic variability of this enzyme remains of paramount importance to the solution of pathological consequences of alcohol abuse.

Human liver ADH exists in multiple molecular forms, and the number and amount of each isoenzyme seems dependent upon the genetic background and health of the donors. However, thus far, only a few of these isoenzymes have been characterize with respect to their biochemical properties.

The recent discovery in our laboratories of a new and unique isoenzyme of human liver ADH, exhibiting kinetic properties markedly different from those of isoenzymes studied previously, may prove to be an important link between the physiology of ethanol metabolism of man and its pathological deviations. This new ADH isoenzyme exhibits a K_m for ethanol which is as much as 100 times greater than that of the other forms; moreover, it is characteristically insensitive to inhibition by pyrazole compounds, potent inhibitors of all mammalian ADH's studied thus far (4). Because of its insensitivity to pyrazole this form of liver ADH has been designated π-ADH (5). We have further succeeded in separating other isoenzymes of human liver ADH consisting of the α and β subunit types, allowing the study and establishment of the functional roles of these various forms.

Characteristically, the mammalian liver alcohol dehydrogen ases exhibit very broad substrates specificity. The substrates known thus far include primary and secondary alcohols, diols, and aromatic alcohols, ω-hydroxylated fatty acids, terpenes and certain hydroxy and keto steroids. Not surprisingly, a number of these compounds - other than ethanol - have been proposed to serve as physiological substrates for this enzyme. Indeed, cer tain horse liver ADH isoenzymes exhibit vastly different kineti properties toward ethanol, on one hand, and some steroid substrates, on the other (10 - 12). In this regard, the discovery that human π-ADH exhibits specificities for certain substrates which differ strikingly from those of other isoenzymes, suggest the possibility that they serve functionally different roles (5). The combination of differences in isoenzyme concentration and their substrate specificities, potentiates the possibility that human liver ADH could serve a more general role in the detoxification and metabolism of a variety of known alcohols and other hydroxylated compounds than has been hitherto suspected.

MOLECULAR HETEROGENEITY OF HUMAN LIVER ALCOHOL DEHYDROGENASE

Vallee and co-workers initially demonstrated the existence of multiple isoenzyme forms of human liver ADH (13), as was confirmed later by several laboratories, which subsequently reported the identification of up to 10 ADH forms, as visualized by starch-gel electrophoresis of certain crude liver extracts (12, 14 - 16). In addition, individual variations in total and specific enzymatic activities as well as changes in the number and the amount of isoenzyme forms were reported in crude tissue extracts. Smith *et al.* (16, 17) proposed a genetic model to explain these observations and postulated the existence of three separate gene loci coding for three different subunit forms, α, β and γ, with polymorphism at the locus. These subunits randomly combine to form the active dimeric enzyme. In a subsequent study (12), dissociation and recombination experiments with isolated specific isoenzymes fully documented that the expected homo- and heterodimeric isoenzymes could be formed from the α, β, and γ chains and that the molecular weight of all isoenzyme forms was identical. Interestingly, the early fetus exhibits only the αα, and in some cases the π-ADH forms. In the neonate and infant, αα, αβ and ββ forms are observed. Isoenzymes containing the γ chains appear at approximately 3 - 6 months of age, giving rise to the complex isoenzyme pattern observed in adults.

Because the genetic model was based almost entirely on studies employing crude homogenates and postmortem specimens without regard to cause of death or the relative stability of the different ADH forms, its validity could not be accepted with certainty. Subsequent studies comparing the activities and electrophoretic patterns of liver biopsy and autopsy samples obtained within 12 hours of death indicated wide variations in the specific activity of ADH dependent upon the health of the donor before death and the tissue storage conditions employed. More importantly, a comparison of the ADH isoenzymes by starch gel electrophoresis revealed the existence of some livers, in addition to the isoenzymes described by Smith *et al.* (17), of a previously unknown activity band with electrophoretic mobility less than that of the αα enzyme form. Until then, the αα molecular form exhibited the least electrophoretic mobility and was the major activity band in premature infant livers. Surprisingly, the newly identified ADH form, initially called the "anodic band," was particularly prevalent in biopsy samples and livers with high activity obtained at autopsy following sudden death, but was seen only faintly or not at all in livers from routine autopsies (18, 19). Therefore, in the past, the existence of this anodic ADH form had not been recognized consistently, because of its lability both *in vivo* and *in vitro* (1920).

It became apparent that both the premortem history and storage conditions of the liver samples influence the specific activity and the isoenzyme composition of ADH obtained at autopsy from human livers. Since the genetic model of ADH isoenzymes does not account for the existence of the anodic band, the molecular basis for the heterogeneity of ADH required re-evaluation.

PURIFICATION OF HUMAN LIVER ALCOHOL DEHYDROGENASE AND ISOENZYMES BY AFFINITY CHROMATOGRAPHY

Previous attempts to purify ADH from human liver using salt fractionation and ion exchange chromatography (12, 13, 15, 21) generally did not yield sufficient amounts of homogenous enzyme to allow determination of its molecular and catalytic properties. We have recently developed a new and effective affinity chromatographic procedure that co-purifies in high yield the major complement of ADH isoenzyme forms, free of contaminating proteins (22). The affinity technique takes advantage of the known inhibitory characteristics of pyrazole and its 4-substituted analogs. The affinity ligand, 4-[3-(3-*N*-6-aminocaproyl)aminopropyl]pyrazole, is immobilized on Sepharose (CapGapp-Sepharose). ADH is absorbed onto the resin in the presence of NAD^+ by forming specific and tight "dead-end" ternary complexes. Elution is accomplished by addition of ethanol which competes with the affinity ligand for the enzyme's active site, and forms the productive ternary enzyme complex (Fig. 1). Due to this dual affinity of the inhibitor and substrate for the enzyme-coenzyme complex, a high degree of both specificity and purification is attained through this method, termed "double-ternary complex affinity chromatography."

This affinity chromatographic technique successfully purifies ADH from human, horse, rat, and rabbit livers (22) (Table I). In marked contrast to the several days required for purification by previous methods, the new procedure results in homogeneous human liver ADH in a remarkably short period, less than 6 hr (Table II). Human ADH purified in this manner encompasses all of the α, β, and γ subunit-containing forms present originally in the liver extracts (5, 20). The molecular and kinetic properties of this enzyme preparation have been reported previously by Lange *et al.* (23).

Since in certain livers with particularly high ADH activity the anodic band appeared to be the major form, isolation of this moiety was attempted by affinity chromatography (Table III Crude liver homogenate supernatants were purified partially

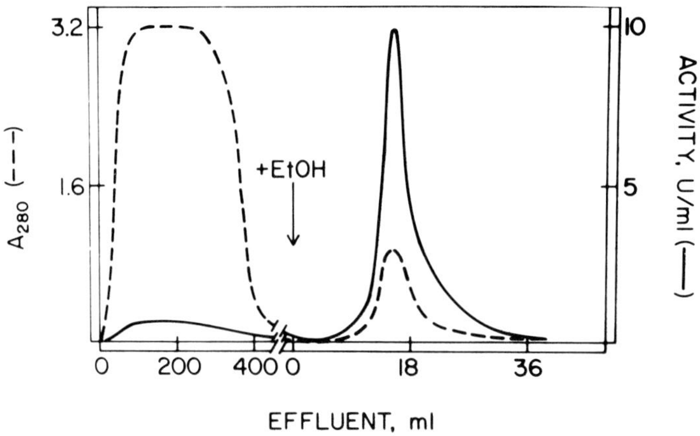

Fig. 1. CapGapp-Sepharose affinity chromatography of human liver ADH. The supernatant from a liver homogenate was first purified on DEAE-cellulose and then applied to a 0.9 × 35 cm column of CapGapp-Sepharose in 50 mM NaPi and 1.2 mM NAD at pH 7.5. ADH activity was determined with 33 mM ethanol, and 2.4 mM NAD^+ in 0.1 M glycine-NAOH at pH 10.0.

TABLE I. Isolation of Liver Alcohol Dehydrogenase from Crude Extracts by Affinity Chromatography[a]

Liver	Initial sp. act.	final sp. act.	Overall yield (%)
Rabbit	0.0124	3.60	78
Rat	0.0048	2.04	69
Horse	0.0320	14.9	88

[a]Rabbit, rat, and horse liver (110, 80, and 50 g wet weight, respectively) were homogenized in cold water and then the DEAE effluent was chromatographed over CapGapp-Sepharose. Specific activity is given as ΔA_{340}/min A_{280} and the yield is calculated based on the amount of alcohol dehydrogenase activity originally present in the crude extract.

TABLE II. Purification of Human Liver Alcohol Dehydrogenase[a]

Step	Volume (mg)	Total Protein (mg)	Activity (U/ml)	Sp. Act. (U/mg)	Yield (%)	Time (hr)
Crude extract	950	100,000	2.2	0.021	100	2.5
DEAE-cellulose	950	10,000	2.2	0.21	100	0.5
Affinity chromatography	245	195	4.5	5.7	65	3.0

[a]Human liver, 606 g (wet weight), was homogenized in 1200 ml of cold H_2O and the mixture was stirred at 4° for 2 hours The strained crude extract was passed through DEAE-cellulose (10 × 15 cm) equilibrated with 0.01 M Tris-Cl, pH 7.9, and the resultant effluent was chromatographed on CapGAPP-Sepharose.

TABLE III. Purification of π-ADH

Step	Total Activity (units)	Specific activity[a] (units/mg)	Yield (%)
1. Supernatant	315	0.06	100
2. DEAE-cellulose	182	0.19	85
3. CapGAPP-Sepharose			
a. π-ADH	30	0.03	10
b. Other forms	95	2.5	30
4. Agarose-hexane-AMP	20	1.3	6

[a]Alcohol oxidizing activity was determined spectrophotometrically in a 3 ml assay at 25° with 0.1 M glycine, pH 10.0, 33 mM ethanol, and 2.4 mM NAD^+.

using batch DEAE-cellulose followed by chromatography on CapGAPP-Sepharose in the presence of NAD^+. Significantly, a major fraction of the activity does not bind to the affinity resin but elutes with the bulk of the protein in the void fractions of the column. As had been observed previously (22), addition of ethanol elutes the remainder of the activity initially placed on the column. The affinity resin is unable to bind a significant portion of the ADH isoenzymes which are remarkably resistant to inhibition by pyrazole compounds. Therefore, the electrophoretic patterns of the enzyme from liver homogenate and the fractions which bind and do not bind to the resin were examined by staining for ethanoloxidizing activity in the presence and absence of 4-methylpyrazole. The enzymatically active band in the homogenate displaying the least cathodic mobility was identified as the anodic band, whereas the remainder were characterized of phenotype ADH_3^2, as described by Smith *et al.* (17). Interestingly, the fractions that did not bind to CapGapp-Sepharose contained solely the anodic isoenzyme form, while all the remaining molecular species remained bound to the resin (Fig. 2). Confirming the earlier suggestion, concentrations of 4-methylpyrazole which completely inhibited all other molecular forms left the anodic enzyme form fully active. Clearly, the insensitivity of the anodic isoenzyme to inhibition by pyrazole accounts for its separation from all other pyrazole-sensitive ADH forms (Fig. 3).

Further purification of this anodic material, baptized π-ADH, was accomplished by affinity chromatography on AMP-Agarose where all the ADH activity applied binds to the resin and is eluted using a linear NADH gradient. As a result of this second affinity column, homogeneous π-ADH is obtained (24). Moreover, the dimeric enzyme is composed of identical 42,000

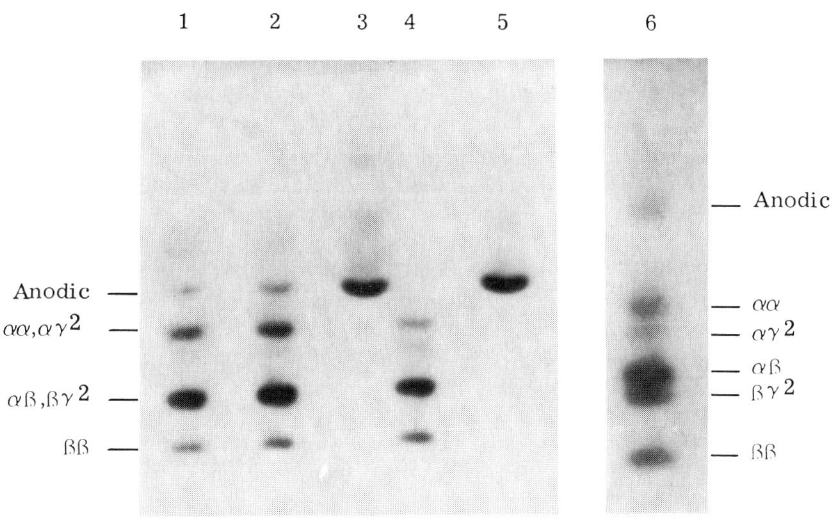

Fig. 2. Starch gel electrophoresis of human liver alcohol dehydrogenase. Electrophoresis was performed at pH 7.7 (samples 1-5) and 8.6 (sample 6), and stained for ethanol oxidizing activity (5). Samples 1 and 6 are homogenate supernatants and samples 2-5 correspond to purification steps, 2, 3a, 4 in Table III. Samples 1, 2, 4, and 6 contain 0.04 units of activity and samples 3 and 5 contain 0.01 units.

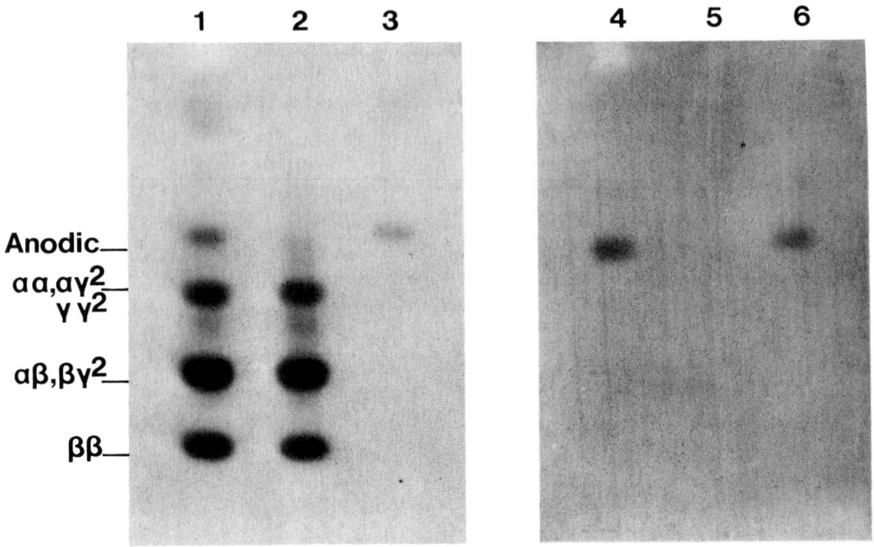

Fig. 3. Identification of the human liver ADH molecular forms in a homogenate supernatant and the enzyme fractions separated by CapGapp-Sepharose. Starch gel electrophoresis of the homogenate supernatant (samples 1 and 4) and the enzyme fraction that bound (samples 2 and 5) and did not bind (samples 3 and 6) to the affinity resin was performed at pH 7.7. Gels were stained at pH 8.6 with 100 mM ethanol in the absence (samples 1 - 3) or presence (samples 4 - 6) of 2 mM 4-methylpyrazole.

dalton subunits, as determined by SDS-polyacrylamide gel electrophoresis and analytical ultracentrifugation (20). The zinc content of 4 g-atoms/mole is identical and the amino acid composition are similar to those of the other human liver ADH isoenzymes (24). Hence, many of the physical and kinetic properties of purified π-ADH are indeed similar to those observed for horse liver ADH (4) and the human liver isoenzymes isolated previously (3).

The presence of π-ADH in biopsy liver samples as opposed to its absence in certain autopsy specimens suggested that π-ADH may be labile *in vivo*. This conclusion is reinforced by the fact that purified π-ADH is also very unstable *in vitro*. Fifty percent of the activity is lost after 24 hr, pH 7.5, 4°. However, the addition of 10^{-2} *M* ethanol stabilizes enzymatic activity effectively for up to 2 weeks (20).

The oxidation of ethanol and the reduction of acetaldehyde by π-ADH follows an ordered BiBi mechanism, as indicated by detailed kinetic studies (24). The optimal pH for ethanol oxidation is above 10.0, and coenzymes NAD^+ and NADH are specific for the enzyme. Similar to other human liver ADH isoenzymes, π-ADH exhibits a broad specificity for alcohol and aldehyde substrates. As the alcohol chain length is increased from 2 to 5 carbons, the K_m decreases from 34 to 0.036 m*M*, while V_{max} remains relatively constant at 0.48 to 0.50 µmol/min/mg (Table IV). The linear relationship between log $1/K_m$ and the log of the octanol-water partition coefficient for these alcohols suggests an important role for hydrophobic binding energies in π-ADH substrate specificity (25).

We have recently separated and purified the αα, αβ, and ββ isoenzymes from infant liver (see footnote) by chromatography on CapGapp Sepharose (22) and carboxymethyl-cellulose (13) (Fig. 4). Preliminary studies indicate that their kinetic properties and sensitivity to inhibition by pyrazole derivatives are similar to those described previously for impure preparations of human ADH isoenzymes (12, 15, 21, 23). However, certain kinetic parameters for π-ADH differ strikingly from those of the purified isoenzymes containing α and β subunits. Thus, the π-ADH form neither oxidizes methanol nor ethylene glycol at concentrations as high as 100 m*M*, in contrast to the others. In addition, the K_m values at pH 7.5 for both ethanol and acetaldehyde, 34 and 30 m*M*, respectively, are as much as 100 times greater than those reported previously.

Moreover, as indicated already, π-ADH is remarkably insensitive to inhibition by 4-methylpyrazole, as revealed both spectrophotometrically and by starch gel activity staining at concentrations that completely inhibit the other forms (5, 20).

However, inhibition of π-ADH is eventually accomplished at values 1000(!)-fold over and above the K_I reported previously for horse and human ADH preparations (21, 26). Under these conditions, the mode of inhibition is no longer competitive with

TABLE IV. Substrate Specificity of π-ADH[a]

Alcohol	K_m *(µmol/min/mg)*	K_m *(mM)*
Methanol	*0*	-
Ethanol	*0.50*	*34*
Butanol	*0.50*	*0.14*
Pentanol	*0.48*	*0.036*
3-Pyridylcarbinol	*0.49*	*0.24*
Ethylene glycol	*0*	-

[a] *0.1 M NaPi, pH 7.5, 2.4 mM* NAD^+*, 25°.*

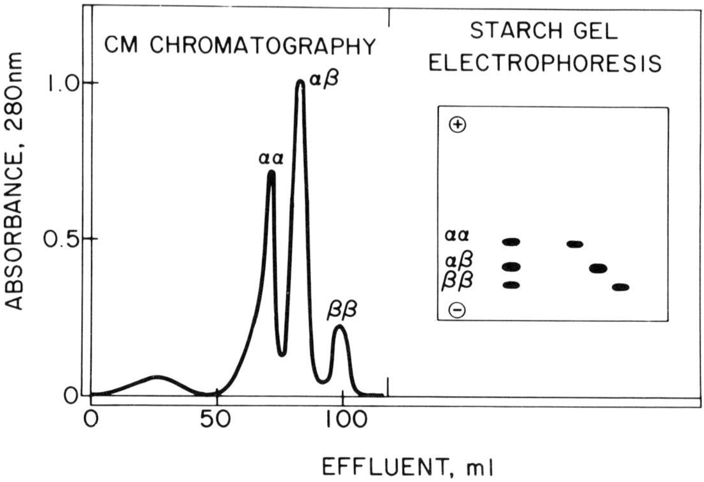

Fig. 4. Isoenzyme separation of infant liver ADH. The supernatant from a liver homogenate was first purified on DEAE-cellulose and CapGapp-Sepharose and then applied to a 0.9 × 25 c column of CM-cellulose in 5 mM Tris, 1 mM NAD⁺, 1 mM DTT, pH 7. at 4°. The isoenzymes were eluted with a linear gradient of 0 to 0.1 M NaCl beginning at 50 ml of effluent. Starch gel electrophoresis of the liver homogenate was performed at pH 7.7, ar the gel was stained with ethanol as substrate.

respect to acetaldehyde. From the increase in slope of the reciprocal plots, K_I values of 30 and 2 m*M* for pyrazole and 4-methylpyrazole were calculated, respectively (24). In contrast the pyrazole-sensitive forms of human liver ADH exhibited K_I values of 2.6 and 0.21 μ*M* (21).

SUBSTRATE SPECIFICITIES OF HUMAN LIVER ALCOHOL DEHYDROGENASE

For many years the physiological role of mammalian liver alcohol dehydrogenase has been the subject of much speculation. Although the liver contains a significant amount of ADH activity, its essential physiological role in normal metabolism has not been apparent. It has been demonstrated that the microbial flora of the gastrointestinal tract is capable of producing substantial amounts of ethanol that is subsequently absorbed into the body (27). Moreover, the existence of ethanol present in the blood obtained from animals raised under sterile conditions, whose gastrointestinal tract is free of bacteria, suggests additional endogenous pathways of ethanol synthesis. These studies have largely eliminated the past uncertainty concerning the natural occurrence of physiologically generated ethanol, albeit in small amounts. However, numerous other alcohols have been shown to serve as substrates for this enzyme and a number of compounds have been proposed as candidates for the true "physiological" substrate.

Arguments for activities toward physiological substrates, other than ethanol, have been based generally on the fact that the K_m values for many of these alcohols are lower than that for ethanol. The behavior of straight chain, primary alcohol substrates for the human liver enzyme illustrates this point. The K_m value falls from 2.7 m*M* for ethanol to 0.04 m*M* for 1-pentanol (23) at pH 10. Overall, the very number of potentially important substrates for alcohol dehydrogenase suggests that the enzyme's physiological action is not limited to ethanol and performs a more general role in metabolism and detoxification, acting on a variety of alcohols and aldehydes.

Previous studies on the substrate specificity of alcohol dehydrogenase have been limited largely to a small number of specific alcohols and sterols and their analogs. The literature gives little evidence that compounds other than aliphatic alcohols and sterols have been given serious consideration as substrates for ADH, based on other structural or conformational properties.

Work in our laboratory using various polyol and polyol-like substrates have recently revealed hitherto unrecognized activities for ADH. Certain polyols, e.g., *D*-ribose and *D*-sorbitol are substrates for human liver ADH. In addition, the enzyme also oxidizes several other sugars. The K_m values range from 7 m*M* for *D*-sorbitol to >100 m*M* for glycerol and *D*-ribose, all K_m values greater than that for ethanol (Table V). Significantly, relative to the maximal activity for ethanol, the k_{cat} values for glycerol and these sugar substrates range from 32 to 88%. In contrast, the activities for the majority of substrates for ADH thought to be potentially physiological are less than 10% of the rate of ethanol (10 - 12). Although numerous naturally occurring sugars were studied as potential substrates, including *D*-galactose, *D*-mannose, and sucrose, ADH catalyzed the oxidation of only a relatively small number of these.

The activity of human liver ADH toward sugars is particularly intriguing. The increased binding of the deoxy sugars when compared to *D*-ribose supports the known hydrophobic character of the ADH site.

In addition, a comparison of the solution conformation of

TABLE V. Catalytic Versatility of Human Liver Alcohol Dehydrogenase[a]

Substrate	k_{cat}[b]	K_m (mM)
Ethanol	100	2.7
Glycerol	32	>100
D-Sorbitol	2	7
D-Digitoxose	88	20
2-Deoxy-D-ribose	40	20
D-Ribose	35	>100

[a]Alcohol oxidizing activity was determined as in Table III.
[b]Relative to ethanol.

sugar substrates and the ring system configuration of steroid substrates may ultimately assist in mapping the requirements of the substrate's binding site. Despite the obviously reduced binding of the sugars when compared to ethanol, the potential for ADH playing an important role in the conversion of saccharide intermediates in cellular metabolism requires further investigation. Although the particular examples of sugar oxidation noted are not thought to be significant to mammalian metabolism, studies of the features common to these activities may reveal hitherto unknown metabolic functions of the enzyme which we continue to examine.

CONCLUSIONS

The understanding of normal human alcohol metabolism shoul benefit from the discovery of π-ADH as a functionally distinct enzyme form. In the past, the failure of pyrazole compounds to inhibit ADH activity at high ethanol concentrations has been cited as evidence for non-ADH catalyzed pathways of ethanol metabolism. The pyrazole insensitivity of π-ADH demonstrates that alternate pathways in humans cannot be inferred exclusively from the effects of these compounds.

The genetic control of molecular heterogeneity of liver AD and alcoholism in some individuals has already been described. The possibility that the presence or absence of π-ADH or any of the other isoenzyme forms provide a biochemical link to alcoholism is provocative, and has been discussed (5). The question o alternatives in the relative distribution and amount of π-ADH and other isoenzyme forms resulting from disease, malnutrition or chronic alcohol consumption are particularly intriguing possibilities. Variations in substrate specificity of isoenzymes have been mentioned previously as pertinent to human liver π-ADH. In this regard, further investigation of the activity of ADH toward metabolically important substrates as a function of isoenzyme composition might establish different physiologically important functions of ADH. It is reasonable to predict that the various isoenzymes of human liver ADH will similarly exhibit different substrate specificities.

ACKNOWLEDGMENTS

We are grateful to Dr. Virginia Swanson-Kamen for her assistance in this work.

Supported by the National Institute of Health Grant GM 15003 and U.S. Public Health Service Grant AA 02342.

REFERENCES

1. Partanen, J., Brunn, K., and Markkanen, T. (1966). "Inheritance of Drinking Behavior." Rutgers University Center of Alcohol Studies, New Brunswick, New Jersey.
2. Goodwin, D. W., Schulsinger, F., Moller, N., Hermansen, L., Winokur, G., and Guze, S. B. (1974). *Arch. Gen. Psychiat. 31,* 164-169.
3. Li, T. K. (1977). *Advan. Enzymol. 45,* 427-483.
4. Brändén, C. I., Jörnvall, H., Eklund, H., and Furugren, B. (1975). *In* "The Enzymes" (Paul D. Boyer, ed.), 3rd ed., Vol. XI, Part A, pp. 204-290. Academic Press, New York.
5. Li, T. K., Borson, W. F., Dafeldecker, W. P., Lange, L. G., and Vallee, B. L. (1977). *Proc. Natl. Acad. Sci. 74,* 4378-4381.
6. Sund, H. and Theorell, H. (1963). *In* "The Enzymes" (P. D. Boyer, H. A. Lardy, and K. Myrback, eds.), Vol. 7, p. 25. Academic Press, New York.
7. Brändén, C. I., Jörnvall, H., Eklund, H., and Furugren, B. (1975). *In* "The Enzymes" (P. D. Boyer, ed.), Vol. 11, p. 203. Academic Press, New York.
8. Björkhem, I. (1972). *Eur. J. Biochem. 30,* 441.
9. Pietruszko, R., Crawford, K., and Lester, D. (1973). *Arch. Biochem. Biophys. 159,* 50.
10. Pietruszko, R., Clark, A., Graves, J. M. H., and Ringold,

J. J. (1966). *Biochem. Biophys. Res. Commun. 23,* 526-534.
11. Theorell, H., Taniguchi, S., Åkeson, Å., and Skursky, L. (1966). *Biochem. Biophys. Res. Commun. 24,* 603-610.
12. Pietruszko, R., Theorell, H., and deZalenski, C. (1972). *Biochem. Biophys. 153,* 279-293.
13. Blair, A. H. and Vallee, B. L. (1966). *Biochemistry 5,* 2026-2034.
14. von Wartburg, J. P., Bethune, J. L., and Vallee, B. L. (1964). *Biochemistry 3,* 1775-1782.
15. Schenker, T. M., Teeple, L. J., and van Wartburg, J. P. (1971). *Eur. J. Biochem. 24,* 271-279.
16. Smith, M., Hopkinson, D. A., and Harris, H. (1971). *Ann. Human Genet. 34,* 251-271.
17. Smith, M., Hopkinson, D. A., and Harris, H. (1972). *Ann. Human Genet. 35,* 243-253.
18. Azevedo, E., Smith, M., Hopkinson, D. A., and Harris, H. (1974). *Ann. Human Genet. 38,* 31-37.
19. Li, T. K. and Magnes, L. J. (1975). *Biochem. Biophys. Res. Commun. 63,* 202-208.
20. Bosron, W. F., Li, T. K., Lange, L. G., Dafeldecker, W. P., and Vallee, B. L. (1977). *Biochem. Biophys. Res. Commun. 74,* 85-91.
21. Li, T. K. and Theorell, H. (1969). *Acta Chem. Scand. 23,* 892-902.
22. Lange, L. G. and Vallee, B. L. (1976). *Biochemistry 15,* 4681-4686.
23. Lange, L. G., Sytkowski, A. J., and Vallee, B. L. (1976). *Biochemistry 15,* 4687-4693.
24. Bosron, W. F., Li, T. K., Dafeldecker, W. P., and Vallee, B. L. (1979). *Biochemistry 18,* 1101-1105.
25. Hansch, C., Schaeffer, J., and Kerley, R. (1972). *J. Biol. Chem. 247,* 4703-4710.
26. Theorell, H., Yonetani, T., and Sjöberg, B. (1966). *Acta Chem. Scand. 23,* 255-260.
27. Krebs, H. A. and Perkins, J. R. (1970). *Biochem. J. 118,* 635.
28. Dafeldecker, W. P., Parés, X., Vallee, B. L., Bosron, W. F., and Li, T.-K. (1981). *Biochemistry 20,* 856-851.
29. Dafeldecker, W. P., Meadow, P. E., Parés, X., and Vallee, B. L. (1981). *Biochemistry 20,* 6729-6734.
30. Parés, X. and Vallee, B. L. (1981). *Biochem. Biophys. Res. Commun. 98,* 122-130.
31. Dafeldecker, W. P. and Vallee, B. L. (1982). *J. Prot. Chem.* (in press).

Note Added in Proof

The differential affinity chromatography procedure, employed here to isolate and purify π-ADH from human liver, was also used to separate the alcohol dehydrogenases of squirrel monkey (28) and rhesus monkey liver (29) into two classes of isoenzymes. The inhibition of these enzyme fractions by 4-methylpyrazole and their substrate specificities are analogous to those observed for the corresponding molecular forms of human liver. However, their physicochemical and compositional characteristics, i.e., molecular weight, zinc content, and dimeric structure, are all virtually identical with those of other mammalian liver alcohol dehydrogenases studied thus far. The presence of two functionally distinct classes of liver ADH in these monkeys suggests that the origin and variability of enzymes involved in human ethanol metabolism may be susceptible to experimental delineation in these primates.

Past investigations of human liver isoenzymes have been limited to those exhibiting cathodic mobility on starch gel electrophoresis at pH 7.7 to 8.6. Even the "anodic band" of Li and Magnes (19), now known as π-ADH (5) and described here, moves toward the cathode, although it has the least mobility of all the cathodic molecular forms. Until now, no isoenzymes of liver alcohol dehydrogenase had been described that actually move toward the anode. However, recently we reported the detection, purification, and initial characterization of two human liver ADH forms that do migrate toward the anode on starch gel electrophoresis and that stain with 1-pentanol but not with ethanol (30). So far these two anodic bands, termed χ-ADH, have been found in all human liver homogenates examined. Importantly, these anodic isoenzymes were also detected and isolated from rhesus monkey (29) and horse liver (31). While the physical properties of human, simian, and equine liver χ-ADH are remarkably similar to other mammalian alcohol dehydrogenases, their substrates specificities and kinetic characteristics differ significantly. Thus, none of them are active toward methanol and cyclohexanol nor are they saturated by ethanol, and 4-methylpyrazole inhibits them only at high concentrations. Moreover, the distinctive enzymatic specificity of these χ-ADH variants toward long chain aliphatic alcohols and ω-hydroxy fatty acids is particularly intriguing and may suggest important physiological functions for these ADH variants.

The present data again underscore both the complexity and diversity of mammalian liver alcohol dehydrogenases. Whatever their origin, inherited or induced, their multiplicity and functional variety must play a significant role in the normal and abnormal metabolism of important metabolic intermediates. The functional and structural features of these isoenzymes are clearly much more complex than had been anticipated.

INTRINSIC BINDING ENERGY, ENZYMATIC CATALYSIS, AND COUPLED VECTORIAL PROCESSES

William P. Jencks

Graduate Department of Biochemistry
Brandeis University
Waltham, Massachusetts 02154

The hypothesis that enzymes derive much of their catalytic activity from their ability to express the binding energy of nonreacting parts of their specific substrates in the transition state, but not in the enzyme-substrate complex, is examined here with particular reference to the interactions of succinyl-CoA - acetoacetate-CoA transferase with the coenzyme A moiety of its substrate. The expression of binding energy of ligands in one state, but not another state, is utilized by coupled vectorial systems that bring about mechanical or osmotic work in order to maintain the different states of the system at comparable concentrations so that turnover may occur rapidly under physiological conditions. Thus, both types of enzymes may be regarded as machines that work as a consequence of their ability to express the intrinsic binding energy of their substrates in one state, but not another. The coupling in vectorial systems results from the additional ability of these systems to change their substrate specificity in different states; these changes in specificity are defined by a set of coupling rules.

 ISBN 0-12-397580-8

INTRODUCTION

One hypothesis to explain the enormously greater catalytic activity of enzymes compared with most chemical catalysts is that enzymes, but not the chemical catalysts, can use the energy derived from binding interactions with nonreacting groups of their specific substrates to increase reaction rates. This hypothesis describes one of the oldest and most obvious mechanisms to account for enzymatic catalysis, but its implications with respect to the catalytic process are only now beginning to be explored experimentally. In particular, the efficient utilization of this binding energy requires that it not be expressed as very tight binding of specific substrates to enzymes in enzyme - substrate complexes, but that it should be expressed maximally only in the transition state of the catalyzed reaction (1 - 4).

The enzyme succinyl-CoA - acetoacetate-CoA transferase provides one example of the association of the binding energy of a specific substrate with catalysis (5, 6). The enzyme reacts with its specific substrate, succinyl-CoA, more rapidly than with the nonspecific substrate, succinyl-SR, by a factor of 3×10^{12}.

$$\text{EnzCOO}^- + \text{Succ-S}\boxed{\text{CoA}} \qquad k = 9.1 \times 10^{5}\ M^{-1}\ \text{min}^{-1}$$

$$\text{EnzCOO}^- + \text{Succ-SR} \qquad k = 2.9 \times 10^{-7}\ M^{-1}\ \text{min}^{-1}$$

The thiol group in the nonspecific substrate is methyl mercaptopropionate, which has a *pK* similar to that of coenzyme A, and the nonspecific substrate has the same chemical reactivity as the specific substrate toward alkaline hydrolysis. Thus, it is an experimental fact that the nonreacting groups of the large coenzyme A moiety of the specific substrate bring about a rate increase of 10^{12}, corresponding to a decrease in the Gibbs free energy of activation of 17 kcal mol^{-1}, compared with the short side chain of the nonspecific substrate. This is almost as large as the overall rate increase of 3×10^{13} for the enzymatic reaction compared to the corresponding nonenzymatic reaction of acetate with succinyl-CoA (6).

$$\text{EnzCOO}^- + \text{Succ-SCoA} \qquad k = 9.1 \times 10^{5}\ M^{-1}\ \text{min}^{-1}$$

$$\text{CH}_3\text{COO}^- + \text{Succ-SCoA} \qquad k = 3.4 \times 10^{-8}\ M^{-1}\ \text{min}^{-1}$$

The hypothesis states that the binding interactions between coenzyme A and the enzyme are directly responsible for catalysis by an amount corresponding to a stabilization of the transition state by 17 kcal mol^{-1}.

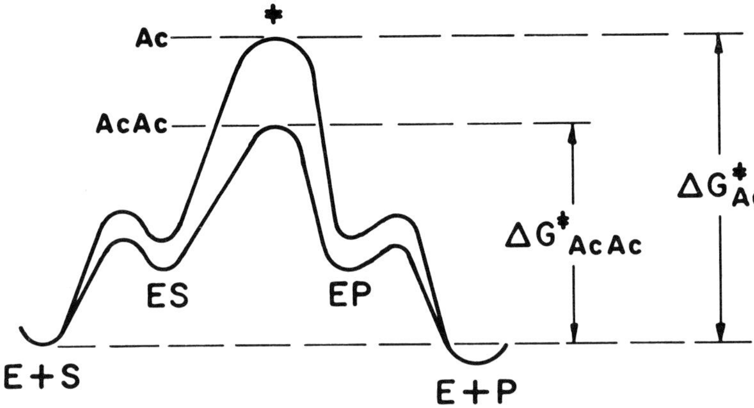

Fig. 1. Specificity and catalysis.

It may be conceivable that the effect of the coenzyme A in this reaction is simply to prevent nonproductive binding or to exert a control function by changing the enzyme from an inactive to an active conformation by an induced fit mechanism, but this does not seem likely. It is more interesting to explore the hypothesis that the interaction of the enzyme with coenzyme A is directly responsible for much of the catalytic process, as would be expected from the development of catalytic activity through evolution. Even if the nonspecific substrate were to bind nonproductively, there would have to be some major barrier to prevent its productive binding by a factor of 10^{12}; this barrier would then have to be overcome for the specific substrate by utilizing the binding energy of the coenzyme A.

Large rate increases may also be brought about by interactions with the active site of small groups on the substrate. The second-order rate constant, k_{cat}/K_m, for acetoacetate is larger by a factor of 2×10^5 than that for acetate, a smaller, nonspecific substrate of CoA transferase (5, 7). This corresponds to a stabilization of the transition state by 7.2 kcal mol^{-1} that arises from noncovalent interactions with the CH_3CO-group of acetoacetate (Fig. 1). Some of this stabilization (2.4 kcal mol^{-1}) represents a more favorable K_s for the specific substrate, but most of it (4.9 kcal mol^{-1}) reflects a faster reaction of the ES complex, k_{cat}. Just the substitution of a carbonyl oxygen atom for two hydrogen atoms at the β position of butyrate causes a stabilization of the transition state by 6.5 kcal mol^{-1}.

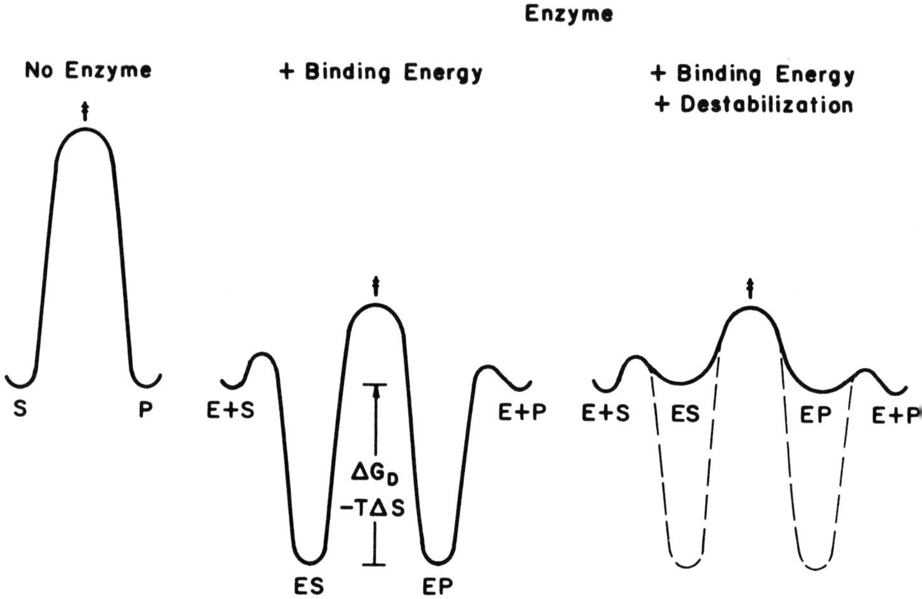

Fig. 2

There are many examples of such rate accelerations with specific substrates, but few have been carefully examined experimentally. The most thoroughly studied from the point of view of specific interactions with parts of the substrate is the 10^{10} faster reaction of the phosphoenzyme form of phosphoglucomutase with the hydroxyl group of its specific substrate, glucose 1-phosphate, than with the hydroxyl group of water, but even in this case the reasons for the rate increase are not completely identified (4).

In order to obtain a large catalytic advantage from the utilization of binding energy it is essential that this energy not be expressed in the enzyme - substrate complex (1 - 4). Suppose that CoA transferase had evolved so as to bind coenzyme A more tightly than the small R group by 17 kcal mol^{-1} in the transition state and that the same difference in binding energies existed in the enzyme - substrate complex. This would be a useless enzyme (Fig. 2) because the barrier for the reaction of the ES complex would be just as large as that for the nonenzymatic reaction, as shown by the first two energy diagrams in the figure. The enzyme will work only if the Gibbs energy of the ES complex is increased so that it is close to that of the transition state and the substrate cannot fall into a deep energy well in the ES complex, as shown in the right hand diagram of the figure. This increase in Gibbs energy can be brought about by some kind of destabilization mechanism or by decreasing the entropy in the ES complex.

Thus, the role of destabilization mechanisms and entropy loss is to make it easier to reach the transition state from the bound substrate and to prevent excessively tight binding of a specific substrate. The introduction of such destabilization and entropy loss does not directly increase the second-order rate constant k_{cat}/K_m for catalysis of reactions of very dilute substrates but, as illustrated in Fig. 2, it is essential that it exist in order to keep the K_m value large enough to prevent the substrate from getting stuck in the low energy well of a stable ES complex (2, 3). The addition to a substrate of a substituent that binds to the enzyme in the transition state will, of course, increase k_{cat}/K_m.

This destabilization may be so large that binding of some group on the substrate, or even the whole substrate molecule, is energetically disfavored and does not occur to more than a very small extent under physiological conditions. Thus, it may not be possible to observe a destabilization mechanism directly in an ES complex, because the destabilized group is hanging out in space. The important point is that the destabilization should exist and it makes little difference whether the balance of energies is such that actual binding of a group is observed (Fig. 3, $K_z > 1$) or not observed ($K_z < 1$); indeed, the most efficient way of dealing with the binding energy may be to have sufficient destabilization that binding is not observed in the ES complex (2, 3).

This situation may well hold for the binding of sugars to the D subsite of lysozyme. There is only a very small increment of observed binding energy, -0.6 to -0.9 kcal mol^{-1}, upon adding a sugar that is expected to bind to the D subsite to a specific trisaccharide, and the very similar binding of the glucose and xylose derivatives suggests that there is little or no direct destabilization of the hydroxymethyl group of bound glucose in the ES complex (8). However, the second-order rate constant for the hydrolysis of a dinitrophenyl trisaccharide is at least 1300 times faster with a glucose than with a xylose residue at the terminal cleavage site, so that the hydroxymethyl group of glucose must provide at least an additional 4.3 kcal mol^{-1} of binding energy in the transition state that is not manifested in the ground state (9).

In CoA transferase the binding energy of the coenzyme A is manifested only very weakly in the ES complex, although it is 17 kcal mol^{-1} tighter than that of SR in the transition state. Thus, coenzyme A is only a weak inhibitor of the enzyme (10), the binding of acetoacetyl-CoA is only twofold tighter (0.4 kcal mol^{-1}) than that of acetoacetate (5) and noncovalent binding interactions in the enzyme-CoA intermediate provide little or no net stabilization, as shown by the fact that the free energy of

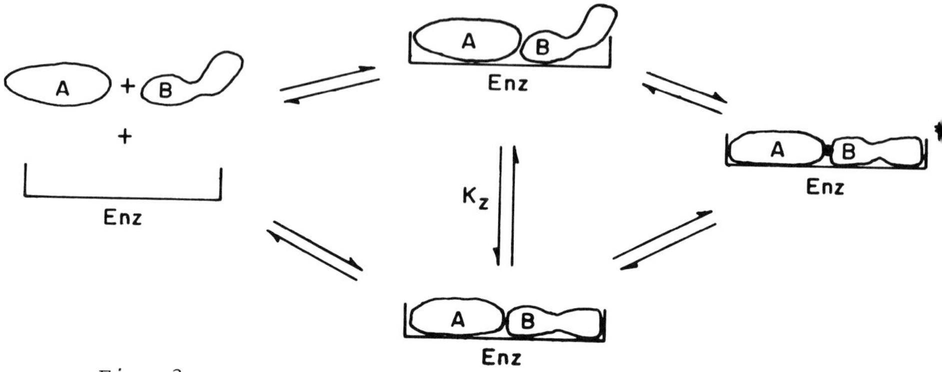

Fig. 3

hydrolysis of enzyme-CoA (a γ-glutamyl thiol ester) is almost the same as that of succinyl-CoA (5).

The important point here, and in related systems, is a matter of economics - you must pay for what you get. This has caused much confusion in this field, and elsewhere. In order to get an increase in k_{cat} from the utilization of binding energy it is necessary to <u>use up</u> that binding energy so that it is not manifested as observed binding. The observed binding energy of coenzyme A, for example, is only what is left over after this energy has been utilized when binding occurs, or what is observed from the binding of some parts of the substrate if the destabilization is sufficient to prevent completely the binding of other parts of the specific substrate. This matter of economics is responsible for the fact that specific substrates often bind no more tightly than nonspecific substrates, and sometimes bind even less tightly. The total binding energy that could be obtained from a particular substrate, if all of it were directly expressed as binding, can be called the <u>intrinsic binding energy</u>; this will generally be much larger than the observed binding energy (2).

These considerations may be summarized by the statement that an enzyme is a machine that works because it expresses the binding energy of a specific substrate in one state, the transition state, but not in another, the enzyme - substrate complex.

State		*Intrinsic binding energy*
1	*ES complex*	*Not expressed*
2	*Transition state*	*Expressed*

MECHANISMS

The real problem is to determine the mechanisms by which binding energies are not allowed to be manifested in the enzyme - substrate complex but are instead utilized to bring about catalysis. We do not know these mechanisms for CoA transferase, or for any other enzyme, but there are some hints and there is a large body of work in nonenzymatic systems that demonstrates the existence and possible magnitude of catalytic mechanisms of this kind (2, 3).

There are three kinds of destabilization mechanisms that can make it easier for an ES complex to reach the transition state:

1. *Geometric destabilization* by distortion or strain in the ES complex is the best known such mechanism, but is subject to the limitation of the force constants that can be developed for pushes, pulls, and twists in the active site of an enzyme. This type of destabilization is better represented at the present time by the numerous chemical reactions that are accelerated by strain than by enzymatic reactions.

2. *Destabilization by desolvation* of charged or polar groups on the substrate or the enzyme has the potential of bringing about very large rate accelerations because of the enormous solvation energies of ions in water. This mechanism is illustrated by the increases in the range of 10^4 - 10^8 for the rates of decarboxylation of a pyruvate-thiazolium adduct and 3-carboxybenzisoxazoles in organic solvents compared to water (11). Its relevance to enzymatic catalysis is supported by evidence from dye-binding studies that the active site of pyruvate decarboxylase has hydrophobic character and by the extremely tight binding to pyruvate dehydrogenase ($K_i < 10^{-10}$ *M*) of an analog of thiamin pyrophosphate in which the nitrogen atom of the thiazole ring does not have a positive charge (12).

3. *Electrostatic destabilization* by repulsion of like charges accounts for a change of 4.7 units in the pK of an ε-amino group of lysine in the active site of acetoacetate decarboxylase (13) and should be able to bring about similar changes in the energy of ES complexes in other systems.

The other way of increasing the Gibbs free energy of the ES complex is to decrease the entropy of the bound substrate so that little or no additional loss of entropy is required to reach the transition state (2, 14). The possible role of this entropy loss is perhaps most easily seen by consideration of Eq. (1). Suppose that

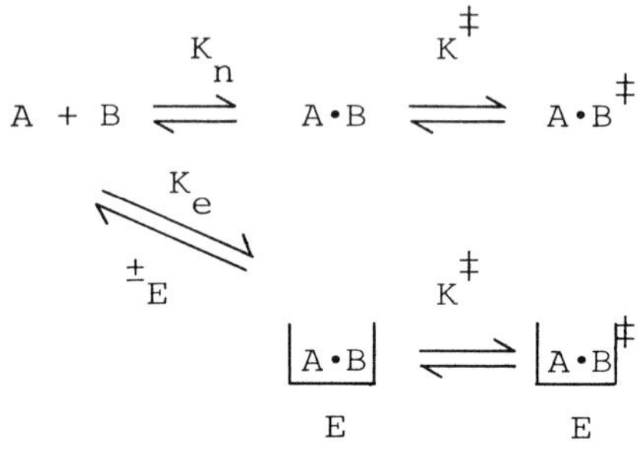

(1)

it were possible to separate a nonenzymatic reaction into two stages: (1) an initial bringing together of A and B into a complex with loss of their translational and rotational entropy so that they are in exactly the right position to react with each other with no further loss of entropy and (2) the activation of this complex to the transition state. Now suppose that the enzyme does not make the activation process easier, so that $K^{\ddagger}$ is the same for the enzymatic and nonenzymatic reactions, but is able to bind A and B at the active site so that they are properly positioned for reaction, as in the nonenzymatic reaction complex. The enzyme <u>uses</u> the energy of its binding interactions with specific substrates to overcome the loss of entropy that occurs on formation of this complex and to give the observed binding. The advantage for the enzymatic reaction is then given simply by the ratio of the equilibrium constants for formation of the enzymatic and nonenzymatic complexes, K_e/K_n (K_e, a binding constant, is the reciprocal of K_s). Since K_n can be as small as 10^{-8} M^{-1} and binding constants for substrates are commonly on the order of 10^4 M^{-1}, the potential advantage to the enzyme from this kind of utilization of binding energy can be as large as $10^4 \times 10^4/10^{-8} = 10^{16}$ M^{-1}. In a simpler example (Eqs. 2 and 3), which avoids the units of "effective molarity" in this ratio, the comparison

$$\mathrm{A + B} \underset{}{\overset{K_n}{\rightleftharpoons}} \mathrm{A{\cdot}B} \overset{K^{\ddagger}}{\rightleftharpoons} \mathrm{A{\cdot}B} \qquad (2)$$

$$\mathrm{A + E{-}B} \overset{K_e}{\rightleftharpoons} \underset{\mathrm{E}}{\mathrm{A{\cdot}B}} \overset{K^{\ddagger}}{\rightleftharpoons} \underset{\mathrm{E}}{\mathrm{A{\cdot}B}} \qquad (3)$$

is between a group B on the enzyme that reacts with the substrate (such as the carboxylate group of CoA transferase) and a comparable group in solution (such as acetate). A binding constant of 10^4 for A then gives a maximal advantage of

$K_e/K_n = 10^4/10^{-8} = 10^{12}$.

The actual advantage from this mechanism in a real situation will be smaller than this by (1) the amount that the necessary loss of entropy is not complete in the ES complex, (2) the amount that the requirement for entropy loss in the nonenzymatic reaction is less stringent than indicated by the value of $K_n = 10^{-8}\ M^{-1}$, and (3) the amount that the observed binding constant is smaller than $10^4\ M^{-1}$. A diffusion-controlled reaction or a reaction with a very loose transition state could have a value of K_n near $0.1\ M^{-1}$; large effects can only occur in reactions with tight transition states, such as those that involve the formation of a new covalent bond in a tetrahedral addition intermediate (very little entropy remains in the stretching vibration of an ordinary covalent bond). A maximal loss of entropy requires very tight fixation of the substrate in the ES complex, even tighter than in a molecular solid which can have 30 units of remaining entropy (2, 15). Thus, it is necessary to compress the reacting groups together in order to obtain the maximal advantage from entropy loss, so that in a real situation maximal entropy loss will be inseparable from the development of strain. It may not be unreasonable to expect that enzymes could learn how to do this over several billion years of evolution.

It is not necessary that binding energy be utilized to bring about entropy loss in the ES complex in order to gain the maximal advantage from such utilization; all that is necessary is that it be utilized to freeze those motions that need to be frozen by the time the transition state is reached. As described earlier for a simple destabilization mechanism, if the available binding energy is slightly too small to force the substrate completely into the active site in the ES complex, some of it will be incompletely bound in this complex and maximal binding and entropy loss will occur only when the transition state is reached. In such cases it may be useful to think of a second complex, ES^*, in which this loss of entropy and associated destabilization processes have taken place

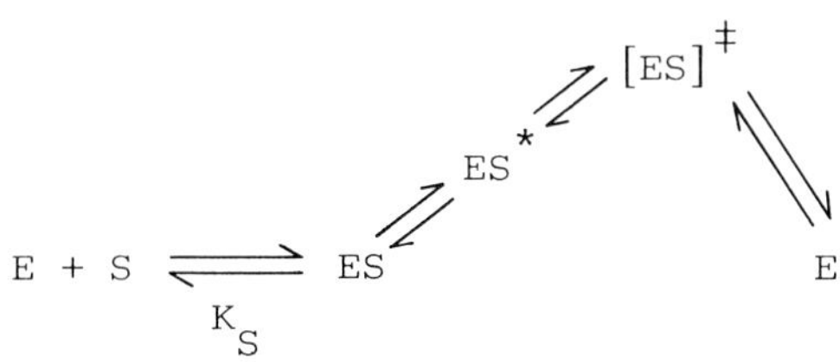

(4)

and have been paid for by the utilization of binding energy. The Gibbs free energy of this complex will then be higher than that of the first complex, so that it will not be populated to a significant extent in the ground state and no evidence for strain and entropy loss will be seen upon physical examination of the predominant ES complex. The arguments described above may then be applied to this ES* complex and the utilization of a given amount of binding energy for catalysis will be just as efficient as it would be if the entropy loss had occurred in the ES complex. Transition state and bisubstrate analogs will also bind as well as they would if destabilization and entropy loss had occurred in the ES complex. This kind of situation might well hold, for example, in lysozyme, in which the binding energy of the hydroxymethyl group of specific substrates could bring about entropy loss as well as strain in the ES* complex and transition state and thereby account for the much faster reaction of glucose than of xylose derivatives, but not fit tightly into the observed ES complex.

In CoA transferase there are three indications of how the binding energy of the coenzyme A may be utilized for catalysis. First, the enzyme-CoA complex is inactivated by reaction of a sulfhydryl group with 5,5'-dithiobis(2-nitrobenzoate) (DTNB) with a second-order rate constant that is 100 times larger than that for the free enzyme (5). This suggests that the thiol group is not easily accessible to DTNB in the free enzyme, but that the coenzyme A forces the enzyme into an open form in which the thiol group can be attacked more easily (Fig. 4). A comparison of the equilibrium constants for formation of the E - SCoA and the E - SR complexes suggests that the energy that is required to force the enzyme to change from the closed to the open form is on the order of 5.7 kcal mol^{-1} (6).

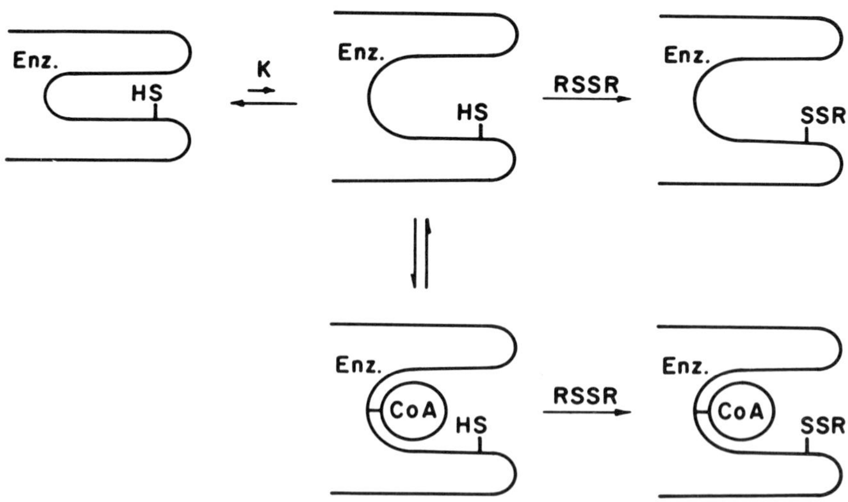

Fig. 4

Second, the coenzyme A group increases the chemical reactivity of the thiol ester in the enzyme-CoA intermediate by a factor of 10^2 to 10^3. This activation may represent a destabilization of the thiol ester relative to the transition state that is brought about by coenzyme A. The reactivity of the thiol ester group of E-SR toward hydrolysis and borohydride reduction is approximately the same as that of thiol esters of acetic acid, i.e., it behaves like a normal thiol ester in its chemical reactivity. However, the reactivity of E-CoA is greater than that of E-SR toward hydrolysis, borohydride reduction, and reaction with thiol anions by factors of approximately 350, 400, and 180, respectively (6).

Third, the reaction rate is certainly increased by the increased probability, or decreased requirement for loss of entropy, in the enzymatic reaction that results from bringing reacting groups together at the active site. We do not know how large this contribution to catalysis is in the enzymatic reaction, but limits to it can be set from the rate accelerations that are observed in intramolecular model reactions. The enzymatic reaction involves the attack of a carboxylate group on a thiol ester, presumably to form an intermediate anhydride and then a new thiol ester (5). The corresponding intramolecular reaction of a thiol ester of succinate (Eq. (5) (6)) is faster than the

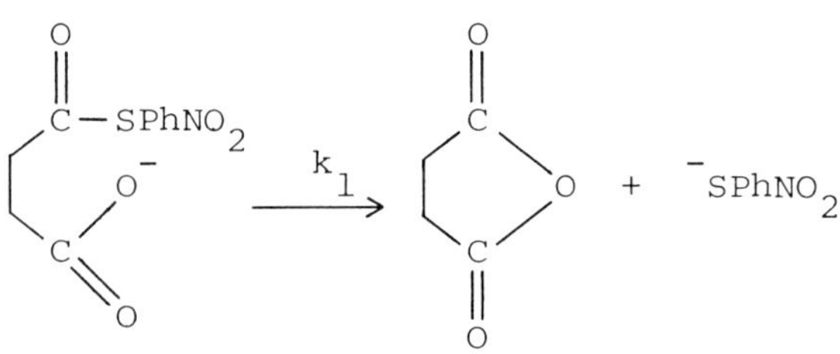

(5)

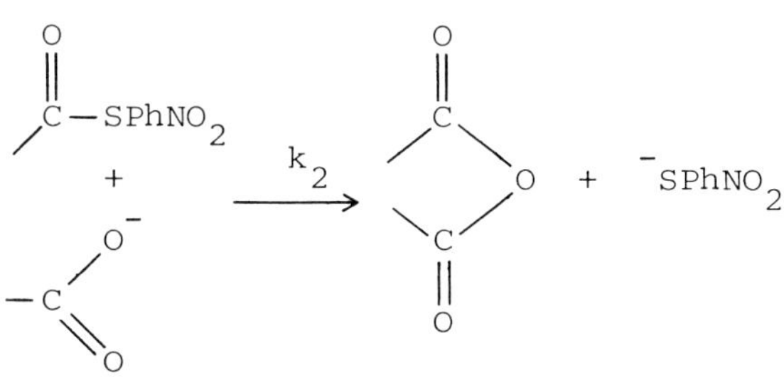

(6)

corresponding bimolecular reaction of 1 *M* acetate with the acetate thiol ester (Eq. 6) by a factor of 6×10^5. Similar rate accelerations have been observed for a number of other intramolecular reactions, including reactions of oxygen esters of succinate (16). Since the succinate molecule is free to rotate around three bonds, this large rate acceleration cannot be attributed to strain or desolvation of the reacting groups; in fact, it is likely that the attacking carboxylate group spends most of its time rotating away from the ester group. This result means that if the enzyme can hold the carboxylate and thiol ester groups in as close a relationship to one another as they are held in succinate, it can bring about a rate acceleration of 6×10^5 *M*, corresponding to 27 e.u. or 8 kcal mol^{-1}, by increasing the probability of reaction. If the three rotations of succinate were frozen with the carboxylate group in the correct position for reaction, an additional factor of about 10^3 could be expected (approximately 10 for each rotation (14), giving a total rate increase of 6×10^8 *M*. It is unlikel that the enzyme can hold the reacting groups rigidly enough relative to each other to give this maximal rate acceleration, but the magnitude of this upper limit for the probability effect suggests that its contribution to catalysis is likely to be large.

These three factors represent contributions to catalysis of 5.7, 3.4, and 8 kcal mol^{-1}, respectively, that are brought about by utilization of the binding energy of coenzyme A. These numbers are crude estimates, at best, and may overlap with one another. They do indicate, however, that it is not altogether unreasonable that noncovalent interactions of the enzyme with the coenzyme A moiety of the substrate could bring about a decrease in the Gibbs energy of activation by 17 kcal mol^{-1}. We might speculate that the enzyme exists initially in a closed form and that the binding energy of coenzyme A is used to force the active site open. The thiol ester would then be firmly held in the correct position for reaction against a desolvated carboxylate group, with compression by the "jaws" of the active site that serves both to decrease the entropy of the bound substrate and to overcome some van der Waals repulsion of the reacting atoms.

COUPLED VECTORIAL PROCESSES

The binding interactions with specific substrates that are utilized by enzymes for catalysis are utilized in coupled vectorial processes, such as muscle contraction and ATP-driven ion transport, in order to balance the energy of different states of the system, so that turnover can proceed at a useful rate. It may be useful to summarize here, for comparison with other enzymes, some ways of looking at these systems that have been described elsewhere in greater detail (2, 17). It is important to emphasize that most of the points made here have been made previously by others, either explicitly or implicitly (18 - 20); however, the author has found these approaches helpful in thinking about these difficult systems and hopes that they may be helpful to others.

One of the remarkable properties of enzymes that carry out coupled vectorial processes is their ability to convert reactions that are strongly unfavorable in solution, such as the synthesis of ATP, to readily reversible reactions at the active site of the enzyme. For example, the equilibrium constant K_2 for the synthesis of ATP from ADP and phosphate at a standard state of 1 *M* in solution is $\sim 10^{-7}\ M^{-1}$ (Eq. 7), whereas the equilibrium constant K_1 for the same reaction at the active site of myosin (Eq. 8) is approximately 0.1, more favorable by a factor of 10^6 *M* (21).

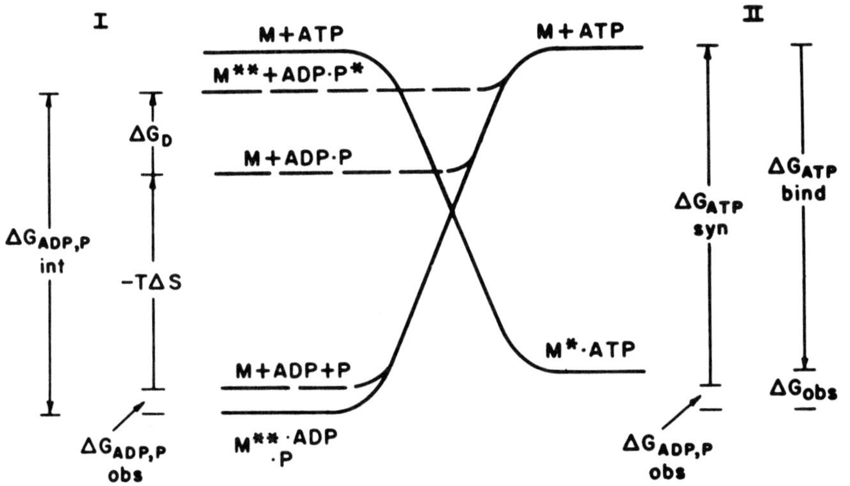

Fig. 5. Gibbs free energy diagram to illustrate how the intrinsic binding energy of ADP and P_i is utilized to decrease the entropy and destabilize the reactants in order to facilitate ATP synthesis (I) and the intrinsic binding energy of ATP is expressed directly as binding to overcome the unfavorable free energy of ATP synthesis (II) (17).

$$\mathrm{ADP} + \mathrm{P_i} \underset{}{\overset{K_2}{\rightleftharpoons}} \mathrm{ATP} \tag{7}$$

$$\frac{\mathrm{ADP \cdot P_i}}{\mathrm{Myosin}} \overset{K_1}{\rightleftharpoons} \frac{\mathrm{ATP}}{\mathrm{Myosin}} \tag{8}$$

How can the ready reversibility of such reactions at the enzyme active site be explained? One might say that the Gibbs energy of ADP and phosphate is somehow increased so that they can form ATP easily, or that the energy of ATP is reduced relative to that of ADP and phosphate so that its synthesis is easy. Taking the former approach, we can ask how the ADP and phosphate are altered at the active site so as to facilitate ATP synthesis.

The situation may be described by the diagram of Fig. 5. The Gibbs energy of enzyme-bound ATP is only slightly higher than that of ADP and phosphate, but the energy of free ATP is much higher than ADP and phosphate. Now, one thing that the enzyme certainly does is to bring the ADP and phosphate together so that they can form ATP with a higher probability and smaller loss of entropy. If ADP and phosphate are brought together in the absence of enzyme into the correct position for reaction with a corresponding loss of entropy, as shown on the left side of the diagram, their Gibbs energy will be higher and ATP synthesis will be easier. This effect alone could account for the observed increase in the equilibrium constant for ATP synthesis of 10^6 *M*, because the maximum entropy effect is $\sim 10^8$ *M*, but it is virtually certain that the enzyme - ADP - P_i complex is also destabilized or strained in order to hold the ADP and phosphate tightly together. This additional destabilization energy is indicated as ΔG_D in Fig. 5. The sum of the unfavorable loss of entropy, $-T\Delta S$, and destabilization energy, ΔG_D, brings the Gibbs energy of this low-entropy, strained M** and ADP·P* close to that of ATP so that ATP synthesis can occur easily.

The binding to the enzyme of ATP is extremely strong (22). The binding of ADP·P* to M** is also very strong ($\Delta G_{ADP,P,int}$), because the binding of ADP·P* requires a relatively small loss of entropy and M** is complementary to ADP·P*. It is reasonable that these two binding energies should be large and comparable, because the same adenine, ribose, and phosphate binding sites are present in both systems. However, the *observed* binding of ADP and phosphate to myosin is very weak ($\Delta G_{ADP,P,obs}$). This is because the binding requires loss of entropy of both ADP and phosphate and overcoming the destabilization energy, ΔG_D. The binding of ATP is strong, as shown on the right side of Fig. 5, because it requires the loss of entro

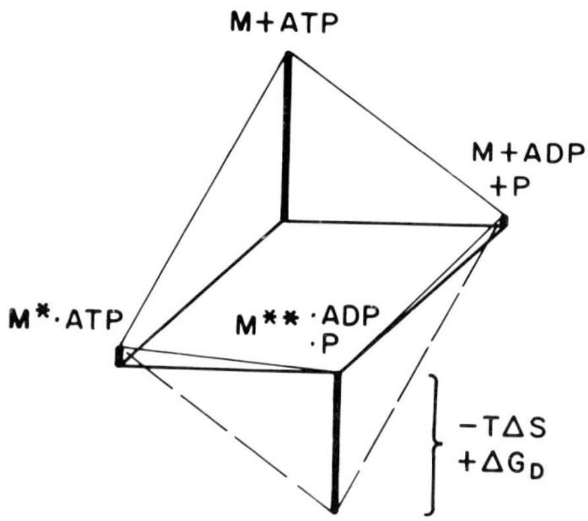

*Fig. 6. Bar graph of Gibbs energies to illustrate how the entropy loss and destabilization of the M**·ADP·P_i complex make ATP synthesis reversible (17).*

py of only one molecule and does not require overcoming ΔG_D, because the active site is complementary to ATP.

The situation is exactly analogous to the binding of substrates and transition states by enzymes. In the binding of ADP and phosphate, or enzyme substrates generally, the intrinsic binding energy of the substrate is used up to make the reaction more probable by overcoming entropy loss and bringing about binding in a destabilized, high energy state. The observed binding energy is only what is left over after this utilization of energy. The binding of ATP or transition states is very much tighter and facilitates the reaction, because the intrinsic binding energy is expressed directly as binding. Thus, we can regard myosin as a machine that makes ATP hydrolysis reversible by existing in two states: in state 1, the M*·ATP complex, the intrinsic binding energy between the enzyme and substrate is expressed, and in state 2, the M**·ADP·P_i complex, is not expressed.

State		*Intrinsic binding energy*
1	*M*·ATP*	*Expressed*
2	*M**·ADP·P*	*Not expressed*

In the contraction cycle of muscle, this balancing of energies maintains the Gibbs energy of hydrolysis of ATP almost intact through the hydrolytic step; it is only dissipated when ADP and phosphate are allowed to dissociate from the enzyme into dilute solution.

The same situation may be illustrated by the bar graph of Gibbs free energies in Fig. 6. The energy of unbound ATP is much higher than that of ADP and phosphate, but this energy difference becomes small when the reactants are bound to the active site of myosin. This change in relative energies reflects the very tight binding of ATP to the enzyme and the relatively much weaker binding of ADP and phosphate at some standard state. Now suppose that the same intrinsic binding energy of the adenine, ribose, and phosphate groups were expressed as observed binding energy in both cases. The diagram would then be as described by the dashed lines, with the connecting lines parallel and a plane passing through the energies of the four species. Such a system would not work, because all of the enzyme would fall into the low energy hole of myosin·ADP·P_i and this species would pile up. The reversible synthesis of ATP and the ready dissociation of ADP and phosphate are possible only because the Gibbs energy of the M**·ADP·P_i complex is increased by destabilization and entropy loss of the reactants, so that relatively little of the intrinsic binding energy of the reactants is expressed. This increase in Gibbs energy corresponds to the interaction or destabilization energy described by Wyman, Weber, Hill, and others (18); these diagrams provide a way of visualizing the energy relationships of a coupled or linked system of this kind.

The currently accepted mechanism of muscle contraction (19) consists of a sequence of three steps in which binding energies are balanced against each other and either expressed or not expressed (Eqs. 9 - 11). Starting with actomyosin, the tight binding of actin is displaced by the even tighter binding of ATP to give M*·ATP (Eq. 9). The

$$A_1 \cdot M + ATP \rightleftharpoons A_1 + M^* \cdot ATP \quad (9)$$

$$M^* \cdot ATP \rightleftharpoons M^{**} \cdot ADP \cdot P \quad (10)$$

$$M^{**} \cdot ADP \cdot P + A_2 \rightleftharpoons A_2 \cdot M + ADP + P \quad (11)$$

system works because there is a destabilization energy in the A·M·ATP complex such that the full binding energy of actin is not expressed when ATP binds and vice versa; without this interaction energy the system would pile up as the A·M·ATP complex. The second step (Eq. 10) makes ATP synthesis reversible because of the destabilization energy in the M**·ADP·P_i complex, as described above. In the final step (Eq. 11) the large binding energy of actin overcomes the small binding energy of ADP and phosphate so that the process is strongly downhill. In an intact muscle this step can be coupled to the performance of mechanical work if the myosin is constrained to react with an actin site, A_2, that is different from that with which it was originally combined, A_1.

The calcium-transporting ATPase of sarcoplastic reticulum has the remarkable property of binding inorganic phosphate and converting it to an acyl phosphate in a readily reversible reaction (20). The enzyme also reacts reversibly with ATP to form what appears to be the same acyl phosphate. Knowles and Racker have allowed the phosphoenzyme to form spontaneously from inorganic phosphate and then added calcium and ADP to obtain a net synthesis of ATP (23); an analogous result has been obtained with the sodium enzyme (24). How can the energy balances in these unusual reactions be rationalized?

Since formation of the acyl phosphate at the active site of the enzyme from bound inorganic phosphate occurs spontaneously (K_1, Eq. 12), the equilibrium constant is close to 1.0, which is more favorable than the formation of acyl phosphate from 1 *M* acetate and phosphate in solution by a factor of about 10^8 ($K_2 \cong 10^{-8}\ M^{-1}$; Eq. 13) (25).

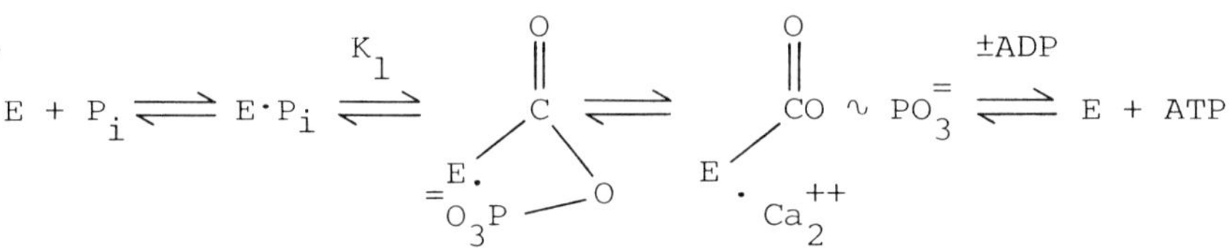

(12)

$$CH_3COO^- + P_i \overset{K_2}{\rightleftharpoons} CH_3CO \sim PO_3^{=} \qquad (13)$$

It is useful to consider the enzymatic reaction as occurring in two steps (Fig. 7). In the first step the acyl phosphate is formed from inorganic phosphate without allowing any noncovalent binding interactions to develop between the phosphate group and the enzyme; the phosphate group might extend into the solvent. In the second step the noncovalent interactions between the phosphate and the active site are allowed to develop, as indicated by the dot in Fig. 7. This shows how the intrinsic binding energy from the noncovalent binding interactions of the phosphate group with the metal-containing active site, $\Delta G_P(\text{int})$, provides the driving force for the spontaneous synthesis of the acyl phosphate from 1 *M* inorganic phosphate in solution. This binding energy must be very large, because it accounts not only for the thermodynamically unfavorable synthesis of an acyl phosphate, $\Delta G_{hyd} \sim P \cong -11$ kcal mol^{-1}, but also for the observed favorable equilibrium constant of the

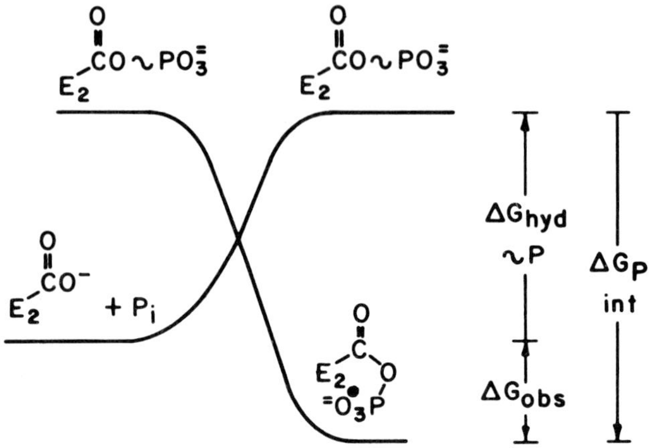

Fig. 7. Gibbs energy diagram to show that the formation of an acyl phosphate is unfavorable in the absence of noncovalent interactions with the enzyme and that the reaction becomes favorable when these interactions develop, as indicated by the dot (17).

reaction (26) corresponding to a Gibbs energy change of -4 kcal mol^{-1}.

In this state of the enzyme, which can be called E_2-P, the intrinsic binding energy of phosphate is expressed so that acyl phosphate formation is spontaneous. However, most of this binding energy is not expressed in the $E_2 \cdot P_i$ complex. If it were, acyl phosphate formation would be strongly unfavorable and would not occur. The system works, as in the case of ATP synthesis, because the binding energy of the phosphate is used up to overcome the necessary loss of entropy, -TΔS, and the destabilization energy, ΔG_D, for the formation of the $E_2 \cdot P_i$ complex. It is only expressed when the acyl phosphate bond is formed, and thereby makes the process spontaneous.

Another form of the enzyme, which can be called E_1~P, is at equilibrium with ATP and is, therefore, a high-energy state of the enzyme. This state can be viewed as one in which most of the binding energy of the phosphate is not expressed, so that the enzyme - phosphate behaves more nearly like a normal high-energy acyl phosphate. Thus, we can summarize these states of the system as follows:

State		*Intrinsic binding energy*
1	$E_2 \cdot P_i$	*Not expressed*
2	E_2-P	*Expressed*
3	E_1~P	*Not expressed*

In order for the calcium-transporting enzyme to work at a reasonable rate, it is necessary not only that there be one state of the phosphoryl enzyme that is in readily reversible equilibrium with inorganic phosphate and another that is at equilibrium with ATP, but also that the state in which the calcium binding site is exposed to the outside of the vesicle should have a very high affinity for calcium, so that it can bind calcium and transport it into the vesicle at external concentrations as low as 10^{-8} *M*, and that the state in which this binding site is exposed to the inside of the vesicle should have a low affinity for calcium so that the transported calcium can be discharged. If these requirements were not met, almost all of the enzyme would pile up in one state, as $E_1{\sim}P{\cdot}Ca_2$ for example, so that rapid turnover would not be possible.

These requirements are nicely balanced against each other in the enzyme so that the reaction can proceed rapidly and reversibly, as illustrated in Fig. 8. The $E_1{\sim}P$ form, at equilibrium with ATP, is of much higher energy than the E_2-P form, at equilibrium with inorganic phosphate, and the direct interconversion of these states is forbidden. Calcium binding to the $E_1{\sim}P$ (outside) state is strongly favorable, downhill, whereas calcium binding to the E_2-P state is much less favorable and is uphill at some standard state concentration on the order of 10^{-4} *M*. The difference between these two energies for the binding of two calcium ions brings the two bound species to comparable energy levels, so that they are readily interconvertible. Thus, the binding energies of phosphate and calcium are expressed in the two states in a complementary manner (see tabulation on next page). Under physiological conditions the difference in the calcium concentrations of the two sides of the membrane will tend to equalize the concentrations of the two states of both the free and the calcium-bound forms of the enzyme.

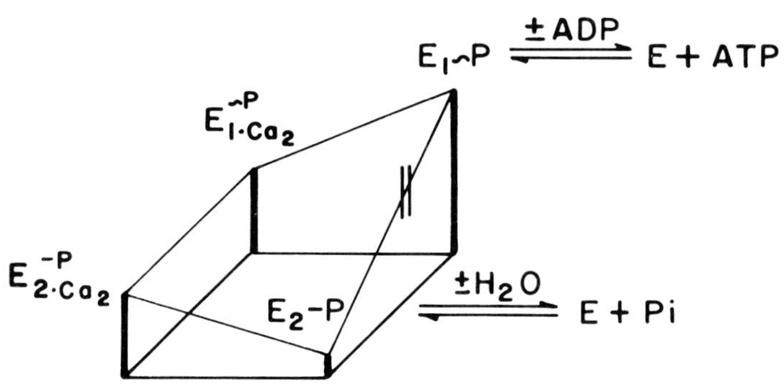

Fig. 8. Bar graph to illustrate the balance of binding energies in the calcium ATPase (17).

Intrinsic binding energy	*State* 1	*State* 2
P_i	−	+
Ca^{2+}	+	−

RULES FOR COUPLING

The selective utilization of the intrinsic binding energies of substrates and ions in the different states of the enzyme serves the important function of balancing the Gibbs energies of the different states of coupled vectorial systems under physiological conditions, so that the system can turn over at a useful rate and the enzyme does not pile up in a single state. However, these binding energies, in themselves, do not account for coupling, movement, or transport in these systems. The coupling process itself is defined by a set of rules, which are implicit in most of the models that have been proposed for the action of coupled processes but deserve examination in their own right.

A typical model for the coupled transport of calcium by the calcium ATPase is shown in Eq. (14). According to this mechanism, ATP and calcium on the outside of the vesicle bind

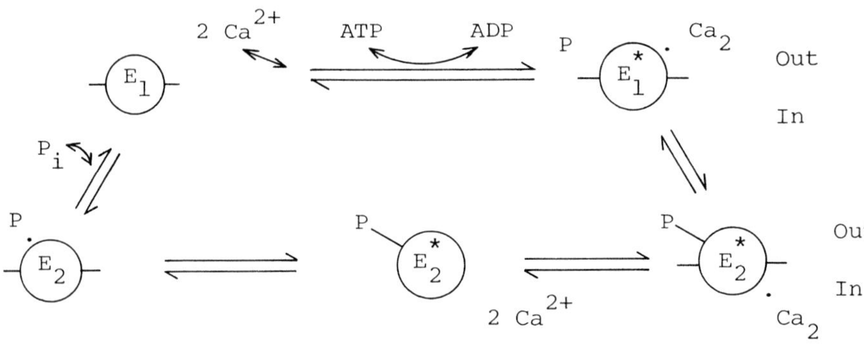

(14)

to the E_1 state of the enzyme to give the high-energy phosphorylated state $E_1^{*}\sim P\cdot Ca_2$. The enzyme then changes to the E_2^{*}-$P\cdot Ca_2$ state, in which the calcium is exposed to the inside of the vesicle, and then discharges calcium, cleaves and discharges phosphate, and returns to the E_1 state. The reverse process results in the synthesis of ATP from ADP and phosphate in the presence of a calcium gradient. This model works as a consequence of the following rules:

1. $E_1{\sim}P$ shall react reversibly with ADP, not HOH.
2. E_2-P shall react reversibly with HOH, not ADP.
3. E* shall transfer in-out (between states E_1 and E_2) only when bound to two calcium ions.
4. E shall transfer in-out only when not bound to calcium.

Rule 1 is required because if $E_1{\sim}P$ could react with water the enzyme would be an uncoupled ATPase. Similarly, if E_2-P could react with ADP to give ATP, the reverse reaction would lead to uncoupled ATPase activity. Rules 3 and 4 are required to prevent leakage, which would also result in uncoupled ATPase activity.

These rules represent changes in the specificity of the enzyme as it changes from one state to another. The change from the E_1 to the E_2 state represents a change in specificity between ADP and water as phosphate acceptor (or between ATP and inorganic phosphate as phosphate donor) and the change from the E* to the E state represents a change in ion-binding specificity. In the case of the calcium enzyme the E* state corresponds to the phosphorylated enzyme.

The specificity of enzymes, such as the 10^{10} faster reaction of phosphorylated phosphoglucomutase with the hydroxyl group of glucose 1-phosphate than with the hydroxyl group of water (4), is well known, if not well understood. The remarkable and characteristic property of enzymes that carry out coupled vectorial processes is that they exhibit at least two different specificities in different states - their activity is made possible by the change in substrate specificity as they change from one state to the other. The vectorial nature of the reaction is a consequence of the coupling of this change in substrate specificity to some kind of movement, such as the alternate exposure of a calcium binding site to the inside or the outside of a vesicle.

The important rule that makes it possible for muscle to perform work with a high efficiency according to the model of Eqs. (9 - 11) is that the M**·ADP·P complex shall combine only with A_2, an actin molecule different from that with which the myosin was originally combined. Since the M*·ATP complex must be able to combine with the actin from which it was formed, by microscopic reversibility, this rule demands that the critical conformation change that gives rise to the change in specificity for combination with A_1 or A_2 must occur in the hydrolysis step in which M*·ATP is converted to M**·ADP·P. Another rule for coupling is that the actin-myosin complex shall not catalyze the hydrolysis of ATP. It is not clear at this time whether this rule is followed as a consequence of a change in catalytic activity of the enzyme between myosin and actomyosin or whether the dissociation of actomyosin occurs fast enough in the presence of ATP so that the rule is enforced by a kinetic specificity of the system.

For several years it has appeared that we are on the verge of understanding how the chemical energy of ATP is transformed into other types of energy in coupled vectorial processes and it is time to ask what questions need to be answered in order that we can say that we do understand how such a system works. Two such questions are, first, how is the intrinsic binding energy of the enzyme for its ligands utilized in the different states of the system and, second, what are the rules that make the system coupled? These questions are far from solution on the microscopic and structural level, but our knowledge of the differences in interaction energies with ligands and the different catalytic specificities of these systems in different states is sufficient that we can say at this time that the basic mechanisms by which several of these systems work are already understood.

ACKNOWLEDGMENT

Publication No. 1213 of the Graduate Department of Biochemistry, Brandeis University. Supported in part by grants from the National Science Foundation (BM471-01501) and the National Institute of General Medical Sciences of the National Institutes of Health (GM 20888).

REFERENCES

1. Haldane, J. B. S. (1930). "Enzymes." Longmans, Green, London; Pauling, L. (1946). *Chem. Eng. News 24,* 1375; Jencks, W. P. (1966). *In* "Current Aspects of Biochemical Energetics" (N. O. Kaplan and E. P. Kennedy, eds.), p. 273. Academic Press, New York; Wolfenden, R. (1972). *Acc. Chem. Res. 5,* 10; Lienhard, G. E. (1973). *Science 180,* 149.
2. Jencks, W. P. (1975). *Advan. Enzymol. 43,* 219.
3. Fersht, A. R. (1974). *Proc. Roy. Soc. London, Ser. B, 187,* 397; Fersht, A. R. (1977). "Enzyme Structure and Mechanism." Freeman, San Francisco; Albery, W. J. and Knowles, J. R. (1977). *Angew. Chem. Int. Ed. Engl. 16,* 285.
4. Ray, W. J., Jr., and Long, J. W. (1976). *Biochemistry 15,* 3993; Ray, W. J., Jr., Long, J. W., and Owens, J. D. (1976). *Ibid.,* 4006.

5. White, H. and Jencks, W. P. (1976). *J. Biol. Chem. 251,* 1688; White, H., Solomon, F., and Jencks, W. P. (1976). *Ibid.,* 1700.
6. Moore, S. (1978). Doctoral dissertation, Brandeis University, Waltham, Massachusetts.
7. The rate constants have been corrected for the effect of the difference in pK of the two substrates on catalytic activity (5).
8. Schindler, M., Assaf, Y., Sharon, N., and Chipman, D. M. (1977). *Biochemistry 16,* 423.
9. Capon, B. and Dearie, W. M. (1974). *J. Chem. Soc., Chem. Comm. 370;* Ballardie, F. W., Capon, B., Cuthbert, M. W., and Dearie, W. M. (1977). *Bioorg. Chem. 6,* 483.
10. Blair, J. B. (1969). *J. Biol. Chem. 244,* 951.
11. Kemp, D. S. and Paul, K. (1970). *J. Amer. Chem. Soc. 92,* 2553; Crosby, J., Stone, R., and Lienhard, G. E. (1970). *Ibid,* 2891; Kemp, D. S. and Paul, K. G. (1975). *Ibid., 97,* 7305; Kemp, D. S., Cox, D. D., and Paul, K. G. (1975). *Ibid., 97,* 7312.
12. Ullrich, J., Ostrovsky, Y. M., Eyzaguirre, J., and Holzer, H. (1970). *Vitamins Hormones 28,* 365; Wittorf, J. H. and Gubler, C. J. (1970). *Eur. J. Biochem. 14,* 53; Gutowski, J. A. and Lienhard, G. E. (1976). *J. Biol. Chem. 251,* 2863.
13. Frey, P. A., Kokesh, F. C., and Westheimer, F. H. (1971). *J. Amer. Chem. Soc. 93,* 7266; Kokesh, F. C. and Westheimer, F. H. (1971). *Ibid,* 7270.
14. Page, M. I. and Jencks, W. P. (1971). *Proc. Natl. Acad. Sci. U.S. 68,* 1678.
15. Becka, L. N. and Cruickshank, D. W. J. (1963). *Proc. Roy. Soc. London Ser. A273,* 435, 455; Westrum, E. F., Jr. and McCullough, J. P. (1963). *In* "Physics and Chemistry of the Organic Solid State" (D. Fox, M. M. Labes, and A. Weissberger, eds.), Vol. 1, p. 1. Wiley (Interscience), New York; Schrader, B., Meier, W., Gottlieb, K., Agatha, H., Barentzen, H., and Bleckmann, P. (1971). *Ber. Bunsenges. Phys. Chem. 75,* 1263; Bleckmann, P., Schrader, B., Meier, W., and Takahashi, H. (1971). *Ibid.,* 1279.
16. Bender, M. L. and Neveu, M. C. (1958). *J. Amer. Chem. Soc. 80,* 5388; Gaetjens, E., and Morawetz, H. (1960). *J. Amer. Chem. Soc. 82,* 5328; Bruice, T. C. and Pandit, U. K. (1960). *Ibid,* 5858; Bruice, T. C. and Pandit, U. K. (1960). *Proc. Natl. Acad. Sci. U.S. 46,* 402; Bruice, T. C. and Turner, A. (1970). *J. Amer. Chem. Soc. 92,* 3422.
17. Jencks, W. P. (1980). *Advan. Enzymol. 51,* 75.
18. See, for example, Wyman, J., Jr. (1948). *Advan. Protein*

Chem. 4, 407; Wyman, J., Jr. (1964). *Ibid. 19,* 223; Weber, G. (1972). *Proc. Natl. Acad. Sci. U.S. 69,* 3000; Weber, G. (1975). *Adv. Protein Chem. 29,* 1; Hill, T. L. (1975). *Biochemistry 14,* 2127; Hill, T. L. and Simmons, R. M. (1976). *Proc. Natl. Acad. Sci. U.S. 73,* 95; Hill, T. L. (1977). "Free Energy Transduction in Biology." Academic Press, New York.

19. Taylor, E. W. (1973). *Current Top. Bioenerg. 5,* 201; Trentham, D. R., Eccleston, J. F., and Bagshaw, C. R. (1976). *Quart. Rev. Biophys. 9,* 217.
20. Hasselbach, W. (1974). *The Enzymes,* 3rd ed., Vol. *10,* 431
21. Alberty, R. A. (1969). *J. Biol. Chem. 244,* 3290; Bagshaw, C. R. and Trentham, D. R. (1973). *Biochem. J. 133,* 323.
22. Wolcott, R. G. (1974). *Fed. Proc. 33,* 1294.
23. Knowles, A. F. and Racker, E. (1975). *J. Biol. Chem. 250,* 1949.
24. Taniguchi, K. and Post, R. L. (1975). *J. Biol. Chem. 250,* 3010.
25. Stadtman, E. R. (1954). *In* "The Mechanism of Enzyme Action" (W. D. McElroy and B. Glass, eds.), p. 581. Johns Hopkins Press, Baltimore; Gerstein, J. and Jencks, W. P. (1964). *J. Amer. Chem. Soc. 86,* 4655.
26. Masuda, H. and de Meis, L. (1973). *Biochemistry 12,* 4581; Kanazawa, T. and Boyer, P. D. (1973). *J. Biol. Chem. 248,* 3163.

OXYGEN INACTIVATION AND STABILIZATION OF *Azotobacter* NITROGENASE

Alexander M. Klibanov
Nancy M. Weare

Department of Chemistry
University of California at San Diego
La Jolla, California 92093

This chapter is dedicated to Martin D. Kamen. The many outstanding scientific achievements of Professor Kamen are well known in a wide range of fields from atomic physics to biochemistry. In the latter his contributions include the discovery of N_2 fixation (1) and H_2 evolution (2) in photosynthetic bacteria, as well as the discovery and extensive investigation of different types of bacterial cytochromes (3) and high potential iron-sulfur proteins (4). Over the past few years, M. D. Kamen has devoted much of his attention to applied bioenergetics, namely, to solar energy bioconversion. The pioneering research (5) by Kamen's and Kaplan's groups on the application of a chloroplast-hydrogenase system to the biophotolysis of water is considered to be a major breakthrough in the field. This study proved that it is possible to use a biochemical reactor for the biophotolysis of water but raised questions about the feasibility of this system since its main components, chloroplasts and hydrogenase, are very labile (to light and oxygen, respectively). This work, in turn, has initiated a number of investigations on the stabilization of both chloro-

 ISBN 0-12-397580-8

plasts and hydrogenase [see (6) for a review]. These studies have led not only to the stabilization of hydrogenase per se but also to the establishment of some general principles of enzyme stabilization against O_2 inactivation (7, 8). The present work continues this line of research.

Nitrogenase is the enzyme complex responsible for biological nitrogen fixation, the reduction of molecular N_2 to ammonia Since the above process is of great importance for life on earth, the studying of nitrogenase and its reactions is a vigorously growing field of biochemistry (9 - 11). The problem of nitrogen fixation also has a significant practical aspect due t shortages in the world's food and energy supplies. The industrial and biological fixation of N_2, both crucial for meeting world-wide agricultural demands, are highly energy-consumptive processes. However, many N_2-fixing organisms can use solar energy directly to satisfy all or a part of these energy needs. Moreover, these organisms can produce H_2 (12) and ammonia (13), an important N fertilizer, in nitrogenase related reactions.

An extreme oxygen lability of nitrogenase is the main obstacle for its use in practical applications. Both functional components of nitrogenase, Fe protein and Mo-Fe protein, are irreversibly inactivated by oxygen after exposure to air for minutes. Since many N_2-fixing bacteria are aerobes, the question arises: How do they manage to combine nitrogenase-catalyzed nitrogen fixation and the aerobic habit? It turns out, that aerobic N_2-fixing bacteria have developed various means of protecting their nitrogenase systems from interference by oxygen (16). For example, there are at least two distinct mechanisms which enable obligatory aerobic, N_2-fixing bacteria of the genus *Azotobacter* to preserve a functional nitrogenase in an O_2-containing environment. When the organism is actively fixing N_2, oxygen is scavenged from sites of nitrogenase activity by a highly active and specialized respiratory system. This mechanism is termed "respiratory protection" (17). In the second mechanism, "conformational protection" (18), nitrogenase is protected against oxygen inactivation by its spatial location either from access of oxygen or from damage consequent on such access.

Thus, we know in principle (although, not in detail) how Nature stabilizes nitrogenase against O_2 inactivation. It is attractive to apply this knowledge to biotechnology. In this chapter we elucidate some aspects of oxygen inactivation and the natural stabilization of nitrogenase from *Azotobacter vinelandii* as well as describe our attempts to enhance and optimize this stabilization by artificial means.

EXPERIMENTAL PROCEDURE

Azotobacter vinelandii OP was grown in a modified Burk's N-free medium as described in (19). The cells were suspended in 25 m*M* Tris·HCl (pH 7.3) and disintegrated by passage through French Press cell at a pressure of about 20,000 psi. Whole cells and cell debris were removed by centrifugation at 18,000 *g* for 30 min. The subcellular particles containing active nitrogenase were obtained from the supernatant by successive centrifugation at 120,000 *g* for 1 hr (sedimented material descarded) and at 160,000 *g* for 5 hr (sedimented particles saved) (20). These particles were termed "particulate nitrogenase." The particulate nitrogenase was further purified by DEAE-cellulose chromatography under anaerobic conditions (21, 22). The Mo-Fe protein appeared in the 0.25 *M* NaCl fraction, and the Fe protein in the 0.50 *M* NaCl fraction.

To assay nitrogenase, we used ATP-driven enzyme-catalyzed H_2 evolution with sodium dithionate as the electron donor (23). The rate of hydrogen gas evolution was determined with a Clark electrode calibrated against H_2-saturated aqueous solutions (7).

The time course of air (oxygen) inactivation of nitrogenase was measured as follows: A small volume of the enzyme (particulate nitrogenase, purified Mo-Fe protein, purified Fe protein, or purified Mo-Fe protein together with purified Fe protein) was stirred at room temperature in an open beaker and aliquots were withdrawn at various times for enzyme assay. When the O_2 stability of a single purified nitrogenase component (Fe protein or Mo-Fe protein) was studied separately, this component was stirred under air and then aliquots were withdrawn at various times and mixed with the second component (which had been stored under argon) in the assay system under anaerobic conditions.

RESULTS AND DISCUSSION

It is known from the literature that whereas particulate *Azotobacter* nitrogenase is fairly oxygen stable (24), the purified components of the enzyme are very O_2 labile (14, 15). However, mainly qualitative information of this sort is available. Therefore, we have studied in detail the kinetics of air (oxygen) inactivation of different species of nitrogenase from *A. vinelandii* (Fig. 1).

As shown by curve a in Fig. 1 the particulate nitrogenase only slowly inactivates under air: even after 19 hr of such

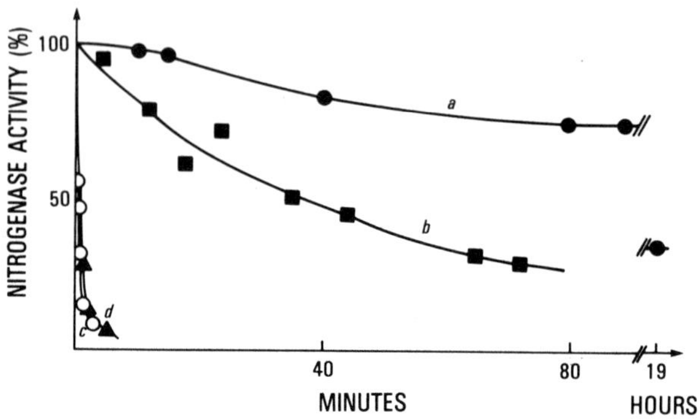

Fig. 1. Air (oxygen) inactivation of different species of nitrogenase system from Azotobacter vinelandii. (a) The particulate enzyme (100 mg/ml); (b) Mo-Fe protein; (c) Fe protein; (d) Fe protein in the complex with Mo-Fe protein. 25 mM Tris·HCl, pH 7.3, 20°C.

aerobic incubation the residual activity is still about 40%. The Mo-Fe protein of the nitrogenase system is much more sensitive to air (Fig. 1, curve b). Finally, the Fe protein is extremely labile (Fig. 1, curve c) with a half-life under air of less than 30 sec. It should be pointed out that under argon, nitrogen, or hydrogen all the species mentioned above are stable for many hours. Therefore, "air inactivation" of nitrogenase is essentially its oxygen inactivation.

The striking feature of the data presented in Fig. 1 is that the sharp destabilization of nitrogenase (more than 2000 times if one compares half-lives of the particulate nitrogenase curve a, and the purified Fe protein, curve c), occurs during just a single step of purification, DEAE-cellulose chromatography. The question arises: Is an essential factor lost during this step? The obvious candidate for this missing factor is the Mo-Fe protein, which is necessary for nitrogenase activity. Therefore, we determined the oxygen stability of purified Fe protein when it is complexed with purified Mo-Fe protein. However, as seen from Fig. 1, curve d, association with Mo-Fe protein during exposure to oxygen does not stabilize the Fe protein. Since the enzymatic activities of the particulate and the reconstituted (from purified Fe protein and Mo-Fe protein) nitrogenase are about the same, one can conclude that the particulate enzyme contains some factor X [probably, Veeger's iron sulfur protein II (25, 26)], which does not affect the nitrogenase activity but is crucial for its oxygen stability.

It is important to know how this stabilizing factor X is bound to the catalytic components. Therefore, we studied the effect of various effectors on the oxygen stability of the particulate nitrogenase. Yates (27) showed that chelating agents sharply destabilized the particulate nitrogenase from *A. chroococcum* by solubilizing Fe protein. We demonstrated that chelating agents have the same influence on *A. vinelandii* particulate nitrogenase: both EDTA and the chelating resin, Chelex, substantially reduced its oxygen stability (Fig. 2, curves c and d, respectively). We also found that 1 *M* NaCl drastically decreased oxygen stability of the particulate nitrogenase (Fig. 2, curve b). In addition, as in the case of the *A. chroococcum* enzyme, simple dilution of the particulate nitrogenase from *A. vinelandii* reduces its stability (Fig. 2, curve e). The effect of the effectors described above might be expressed by the following scheme:

$$\text{Particulate nitrogenase} \xrightarrow[\text{dilution}]{\text{high salt concentration, chelating agents}} \text{Stabilizing factor X + Fe protein + Mo-Fe protein}$$

$$\text{Fe protein} \xrightarrow{\text{oxygen}} \text{Inactive protein}$$

Comparison of the data in Figs. 1 and 2 suggests that the first stage in this scheme is rate-limiting for the inactivation process because the decay of activity of the particulate nitrogenase in the presence of the effectors tested (Fig. 2, curves b - e) is much slower than that of Fe protein (Fig. 1, curve c)

The above results indicate that, first, Fe protein is relatively loosely bound to the stabilizing factor X (since the former could be dissociated from the latter by simple dilution); and, second, Fe protein is bound to the stabilizing factor X through ionic bonds and Me^{2+} (since high ionic strength and chelating agents, respectively, dissociate the former from the latter). The structure of the particulate enzyme is schematically shown in Fig. 3. Apparently, it is the stabilizing factor X which creates the "conformational protection" of Postgate (18, 14).

According to the above scheme, inactivation of nitrogenase follows dissociation of the particulate enzyme. Therefore, preventing or hindering dissociation should obviously have a stabilizing effect. It is evident from Fig. 3 that the dissociation might be prevented by firm (covalent) attachment of Fe protein to the stabilizing factor X, for example, by cross-linking of the particulate nitrogenase. We experimentally proved the validity of this idea. As it is shown in Fig. 4, covalent cross-linking of the particulate nitrogenase via either amino groups (using diimido esters) or carboxy groups (using

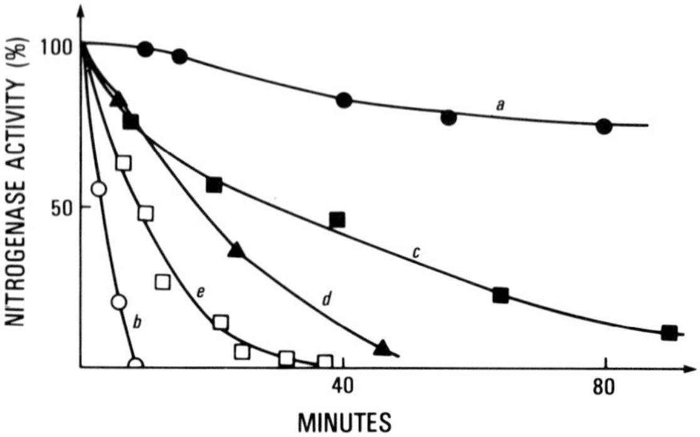

Fig. 2. The effect of different factors on air (oxygen) inactivation of the particulate nitrogenase from Azotobacter vinelandii. (a) The enzyme (100 mg/ml) in buffer; (b) the enzyme (100 mg/ml) in the presence of 1 M NaCl; (c) the enzyme (100 mg/ml) in the presence of 10 mM EDTA; (d) the enzyme (100 mg/ml) in the presence of chelating resin Chelex (0.2 g/ml); (e) the enzyme (10 mg/ml in buffer. 25 mM Tris·HCl, pH 7.3, 20°C.

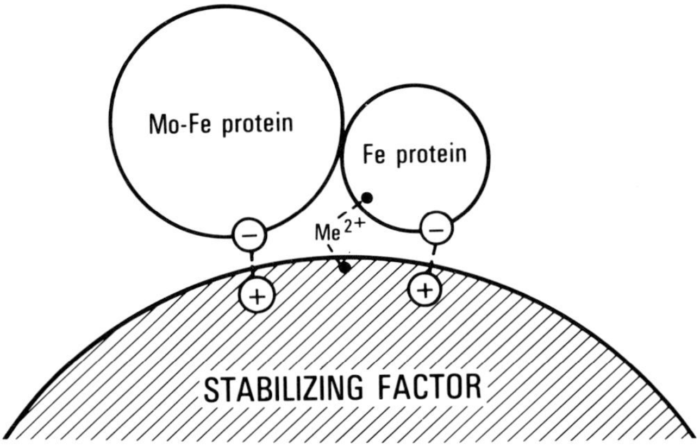

Fig. 3. Schematic representation of the structure of the particulate nitrogenase from Azotobacter vinelandii.

activation with carbodiimide followed by treatment with 1,6-diaminohexane) substantially improves its stability against oxygen inactivation brought about by dilution.

Although the stabilization effect (Fig. 4) is not very high these data might be considered as promising. It is hoped that further investigations of nitrogenase stabilization including some already developed principles of enzyme stabilization (28) will result in enzyme preparations with the required

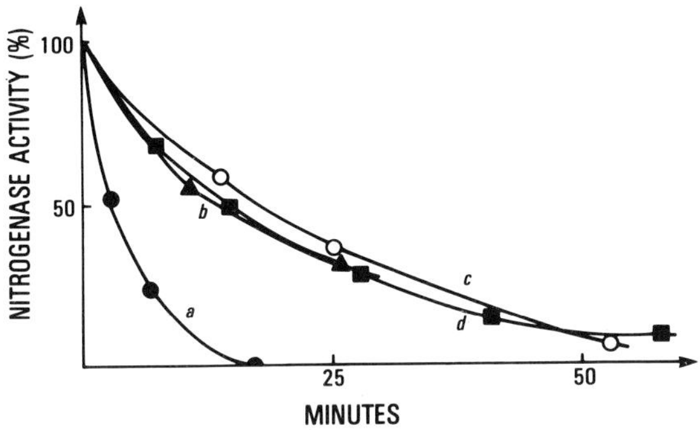

Fig. 4. The effect of covalent cross-linking on air (oxygen) inactivation of the particulate nitrogenase (5 mg/ml) from Azotobacter vinelandii. (a) Control; (b) the enzyme cross-linked by dimethyl suberimidate; (c) the enzyme cross-linked by dimethyl adipimidate; (d) the enzyme treated with 1-ethyl-3-(3-dimethylaminopropyl)carbodiimide followed by cross-linking by 1,6-diaminohexane. 25 mM Tris·HCl, pH 7.3, 20°C.

stability for technological application. Such preparations could eventually be coupled with stabilized chloroplasts and employed to fix N_2 using solar energy.

ACKNOWLEDGMENT

This work was supported by a grant from Solar Energy Research Institute (EH-8-1214-1).

REFERENCES

1. Kamen, M. D. and Gest, H. (1949). *Science 109*, 560.
2. Gest, H. and Kamen, M. D. (1949). *Science 109*, 558-559.
3. Kamen, M. D. and Horio, T. (1970). *Annu. Rev. Biochem. 39*, 673-700.
4. Bartsch, R. G. (1978). *Methods Enzymol. 53*, 329-340.
5. Benemann, J. R., Berenson, J. A., Kaplan, N. O., and Kamen, M. D. (1973). *Proc. Natl. Acad. Sci. U.S. 70*, 2317-2320.

6. Weaver, P., Lien, S., and Seibert, M. (1979). "Photobiological Production of Hydrogen-A Solar Energy Conversion Option." SERI, Golden, Colorado.
7. Klibanov, A. M., Kaplan, N. O., and Kamen, M. D. (1978). *Proc. Natl. Acad. Sci. U.S. 75,* 3640-3643.
8. Klibanov, A. M., Kaplan, N. O., and Kamen, M. D. (1979). *Biochim. Biophys. Acta 547,* 411-416.
9. Burns, R. C. and Hardy, R. W. F. (1975). "Nitrogen Fixation in Bacteria and Higher Plants." Springer-Verlag, Berlin.
10. Newton, W., Postgate, J. R., and Rodrigues-Barrueco, C. (eds.) (1977). "Recent Developments in Nitrogen Fixation." Academic Press, London.
11. Hardy, R. W. F. (ed.). (1977 - 1978). "A Treatise on Dinitrogen Fixation." Wiley, New York.
12. Benemann, J. R. and Weare, N. M. (1974). *Science 184,* 174-175.
13. Weare, N. M. (1978). *Biochim. Biophys. Acta 502,* 486-494.
14. Eady, R. R. and Postgate, J. R. (1974). *Nature (London) 249,* 805-810.
15. Zumft, W. G. (1976). *Structure Bonding 29,* 1-65.
16. Postgate, J. R. (1974). *In* "The Biology of Nitrogen Fixation" (A. Quispel, ed.), pp. 663-686. North-Holland, Amsterdam.
17. Dalton, H. and Postgate, J. R. (1969). *J. Gen. Microbiol. 54,* 463-473.
18. Drozd, J. and Postgate, J. R. (1970). *J. Gen. Microbiol. 63,* 63-73.
19. Strandberg, G. W. and Wilson, P. W. (1968). *Can. J. Microbiol. 14,* 25-34.
20. Wong, P. P. and Burris, R. H. (1972). *Proc. Natl. Acad. Sci. U.S. 69,* 672-675.
21. Shah, V. K. and Brill, W. J. (1973). *Biochim. Biophys. Acta 305,* 445-454.
22. Emerich, D. W. and Burris, R. H. (1978). *Methods Enzymol. 53,* 314-329.
23. Burris, R. H. (1972). *Methods Enzymol. 24,* 415-431.
24. Bulen, W. A., Burns, R. C., and LeComte, J. R. (1965). *Proc. Natl. Acad. Sci. U.S. 53,* 532-535.
25. Haaker, H. and Veeger, C. (1977). *Eur. J. Biochem. 77,* 1-10.
26. Scherings, G., Haaker, H., and Veeger, C. (1977). *Eur. J. Biochem. 77,* 621-630.
27. Yates, M. G. (1972). *Eur. J. Biochem. 29,* 386-392.
28. Klibanov, A. M. (1979). *Anal. Biochem. 93,* 1-25.

Regulatory Mechanisms and Metabolic Miscellany

HORMONES OF THE ANTERIOR PITUITARY GLAND

Choh Hao Li

Hormone Research Laboratory
University of California
San Francisco, California 94143

Of all the endocrine glands, the anterior lobe of the hypophysis occupies the prime place. In addition to its influence on general body growth and carbohydrate metabolism, the adenohypophysis exercises direct control on the functions of the adrenal cortex, the thyroid, and the gonads. Hence, the hormones secreted by the anterior pituitary (except the growth hormone and prolactin) are called "tropic" hormones; i.e., their function is the stimulation of a specific target organ, an endocrine gland, whose activity is thereby either enhanced or awakened.

After nearly 40 years of intense work on the chemistry and biology of anterior pituitary hormones, all of the ten hormones have been completely purified and characterized (1). Their primary structures are known (see Figs. 1 - 3). Six of them have been either partially or totally synthesized in the laboratory. From the studies on their primary structures, it is possible to divide them into three groups having certain structural features in common. Because of these common structures, they exhibit overlapping biological activities. A high degree of structural

 ISBN 0-12-397580-8

5 10 15
H-Glu-Leu-Thr-Gly-Gln-Arg-Leu-Arg-Gln-Gly-Asp-Gly-Pro-Asn-Ala-

20 25 30
Gly-Ala-Asp-Asp-Gly-Pro-Gly-Ala-Gln-Ala-Asp-Leu-Glu-His-Ser-

35 40 45
Leu-Leu-Val-Ala-Ala-Glu-Lys-Lys-Asp-Glu-Gly-Pro-Tyr-Arg-Met-

50 55 60
Glu-His-Phe-Arg-Trp-Gly-Ser-Pro-Pro-Lys-Asp-Lys-Arg-Tyr-Gly-

65 70 75
Gly-Phe-Met-Thr-Ser-Glu-Lys-Ser-Gln-Thr-Pro-Leu-Val-Thr-Leu-

80 85 89
Phe-Lys-Asn-Ala-Ile-Ile-Lys-Asn-Ala-Tyr-Lys-Lys-Gly-Glu-OH

The Amino Acid Sequence of Human β-Lipotropin

5 10 15
H-Ser-Tyr-Ser-Met-Glu-His-Phe-Arg-Trp-Gly-Lys-Pro-Val-Gly-Lys-

20 25 30
Lys-Arg-Arg-Pro-Val-Lys-Val-Tyr-Pro-Asn-Gly-Ala-Glu-Asp-Glu-

35 39
Ser-Ala-Glu-Ala-Phe-Pro-Leu-Glu-Phe-OH

The Amino Acid Sequence of Human ACTH

5 10 13
Ac-Ser-Tyr-Ser-Met-Glu-His-Phe-Arg-Trp-Gly-Lys-Pro-Val-NH_2

The Amino Acid Sequence of Human α-MSH

5 10 15
H-Ala-Glu-Lys-Lys-Asp-Glu-Gly-Pro-Tyr-Arg-Met-Glu-His-Phe-Arg-Trp-Gly-

22
Ser-Pro-Pro-Lys-Asp-OH

The Amino Acid Sequence of Human β-MSH

resemblance has been found to exist among the glycoprotein hormones: thyrotropin, lutropin and follitropin. Somatotropin and prolactin form the second group, exhibiting considerable structural homology as well as common biological properties. Corticotropin, the melanotropins, and the lipotropins fall into the third group of structurally and biologically related molecule.

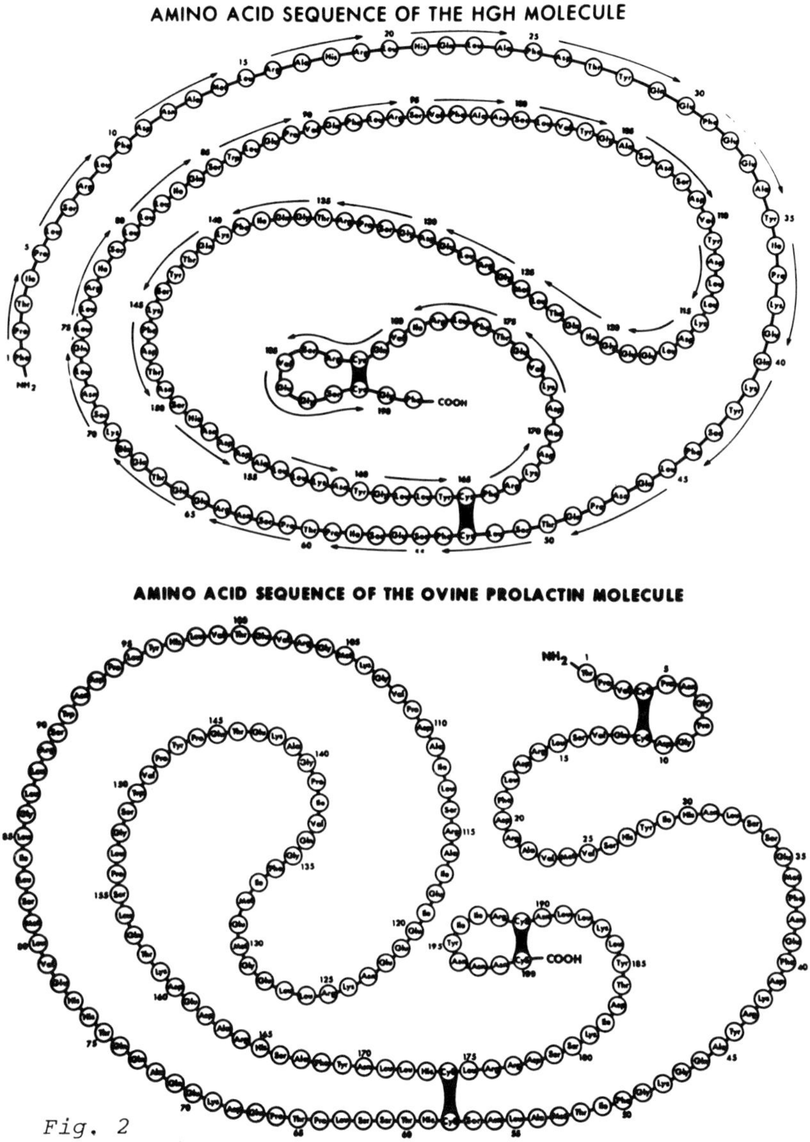

Fig. 2

GLYCOPROTEIN HORMONES

In 1927, Smith and Engle (2) showed that the pituitary gland profoundly affected the various reproductive processes in the rat. Subsequently, Fevold *et al.* (3) clearly demonstrated that there were two pituitary factors involved: follitropin (FSH) and lutropin (LH, ICSH). In the female, FSH stim-

5 10 15
H-Val-Gln-Asp-Cys-Pro-Glu-Cys-Thr-Leu-Gln-Glu-Asn-Pro-Phe-Phe-

20 25 30
Ser-Gln-Pro-Gly-Ala-Pro-Ile-Leu-Gln-Cys-Met-Gly-Cys-Cys-Phe-

35 40 45
Ser-Arg-Ala-Tyr-Pro-Thr-Pro-Leu-Arg-Ser-Lys-Lys-Thr-Met-Leu-

CHO 50 55 60
Val-Gln-Lys-Asn-Val-Thr-Ser-Glu-Ser-Thr-Cys-Cys-Val-Ala-Lys-

65 70 CHO 75
Ser-Tyr-Asn-Arg-Val-Thr-Val-Met-Gly-Gly-Phe-Lys-Val-Glu-Asn-

80 85 89
His-Thr-Ala-Cys-His-Ser-Cys-Thr-Cys-Tyr-Tyr-His-Lys-Ser-OH

Amino Acid Sequence of Human TSH-α

5 10 15
H-Phe-Cys-Ile-Pro-Thr-Glu-Tyr-Met-Thr-His-Ile-Glu-Arg-Arg-Glu-

20 CHO 25 30
Cys-Ala-Tyr-Cys-Leu-Thr-Ile-Asn-Thr-Thr-Ile-Cys-Ala-Gly-Tyr-

35 40 45
Cys-Met-Thr-Arg-Asp-Ile-Asn-Gly-Lys-Leu-Phe-Leu-Pro-Lys-Tyr-

50 55 60
Ala-Leu-Ser-Gln-Asp-Val-Cys-Thr-Tyr-Arg-Asp-Phe-Ile-Tyr-Arg-

65 70 75
Thr-Val-Glu-Ile-Pro-Gly-Cys-Pro-Leu-His-Val-Ala-Pro-Tyr-Phe-

80 85 90
Ser-Tyr-Pro-Val-Ala-Leu-Ser-Cys-Lys-Cys-Gly-Lys-Cys-Asp-Thr-

95 100 105
Asp-Tyr-Ser-Asp-Cys-Ile-His-Glu-Ala-Ile-Lys-Thr-Asn-Tyr-Cys-

110
Thr-Lys-Pro-Gln-Lys-Ser-Tyr-OH

Amino Acid Sequence of Human TSH-β

H-Val-Gln-Asp-Cys-Pro-Glu-Cys-Thr-Leu-Gln-Glu-Asn-Pro-Phe-Phe-
1 5 10 15

Ser-Gln-Pro-Gly-Ala-Pro-Ile-Leu-Gln-Cys-Met-Gly-Cys-Cys-Phe-
20 25 30

Ser-Arg-Ala-Tyr-Pro-Thr-Pro-Leu-Arg-Ser-Lys-Lys-Thr-Met-Leu-
35 40 45

CHO
Val-Gln-Lys-Asn-Val-Thr-Ser-Glu-Ser-Thr-Cys-Cys-Val-Ala-Lys-
50 55 60

CHO
Ser-Tyr-Asn-Arg-Val-Thr-Val-Met-Gly-Gly-Phe-Lys-Val-Glu-Asn-
65 70 75

His-Thr-Ala-Cys-His-Ser-Cys-Thr-Cys-Tyr-Tyr-His-Lys-Ser-OH
80 85 89

Amino Acid Sequence of Human Lutropin-α

H-Ser-Arg-Glu-Pro-Leu-Arg-Pro-Trp-Cys-His-Pro-Ile-Asn-Ala-Ile-
1 5 10 15

CHO
Leu-Ala-Val-Glu-Lys-Glu-Gly-Cys-Pro-Val-Cys-Ile-Thr-Val-Asn-
20 25 30

Thr-Thr-Ile-Cys-Ala-Gly-Tyr-Cys-Pro-Thr-Met-Met-Arg-Val-Leu-
35 40 45

Gln-Ala-Val-Leu-Pro-Pro-Leu-Pro-Gln-Val-Cys-Thr-Tyr-Arg-Asp-
50 55 60

Val-Arg-Phe-Glu-Ser-Ile-Arg-Leu-Pro-Gly-Cys-Pro-Arg-Gly-Val-
65 70 75

Asp-Pro-Val-Val-Ser-Phe-Pro-Val-Ala-Leu-Ser-Cys-Arg-Cys-Gly-
80 85 90

Pro-Cys-Arg-Arg-Ser-Thr-Ser-Asp-Cys-Gly-Gly-Pro-Lys-Asp-His-
95 100 105

Pro-Leu-Thr-Cys-Asp-Pro-Gln-His-Ser-Gly-OH
110 115

Amino Acid Sequence of Human Lutropin-β

Fig. 3

ulates the development of immature ovarian follicles, bringing them to maturity. LH, in turn, stimulates the mature follicles to form estrogens and gives the follicle impetus for further development as a corpus luteum. The growing corpus luteum is then brought to maturity by LH. The secretion of progesterone from the functional corpora lutea is under the control of LH and prolactin.

Thyrotropin (TSH) is the principal stimulatory agent in the physiological regulation of thyroid gland function. The earliest response that has been observed in the thyroid gland after TSH administration is the activation of the mechanism in the follicular cells for the resorption of thyroglobulin from follicular lumen depots and the release of thyroid hormones into the bloodstream.

The three glycoprotein hormones consist of two subunits (α and β). In solutions of acidic pH and denaturants (8 *M* urea or 6 *M* guanidine-HCl), the native hormones dissociate into α and β subunits which can be isolated by countercurrent distribution or chromatographic procedures (4, 5). The amino acid sequences of the α subunit in FSH, LH, and TSH are nearly identical whereas the primary structures of the β subunit of these three hormones are different (see Fig. 3). Apparently, the β subunit is hormonally specific. When the β subunit of TSH recombines with FSH-α, it generates thyrotropic activity. Similarly, when TSH-α recombines with FSH-β, it generates follitropic activity. The complete amino acid sequences of human FSH (6) and TSH (7) have recently been determined.

SIMPLE PROTEINS

The first indication that a hormone from the anterior pituitary might bring about lactation in mammals came from early experiments of Stricker and Grueter (8) in 1928, which showed that lactation could be elicited in the pseudopregnant rabbit, before and after spaying, by the injection of bovine anterior pituitary extract. These observations have been confirmed and extended by a number of investigators. In 1932, it was demonstrated by Riddle *et al.* (9) that macroscopically observable growth of the crop glands of the pigeon occurs after the administration of pituitary extracts. Subsequent work has indicated that this effect in the bird is probably due to the same factor that induces lactation in mammals.

Lactogenic activity has since been reported to be present in the pituitary extracts of various species, including humans, sheep, pigs, horses, whales, rabbits, cats, rats, guinea pigs, mice, fish, amphibians, reptiles, and birds. Since 1937, highly purified prolactin preparations have been obtained from cattle, sheep, rat, human, and fish pituitary glands. The primary structure of prolactin from sheep pituitary gland was finally shown in 1969 to consist of 199 amino acids with a linear sequence (10) (see Fig. 2).

From clinical observations of disturbances in the growth process, which are possibly the most obvious to perceive of all endocrine results, it was evident very early that the pituitary gland bears a direct relationship to body growth. Early specific experiments confirmed these general clinical observations, offering a convincing argument that the anterior pituitary secretes a hormone which possesses growth-promoting activity (11). Final proof came with the isolation, in highly purified form from beef pituitary glands (12) of a hormone known as growth hormone (from its obvious effects on growth) or somatotropin (from soma, the Greek word for body, since the sphere of the hormone's influence is general to body growth).

Although the effects of the bovine growth hormone were readily evident in the rat and mouse, it seemed to be ineffective in humans. Evidence was gathered by a number of investigators which pointed to an important connection between the biologic effectiveness of the hormone and the species from which it was derived. For example, it was found that the bovine hormone is inactive not only in humans but also in monkeys, whereas the hormone preparation from monkey glands is effective in that particular species. In the same way, a growth hormone concentrate prepared from fish pituitaries is effective in fish but not in rats. The bovine hormone, however, is effective in both fish and rats, although it has also been reported to have no effect in the guinea pig. All this evidence suggested that the growth hormones isolated from various species may not be chemically identical.

This was borne out in 1956, when highly purified preparations of growth hormone were successfully isolated from both human and monkey pituitaries (13). Chemical studies on these preparations have clearly demonstrated a difference between the human growth hormone (HGH) and the bovine hormone; for example, HGH is a monomer in neutral solutions, whereas the bovine hormone is a dimer.

HGH is a globular protein with a molecular weight of 21,500, isoelectric point at pH 4.9, and α-helical content of 55%. It contains 191 amino acids with a single tryptophan residue and two disulfide bridges (14) as shown in Fig. 2. A comparison of the primary structure of HGH with that of ovine prolactin (see Fig. 2) indicates a high degree of homology and suggests that these two molecules must have evolved from a common ancestor.

SIMPLE PEPTIDES

The removal of the anterior pituitary from almost all the species that have been studied brings about a pronounced atrophy of the cortex of the adrenal glands, whereas the medulla is scarcely affected (15). On the other hand, implantation of anterior pituitary tissue of injection of pituitary extracts into animals that are normal produces as its chief effect a marked hypertrophy of the adrenal cortex. Hence, the name adrenocorticotropic hormone (ACTH) has been employed to designate the active principle in pituitary extracts that accomplishes repair of adrenocortical atrophy following hypophysectomy. In both experimental animals and human subjects, injections of ACTH accelerate the synthesis and secretion of adrenal hormones including cortisol and aldosterone.

ACTH has been isolated in highly purified form from sheep, pig, cattle, human, fish, whale, and ostrich (16). Although all ACTHs have similar biologic properties, they differ slightly in amino acid composition and in chemical structure. They are single-chain polypeptides composed of 39 amino acids with serine and phenylalanine as NH_2-terminal and COOH-terminal residues, respectively (see Fig. 1). Human ACTH has been synthesized by both solution (17) and solid-phase (18) methods.

Two melanotropins are present in pituitary extracts (19), namely, α-MSH and β-MSH. α-Melanotropin consists of 13 amino acid residues and its sequence is identical to the first 13 residues of ACTH, except that the amino group of serine in position 1 is acetylated and the carboxyl group of valine in position 13 is present as an amide (see Fig. 1). Synthesis of α-MSH was achieved by both solution (20) and solid-phase (21) methods.

β-Melanotropins from various species consist of 18 amino acids except that the human hormone consists of 22 amino acids (see Fig. 1). Human β-MSH has been synthesized by both solution (22) and solid-phase (23) methods.

The discovery of lipotropins (LPHs) occurred only 14 years ago. During the course of improving the yield of ACTH from sheep pituitary extracts, a new component was isolated and characterized to be different chemically from all known pituitary hormones (24, 25). It was first assayed for lipolytic activity in rabbit fat pad and shown to be a potent lipotropic agent. There are two lipotropins: β-LPH (26) and γ-LPH (27). β-Lipotropin consists of 91 amino acids with the sequence as shown in Fig. 1 (26, 28). γ-Lipotropin has a structure corresponding to the 58 NH_2-terminal residues of β-LPH. Total synthesis of sheep β-LPH was recently accomplished by the solid-phase method (29).

In the search for β-lipotropin from camel pituitary glands,

a new peptide with 31 amino acids was discovered and its sequence (see Fig. 1) was identical to the COOH-terminal 31 residues of β-LPH (30). It possesses significant opiate activity *in vitro* (30, 31) and *in vivo* (32). This untriakontapeptide was designated β-endorphin (β-EP) (30). Both camel (33) and human (34) β-endorphins have been synthesized by the solid-phase method.

CONCLUDING REMARKS

The first anterior pituitary hormone used clinically was adrenocorticotropin. Since 1949, ACTH has been employed for remission of rheumatoid arthritis and other metabolic disorders as well as for a diagnostic agent in tests of adrenal functions. After the isolation and partial characterization of HGH in 1956, it has been shown by various investigators (for a review, see 35, 36) that the hormone is effective in promoting growth in children with growth problems. Recent studies show that HGH exhibits beneficial effects on patients with bleeding ulcers (37). Clinical usefulness of HGH has not been fully explored due to the limited supply of the hormone. It is hoped that a synthetic product with full HGH activity will be developed in the future for extensive evaluation of HGH potential in clinical medicine.

In addition to the profound analgesic potency of β-endorphin, the peptide exhibits behavioral activity in experimental animals (for a review, see 38). Preliminary studies (39) have shown that β-endorphin exerts short-term improvements in depressed patients and longer duration improvements in some chronic schizophrenics after intravenous injections. In addition, β-endorphin has narcotic abstinence-suppressing properties in man (40, 41). If future studies confirm these preliminary clinical results, β-endorphin may play an important role in treatment of narcotic withdrawal syndrome, schizophrenia, and other forms of mental disease.

ACKNOWLEDGMENT

I thank my co-workers (cited in the references) for their contributions to these investigations. The experimental work was supported in part over the years by grants from the National Institutes of Health (GM-2907, AM-6097, AM-18677), the National Institute of Mental Health (MH-30245), the American Cancer Society, and the Hormone Research Foundation.

REFERENCES

1. Li, C. H. (1972). *Proc. Amer. Phil. Soc. 116,* 365.
2. Smith, P. E. and Engle, E. T. (1927). *Amer. J. Anat. 40,* 159.
3. Fevold, H. L., Hisaw, F. I., and Lenard, S. L. (1931). *Amer. J. Physiol. 97,* 291.
4. Papkoff, H., Ryan, R. J., and Ward, D. N. (1977). *In* "Frontiers in Reproduction and Fertility Control," (R. O. Greep and M. A. Koblinsky, eds.), pp. 1-10. MIT Press, Boston.
5. Pierce, J. G., Liao, T. H., and Carlsen, R. B. (1973). *In* "Hormonal Proteins and Peptides," (C. H. Li, ed.), Vol. 1, pp. 17-57. Academic Press, New York.
6. Shome, B. and Parlow, A. F. (1974). *J. Clin. Endocrinol. Metab. 39,* 199.
7. Sairam, M. R. and Li, C. H. (1977). *Can. J. Biochem. 55,* 755.
8. Stricker, P. and Grueter, F. (1928). *C. R. Soc. Biol. (Paris) 99,* 1978.
9. Riddle, O., Bates, R. W., and Dyshorn, S. W. (1932). *Proc. Soc. Exp. Biol. Med. 29,* 1211.
10. Li, C. H., Dixon, J. S., Lo, T. B., Pankov, Y. A., and Schmidt, K. D. (1969). *Nature (London) 224,* 695; Li, C. H. (1976). *Int. J. Peptide Protein Res. 8,* 205.
11. Evans, H. M. and Long, J. A. (1921). *Anat. Rec. 21,* 62.
12. Li, C. H. and Evans, H. M. (1944). *Science 99,* 183.
13. Li, C. H. and Papkoff, H. (1956). *Science 124,* 1293; Li, C. H. (1957). *Fed. Proc. 16,* 775.
14. Li, C. H., Dixon, J. S., and Lin, W. K. (1969). *Arch. Biochem. Biophys. 133,* 70.
15. Smith, P. E. (1930). *Amer. J. Anat. 45,* 205.
16. Li, C. H., Chung, D., Oelofsen, W., and Naudé, R. J. (1978). *Biochem. Biophys. Res. Commun. 81,* 900.
17. Sieber, P., Rittal, W., and Riniker, B. (1972). *Helv. Chim. Acta 55,* 1243.
18. Yamashiro, D. and Li, C. H. (1973). *J. Amer. Chem. Soc. 95,* 1310.
19. Lee, T. H. and Lerner, A. B. (1956). *J. Biol. Chem. 221,* 943.
20. Guttmann, S. and Boissonnas, R. A. (1959). *Helv. Chim. Acta 42,* 1257.
21. Blake, J. and Li, C. H. (1971). *Int. J. Protein Res. 3,* 185.
22. Yajima, H., Okada, Y., Kimomura, Y., and Minami, H. (1968). *J. Amer. Chem. Soc. 90,* 527.
23. Wang, K. T., Blake, J., and Li, C. H. (1973). *Int. J. Peptide Protein Res. 5,* 33.

24. Li, C. H. (1964). *Nature (London) 201,* 924.
25. Birk, Y. and Li, C. H. (1964). *J. Biol. Chem. 239,* 1048.
26. Li, C. H., Barnafi, L., Chrêtien, M., and Chung, D. (1966) *Excerpta Med. Found. Int. Congr. Ser. 112,* 334-364.
27. Chrêtien, M. and Li, C. H. (1967). *Can. J. Biochem. 45,* 1163.
28. Li, C. H. and Chung, D. (1976). *Nature (London) 260,* 622.
29. Yamashiro, D. and Li, C. H. (1978). *J. Amer. Chem. Soc., 100,* 5174-5179.
30. Li, C. H. and Chung, D. (1976). *Proc. Natl. Acad. Sci. U.S. 73,* 1145.
31. Cox, B. M., Goldstein, A., and Li, C. H. (1976). *Proc. Natl. Acad. Sci. U.S. 73,* 1821.
32. Loh, H. H., Tseng, L.-F., Wei, E., and Li, C. H. (1976). *Proc. Natl. Acad. Sci. U.S. 73,* 2895.
33. Li, C. H., Lemaire, S., Yamashiro, D., and Doneen, B. A. (1976). *Biochem. Biophys. Res. Commun. 71,* 19.
34. Li, C. H., Yamashiro, D., Tseng, L.-F., and Loh, H. H. (1977). *J. Med. Chem. 20,* 325.
35. Tanner, J. M. (1972). *Nature (London) 237,* 433.
36. Escamilla, R. F. (1975). *In* "Hormonal Proteins and Peptides" (C. H. Li, ed.), Vol. 3, pp. 147-190. Academic Press, New York.
37. Vanamee, P., Sherlock, P., and Winawer, S. J. (1972). *Gastroenterology 62,* 824.
38. Li, C. H. (1977). *Arch. Biochem. Biophys. 183,* 592.
39. Kline, N., Li, C. H., Lehmann, H. E., Lajtha, A., Lashi, E., and Cooper, T. (1977). *Arch. Gen. Psych. 34,* 1111.
40. Catlin, D. H., Hui, K. K., Loh, H. H., and Li, C. H. (1977). *Commun. Psychopharmacol. 1,* 492.
41. Su, C. Y., Lin, S.-H., Wang, Y.-T., Li, C. H., Hung, L. H., Lin, C. S., and Lin, B. C. (1978). *J. Formosan Med. Assoc. 77,* 133.

SAINT-AGNÈS, MUSIC, AND SPORULATION

Jekisiel Szulmajster

Laboratoire d'Enzymologie du C.N.R.S.
91190 Gif-sur-Yvette, France

It was a delightful summer evening in 1959* when Sol Spiegelman, on one of his frequent visits in Paris, asked Huguette and myself to meet him in a café on rue Soufflot. He was in the company of Martin Kamen. I had, of course, heard of Martin many years before and had attended his first seminar after the war at the Pasteur Institute in 1956, when he was permitted to go abroad after the sad MacCarthy period. I do not remember in what bistro or restaurant we ended up that evening but I do remember that since that evening we have been great friends and have never missed the opportunity to meet each other during his visits to Paris.

In 1966, Martin Kamen was appointed for a period of 3 years, as Directeur de Recherche at the C.N.R.S. As a scientific advisor to the Laboratoire de Photosynthèse he helped Dr. Lavorel, the newly promoted director of this laboratory,

**Some of the dates mentioned here are probably wrong, but this does not matter, as the instances are more relevant.*

 ISBN 0-12-397580-8

to initiate a few specific projects on *Rhodopseudomonas spheroides* and *Rhodospirillum rubrum*.

This brought Martin Kamen frequently to Gif for relatively long periods of time. He lived then in a comfortable apartment at the Chateau at Gif (digne d'un Grand Directeur!). The occasions of Martin's visits to the Laboratoire d'Enzymologie were a time of fruitful discussion and many friendly and enjoyable dinners either at home or in some charming Parisian restaurant.

In going back in my mind to the marvelous time and "la joie de vivre" we shared with Martin, I recall more vividly the first visit, in August of 1969, of Martin and Virginia to our reconstructed ruin in Sainte-Agnés. Sainte-Agnés is a small medieval village located 12 k from Menton, 2200 feet above the level of the Mediterranean sea. When I picked up Martin and his wife at the railway station at Mention, I noticed that he had brought with him his wonderful toy: an Albany viola. He almost took as much care of it as of Virginia, whom I met for the first time. It is with a particular pleasure and deep nostalgia that I remember the first breakfast we shared together in Sainte-Agnès. Huguette and myself arose early in the morning to prepare breakfast and to watch the sun rise from our terrace behind our house, which faces, in the background, the high mountains of the Italian border. Just in front of us was the sharply outlined steeple of a typical country church. With the smell of the coffee coming from the house, suddenly wonderful music reached our ears. Entering the house was Martin, in his white and blue striped morning dress, playing a Mozart concerto. The acoustics in this old house, with its 1-meter thick walls was fantastic and the amplification great. Virginia with her charming smile, followed by our daughters, Anne and Claire, full of admiration, came downstairs and all of us, touched by this music, sat down for breakfast. What an exceptional morning!

Martin's second visit to Sainte-Agnès was in August 1973, about 2 months before the International Spore Meeting that I (in collaboration with P. Schaeffer and J. P. Aubert) was organizing in Gif (1). I was then preparing my paper and was happy to discuss the recent developments in sporulation problems with him. Discussions of more general interest went on during our promenades on the road leading to the "Col de la Banquette," where we collected dried pine cones for barbecue. I was delighted to let Martin read my paper for the Spore Meeting. He made a very kind remark that the manuscript was conceived in Yiddish, expressed in French, and translated into English! He polished it for presentation.... I am not sure of the scientific value of this paper but it is surely written in good English. Martin was interested in sporulation not only because his friend was making a living on this problem, but mainly because his enthusiasm for science was unlimited. He was also interested in sporulation because it represents a biological phenomenon of cryptobiosis, in that spores are endowed with extraordinary properties of resistance against destruction by heat and other deleterious agents. Furthermore, the transition from the fragile vegetative cell to the dormant form constitutes an example of unicellular differentiation unique in the biological world (2).

It is worthwhile to give a very general outlook on the sporulation problem with particular emphasis on the progress made since that meeting in 1973 on the control of sporulation at the transcription level.

In 1973, the confusion of this particular problem was great and tenacious, although confusion in science is a frequent, and maybe an unavoidable, phenomenon. The attention of the sporologists was mainly focused on the mystery of behavior of the DNA-dependent RNA polymerase (RPase) during sporulation. The structure and mode of action of this enzyme during spore formation have been and to some extent still are the subjects of a number of studies with contradictory results (3). This is not unexpected in view of the complexity of the structure and function of this enzyme.

For many years bacteriophage-infected cells were taken as a model for studying the mechanisms of regulation, which govern sequential gene expression at the transcriptional level in prokaryotic developmental system. In general, the synthesis of different RNA species during the programmed transformation of one defined type of cell to another can be accomplished by two types of changes at the RNA polymerase level: (1) the first possibility would be the complete replacement of the bacterial polymerase by a *de novo* polymerase. (2) The second mechanism is a substantial alteration in the structure in preexisting host enzymes.

In phage development, changes in transcription patterns are indeed correlated with these two types of molecular mechanisms.

After infection of *E. coli* by T3 and T7, new phage-specific RPases have been identified (4). In this case, the enzyme of the host cell ($\beta\beta'\alpha_2\sigma$, MW = 450,000) is replaced by a new phage-coded RPase which is composed of one polypeptide chain of a molecular weight of 110,000 daltons necessary for transcription of late phage genes. These RPases recognize phage-specific promotors and do not require sigma factor for activity.

An example of the second mechanism is the modification of the existent host RPase as exemplified by the SP01 and SP82 phages after infection of *B. subtilis* (5, 6). Here, phage-specific polypeptides are synthesized which are associated with the host core enzyme to form an enzyme with an altered template specificity, which no longer recognizes host promotor sites. In both cases the RNA polymerase retains the core enzyme structure ($\alpha_2\beta\beta'$) losing the vegetative sigma factor.

It was disappointing for the sporologist to realize after several years of effort that extrapolation of data derived from phage studies to sporulation problems was leading them in the wrong direction. Initially the ideas of similarity of events in phage-infected cells and during spore development were strengthened by studies demonstrating that the transition of a vegetative cell to the sporulating state was associated with two specific molecular events which are expressed *in vivo* and *in vitro*: change in template specificity and a structural modification of RPase (3, 7 - 9). It was observed that when vegetative cells of *B. subtilis* were infected by the virulent phage ϕe, they give rise to a large number of progeny phage. However, when sporulating cells were infected by the same phage, the bacteria no longer supported phage multiplication. This behavior of the phage was explained by means of *in vitro* experiments using partially purified RPase isolated from both exponential phase and sporulating cells. It was observed that the vegetative enzyme was able to transcribe either ϕe DNA or poly d(AT) whereas the RNA polymerase from the sporulating cells was able to transcribe only poly d(AT) but not the phage DNA. As a consequence of the change in template specificity it was also found that during sporulation, one of the β subunits of the RPase was modified and replaced by a polypeptide with a lower molecular weight.

This nice picture of events was conceptually and intellectually satisfactory but experimentally wrong. The disenchantment started from the moment when it was shown that the structural modification of RNA polymerase is an artefact due to the treacherous activity of some clandestine proteolytic enzymes appearing during sporulation and, hence, present in the extracts of these cells. By protecting the RNA polymerase preparations by protease inhibitors (PMSF, DFP, and EDTA) no modification was noted in any of the subunits of the RPase of *B. subtilis* and the enzyme obtained was similar to that isolate from vegetative cells (10).

Concerning the hypothesis of alteration of the template specificity of RPase which, per se, was an attractive type of control mechanism, it was usually based only on the observed decrease in the capacity of the spore enzyme to transcribe ϕe DNA *in vitro*. This observation still remains controversial and different suggestions have been made to explain this decrease in the activity toward phage ϕe DNA: stationary phage RPase (10), dependence on growth medium (11, 12), presence of an inhibitor (3, 13) or interference by a specific nuclease (14) responsible for the degradation of the RNA synthesized on the ϕe DNA template.

It is obvious that before changes in specificity of the enzyme can be considered to be established as a significant control mechanism, it must be shown that the spore RPase is able to synthesize RNA transcripts *in vitro* that are not produced by the vegetative enzyme. Unfortunately, this is not the case at present.

From an historical point of view, it is worthwhile to mention that the concept of sigma factor function was extremely attractive as a basis for explanation of positive control involved in regulation mechanisms of gene expression. One could indeed assume that sporulating cells are synthesizing different sigma-like factors each time before a new gene or set of genes has to be transcribed. Unfortunately, the solution was not to be so. Indeed, we now know that only one and the same sigma factor is found in the RPase isolated from vegetative and sporulating cells, although the sporulation sigma factor appears to be weakly bound to the RPase core enzyme. The vegetative RPase core appears also to be preserved unchanged throughout the sporulation process since sporulating cells like vegetative ones remain sensitive to rifampicin (antibiotic affecting specifically the DNA-dependent RNA polymerase) during the entire life cycle.

What seems now to be well established is that a control of sporulation at the level of transcription does take place. This assessment is based on two types of evidence (15, 16): (1) Mutants of *B. subtilis* have been isolated which by virtue of a single mutational event have gained simultaneously resistance to rifampicin *in vivo* and *in vitro* and have lost the ability to sporulate. (2) DNA - RNA hybridization-competition studies with RNA from sporulating cells have revealed that new species of mRNA appear sequentially during spore development. These studies have surprisingly shown that 70 - 80% of the vegetative genes continue to be expressed during the sporulation process (15, 16). This situation makes the "simple" model of spore differentiation more complex than was supposed some years ago.

In spite of the existence of spore-specific mRNA's, a regulatory function of the RPase has not yet been clearly established. The latest studies are mainly focused on the modulation of RPase by certain polypeptides of different molecular weights specifically associated with core or holoenzyme and appearing at various stages of sporulation (15). However, the role of these polypeptides as control elements in the sporulation process still remains to be established.

While the regulatory function of the sporulation RPase still remains a subject of intensive investigation, more attention is now being drawn to a new approach of studying gene expression during spore morphogenesis. This approach is taking advantage of the powerful tool of molecular cloning of recombinant DNA, which permits the isolation of chromosome segments bearing specific genes and amplification of the segments in sufficient quantities for *in vitro* transcription and translation. Although a few encouraging preliminary results were reported at the last Spore Meeting in Madison (17), several difficulties have to be eliminated to make rapid and further progress in this field. The main difficulty now is to find a suitable vector for cloning *B. subtilis* DNA fragments. Although great progress has been made during the last few years in the studies of plasmids and bacteriophages as vectors for cloning DNA molecules, there was not until very recently such a vector available for studies of genetic recombinations *in vitro* with *B. subtilis* DNA. All the preliminary studies were therefore carried out using *E. coli* plasmids. The amplification of these types of recombinant plasmids in *E. coli* and their expression in *B. subtilis* presents some difficulties which make the experimentation very laborious and time consuming. The discovery of *B. subtilis* plasmid (18) bearing antibiotic markers (kan. or neomycin) will make further studies easier and more likely to be successful.

Now, in considering "the future of the present" one may ask what will be the next step in spore research. In assuming that all the cloning difficulties will be overcome and that many spore genes will be cloned in *B. subtilis*, the goal to reach is the *in vitro* transcription and translation of these genes in order to find out the type of proteins synthesized under the control of these spores genes. Fortunately, the number of gene products in *B. subtilis* having an authentic function in the formation of the spore is limited to the spore coat proteins and to the enzymes involved in the synthesis of the cortical peptidoglycan and dipicolinic acid.

The identification of these products can be carried out by immunological assays using antibodies for some of the above macromolecules now available in our laboratory as well as in others.

Studies by this approach open new avenues to explore and many obstacles to cross. It is to be hoped, however, that for Martin's next celebration we will take a quantum jump in our understanding of control mechanism operating during spore development. (Maybe!)

I tried in this chapter to present a brief summary of the progress made in the last few years on one aspect of spore research, i.e., the one I am most interested. This very subjective presentation should not put into the shadows other important aspects of sporulation and, in particular, those concerning the basis and mechanisms of bacterial spore resistance. As mentioned above bacterial spores exemplify to the extreme degree the general biological phenomenon of cryptobiosis. Despite the practical and theoretical implications and in more than a century of research no fundamental explanation of spore resistance has been found. Here too, efforts are being made in recent years with new experimental approaches and biophysical methods that might be applied to sporology (19, 20).

REFERENCES

1. "Régulation de la Sporulation Microbienne" (1973). Colloque Int. C.N.R.S., (J. P. Aubert, P. Schaeffer, and J. Szulmajster, eds.). No. 227.
2. Szulmajster, J. (1973). *Symp. Soc. Gen. Microbiol. 23*, 45.
3. Szulmajster, J. (1973). *In* "Regulation de la Sporulation Microbienne" (J. P. Aubert, P. Schaeffer, and J. Szulmajster, eds.). Colloque Int. C.N.R.S., no. 227.
4. Bautz, E. K. F. (1976). *In* "RNA Polymerase" (R. Losick and M. Chamberlin, eds.), p. 273. C.S.H.
5. Petrusek, R., Duffy, J. J., and Geiduschek, E. P. (1976). *In* "RNA Polymerase" (R. Losick and M. Chamberlin, eds.), p. 567. Cold Spring Harbor Lab., New York.
6. Whiteley, H. R., Spiegelman, G. B., Lawrie, J. M., and Hiatt, W. R. (1976). *In* "RNA polymerase" (R. Losick and M. Chamberlin, eds.), p. 587. Cold Spring Harbor Lab., New York.
7. Losick, R. (1973). *In* "Régulation de la Sporulation Microbienne" (J. P. Aubert, P. Schaeffer, and J. Szulmajster, eds.). Colloque Int. C.N.R.S., no. 227.
8. Losick, R., Shorenstein, R. G., and Sonenshein, A. L. (1970). *Nature (London) 227*, 910.
9. Kerjan, P. (1973). *In* "Régulation de la Sporulation Microbienne" (J. P. Aubert, P. Schaeffer, and J. Szulmajster, eds.). Colloque Int. C.N.R.S., no. 227, p. 41.
10. Orrego, C., Kerjan, P., Manca de Nadra, M. C., and Szulmajster, J. (1973). *J. Bacteriol. 116*, 636.
11. Strauss, N. (1973). *In* "Régulation de la Sporulation Microbienne" (J. P. Aubert, P. Schaeffer, and J. Szulmajster, eds.). Colloque Int. C.N.R.S., no. 227, p. 11.
12. Nishimoto, H. and Takahashi, I. (1973). *In* "Régulation de la Sporulation Microbienne" (J. P. Aubert, P. Schaeffer, and J. Szulmajster, eds.). Colloque Int. C.N.R.S., no. 227, p. 9.
13. Segall, J., Tjian, R., Pero, J., and Losick, R. (1974). *Proc. Natl. Acad. Sci. U.S. 71*, 4860.
14. Kerjan, P. and Szulmajster, J. (1976). *Biochimie 58*, 533.
15. Doi, R. H. (1977). *Bacteriol. Rev. 41*, 568.
16. Bonamy, C., Manca de Nadra, M. C., and Szulmajster, J.

(1976). *Eur. J. Biochem. 63,* 53.
17. Bonamy, C., Keryer, E., and Szulmajster, J. (1978). *In* "Spores VII" (G. Chambliss and J. C. Vary, eds.), p. 139. Amer. Soc. Microbiol.
18. Gryczan, T. J. and Dubnau, D. (1978). *Proc. Natl. Acad. Sci. U.S. 75,* 1428.
19. Gould, G. W. (1977). *J. Appl. Bacteriol. 42,* 297.
20. "Spores VII" (1978). (G. Chambliss and J. C. Vary, eds.). Amer. Soc. Microbiol.

MOLECULAR MECHANISMS AND FUNCTION OF BIOLUMINESCENCE IN FISHES

Frederick I. Tsuji

Marine Biology Research Division A002
University of California at San Diego
La Jolla, California 92093

Brentwood Veterans Hospital
Los Angeles, California 90073

Department of Biological Sciences
University of Southern California
Los Angeles, California 90073

One of the most striking features of oceanic life is the widespread occurrence of bioluminescence among marine organisms, especially among fish, crustaceans, and cephalopods (1). Sunlight penetrates to a depth of around 1000 m, above and below which many of these organisms live. Another distinct feature of oceanic life is the diurnal vertical migration that many of these organisms undertake. They migrate in large numbers to the surface at night and descend to lower depths during the day. They presumably migrate in search of food in the upper photic zone, where plant and animal life thrive. Among the most important participants in this migration are the deep-sea fish (1, 2). It is an interesting fact that luminescent fishes are all marine and they represent the highest form of life known to be bioluminescent. The prevalence of bioluminescence in deep-sea fish is shown by a taxonomic study (3) carried out over a 9-year period involving more than 1500 net hauls and 115,747 individuals which indicates that "Luminosity is present as follows: in 50% of the orders, 39% of the families, 81% of the genera, 66% of the species and (thanks again to *Cyclothone* and

 ISBN 0-12-397580-8

Myctophidae) to 96.5% of all individuals." *Cyclothone* and myctophids or lantern fishes are among the most numerous of deep-sea fish. The ability to produce light, however, is not confined only to deep-sea fish, since many shallow-water fishes are also luminous (1, 2, 4, 5). They are found in large numbers in the shallow, coastal waters of the tropical and subtropical Indo-West Pacific.

Fish produce bioluminescence in luminous organs or photophores by one of two methods: (1) by a bacterial reaction involving symbiotic luminous bacteria or (2) by a luciferin-luciferase reaction involving the oxidation of a luciferin (substrate) by molecular oxygen, catalyzed by luciferase (enzyme). The two methods of light production do not occur together in a single fish, with the exception of the angler fish, *Linophryne* and possibly *Himantolophus* (6, 7). In many fish, the luminous organ is highly developed morphologically, approaching that of the eye. It is usually located on the ventral or ventrolateral surface of the body or within the body cavity. The organ may have accessory structures associated with it to control the intensity and color of light. A luminous organ commonly consists of a central core of photogenic cells, a reflector, a lens, and a melanophore or pigment layer to filter the light. Dermal organs are often of this type. Luminous organs located internally are almost always connected to the gut by a duct. In a young fish, this duct may serve as a route through which initial infection of the organ by symbiotic luminous bacteria takes place and subsequently serves as a means of getting rid of dead bacteria. In some shallow-water luminescent fish, e.g., apogonids, the duct appears to serve as a route through which luciferin derived from the ingestion of luminescent organisms enters the luminous organ. In apogonids, the luminous organ is a small outcropping of the gut which is used to store luciferin and luciferase and which is continuously luminous. In deep-sea fish, on the other hand, a duct connection is usually absent. Instead, the luminous organ is embedded in the dermis and the organs are arranged, often serially, in species-specific patterns over the head and trunk. In others, the luminous organ may be a specialized structure, such as the luminous lure or esca of the angler fish.

BACTERIAL METHOD OF LIGHT PRODUCTION

This method of light production appears to be more common among shallow-water than deep-sea fishes. Best known examples are the 20 or more species of *Leiognathus* (5), *Photoblepharon*

(8 - 11), and *Anomalops* (8, 12) (Fig. 1). Light and electron microscopic studies of the organs show the presence of numerous bacteria and cell-free extracts of the organ contain bacterial luciferase. In *Leiognathus* the luminous organ consists of a ring of luminous tissue which encircles the distal end of the esophagus at its juncture with the stomach. The organ has two ducts leading into the esophagus. The light from the organ enters the swimbladder chamber and is reflected out through the translucent ventral abdominal wall. The bacterium responsible for the light emission is *Photobacterium leiognathi* (13). Between 10 - 25% of the mass of the organ is made up of viable luminescent bacteria (14). In contrast, *Photoblepharon* and *Anomalops* possess a pair of large, elliptical organs located beneath the eyes that shine continuously. The organ is attached to the body at its dorsoanterior edge by a small piece of cartilaginous tissue. *Photoblepharon* occludes its light by raising a black fold over the organ, whereas *Anomalops* rotates its luminous organ downward along its long axis thereby turning the luminous face out of view and displaying the opposite black pigmented face. The light of the two fish is produced by symbiotic luminous bacteria which thus far have been found to be nonculturable.

Fig. 1. Photoblepharon palpebratus (above), approximately 80 mm long, and Anomalops katoptron (below), approximately 94 mm long, showing suborbital luminous organs (LO).

Only four species of luminous bacteria are known: *Photobacterium fischeri, P. phosphoreum, P. leiognathi,* and *Beneckea harveyi* (15, 16), with only the first three having been found in the light organs of fish. The bacterial reaction responsible for light emission is a monooxygenase reaction. *In vitro* studies have shown that the reaction involves the oxidation of $FMNH_2$ and a long-chain aldehyde by molecular oxygen, catalyzed by bacterial luciferase (17):

$$FMNH_2 + RCHO + O_2 \xrightarrow[\text{luciferase}]{\text{Bacterial}} h\nu_{\lambda_{max}}\ 496\ nm + FMN + H_2O + RCOOH$$

A long-lived oxygen-containing intermediate is formed, which appears to be the 4a-peroxy adduct of $FMNH_2$. The intermediate reacts with aldehyde to yield several additional intermediates, one of which serves as the emitter in the reaction. Aldehydes with 8 - 14 carbon atoms are active in the reaction, with tetradecanal apparently being the natural aldehyde (18). Where the *in vitro* reaction normally yields an emission peak at around 496 nm, the emission peak in the case of the *in vivo* reaction is usually at a shorter wavelength. For example, in *P. phosphoreum,* it is 476 nm. The *in vivo* emitter has not been identified, but according to recent reports may be a chromophore attached to a protein which acts as a sensitizer of the bioluminescence reaction (19 - 22).

LUCIFERIN-LUCIFERASE METHOD OF LIGHT PRODUCTION

This method of light production is used by a large number of fish and, in many, the reaction appears to be similar, if not identical, to the reaction in the small, marine ostracod crustacean, *Cypridina hilgendorfii.* Luminous species of *Cypridina* are widely distributed in the waters of the world, with the most studied species being *C. hilgendorfii* (1, 23). The organism is about 3 mm long and is found in the south coastal waters of Japan. It can be readily collected and the luciferin and luciferase extracted and purified (1, 24). When disturbed, *Cypridina* ejects luciferin and luciferase from separate glands into sea water. A bright blue cloud is produced as a result of the oxidation of luciferin by molecular oxygen, catalyzed by luciferase (Fig. 2) (1, 25, 26). The reaction has been studied in considerable detail and it appears to involve the formation of a dioxetanone intermediate during the reaction which decomposes to yield light, carbon dioxide, and oxyluciferin (27 - 31). The emitter in the reaction is the

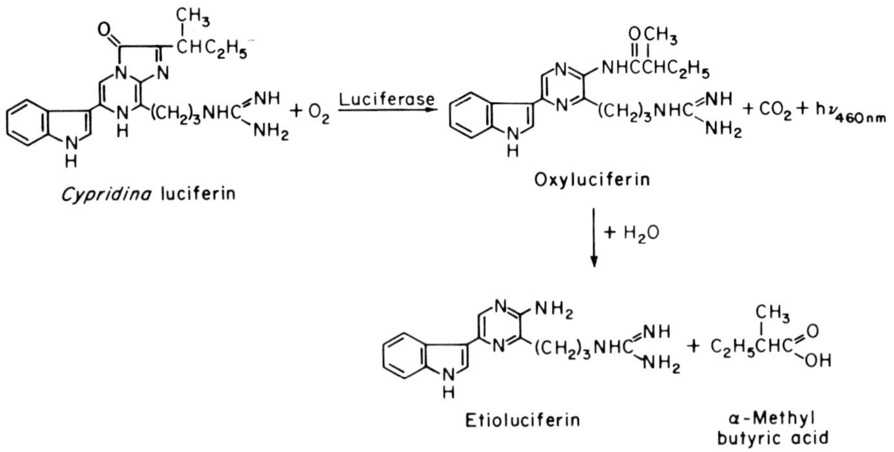

Fig. 2. Scheme for the Cypridina bioluminescence reaction.

oxyluciferin - luciferase complex (32).

Both *Cypridina* luciferin and luciferase give reciprocal light-emitting cross-reactions with the luciferins and luciferases of a variety of fish. They include the luciferins and luciferases of the apogonid fishes, *Apogon ellioti* (33 - 36), *A. striata* (37), *Archamia fucata* (37), *A. zosterophora* (37), *A. lineolata* (37), and *Rhabdamia cypselura* (37) and with those of the pempherid fish, *Parapriacanthus ransonneti* (38 - 40). Further, the chemical properties of *Apogon* and *Parapriacanthus* luciferins are similar, if not identical, to those of *Cypridina* luciferin (35, 39). *Apogon* luciferase and *Cypridina* luciferase, however, differ from each other in their immunochemical and chromatographic properties (36). Since the luminous organs are connected to the gut by a duct and since *Cypridina* organisms are found in the same waters as the fish, it has been suggested that the luciferins originate from the ingestion of luminescent *Cypridina* organisms (41, 42), whereas the luciferases are either synthesized *de novo* or are obtained through the diet and used in modified form (36). Apogonid and pempherid luciferins and luciferases also give reciprocal light-emitting cross-reactions (34, 37, 40, 42).

Other cross-reactions have been demonstrated between the light-emitting components of *Cypridina* and those of the coastal fishes, *Porichthys porosissimus* (43) and *P. notatus* (42), the deep-sea myctophid or lantern fishes, *Diaphus elucens* Brauer (Fig. 3) (44) and *Stenobrachius leucopsarus* (44), and the deep-sea hatchet fishes, *Polyipnus spinosus* (44) and *Sternoptyx* diaphana (44). *Porichthys* luciferin also cross-reacts with apogonid and pempherid luciferases (42). A compound with an

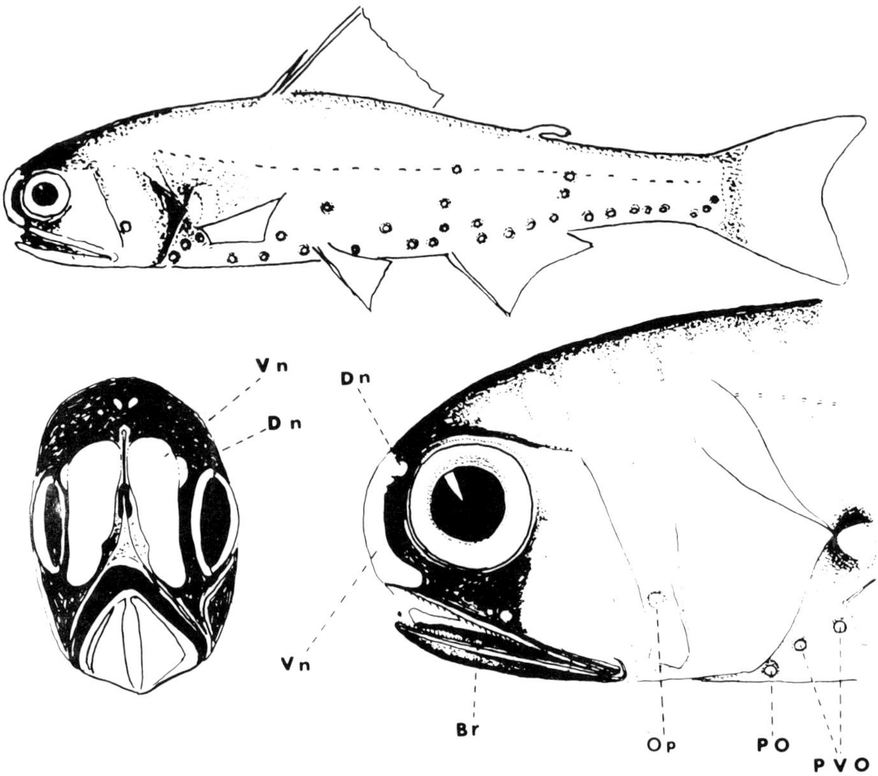

Fig. 3. Deep-sea lantern fish, Diaphus elucens Brauer, approximately 120 mm long, showing distribution of luminous organs: Vn, ventro-nasal; Dn, dorso-nasal; Br, branchiostegal; Op, opercular; PO, pectoral; and PVO, subpectoral.

imidazopyrazine structure similar to *Cypridina* luciferin has also been isolated from the liver of *Neoscopelus microchir*, a deep-sea fish belonging to a family closely related to the myctophid family (45). The structure of this compound is apparently identical to the structure proposed for the luciferin of two decapod shrimps, *Oplophorus spinosus* and *Heterocarpus laevigatus* (Fig. 4, I) (46). Since *Neoscopelus* and the shrimps are found in the same waters, the possibility exists that the compound in *Neoscopelus* arises through the ingestion of these shrimps. Interesting enough, the proposed structure is also identical to the structure of *Renilla reniformis* (coelenterate) luciferin (Fig. 4, II) (47, 48) and basically the same as the structure proposed for the luciferin of the Japanese firefly-squid, *Watasenia scintillans* (Fig. 4, III) (49). The imidazopyrazine structure itself appears to be unique in

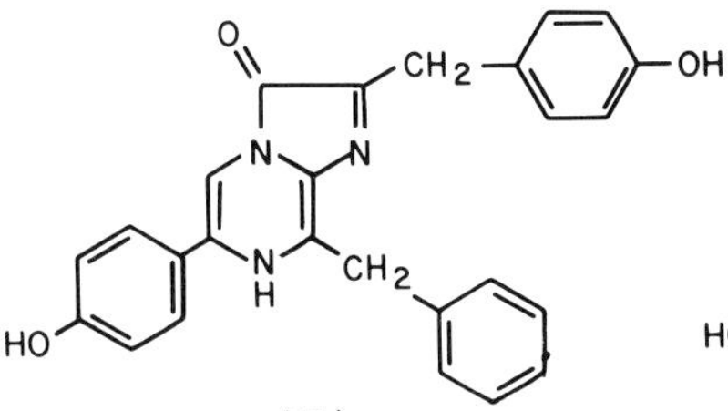

(I)

Oplophorus luciferin

(Crustacean)

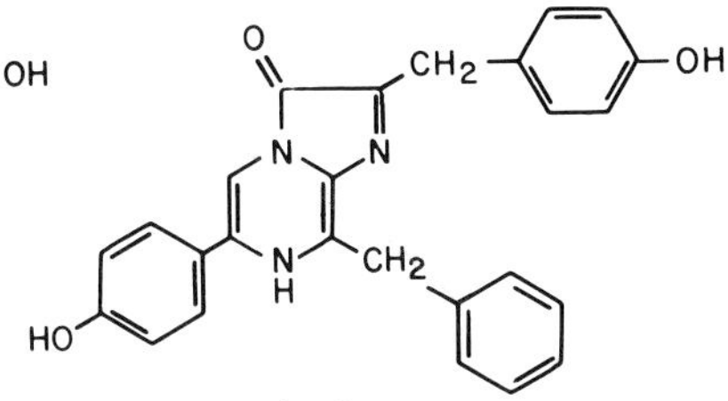

(II)

Renilla luciferin

(Coelenterate)

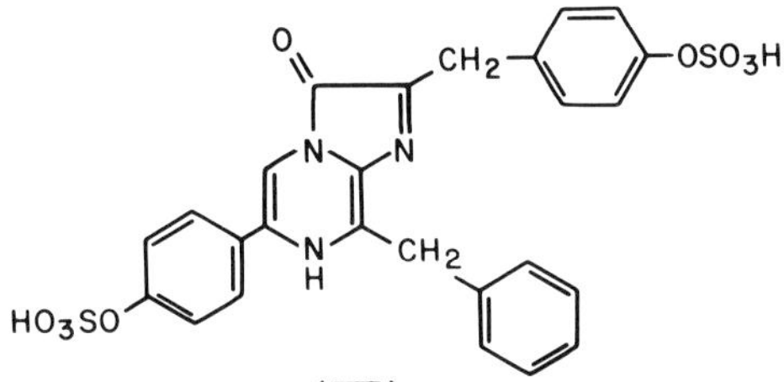

(III)

Watasenia luciferin

(Cephalopod)

Fig. 4. Structure of Renilla luciferin and proposed structures for Oplophorus and Watasenia luciferins.

that it has not been found elsewhere. The fact that the structure appears in the luciferins, either known or proposed, of such a wide range of organisms suggests that the luciferins may have a common evolutionary origin or alternatively that they arise from a common food source such as a group of luminescent planktons in the food chain.

The question as to whether luminescent fish synthesize luciferin *de novo* or obtain it by ingesting luminescent organisms has been studied in most detail in the California midshipman fish, *Porichthys notatus* (Fig. 5). This fish, commonly found along the Pacific Coast of North America, possesses more than 700 dermal photophores (1-mm diameter) arranged in characteristic patterns over the head and trunk. There are no duct connections between the gut and the photophores. Bright light is emitted from the photophores following subcutaneous or intraperitoneal injection of epinephrine, norepinephrine, amphetamine, or 6-hydroxydopamine (50 - 56). The bioluminescence reaction involves the oxidation of luciferin by molecular oxygen, catalyzed by luciferase (42, 43). The reaction is very similar, if not identical, to the reaction in *Cypridina* (Fig.

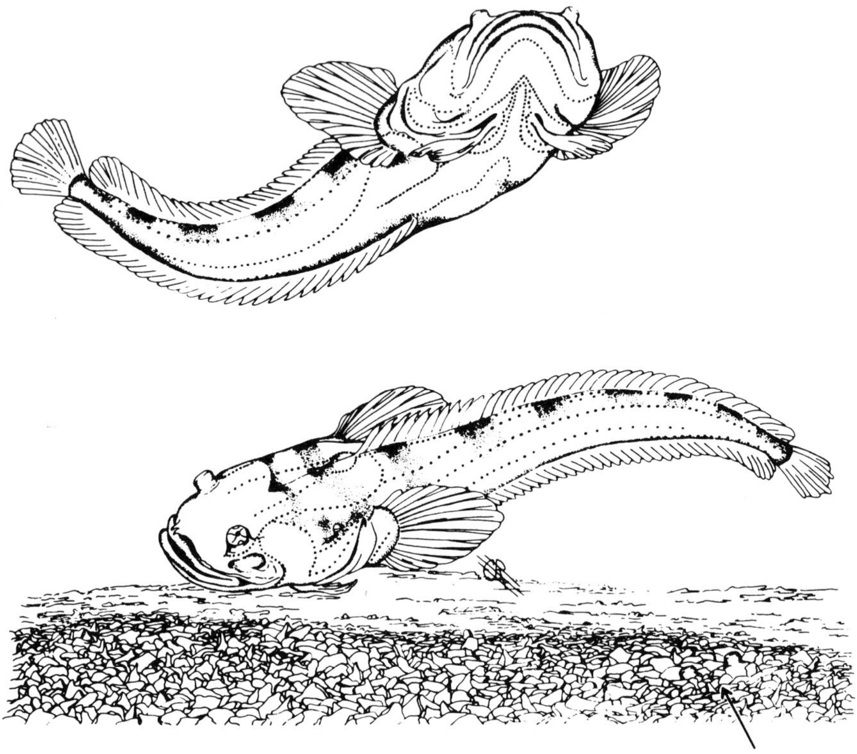

Fig. 5. Porichthys notatus swimming in aquarium showing serial photophores on head and trunk. Adults measure between 175 - 350 mm in length. A third Porichthys, buried at the bottom right, is shown with a pair of eyes protruding (arrow).

2). In contrast, specimens of *P. notatus* collected in Puget Sound, Washington, are nonluminous, even though ultrastructurally their photophores do not differ from those of the luminous California fish (52, 57). Luminescence capability may be induced in the Puget Sound fish by (a) forced feeding of dried *Cypridina hilgendorfii* organisms placed in a gelatin capsule, (b) intraperitoneal injection or delivery to the stomach by intramedic tubing of highly purified *C. hilgendorfii* luciferin, or (c) intraperitoneal injection of synthetic *C. hilgendorfii* luciferin (58 - 60). After 4 - 7 days, luminescence may be elicited by an injection of norepinephrine. A single dose of either *Cypridina* organisms or luciferin is all that is required to induce luminescence capability and, once developed, the fish appears to retain its ability to luminesce almost indefinitely. The minimum effective dose is about 10 μg of *Cypridina* luciferin. The development of luminescence capability is paralleled

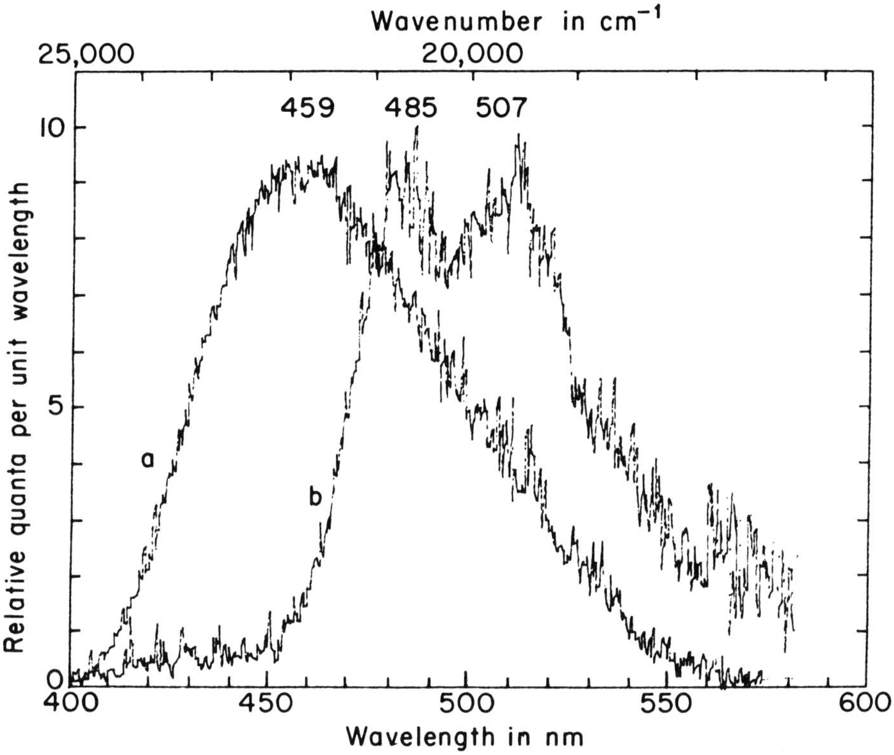

Fig. 6. Bioluminescence emission spectra of (a) Cypridina reaction, measured in vitro, and (b) California Porichthys notatus, measured in vivo.

by the appearance of an intense greenish-yellow fluorescence of the photophores. Since *Cypridina* luciferin also fluoresces greenish-yellow (535 nm) in near uv light, the fluorescence is apparently due to the accumulation of luciferin in the photophores.

The *in vivo* bioluminescence emission spectrum produced by the induced Puget Sound fish is indistinguishable from that of the naturally luminous California fish (60). In both, the emission spectrum extends from around 400 to 600 nm, with maxima at 485 and 507 nm (Fig. 6). In contrast, the *in vitro* spectrum of the *Porichthys* luciferin x *Cypridina* luciferase reaction has only a single maximum, which is close to 460 nm, the same as in *Cypridina* (Fig. 6). The bimodal nature of the emission spectrum suggests that two emitters are involved, possibly involving energy transfer from a primary emitter (an electronically excited state of an oxyluciferin-luciferase complex) to a second emitting species. This is not unreasonable since energy transfer to the chromophore of a second protein molecule has been demonstrated in the case of coelenterate bioluminescence (61 - 74) and a similar process appears to be implicated in bacterial bioluminescence (19 - 22). Such a process could account for a discrepancy previously observed in which bioluminescence emission spectra of different strains of bacteria show *in vivo* maxima ranging from 476 to 505 nm, whereas *in vitro* maxima are grouped around 496 nm, with a maximum peak difference for a single strain being about 18 nm (75). In one strain of *Photobacterium fischeri*, the peak of the *in vivo* emission is at 545 nm, whereas the peak of the *in vitro* emission is at 495 nm (19). Another explanation for the bimodal emission in *Porichthys* is that it is due to complete or partial excited state protonation of the *Cypridina* oxyluciferin anion formed during the reaction. It has been shown that the anion of *Cypridina* oxyluciferin and certain of its analogs possess a fluorescence emission maximum near 500 nm, while the neutral species exhibits fluorescence maxima ranging from 420 to 477 nm in aprotic solvents (27, 32, 76, 77). Still another possibility is that the emissions are due to two different luciferases present in the photophores.

The inability of Puget Sound *Porichthys* to luminesce has been correlated with a lack of luciferin in all developmental stages of the animal, whereas the California *Porichthys* possesses luciferin in all stages of development (58 - 60). In both California and Puget Sound *Porichthys*, luciferase is detectable in the larvae around day 29, after which luciferase is found in all stages of development. The persistence of luminescence capability in Puget Sound *Porichthys* is not due to recycling of degradation products of luciferin. Injection of air-oxidized *Cypridina* luciferin, synthetic *Cypridina* oxyluciferin, or *Cypridina* luciferin oxidized with a high concentration of purified *Cypridina* luciferase is ineffective in inducing bioluminescence capability. At present, no compound tested besides luciferin from *Cypridina* and *Porichthys* has been found to induce bioluminescence capability in the Puget Sound fish. Further, *Cypridina* luciferin is a highly labile compound, autoxidizing in the presence of molecular oxygen with loss of activity. The fish, therefore, appears to possess a mechanism for protecting luciferin from autoxidation.

The mechanism of induction of bioluminescence in Puget Sound *Porichthys* is still unknown. The mechanism could involve (1) absorption, transport, concentration, and protection of the administered luciferin or (2) *de novo* luciferin synthesis triggered by luciferin. There is no known process in plants and animals in which *de novo* synthesis of a substrate is induced by the substrate. It does not resemble amplification as has been observed in aggregating cells of *Dictyostelium discoideum* in which cyclic AMP stimulates the release of cyclic AMP (78, 79). The photophores of Puget Sound *Porichthys* do not contain any luciferin. This is shown by a complete lack of light emission when norepinephrine is injected into the fish, by the nonfluorescence of the photophores when exposed to near uv light, and by an inactivity of photophore extracts when mixed with *Cypridina* luciferase. Thus, further understanding of the mechanism of induction will have to await the synthesis of *Cypridina* ^{14}C-labeled luciferin, which is presently underway, and the determination of its specific activity following injection into the nonluminous Puget Sound fish.

FUNCTION OF BIOLUMINESCENCE

This subject has been reviewed in several articles (80 - 82). According to the various hypotheses, bioluminescence is believed to (1) attract prey, (2) deter or distract a predator, (3) match downwelling ambient sunlight in order to obscure or obliterate a fish's silhouette from a predator below, (4) serve as a warning signal to members of the species of the presence of a nearby predator and reveal the predator's presence to other predators, (5) allow members of a species to recognize each other and to attract the opposite sex, (6) see and navigate in the dark, (7) serve as a reference source for spacing out individuals over a given territory, (8) aid in schooling, and (9) mimic other luminescent organisms in order to provide protection against a predator or to approach a prey.

At present, the best evidence that is available in support of a function is for the countershading or camouflage hypothesis. It is a well-known fact that many luminescent fish have their luminous organs or photophores located predominantly on the ventral and ventrolateral surface of the body. Such a placement of photophores has been thought to be for the purpose of emitting light in the ventral direction. If the fish is able to match ambient light correctly, it should be able to obliterate its own silhouette and be less subject to predation from below (83 - 95). This mechanism appears to be especially applicable to *Argyropelecus* or the hatchet fish. This fish is apparently able to match ambient light with luminescent light of just the right intensity, correct spectral composition, and right angular distribution (89, 91, 92, 95). In addition, some luminescent fish appear to be able to adjust their skin coloration to match ambient light conditions by means of changes in their chromatophores (88). Surface-dwelling fish do not have ventral photophores apparently because of the difficulty of producing light of sufficient intensity to match ambient light.

The countershading mechanism has also been suggested for the squid (96 - 101) and shrimp (102). How a fish is able to match ventral light to downwelling ambient light has been a subject of speculation inasmuch as the ventral photophores are out of sight of the fish. Possibly, the fish may see some of its own light as a result of scattering from detritus. A current hypothesis is that fish use the light of orbital photophores as a reference source with which to compare downwelling ambient light (93, 103, 104). Orbital photophore light is also believed to be used to light-adapt the eye before other photophore lights are turned on. A difficulty with both hypotheses is that many lantern fishes do not possess orbital photophores and even for fish with orbital photophores, it is not certain whether the light has any relationship to ventral light intensity. In reality, such a closed-loop or feedback system may not be necessary to explain how matching is accomplished. It is possible that the fish has developed, as a result of evolutionary pressure, an open-loop system in which the fish is able to sense from perceiving ambient light the correct nerve impulse to send to the ventral photophores in order to produce light of matching intensity. A reasonable case can be made in this connection for *Opisthoproctus,* which lives at depths ranging from 300 to 700 m and possesses tubular eyes that are focused upwards (105, 106). *Opisthoproctus* does not have orbital light organs (106). An open-loop system could serve this fish well in adjusting the "window" of its luminous rectal bulb, which produces light by means of symbiotic luminous bacteria.

While nearly all studies support the countershading hypothesis, there are some observations that are difficult to explain. For example, it has been found by use of an echo sounder and by direct visual observation from a submersible vehicle that large swarms of myctophids or lantern fish congregate to form major parts of the deep scattering layer (107, 108). These schools participate in diurnal vertical migration. Fish in such schools appear motionless and randomly oriented, often occupying slightly oblique positions. The fish do not emit light, except when caught in a net attached to the submersible vehicle. It seems doubtful that fish in such large schools can carry out effective countershading. Another example concerns the benthic fish, *Porichthys,* which has a very prominent distribution of ventral photophores (see Fig. 5). It is a nocturnal animal, which spends its day buried in the ocean bottom and comes out to feed at night, rising well into midwater. It is difficult to imagine when the ventral photophores would be used for countershading, unless they are used in countershading for moon- or starlight (42). Thus, it is conceivable that bioluminescence is a mechanism used largely at night. Further, it is interesting to note that the population of *Porichthys* in Puget Sound does not appear to show adverse effects from the loss of bioluminescence capability. The fish is found in large numbers and specimens caught appear perfectly healthy. Functions for which bioluminescence has been proposed in *Porichthys* include courtship display during spawning (109), warning signal on the approach of a predator (110), and mimicking other luminescent organisms (42, 111). A third example concerns the Japanese pony fish, *Leiognathus elongatus,* a moderately deep-water fish with sexually dimorphic organs (5). The luminous organ is extremely large in the male and minute in the female. When viewed in the dark, the exposed female organ presents itself as a tiny point source of light. The organ is so small and the light so weak as to raise serious doubts as to whether effective countershading would be possible. Thus, bioluminescence more likely plays a sexual role in this species.

Other roles for which bioluminescence has received attention include the following. The males of the lantern fish, *Tarlentonbeania,* possess a large luminous organ or "stern chaser" on the dorsal surface of the caudal peduncle, which is absent in the female. It is thought that the organ, which can produce a very bright flash of light, is used by the male to draw away a predator from a female. The female, by turning off all her lights and remaining still, escapes attention while the male makes its getaway (112). However, such tactics may not work against predators with powerful chemosensory receptors for tracking prey. Another hypothesis may be mentioned here by way of connecting the bioluminescence of another organism to fish. Luminescent dinoflagellates, which sometimes multiply very rapidly and form large patches in the ocean and on which plankton filtering animals such as sardines feed, are said to emit bioluminescent flashes to attract secondary predators, e.g., tunas, as a means of self-protection (113). The deep-sea fish *Stomias* and *Chauliodus* possess numerous minute, ball-shaped bodies, in addition to serial ventral photophores, which are embedded in a gelatinous sheath surrounding the fish's body (114). When stimulated, these bodies produce a luminous glow which envelopes the fish's body and reveals clearly the contour of the fish, especially to any predator nearby. This effect is just the opposite from what might be expected under the countershading hypothesis.

Shallow-water fishes such as *Anomalops* and *Photoblepharon* appear to use their luminous organs to see, attract prey, and carry out intraspecific communications (8 - 10). Another specialized organ is the luminous lure or esca of the female deep-sea angler fish. The fish probably uses it in the same manner as its shallow-water relative, *Lophius*, which dangles a nonluminous lure to catch prey (115 - 118).

The relationship of bioluminescence to vision has received some attention (119). That fish can see their own bioluminescence has been demonstrated in a number of cases. The deep-sea fish, *Pachystomias,* which emits red light from large photophores near the eye has its maximal retinal absorption near 575 nm (120). The red light may well be used to hunt for unsuspecting prey. The deep-sea fish, *Aristostomias scintillans,* emits red light from a suborbital organ and green light from a postorbital organ and the absorbance maxima for the two visual pigments are at 551 and 526 nm (121). In *Photoblepharon* the maximum for the *in vivo* bioluminescence spectrum is at 496 nm, which corresponds to the maximum for the spectral sensitivity of the retina (122). For *Porichthys* a good match exists between the *in vivo* bimodal emission spectrum and the spectral sensitivity curve, indicating that the fish can perceive its bioluminescence very well (123). Thus, fish may very well carry out intraspecific communication in addition to interspecific functions.

While luminous bacteria may be of use to fish that use them in their luminous organs, it might be asked of what value could light be to the bacteria that produce it? It has been suggested that a bioluminescence reaction is either a vestigial detoxification reaction previously employed for removing molecular oxygen (124) or an oxygenase reaction which has been secondarily adapted for use in light production (125). The fact that luciferase constitutes 2 - 5% of the protein in a luminescent bacterium (126, 127), which seems like a relatively large amount, suggests that bacterial light is probably not an adventitious phenomenon but has some survival value. It is conceivable that bacterial luminescence serves some interspecific function. For example, while free-living luminescent bacteria in the open ocean may not be luminescent (128), those growing rapidly on dead tissue are often brightly luminescent. Recent studies have shown that bacteria have protozoan and copepod predators (129, 130). Thus, it is possible that such bacteria produce light to deter or mask themselves from predators which otherwise might graze on them.

ACKNOWLEDGMENT

Supported by research grant PCM 75-23042 from the National Science Foundation.

REFERENCES

1. Harvey, E. N. (1952). "Bioluminescence," 649 pp. Academic Press, New York.
2. Herring, P. J. and Morin, J. G. (1978). Bioluminescence in fishes. *In* "Bioluminescence in Action" (P. J. Herring, ed.), pp. 273-329. Academic Press, New York.
3. Beebe, W. (1937). Preliminary list of Bermuda deep-sea fish. *Zoologica 22,* 197-208.
4. Haneda, Y. (1955). Luminous organisms of Japan and the Far East. *In* "The Luminescence of Biological Systems" (F. H. Johnson, ed.), pp. 335-385. Amer. Assoc. Advan. Sci., Washington, D.C.
5. Haneda, Y. and Tsuji, F. I. (1976). The luminescent systems of pony fishes. *J. Morphol. 150,* 539-552.
6. Hansen, K. and Herring, P. J. (1977). Dual bioluminescent systems in the anglerfish genus *Linophryne* (Pisces: Ceratioidea). *J. Zool. (London) 182,* 103-124.
7. Haneda, Y. (1968). Observations on the luminescence of the deep-sea luminous angler fish, *Himantolophus groenlandicus. Sci. Rep. Yokosuka Cy Mus. 14,* 1-6.
8. Haneda, Y. and Tsuji, F. I. (1971). Light production in the luminous fishes *Photoblepharon* and *Anomalops* from the Banda Islands. *Science 173,* 143-145.
9. Morin, J. G., Harrington, A., Nealson, K., Krieger, N., Baldwin, T. O., and Hastings, J. W. (1975). Light for all reasons: versatility in the behavioral repertoire of the flashlight fish. *Science 190,* 74-76.
10. Meyer-Rochow, V. B. (1976). Some observations on spawning and fecundity in the luminescent fish *Photoblepharon palpebratus. Mar. Biol. 37,* 325-328.
11. Kessel, M. (1977). The ultrastructure of the relationship between the luminous organ of the teleost fish *Photoblepharon palpebratus* and its symbiotic bacteria. *Cytobiologie 15,* 145-158.
12. Meyer-Rochow, V. B. (1976). Loss of bioluminescence in *Anomalops katoptron* due to starvation. *Experientia 32,* 1175-1176.
13. Reichelt, J. L., Nealson, K., and Hastings, J. W. (1977). The specificity of symbiosis: pony fish and luminescent bacteria. *Arch. Microbiol. 112,* 157-161.
14. Hastings, J. W. and Mitchell, G. (1971). Endosymbiotic bioluminescent bacteria from the light organ of pony fish. *Biol. Bull. 141,* 261-268.
15. Reichelt, J. L. and Baumann, P. (1973). Taxonomy of the marine, luminous bacteria. *Arch. Mikrobiol. 94,* 283-330.
16. Reichelt, J. L., Baumann, P., and Baumann, L. (1976). Study of genetic relationships among marine species of the

genera *Beneckea* and *Photobacterium* by means of *in vitro* DNA/DNA hybridization. *Arch. Mikrobiol. 110,* 101-120.
17. Hastings, J. W. and Nealson, K. H. (1977). Bacterial bioluminescence. *Annu. Rev. Microbiol. 31,* 549-595.
18. Ulitzur, S. and Hastings, J. W. (1979). Evidence for tetradecanal as the natural aldehyde in bacterial bioluminescence. *Proc. Natl. Acad. Sci. U.S. 76,* 265-267.
19. Ruby, E. G. and Nealson, K. H. (1977). A luminous bacterium that emits yellow light. *Science 196,* 432-434.
20. Gast, R. and Lee, J. (1978). Isolation of the *in vivo* emitter in bacterial bioluminescence. *Proc. Natl. Acad. Sci. U.S. 75,* 833-837.
21. Gast, R., Neering, I. R., and Lee, J. (1978). Separation of a blue fluorescence protein from bacterial luciferase. *Biochem. Biophys. Res. Commun. 80,* 14-21.
22. Lee, J., Small, E. D., Liu, Y.-M., and Sinha, S. (1979). High molecular weight blue fluorescence protein from the bioluminescent bacterium, *Photobacterium fischeri. Biochem. Biophys. Res. Commun. 86,* 1241-1247.
23. Tsuji, F. I., Lynch, R. V. III, and Haneda, Y. (1970). Studies on the bioluminescence of the marine ostracod crustacean *Cypridina serrata. Biol. Bull. 139,* 386-401.
24. Tsuji, F. I. (1978). *Cypridina* luciferin and luciferase. *Methods Enzymol. 57,* 364-372.
25. Kishi, Y., Goto, T., Hirata, Y., Shimomura, O., and Johnson, F. H. (1966). The structure of *Cypridina* luciferin. *In* "Bioluminescence in Progress" (F. H. Johnson and Y. Haneda, eds.), pp. 89-113. Princeton University Press, Princeton, New Jersey.
26. Stone, H. (1968). The enzyme catalyzed oxidation of *Cypridina* luciferin. *Biochem. Biophys. Res. Commun. 31,* 386-391.
27. McCapra, F. and Chang, Y. C. (1967). The chemiluminescence of a *Cypridina* luciferin analogue. *J. Chem. Soc. Chem. Commun.,* pp. 1011-1012.
28. Shimomura, O. and Johnson, F. H. (1971). Mechanism of the luminescent oxidation of *Cypridina* luciferin. *Biochem. Biophys. Res. Commun. 44,* 340-346.
29. Tsuji, F. I., De Luca, M., Boyer, P. D., Endo, S., and Akutagawa, M. (1977). Mechanism of the enzyme-catalyzed oxidation of *Cypridina* and firefly luciferins studied by means of $^{17}O_2$ and $H_2{}^{18}O$. *Biochem. Biophys. Res. Commun. 74,* 606-613.
30. Hart, R. C., Stempel, K. E., Boyer, P. D., and Cormier, M. J. (1978). Mechanism of the enzyme-catalyzed bioluminescent oxidation of coelenterate-type luciferin. *Biochem. Biophys. Res. Commun. 81,* 980-986.
31. Wannlund, J., De Luca, M., Stempel, K., and Boyer, P. D. (1978). Use of ^{14}C-carboxyl-luciferin in determining the mechanism of the firefly luciferase catalyzed reactions.

Biochem. Biophys. Res. Commun. 81, 987-992.
32. Shimomura, O., Johnson, F. H., and Masugi, T. (1969). *Cypridina* bioluminescence: light-emitting oxyluciferin-luciferase complex. *Science 164,* 1299-1300.
33. Haneda, Y., Johnson, F. H., and Sie, E. H.-C. (1958). Luciferin and luciferase extracts of a fish, *Apogon marginatus,* and their luminescent cross-reactions with those of a crustacean, *Cypridina hilgendorfii. Biol. Bull. 115,* 336.
34. Johnson, F. H., Haneda, Y., and Sie, E. H.-C. (1960). An interphylum luciferin-luciferase reaction. *Science 132,* 422-423.
35. Sie, E. H.-C., McElroy, W. D., Johnson, F. H., and Haneda, Y. (1961). Spectroscopy of the *Apogon* luminescent system and of its cross reaction with the *Cypridina* system. *Arch. Biochem. Biophys. 93,* 286-291.
36. Tsuji, F. I. and Haneda, Y. (1966). Chemistry of the luciferases of *Cypridina hilgendorfii* and *Apogon ellioti. In* "Bioluminescence in Progress" (F. H. Johnson and Y. Haneda, Y., eds.), pp. 137-149. Princeton University Press, Princeton, New Jersey.
37. Haneda, Y., Tsuji, F. I., and Sugiyama, N. (1969). Luminescent systems in apogonid fishes from the Philippines. *Science 165,* 188-190.
38. Haneda, Y. and Johnson, F. H. (1958). The luciferin-luciferase reaction in a fish, *Parapriacanthus beryciformis,* of newly discovered luminescence. *Proc. Natl. Acad. Sci. U.S. 44,* 127-129.
39. Johnson, F. H., Sugiyama, N., Shimomura, O., Saiga, Y., and Haneda, Y. (1961). Crystalline luciferin from a luminescent fish, *Parapriacanthus beryciformes. Proc. Natl. Acad. Sci. U.S. 47,* 486-489.
40. Haneda, Y. and Johnson, F. H. (1962). The photogenic organs of *Parapriacanthus beryciformes* Franz and other fish with the indirect type of luminescent system. *J. Morphol. 110,* 187-198.
41. Haneda, Y., Johnson, F. H., and Shimomura, O. (1966). The origin of luciferin in the luminous ducts of *Parapriacanthus ransonneti, Pempheris klunzingeri,* and *Apogon ellioti. In* "Bioluminescence in Progress" (F. H. Johnson and Y. Haneda, eds.), pp. 533-545. Princeton University Press, Princeton, New Jersey.
42. Tsuji, F. I., Haneda, Y., Lynch, R. V. III, and Sugiyama, N. (1971). Luminescence cross-reactions of *Porichthys* luciferin and theories on the origin of luciferin in some shallow-water fishes. *Comp. Biochem. Physiol. 40A,* 163-179.
43. Cormier, M. J., Crane, J. M. Jr., and Nakano, Y. (1967). Evidence for the identity of the luminescent systems of

Porichthys porosissimus (fish) and *Cypridina hilgendorfii* (crustacean). *Biochem. Biophys. Res. Commun. 29*, 747-752.
44. Tsuji, F. I. and Haneda, Y. (1971). Luminescent system in a myctophid fish, *Diaphus elucens* Brauer. *Nature (London) 233*, 623-624.
45. Inoue, S., Okada, K., Kakoi, H., and Goto, T. (1977). Fish bioluminescence. I. Isolation of a luminescent substance from a myctophina fish, *Neoscopelus microchir,* and identification of it as *Oplophorus* luciferin. *Chem. Lett.*, pp. 257-258.
46. Inoue, S., Kakoi, H., and Goto, T. (1976). *Oplophorus* luciferin, bioluminescent substance of the decapod shrimps, *Oplophorus spinosus* and *Hetrocarpus laevigatus*. *J. Chem. Soc., Chem. Commun.*, pp. 1056-1057.
47. Hori, K., Charbonneau, H., Hart, R. C., and Cormier, M. J. (1977). Structure of native *Renilla reniformis* luciferin. *Proc. Natl. Acad. Sci. U.S. 74*, 4285-4287.
48. Inoue, S., Kakoi, H., Murata, M., Goto, T., and Shimomura, O. (1977). Complete structure of *Renilla* luciferin and luciferyl sulfate. *Tetrahedron Lett.*, No. 31, pp. 2685-2688.
49. Inoue, S., Kakoi, H., and Goto, T. (1976). Squid bioluminescence. III. Isolation and structure of *Watasenia* luciferin. *Tetrahedron Lett*. No. 34, pp. 2971-2974.
50. Greene, C. W. and Greene, H. H. (1924). Phosphorescence of *Porichthys notatus,* the California singing fish. *Amer. J. Physiol. 70*, 500-506.
51. Nicol, J. A. C. (1957). Observations on photophores and luminescence in the teleost *Porichthys*. *Quart. J. Microscop. Sci. 98*, 179-188.
52. Strum, J. M. (1969). Photophores of *Porichthys notatus*: ultrastructure of innervation. *Anat. Rec. 164*, 463-477.
53. Case, J. F., Reynolds, G. T., Buck, J., Burns, J., and Halverson, R. (1970). Comparative physiology of neurally controlled luminescence. *Biol. Bull. 139*, 417.
54. Baguet, F. and Case, J. (1971). Luminescence control in *Porichthys* (Teleostei) excitation of isolated photophores. *Biol. Bull. 140*, 15-27.
55. Baguet, F. (1975). Excitation and control of isolated photophores of luminous fishes. *Progr. Neurobiol. 5*, 97-125.
56. Anctil, M. and Case, J. F. (1976). Pharmacomorphological study of denervation induced by 6-hydroxydopamine in *Porichthys* photophores. *Cell Tissue Res. 166*, 365-388.
57. Strum, J. M. (1969). Fine structure of the dermal luminescent organs, photophores, in the fish, *Porichthys notatus*. *Anat. Rec. 164*, 433-461.
58. Tsuji, F. I., Barnes, A. T., and Case, J. F. (1972).

Bioluminescence in the marine teleost, *Porichthys notatus*, and its induction in a non-luminous form by *Cypridina* (ostracod) luciferin. *Nature (London) 237*, 515-516.
59. Barnes, A. T., Case, J. F., and Tsuji, F. I. (1973). Induction of bioluminescence in a luciferin deficient form of the marine teleost, *Porichthys*, in response to exogenous luciferin. *Comp. Biochem. Physiol. 46A*, 709-723.
60. Tsuji, F. I., Nafpaktitis, B. G., Goto, T., Cormier, M. J., Wampler, J. E., and Anderson, J. A. (1975). Spectral characteristics of the bioluminescence induced in the marine fish, *Porichthys notatus*, by *Cypridina* (ostracod) luciferin. *Molec. Cell. Biochem. 9*, 3-8.
61. Johnson, F. H. (1967). Bioluminescence. *Comprehensive Biochem. 27*, 79-136.
62. Hastings, J. W. and Morin, J. G. (1968). Calcium activated bioluminescent proteins from ctenophores *(Mnemiopsis)* and colonial hydroids *(Obelia)*. *Biol. Bull. 135*, 422.
63. Hastings, J. W. and Morin, J. G. (1969). Comparative biochemistry of calcium-activated photoproteins from the ctenophore, *Mnemiopsis* and the coelenterates *Aequorea, Obelia, Pelagia,* and *Renilla*. *Biol. Bull. 137*, 402.
64. Hastings, J. W. and Morin, J. G. (1969). Calcium-triggered light emission in *Renilla*. A unitary biochemical scheme for coelenterate bioluminescence. *Biochem. Biophys. Res. Commun. 37*, 493-498.
65. Morin, J. G. and Hastings, J. W. (1971). Biochemistry of bioluminescence of colonial hydroids and other coelenterates. *J. Cell. Physiol. 77*, 305-311.
66. Morin, J. G. and Hastings, J. W. (1971). Energy transfer in a bioluminescent system. *J. Cell. Physiol. 77*, 313-318.
67. Wampler, J. E., Hori, K., Lee, J. W., and Cormier, M. J. (1971). Structured bioluminescence. Two emitters during both the *in vitro* and the *in vivo* bioluminescence of the sea pansy, *Renilla*. *Biochemistry 10*, 2903-2909.
68. Wampler, J. E., Karkhanis, Y. D., Morin, J. G., and Cormier, M. J. (1973). Similarities in the bioluminescence from the *Pennatulacea*. *Biochim. Biophys. Acta 314*, 104-109.
69. Cormier, M. J., Hori, K., Karkhanis, Y. D., Anderson, J. M., Wampler, J. E., Morin, J. G., and Hastings, J. W. (1973). Evidence for similar biochemical requirements for bioluminescence among the coelenterates. *J. Cell. Physiol. 81*, 291-297.
70. Cormier, M. J., Hori, K., and Anderson, J. M. (1974). Bioluminescence in coelenterates. *Biochim. Biophys. Acta 346*, 137-164.
71. Morise, H., Shimomura, O., Johnson, F. H., and Winant, J.

(1974). Intermolecular energy transfer in the bioluminescent system. *Biochemistry 13,* 2656-2662.

72. Ward, W. W. and Cormier, M. J. (1976). *In vitro* energy transfer in *Renilla* Bioluminescence. *J. Phys. Chem. 80,* 2289-2291.
73. Ward, W. W. and Cormier, M. J. (1978). Energy transfer via protein-protein interaction in *Renilla* bioluminescence *Photochem. Photobiol. 27,* 389-396.
74. Ward, W. W. and Cormier, M. J. (1979). An energy transfer protein in coelenterate bioluminescence. *J. Biol. Chem. 254,* 781-788.
75. Seliger, H. H. and Morton, R. A. (1968). A physical approach to bioluminescence. *In* "Photophysiology" (A. C. Giese, ed.), pp. 253-314. Academic Press, New York.
76. Goto, T., Inoue, S., and Sugiura, S. (1968). *Cypridina* bioluminescence. IV. Synthesis and chemiluminescence of 3,7-dihydroimidazo (1,2-α)pyrazin-3-one and its 2-methyl derivative. *Tetrahedron Lett.* No. 36, pp. 3873-3876.
77. Goto, T., Inoue, S., Sugiura, S., Nishikawa, K., Isobe, M. and Abe, Y. (1968). *Cypridina* bioluminescence. V. Structure of emitting species in the luminescence of *Cypridina* luciferin and its related compounds. *Tetrahedron Lett.* No. 37, pp. 4035-4038.
78. Roos, W., Nanjundiah, V., Malchow, D., and Gerisch, G. (1975). Amplification of cyclic-AMP signals in aggregating cells of *Dictyostelium discoideum. FEBS Lett. 53,* 139-142.
79. Roos, W. and Gerisch, G. (1976). Receptor-mediated adenylate cyclase activation in *Dictyostelium discoideum. FEBS Lett. 68,* 170-172.
80. Nicol, J. A. C. (1969). Bioluminescence. *In* "Fish Physiology," (W. S. Hoar and D. J. Randall, eds.), Vol. 3, pp. 355-400. Academic Press, New York.
81. Herring, P. J. (1977). Bioluminescence of marine organisms. *Nature (London) 267,* 788-793.
82. Buck, J. B. (1978). Functions and evolutions of bioluminescence. *In* "Bioluminescence in Action" (P. J. Herring, ed.), pp. 419-460. Academic Press, New York.
83. Rauther, M. (1927). Echte Fische, Teil 1. Anatomie, Physiologie und Entwicklungsgeschichte. *In* "Klassen und Ordnungen des Tier-Reichs," (H. G. Bronn, ed.), Vol. 6, pp. 1-1050. Akademische Verlagsgesellschaft, Leipzig.
84. Jerzmańska, A. (1960). The structure and the biological significance of light organs in Teleostei. *Przegl. Zool. 4,* 112-118.
85. Fraser, J. (1962). "Nature Adrift. The Story of Marine Plankton," 178 pp. G. T. Foulis, London.
86. Clarke, W. D. (1963). Function of bioluminescence in mesopelagic organisms. *Nature (London) 198,* 1244-1246.

87. McAllister, D. E. (1967). The significance of ventral bioluminescence in fishes. *J. Fish. Res. Bd. Can. 24,* 537-554.
88. Badcock, J. (1969). Colour variations in two mesopelagic fishes and its correlation with ambient light conditions. *Nature (London) 221,* 383-385.
89. Denton, E. J. (1970). On the organization of reflecting surfaces in some marine animals. *Phil. Trans. Roy. Soc. B258,* 285-313.
90. Hastings, J. W. (1971). Light to hide by: ventral luminescence to camouflage the silhouette. *Science 173,* 1016-1017.
91. Denton, E. J. (1971). Reflectors in fishes. *Sci. Amer. 224,* 64-72.
92. Denton, E. J., Gilpin-Brown, J. B., and Wright, P. G. (1972). The angular distribution of the light produced by some mesopelagic fish in relation to their camouflage. *Proc. Roy. Soc. B182,* 145-158.
93. Lawry, J. V. Jr. (1974). Lantern fish compare downwelling light and bioluminescence. *Nature (London) 247,* 155-157.
94. Case, J. F., Warner, J., Barnes, A. T., and Lowenstine, M. (1977). Bioluminescence of lantern fish (Myctophidae) in response to changes in light intensity. *Nature (London) 265,* 179-181.
95. Denton, E. J. and Herring, P. J. (1978). On the filters in the ventral photophores of mesopelagic animals. *J. Physiol. (London) 284,* 42P.
96. Dahlgren, U. (1916). Production of light by animals. *J. Franklin Inst. 181,* 525-556.
97. Young, R. E. (1973). Information feedback from photophores and ventral countershading in midwater squid. *Pac. Sci. 27,* 1-7.
98. Young, R. E. and Roper, C. F. E. (1976). Bioluminescent countershading in midwater animals: evidence from living squid. *Science 191,* 1046-1048.
99. Young, R. E. and Roper, C. F. E. (1977). Intensity regulation of bioluminescence during countershading in living midwater animals. *Fish. Bull. U.S. 75,* 239-252.
100. Young, R. E. (1978). Ventral distribution and photosensitive vesicles of pelagic cephalopods from Hawaiian waters. *Fish. Bull. U.S. 76,* 583-615.
101. Young, R. E., Roper, C. F. E., and Walters, J. F. (1979). Eyes and extraocular photoreceptors in midwater cephalopods and fishes: their roles in detecting downwelling light for counterillumination. *Mar. Biol. 51,* 371-380.
102. Warner, J. A., Latz, M. I., and Case, J. F. (1979). Cryptic bioluminescence in a midwater shrimp. *Science 203,* 1109-1110.
103. Nicol, J. A. C. (1967). The luminescence of fishes.

Symp. Zool. Soc. London No. 19, pp. 27-55.
104. Herring, P. J. (1977). Oral light organs in *Sternoptyx*, with some observations on the bioluminescence of hatchet-fishes. *In* "A Voyage of Discovery: George Deacon 70th Anniversary Volume" (M. Angel, ed.), pp. 553-567. Pergamon, Oxford.
105. Bertelsen, E. and Munk, O. (1964). Rectal light organs in the argentinoid fishes *Opisthoproctus* and *Winteria*. Dana-Report No. 62, 17 pp. Carlsberg Foundation, Copenhagen.
106. Herring, P. J. (1975). Bacterial bioluminescence in some argentinoid fishes. *Proc. 9th Europ. Mar. Biol. Symp.*, pp. 563-572. Aberdeen University Press, Aberdeen.
107. Barham, E. G. (1966). Deep scattering layer migration and composition: observations from a diving saucer. *Science 151*, 1399-1403.
108. Backus, R. H., Craddock, J. E., Haedrich, R. L., Shores, D. L., Teal, J. M., Wing, A. S., Mead, G. W., and Clarke, W. D. (1968). *Ceratoscopelus maderensis*: peculiar sound scattering layer identified with this myctophid fish. *Science 160*, 991-993.
109. Crane, J. M. Jr. (1965). Bioluminescent courtship display in the teleost *Porichthys notatus*. *Copeia*, pp. 239-241.
110. Lane, E. D. (1967). A study of the Atlantic midshipman, *Porichthys porosissimus*, in the vicinity of Port Aransas, Texas. *Contrib. Mar. Sci. 12*, 1-53.
111. Springer, V. G. (1957). Mysterious midshipman ... little known curiosity of the sea. *Tex. Game Fish*, pp. 6-7.
112. Bolin, R. L. (1961). The function of the luminous organs of deep-sea fishes. *Proc. 9th Pac. Sci. Congr. 1957 10*, 37-39.
113. Burkenroad, M. D. (1943). A possible function of bioluminescence. *J. Mar. Res. 5*, 161-164.
114. O'Day, W. T. (1973). Luminescent silhouetting in stomiatoid fishes. *Contr. Sci.* No. 246, 8 pp.
115. Waterman, T. H. (1948). Studies of deep-sea angler-fishes (Ceratioidea). III. The comparative anatomy of *Gigantactis longicirra* Waterman. *J. Morphol. 82*, 81-149.
116. Bertelsen, E. (1951). The ceratioid fishes. Dana-Report No. 39, 276 pp. Carlsberg Foundation, Copenhagen.
117. Wilson, D. P. (1937). The habits of the angler-fish, *Lophius piscatorius L.*, in Plymouth aquarium. *J. Mar. Biol. Assoc. U.K. 21*, 477-496.
118. Gudger, E. W. (1945). The angler-fishes, *Lophius piscatorius et Americanus*, use the lure in fishing. *Amer. Natur. 79*, 542-548.
119. Nicol, J. A. C. (1978). Bioluminescence and vision. *In* "Bioluminescence in Action" (P. J. Herring, ed.), pp.

367-398. Academic Press, New York.

120. Denton, E. J., Gilpin-Brown, J. B., and Wright, P. G. (1970). On the "filters" in the photophores of mesopelagic fish and on a fish emitting red light and especially sensitive to red light. *J. Physiol. (London) 208,* 72P-73P.

121. O'Day, W. T. and Fernandez, H. R. (1974). *Aristostomias scintillans* (Malacosteidae): a deep-sea fish with visual pigments apparently adapted to its own bioluminescence. *Vision Res. 14,* 545-550.

122. Girsch, S. J. (1976). Interfacing of vision and bioluminescence in the flashlight fish, *Photoblepharon palpebratus. Abstr. 4th Annu. Meet. Amer. Soc. Photobiol., Denver, Colorado, Feb., 1976,* p. 41.

123. Fernandez, H. R. and Tsuji, F. I. (1976). Photopigment and spectral sensitivity in the bioluminescent fish *Porichthys notatus. Mar. Biol. 34,* 101-107.

124. McElroy, W. D. and Seliger, H. H. (1962). Origin and evolution of bioluminescence. *In* "Horizons in Biochemistry" (M. Kasha and B. Pullman, eds.), pp. 91-101. Academic Press, New York.

125. Seliger, H. H. (1975). The origin of bioluminescence. *Photochem. Photobiol. 21,* 355-361.

126. Hastings, J. W., Riley, W. H., and Massa, J. (1965). The purification, properties, and chemiluminescent quantum yield of bacterial luciferase. *J. Biol. Chem. 240,* 1473-1481.

127. Kuwabara, S., Cormier, M. J., Dure, L. S., Kreiss, P., and Pfuderer, P. (1965). Crystalline bacterial luciferase from *Photobacterium fischeri. Proc. Natl. Acad. Sci. U.S. 53,* 822-828.

128. Booth, C. R. and Nealson, K. H. (1975). Light emission by luminous bacteria in the open ocean. *Biophys. J. 15,* 56a.

129. Curds, C. R. and Bazin, M. J. (1977). Protozoan predation in batch and continuous culture. *In* "Advances in Aquatic Microbiology" (M. R. Droop and H. W. Jannasch, eds.), pp. 115-177. Academic Press, New York.

130. Rieper, M. (1978). Bacteria as food for marine harpacticoid copepods. *Mar. Biol. 45,* 337-345.

REGULATION OF YEAST HEXOKINASE ACTIVITY

Sidney P. Colowick
Frances C. Womack

Department of Microbiology
Vanderbilt University School of Medicine
Nashville, Tennessee

In eukaryotic cells, the first step in glucose utilization is its transport into the cell by a specific carrier system. The intracellular glucose is then converted to glucose 6-phosphate by the hexokinase reaction:

$$Mg \cdot ATP + glucose \rightarrow glucose\ 6\text{-}phosphate + Mg \cdot ADP$$

It has been known since the classical work of Crane and Sols (1) in 1954 that the hexokinase in mammalian cells is strongly inhibited by the reaction product, glucose 6-phosphate. They presented evidence that the inhibition is due to the action of glucose 6-phosphate at a site on the enzyme different from the catalytic site. This was probably the first clear statement of the concept of allosteric regulation of enzyme activity.

It is now generally believed that this inhibition is an important factor in the regulation of glucose utilization by mammalian cells. The concentration of glucose 6-phosphate is in turn regulated by its rate of removal via conversion to

 ISBN 0-12-397580-8

fructose 6-phosphate and then to fructose 1,6-diphosphate by the phosphofructokinase reaction:

Mg·ATP + fructose 6-phosphate → fructose-1,6-diphosphate + Mg·ADP

The experiments of Lowry and Passoneau (2) showed that the Pasteur Effect in the brain, i.e., the suppression of glucose utilization in the presence of oxygen, was entirely accounted for by changes in concentrations of metabolites which are allosteric regulators of phosphofructokinase activity. Thus, oxygen, by altering metabolite concentrations in such a way as to suppress phosphofructokinase activity, would suppress glycolysis. However, in order to explain a diminished utilization of glucose, one must also assume that the suppression of phosphofructokinase leads to an accumulation of glucose 6-phosphate which in turn inhibits hexokinase activity.

In the case of yeast, this explanation for the Pasteur Effect is not tenable, because the hexokinase activity of yeast does not show the marked sensitivity to glucose 6-phosphate exhibited by the mammalian enzyme. We will describe here some regulatory effects of metabolites other than glucose 6-phosphate on one of the yeast hexokinase isozymes. Although these observations do not provide an explanation for the Pasteur Effect or glucose phosphorylation in yeast, they are presumably of physio logical significance and, in addition, raise some interesting questions about the molecular mechanism of the effects observed.

The first observation of regulatory effects on yeast hexokinase was made by Kosow and Rose (3), who showed that one of the isoenzymes (form P-II) could be activated by a variety of metabolites, including citrate, malate, and 3-phosphoglycerate, provided that the tests were carried out at suboptimal pH values of 7 or below, instead of at the optimum of pH 8. When we tried to repeat their observations, we found that we could only do so by using the same buffer which they used, i.e., one of the sulfonated buffers known as MES or TES. When we used imidazole or Tris buffer, we found that the activity was not suppressed by lowering the pH to 7, so that no activation was seen with citrate, etc.

We then found (4, 5) that if we added low concentrations of adenosinediphosphate to the imidazole or Tris-buffered system, we could obtain the same inhibited state at pH 7 which Kosow and Rose had observed in MES or TES buffer. This is illustrated in Fig. 1. The ADP-inhibited state shows biphasic kinetics ("negative cooperativity") with respect to ATP concentration. The inhibition is also seen with GDP but not with pyrimidine nucleoside diphosphates. When the concentration of ADP or GDP is varied, the maximum inhibition obtainable is around 70%, and half-maximal inhibition is obtained with about 0.1 m*M* of either ADP or GDP (Fig. 2).

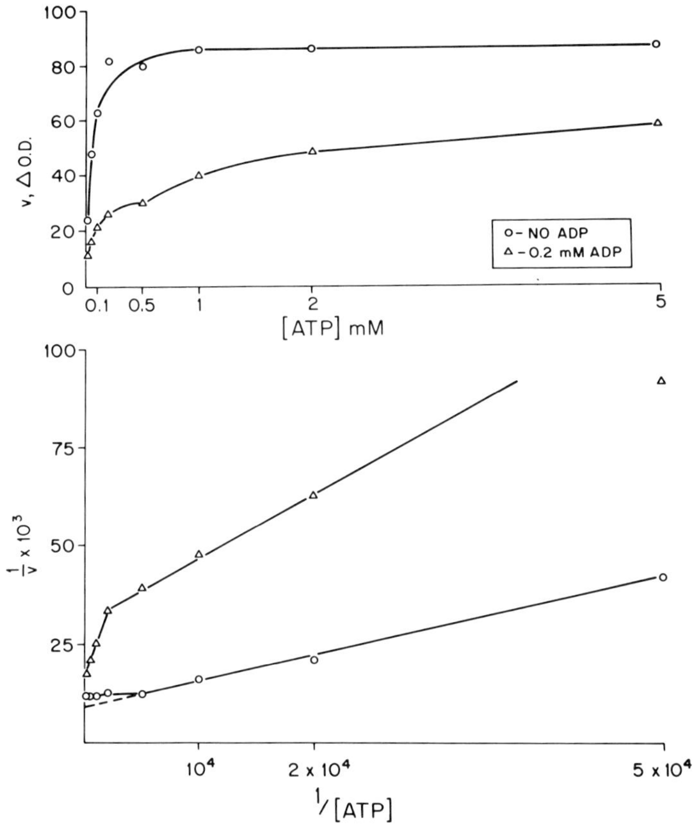

Fig. 1. Inhibition of hexokinase S-II by ADP at varying ATP concentration. Buffer is 50 mM Tris, pH 7. From (5). Hexokinase S-II is a protease-modified form of P-II.

When the pH is varied, one finds that the inhibition by GDP is absolutely pH dependent: there is no inhibition at pH 8, but marked inhibition at pH 7 or below (Fig. 3). There is a similar pH dependence for ADP inhibition, but in this case there is some inhibition even at pH 8 (Fig. 3); this is presumably related to the fact that ATP is a good substrate for hexokinase, while GTP is practically nonreactive, so that ADP inhibition at pH 8 is presumably attributable to back-reaction rather than regulation.

Figure 3 also shows that the pH-dependent inhibition by purine nucleoside diphosphates is completely reversed by the presence of an activator such as citrate, in confirmation of Kosow and Rose (3). Figure 4 shows concentration curves for the reversal of ADP inhibition by various activators. In ad-

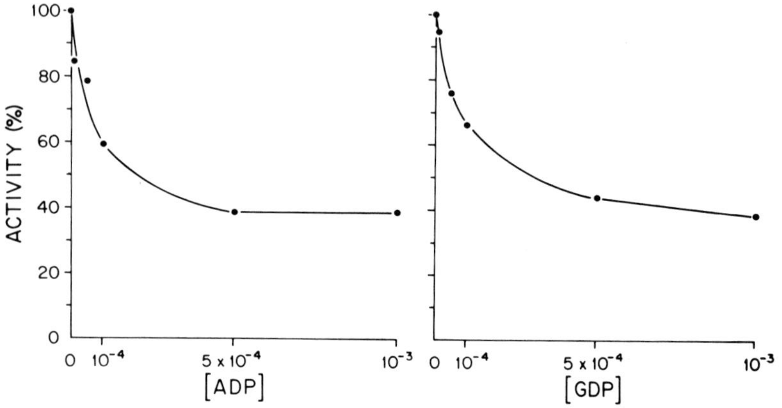

Fig. 2. Effect of ADP and GDP concentration on hexokinase P-II activity. Buffer is 50 mM imidazole, pH 6.9; ATP, 0.5 mM. From (5).

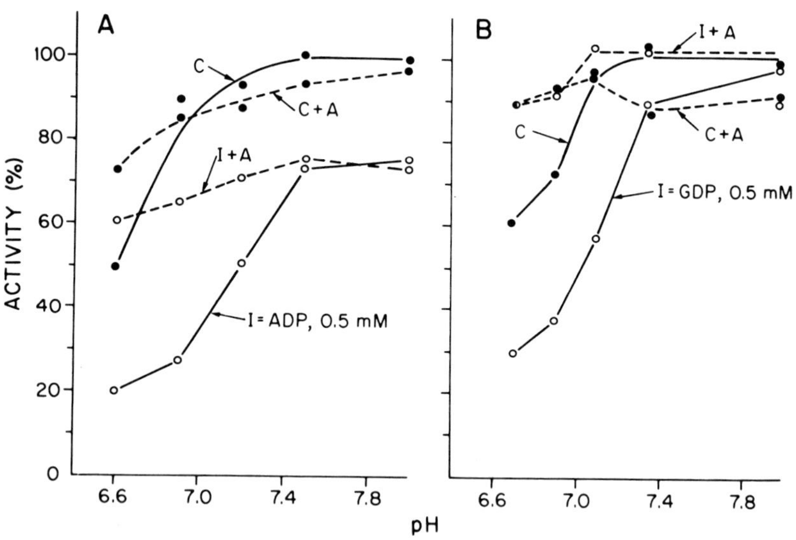

Fig. 3. The pH-dependent regulation of hexokinase P-II. Buffer is 50 mM imidazole. ATP, 0.5 mM; C, control; A, activator (citrate, 0.5 mM); I, inhibitor (ADP or GDP, 0.5 mM). From (5).

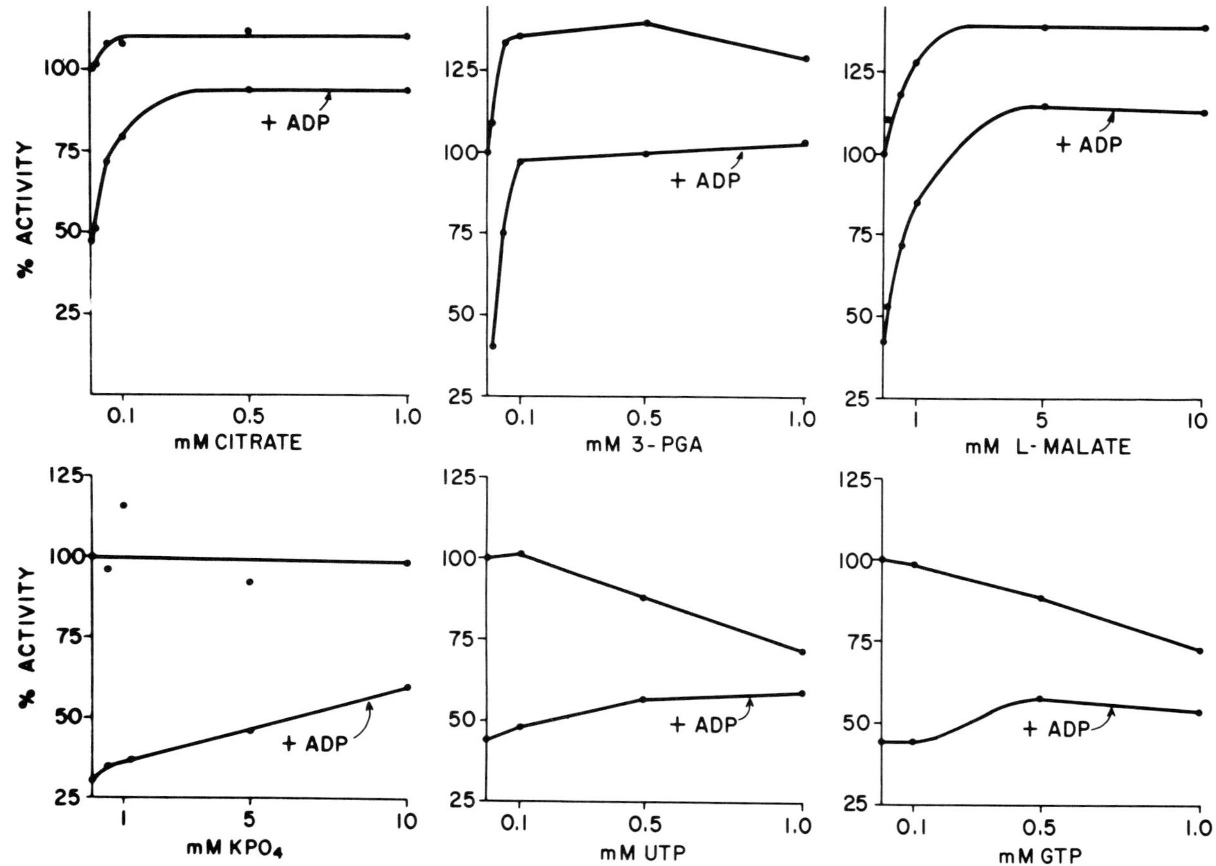

Fig. 4. Effect of activator concentration on reversal of ADP inhibition of hexokinase P-II. pH 6.9. From (5).

dition to the Kosow-Rose activators, we have found that catecholamines at 1 mM concentration can reverse ADP inhibition. Catecholamines had been reported earlier to activate yeast hexokinase at low pH (6).

Figure 5 shows that the effects observed are much less pronounced when the isoenzyme P-I is used. This suggests that the isoenzymes P-I and P-II may serve different functions in the cell. Kosow and Rose (3) have suggested that citrate, which is a known inhibitor of phosphofructokinase, would, by activating hexokinase P-II, promote a pathway for storing glucose as glycogen when glycolytic energy is not required. If the same concept is extended, ADP would function to block this glycogen storage pathway while stimulating glycolysis and respiration.

It would be of interest to determine whether mutants lacking isoenzyme P-II are defective with respect to glycogen storage. However, it is unlikely that the different hexokinases serve strictly delimited functions in the cells, since Maitra and Lobo (7) have constructed mutants of *Saccharomyces* which have only P-I, or only P-II, or only glucokinase. All of these mutants are capable of growth on glucose, and all exhibit a Pasteur Effect. Thus, it seems clear that regulation of a particular hexokinase is not involved in the Pasteur Effect. More likely, control is exerted at the level of membrane transport of glucose.

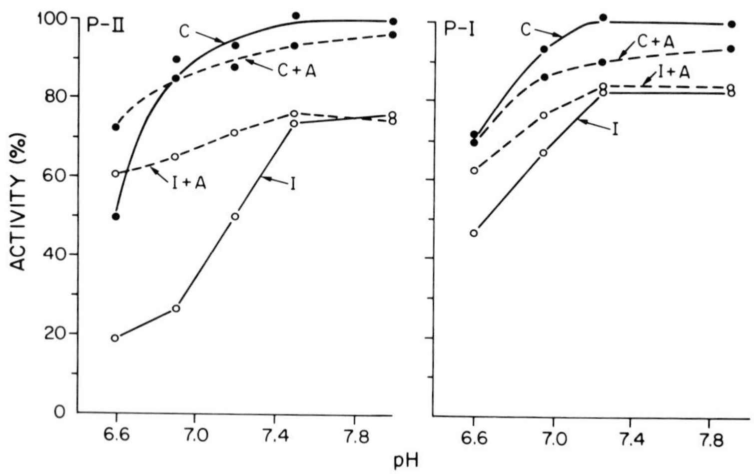

Fig. 5. Comparison of regulatory effects on P-I and P-II. Imidazole buffer; C, control; A, activator (citrate 0.5 mM); I, inhibitor (ADP, 0.5 mM). From (5).

Finally, a few words about the molecular basis of the regulation of P-II activity by nucleoside diphosphates. The fact that the inhibited state is characterized by negative cooperativity with respect to ATP concentration should not lead to the conclusion in this case that interaction between subunits is involved. We find that the same phenomena of inhibition and activation are observed when we use native P-II or a trypsin-modified form (S-II) which exists predominantly as the monomer (5). We feel therefore that the shift between a low-activity and a high-activity state does not involve a shift between monomeric and dimeric forms. Rather, we agree with Neet and his co-workers (8) that what is involved here is a shift between active and inactive states of a monomer, and that the cooperativity observed is actually a "temporal cooperativity" resulting from the slowness of interconversion between the two states. Neet's group has been able to demonstrate this slow transition directly at low pH, where an initial burst of activity is followed by a slower steady state rate.

We present in Fig. 6 a scheme modified from that of Kosow and Rose (3) to illustrate our view of the changes in the conformation and activity of the monomer. Protonation of the active form can occur without change in activity (e.g., when pH is changed from 8 to 7). This active protonated form can be converted to a low-activity form either by combining with a nucleoside diphosphate or with a second proton, or it can be stabilized in the high activity state by combining with an activator.

A major question which remains unresolved is whether the binding of inhibitory nucleoside diphosphates or of activators

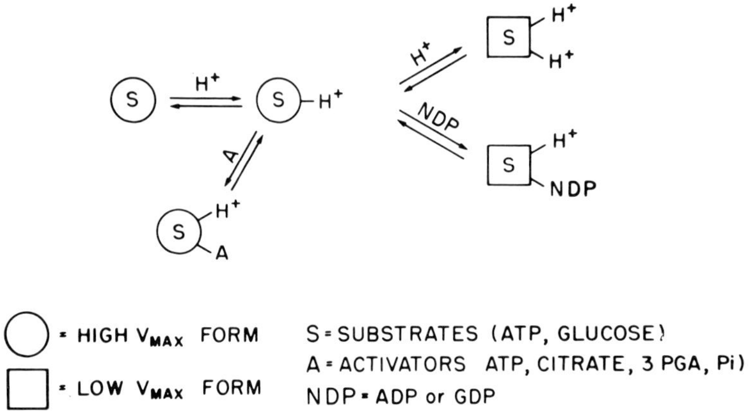

Fig. 6. Regulation of hexokinase P-II, a modification of Kosow and Rose's scheme (3). From (5).

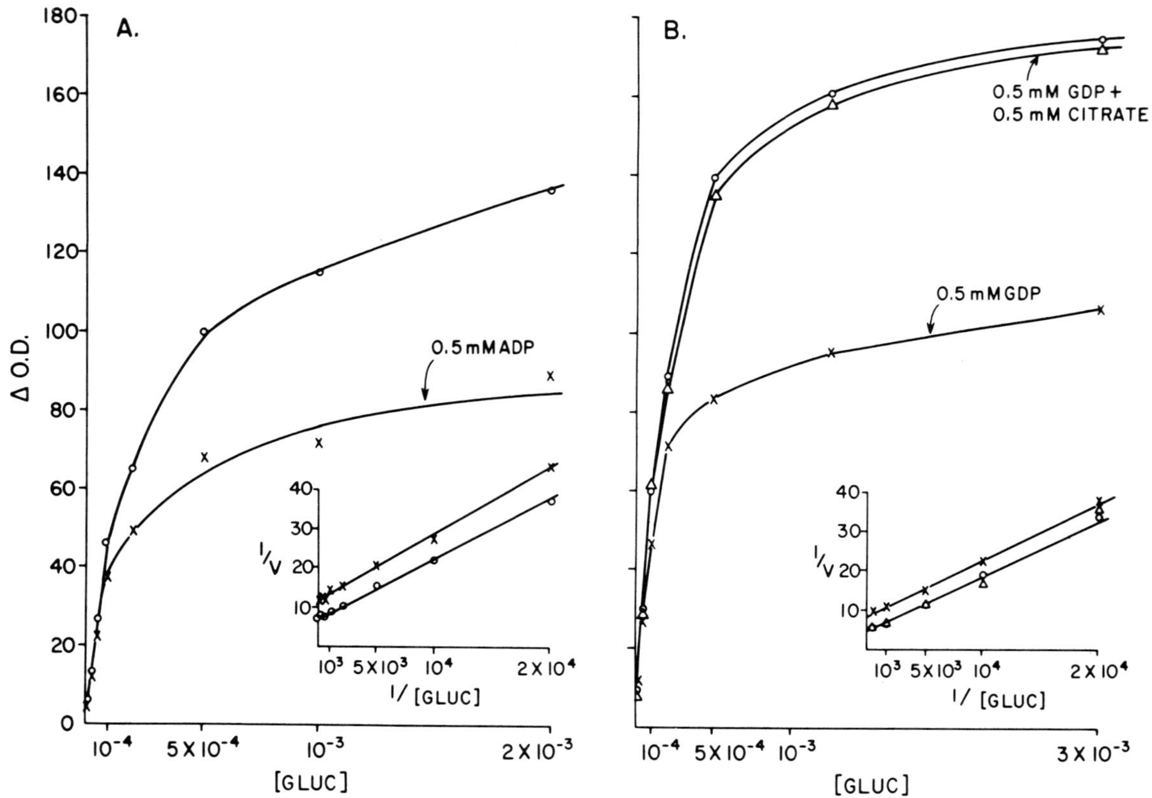

Fig. 7. Inhibition of hexokinase P-II by ADP or GDP at varying glucose concentration. Buffer pH 6.9. ATP, ADP, and GDP, all 0.5 mM. From (5).

such as citrate occurs at the catalytic site or at a separate regulatory site. In our measurements of nucleoside diphosphate binding, we have found only 1 mol of ADP bound per mole of monomer (9) and no GDP binding (5); the ADP binding was dependent on formation of a ternary complex with a glucose analog (D-lyxose), and presumably occurred at the active site. Although kinetic studies show that inhibition by ADP or GDP is promoted by high glucose concentrations (Fig. 7) there is no evidence as yet for sugar-dependent binding of ADP or GDP at a regulatory site, or is there any evidence for citrate binding to hexokinase. The Steitz laboratory, which has done excellent X-ray crystallographic studies of hexokinase P-II, finds an extra nucleotide binding site in a space between the two subunits of the dimer (10). Perhaps this represents the regulatory site for nucleoside diphosphates. If so, it is necessary to postulate that binding of nucleoside diphosphates can also occur at the inter-subunit site even when the site is disrupted by dissociating into monomers.

REFERENCES

1. Crane, R. K. and Sols, A. (1954). *J. Biol. Chem. 210,* 597.
2. Passonneau, J. V. and Lowry, O. H. (1962). *Biochem. Biophys. Res. Commun. 7,* 10.
3. Kosow, D. P. and Rose, I. A. (1971). *J. Biol. Chem. 246,* 2618-2625.
4. Womack, F. C. and Colowick, S. P. (1971). *Fed. Proc. 30,* 1059.
5. Womack, F. C. and Colowick, S. P. (1978). *Arch. Biochem. Biophys. 191,* 748-755.
6. Harrison, W. H. and Gray, R. M. (1972). *Arch. Biochem. Biophys. 151,* 357-360.
7. Lobo, Z. and Maitra, P. K. (1977). *Arch. Biochem. Biophys. 182,* 639-645.
8. Shill, J. P. and Neet, K. E. (1974). *Biochemistry 13,* 3864-3871.
9. Colowick, S. P., Womack, F. C., and Nielsen, J. (1969). *In* "The Role of Nucleotides for the Function and Conformation of Enzymes" (H. M. Kalckar, ed.), pp. 15-43. Munksgaard, Copenhagen.
10. Steitz, T. A., Anderson, W. F., Fletterick, R. J., and Anderson, C. M. (1977). *J. Biol. Chem. 252,* 4494-4500.

Note Added in Proof

Since this paper was presented, we have found that the apparent regulatory effects described here are entirely accounted for by the presence of traces of aluminum in the nucleotides used (ATP, ADP, GDP). A pH-dependent inhibition by Al·ATP ($K_i = 10^{-7}M$) is reversed by citrate, 3-phosphoglycerate, etc., by virtue of their ability to chelate aluminum. For further information, see:

Womack, F. C. and Colowick, S. P. (1979). *Proc. Natl. Acad. Sci. U.S. 76*, 5080-5084.

Viola, R. E., Morrison, J. F., and Cleland, W. W. (1980). *Biochemistry 19*, 3131-3137.

Neet, K. E., Furman, T. C., and Hueston, W. J. (1982). *Arch. Biochem. Biophys. 213*, in press.

DEACYLATION OF CELLULAR LIPIDS: AN ASSAY FOR TUMOR PROMOTERS

Lawrence Levine

Department of Biochemistry
Brandeis University
Waltham, Massachusetts 02254

INTRODUCTION

Unsaturated fatty acids do not exist free in cells. They are found in the form of phosphoglycerides and triglycerides and must be deacylated to provide substrate for the lipoxygenases and cyclooxygenases (Fig. 1). It is the membrane phospholipids that are the richest source of these precursor polyunsaturated fatty cells, and it has been postulated that the phospholipases of the cell are part of the sequence of events involved in prostaglandin biosynthesis (1 - 3).

The arachidonic acid metabolic pathway is shown in Fig. 2. In this scheme arachidonic acid, which is the unsaturated fatty acid found in highest concentration in membrane phospholipids, is shown as the fatty acid undergoing metabolism. The endoperoxides (PGG_2 and PGH_2) are the immediate products of the cyclooxygenase (4, 5); they then undergo enzymatic and nonenzymatic transformations to form the thromboxanes (TXA_2 and TXB_2) (6, 7), the prostacyclins (PGI_2) (8), and the prostaglan-

 ISBN 0-12-397580-8

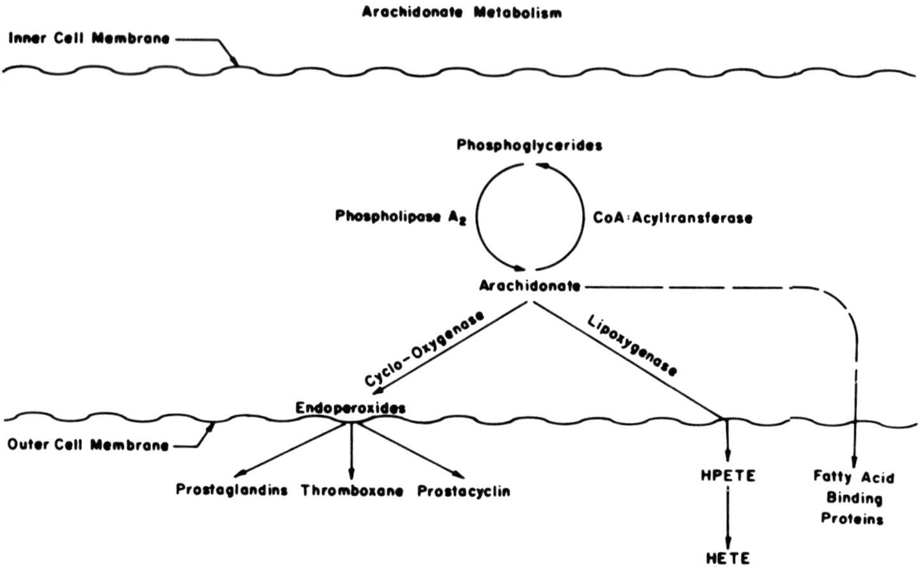

Fig. 1. Schematic representation of the components and characteristics of arachidonic acid utilization.

dins (PGE_2, $PGF_{2\alpha}$, and PGD_2). The products formed from the endoperoxides, as well as the endoperoxides themselves, are potent pharmacologically and their activities, qualitatively and quantitatively, depend on the nature of the enzymatic (and nonenzymatic) transformations. Arachidonic acid also can be converted by a lipoxygenase to 12-hydroperoxy arachidonate and 12-L-hydroxy arachidonate (9).

METHODS

The arachidonic acid metabolic products have been identified and quantified by radioimmunoassay (10). Radioimmunoassay for prostaglandins offers distinct advantages over other analytical procedures (11). Chromatographic methods including thin-layer and gas-liquid chromatography are among the most widely used analytical tools to study the metabolism of the prostaglandins, especially when labeled compounds are available to aid in the identification and quantitation of the parent compound and its metabolites. Gas - liquid chromatography is particularly effective when combined with mass spectrometry (12). With such systems, sensitivity down to the picomole levels is not uncommon. These chromatographic methods normally

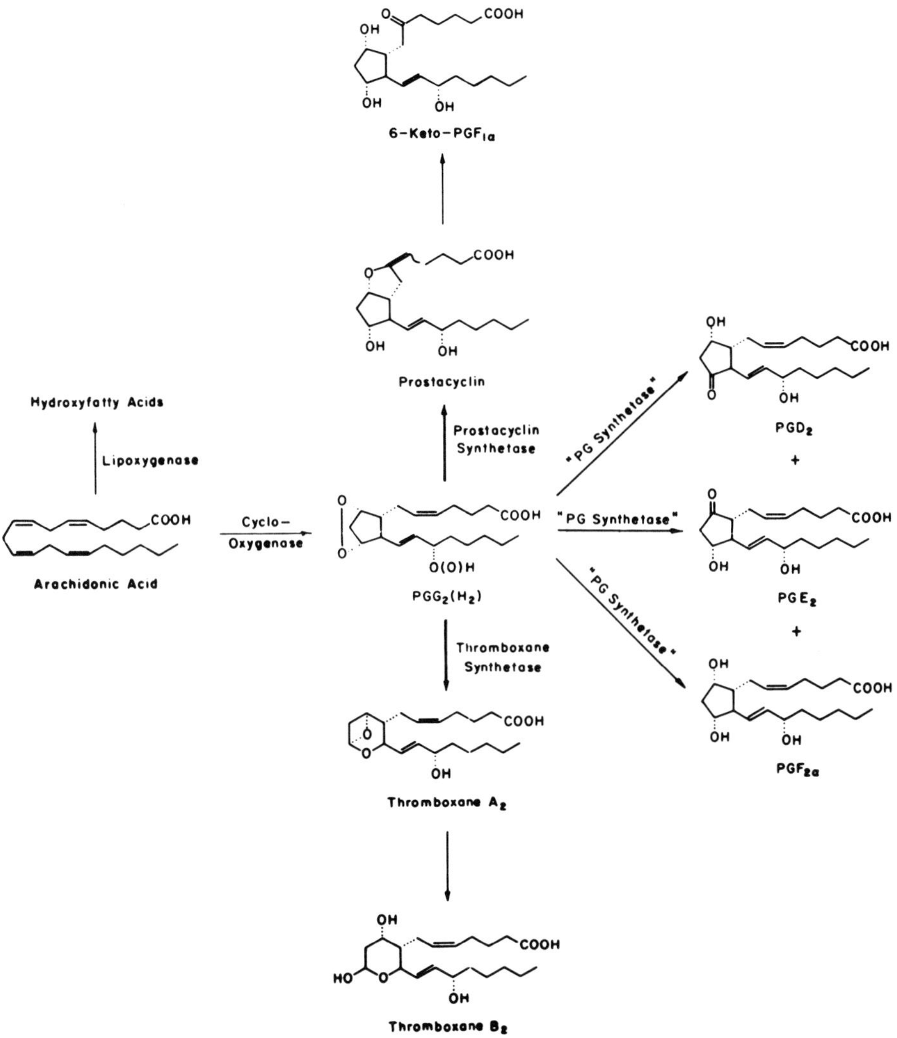

Fig. 2. Some biochemical pathways of arachidonic acid metabolism.

require that samples be extracted and separated from substances that might interfere with the assays. The sensitivity of radioimmunoassays equals or surpasses that of the available chromatographic and spectral methods. The added dimension of antigen-antibody specificity with most prostaglandin immune systems allows radioimmunoassay to be used for the direct analysis of compounds in culture fluids.

In our experiments, a dog kidney cell line (MDCK) was used. Exponentially growing cells were treated with trypsin and seeded at 2×10^5 cells/60-mm Falcon tissue culture dish in 4 ml of Eagle's minimal essential medium containing 2 m*M* l-glutamate and supplemented with 10% (v/v) fetal bovine serum and penicillin-streptomycin and incubated for 24 hr. The cells were washed twice with 2 ml of the medium lacking fetal bovine serum and incubated for the desired time with and without the reagents used in the particular experiment. Aliquots of the medium were used for the serologic measurements of prostaglandins.

So that many culture fluids can be measured for arachidonic acid metabolic products, we have produced antibodies directed toward PGE_2, $PGF_{2\alpha}$, $PGF_{2\beta}$, 13,14-dihydro-PGE_2, 15-keto-$PGF_{2\alpha}$, 13,14-dihydro-15-keto-$PGF_{2\alpha}$, 13,14-dihydro-15-keto-PGE_2, 6-keto-$PGF_{1\alpha}$, thromboxane B_2, 5,6-dihydro-PGI_1, and endoperoxide analogs. While some of these antibodies can be used to measure cyclooxygenase products in biological fluids without prior extraction and separation, some cannot. Therefore, to obtain accurate specific activities of these metabolites, we have developed procedures for analyses of many of these compounds by a combination of high pressure liquid chromatography and radioimmunoassay (13). In addition, stimulation of deacylation of cellular lipids and prostaglandin production can be demonstrated by release of radioactive compounds from the cells labeled with [^{3}H]arachidonic acid. The distribution of [^{3}H]-arachidonic acid among the lipids of the cells depends on the length of labeling. Incorporation of [^{3}H]arachidonic acid into phosphatidylcholine is rapid and reaches a steady state in 5 hr. Incorporation of radioactivity into phosphatidylethanolamine and phosphatidylserine is slower but increases up to 20 hr. Radioactivity, found in triglycerides, reaches a maximum in 5 hr and decreases afterward. After 2 hr, phosphatidylcholine is the major radioactive component (60%) followed by triglycerides (25%). Only low levels of radioactivity are found in phosphatidylethanolamine (10%). After longer periods of labeling (20 hr), a relatively high level of radioactivity is found in the phosphatidylethanolamine fraction (33%), and the levels of label in the triglyceride fraction have decreased (8%) (14).

RESULTS AND DISCUSSION

Various tumors contain more prostaglandins than normal tissue (15). In mice bearing the prostaglandin-producing $HSDM_1$ fibrosarcoma, plasma concentrations of 13,14-dihydro-15-keto-PGE_2 are elevated and the magnitude of the rise is greater than that of PGE_2. When hydrocortisone, which inhibits synthesis of PGE_2 by $HSDM_1$ cells in culture, is administered to tumor-bearing mice, the steroid hormone prevents the rises in plasma PGE_2 metabolite (16). In rabbits bearing the prostaglandin-producing VX_2 carcinoma, the plasma concentrations of 13,14-dihydro-15-keto-PGE_2 are elevated within 1 week after tumor implantation (17, 18). Some studies of cells growing in tissue culture suggest that increased prostaglandin production may be associated with neoplasia. Polyoma virus transformation of baby hamster kidney cells leads to a large increase in PGE_2 production by the transformed lines (19). Cell lines transformed from 3T3 spontaneously, by radiation, or by treatment with chemical carcinogens, polyoma- and SV40 virus, produce up to 5 times more prostaglandins than their parent line (20). On the other hand, there are virus and chemically transformed cell lines that produce little or no prostaglandins (20).

The cocarcinogenic role of environmental factors in human cancer is well recognized. Cocarcinogenesis as used here is a general term referring to the function of any agent or treatment that may modify carcinogenesis to affect tumor yield. However, when referring to the phorbol diesters, the term promoters may be used. Carcinogenesis may be a multistep process (21) and it is probable that various environmental factors act at different steps in this process. One of the best-studied models is the so-called "2-stage carcinogenesis" system in mouse skin (22). In this system, two distinct stages in the carcinogenic process have been identified, the first designated "initiation" and the second "promotion."

The essential features of this model are: (1) one application of the initiator alone to mouse skin will produce few or no tumors; (2) many applications of the promoting substance alone will produce inflammation followed by epithelial hyperplasia, but will not produce tumors; and (3) one application of the initiator followed by many applications of the promoter results in both benign and malignant tumors. The "promoting stage" has been extended into at least two separate stages.

In formulating the "simple" two-stage model in mouse skin (22) the promoting agent used was oil (croton oil) from the seeds of *Croton tiglium*. The promoting fractions from the croton oil have been purified and characterized by Hecker (23) and Van Duuren (24) and their colleagues. The most extensively studied promoters are the tumor-promoting phorbol diesters purified from croton oil; the most biologically active of which is the 12-*O*-tetradecanoyl phorbol-13-acetate (TPA).

Two tumor-promoting phorbol-12,13-diesters, 12-*O*-tetradecanoyl phorbol-13-acetate (TPA) and phorbol-12,13-didecanoate (PDD), at concentrations of 10^{-9} to 10^{-10} *M*, stimulate prostaglandin production by dog kidney (MDCK) cells (Table I). The

TABLE I. Effect of Phorbol-12,13-Diesters on Prostaglandin Production by MDCK Cells[a]

Effector	ng/ml	PGE_2[c]	$PGF_{2\alpha}$[c]
No addition	--	3.2	10.5
TPA	1.0[b]	115.0	131.0
	0.4	35.6	58.8
	0.1	9.8	18.5
	0.08	6.2	12.4
	0.02	4.1	9.0
	0.01	3.3	11.0
PDD	1.0[b]	30.1	45.4
	0.4	13.6	21.1
	0.1	7.7	20.1
	0.08	3.8	9.6
	0.02	3.6	18.8
	0.01	5.2	11.0
4-PDD	100.0	4.5	9.7
	10.0	4.2	12.2

[a]*MDCK cells, 2 × 10^5 per 60 mm dish, were incubated with increments of the phorbol-12,13-diesters in serum-supplemented medium for 24 hr, washed twice, and incubated for a second 24 hr in serum-supplemented medium lacking the effectors. The medium from the second 24-hr incubation was assayed serologically. At the end of the second 24-hr incubation, the MDCK cells were removed from the dish after treatment with trypsin, counted, and tested for viability.*

[b]*Viability, as judged by trypan blue exclusion, was 98%.*

[c]*Data in fg/cell.*

nontumor-producing phorbol diester, 4α-phorbol-12,13-didecanoate (4α-PDD) at a concentration of 10^{-7} *M* has no effect (25, 26); TPA and/or PDD are not required throughout the incubation. If the cells are incubated at 2 × 10^5/dish for 60 min, washed twice, and incubated in medium in the absence of these reagents for a further 24 hr, the cells continue to produce increased levels of prostaglandins. The levels of PGE_2 in the serum-supplemented medium, after treatment of the cells with TPA and PDD for 60 min and after removal of the phorbol diesters, increase with time of incubation. Stimulation of prostaglandin production is a slow process and at low concentrations of promoters the reaction(s) is not complete even after 24 hr. At all levels of the phorbol promoters there is a pronounced lag time, the length of which depends on the amount of promoter used during the initial incubation.

Indomethacin, which blocks conversion of arachidonic acid to endoperoxide (27), is an effective inhibitor of the TPA-stimulated prostaglandin synthesis; 50% inhibition is obtained at a 8×10^{-8} *M* concentration. Hydrocortisone (1 μg/ml), which inhibits expression of phospholipase activity in MC5-5 cells, does not inhibit prostaglandin synthesis of MDCK cells even with a 2 hr preincubation and continuous treatment after TPA removal. Cycloheximide and puromycin, eukaryotic protein synthesis inhibitors, also block prostaglandin synthesis. The inhibition of the expression of prostaglandin synthesis after TPA treatment is marked (50 - 60%) even with 0.05 μg/ml of cycloheximide. The inhibition by cycloheximide is not observed when the cells are treated with cycloheximide for 1 hr during the TPA treatment or 2 hr before the TPA treatment if the cycloheximide then is removed with the residual TPA. Actinomycin D, 0.01 - 0.1 μg/ml, also inhibits the expression of prostaglandin synthesis after TPA treatment but many of the cells become detached from the culture dish.

The increase in the availability of the substrate for prostaglandin synthesis can be demonstrated by measuring the release of radioactive compounds from cells labeled with [^{3}H]-arachidonic acid. MDCK cells, treated with TPA (1 ng/ml), release about 2.5 times more radioactive materials than cells stimulated by serum alone. Hydrocortisone (1.0 μg/ml) has no effect on this TPA stimulation, but cycloheximide (0.5 μg/ml) inhibits it 76%. Indomethacin (0.5 μg/ml) does not inhibit the early release of radioactive materials by the TPA-treated cells, but inhibits later release of radioactivity. Most of the extracted radioactive materials released from the cells incubated in serum-supplemented medium chromatograph as arachidonic acid. Only small amounts of this arachidonic acid are converted to PGE_2 and $PGF_{2\alpha}$. The TPA-treated cells release 2.4 times more arachidonic acid, much of which is converted to PGE_2 and $PGF_{2\alpha}$. In the presence of indomethacin the conversion of the released arachidonic acid into PGE_2 and $PGF_{2\alpha}$ is inhibited almost completely. Indomethacin, at 0.5 μg/ml, appears to be inhibiting a reaction in addition to the cyclooxygenase step, possibly phospholipase activity, since the overall stimulation of released radioactivity is only 1.7 instead of 2.4-fold. Hydrocortisone, which does not inhibit prostaglandin production by MDCK cells, does not affect the stimulated release of radioactive compounds by TPA either qualitatively or quantitatively. Cycloheximide (0.5 μg/ml) inhibits both the release of radioactive materials stimulated by TPA and the conversion of the released arachidonic acid into prostaglandins.

Treatment of MDCK cells with TPA, at 1 ng/ml, causes a change in cell morphology. Many of the treated cells have long, thin processes and appear longer and thinner than normal cells. The treated cells appear to have much less contact with neighboring cells, and occasionally giant cells with multiple processes are seen. Cells treated with PDD (1.0 ng/ml) have the same altered morphology as the TPA treated cells. Cells treated with 4α-PDD, at 100 ng/ml, or cells treated with TPA (1 ng/ml) and cycloheximide (0.5 μg/ml) can not be distinguished morphologically from normal cells. The alteration of morphology after TPA treatment takes time, i.e., after 2 hr of incubation (following the TPA treatment) slight changes in cell shape can be seen in some but not all cells; after 4 hr, the changes become more prominent and around 12 hr after TPA treatment, many cells are longer and thinner than control cells; from 12 to 24 hr, most of the cells have altered morphology. Indomethacin, 0.5 μg/ml, does not affect the morphological changes induced by TPA as judged by microscopic examination after 24 hr of incubation.

The stimulation of prostaglandin production by TPA is reversible. When MDCK cells are treated with TPA (1 ng/ml) for 60 min, washed to remove residual TPA, and incubated in serum-supplemented medium for a second 24 hr, washed again and incubated in fresh medium for a third 24 hr, the degree of stimulation decreases after the first 24-hr incubation. During the third 24-hr incubation, prostaglandin production by cells treated for 1 hr with TPA decreases to almost that of control cells. In another experiment, when the cells were treated with 0.5 ng of TPA for 60 min, a 4.2-fold stimulation was seen in the first 24-hr incubation (0.45×10^6, control; and 0.26×10^6 cells/dish, TPA), a 2.6-fold stimulation was found in the second 24-hr incubation (0.59×10^6, control, and 0.58×10^6 cells/dish, TPA), but the stimulation in the third 24-hr incubation cultures was insignificant (1.39×10^6, control, and 1.35×10^6 cells/dish, TPA).

Purified epidermal growth factor (EGF) also stimulates MDCK cells to produce prostaglandins (Fig. 3). During the incubation there is no appreciable growth of MDCK cells. As little as 0.5 ng/ml (8×10^{-11} *M*) of EGF stimulates production of both $PGF_{2\alpha}$ and PGE_2 after 12 hr of incubation (1.7 ng of $PGF_{2\alpha}$ and 0.2 ng PGE_2/ml in the presence of EGF compared to 1.2 ng $PGF_{2\alpha}$ and 0.16 ng of PGE_2/ml in its absence) (28). Indomethacin (0.01 μ*M*) blocks the EGF-stimulated prostaglandin production by MDCK cells. Incubation of EGF with MDCK cells in which [^{3}H]arachidonic acid has been incorporated into cellular lipids also results in stimulated release of radioactive materials from the cells into the medium - even as early as 2 hr after addition of the EGF. When the radioactive materials released into the medium after 8 hr of incubation with EGF are extracted with ethyl acetate and analyzed by thin-layer chromatography, increased levels of radioactive prostaglandins and

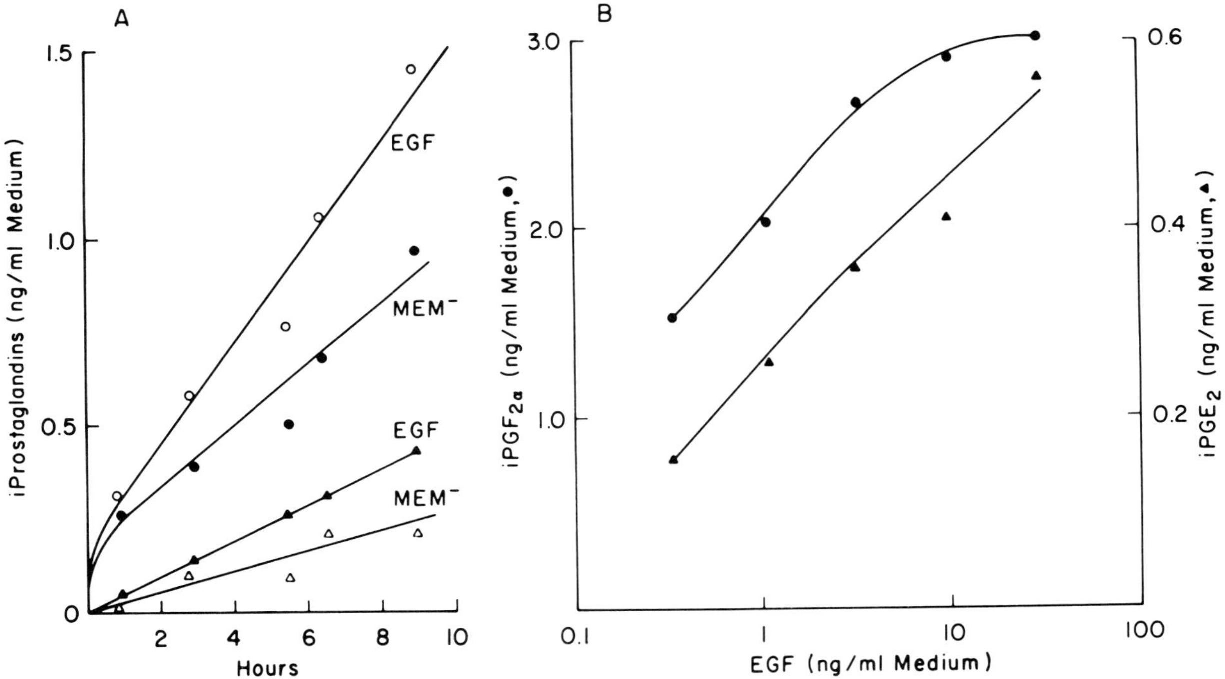

Fig. 3. Effect of EGF on prostaglandin production by MDCK cells. MDCK cells at 2×10^5 cells/dish were incubated with MEM⁻ or EGF (30 ng/ml) for various periods of time (A) or with increments of EGF for 12 hr (B). Duplicate dishes were used for each time point and each EGF concentration. The media were collected and assayed (A) for $PGF_{2\alpha}$ in the presence (○) and absence of EGF (●) and PGE_2 in the presence (▲) and absence of EGF (△). In the presence of increments of EGF (B), the media were assayed for $PGF_{2\alpha}$ (●) and PGE_2 (▲). In the absence of EGF, 1.2 ng/ml of $PGF_{2\alpha}$ and 0.16 ng/ml PGE_2 were produced. The values obtained for the media in the duplicate dishes agreed within 10% of the mean values.

TABLE II. Effect of Retinoids and TPA on Prostaglandin Production and the Release of Radioactivity from [^{3}H]Arachidonic Acid-Labeled MDCK Cells[a]

Treatment[b]	Radioactivity (cpm × 10^{-3}/dish)	Prostaglandins produced (ng/ml)[c]		
		PGE_2	$PGF_{2\alpha}$	6-keto-$PGF_{1\alpha}$[d]
None	19.9 ± 1.0	0.11 ± 0.04	0.66 ± 0.09	0.47 ± 0.12
+ TPA	93.3 ± 6.4*	2.47 ± 0.35*	6.50 ± 0.66*	9.03 ± 0.93*
Cis-Retinoic acid	36.6 ± 0.6*	0.38 ± 0.05**	1.15 ± 0.05**	2.13 ± 0.40*
+ TPA	101.6 ± 1.7*	2.32 ± 0.21*	4.83 ± 0.59*	10.10 ± 1.77*
TMMP Retinoic acid	43.2 ± 1.3*	0.64 ± 0.07*	2.60 ± 0.35*	4.23 ± 0.75*
+ TPA	172.3 ± 1.4*§	5.20 ± 0.58*§§	12.67 ± 0.58*§	22.67 ± 0.58*§

[a]The values are the mean ± S.D. of triplicate dishes. The differences between None and cis-retinoic acid or TMMP retinoic acid are statistically significant (*$p < 0.001$, **$p < 0.005$). The differences between none and TPA, cis-retinoic acid and cis-retinoic acid plus TPA, and TMMP retinoic acid and TMMP retinoic acid plus TPA are statistically significant ($p < 0.001$). The differences between TPA and cis-retinoic acid plus TPA are not statistically significant. The differences between TMMP retinoic acid plus TPA and TPA or cis-retinoic acid plus TPA are statistically significant (§$p < 0.001$, §§$p < 0.005$).

[b]MDCK cells were treated for 22 hr with TPA (0.0005 μg/ml), cis-retinoic acid (2 μg/ml), TMMP retinoic acid (2 μg/ml) or their mixture.

[c]Analyzed by radioimmunoassay.

[d]Production of 6-keto-$PGF_{1\alpha}$ by MDCK cells was verified by high pressure liquid chromatographic analyses (13).

arachidonic acid as well as some unidentified compounds are found.

Vitamin A and its analogs have exhibited cocarcinogenic and anti-carcinogenic activities (32). MDCK cells incubated in the presence of the trimethylmethoxyphenyl (TMMP) analog of retinoic acid, like the tumor-promoting TPA and cocarcinogen EGF, produce increased levels of PGE_2 and $PGF_{2\alpha}$ (29). This stimulation is observed with as little as 0.1 μg/ml (2.7×10^{-7} *M*) TMMP retinoic acid. With the exception of retinyl palmitate, all of the retinoids tested (retinol, retinal, *cis*- and *trans*-retinoic acids, and the TMMP retinoic acid) stimulate prostaglandin production. Retinyl palmitate, at levels greater than 5.0 μg/ml, inhibits prostaglandin production, but this inhibition could be accounted for by its content of the phenolic antioxidant, butylated hydroxyanisole, used as preservative. Stimulation of deacylation of the cellular lipids by retinoids also was demonstrated by the release of radioactive materials from [^{3}H]arachidonic acid-labeled cells treated with *cis*-retinoic acid or TMMP retinoic acid (2.0 μg/ml) (Table II). Levels of PGE_2, $PGF_{2\alpha}$, and the prostacyclin nonenzymatic product, 6-keto-$PGF_{1\alpha}$, also are increased. The tumor promoter, TPA, is more effective than the retinoids at stimulating release of radioactive materials and enhancing prostaglandin and prostacyclin production. The stimulating effects of both TPA and TMMP retinoic acid are more than additive; they act in a synergistic manner. The effect of *cis*-retinoic acid and TPA, however, is not synergistic; in fact, their combined effect is slightly less than additive.

Several stimulators of prostaglandin production and deacylation of cellular lipids, in addition to the tumor-promoting phorbol diesters, epidermal growth factor, and retinoids, such as estrogens, anthralin, polycyclic hydrocarbons, iodoacetate, colcimid, and uv radiation (Table III) have been demonstrated to be cocarcinogenic *in vivo* or have been associated with carcinogenesis (30 - 33).

Among the inhibitors of prostaglandin production (Table IV) (34) several compounds including indomethacin, anti-inflammatory steroids, and butylated hydroxyanisole have been shown to be anti-carcinogenic *in vivo* (35 - 37). Thus, there appears to be coincidence between both prostaglandin production and deacylation of phospholipids and cocarcinogenesis. One possible conclusion to be reached from this coincidence and the increased prostaglandin content of tumors is that some arachidonate metabolites give the tumor cells a proliferative advantage - perhaps by inhibiting local immunologic surveillance. Recently it has been reported that prostaglandins can act as cocarcinogens *in vivo* (38). If arachidonic acid metabolism accompanies cocarcinogenic events, prostaglandin production by several cell lines growing in culture could be used to screen for environmental

TABLE III. Some Stimulators of Prostaglandin Production by Cells in Culture

Tumor-promoting phorbol diesters[a,b]
Epidermal growth factor[a,b]
Adriamycin
17β-Estradiol[b]
Anthralin[b]
Retinoids[a,b]
Polycyclic hydrocarbons[a,b]
Melittin[a]
Iodoacetate[b]
Colcimid[b]
Thrombin[a]
Bradykinin[a]
uv Radiation[b]

[a]*Stimulates release of arachidonic acid.*
[b]*Cocarcinogenic or cocarcinogen suspect.*

TABLE IV. Some Inhibitors of Prostaglandin Production by Cells in Culture

Nonsteroidal antiinflammatory drugs[a]
Indomethacin[b]
Aspirin
Steroidal antiinflammatory drugs[b]
Dexamethasone
Hydrocortisone[b]
Some Anti-oxidants
Butylated hydroxyanisole[b]
α-Tocopherol[d]

[a]*Inhibits prostaglandin production by all cultured cells tested.*
[b]*Anti-carcinogenic.*
[c]*Does not affect prostaglandin production by all cells.*
[d]*Inhibits TPA-stimulated prostaglandin production.*

cocarcinogens. A battery of different cell lines would have to be tested because putative cocarcinogens such as saccharin, cyclamates or bile acid may stimulate prostaglandin production by bladder cells, for example, but not that by fibroblasts or kidney cells. Cocarcinogens in cigarette condensates may stimulat prostaglandin production by lung cells, for example, but not that by bladder cells, kidney cells or fibroblasts.

ACKNOWLEDGMENTS

This research has been supported by grants from the National Institutes of Health. The detailed studies from which this summary was abstracted were performed in collaboration with Iftekhar Alam, Aviv Hassid and Kazuo Ohuchi.

REFERENCES

1. Lands, W. E. M. and Samuelsson, B. (1968). Phospholipid precursors of prostaglandins. *Biochim. Biophys. Acta 164,* 426.
2. Vonkeman, H. and Van Dorp, D. A. (1968). The action of prostaglandin synthetase on 2-arachidonyl-lecithin. *Biochim. Biophys. Acta 164,* 430.
3. Kunze, H. and Vogt, W. (1971). Significance of phospholipase A for prostaglandin formation. *Ann. N.Y. Acad. Sci. 180,* 123.
4. Nugteren, D. H. and Hazelhof, E. (1973). Isolation and properties of intermediates in prostaglandin biosynthesis. *Biochim. Biophys. Acta 326,* 448.
5. Hamberg, M. and Samuelsson, B. (1973). Detection and isolation of an endoperoxide intermediate in prostaglandin biosynthesis. *Proc. Natl. Acad. Sci. U.S. 70,* 899.
6. Hamberg, M., Svensson, J., and Samuelsson, B. (1975). Thromboxanes: A new group of biologically active compounds derived from prostaglandin endoperoxides. *Proc. Natl. Acad. Sci. U.S. 72,* 2994.
7. Needleman, P., Moncada, S., Bunting, S., Vane, J. R., Hamberg, M., and Samuelsson, B. (1976). Identification of an enzyme in platelet microsomes which generates thromboxane A_2 from prostaglandin endoperoxides. *Nature (London) 261,* 558.
8. Moncada, S. and Vane, J. R. (1977). The discovery of prostacyclin - A fresh insight into arachidonic acid metabolism. *In* "Biochemical Aspects of Prostaglandins and Thromboxanes" (N. Kharasch and J. Fried, eds.), pp. 155-177. Academic Press, New York.
9. Hamberg, M. and Samuelsson, B. (1974). Prostaglandin endoperoxides. Novel transformations of arachidonic acid in human platelets. *Proc. Natl. Acad. Sci. U.S. 71,* 3400.
10. Levine, L., Gutierrez-Cernosek, R. M., and Van Vunakis, H. (1971). Specificities of prostaglandin B_1, $F_{1\alpha}$ and $F_{2\alpha}$ antigen-antibody reactions. *J. Biol. Chem. 246,* 6782.
11. Langone, J. J., Van Vunakis, H., and Levine, L. (1975).

Antibodies: Analytical tools to study pharmacologically active compounds. *Accounts Chem. Res. 8,* 335.
12. Green, K. (1973). Methods for quantitative determination of prostaglandins using gas chromatography-mass spectrometry. *Les Prostaglandines, 1973,* Inserm, Paris.
13. Alam, I., Ohuchi, K., and Levine, L. (1979). Determination of cyclooxygenase products and prostaglandin metabolites using high pressure liquid chromatography and radioimmunoassay. *Analytical Biochemistry 93,* 339-345.
14. Pong, S. S., Hong, S. L., and Levine, L. (1977). Prostaglandin production by methylcholanthrene transformed mouse BALB/3T3: Requirement for protein synthesis. *J. Biol. Chem. 252,* 1408.
15. Jaffe, B. M. (1974). Prostaglandins and cancer: An update. *Prostaglandins 6,* 453.
16. Tashjian, A. H., Jr., Voelkel, E. F., and Levine, L. (1977). Effects of hydrocortisone on the hypercalcemia and plasma 13,14-dihydro-15-keto-PGE_2 in mice bearing the $HSDM_1$ fibrosarcoma. *Biochem. Biophys. Res. Commun. 74,* 199.
17. Seyberth, H. W., Hubbard, W. C., Oelz, O., Sweetman, B. J., Watson, J. T., and Oates, J. A. (1977). Prostaglandin-mediated hypercalcemia in the VX_2 carcinoma-bearing rabbit. *Prostaglandins 14,* 319.
18. Tashjian, A. H., Jr., Voelkel, E. F., and Levine, L. (1978). Plasma concentrations of 13,14-dihydro-15-keto-prostaglandin E_2 in rabbits bearing the VX_2 carcinoma: Effects of hydrocortisone and indomethacin. *Prostaglandins 14,* 309.
19. Hammarstrom, S., Samuelsson, B., and Bjursell, G. (1973). Prostaglandin levels in normal and transformed baby-hamster-kidney fibroblasts. *Nature New Biology 243,* 50.
20. Hong, S. L., Wheless, C. M., and Levine, L. (1977). Elevated prostaglandin synthetase activity in methylcholanthrene-transformed mouse BALB/3T3. *Prostaglandins 13,* 271.
21. Boutwell, R. K. (1974). The function and mechanism of promoters of carcinogenesis. *CRC Crit. Rev. Toxicol. 2,* 419.
22. Berenblum, I. (1969). A re-evaluation of the concept of cocarcinogenesis. *Progr. Exp. Tumor Res. 11,* 31.
23. Hecker, E. (1968). Cocarcinogenic principles from the seed oil of *Croton tiglium* and from other *Euphorbiaceae. Cancer Res. 28,* 2338.
24. Van Duuren, B. L. (1969). Tumor promoting agents in two-stage carcinogenesis. *Progr. Exp. Tumor Res. 11,* 31.
25. Levine, L. and Hassid, A. (1977). Effects of phorbol-12, 13-diesters on prostaglandin production and phospholipase activity in canine kidney (MDCK) cells. *Biochem. Biophys.*

Res. Commun. 79, 477.

26. Ohuchi, K. and Levine, L. (1978). Stimulation of prostaglandin synthesis by tumor-promoting phorbol-12,13-diesters in canine kidney (MDCK) cells: Cycloheximide inhibits the stimulated prostaglandin synthesis, deacylation of lipids and morphological changes. *J. Biol. Chem. 253,* 4783.
27. Vane, J. R. (1971). Inhibition of prostaglandin synthesis as a mechanism of action for aspirin-like drugs. *Nature New Biology 231,* 232.
28. Levine, L. and Hassid, A. (1977). Epidermal growth factor stimulates prostaglandin biosynthesis by canine kidney (MDCK) cells. *Biochem. Biophys. Res. Commun. 76,* 1181.
29. Levine, L. and Ohuchi, K. (1978). Retinoids, like tumor-promoting phorbol diesters, enhance deacylation of cellular lipids and prostaglandin production in MDCK cells. *Nature (London) 276,* 274-275.
30. Berenblum, I. (1975). Sequential aspects of chemical carcinogenesis: Skin. *In* "Cancer I" (F. F. Becker, ed.). Plenum, New York.
31. Rose, S. P., Stahn, R., Passovoy, D. S., and Herschman, H. (1976). Epidermal growth factor enhancement of skin tumor induction in mice. *Experientia 32,* 913.
32. Sporn, M. B., Dunlop, N. M., Newton, D. L., and Smith, J. M. (1976). Prevention of chemical carcinogenesis by Vitamin A and its synthetic analogs (retinoids). *Fed. Proc. 35,* 1332.
33. Heddle, J. A. and Bruce, W. R. (1977). Comparison of tests for mutagenicity or carcinogenicity using assays for sperm abnormalities, formation of micronuclei, and mutations in *Salmonella*. *In* "Origins of Human Cancer" (H. H. Hiatt, J. D. Watson, and J. A. Winsten, eds.), pp. 1549-1577. Cold Spring Harbor Laboratory, New York.
34. Hong, S. L. and Levine, L. (1976). Inhibition of arachidonic acid release from cells as the biochemical action of anti-inflammatory corticosteroids. *Proc. Natl. Acad. Sci. U.S. 73,* 1730.
35. Wattenberg, L. W., Loub, W. D., Lam, L. K., and Speier, J. L. (1976). Dietary constituents altering the responses to chemical carcinogens. *Fed. Proc. 35,* 1327.
36. Belman, S. and Troll, W. (1972). The inhibition of croton-oil promoted mouse skin tumorigenesis by steroid hormones. *Cancer Res. 32,* 450.
37. Viaje, A., Slaga, T. J., Wigler, M., and Weinstein, I. B. (1977). Effects of anti-inflammatory agents on mouse skin tumor promotion, epidermal DNA synthesis, phorbol ester-induced cellular proliferation and production of plasminogen activators. *Cancer Res. 37,* 1530.
38. Lupulescu, A. (1978). Enhancement of carcinogenesis by prostaglandins. *Nature (London) 272,* 634.

METABOLISM OF CYCLIC NUCLEOTIDES AND RELATED COMPOUNDS IN DIFFERENT LAYERS OF THE RETINA

Oliver H. Lowry
Sosamma J. Berger
Gerald W. DeVries
Joyce G. Carter
Demoy W. Schulz
James A. Ferrendelli

Departments of Pharmacology and Neurology
and Neurological Surgery
Washington University School of Medicine
St. Louis, Missouri 63110

The function of guanosine-3',5'-monophosphate (cGMP) in biological material is not understood. cGMP and adenosine-3', 5'-monophosphate (cAMP) respond differently, often in opposite directions, to particular stimuli. There is a growing belief that the mode of action of the two nucleotides may possibly be quite different.

The retina is uniquely rich in cGMP (1) as well as in the enzymes which make and degrade it: guanylate cyclase (2) and cGMP phosphodiesterase (PDE) (3). Moreover, the cGMP levels in retina are reduced by light and this is quite surely due to activation of the PDE (4 - 7). For these reasons, the retina should be an excellent place to study cGMP physiology - and, as will be seen, for studying metabolism of guanine nucleotides in general.

Another feature that makes the retina an attractive place to study almost anything associated with the nervous system, is that it is a discretely layered structure, with different neuronal elements segregated and easily separable for analysis. Furthermore, from the first there was evidence that the high

 ISBN 0-12-397580-8

levels of cGMP and related enzymes were not shared by all parts of the retina.

Because of these considerations, we decided to join in a collaborative study of other guanylate compounds and enzymes in the retina and their distribution among the several layers.

METHODS

To analyze the retinal layers separately, the eye is quick frozen and 5-μm frozen sections are cut parallel to the surface. Therefore, layers that are actually only 25-30 μm thick are spread out 10-fold or more in the sections. The frozen sections are dried under vacuum at -40°C, after which it is a simple matter to dissect (at room temperature) pure samples of each layer which weigh 20 - 50 ng (8).

The weights are determined on a simple quartz "fish pole" balance (8) and analyzed either by radioimmune assay [cGMP itself (9)] or by enzymatic methods, which will be published separately. Initial steps are carried out in volumes as small as 0.1 μl. Evaporation is prevented by working under oil (the "oil well technique") (8).

The enzymatic methods all terminate in a pyridine nucleotide step. In some cases, there was a sufficient amount of pyridine nucleotide oxidized or reduced to measure directly by fluorometry. In other cases amplification was needed. One example should suffice:

To measure the sum of GTP plus GDP, the sample was heated in NaOH to destroy enzymes and then a reagent added to carry out the following cyclic reaction originally devised by Cha and Cha (10) and exploited by Goldberg *et al.* (11).

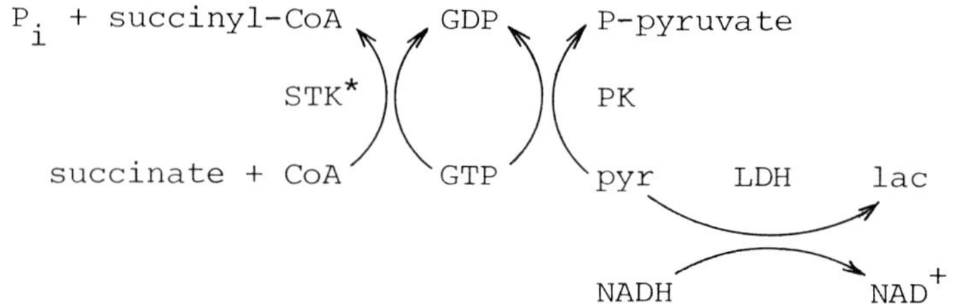

**Key to abbreviations:* *STK, succinic thiokinase; PK, pyruvate kinase; LDH, lactic dehydrogenase; pyr, pyruvate; lac, lactate.*

With sufficient STK and PK, 1000 molecules or more of pyruvate (and hence NAD^+) are formed for each molecule of GTP or GDP. After stopping the reaction with HCl (which destroys the excess NADH), the NAD^+ is converted with 6 *N* NaOH into a highly fluorescent compound which is measured in the fluorometer.

RESULTS

cGMP and cAMP

Figure 1 shows the distribution of cGMP in the rabbit retina in dark and light adaptation. The highest levels are in the outer segment layer where, with dark adaptation, the concentration is 300 times that in average brain. The levels are

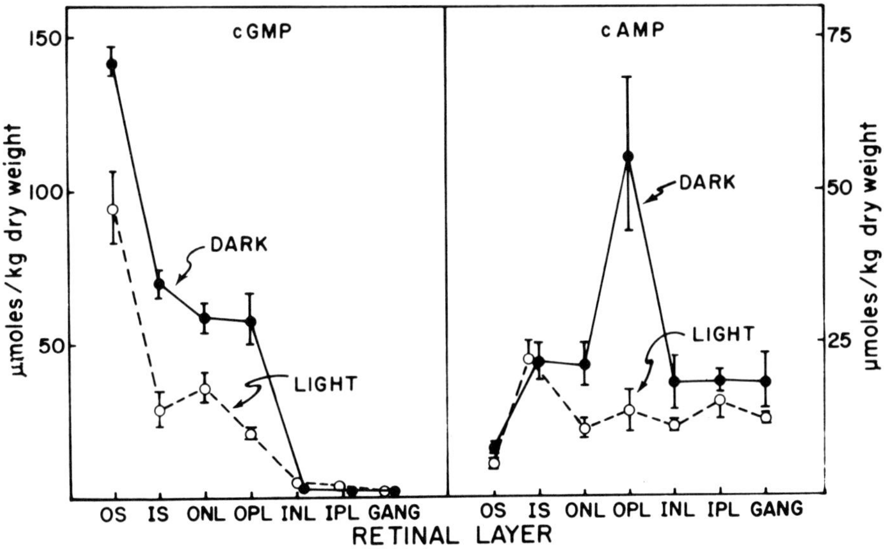

Fig. 1. cGMP and cAMP in the layers of light- and dark-adapted rabbit retina. The eyes were frozen after about 1 min of ischemia. Each point is the average for three eyes ± SEM, except for the dark-adapted OPL layer (cGMP and cAMP), where the average of two eyes is presented. Abbreviations: OS, outer segment; IS, inner segment; ONL, outer nuclear layer; OPL, outer plexiform layer; INL, inner nuclear layer; IPL, inner plexiform layer; GANG, ganglion cell layer. From Orr et al. (7).

lower but still very high in deeper layers of the photoreceptor cells, but cGMP then falls to very low levels in the remainder of the retina. In great contrast, cAMP, which is an order of magnitude higher than cGMP in brain is, in the outer segment layer, an order of magnitude lower than cGMP, and the distributions of the two nucleotides in the rest of the retina differ completely (1).

Light causes levels of cGMP to fall in all layers of the photoreceptor cells. Again the response of cAMP to light adaptation is quite different. There is a dramatic fall in the outer plexiform layer and a substantial fall in the outer nuclear layer. Otherwise there are no significant changes. The difference in the location of the changes in the two nucleotide raises the possibility that cGMP is involved more directly in the transduction or transmission process, whereas cAMP may be more concerned with secondary metabolic events.

cGMP cycle

The data will be presented in relation to the "cGMP cycle:"

$$\text{GTP} \longrightarrow \text{cGMP} \longrightarrow \text{GMP} \longrightarrow \text{GDP} \longrightarrow \text{GTP}$$

Data have been obtained for the distribution of all the enzymes concerned and all the nucleotides except GMP. (The levels of GDP are low compared to ATP, and these will not be reported here.)

GTP + GDP and ATP + ADP

In all the photoreceptor cell layers of monkey retina, GTP + GDP levels are almost equal to those of ATP + ADP (Fig. 2). Similar results have been obtained in the rabbit. As far as we know, this is the only place in the body where GTP even approaches ATP in concentration. The ratio of 1:3 or 1:4 seen here in the deeper retinal layers is typical of everage brain. These results confirm the finding of Klethi *et al.* (12) who found that in whole retina the ratio of total guanylates to total adenylates was about 2:3.

Guanylate Cyclase

The level of this enzyme has been found to be highest by far in the outer segment layer in the ground squirrel (a predominantly cone retina), in the rabbit (a predominantly rod

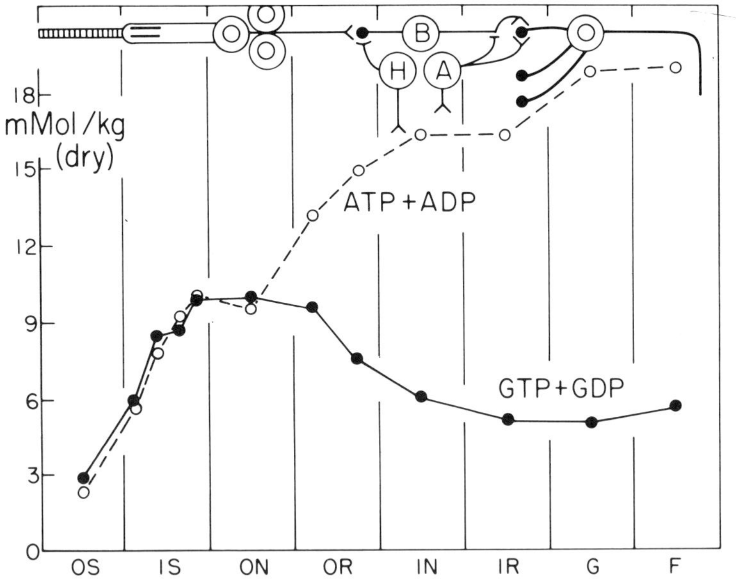

Fig. 2. Distribution of ATP plus ADP, and GTP plus ADP in monkey retina. Abbreviations are outer segments (OS), inner segments (IS), outer nuclear (ON), outer reticular (OR), inner nuclear (IN), inner reticular (IR), ganglion cell layer (G), fiber layer (F). In each layer the bulk of each nucleotide was present as the triphosphate.

retina) and in the monkey (a mixed rod and cone retina). In other layers of the photoreceptor cells, and in deeper retinal layers, levels were only 5 or 10% as high (not shown).

Phosphodiesterase (PDE)

Bitensky's group has shown that the cyclic nucleotide PDE in retina has special properties quite different from the PDE's of brain (13). It is active toward both cAMP and cGMP with about the same maximum velocity, but the Michaelis constant is more favorable for cGMP by a factor of almost 50. The distribution of this enzyme was measured in the retina with a reagent containing protamine and guanylylimido diphosphate which brings out full activity whether the retina has been light- or dark-adapted (13).

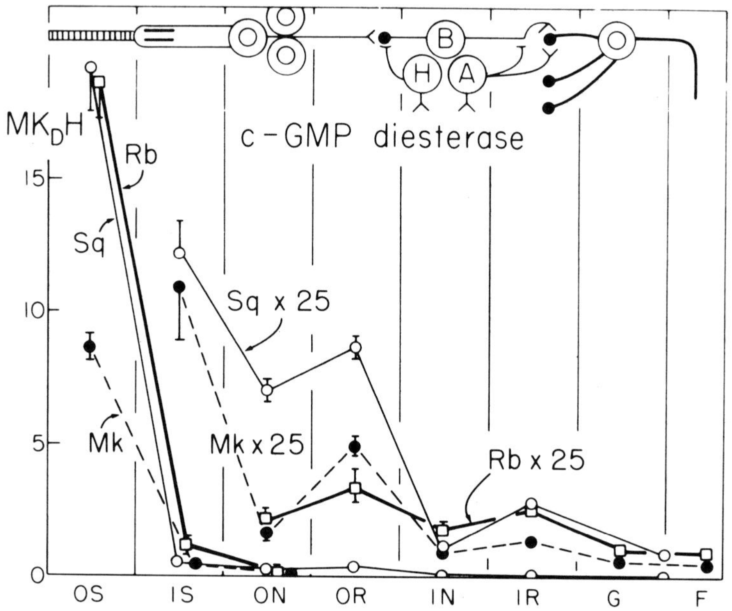

Fig. 3. The distribution of cGMP phosphodiesterase in the retina of monkey (Mk), rabbit (Rb), and ground squirrel (Sq). The values in the deeper layers are also shown multiplied by a factor of 25 (25). Activity is recorded as mol kg^{-1} (dry) hr^{-1} (MK_DH). Other abbreviations are the same as in Fig. 2.

The distributions of PDE activities (Fig. 3) are very similar for the rabbit (Rb), the ground squirrel (Sq), and the monkey (Mk). In all cases the enzyme is essentially restricted to the outer segment layer. The levels are about 100 times higher in this layer than in the nonphotoreceptor layers.

Guanylate Kinase

The next enzyme in the cGMP cycle, 5'-GMP kinase, like the PDE, is nearly confined to the photoreceptor cells (Fig. 4), but is shifted to the deeper layers, and is actually very low in the outer segment layer. Therefore, when cGMP is hydrolyzed

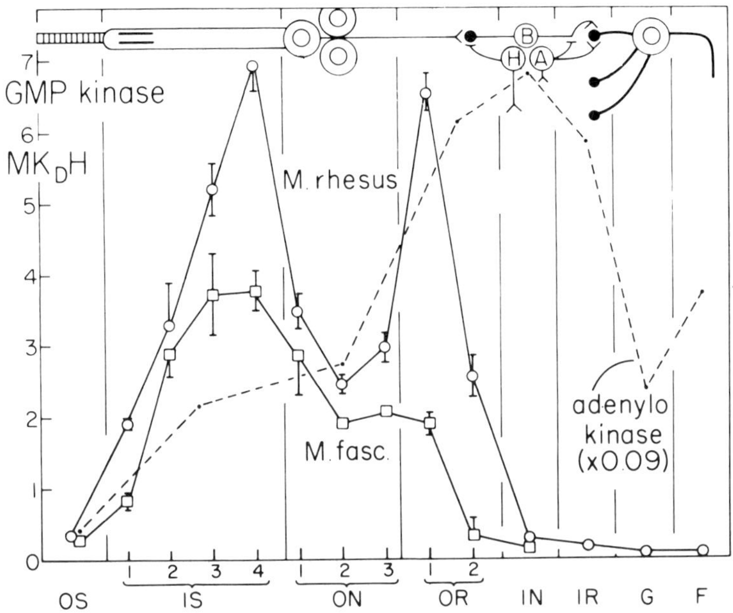

Fig. 4. The distribution of guanylate kinase (GMP kinase) in the retina of two species of monkey. The adenylokinase data are for M. rhesus and are taken from an earlier publication (14). Abbreviations are the same as in Fig. 2 and 3.

it presumably must diffuse from the outer to the inner segments before removal. A similar pattern for GMP kinase distribution was found in rabbit retina.

Figure 4 also shows the activity of the homologous enzyme adenylokinase, which has a completely different distribution from that of guanylate kinase. The ratio between the two enzymes is 50 times greater in the inner nuclear and inner reticular layers than in the inner segment layer.

Guanosine Diphosphate Kinase

The last step in the cGMP cycle, the conversion of GDP to GTP, could occur directly at the succinic thiokinase step in

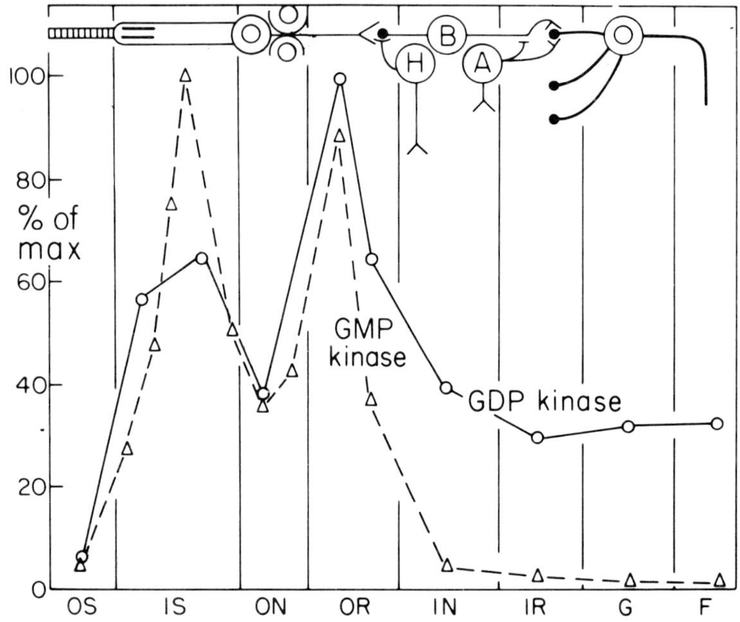

Fig. 5. Guanosine diphosphate kinase (GDP kinase) distribution in monkey retina. For comparison, GMP kinase is repeated from Fig. 4. Abbreviations are the same as in Fig. 2. The peak activity of GDP kinase was 12 mol kg^{-1} (dry) hr^{-1}.

the citric acid cycle. There is a dense concentration of mitochondria in the inner segments at the end nearest to the outer segments. Another possibility is formation of GTP by phosphate transfer from ATP by the action of GDP kinase (nucleoside diphosphokinase). This enzyme is distributed similarly to GMP kinase among the four layers of the photoreceptor cell, but it does not fall to low levels in the deeper retinal layers (Fig. 5). This difference in distribution is reasonable since GDP kinase would be required by all cells with the citric acid cycle, and would have to be active in all cells which have a high metabolic rate.

DISCUSSION

From the data presented concerning the enzyme distributions it seems clear that cGMP is formed and broken down to GMP in the outer segments, but that the resynthesis of GTP probably takes place mainly in the other parts of the photoreceptor cell, i.e., GMP diffuses from outer to inner segments and GTP diffuses back. Whether GTP is formed directly in the citric acid cycle or indirectly from ATP, the high-energy phosphate bond is probably generated at the tip of the inner segments where the mitochondria are located.

One other enzyme is involved in the cGMP cycle - this is inorganic pyrophosphatase which is needed to hydrolyze the PP_i liberated when cGMP is formed. This enzyme is also low in outer segments but unusually rich in the rest of the photoreceptor cell, where the levels are 4 to 8 times higher than in the deeper layers of the retina and in brain (14). Thus PP_i probably also diffuses out of the outer segments for hydrolysis and resynthesis into GTP.

The question arises, if cGMP is involved in vision, is it concerned with a very rapid event such as transduction or a slower event such as light or dark adaptation. A clue to this can be obtained from the absolute nucleotide levels and the enzyme activities and their kinetic constants. Peak GTP levels exceed peak cGMP levels by a factor of 100 (Table I) and are probably high enough to saturate guanylate cyclase. However, the peak cyclase activity is low in absolute terms, and if the cGMP level were somehow reduced to one-half it would take perhaps 3000 msec to restore it back to normal. On the other hand, the cGMP diesterase is several orders of magnitude more active than the cyclase (Table I) and could reduce cGMP to one-half in as little as 25 msec or to 90% in 3 or 4 msec; i.e., changes could occur in the right time frame to play a role in transduction. The restoration of GMP back to GTP is even faster - with half-times for the two steps that are in the order of a very few milliseconds (Table I).

ACKNOWLEDGMENT

Supported in part by grants from the American Cancer Society (BC-4Q) and the U.S. Public Health Service (NS-08862 and E.Y.02294).

TABLE I. Highest Levels of Guanylates and Guanylate Enzumes among the Layers of the Photoreceptor Cell[a]

	Concentration ($mmol\ kg^{-1}$)	V_{max} ($mmol\ kg^{-1} sec^{-1}$)	$t_{1/2}$ (msec)
GTP	2.		
Guanylate cyclase		0.005	3000[b]
cGMP	0.03		
cGMP PDE		1.8	25
GMP kinase		1.0	4
GDP kinase		3.	5

[a]*The half times are rough estimates from the activities and kinetic constants at 25° and would probably be somewhat shorter at 38°. V_{max} activities are calculated per kilogram wet weight.*

[b]*Time for guanylate cyclase to raise cGMP from 0.015 to 0.03 mmol/kg.*

REFERENCES

1. Goridis, C., Virmaux, N., Cailla, H. L., and DeLaage, M. A. (1974). *FEBS Lett. 49,* 167-169.
2. Goridis, C., Virmaux, N., Urban, P. F., and Mandel, P. (1973). *FEBS Lett. 30,* 163-166.
3. Pannbacker, R. G., Fleischman, D. E., and Reed, D. W. (1972). *Science 175,* 757-758.
4. Miki, N., Keirns, J., Marcus, F., and Freeman, J. (1973). *Proc. Natl. Acad. Sci. U.S. 70,* 3820-3824.
5. Chader, G. J., Herz, L., and Fletcher, R. T. (1974). *Biochim. Biophys. Acta 347,* 491-493.
6. Goridis, C. and Virmaux, N. (1974). *Nature (London) 248,* 57-58.
7. Orr, H. T., Lowry, O. H., Cohen, A. I., and Ferrendelli, J. A. (1976). *Proc. Natl. Acad. Sci. U.S. 73,* 4442-4445.
8. Lowry, O. H. and Passonneau, J. V. (1972). "A Flexible System of Enzymatic Analysis." Academic Press, New York.
9. Ferrendelli, J. A., Rubin, E. H., Orr, H. T., Kinscherf, D. A., and Lowry, O. H. (1977). *Anal. Biochem. 56,* 392-407.
10. Cha, S. and Cha, C.-J. M. (1965). *Molec. Pharmacol. 1,* 178-189.
11. Goldberg, N. D., Dietz, S. B., and O'Toole, A. G. (1969). *J. Biol. Chem. 244,* 4458-4466.

12. Klethi, J., Urban, P. F., and Mandel, P. (1970). *FEBS Lett. 8*, 119-121.
13. Bitensky, M. H., Miki, N., Keirns, J. J., Keirns, M., Baraban, J. M., Freeman, J., Wheeler, M. A., Lacy, J., and Marcus, F. R. (1975). *In* "Advances in Cyclic Nucleotide Research" (G. I. Drummond, P. Greengard, and G. A. Robinson, eds.), Vol. 5, pp. 213-240. Raven Press, New York.
14. Lowry, O. H. (1964). *In* "Morphological Correlates of Neural Activity" (M. M. Cohen and R. S. Snider, eds.), pp. 178-191. Harper and Row, New York.

BIOLOGICAL FUNCTIONS OF LACTATE DEHYDROGENASE

Johannes Everse

Department of Biochemistry
Texas Tech University Health Sciences Center
Lubbock, Texas 79430

*Nathan O. Kaplan**

Department of Chemistry, Q-058
University of California at San Diego
La Jolla, California 92093

**Much has been said about Martin during the course of these meetings. I too, would like to express my own personal affection for him. I met Martin approximately 40 years ago when I was a student at Berkeley. I have great admiration for him as a scientist. His friendship and humanity have been a source of inspiration to me. His compassion as a colleague at Brandeis was one of constant support not only to me but the other members of the Brandeis community as well. I suspect many of us who are here at UCSD were attracted largely by Martin's presence. I hope the future will bring him many more productive and happy years, and that we will be able to share them with him.*

I think it fitting to present as a dedicatory paper in Martin's honor some aspects of studies on isozymes. It was in those early years at Brandeis that we first became involved in this field. Many discussions with Martin were of great benefit to my associates and myself in pursuing this line of endeavor.

 ISBN 0-12-397580-8

INTRODUCTION

About 20 years ago the first attempt was made to explain the biological role of the lactate dehydrogenase (LDH) isoenzymes (1). The hypothesis advanced at that time states that the LDH isoenzymes have evolved in order to fulfill specific metabolic purposes. This has become known as the aerobic-anaerobic theory. The hypothesis has proved to be of value when additional information regarding the distribution and the kinetic properties of the LDH isoenzymes became available. As a result, this theory has recently been discussed in various biochemical textbooks.

As indicated by several recent communications in TIBS (2 - 6), however, there appears to be some confusion concerning the scope and interpretations of the theory. It may be appropriate, therefore, to review the biological functions of the LDH isoenzymes in some detail. We hope this will help to clarify those aspects of the aerobic-anaerobic theory that appear to be subject to misinterpretation.

BIOLOGICAL FUNCTION OF LACTATE DEHYDROGENASE

Lactate dehydrogenase plays a unique role in intermediary metabolism because it serves various functions in the overall metabolism of the mammalian cell. These functions can be distinguished by considering the role the enzyme plays in the flow of metabolites through the metabolic pathways. To illustrate this point we will discuss three examples in which LDH serves different metabolic functions.

In its most simple function, LDH acts as the final enzyme in the glycolytic pathway. As such, it serves two purposes: (1) it reoxidizes NADH generated by the triose phosphate dehydrogenase reaction, and (2) the large negative ΔG of the LDH reaction, together with that of the pyruvate kinase reaction, assures a rapid and effective production of ATP from glucose. As a result, in cells that require rapid production of ATP and are entirely dependent on glycolysis for its generation, LDH functions entirely as a pyruvate reductase.

In cells that produce ATP via oxidative phosphorylation, LDH catalyzes a "dead end" side reaction, and does little more than provide for storage of excess reducing power (Fig. 1). This function is not unlike the function of creatine kinase, which catalyzes a similar "dead end" reaction. For LDH, however, the stored energy (lactate) can leave the cell and be reutilized in a different cell or even a different organ.

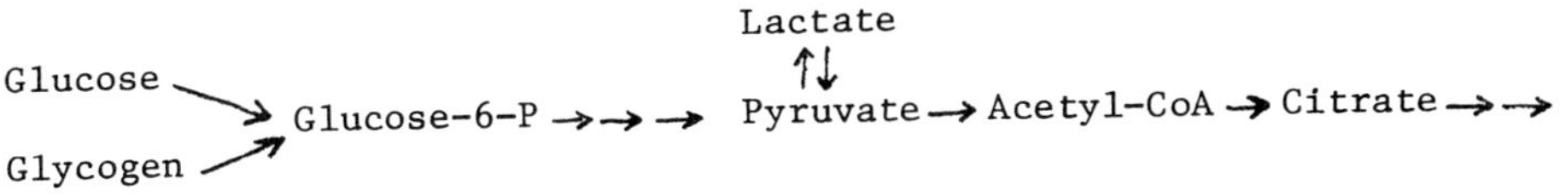

Fig. 1. Schematic diagram of pyruvate metabolism in aerobic tissues.

A third function of LDH is found in cells that do not produce either chemical or mechanical energy as their primary function. These cells possess the glycolyte enzymes, but the primary function of the pathway is not to produce ATP, but rather to provide metabolites and reducing power for other purposes. For example, the primary function of the glycolytic pathway in erythrocytes is probably to provide sufficient levels of 2,3-diphosphoglycerate (DPG), as well as to maintain sufficient levels of NADH needed for an efficient reduction of methemoglobin. In this type of cell the kinetic properties of LDH must satisfy the need for relatively high levels of 2,3-DPG and NADH.

DISTRIBUTION OF LDH ISOENZYMES

Experiments have shown that the M-type LDH is ideally suited to serve as a pyruvate reductase in anaerobic tissues. Considerable confusion, however, appears to exist concerning the function of LDH in tissues that possess oxidative metabolism, such as heart and liver, and the type of LDH isoenzyme found in these tissues. Pyruvate metabolism in both tissues may be represented in part by the schematic diagram in Fig. 1.

Metabolism in heart muscle is directed entirely toward the production of ATP via oxidative phosphorylation, and the main sources of energy for the heart are fatty acids, glucose, and lactate. The oxidation of 1 mole of lactate produces 1 mole of NADH in a single enzyme reaction and this NADH can immediately be transported into the mitochondria to produce ATP. It should, therefore, be apparent that lactate would be the preferred substrate over glucose. This preference for lactate has indeed been demonstrated with intact hearts (7 - 9). The heart would, therefore, require an LDH that readily catalyzes the oxidation of lactate. On the other hand, when glucose is used as the energy source the reduction of pyruvate should be inhibited so that pyruvate may enter the Krebs cycle, and the NADH may be reoxidized via the mitochondria.

It is well known that the H-type has kinetic properties that are consistent with these expectations (10 - 13).

If one considers only cell types with a metabolism that is geared to energy production, the aerobic-anaerobic theory can readily be applied to explain the experimental observations. However, in order to apply the theory to tissues with different kinds of metabolism, e.g., liver and erythrocytes, other factors need to be considered.

Pyruvate metabolism in liver cells is in accord with the scheme presented in Fig. 1. But the liver, unlike the heart, converts large amounts of lactate into glycogen via pyruvate. Both the heart and liver are effective scavengers of blood lactate; the difference is that the heart uses the lactate as an energy source whereas the liver converts lactate to glycogen or glucose for future energy needs. Since the heart is dependent on lactate for immediate energy needs, one would predict that the heart would have a more effective mechanism for the uptake of blood lactate than the liver, assuming other factors to be equal. As we have pointed out previously in detail (13), the difference in K_m for lactate of the H- and M-type LDH provides for this difference in effectiveness. When the blood lactate level decreases, the higher K_m of the liver enzyme will result in a decrease in the rate of lactate uptake by the liver, while at the same time the lower K_m of the heart enzyme will still allow the heart to utilize blood lactate efficiently as a fuel (see Fig. 2). It would not be to the advantage of most mammalian species if the heart and the liver were competing for the available lactate on an equal basis!

Exceptions may be found in mammals that depend on a rapid removal of lactate from their blood. This may be essential when an animal produces large amounts of lactate over short periods of time, for example, when an animal needs to run for long distances. In these cases, the presence of the H-type LDH in their livers may provide a definite advantage (13). It is of interest that most birds have a high percentage of H-type enzyme in their livers which is commensurate with the large amounts of lactate that are produced in their skeletal muscles (14).

A similar situation is encountered in the kidney. Most of the energy-requiring transport processes take place in the cortex during filtration and reabsorption. Concentration of the urine, associated with the transport of Na ions, occurs in the medulla whereas no energy-requiring process occur in the papilla. The cortex, which has a high oxidative metabolism, contains the H-type LDH almost exclusively. In contrast, the papilla has almost all M-type LDH, whereas the medulla contains about 50% of the H-type isoenzyme (15). Thus in the kidney, where we have a sustained demand for energy, the percentage H-type LDH correlates directly with the amount of energy required for tissue

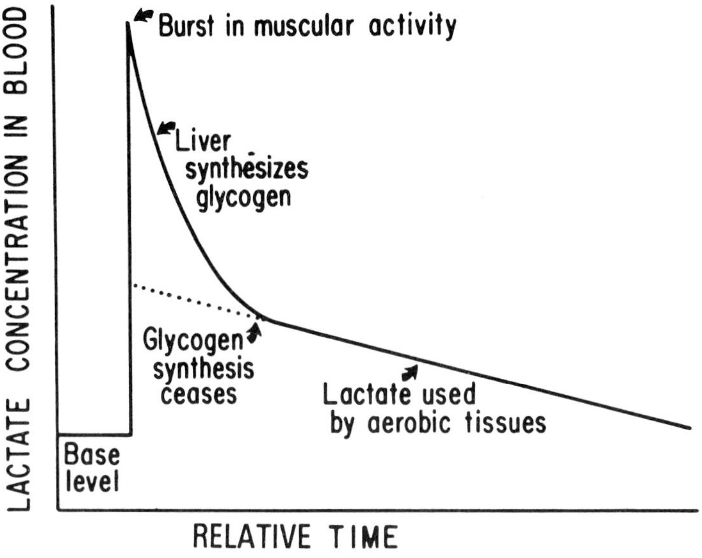

Fig. 2. Hypothetical scheme illustrating the removal of lactate from the blood by muscle and liver (glycogen synthesis) and by aerobic oxidation.

functioning.

Inhibition of the LDH in erythrocytes would result in increased levels of 2,3-DPG and of NADH. Hence, one would predict that mammalian erythrocytes contain the H-type isoenzyme. This isoenzyme is, indeed, found in the erythrocytes of some mammals (human, rabbit). However, other mammals have predominantly the M-type LDH in their erythrocytes (rat, mouse) (13). These findings suggest to us that other factors besides the levels of 2,3-DPG and NADH have to be considered in erythrocyte metabolism.

H-TYPE LDH DEFICIENCY

An H-type deficiency was recently discovered in humans (16) and this has raised several queries (5), particularly with regard to the possible advantages that the H-type LDH may provide to the organism. We feel that the presence of the H-type LDH provides certain metabolic advantages to the organism as outlined above, but this does not mean that an absence of these advantages is necessarily lethal. For example, the absence of the H-type LDH in heart tissues prevents the heart from using lactate as an energy source. That this does not lead to serious consequences is clear from the fact that in diabetes the heart uses fatty acids almost exclusively as the energy source (17). It is our view that the lack of the H-type isoenzyme may not have serious consequences as long as it is a true enzyme deficiency. On the other hand, one would expect that a lack of the M-type in muscle, or replacing the H-type with the M-type LDH in heart, would be lethal.

HYBRIDS OF THE LDH ISOENZYME

The various functions of LDH described above are obviously illustrations of extreme cases, that is, where the tissue metabolism is fully aimed toward a single goal (e.g., energy production in heart muscle). Many tissues will have metabolic features supporting a variety of these goals. Even though the extreme cases would call for either the H-type or the M-type LDH, most tissues would be expected to have a mixture of the two isoenzymes with perhaps a predominance of one of the two types. Since the subunits of LDH are catalytically independent of each other, the presence of hybrid isoenzymes would not affect their individual functioning.

RECENT RESULTS IN SUPPORT OF THE THEORY

No theory has scientific value unless its validity can be experimentally tested. Our theory is based on the premise that the M-type LDH operates *in vivo* as an unregulated enzyme whereas the H-type LDH is subject to metabolic control. The theory predicts that the H-type LDH in tissues, such as heart muscle, should be present in a substantially inhibited form in order to promote the aerobic oxidation of pyruvate. The M-type in white skeletal muscles, on the other hand, should be present as a fully active enzyme.

Many *in vitro* experiments have been done to demonstrate that the H-type LDH becomes inhibited in the presence of physiological concentrations of NAD^+ and pyruvate, whereas the M-type enzyme is inhibited to a much lesser extent or not at all under these conditions. However, this does not prove that a similar situation exists *in vivo*.

More direct evidence would be obtained if one demonstrates that heart muscle, unlike skeletal white muscle, contains an inhibited form of the enzyme. This is difficult experimentally because the inhibited enzyme (ternary complex) starts dissociating (and thus reactivating) at the moment that a cell is broken, simply because of mass action:

$$\underset{\text{(Ternary Complex)}}{(\text{LDH} \cdot \text{NAD} \cdot \text{pyruvate})} \longrightarrow \text{LDH} + \text{NAD}^{+} + \text{pyruvate}$$

The dissociation is rather slow, however, and we were, therefore, able to demonstrate the presence of the inhibited enzyme *in vivo* using a stopped-flow apparatus. We already reported the results of such experiments using heart and muscle tissues from chicken (12, 13). More recently we have done similar experiments using mammalian tissues such as hearts, muscles, and livers from rats, rabbits, and dogs. These experiments yielded results similar to those obtained with the chicken tissues: an estimated 70 - 80% of the H-type LDH in heart muscle is in the inhibited form whereas the M-type LDH was not inhibited (Fig. 3). (Details of these experiments will be presented elsewhere). These experimental results provide strong evidence for the concept that the LDH isoenzymes operate *in vivo* in the manner outlined earlier in this review.

MISCONCEPTIONS CONCERNING SCOPE OF THE THEORY

When the aerobic-anaerobic theory was first proposed (1) we wanted to show that the different catalytic properties of the H- and M-type LDH may have a definite metabolic significance. The theory was illustrated by comparing the (anaerobic) metabolism of striated muscle with that of (aerobic) heart muscle. We certainly did not wish to suggest that the biological significance of the isoenzyme was limited only to muscle tissue. Our intention was to state, as we have done more explicitly in this and other articles (12, 13), that a careful consideration of the metabolic processes that take place in a given tissue or cell type leads one to the conclusion that, first, LDH plays different roles in different cell types, and second, that these roles are hard to reconcile with a single enzyme activity but are consistent with separate roles for the two isoenzymes. The association of the M-type LDH with anaerobic metabolism and of the H-type LDH with aerobic metabolism

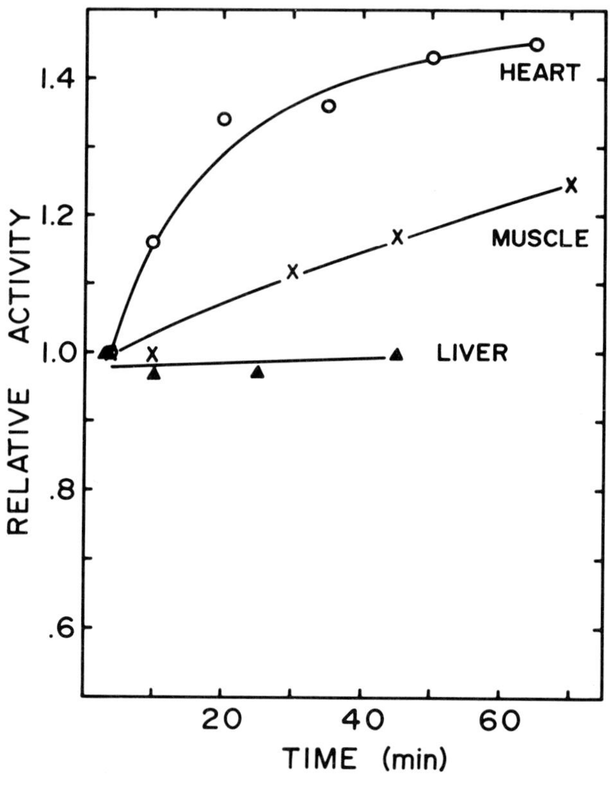

Fig. 3. Increases in activity rates of LDH from dog hearts, skeletal muscle, and liver following disruption of the tissue. Four grams of fresh heart or muscle tissue (1 g of fresh liver tissue) were minced in a Polytron homogenizer for 30 sec with 20 ml of 0.1 M phosphate buffer, pH 7.4. The homogenate was filtered over a Büchner funnel and the filtrate was transferred to one of the syringes of an Aminco-Morrow stopped-flow apparatus. The other syringe contained 0.28 mM NADH and 1 mM pyruvate in the same buffer. The time the homogenization was started is indicated as zero minutes, and the activity during the first assay was arbitrarily set at 1.0. Dog hearts, muscle, and liver contains 90, 45, and 10% H-type LDH, respectively.

was valid for the examples that we chose, but we did not suggest that this was universal. In fact, we were well aware at that time of the exception presented by the liver (1). Similarly, the designation of the M-type LDH as a pyruvate reductase and of the H-type as a lactate dehydrogenase is only valid for the muscles that we used as examples and should not be taken as a general rule.

Most of the confusion, it seems to us, has arisen from the use of these terms in a more general sense than we intended, and especially from the fact that our theory became known as the aerobic-anaerobic theory, and was overinterpreted. We certainly did not have any idea that our suggestions would become a dogma but rather a concept which might predict the levels of various LDH isozymes in a given cell type, and which would be of use in explaining the metabolic activities of different cell types.

SUMMARY

In this chapter we have attempted to correct a number of misconceptions and misinterpretations that have arisen from the use of the theory concerning the significance of the LDH isoenzymes, commonly referred to as the aerobic-anaerobic theory. The concept that the H-type LDH is to be found in tissues with an aerobic metabolism whereas the M-type LDH is associated exclusively with anaerobic tissues is wrong, even though the results may be confirmatory in most cases. Similarly, classifying the M-type-containing liver as an exception to the aerobic-anaerobic theory is a misinterpretation of the theory. In attempts to predict the presence of a given LDH isoenzyme in a certain tissue one has to consider the metabolic function of LDH in that tissue. If there are reasons why the reduction of pyruvate should not occur *ad libitum* or if there is any reason why it would be advantageous to the cell to control the activity of the LDH one would expect to find the H-type isoenzyme. In all other cases, one would predict the M-type LDH to be present. Such considerations are consistent with the concepts of our theory.

It is unfortunate that the theory became known as the aerobic-anaerobic theory since this name is definitely misleading. The concept of the theory deals with the metabolic function of the isoenzymes in a cell, not with the aerobicity of the cell's metabolism. We hope that the information presented in this chapter will help to clarify our ideas concerning the importance of the LDH isoenzymes in cellular metabolism.

REFERENCES

1. Cahn, R. D., Kaplan, N. O., Levine, L., and Zwilling, E. (1962). *Science 136*, 962-969.
2. Flores, R. (1979). *Trends Biochem. Sci. 4*, N32-N33.
3. Huijing, F. (1979). *Trends Biochem. Sci. 4*, N132.
4. Heffron, J. J. A. (1979). *Trends Biochem. Sci. 4*, N133.
5. Svasti, J. (1979). *Trends Biochem. Sci. 4*, N133.
6. Masters, C. J. (1979). *Trends Biochem. Sci. 4*, N276.
7. Bing, R. J., Siegel, A., Vitale, A., Balboni, F., and Sparks, E. (1953). *J. Clin. Invest. 32*, 556.
8. Bing, R. J. (1965). *Physiol. Rev. 45*, 171-213.
9. Krasnow, E. M., Neill, W. A., Messer, J. V., and Gorlin, R. (1962). *J. Clin. Invest. 41*, 2075-2085.
10. Everse, J., Barnett, R. E., Thorne, C. J. R., and Kaplan, N. O. (1971). *Arch. Biochem. Biophys. 143*, 444-460.
11. Everse, J., Berger, R. L., and Kaplan, N. O. (1970). *Science 169*, 1236-1238.
12. Everse, J., and Kaplan, N. O. (1973). *Advan. Enzymol. 37*, 61-133.
13. Everse, J., and Kaplan, N. O. (1975). *In* "Isozymes II, Physiological Function" (C. L. Markert, ed.), pp. 29-43. Academic Press, New York.
14. Fine, I. H., Kaplan, N. O., and Kuftinec, D. (1963). *Biochemistry 2*, 116-121.
15. Dawson, D. M., Goodfriend, T. L., and Kaplan, N. O. (1964). *Science 143*, 929-933.
16. Kitamura, M., and Hishina, J. (1975). *In* "Isozymes II, Physiological Function" (C. L. Markert, ed.), pp. 97-111. Academic Press, New York.
17. Schlant, R. C. (1978). *In* "The Heart" (J. W. Hurst, R. B. Logue, R. C. Schlant, and N. K. Wenger, eds.), pp. 107-118. McGraw-Hill, New York.

ADDENDUM

It is of some interest to note that recently Anderson and Kovacik (*Proc. Natl. Acad. Sci. U.S. 78,* No. 5, 3209-3213, May, 1981) have reported that a new form of LDH exists in human tumors which may relate to the rat-specific gene product of Kirsten murine sarcoma virus. The enzyme has been reported to work only at low oxygen concentrations. It has properties similar to M_4LDH. Much further work is required before a definite conclusion can be made that this enzyme is a novel entity and particularly characteristic of neoplastic tissues.

SOME OBSERVATIONS ON THE EVOLUTION OF THE REGULATION OF THE *de novo* PATHWAY FOR UMP BIOSYNTHESIS IN PROKARYOTES

Mary Ellen Jones
Richard I. Christopherson

Department of Biochemistry
University of North Carolina
Chapel Hill, N.C. 27514

Dr. Jones' interest in the regulation of UMP biosynthesis was a rather direct result of her years (1957 - 1966) with the Graduate Department of Biochemistry at Brandeis University and the exciting intellectual interchange that existed in that new department because of the leadership of Nathan Kaplan and Martin Kamen.

There was a lively interest in the department during those years in the evolutionary aspects of biochemistry: Martin Kamen was investigating photosynthesis as it occurred in a number of species; Nathan Kaplan was looking at the isozymes of lactic dehydrogenase of various tissues and species; Julius Marmur was comparing the (G+C)/(A+T) ratios of DNAs from various species and Allan Wilson was advising all of us on the species and their interrelationships to one another.

Dr. Jones' interests were really not in this direction. Rather, she was concerned with the mechanism of the carbamyl-P synthetases (1 - 6). However, a fellow from Israel, Dr. Joseph Neumann, who had trained as a botanist, was anxious to learn enzymology. He originally selected to work on the regulatory

 ISBN 0-12-397580-8

TABLE I. Prokaryotic ATCases[a]

	Physical characteristics			
	Stokes radius (ref. 9)	Molecular size of the		
Class	(mμ)	Holo-enzyme	Catalytic polypeptide[b]	Regulatory polypeptide
Organisms with ATCase				
A	8.5	360,000	180,000(2)	-
B	6.5	300,000	33,000(6)	17,000(6)
C	4.0	130,000	32,500 (3 or 4)[d]	None
		100,000	33,500(3)	None
Organisms with Two ATCases				
B	6.5	250,000	N.D.	N.D.
C	4.0	93,000	N.D.	N.D.

[a]The quantitative values listed above are derived from studies on the first named organism or the organism listed without brackets. The organisms listed in brackets were placed in the category indicated by the studies of Bethell and Jones (9), and the limited kinetic data available also places them in the indicated category.

Regulated by nucleotides	Kinetic characteristics[c] Asp. (K_m, mM)[b]	CAP (K_m, μM)[b]	% G + C (ref.10)	Organisms	Ref.
+	H (3)	H (14)	60 - 62	Ps.fluorescens	8,9
+				(Az.vinelandii)	11
+	S (6)	S (200)	50 - 62	E. coli (S.typhimurium)	9,12-15
+					
-	H (10)	H (8)	34 - 36	S.faecalis	9,16-18
-	H (5.7)	H (500)	42 - 44	B.subtilis	8,9,19
+	H (20)	H (70)	50 - 52 (38 - 40	C.freundii P.vulgaris)	9,20
+	H (20)	H (40)			

[b]The bracketed values represent the number of polypeptide chains present in the holoenzyme. The bracketed values under the ASP and CAP represent the K_m values for these two substrates in mM or μM terms, respectively.

[c]Hyperbolic (H) or Sigmoid (S) substrate curve.

[d]The tetramer structure of the S. faecalis enzyme rests mainly on the determination of the molecular weight of the holoenzyme (17) if this value is in error it is possible that this enzyme is a trimer, i.e. C_3.

properties of lettuce aspartate transcarbamylase (ATCase). When the characteristics of the lettuce enzyme were found to be different from those of the *Escherichia coli* enzyme (7), he decided to make a survey of various microorganisms to see the type of variability that could occur. Dr. Neumann's studies (8) and Dr. Martha Bethell's thesis work at Brandeis (9) showed that the bacteria could be divided into at least three major groups (A, B, or C) on the basis of the size of their ATCases (Table I). The class C ATCases consist of a single catalytic trimer (C_3) or tetramer (C_4).* The higher molecular weight of the *Enterobacteriacae* (class B) ATCases is due to the presence, in two instances (*E. coli* and *Salmonella typhimurium*), of two catalytic trimers (C_3) and three regulatory dimers (R_2) to yield a holoenzyme containing twelve polypeptides $(C_3)_2(R_2)_3$. A regulatory subunit has been observed only in the ATCases of the Enterobacteriacae (8, 9, 12, 13). The class A enzymes contain two polypeptides having a molecular weight of 180,000. At least one of these must contain the catalytic site and if the polypeptides are chemically identical then the regulatory and catalytic sites would be on the same polypeptide chain to yield an $(R\text{-}C)_2$ structure.

There are many kinetic similarities between the similarly sized catalytic subunits of the class B and C organisms, i.e., the nature and detailed structure of the dicarboxylic acids which are inhibitors (21, 22) (Table II), and the strong sensitivity to the inhibitor, inorganic pyrophosphate (21, 24). However, the unregulated class C enzyme from *Streptococcus faecalis* differs markedly from the *E. coli* holoenzyme (class B) which is the classic example of an enzyme which has sigmoid substrate curves that can be modulated by allosteric effectors (12, 13). The aspartate concentration curve for *S. faecalis* ATCase appears hyperbolic (17) unless the anion concentration is held very low (about 1 m*M*) in which case the velocity as the aspartate concentration is raised does show positive cooperativity (i.e., a sigmoid inflection at subsaturating aspartate concentrations); however, a marked inhibition by high concentrations of aspartate is always apparent with this enzyme (16, 17). Studies of the anion effects (17, 18) indicate that the active site of this catalytic subunit can bind an anion activator and that the order of the substrate addition changes (Fig. 1) if such an activator is present. There is reason to believe that the substrate, aspartate, may be the natural anion activator (17); in this case aspartate would bind twice to the

**The tetramer structure of the S. faecalis enzyme rests mainly on the determination of the molecular weight of the holoenzyme (17) if this value is in error it is possible that this enzyme is a trimer, i.e.* C_3.

TABLE II. Comparison of the Effects of Dicarboxylic Acids as Competitive Inhibitors of S. faecalis and E. coli ATCases

	K_m or K_i values (in mM) for enzyme from		
Dicarboxylic Acid	*S. faecalis*[a]	*E. coli*[b]	
Aspartate	10	20	
Maleate	1.3	4.3	
Succinate	3.4	3.5	
Acetylene dicarboxylate	110	-	
Fumarate	$\geq$ 300	145	
D-Malate	20	-	(11)
L-Malate	60	60	(12)
meso-Tartarate	11	2.7	
Oxalate	> 150	-	
Malonate	50	-	
Succinate	3.4	3.5	
Glutarate	N.O.[c]	260	
L-Glutamate	> 150	120	

[a]*These values were determined by using two L-aspartate concentrations, 4 and 8 mM, with sufficient potassium acetate (800 mM) to fill the anion activator site for this enzyme with acetate. The dicarboxylic acid (I) was varied and the place where the two lines intersected in the -I quadrant of the Dixon plot was taken as the K_i value. The Tris buffer was pH 8.5 and the carbamyl-P was 10 mM. Samples were incubated at 37°C for 10 min. Carbamyl aspartate was determined colormetrically (21).*

[b]*Values from references 22 and 23. Values in parenthesis are for pH 6.9; all other values are for pH 7.9 or 8.*

[c]*N.O. = no inhibition observed.*

enzyme, as in Scheme b of Fig. 1. There is no specific effect of nucleotides on this enzyme or another class C enzyme from *Bacillus subtilis* which has very similar kinetic properties. The subunits of the *S. faecalis* and *B. subtilis* enzymes are quite similar in size to the catalytic polypeptides of the *E. coli* enzyme but the class C enzymes do not cross-react with antibody to the *E. coli* catalytic subunit (17, 19). Since the *E. coli* catalytic subunit does not show the sigmoid substrate

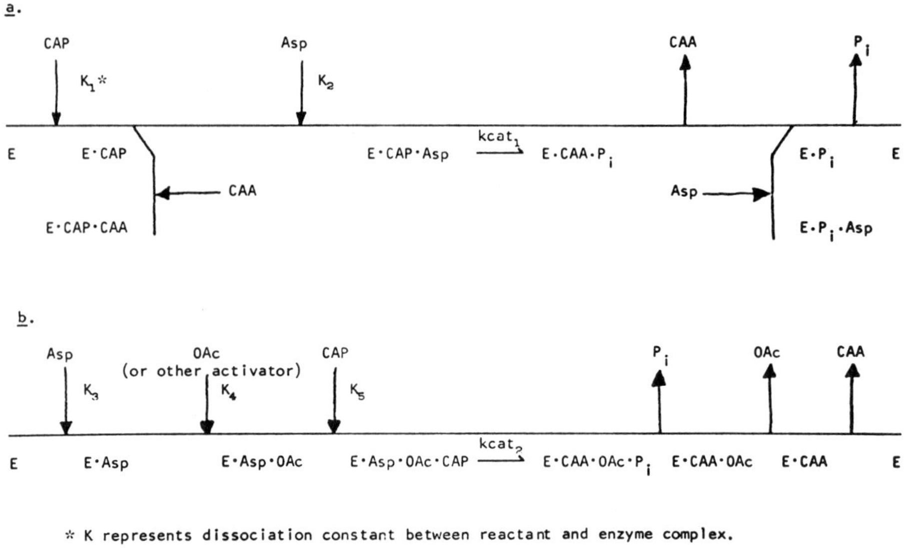

Fig. 1. Order of substrate addition for S. faecalis ATCase. (a) Order of substrate addition when total anion concentration is low, (b) order of substrate addition when total anion concentration is high and the anion activator site is occupied. Abbreviations used: E, S. faecalis ATCase; CAP, carbamyl phosphate; CAA, carbamyl L-aspartate; Asp, L-aspartate P_i, orthophosphate; OAc, acetate.

curve characteristic of the holoenzyme, it is possible that there is a large degree of similarity between the catalytic sites of the ATCases from these organisms. The subsequent development of the regulatory subunit would confer tremendous advantage on such organisms as the Enterobacteriacae whose existence in the lower intestine of mammals is subject to a spasmodic input of nutrients. The *in vivo* control of the class C organisms, such as *S. faecalis*, is still unknown and such knowledge would require whole cell studies.

The survey of the bacterial ATCases raised two other interesting problems in the regulation of pyrimidine biosynthesis for microorganisms. *Proteus vulgaris* and *Citrobacter freundii* have two ATCases (see Table I) (9). Only the *Citrobacter* enzymes have been studied further (20) and even here our studies are still preliminary. These studies suggest, however, that the smallest *Citrobacter* enzyme, which sieves as if it had a MW of 93,000, is the major form of the enzyme present in these cells during logarithmic growth of this organism. The larger enzyme, sieving as if its molecular weight were 250,000, is the only enzyme present during the stationary phase of cell growth.

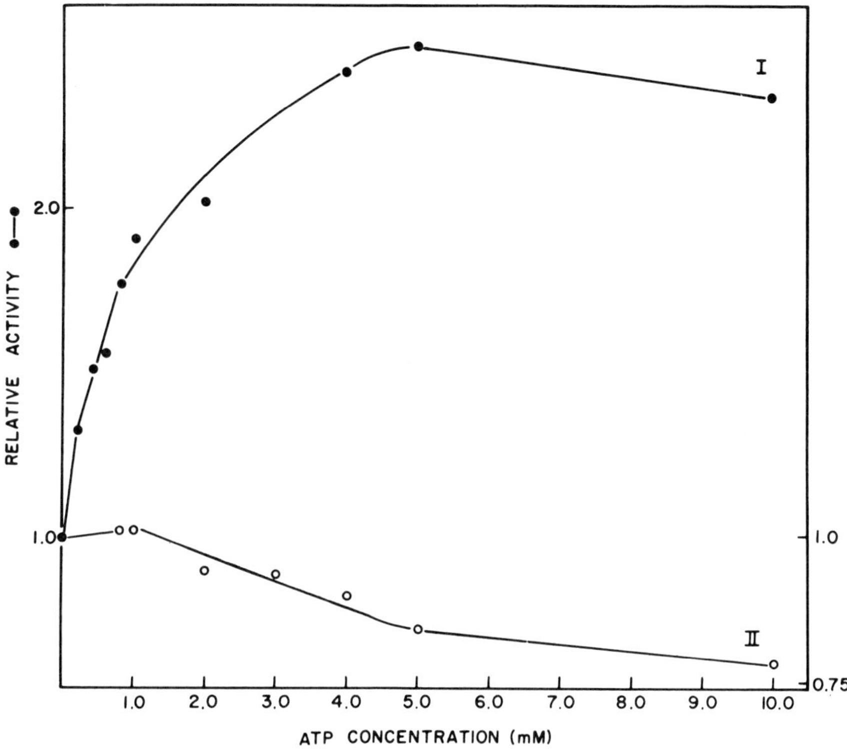

Fig. 2. Effect of ATP on ATCase I (MW of 250,000) and ATCase II (MW of 93,000) when aspartate concentrations were low (15 mM). Experimental details are in reference (20).

The two enzymes have almost identical kinetic behavior except that the large enzyme is activated by ATP (Fig. 2). This strong ATP activation suggests that the enzyme that predominates in stationary phase of culture might be sensitive to adenylate charge (25) and therefore become more active when renewed nucleotide or nucleic acid biosynthesis can occur, but it may be inactive when the adenylate charge is low. *In vitro* the large enzyme can be converted to the small enzyme (20) when it is subject to a large dilution. One can also ask if this conversion could also occur *in vivo* when the cell culture is diluted with fresh medium to induce logarithmic growth?

The possibility has been raised that the small enzyme is always a proteolytic product of the larger enzyme (26). Only experiments, as yet not done, conducted on homogenates prepared in the presence and absence of protease inhibitors, could clarify this possibility. Prokaryotes have low levels of proteolytic enzymes particularly during logarithmic growth (27, 28); the level of such activity increases in the stationary phase (29, 30). It, therefore, might be predicted that if proteolysis of a large enzyme were the reason for the appearance of a small enzyme that it should be observed in the cells harvested in stationary phase. Since the opposite is the case, i.e., the small enzyme is observed in logarithmic growth while the large enzyme occurs in the stationary phase, we are inclined to believe that the appearance of the small enzyme in logarithmic cells and the appearance of the large enzyme in the stationary phase do not represent experimental artifacts but are related to the metabolic events in these growth phases of the *Citrobacter* cells.*

The exact molecular weight of these two enzymes also requires more detailed studies since the sieving experiments could or could not be compatible with a monomer - dimer interconversion. The ability to isolate a rather stable "dimer," i.e., the larger ATCase, would not be compatible with a rapidly equilibrating monomer - dimer interconversion and therefore might indicate that the presence of a special small protein is essential for creation of the stable "dimer," or alternatively that formation of covalent disulfide bridges between subunits might account for the aggregation.

The third group of bacterial ATCases (class A) is found in *Pseudomonas* and *Azotobacter* (9, 11). This ATCase is most unusual for it is composed of two very large subunits with an apparent molecular weight of 180,000 each. The large subunit size suggests that this protein might be multienzymatic (31, 32), i.e., have more than one enzyme activity per polypeptide chain. The simplest eukaryotes, yeast and *Neurospora*, have a large subunit with ATCase activity. The polypeptide contains two enzyme centers, namely a pyrimidine-specific carbamyl-P synthetase (CPSase) and an ATCase (33, 34). In mammals, also, a large subunit of 200,000 daltons may contain three enzyme centers, a pyrimidine-specific carbamyl-P synthetase, ATCase, and dihydroorotase (35 - 39). In our studies on the *Pseudomonas* enzyme, we tested to see if the CPSase activity of the cell copurified with the ATCase activity. Since it did not do so, Drs. Adair and Jones (11) did not consider this a multienzymatic protein; however, they did not test to see if the synthetase was enriched in another fraction. The possibility remains, therefore, that the synthetase subunit is larger because it is a multienzymatic protein which contains a CPSase active

**When undiluted fractions containing the large ATCase I were held at 4°C for 10 days no ATCase II (the small class C) enzyme was formed.*

site plus an ATCase active site. The active site of the mammalian CPSase is readily denatured even though the 200,000 dalton subunit retains its size and also retains most of the ATCase and dihydroorotase activities (35, 37, 40). This rapid and selective loss of the CPSase activity was the major reason that the glutamine-dependent carbamyl-P synthetase of mammals was not discovered for many years (37).

The *Pseudomonas* enzyme has two inhibitory sites for nucleotides. When carbamyl-P concentrations are near the K_m value (14 μM), phosphate compounds, particularly inorganic pyrophosphate and trinucleotides, can bind competitively versus carbamyl-P. A second characteristic usually observed when nucleotides can bind at the carbamyl-P sites is that the nature of the nucleotide base is relatively unimportant (11, 20, 52). The K_i values for trinucleotide versus carbamyl-P are near 8 μM and are, therefore, low enough that this competition could be important *in vivo* unless this ATCase is part of a multienzymatic protein possessing a CPSase so that the carbamyl-P formed by the synthetase does not have to enter the cytosol but is retained at a high local concentration in the small space between a synthetase and an ATCase site. Such channeling of carbamyl-P has been observed with the yeast CPSase: ATCase multienzymatic protein (33) and is known to occur *in vivo* in *Neurospora* (34), but does not occur in *E. coli* since the enzymes are distinct. The second nucleotide binding site for the *Pseudomonas* enzyme can be observed when the solvent carbamyl-P concentration is high enough to keep the carbamyl-P site saturated. Under this condition and at low (near K_m or below) aspartate concentrations, nucleotides inhibit noncompetitively with aspartate. There is now some selection for the nucleotide base, i.e., GTP and CTP (or CDP) are more effective than UTP and ATP.

A tremendous vacancy exists in the area of regulation of pyrimidine biosynthesis in that whole cell studies need to be done to determine which of the kinetic features observed with the isolated enzymes for *de novo* synthesis of pyrimidine nucleotides are important *in vivo* [despite the fact that this was the thrust of the early Yates and Pardee experiments (41)]. Figure 3 summarizes our current knowledge of the regulation of pyrimidine biosynthesis in the *Enterobacteriacae*. The rate of synthesis of pyrimidine nucleotides may be regulated in three main ways. Feedback repression by pyrimidine end products on the syntheses of enzymes of the pathway provides a slow, coarse form of control, but this slow control is most energy conserving. Studies using growing cells of appropriate mutants have enabled the corepressors to be identified either as uracil or cytosine compounds (Fig. 3), but the actual active corepressors have not been determined (42 - 44). Regulation of enzymatic activity by changing concentrations of substrates and ef-

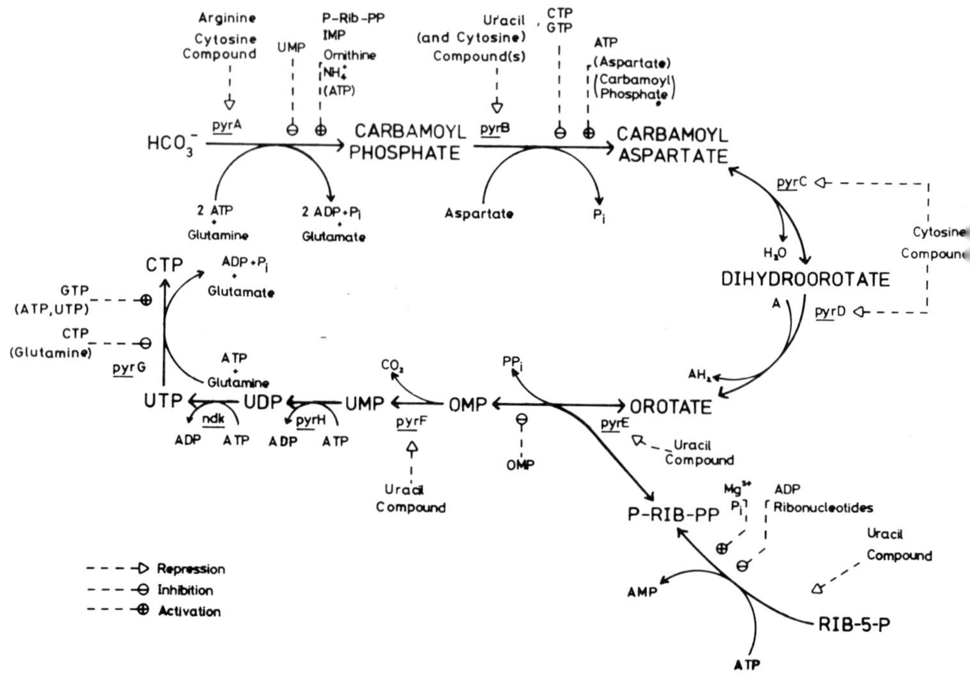

Fig. 3. Regulation of the pyrimidine biosynthetic pathway of E. coli and S. typhimurium. The synthesis of P-Rib-PP by P-Rib-PP synthetase has been included because the supply of P-Rib-PP may limit the flux through the pathway under some conditions. Genetic symbols for the enzymes are shown and the gene designations are as follows: pyrA, carbamoyl phosphate synthetase (EC 2.7.2.9); pyrB, aspartate carbamoyltransferase (EC 2.1.3.2); pyrC, dihydro-orotase (EC 3.5.2.3); pyrD, dihydro-orotate dehydrogenase (EC 1.3.3.1); pyrE, orotate phosphoribosyltransferase (EC 2.4.2.10); pyrF, OMP decarboxylase (EC 4.1.1.23); pyrH, UMP kinase (EC 2.7.4.4); ndk, nucleosidediphosphate kinase (EC 2.7.4.6); pyrG, CTP synthetase (EC 6.3.4.2); to our knowledge, the genetic locus for ribose phosphate pyrophosphokinase (EC 2.7.6.1) has not been mapped. The effects of metabolites as corepressors regulating the expression of gene loci ---▷ and as activators ---⊕ or inhibitors ---⊖ of enzymatic activity are shown. Compounds in brackets are substrates for that reaction exhibiting positive ---⊕ or negative ---⊖ homotropic effects. The role of a cytosine compound as a corepressor in the expression of pyrB remains uncertain at this time.

fector molecules (pyrimidines and related metabolites) provides an immediate fine control on the flux through the pathway. The mechanisms proposed to regulate the pathway *in vivo* (Fig. 3) have been determined from studies performed with isolated enzymes *in vitro*. Some work has been done using intact cells to determine which of these proposed mechanisms actually operate *in vivo*. Such work is difficult because most of the proposed effectors shown in Fig. 3 are impermeable to the intact cell and/or are degraded in the periplasmic space prior to assimilation. A valid procedure for looking at regulation of the pathway by effectors *in vivo* is to perturb the pathway by addition of an appropriate compound and simultaneously measure the concentrations of the effectors and intermediates of the pathway and the relative activity of the enzyme under investigation. Changes in the total flux through a particular enzymatic step, as indicated by changes in the concentrations of substrates and products may then be correlated with any changes observed in the concentration of the effector of interest. Such correlations must be made cautiously, however, because of the undefined, heterogeneous nature of the cytosol of a growing cell.

Such a study has been carried out in the laboratory of Dr. Lloyd Finch of the Biochemistry Department, University of Melbourne, Australia by one of us (45). Cultures of *Escherichia coli* K12 were grown exponentially in glucose-minimal salts medium, and then adenine or uracil added to yield a final concentration of 0.1 m*M*. Samples were taken at appropriate times relative to the adenine or uracil addition for determination of carbamyl phosphate, aspartate, and orotate concentrations as well as the CPSase, ATCase, and orotate phosphoribosyltransferase (OPRTase) activities.

Earlier work by Bagnara and Finch (46) had shown that addition of adenine to a culture of *E. coli* K12 resulted in a rapid depletion of P-Rib-PP to 5% of its initial intracellular concentration of approximately 500 μ*M* and also caused decreases in UTP and CTP levels and increases in ATP and GTP levels. The drastic decrease in R-Rib-PP resulted in inhibition of the pyrimidine pathway, as indicated by depletion of UTP and CTP, which could occur at two possible sites: P-Rib-PP is a positive effector of CPSase (47) and a substrate for the synthesis of OMP. Christopherson and Finch (49) found that after addition of adenine, orotate accumulated in the cells and subsequently leaked into the culture medium, demonstrating that when a limitation of P-Rib-PP as a substrate for the synthesis of UMP is created, that the synthesis of OMP is the primary site of inhibition of pyrimidine biosynthesis. After addition of adenine, ATCase and OPRTase were derepressed to 3.05 and 1.90 times their initial activities, respectively, while CPSase showed only a slight increase of activity. The increases in these enzyme activities coincided with decreased intracellular concentrations of UTP and CTP and presumably, therefore, with a decreased concentration of the active corepressors of these enzymes (Fig. 3).

After addition of uracil to a culture of *E. coli*, the cellular doubling time decreased from 65 to 53 min, P-Rib-PP decreased to 60% of its initial concentration and then oscillated, UTP and CTP levels increased, while ATP and GTP levels decreased (46). Carbamyl phosphate repidly decreased to 20% of its initial intracellular concentration of approximately 840 μ*M* and continued to decrease for the 40-min duration of the experiment. This latter effect may be attributed to elevated concentrations of UMP synthesized from uracil since CPSase is progressively inhibited by UMP concentrations in the range of 5 to 1.6 m*M* (48, Fig. 3).

Anderson and Meister found that CPSase shows a sigmoidal kinetics with respect to ATP concentration and can be maximally sensitive to changes in ATP about a concentration of 4 m*M* (48), which approximates the *in vivo* intracellular ATP level (46). The decrease in carbamyl phosphate observed after addition of uracil is also consistent with a decrease in ATP (46), a substrate for its synthesis. After addition of adenine, there is an initial increase in carbamyl phosphate which coincides with an increase in the concentration of ATP (46). Thus the sigmoidal kinetics of CPSase observed *in vitro* seem to apply *in vivo* resulting in synthesis of carbamyl phosphate according to the energy status of the cell.

Under conditions of balanced growth in minimal medium, the intracellular concentration of carbamyl phosphate was 4.1 μmol/g protein (approximately 840 μ*M*) and the total level of intracellular and extracellular orotate was 61 μmol/g protein [approximately 3.8 m*M* intracellular, 2.3 μ*M* extracellular (49)]. Both carbamyl aspartate and dihydro-orotate accumulate to double the levels of orotate in cultures grown in the presence of ^{14}C-aspartate (50), while OMP and UMP were not detectable by the assay procedures used. The cellular growth rate is stimulated equally by exogenous provision of uracil or aspartate, but addition of both compounds together provides no additional stimulation. Exogenous provision of uracil spares aspartate from utilization in *de novo* biosynthesis of pyrimidine nucleotides and results in an initial elevation in the concentration of aspartate (49). The overall effect of diversion of aspartate to orotate unchecked by an effective feedback inhibition from pyrimidine nucleotide end products (Fig. 3) would be an unbalanced demand on the total pool of dicarboxylic acids resulting in limitation of cellular growth by a wastage of precursors related to aspartate. After addition of uracil, the activity of CPSase and ATCase increased considerably and there was a slight increase in OPRTase. In contrast to the result with adenine, these increases in enzyme levels coincide with increases in UTP and CTP and presumably the active corepressors of these enzymes (Fig. 3). A possible explanation for this unexpected observation would be that aspartate is spared from *de novo* pyrimidine biosynthesis resulting in an increased availability of some amino acid(s) for protein synthesis.

From these studies, Christopherson and Finch (49) concluded that the OPRTase step is rate limiting to the pathway during normal growth of *E. coli* K12 in the laboratory and that exogenous provision of adenine results in a rapid consumption of P-Rib-PP (46) whose availability as a substrate for OPRTase then limits the rate of pyrimidine biosynthesis. The adaptive responses of the pyrimidine pathway of *E. coli* K12 to addition of adenine or uracil were consistent with some of the regulatory mechanisms proposed from *in vitro* work (Fig. 3). Only qualitative correlations could be made because of the difficulty in transposing results obtained in isolation in the test tube to events occurring in a growing cell. Despite these difficulties it seems worthwhile to make the effort to test the relative importance of regulatory sites discovered in studies of the isolated enzymes in whole cells.

The diversity of the model of regulation of the pyrimidine pathway of the Enterobacteriacae (Fig. 3) evolved in response to a strong selective pressure for an organism, which can rapidly utilize preformed nutrients as they become available in the host's diet, and grow at the maximal possible rate. When the supply of nutrients is low, these cells must be able to synthesize their own cellular components from simple precursors. Experiments where *E. coli* was grown in minimal medium containing radiolabeled aspartate as a precursor for *de novo* synthesis of pyrimidines and radiolabeled uracil was subsequently added, showed that 30 min after addition of uracil only 20% of the acid-soluble UTP and CTP was derived from *de novo* synthesis and 80% was formed from exogenous uracil (49). The total UTP and CTP contents were similar to those at the time of uracil addition. It is clear from these experiments that *E. coli* K12 has the capacity to rapidly switch from making pyrimidine nucleotides *de novo* to synthesizing them via salvage pathways from exogenous preformed bases.

From the survey made of bacterial ATCases (Table I) we may speculate as to how this molecular diversity arose during the course of evolution. The ATCases of class C may be most closely related to a primordial catalytic subunit. For the class B ATCases of the Enterobacteriacae, a pair of catalytic trimers of ATCase may have subsequently acquired three regulatory dimers which bound ATP and CTP and modulated the ATCase activity. Later acquisition of an ancillary subunit has also been postulated for the glutaminase subunit of the CPSase of *E. coli* (51).

The ATCases of class A may have a nucleotide regulatory site as a portion of the catalytic subunit. Although the nucleotide inhibition observed at low carbamyl-P concentrations might represent nucleotide binding in place of carbamyl-P at the active site, the nucleotide inhibition observed at high concentrations of this substrate probably does not occur at the active site. Therefore, a special nucleotide binding site may exist. One explanation for the increased size of the class A subunit would represent initial fusion of a R gene and a C gene to give rise to a single R-C gene followed by secondary fusion of duplicate R-C genes to form the large subunit observed (11). A second possibility would be the fusion of genes for the ATCase catalytic subunit with another or more than one other enzyme of the *de novo* pathway for pyrimidine biosynthesis. This second (or third, etc.) center could contain the nucleotide binding site which might control both active centers on the protein produced from the fused gene. A detailed study of the molecular and chemical characteristics of these ATCases and ultimately, a direct comparison of the degree of homology of their amino acid sequences might well answer these questions and provide new criteria for bacterial taxonomy at a fundamenta level.

APPENDIX

I (MEJ) first met Martin Kamen in the months when negotiations were proceeding for the creation of the Graduate Department of Biochemistry at Brandeis University, and it was my good fortune to have 9 years in the stimulating intellectual environment of this entirely new department, and to be part of its growth. The first four of these 9 years Martin Kamen was part of our faculty and Martin Kamen's personal attributes, namely a cheerful, positive, and forward-moving outlook typified the department as a whole. His attitude that a fellow professor's problem was nearly as important as his own helped to create the unified department achieved at Brandeis during these years. He never seemed too busy to listen, and he usually had a prompt and positive suggestion for a solution to a problem. Alternatively, he could persuade one that maybe the problem was not that serious. These characteristics have created as one of his major contributions to biochemistry in this country an important role in the building of a newly formed department (Brandeis) or newly arranged departments (at La Jolla and at Southern California). The credit one receives for the large number of hours spent in such service is usually small; the friendships and gains are, however, numerous and lasting. Martin deserves special citation at each of these Universities for the unique role he has played. During all this time he has managed to maintain a creative research program as well as his beloved musical skill. He is a man for all seasons and moods.

ACKNOWLEDGMENT

Research in the laboratory of Dr. Jones was supported by grants generously provided by the National Institute of Child Health Development of the NIH and from the NSF. The research from Dr. Finch's laboratory was supported by the Australian Research Grants Committee and a Commonwealth Postgraduate Research Award to R.I.C.

REFERENCES

1. Jones, M. E. and Spector, L. (1960). *J. Biol. Chem. 235,* 2897-2901.
2. Allen, C. M., Jr. and Jones, M. E. (1966). *Arch. Biochem. Biophys. 114,* 115-122.
3. McKinley, S., Anderson, C., and Jones, M. E. (1967). *J. Biol. Chem. 242,* 3381-3390.
4. Jones, M. E. (1959). *J. Cell. Comp. Physiol. 54* (Suppl. 1), 68-69.
5. Jones, M. E. (1965). *Annu. Rev. Biochem. 34,* 381-418.
6. Jones, M. E. (1976). *In* "The Urea Cycle" (S. Grisolia, R. Baguena, and F. Mayor, eds.), pp. 107-122. Wiley, New York.
7. Neumann, J. and Jones, M. E. (1962). *Nature (London) 195,* 709-710.
8. Neumann, J. and Jones, M. E. (1964). *Arch. Biochem. Biophys. 104,* 438-447.
9. Bethell, M. R. and Jones, M. E. (1969). *Arch. Biochem. Biophys. 134,* 352-365.
10. Marmur, J., Falkow, S., and Mandel, M. (1963). *Annu. Rev. Microbiol. 17,* 329-372.
11. Adair, L. B. and Jones, M. E. (1971). *Biochem. Biophys. Res. Commun. 45,* 796-802.
12. Gerhart, J. C. and Pardee, A. B. (1962). *J. Biol. Chem. 237,* 891-896.
13. Bethell, M. R., Smith, K. E., White, J. S., and Jones, M. E. (1968). *Proc. Natl. Acad. Sci. U.S. 60,* 1442-1449.

14. Gerhart, J. C. and Schachmann, H. K. (1965). *Biochemistry 4,* 1054-1062.
15. O'Donovan, G. A., Holoubek, H., and Gerhart, J. C. (1972). *Nature New Biol. 238,* 264-266.
16. Prescott, L. M. and Jones, M. E. (1970). *Biochemistry 9,* 3783-3793.
17. Chang, Ta-Yuan and Jones, M. E. (1974). *Biochemistry 13,* 629-638.
18. Chang, Ta-Yuan and Jones, M. E. (1974). *Biochemistry 13,* 638-645.
19. Brabson, J. S. and Switzer, R. L. (1975). *J. Biol. Chem. 250,* 8664-8669.
20. Coleman, M. S. and Jones, M. E. (1971). *Biochemistry 10,* 3390-3396.
21. Jones, M. E. (1962). *Methods Enzymol.,* 903-925.
22. Davies, G. E., Vanaman, T. C., and Stark, G. R. (1970). *J. Biol. Chem. 245,* 1175-1179.
23. Porter, R. W., Modebe, M. O., and Stark, G. R. (1969). *J. Biol. Chem. 244,* 1846-1859.
24. Kleppe, K. (1966). *Biochim. Biophys. Acta 122,* 450-461.
25. Atkinson, D. E. (1968). *Biochemistry 7,* 4030-4034.
26. Makoff, A. J. and Radford, R. (1978). *Microbiol. Rev. 42,* 307-328.
27. Hogness, D. S., Cohn, M., and Monod, J. (1955). *Biochim. Biophys. Acta 16,* 99-116.
28. Willetts, N. S. (1967). *Biochem. J. 103,* 462-471.
29. Borek, E., Ponticorvo, L., and Rittenberg, D. (1958). *Proc. Natl. Acad. Sci. U.S. 44,* 369-373.
30. Willetts, N. S. (1967). *Biochem. J. 103,* 453-461.
31. Kirschner, K. and Bisswanger, A. (1976). *Annu. Rev. Biochem. 47,* 143-166.
32. Traut, T. W. and Jones, M. E. (1978). *In* "Advances in Enzyme Regulation," (G. Weber, ed.), Vol. 16, pp. 21-421 Pergamon, New York.
33. Lue, P. F. and Kaplan, J. G. (1970). *Biochim. Biophys. Acta 220,* 365-372.
34. Williams, L. G., Bernhardt, S. A., and Davis, R. H. (1970). *Biochemistry 9,* 4329-4335.
35. Shoaf, W. T., III, and Jones, M. E. (1973). *Biochemistry 12,* 4039-4051.
36. Jones, M. E. (1971). *In* "Advances in Enzyme Regulation," (G. Weber, ed.), Vol. 9, pp. 19-49. Pergamon, New York.
37. Jones, M. E. (1972). *Curr. Topics Cell. Regul. 6,* 227-265.
38. Coleman, P. F., Suttle, D. P., and Stark, G. R. (1977). *J. Biol. Chem. 252,* 6379-6385.
39. Mori, M. and Tatibana, M. (1978). *Eur. J. Biochem. 86,* 381-388.
40. Tatibana, M. and Ito, K. (1969). *J. Biol. Chem. 244,*

5403-5413.

41. Yates, R. A. and Pardee, A. B. (1956). *J. Biol. Chem. 221*, 757-770.
42. Piérard, A., Glansdorff, N., Gigot, D., Crabeel, M., Halleux, P., and Thiry, L. (1976). *J. Bacteriol. 127*, 291-301.
43. Schwartz, M. and Neuhard, J. (1975). *J. Bacteriol. 121*, 814-822.
44. Ingraham, J. L. and Neuhard, J. (1972). *J. Biol. Chem. 247*, 6259-6264.
45. Christopherson, R. I. (1976). Ph.D. Thesis, University of Melbourne, Australia.
46. Bagnara, A. S. and Finch, L. R. (1974). *Eur. J. Biochem. 41*, 421-430.
47. Abdelal, A. T. H. and Ingraham, J. L. (1975). *J. Biol. Chem. 250*, 4410-4417.
48. Anderson, P. M. and Meister, A. (1966). *Biochemistry 5*, 3164-3169.
49. Christopherson, R. I. and Finch, L. R. (1978). *Eur. J. Biochem. 90*, 347-358.
50. Christopherson, R. I. and Finch, L. R. (1977). *Anal. Biochem. 80*, 159-167.
51. Trotta, P. P., Burt, M. E., Haschemeyer, R. H., and Meister, A. (1971). *Proc. Natl. Acad. Sci. U.S. 68*, 2599-2603.
52. Suter, P. and Rosenbusch, J. P. (1977). *J. Biol. Chem. 252*, 8136.

THE PURINE NUCLEOTIDE CYCLE IN BRAIN

John M. Lowenstein

Graduate Department of Biochemistry
Brandeis University
Waltham, Massachusetts 02154

Work carried out over 50 years ago showed that slices of brain produce ammonia when incubated under suitable conditions (1). This observation has been confirmed on many occasions both *in vivo* and *in vitro*. For example, ammonia appears during ischemia of the brain and electrical stimulation of peripheral nerves. Ammonia disappears from the brain during sleep and anesthesia (2, 3). For many years the explanation commonly given to account for ammonia formation involved transamination of amino acids with α-ketoglutarate to form glutamate, and the subsequent deamination of glutamate by glutamate dehydrogenase.

It was shown previously that cell-free extracts of rat brain contain the enzymes of the purine nucleotide cycle and that the operation of the cycle can be demonstrated in such extracts under suitable conditions (4). Experiments are described below in which transient changes in energy metabolism of the whole brain of rat are elicited with electrical shock treatment for 10 seconds. This causes massive neuronal firing and membrane depolarization. The ionic gradients are then re-

 ISBN 0-12-397580-8

established by sodium-potassium ATPase. The resulting energy drain is accompanied by changes in the contents of adenine and hypoxanthine nucleotides and nucleosides as well as ammonia and adenylosuccinate. The results demonstrate the operation of the purine nucleotide cycle in brain.

MATERIALS AND METHODS

Male rats of the Sprague-Dawley strain weighing 225 to 250 g were stimulated electrically as described previously (5). Brains were sampled by the blow-and-rapid-freeze technique of Veech *et al.* (6). Metabolite analyses were performed using enzymatic methods to produce a change in absorption in the metabolite itself or, where this was not possible, by linking the enzyme reaction to formation or removal of NADH either directly or via coupling enzymes (5). Absorbance changes were measured on a Perkin Elmer double-beam spectrophotometer (model 356); this instrument permits the measurement of 0.1 to 0.2 nmol of metabolite per milliliter of tissue extract.

RESULTS AND DISCUSSION

Electrical shock treatment for 10 seconds produced rapid changes in the metabolite content of rat brain. Some changes reached a maximum during the first 10 seconds, and most reached a maximum during the first minute. Ammonia increased threefold within 1 minute after starting the stimulus and remained elevated for 5 minutes (Fig. 1). Its formation coincided with a large increase in AMP content and a somewhat smaller increase in IMP content. Ammonia appearance stopped as AMP levels began to decline.

The fastest response detected by metabolite measurements was the rapid disappearance of high energy phosphate. Figure 2 shows that the content of ATP and creatine phosphate reached their lowest values 10 seconds after starting the stimulation, the earliest time point that was analyzed. Creatine phosphate disappeared at an initial rate of 2.95 μmol/g of fresh weight/10 seconds or 17.7 μmol/g of fresh weight/minute. ATP disappeared at an initial rate of 0.84 μmol/g of fresh weight/10 seconds. Figures 1 and 2 show that the ATP lost was converted into ADP and into AMP plus IMP at approximately equal rates. Thus, the rate of dephosphorylation of ATP corresponds to phosphoric anhydride utilization of $0.84 + 0.84 \times 0.5 = 1.26$ μmol/g

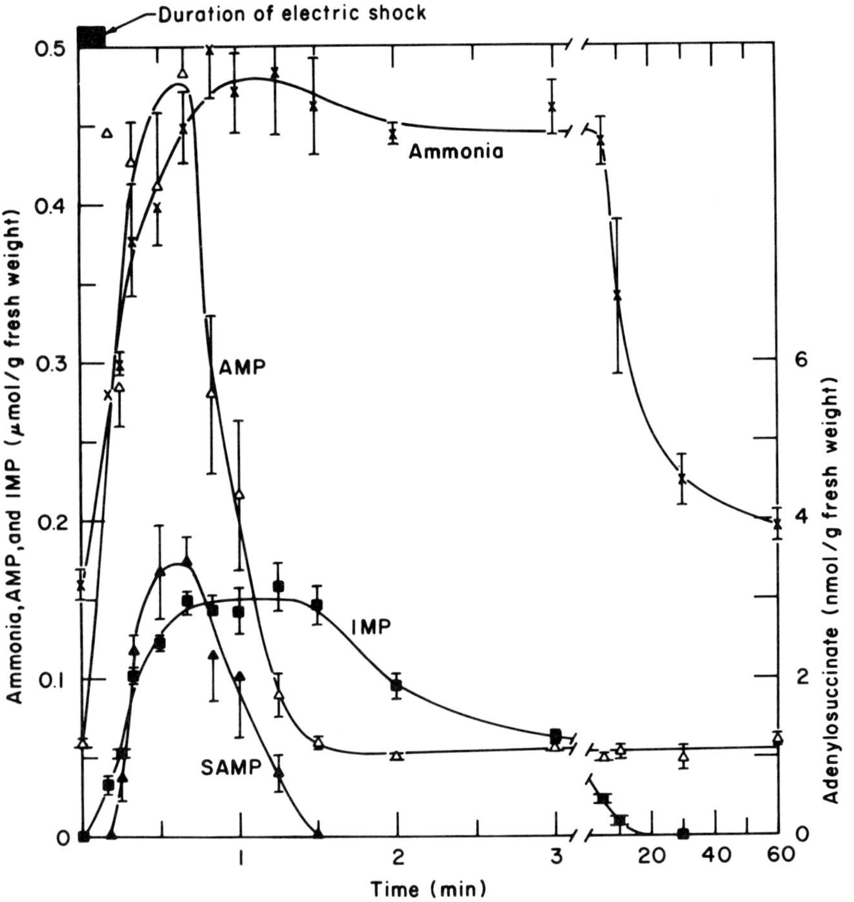

Fig. 1. Changes in purine nucleotide cycle metabolites of rat brain following electrical shock. Rats were stimulated for 10 seconds with monophasic pulses of 8.25 mseconds duration with a pulse repetition rate of 16/second. The voltage was constant at 100 V during each pulse. Current was delivered through silver electrodes applied to the shaved foreheads of the rats. Electrical stimulation was begun at zero time. Brains were removed and frozen rapidly by the method of Veech et al. (5) at the times indicated by the experimental points. In the case of control values shown at zero time, the electrodes were held in place for 10 seconds but no stimulus was applied. Each point shows the mean ± S.E. for four to eight rats, except for the 10 second points which show the average for two rats. Where standard error bars overlap, only half error bars are drawn for clarity. x, ammonia; Δ, AMP; ■ , IMP; and ▲ adenylosuccinate (SAMP). (Reproduced by permission from the Journal of Biological Chemistry.)

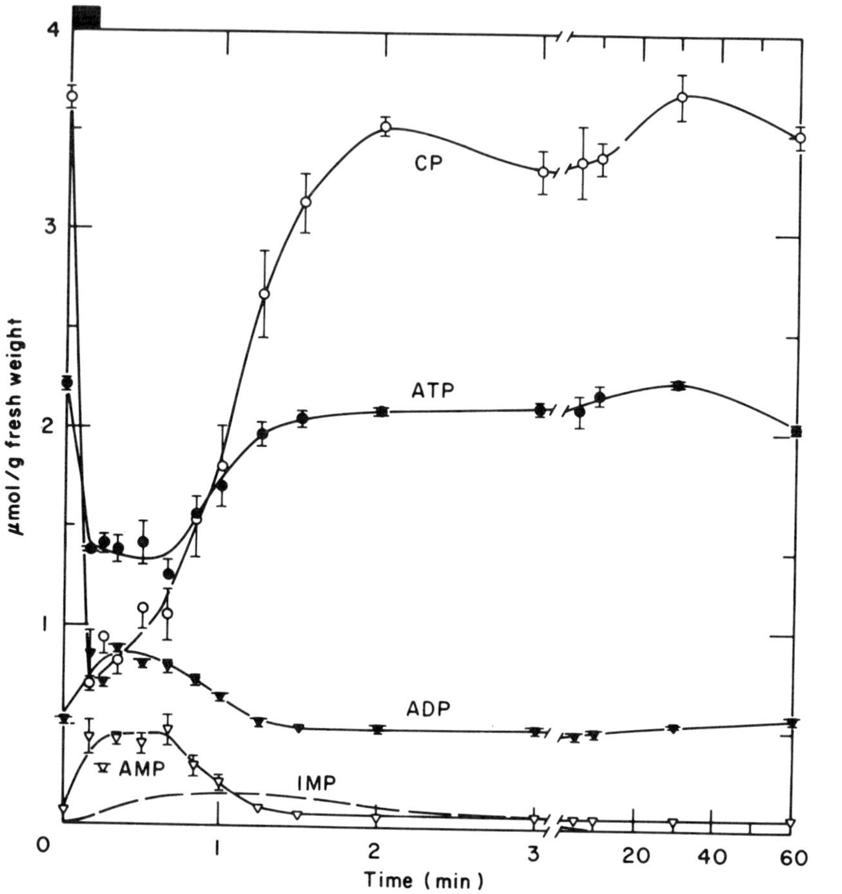

Fig. 2. Changes in adenine mononucleotides and creatine phosphate contents of rat brain following electrical shock. The experimental conditions were as described in Fig. 1. o, creatine phosphate (CP); ●, ATP; ▼, ADP; ▽, AMP. The experimental points for IMP (broken curve) are shown in Fig. 1. (Reproduced by permission from the Journal of Biological Chemistry.)

of fresh weight/10 seconds or 7.56 µmol/g/minute. The rate of dephosphorylation measured by the sum of the rates of loss of ATP and creatine phosphate was, therefore, 25.3 µmol/g of fresh weight/minute. The lactate content increased from 1.04 to 2.62 µmol/g of fresh weight during the first 10 seconds or at a rate of 9.48 µmol/g/minute. If it is assumed that this lactate was derived from glucose, then its production would have generated 9.48 µmol of ATP; if it was produced from glycogen it would have generated 14.2 μmol of ATP. Thus, the total ATP equivalents dephosphorylated fall between 34.8 and 39.5 μmol/g of fresh weight/minute. Rephosphorylation began immediately after the electrical stimulus was turned off and control levels were reached again after 1½ to 2 minutes. The highest net rate of rephosphorylation, measured by the sum of the rates of creatine phosphate and ATP formation, was about 4.6 μmol/g of fresh weight/minute.

The time course of the appearance of IMP closely paralleled the appearance of ammonia, reaching a maximum 1 minute after the beginning of the stimulus and coming to a stop as AMP started to decline (Fig. 1). However, IMP and ammonia did not accumulate in identical amounts. In brain, IMP is removed by two mechanisms, degradation to inosine and hypoxanthine, and conversion to AMP via adenylosuccinate. The maximum rate of IMP removal during the recovery period was 0.12 μmol/g of fresh weight/minute. This would be the rate of operation of adenylosuccinate synthetase if all of the inosine and hypoxanthine formed were derived from adenosine. The sum of the rates of appearance of inosine and hypoxanthine was 0.028 μmol/g of fresh weight/minute. If all of the inosine and hypoxanthine were derived from IMP, the maximum rate of adenylosuccinate synthetase would be 0.092 μmol/g of fresh weight/minute. These rates compare with a maximum activity of this enzyme *in vitro* of 0.058 μmol/g fresh weight/minute (4). *In vitro* measurements in crude homogenates tend to underestimate the activity of adenylosuccinate synthetase, and the agreement between these two rates can be considered to be reasonably good.

Adenylosuccinate reached a maximum of 3.5 nmol/g of fresh weight 40 seconds after starting the shock treatment (Fig. 1). Note that adenylosuccinate declines again at a time when the IMP level is still high, and when creatine phosphate and ATP have been restored to control values. The activity of adenylosuccinase measured *in vitro* is about 5 times that of adenylosuccinate synthetase (4). Perhaps adenylosuccinase requires time to become fully active and this permits a transitory accumulation of adenylosuccinate.

Adenosine, inosine, and hypoxanthine are also formed in response to the electrical shock treatment. Adenosine reached a maximum of 33 nmol/g fresh weight after 1 minute, while inosine and hypoxanthine reached maximums of 41 and 74 nmol/g fresh weight, respectively, after 3 to 5 minutes. The sum of these three substances was just under 0.1 μmol/g fresh weight 5 to 10 minutes after the shock treatment.

The total content of AMP, ADP, ATP, and adenosine decreased during the first minute following stimulation. The decrease was exactly balanced by the appearance of IMP plus inosine plus hypoxanthine. Similarly, during the recovery period, the reappearance of adenine-containing compounds was

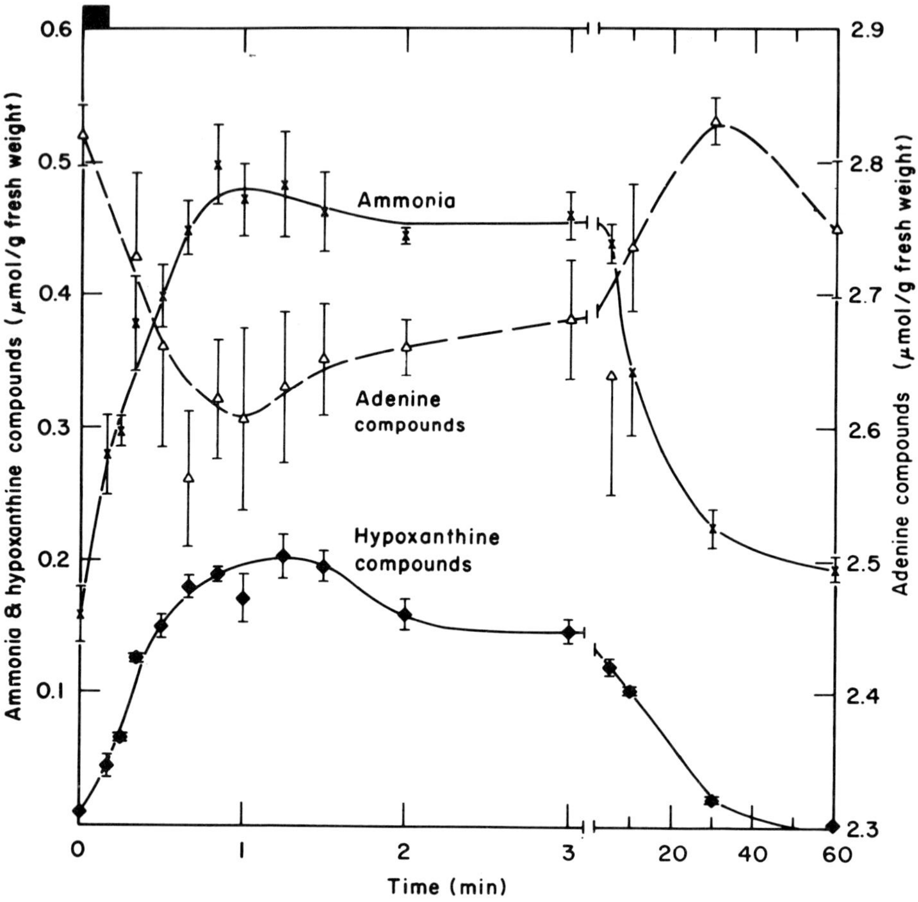

Fig. 3. Comparison of changes in adenine compounds (ATP + ADP + AMP + adenosine) with hypoxanthine compounds (IMP + inosine + hypoxanthine) and ammonia of rat brain following electrical shock. Note that the scale on the right hand ordinate is displaced by 2.3 but is otherwise the same as the scale on the left hand ordinate. "Hypoxanthine compounds" include xanthine since hypoxanthine measurements were made with xanthine oxidase. (Reproduced by permission from the Journal of Biological Chemistry.)

balanced by the disappearance of hypoxanthine-containing compounds (Fig. 3). The accompanying rise in ammonia exceeded the loss of adenine compounds into hypoxanthine compounds by only about 50%. However, this near correspondence may be fortuitous since ammonia was probably being converted to glutamine and was also diffusing from the brain during the period covered in Fig. 3.

α-Ketoglutarate decreased by 60% during the first 10 seconds of shock treatment and then recovered slowly (Fig. 4). Glucose 6-phosphate followed a similar time course, decreasing to 50% of control values after 10 seconds. It then increased rapidly to reach 1.8 times control values at 3 minutes. Thereafter, its return to control levels was slow. The rapid decline in glucose 6-phosphate reflects a sudden increase in the rate of operation of phosphofructokinase, in agreement with previous observations in mice and rats (7, 8). This is followed by an increased rate of glucose uptake (8, 9) which allows the hexokinase reaction to make up, and in this case to overshoot, the previous level of glucose 6-phosphate. Glycogen was not measured in this experiment. However, previous studies showed that glycogen decreases during seizures (8, 9) and it has been suggested that this might contribute to the accumulation of glucose 6-phosphate (7).

The [NADH]/[NAD] ratio for the intramitochondrial space was calculated from the glutamate, α-ketoglutarate, and ammonia contents. It was assumed that the intracellular water accounts for 50% of the fresh weight, that glutamate and α-ketoglutarate are evenly distributed throughout the intracellular water, and that ammonia is evenly distributed throughout the total brain water. The latter was taken to be 78% of the wet weight (10). The total glutamate content was 10.0 μmol/g of fresh weight and remained approximately constant. The results of these calculations show that there was no change in the intramitochondrial [NADH]/[NAD] ratio calculated from the metabolites of glutamate dehydrogenase during the first minute after applying electrical shock (Table I), the period during which most of the ammonia is liberated. The oxidation-reduction state became somewhat more oxidized during the recovery period, perhaps because additional NADH was used to remove ammonia as glutamate.

Following electrical stimulation, the decrease in adenine compounds is exactly balanced by the increase in hypoxanthine compounds, but the amount of ammonia that accumulates (0.31 μmol/g of fresh weight) exceeds the amount of hypoxanthine compounds that accumulates (0.20 μmol/g of fresh weight) by 55% (Fig. 3). As is pointed out above, the maximum rate of operation of adenylosuccinate synthetase is between 0.09 and 0.12 μmol/g of fresh weight/minute. If the enzyme is operating at its maximum rate over the minute during which 0.31 μmol of ammonia accumulates, then this could account for the IMP "missing" from the stoichiometry (0.20 + 0.09 = 0.29 μmol/g). Thus, all of the observed increase in ammonia content could arise via the adenylate deaminase reaction.

The question remains if, and to what extent, the glutamate dehydrogenase reaction contributes to ammonia production. Put

TABLE I. Intramitochondrial [NADH]/[NAD] calculated from Metabolites of Glutamate Dehydrogenase Reaction[a]

Time (min)	*[NADH]/[NAD]*
Control	1.25 ± 0.24
0.17	1.46
0.25	1.28 ± 0.06
0.33	1.22 ± 0.19
0.50	1.08 ± 0.07
0.67	1.07 ± 0.07
0.83	0.92 ± 0.12
1.00	0.96 ± 0.07
1.25	0.72 ± 0.05
1.5	0.74 ± 0.07
2	0.68 ± 0.03
3	0.60 ± 0.04
5	0.53 ± 0.02[b]
10	0.63 ± 0.05[b]
30	0.84 ± 0.08
60	0.98 ± 0.05

[a] *The experimental conditions were as described in the legend to Fig. 1. Calculations were made using the equation*

$$\frac{[NADH]}{[NAD]} = \frac{[glutamate] \times K}{[\alpha\text{-}ketoglutarate][ammonia]}$$

where K, the equilibrium constant, equals 3.87×10^{-6} M at pH 7.0 (11). Assumptions concerning metabolite distribution are explained in the text. Results are expressed as means ±S.E. for four to eight rats, except in the case of 0.17 minutes which shows the mean for two animals. (Reproduced by permission from the Journal of Biological Chemistry.)

[b] *Significantly different from controls ($P < 0.05$) by paired t test.*

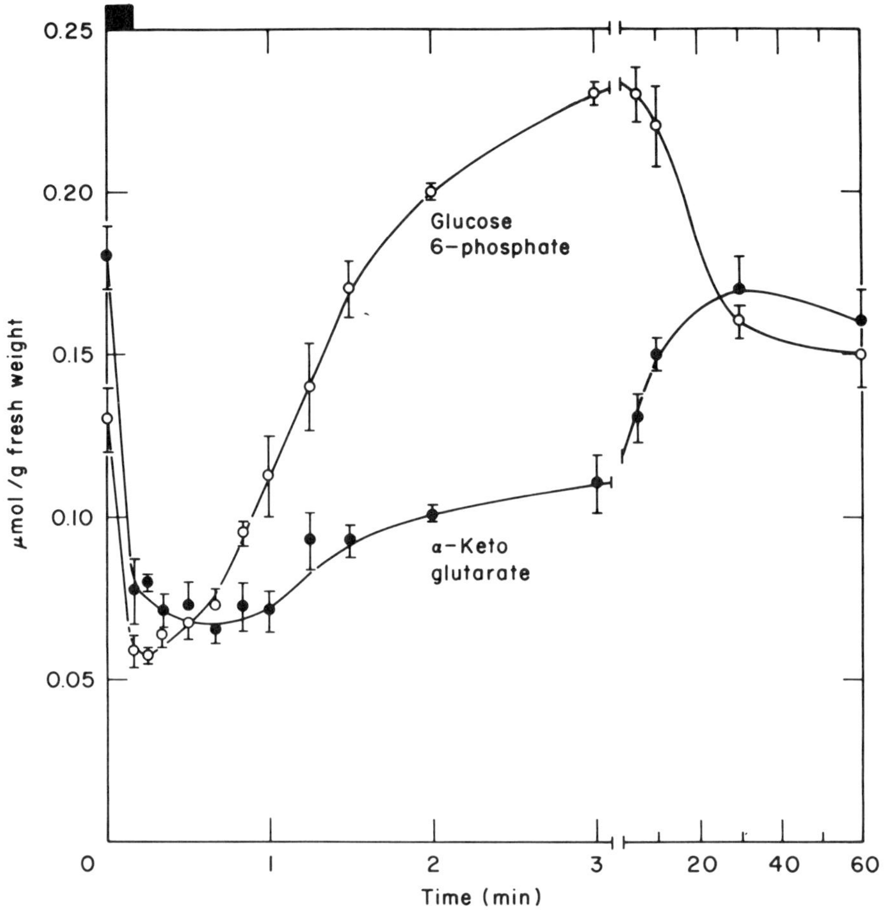

Fig. 4. Changes in glucose 6-phosphate and α-ketoglutarate contents of rat brain following electrical shock. The experimental conditions were as described in Fig. 1. o, glucose 6-phosphate; ●, α-ketoglutarate. (Reproduced by permission from the Journal of Biological Chemistry.)

differently, while our data show clearly that a large portion of the ammonia arises via the purine nucleotide cycle, they do not rule out some contribution by the glutamate dehydrogenase reaction. Ammonia production is associated not only with a rapid rise in IMP from <0.005 to 0.15 μmol/g of fresh weight (Fig. 1), but also with a rapid fall in α-ketoglutarate from 0.18 to 0.07 μmol/g of fresh weight (Fig. 4). A similar fall has been reported to occur in response to electrically induced convulsions in mice (7). The fall in α-ketoglutarate content can be interpreted in at least two ways. It may be due to the sudden rise in ammonia which leads to a readjustment of the glutamate dehydrogenase equilibrium in favor of glutamate formation. In other words, α-ketoglutarate falls because the ammonia produced by other reactions is partially converted to glutamate. Alternatively, the glutamate dehydrogenase equilibrium may be shifted in the direction of ammonia formation by the removal of α-ketoglutarate by some other reaction, such as the α-ketoglutarate dehydrogenase reaction. The latter alternative has been advanced in support of ammonia production via the glutamate dehydrogenase reaction in brain (12) and kidney (13). A shift in the ratio [reduced pyridine nucleotide]/[oxidized pyridine nucleotide] in the mitochondria would also lead to a readjustment of the equilibrium between glutamate, α-ketoglutarate, and ammonia. There are currently no suitable substrates that can be used to calculate the [NADH]/[NAD] ratio in this compartment other than the substrates of glutamate dehydrogenase, and using these would lead to a circular argument. (The β-hydroxybutyrate dehydrogenase reaction equilibrium is used for calculating intramitochondrial [NADH]/[NAD] in liver, but this is not practical in the case of brain (14).) Assuming that the glutamate dehydrogenase reaction is at or near equilibrium during the experimental period, calculations of the intramitochondrial [NADH]/[NAD] ratio show that no changes occur in this ratio during the period over which most of the ammonia is liberated (Table I). No change in intramitochondrial [NADH]/[NAD] was observed in brains of rats in which seizures were induced by single electrical shocks (15). Similar results were reported for the first minute of sustained epileptic seizures, the period over which ammonia accumulated; thereafter, the ammonia content stabilized, and calculations showed that the intramitochondrial [NADH]/[NAD] then became more oxidized (8).

Ammonia formed during electrical activity can be removed by the glutamate dehydrogenase reaction. The activity of the enzyme is 11 μmol/g of fresh weight/minute at 30°C and pH 7, which is equivalent to 18 μmol/g of fresh weight/minute at 38°C (4). Ammonia can also be removed by the glutamine synthetase reaction. The activity of this enzyme is 2.3 μmol/g of fresh weight/minute at 37°C (16). Ammonia can also be lost by diffusion into the blood, but no quantitative estimate of the rate at which this may occur has been reported.

When mice were stimulated electrically, the glutamate content of 13 μmol/g of fresh weight did not change significantly, whereas the glutamine content increased from 4.5 to 5.5 μmol/g of fresh weight (17). This suggests that in our experiments more ammonia may have been formed in response to the shock treatment that was actually measured and that it was partly converted to glutamine.

Various possible functions of the purine nucleotide cycle were proposed and discussed previously, including (1) regulation of the relative amounts of AMP, ADP, and ATP, during periods of sudden energy drain, by deamination of AMP to IMP; (2) deamination of amino acids via aspartate, thus making available the carbon skeletons of amino acids for energy production; (3) control of phosphofructokinase and hence of glycolysis by adjustment of the AMP concentration; (4) replenishment of citric acid cycle intermediates via the generation of fumarate in the adenylosuccinate synthetase reaction; (5) regulation of phosphorylase *b* by accumulation of IMP, which activates the enzyme under conditions when AMP is low and phosphorylase exists largely in the *b* form (18 - 20). The experimental results presented here show that most if not all of the ammonia produced in brain in response to electrical stimulation results from the operation of the purine nucleotide cycle.

REFERENCES

1. Loebel, R. O. (1925). *Biochem. Z. 161,* 219-239.
2. Weil-Malherbe, H. (1962). *In* "Neurochemistry" (K. A. C. Elliott, I. H. Page, and J. H. Quastel, eds.), p. 323. Thomas, Springfield, Illinois.
3. Vrba, R. (1957). *Rev. Czech. Med. 3,* 1-26.
4. Schultz, V. and Lowenstein, J. M. (1976). *J. Biol. Chem. 251,* 485-492.
5. Schultz, V. and Lowenstein, J. M. (1978). *J. Biol. Chem. 253,* 1938-1943.
6. Veech, R. L., Harris, R. L., Veloso, D., and Veech, E. H. (1973). *J. Neurochem. 20,* 183-188.
7. King, L. J., Carl, J. L., and Lao, L. (1973). *J. Neurochem. 20,* 477-485.
8. Chapman, A. G., Meldrum, B. S., and Siesjö, B. K. (1977). *J. Neurochem. 28,* 1025-1035.
9. King, L. J., Lowry, O. H., Passonneau, J. V., and Venson, V. (1967). *J. Neurochem. 14,* 599-611.
10. Long, C. (1961). "Biochemists' Handbook," p. 640, E. and F. M. Spon, London.
11. Williamson, D. H., Lund, P., and Krebs, H. A. (1967). *Biochem. J. 103,* 514-527.
12. Weil-Malherbe, H. (1974). *Molec. Cell. Biochem. 4,* 31-43.
13. Krebs, H. A. and Vinay, P. (1975). *Med. Clin. North Amer. 59,* 595-603.
14. Miller, A. L., Hawkins, R. A., and Veech, R. L. (1973). *J. Neurochem. 20,* 1393-1400.

15. Howse, D. C. and Duffy, T. E. (1975). *J. Neurochem. 24,* 935-940.
16. Lund, P. (1970). *Biochem. J. 118,* 35-39.
17. King, L. J., Carl, J. L., and Lao, L. (1974). *J. Neurochem. 22,* 307-309.
18. Lowenstein, J. M. (1972). *Physiol. Rev. 52,* 382-414.
19. Aragón, J. J., Tornheim, K., and Lowenstein, J. M. (1980). *FEBS Lett. 117,* K56-K64.
20. Aragón, J. J., Tornheim, K., Goodman, M. N., and Lowenstein, J. M. (1981). *Curr. Top. Cell. Reg. 18,* 131-149.

SIMILARITY ANALYSIS BASED UPON RIBOSOMAL PROTEIN COMPOSITIONS IN RHODOSPIRILLACEAE FAMILY: INITIATORY TRIAL

*Kumiko Kawaguchi**

Department of Biochemistry and Biophysics
Research Institutes for Nuclear Medicine and Biology
Hiroshima University
Hiroshima, Japan

Shigehiko Taniguchi

Department of Biochemistry
Okayama University School of Dentistry
Kasumi 1-2-3
Okayama City, Japan

INTRODUCTION

The phylogenetic status of non-sulfur purple bacteria mostly capable of growing either under the light-anaerobic or dark-aerobic conditions, has currently been under extensive focus in light of its possible ancestral status in diverging prokaryotic photosynthesis and respiration, as revealed in the recent trials of a molecular analysis of evolutionary processes (1). Noteworthy, however, is the fact that a minority group in *Rhodospirillum* genus of so-called brown spirilla species comprising *R. fulvum*, *R. molischianum*, and *R. photometricum*, all characteristically growing as strict photo-anaerobes, have been classified under the common *Rhodospiril-*

**Present address: Department of Bacteriology, Hyogo College of Medicine, Mukogawa-cho, 1-1, Nishinomiya City, Hyogo, Japan 663.*

 ISBN 0-12-397580-8

laceae family. Taxonomic and phylogenetic problems of this family have recently been reviewed in detail by Pfennig (2) and Saunders (3). Thus, similarity analysis within this family has been considered as an indispensable step and only a practicable methodology for the purpose has been awaited.

Osawa's cochromatographic analysis (4) of ribosomal protein compositions has achieved unique taxonomic success with enteric bacteria. Thus, the authors attempted to extend the method to the *Rhodospirillaceae* family. The advantages of ribosomal protein composition as a unique index for similarity analysis has been confirmed by the facts thus far accumulated. Noteworthy among them are the observations proving significant similarities between *Erwinia* and enteric bacteria (5, 6) and between *R. tenue* and *Rhodocyclus purpurens* (K. Kawaguchi and N. Pfennig, private communication, 1978). The latter result subsequently suggests that *Rhodocyclus* is probably a new species in the genus *Rhodospirillum,* rather than as an entirely independent genus (7). Application of the same method to a genus of *Mycobacterium* from soil origin has revealed a significant correlation between the assessed similarity and the genetic type determined within the genus (K. Kawaguchi, private communication, 1976). A perspective on the ribosomal protein compositions as fruitful indexes for taxonomic or phylogenetic analysis has been reviewed in light of recent progress of the sequence analysis of ribosomal component proteins (8).

A similar attempt has been made using the rRNA(16 S) sequence, intending to relate the *Rhodospirillaceae* family (*Rps. sphaeroides* as a representative) within the prokaryote kingdom, revealed the family as significantly closer to the Vibrio-Enteric group than to *Bacillaceae* or *Cyanophyta* (9, 10). In fact, the family appears to evolve through an independent route diverging in between two branches, each reaching *E. coli* and *Bacillus* spp. (H. Hori and S. Taniguchi, private communication) judging from the base sequence of rRNA(5 S) as determined quite recently. However, no further taxonomic information, particularly of the intra-family status to be detailed in terms of indexes on intergenus, -species, and -strain similarity, has apparently been uncovered by such an approach.

This chapter deals with the authors' "initiatory trials" of a systematic survey of similarity analysis with the eventual intention of extending it to the entire family of *Rhodospirillaceae*.

TABLE I. List of Strains[a]

	Ribosome type	Strain	Obtained from
Rhodospirillum rubrum	Group I	ATCC 11170*	ATCC/USA
		277	ATCC/USA
		9791	ATCC/USA
		17031	ATCC/USA
		17033	ATCC/USA
		19613	ATCC/USA
		G9**	J. W. Newton
		IFO 3986	IFO/JAPAN
		ST	S. Tsuboi
		K	G. Kukuchi
		Ha	N. Pfennig
		2461	N. Pfennig
		6461	N. Pfennig
		9960	N. Pfennig
	Group II	ATCC 17032	ATCC/USA
		17036	ATCC/USA
tenue		2761	N. Pfennig
		3661	N. Pfennig
		3760*	N. Pfennig
		3761	N. Pfennig
fulvum		B(Berkeley)147	N. Pfennig
molischianum		NTHC 131	N. Pfennig
photometricum		2160	N. Pfennig
Rhodopseudomonas sphaeroides	Group I	ATCC 17023*	ATCC/USA
		17026	ATCC/USA
		17027	ATCC/USA
		17028	ATCC/USA
		17029	ATCC/USA
		17030	ATCC/USA
		Sa	M. Higuchi
		Sb	M. Higuchi
		L	
		BG**	J. Lascelles
		Ga	J. Gibson
		Gif	N. Pfennig
		1760-1	N. Pfennig
		O	H. Kitamura
		S	H. Kitamura
		P_2**	H. Kitamura
	Group II	ATCC 17024	ATCC/USA
		17025	ATCC/USA
gelatinosa		ATCC 17011	ATCC/USA
palustris		ATCC 17001	ATCC/USA

MATERIALS AND METHODS

Reagents

Radioactive amino acids were obtained from Radiochemical Centre, Amersham England, with the following specific activities: *L*-[4,5-^{3}H]lysine (250 Mci/mmol), *L*-[U-^{14}C]lysine (342 Mci/mmol), *L*-[4,5-^{3}H]leucine (58 Mci/mmol), and *L*-[U-^{14}C]leucine (311 Mci/mmol).

Bacterial Strains

The bacterial strains subjected to the similarity analysis were mainly those from the *Rhodospirillum* genus and *Rps. sphaeroides* species in *Rhodopseudomonas* genus. They are listed together with their direct sources in Table I.

Media and Culture Conditions

Rhodospirillum rubrum and *Rps. sphaeroides* were grown in the synthetic medium prescribed by Cohen-Bazire *et al.* (11) where casamino acid was replaced by potassium glutamate, potassium acetate, (both 0.1% in final concentration), and 10 m*M* NaCl. Instead of ammonium malate, ammonium succinate was used throughout as major source of carbon and nitrogen. *Rhodospirillum tenue* and three brown spirilla species were grown in "medium R8AH" as described (12), except that the amount of yeast extract was reduced to 0.01% and 0.05% sodium ascorbate was added. Labeled amino acids added to each culture (as described below) were, 100 μCi of [^{3}H]lysine or 5 μCi of [^{14}C]lysine in the case of *R. rubrum* and *R. tenue* together with *Rps. sphaeroides* species, and 100 μCi of [^{3}H]leucine or 5 μCi [^{14}C]leucine in the case of three brown spirilla species.

Unless otherwise noted, bacterial cells were cultured anaerobically in a light intensity of 500 foot-candles at 30°C in two 15-ml screw-cap tubes completely filled with the appropriate medium. Sometimes, instead of media, argon-CO_2(95:5) was bubbled through. For dark-aerobic cultivation, Erlenmeyer flasks (100 ml capacity) wrapped with black tape and containing

(Footnote to Table I) [a]*"Ribosome type" is a taxon tentatively proposed on the basis of present similarity analysis. See Table III and the text. (*) Strain was that chosen as reference strain in the present similarity analysis; (**) mutant of which parental strain is shown in the text. All the others are natural strains.*

30 ml of culture medium were shaken at 30°C in a thermostated shaker operated at 120 strokes per minute. Cell density of starting culture was made up to 10 - 20 Klett units (as determined by Klett-Summerson photometer with No. 66 filter) and cells exponentially grown to the cell density of around 150 Klett units. The exponential cells were collected by centrifugation, washed with 10 m*M* Tris-HCl buffer (pH 7.8) containing 0.1 m*M* magnesium acetate at 5°C, and kept frozen at -20°C until use. Purity of each cell sample was checked either by microscopic observation or by agar plate test.

Isolation of Ribosomal Proteins

Frozen cells were ground with quartz sand and labeled 30 and 50 S ribosomal; subunits were separately prepared by means of sucrose gradient centrifugation described previously (13). The fractions containing the ribosomal subunits were collected and dialyzed against 0.2 m*M* magnesium acetate and then lyophilized.

Carboxymethyl(CM)-Cellulose Column Chromatography

The fractionation of ribosomal proteins by the CM-cellulose column chromatography was carried out according to the method described by Otaka *et al*. (13). A mixture of two lyophilized samples of 50 S, each from one of two strains (one labeled with [^{14}C]lysine and the other with [^{3}H]lysine) to be mutually compared, was digested with a mixture of pancreatic and T_1 PNase to release rRNA components. The treated sample was then applied onto a 0.5 × 40 cm column of CM-cellulose (Whatman's CM 52) and then eluted with a linear gradient established between each 200 ml of 0.05 and 0.5 *M* sodium acetate buffer, pH 5.6, throughout containing 6 *M* urea and 13 m*M* 2-mercaptoethanol. Cochromatography for the 30 S subunit was conducted separately in the same manner. A nonlabeled sample of ribosomal proteins separately prepared from *R. rubrum* was routinely used as a carrier protein for the above cochromatography. Fractions were collected at a flow rate of about 1.8 ml (one tube) per 20 min. Chromatography was conducted in a thermostated room (24°C).

Estimation of the Similarity from the Cochromatographic Elution Profile

Each elution profile subjected to the similarity analysis comprised in total 35 to 38 resolved peaks assumed to represent each individual component. Among them, 17 - 20 and 15 - 18 components were from 50 and 30 S subunits, respectively. As described previously (4), the similarity (%) was estimated as (number of component judged to be common/number of component detected throughout) × 100. The evaluation of the advantages as well as the limitations of the present method of similarity analysis are presented in "Summary and Discussions."

RESULTS

Interspecies and Intergenus Similarity Analysis

1. Ribosomal Protein Compositions as Compared between Light Anaerobically and Dark Aerobically Grown R. rubrum

Figure 1a and b show cochromatograms of 30 or 50 S ribosomal protein compositions from light anaerobically grown and dark aerobically grown *R. rubrum ATCC* 11170. Two cochromatograms of the ribosomal proteins showed almost indistinguishable elution patterns except for a slight difference in the labeled extent with some components, such as those corresponding to peaks 5, 14, and some others. However, no significance can be attached to these differences in similarity assessment because the height of the peaks was found variable in this extent in repeated experiments, although each peak was precisely eluted at the reproducibly same position in the cochromatograms.

A similar experiment was carried out with another representative of facultative photoaerobe, *Rps. sphaeroides* ATCC 17023. Again no significant difference was detected, although the elution profile is not shown. These results indicate, within the resolution capacity of the present method, that the environmental conditions do not seem to affect the composition of ribosomal proteins.

2. Interspecies Comparison within Rhodospirillum genus - R. rubrum Versus other Species

Two cochromatograms (for 50 and 30 S subunits) of each of the species were characteristically different from one another. Cochromatograms obtained with *R. rubrum* ATCC 11170 and *R. tenue* 3760 are illustrated in Fig. 2. The ribosomal protein compositions were significantly different between these two species. The peaks 2, 3, 5, and 7 from the 50 S subunit and 9, 10, 11, and 17 from the 30 S subunit of *R. rubrum* are judged as common. Thus, overall similarity is estimated to be 21 - 26% (see also Table III). Almost comparable or slightly less similarity was

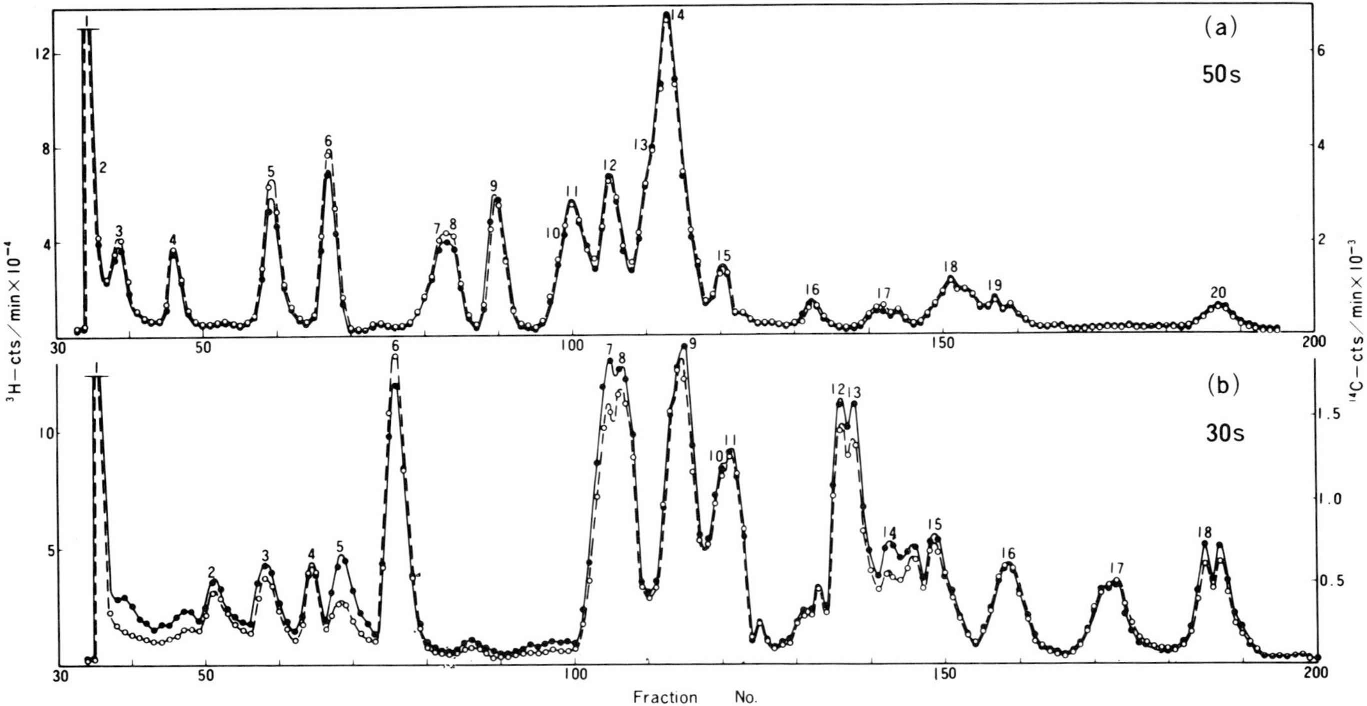

Fig. 1. *Cochromatograms on CM-cellulose columns of the 50 S (a) and 30 S (b) ribosomal proteins from dark aerobically and light anaerobically grown R. rubrum ATCC 11170.* ——●——, *Dark-protein labeled with [^{14}C]lysine.* ------○------, *Light-protein labeled with [^{3}H]lysine.*

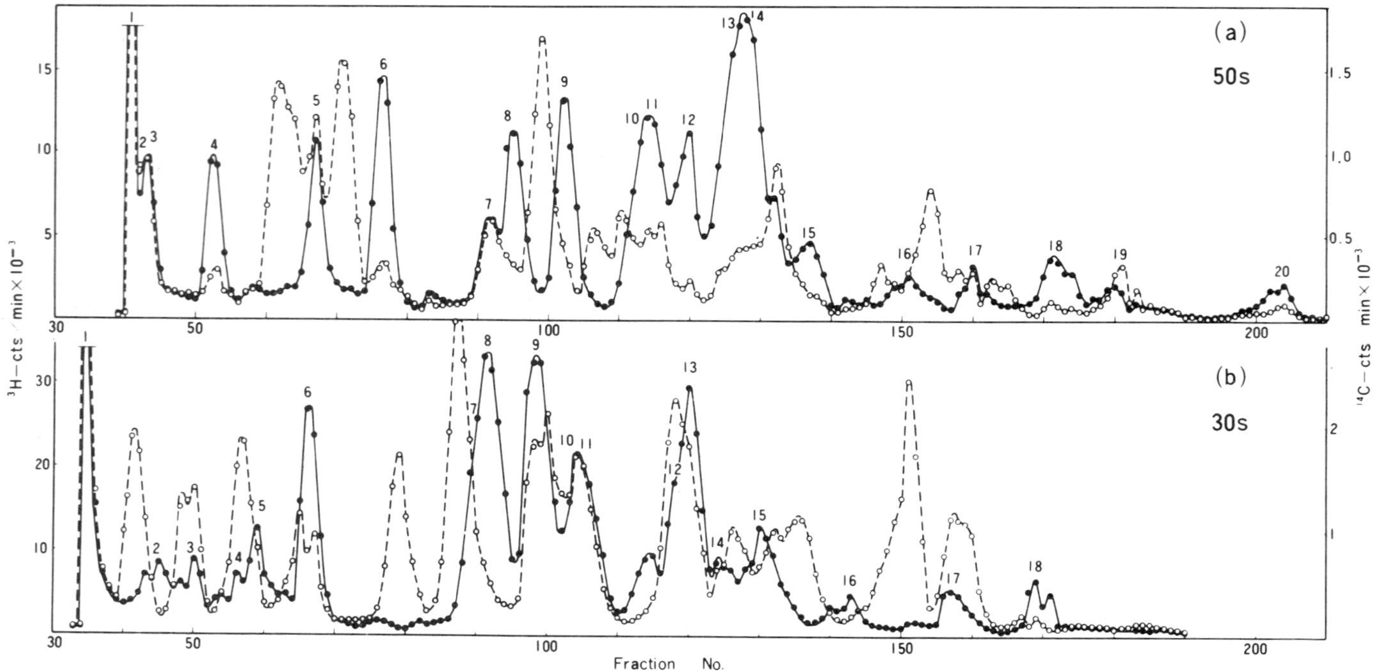

Fig. 2. *Cochromatograms on CM-cellulose columns of the 50 S (a) and 30 S (b) ribosomal proteins from R. rubrum ATCC 11170 and R. tenue 3760.* ——●——, *R. rubrum protein labeled with* $[^{14}C]$*lysine.* ------○------, *R. tenue protein labeled with* $[^{3}H]$*lysine.*

assessed with *R. rubrum* ATCC 11170 versus *R. tenue* 3761(18 - 24%), and *R. tenue* 3661(18 - 21%) and *R. tenue* 2761(11 - 13%).

Since three brown spirilla species do not incorporate lysine but leucine under the present growth conditions, comparison with these species was always performed by using radioactive leucine to label ribosomal proteins. It was preliminarily confirmed with *R. rubrum* ATCC 11170 that distribution of radioactive leucine incorporated in each component of the 50 and 30 S subunits was indistinguishable from that of lysine, as judged from entire overlapping of cochromatograms between [^{14}C]leucine-labeled and [^{3}H]lysine-labeled subunits. Figure 3a and b record the cochromatograms between *R. photometricum* 2160 and *R. rubrum* ATCC 11170. Most of protein components of *R. photometricum* 2160 seem to be distinct from those of *R. rubrum* with the exception of two, corresponding to 19 in 50 S and 6 in 30 S. Taking two more ambiguous components 8 and 17 of 30 S into consideration, the similarity between these two species is estimated to be 5 - 11%.

Much less similarity to *R. rubrum* was observed in cochromatograms with *R. rubrum* ATCC 11170 versus *R. fulvum* B147 or *R. molischianum* NTHC 131 (5 - 8% in both cases). Although the results are not shown here, only three or four peaks were found in common between *R. tenue* 3760 and *R. photometricum* 2160, indicating the similarity as 8 - 11%. In addition, comparison between *R. tenue* 3760 and each one of two other brown spirilla species also indicated the same level of similarity (8 - 11%) for both cases.

3. *Interspecies Comparison within Strictly Photoanaerobic Brown Spirilla Species*

Three species *R. fulvum, R. molischianum,* and *R. photometricum,* characterized by their brown color and inability to grow under aerobic conditions have been classified mainly by cell size and benzoate metabolism (14). Similarity analysis have revealed an unexpectedly higher similarity (70 - 90%). Cochromatograms between *R. fulvum* B147 and *R. molischianum* NTHC 131 are shown in Fig. 4a and b. They indicate that the composition in 30 S is identical between these two species. On the other hand, peaks 2, 12, and 18 all in the 50 S subunit of *R. molischianum* were distinguishable from those in *R. fulvum* (92% similarity).

Comparisons were made between *R. photometricum* 2160 and each one of other two. Although the results are not illustrated in figure, three more peaks in 50 S and five more peaks in 30 S were assigned as distinctive on the cochromatograms. The overall similarity of *R. photometricum* 2160 either to *R. fulvum* B147 or *R. molischianum* NTHC 131 is estimated as 69 - 72%.

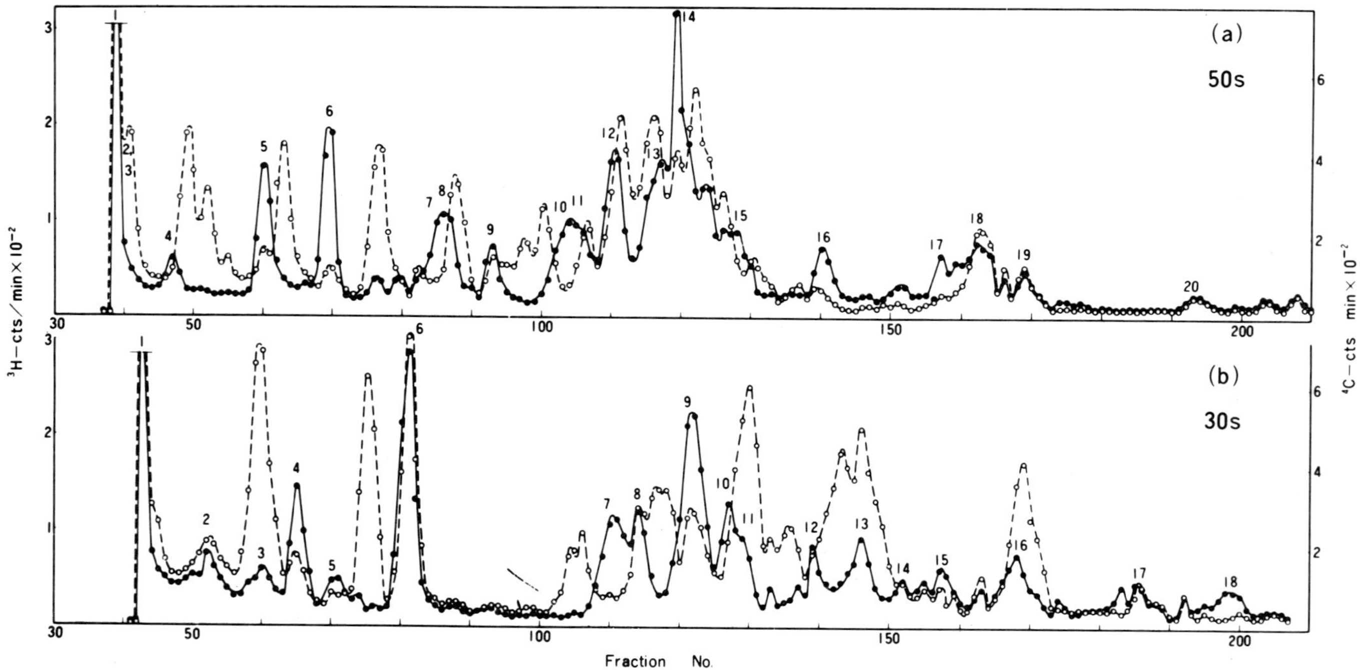

Fig. 3. Cochromatograms on CM-cellulose columns of the 50 S (a) and 30 S (b) ribosomal proteins from *R. rubrum* ATCC 11170 and *R. photometricum* 2160. ———●———, *R. rubrum* protein labeled with [^{14}C]leucine. ------○------, *R. photometricum* 2160 protein labeled with [^{3}H]leucine.

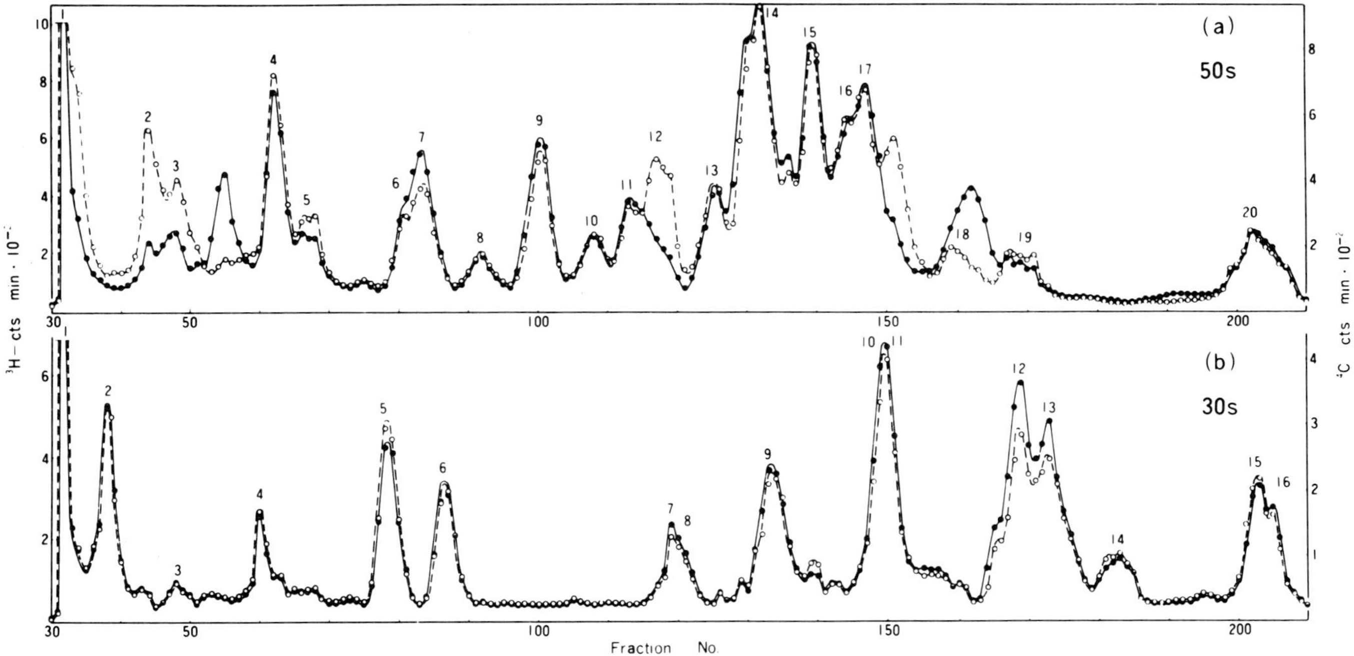

Fig. 4. *Cochromatograms on CM-cellulose columns of the 50 S (a) and 30 S (b) ribosomal proteins from R. fulvum B147 and R. molischianum NTHC 131.* ——●——, *R. fulvum protein labeled with [^{14}C]leucine.* ------○------, *R. molischianum protein labeled with [^{3}H]leucine.*

4. *Interspecies Comparison within Rhodopseudomonas Genus Rps. sphaeroides Versus other Species*

In relation to considerably lower interspecies similarity, as disclosed within *Rhodospirillum* genus, preliminary interspecies analyses were conducted within the *Rhodopseudomonas* genus. In a comparison of the composition between *Rps. sphaeroids* ATCC 17023 and *Rps. gelatinosa* ATCC 17011 and *Rps. palustris* ATCC 17001, although data are not presented, apparently no similarity was detected in any case. The results seem to imply the possibility that the interspecies divergency is more pronounced in the *Rhodopseudomonas* than *Rhodospirillum* genus.

5. *Intergenus Comparison within Rhodospirillaceae Family R. rubrum Versus Rps. sphaeroides*

Rhodopseudomonas sphaeroides and *R. rubrum* are classified under a separate genus because of photopigment and morphological characteristics (14).

Figure 5a and b show the cochromatograms between *Rps. sphaeroides* ATCC 17023 and *R. rubrum* ATCC 11170. At least 20 and 18 peaks in the 50 S and 30 S ribosomal proteins, respectively, were reproducibly resolved in both strains. As illustrated in Fig. 5, all components were eluted out without revealing any noticeable matching on the cochromatograms, indicating 0% similarity between these two groups.

Interstrain Similarity Analysis

1. *Interstrain Comparison with R. rubrum Species*

In contrast with lower inter-similarity thus far found within *Rhodospirillum* genus, few differences in the components were discovered among 16 strains of *R. Rubrum*. From the present interstrain analyses, fourteen strains, with the exception of *R. rubrum* ATCC 17032 and 17036, showed identical ribosomal protein compositions. The carotenoid-less blue green mutant G9 derived from ATCC 11170 strain showed an identical composition to that of the parental strains. Strains of ATCC 17036 and 17032 also showed identical composition with each other. Furthermore, a small, but clear, difference was noted between ATCC 11170 and each one of the other species.

Figure 6a and b show cochromatograms between *R. rubrum* ATCC 17032 and *R. rubrum* ATCC 11170. As indicated by the arrow on the cochromatogram of 50 S, peak 5 of ATCC 11170 and peak 6 of ATCC 17032 were eluted earlier or later than the corresponding peak of other strains, respectively. On the cochromatogram of 30 S, on the other hand, peak 9

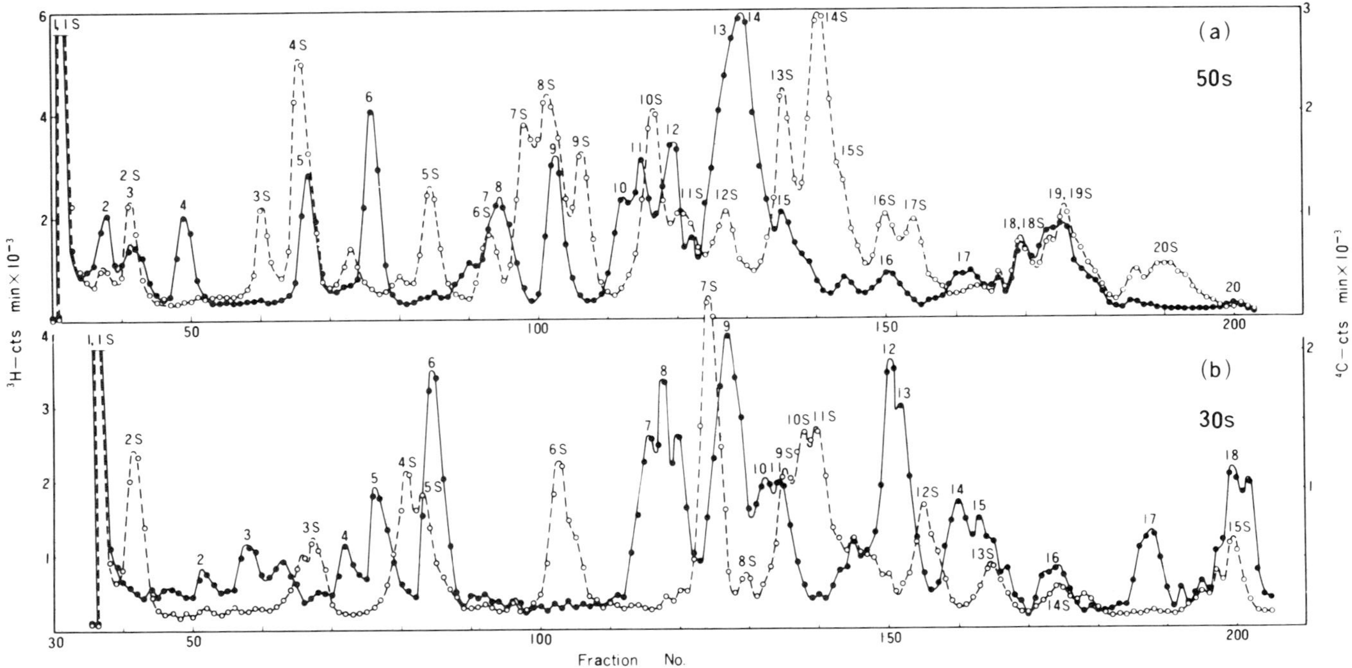

Fig. 5. Cochromatograms on CM-cellulose columns of the 50 S (a) and 30 S (b) ribosomal proteins from R. rubrum ATCC 11170 and Rps. sphaeroides ATCC 17023. ———●———, *R. rubrum protein labeled with [^{14}C]lysine.* ------○------, *Rps. sphaeroides protein labeled with [^{3}H]lysine.*

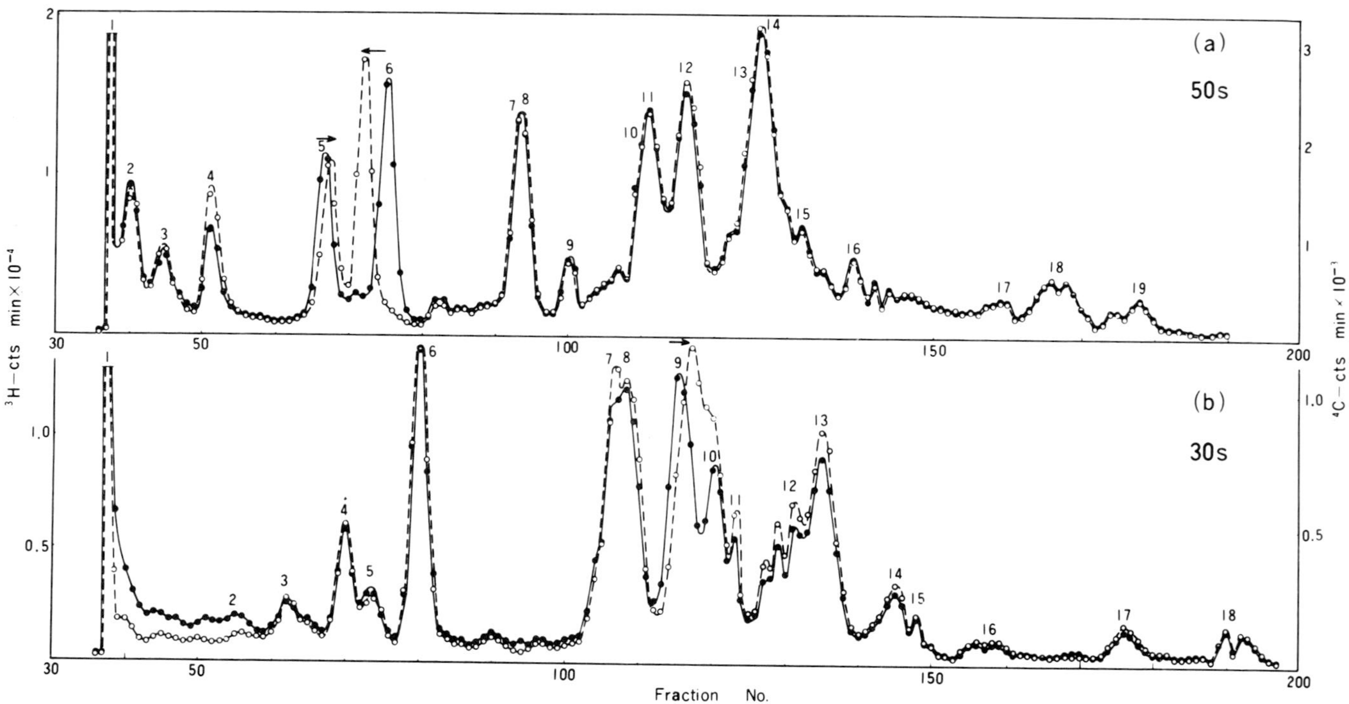

Fig. 6. Cochromatograms on CM-cellulose columns of the 50 S (a) and 30 S (b) ribosomal proteins from *R. rubrum* ATCC 11170 and *R. rubrum* ATCC 17032. ——●——, ATCC 11170 protein labeled with [^{14}C]lysine. ------○------, ATCC 17032 protein labeled with [^{3}H]lysine.

of ATCC 11170 was eluted earlier than the corresponding peak of ATCC 17032. Overall similarity is, therefore, estimated to be 92%.

Thus, fourteen strains compose one major taxon, and strains ATCC 17032 and ATCC 17036 another minor one.

2. *Interstrain Comparison within R. tenue Species*

Interstrain analyses among four strains of *R. tenue* species disclosed a singularity of 2761 strain over the other three. Figure 7a and b show the cochromatograms between *R. tenue* 2761 and *R. tenue* 3760. Components assigned to be common in Fig. 7 are 4, 9, 16, and 17 in 50 S and 4, 8, 12, 13, 16, 17 with two ambiguous components such as 14 and 18 in 30 S. Thus, the similarity is estimated to be 28 - 33%.

Figure 7c and d show only critical differences in elution profiles between *R. tenue* strain 3661 and *R. tenue* 3760, where small but evident displacement in three peaks were noted. On profiles of 3760, peak 5 of 50 S was eluted earlier than the corresponding peak of 3661, whereas in the 30 S cochromatogram, two peaks, 3 and 5, were eluted later than the corresponding peaks of the 3661 strain, respectively. These differences gave 92% similarity between *R. tenue* 3661 and 3760 strain. The last strain *R. tenue* 3761, was also found distinctive from any of other three strains. Although the results are not presented in figure, the cochromatograms between *R. tenue* 3760 and 3761 revealed only one difference in peak 5 in 50 S, but were identical to the corresponding one in *R. tenue* 3661 in a separate cochromatography. However, *R. tenue* 3761 was found to possess identical composition in 30 S with that of 3760. The overall similarity was assessed to be 97% between *R. tenue* 3761 and 3760.

Thus, interstrain comparison among four strains of *R. tenue* species indicates that each strain has its own characteristic composition, but inter-strain similarity is high (>90%) among the three strains, except strain 2761, which was found to be in quite a singular status among those tested.

3. *Interstrain Comparison within Rps. sphaeroides*

Similar to the case of *R. rubrum,* interstrain analyses among 18 strains within *Rps. sphaeroides* species demonstrated that only two strains out of 18 were distinguished from the other 16 strains, thus composing two tentative taxa.

Figure 8a and b show the cochromatographic elution between *Rps. sphaeroides* ATCC 17025 and *Rps. sphaeroides* ATCC 17023. As seen from Fig. 8, peaks 2, 10, 12, 16, and 20 in 50 S and 2, 3, 5, and 11 in 30 S of ATCC 17023 were recognized in the

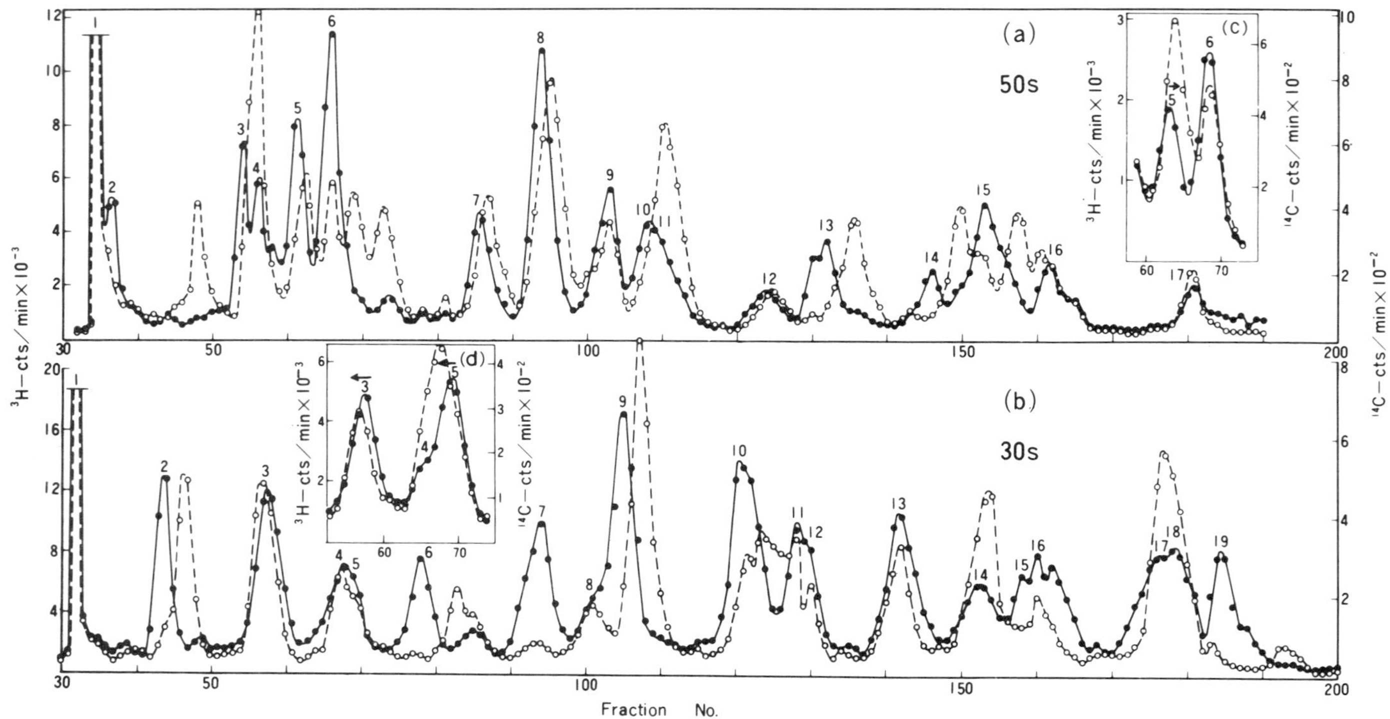

Fig. 7. (c and d) Cochromatograms on CM-cellulose columns of the 50 S (c) and 30 S (d) ribosomal proteins from R. tenue 3760 and R. tenue 3661. ——●——, *3760 protein labeled with [^{14}C]lysine.* ------○------, *3661 protein labeled with [^3H]lysine.*

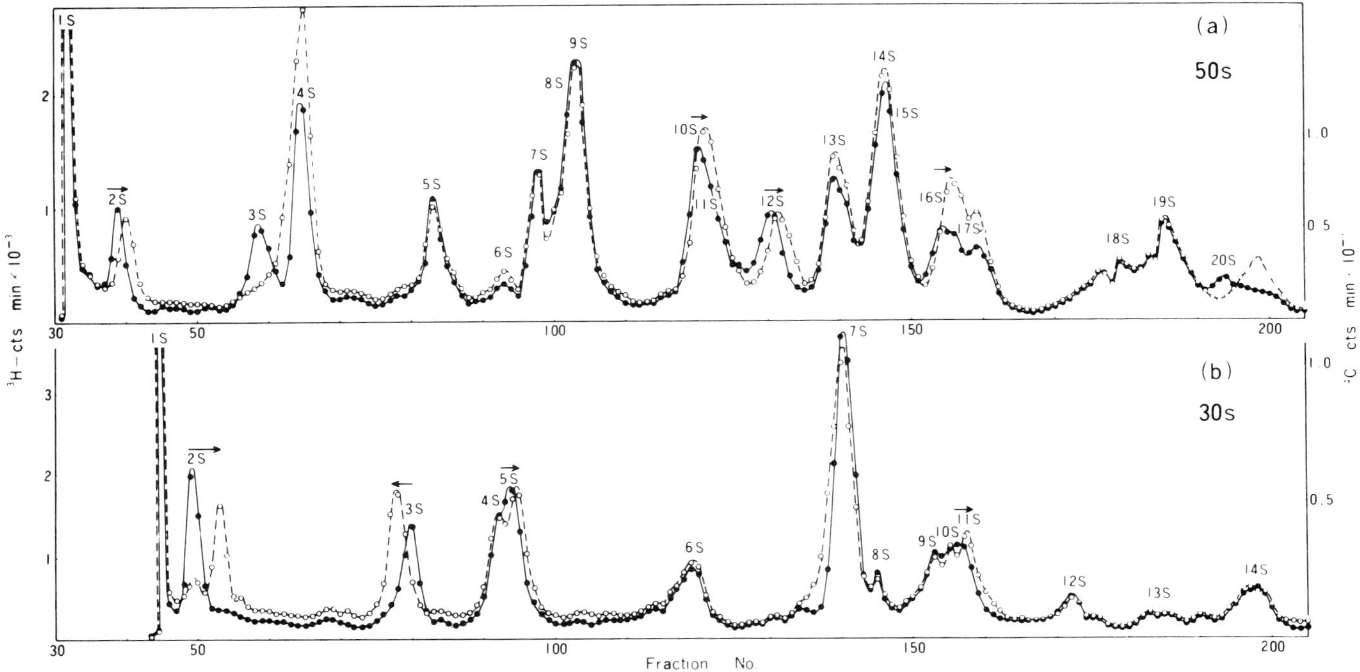

Fig. 8. Cochromatograms on CM-cellulose columns of the 50 S (a) and 30 S (b) ribosomal proteins from Rps. sphaeroides ATCC 17023 and Rps. sphaeroides ATCC 17025. ——●——, ATCC 17023 protein labeled with [^{14}C]lysine. ------○------, ATCC 17025 protein labeled with [^3H]lysine.

profiles of ATCC 17025, with a definite shift as indicated by the arrow. In addition, no match for peak 3 in 50 S of ATCC 17023 is apparently detected in ATCC 17025, thus indicating the overall similarity as 71%.

Only one more strain, ATCC 17024, revealed an identical composition to ATCC 17025. The carotenoidless green mutant BG strain (derived from ATCC 17023), the rapidly growing S strain, and porphyrin-accumulating mutant strain P_1 (derived from S strain) were all of 100% similarity to the reference strain ATCC 17023. Thus, *Rps. sphaeroides* species seem to be also classified into two ribosome groups, as in the case of *R. rubrum*. However, the similarity between these two taxa in *Rps. sphaeroides* was somewhat lower than the case of *R. rubrum* species.

SUMMARY AND DISCUSSION

Evaluation and Prospect of the Methodology

Prior to further phylogenetic examination of the data obtained, of particular pertinence is the confirmation of the methodological principles and characteristics from current points of view.

The choice of ribosomal protein compositions to approximate the extent of overall homology in genome DNA has a number of advantages.

First, the catalogue serving the present similarity analysis comprises the multiple gene products of the structural gene portion of ribosomal gene cluster. This accounts for approximately 2% of the whole genome DNA in the case of *E. coli*. Therefore, the present method is not based upon a single phenotypic expression, as often chosen in hitherto conventional methods, but on 40 or more complexed expressions, all simultaneously being surveyed by a common method. In this context of notable importance is the systematic conduct of overall analysis of whole cellular proteins of *E. coli* employing O'Farell's technique (15), as recently reported by Pedersen *et al.* (16). The catalogue consists of 140 protein species, including approximately 50 ribosomal component proteins thus accounting for the gene products of nearly 5% of the whole genome in *E. coli*. Apparently, this approach has opened the possibility of subjecting a majority of cellular protein species to similarity analysis.

Second, as has already been emphasized by Osawa *et al.* (4), the assumption that overall homology of gene compositions among various bacteria is closely represented by the similarity estimated from their ribosomal protein compositions, has thus far been supported by the following observations with gram-positive bacteria in the *Bacillaceae* family.

Among various bacteria in this family, the ribosomal genes appear to approximate markedly overall DNA homology as judged from the apparent parallelism between the variability in the amino acid sequence of particular protein components (functionally less specific components and probably poorly conserved during the evolutionary process) and those of cellular G+C content (13).

Third, the last characteristic is less variability of ribosomal protein components. A noticeable conservation of ribosomal constitution once settled during the process of cellular evolution seems feasible, in respect to the rarity of the ribosomal mutants, as expected from striking cooperativity inherent to the ribosome assembly (17). In fact, the average rate of substitution of amino acid residues of ribosomal proteins has been estimated as small as that of cytochrome *c* or the rate of nucleotide substitution in rRNA(5 S) (18). Thus, like the case of rRNA(5 S), ribosomal protein highly conserved from genetic variation, appears to be suitable as the catalogue for similarity analysis, particularly because the less variability permits comparison over a wider range in higher taxa. In fact, sequence matching in less variable ribosomal proteins is reported to be feasible even between *E. coli* and yeast (8). As has been discussed (18), amino acid substitution in ribosomal proteins could mostly be inferred to be neutral rather than having resulted from environmental selection.

In addition to above points, is the practical characteristics of the analysis in the subsequent data treatment and interpretation. The present method only indirectly approximates a "true" similarity to be estimated primarily on the basis of the sequence difference. Precise overlapping between two peaks in elution profile has always been counted as a unit numerator over the arithmetically summed up numbers of all detected components as the denominator. Such estimation of the denominator has been substantiated by the reported resemblance of the number and the distribution in molecular weight of ribosomal proteins even among distantly related bacteria (8). This was actually confirmed to be the case in *Rhodospirillaceae* with the ribosome of *Rps. palustris* in an electrophoretic pattern on polyacrylamide gel containing sodium dodecyl sulfate (20).

The error source and the consequent limitation of the methods, when the data are to be interpreted phylogenetically, are summarized in Table II (see also 21). Experimental limitation such as noted in points (3) and (4) in Table II would not bias the results too seriously, since it could reasonably be assumed that the Pauling units are approximately the same among ribosomal proteins (22). However, the way of denominator counting together with points (1) and (2) in Table II undoubtedly mold the unique character of the method to be capable of detecting even interstrain minute differences due to the unique amplification. This is immediately clear when the following two extreme hypothetical cases are considered, assuming, for the sake of simplicity, that each ribosome consists of 40 protein components each composed of 100 amino acid residues. In the first case, when only one component receives a single amino acid substitution causing a charge variation, then the similarity on the sequence basis is 99.975%, whereas that on the component basis is 97.5%. Amplification of dissimilarity becomes pronounced in the next case where 40 substitutions are received by one particular component or where each of 40 components receives a single substitution. The similarity on the component basis is then 97.5% (maximum) in the former case to 0% (minimum in the latter; where as it is 90% on the sequence basis. Such feature of the present method seems to afford the main explanation for the apparent remarkable sensitivity. On the other hand, the results of $G^{+}C$ content analysis, for instance, were reported to be around 65% throughout the *Rhodospirillum* genus (23, 24), indicating that the resolution level is not yet high enough to permit intragenus analysis.

Although the present method is essentially of a tentative and conventional nature, however, this is certainly of particular merit unless or until the sequence data of all component proteins become available. The method, as emphasized, is applicable within a limited size of certain taxa, possibly only of the genus category, as shown in the present study and in the case of *Escherichia* (4). This results solely from the uniqueness of the taxonomic index adopted.

Results from Similarity Analyses and Taxonomic and Phylogenetic Significance

Based upon the fact, confirmed in the present study, of the absence of detectable differences in the elution profile due to varied environmental conditions (light-anaerobic versus dark-aerobic), comparative survey was performed throughout with the photoanaerobically grown cells harvested in the logarithmic phase. The overall results, including those not referred in "Results," are summarized in the form of a similarity matrix as shown in Table III; those of taxonomic and phylogenetic relevance will be discussed.

1. The species of *R. rubrum* (16 strains) and *Rps. sphaeroides* (18 strains) yielded one minor ribosome group as a taxon of 92% (*R. rubrum* ATCC 17032 and 17036) and 70% ((*Rps. sphaeroides* ATCC 17024 and 17025) similarity, respectively.

TABLE II. Error Source and Limitation as Interpreted on Amino Acid Sequence Basis[a]

Matching (in position) in elution profile	Phenotypic difference	Error source and limitation as interpreted on the sequence level
Two peaks matching in identical position	No	1. Uneliminatable possibility of sequence difference due to the variation between uncharged residues
Two peaks matching but with displacement, or assignment of peak matching totally impossible	Yes	2. Total disregard for the extent of difference (more than one mutational change ?)on sequence basis
For any case		Disregard for probably more than one-third of components to be included in comparison due to: 3. Limitation in detecting minute peak or the displacement beyond the resolution level, especially with acid components less sharply eluted 4. Uneliminatable possibility losing some components upon release of ribosomes from the membrane-bound state

[a]In relation to (1), the probability of net charge change due to single amino acid substitution has been theoretically and empirically estimated to be 30 or 40% (21).

TABLE III. Similarity matrix[a]

Reference strain and number of protein components		
	R. rubrum (ATCC 11170)	R. tenue 3760
Grouping	50 S = 20, 30 S = 18	
R. rubrum Group I (14 strains)	———	100 %
Group II (ATCC 17032 & 17036)	50-5,6 30-9	92
R. tenue 3760	50-2,3,5,(6),7 30-(3),9,10,11,17	21-26
R. tenue 3761	50-2,3,(6),7 30-(3),9,10,11,17	18-24
R. tenue 3661	50-2,3,(6),7 30-9,10,11,17	18-21
R. tenue 2761	50-(6),7 30-6,(8),(17)	11-13
R. fulvum B147	50-19 30-6,(17)	5-8
R. molischianum NTHC 131	50-19 30-6,(17)	5-8
R. photometricum 2160	50-19 30-10,11,12	5-11
Rps. sphaeroides Group I (16 strains)		0
Group II (ATCC 17024 & 17025)		0
Rps. gelatinosa ATCC 17011		———
Rps. palustris ATCC 17001		———

[a]The components inside a box are all variants judged as uncommon. Ambiguous components are placed in parentheses.

tenue 3760		R. fulvum B147		Rps. sphaeroides (ATCC 17023)	
S = 17, 30 S - 19		50 S = 20, 30 S = 16		50 S = 20, 30 S = 15	
______	100%				
5	97				
5 3,5	92				
4,9,16,17 4,8,12,13,(14) 16,17,(18)	29-35	50-5,8,(13) 30-7	8-11%		
(6) 7,10,13	8-11	______	(100)		
(6) 7,10,13	8-11	50-2,12,18	92		
(6) 7,10,13	8-11	50-2,(6),10 11,12,18 30-3,5,7,9,16	69-72		
	0			______	100%
	0			50-2,3,10 12,16,20 30-2,3,5,11	71
	______		______		0
	______		______		0

Classification into two taxa remains to be fortified with respect to any other phenotypic character.* On the other hand, a totally dissimilar pattern of composition (estimated as 0%) was observed from intergenus (*R. rubrum* versus *Rps. sphaeroides*) and interspecies (*Rps. sphaeroides* versus *Rps. gelatinosa* and *Rps. palustris*) analyses. These facts essentially substantiate the traditional classification on the above genera and species. However, the total dissimilarity thus clarified from intergenus or inter-species analyses is in characteristic contrast with r-RNA (16S) analyses where similarity indices of *R. rubrum* versus *Rps. sphaeroides* and of *Rps. sphaeroides* versus *Rps. palustris* or *Rps. gelatinosa* were estimated as approximately 43, 40 and 20%, respectively (25). The discrepancy could be largely due to the above-mentioned amplification nature of the present method.

2. In contrast with the lower level of interspecies similarity (11-26%) between *R. rubrum* and *R. tenue*, the representative strain in each of three brown spirilla species revealed a remarkable resemblance to one another with similarity around 90% - as high as usually seen in an interstrain level. On the other hand, much less similarity was estimated by interspecies analyses with brown spirilla species versus *R. tenue* (10%) and versus *R. rubrum* (10%). These facts suggest a requirement for reclassification of these brown species under a separate taxon, despite the difference in cell size which has been used as a criterion for the classification (14). The extension of the similarity analysis to a wider range of strains in brown spirilla species remains to be conducted.

3. Unlike in the case of the *Escherichia* genus, it is rather remarkable that the interspecies similarity assessed within two genera of *Rhodospirillum* and *Rhodopseudomonas* is, except the case of brown spirilla species previously mentioned, so low, particularly within the latter genus. The results could be explained by ascribing it mainly to a more solidly

**R. rubrum ATCC 17032 and 17036 strains were isolated at the Hopkins Marine Station by Dr. F. M. Muller in 1930 and by Dr. C. B. van Niel in 1939, respectively. Some minor morphological differences were noted between strain ATCC 17032 and others in the same species, but were not striking enough to consider it as a distinct species. With the particular strain of ATCC 17036, isolated from ethanol-enriched culture, the effect of Ca ions on the motility was observed by Dr. J. W. Foster.*

Rps. sphaeroides ATCC 17024 and 17025 were both isolated by Dr. C. B. van Niel at Delft, Holland in 1935. (Private communication from Dr. C. B. van Niel, 1971.)

constructed network taxonomy of enteric bacteria fortified by criteria such as serological typing.

4. *R. tenue* 2761 from Göttingen, one among four strains tested in this species, revealed a similarity around 30% to each of the other strains, whereas the intersimilarity among the latter three was more than 90%. The 30% similarity with 2761 is considered far lower than that to be seen at inter-strain level. Phenotypic relevance to this particular strain is, uniqueness in the carotenoid compositions known to be distinct from any of other three, namely, possessing anhydrorhodovibrin and rhovibrin but lycopenal, rhodopinol, and rhodopinal being undetected (26). Like other *R. tenue* strains, however, 2761 strain was proved to be also distantly related both to *R. rubrum* and to the brown spirilla species with approx. 10% similarity in both cases.

5. Aside from the above taxonomic problem of the *R. tenue* 2761 strain, all the other three *R. tenue* strains are related to *R. rubrum* with 19 - 26% similarity,* whereas a far lower similarity (9 - 12%) was estimated with these strains versus the brown spirilla species. Furthermore, the distant relatedness to brown spirilla species is more pronounced in the case of *R. rubrum* species (ca. 5% similarity). These facts seem to support the placement of *R. tenue* species as a possible phylogenetic intermediate between the facultative photoanaerobe, *R. rubrum,* and the strictly photoanaerobic brown spirilla species. This view appears to be more probable when we consider the morphological characteristics of the intracytoplasmic membrane in *R. tenue* species, which could again be assignable to an intermediate stage in membrane differentiation between the lamella and the vesicle type, as observed in brown spirilla species and in *R. rubrum,* respectively (27).

The uncommon status of the *R. tenue* species has already been recognized in respect to the characteristic cytochrome pattern, which is singular among the *Rhodospirillum* genus (28).

Any further resolution of the problematic status of the *R. tenue* species, requires the finding of a sufficient number of strains of intermediate similarity, thus consistently fractionating the divergency span.

Further analysis is in progress and has been extended to the *Rps. capsulata* species of the *Rhodopseudomonas* genus in conjunction with the gene transfer system (29) thus far visualized only among this species in *Rhodospirillaceae*.

**It is noteworthy that the similarity between R. rubrum and R. tenue estimated in the present paper is in good agreement with that assessed from analysis of r-RNA (16S) (Approx. 20%) (25).*

ACKNOWLEDGMENT

The authors would like to gratefully acknowledge Prof. S. Osawa, Dr. E. Otaka, and all members of the Department of Biochemistry and Biophysics, for their help and encouragement throughout our research. The authors are deeply indebted to Prof. Norbert Pfennig (Göttingen University) for his kindness in providing bacterial strains and his valuable discussions, which made this research possible, and to Prof. C. B. van Niel (Hopkins Marine Station) for his kindness in informing us of his unpublished observations on his strains. Thanks are due to all those who donated their strains (as listed in Table II).

Finally, our most sincere thanks go to Prof. M. D. Kamen who unfailingly supported us from the stage of gathering.

This research was supported in part by a scientific research fund (No. 034059, 1974-1976 to S.T.) from the Japanese Ministry of Education, Science and Culture and by the research grant (1973 to K.K.) provided by the Yakult Central Research Institute.

REFERENCES

1. Dickerson, R. E., Timkovich, R., and Almassy, R. J. (1976) The cytochrome fold and the evolution of bacterial energy metabolism. *J. Mol. Biol. 100,* 473-491.
2. Pfennig, N. (1977). Phototrophic green and purple bacteria: a comparative systematic survey. *Annu. Rev. Microbiol. 31,* 275-290.
3. Saunders, V. A. (1978). Genetics of *Rhodospirillaceae*. *Microbiol. Rev. 42,* 357-384.
4. Osawa, S., Itoh, T., and Otaka, E. (1971). Differentiation of the ribosomal protein compositions in the genus *Escherichia* and its related bacteria. *J. Bacteriol. 107,* 168-178.
5. Kado, C. I., Schaad, N. W., and Heskett, M. G. (1972). Comparative gel-electrophoresis studies of ribosomal proteins of *Erwinia* species and members of the *Enterobacteriaceae*. *Phytopathology 62,* 1077-1082.
6. Schaad, N. W. (1974). Comparative immunology of ribosomes and disc gel-electrophoresis of ribosomal proteins from Erwiniae, Pectobacteria, and other members of the family Enterobacteriaceae. *Intern. J. System. Bacteriol. 24,* 42-53.
7. Pfennig, N. (1978). *Rhodocyclus purpureus* gen. nov. and sp. nov., a ring-shaped, vitamin B_{12}-requiring member of

the family *Rhodospirillaceae*. *Intern. J. System. Bacteriol. 28,* 283-288.

8. Wittmann, H. G. (1976). The seventh Sir Hans Krebs lecture: structure, function and evolution of ribosomes. *Eur. J. Biochem. 61,* 1-13.
9. Zablen, Z. and Woese, C. R. (1975). Prokaryote phylogeny IV: Concerning the phylogenetic status of a photosynthetic bacterium. *J. Mol. Evol. 5,* 25-34.
10. Woese, C. R., Pribula, C. R., Fox, G. E., and Zablen, L. B. (1975). The nucleotide sequence of the 5S ribosomal RNA from a photobacterium. *J. Mol. Evol. 5,* 35-40.
11. Cohen-Bazire, G., Sistrom, W. R., and Stanier, R. Y. (1957). Kinetic studies of pigment synthesis by non-sulfur purple bacteria. *J. Cell. Comp. Physiol. 49,* 25-68.
12. "American Type Culture Collection" (1972). Catalogue of Strains, 10th ed., p. 256.
13. Otaka, E., Itoh, T., and Osawa, S. (1968). Ribosomal proteins of bacterial cells: strain- and species-specificity. *J. Mol. Biol. 33,* 93-107.
14. Pfennig, N. and Trüper, H. G. (1974). The phototrophic bacteria. *In* "Bergey's Manual of Determinative Bacteriology" (R. E. Buchanan and N. E. Gibbons, eds.), 8th ed., pp. 23-34. Williams & Wilkins, Baltimore, Maryland.
15. O'Farrell, P. H. (1975). High resolution two-dimensional electrophoresis of proteins. *J. Biol. Chem. 250,* 4007-4021.
16. Pedersen, S., Bloch, P. L., Reeh, S., and Neidhardt, F. C. (1978). Patterns of protein synthesis in *E. coli*: a catalogue of the amount of 140 individual proteins at different growth rates. *Cell 14,* 179-190.
17. Davies, J. and Nomura, M. (1972). The genetics of bacterial ribosomes. *Annu. Rev. Genet. 6,* 203-234.
18. Hori, H., Higo, K., and Osawa, S. (1977). The rates of evolution in some ribosomal components. *J. Mol. Evol. 9,* 191-201.
19. Sun, T. T., Bickle, T. A., and Traut, R. R. (1972). Similarity in the size and number of ribosomal proteins from different prokaryotes. *J. Bacteriol. 111,* 474-480.
20. Bhatnager, Y. M. and Stachow, C. S. (1972). Ribosomal proteins of *Rhodopseudomonas palustris*. *J. Bacteriol. 109,* 1319-1321.
21. Nei, M. and Chakraborty, R. (1973). Genetic distance and electrophoretic identity of proteins between taxa. *J. Mol. Evol. 2,* 323-328.
22. Hori, H. and Osawa, S. (1978). Evolution of ribosomal proteins in Enterobacteriaceae. *J. Bacteriol. 133,* 1089-1095.

23. Silver, M., Friedman, S., Couture, J., and Tanguay, R. (1971). Base composition of deoxyribonucleic acid isolated from *Athiorhodaceae*. *J. Bacteriol. 107,* 368-370.
24. Mandel, M., Leadbetter, E. R., Pfennig, N., and Trüper, H. G. (1971). Deoxyribonucleic acid base compositions of phototrophic bacteria. *Intern. J. System. Bacteriol. 21,* 222-230.
25. Woese, C. R., Gibson, J., and Fox, G. E. (1980). Do genealogical patterns in purple photosynthetic bacteria reflect interspecific gene transfer? *Nature 283,* 212-214.
26. Schmidt, K. (1978). *In* "Photosynthetic Bacteria" (R. K. Clayton and W. R. Sistrom, eds.). Plenum, New York.
27. Oelze, J. and Drews, G. (1972). Membranes of photosynthetic bacteria. *Biochim. Biophys. Acta 265,* 209-239.
28. Ambler, R. P. (1977). Cytochrome *c* and copper protein. *Symp. Evolution Metalloproteins, University of Sussex,* July 1.
29. Marrs, B. R. (1978). Mutations and genetic manipulations as probes of bacterial photosynthesis. *In* "Photosynthetic Bacteria" (R. K. Clayton and W. R. Sistrom, eds.). Plenum, New York.

NUCLEOTIDE EXCISION REPAIR OF DAMAGED DNA

L. Grossman
S. Riazuddin

Department of Biochemistry
The Johns Hopkins University
School of Hygiene and Public Health
Baltimore, Maryland

*W. Haseltine**
K. Lindan

Sidney Farber Cancer Research Institute
Harvard School of Medicine
Boston, Massachusetts

The ability of a cell to survive the effects of DNA modifications by a variety of noninstructional or non-hydrogen binding changes must be *repaired* prior to or post replication events. The model which can be invoked for studying repair of the aforementioned species of damage has, in principle, been derived from photobiology. The facile identification of pyrimidine-pyrimidine dimers, their identifiable chromatographic properties, and the ease of isolation has allowed for the temporal analysis of their effects during postirradiation periods.

The best understood pathway of DNA repair involves excision either of nucleotides, first described by Carrier and Setlow (1) and Boyce and Howard-Flanders (2), or of bases described by Lindahl (3). The initial recognition by site-specific nucleases which act on the primary, chemical or photochemical damage is the rate-limiting step in the excision process. The other path-

**W. Haseltine is an American Cancer Society Faculty Research Awardee.*

 ISBN 0-12-397580-8

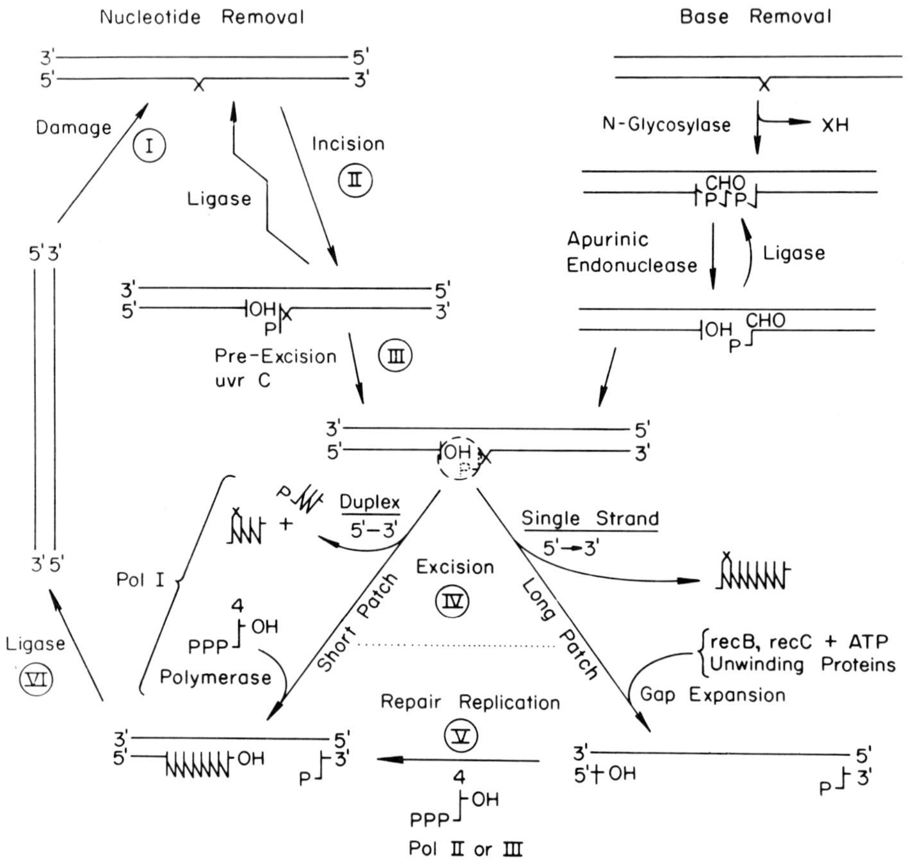

Fig. 1. A proposed DNA excision repair cycle.

ways involving *N*-glycosylic bond hydrolysis by enzymes recognizing unique kinds of chemical changes have recently received a great deal of attention (see I. R. Lehman, this volume).

NUCLEOTIDE EXCISION - DAMAGE-SPECIFIC ENDONUCLEASES (Fig. 2)

There are three damage-specific endonucleases which act on pyrimidine dimer containing DNA. The T4 bacteriophage $\underline{V^{+}}$ gene-coded enzyme, the enzymes controlled by the *uvrA*, *uvrB*

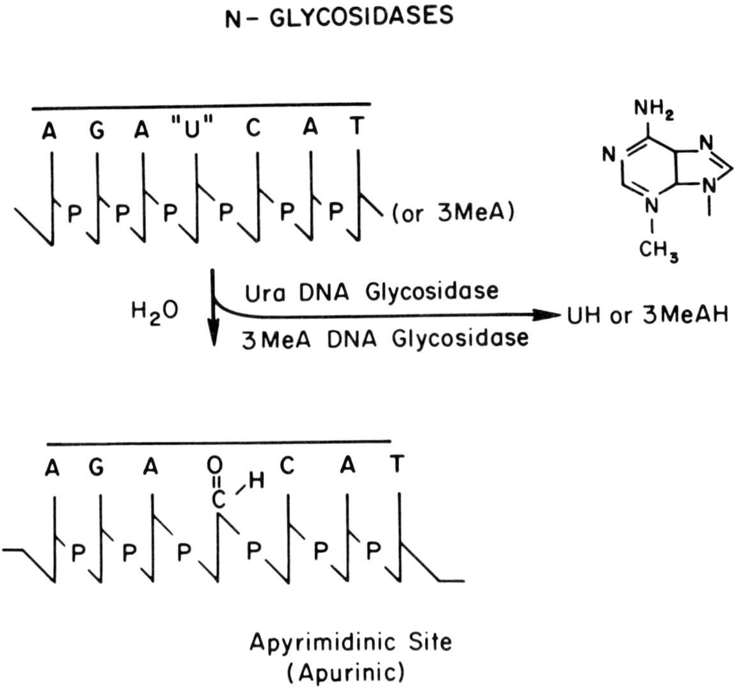

Fig. 2. Mechanism of DNA N-Glycosylase reaction.

structural genes, and the enzymes isolated from *Micrococcus luteus* behave in a similar manner. These enzymes are specific for the damaged strand resulting in phosphodiester bond hydrolysis 5' to the damaged site generated 3'-hydroxyl and 5'-phosphoryl groups. Such sites are removed with polynucleotide ligase by virtue of the juxtaposed 3'-hydroxyl and 5'-phosphoryl group. Although polynucleotide ligase has been used diagnostically in identifying these intermediate structures, it is possible that this "reversibility" may play an important role *in vivo* in which this putative step is under control by the *uvrC* gene.

It is unclear why a controlling step is necessary at this point in repair. However, definitive binding data at such sites by competing enzymes such as DNA polymerase I and ligase is needed. From preliminary information concerning the competition of these two enzymes for nicked or incised sites it can be suggested that ligase can limit polymerization or strand displacement reactions catalyzed by DNA polymerase I.

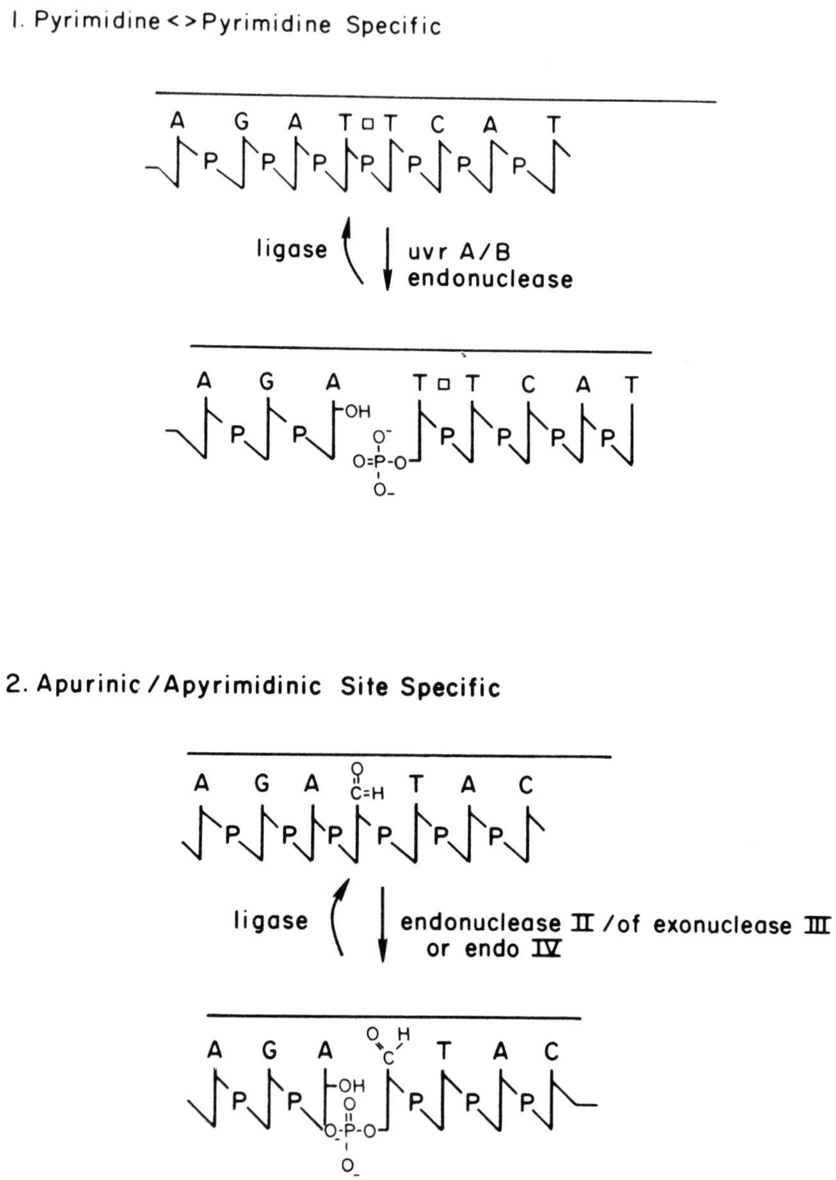

Fig. 3. Mechanisms of damage-specific endonucleases in repair.

DNA N-GLYCOSYLASES (Fig. 3)

Lindahl and co-workers (3, 4) recently discovered a series of bacterial enzymes that recognizes a modified nucleotide and removes the modified purine or pyrimidine base in the absence of phosphodiester bond hydrolysis. The general hydrolytic mechanisms catalyzed by these enzymes result in the formation of a deoxyribose moiety lacking an aglyclone. The two enzymes that have been identified are from *Escherichia coli.* One of these enzymes recognizes the uracil which replaces thymine in the PBS II phage DNA resulting in an A:U pair, or in deaminated DNAs with resultant U:G non-hydrogen bonded base pairs. This enzyme, ura-*N*-glycosylase, is, therefore, specific for the uracil residue generating an apyrimidinic site.

A similar-type enzyme specific for 3-methyladenine and hypoxanthine-containing DNA and yielding an apurinic site has recently been highly purified. These unique sites in DNA are sensitive to β-elimination reactions and hence are alkali labile. Unlike the nucleoside phosphorylases and the NADases, the DNA *N*-glycosylase reactions are neither phosphorolytic nor arsenolytic and are irreversible. The role of these enzymes in metabolic repair has been amply discussed by Dr. I. R. Lehman (this volume).

An endonucleolytic step is required to act specifically at sites generated by the DNA *N*-glycosylases. These apyrimidinic or apurinic sites act as substrates for the action of a number of specific endonucleases. In *Escherichia coli,* exonuclease III, which with its associated endonuclease activity (5), is able to act at such sites generating 5'-phosphoryl and 3'-hydroxyl termini 5' to the deoxyribose moiety (6). It may be suggested that the role of the exonuclease activity in repair is for removal of a few juxtaposed nucleotides thereby limiting the action of ligase and thus <u>abortive repair</u>.

EXCISION OF DAMAGED NUCLEOTIDES (Fig. 4)

DNA Polymerase-Associated Excision Enzymes

The bacterial polymerases with their associated exonucleolytic activities have the capabilities for involvement in a number of important biological functions in addition to polymerization of nucleotides. Most notable is the 5'-exonuclease activity specific for double-stranded DNAs, which acts at nicks in an extremely efficient manner during concomitant polymerization. This <u>nick translational</u> activity or polymerase I functions by

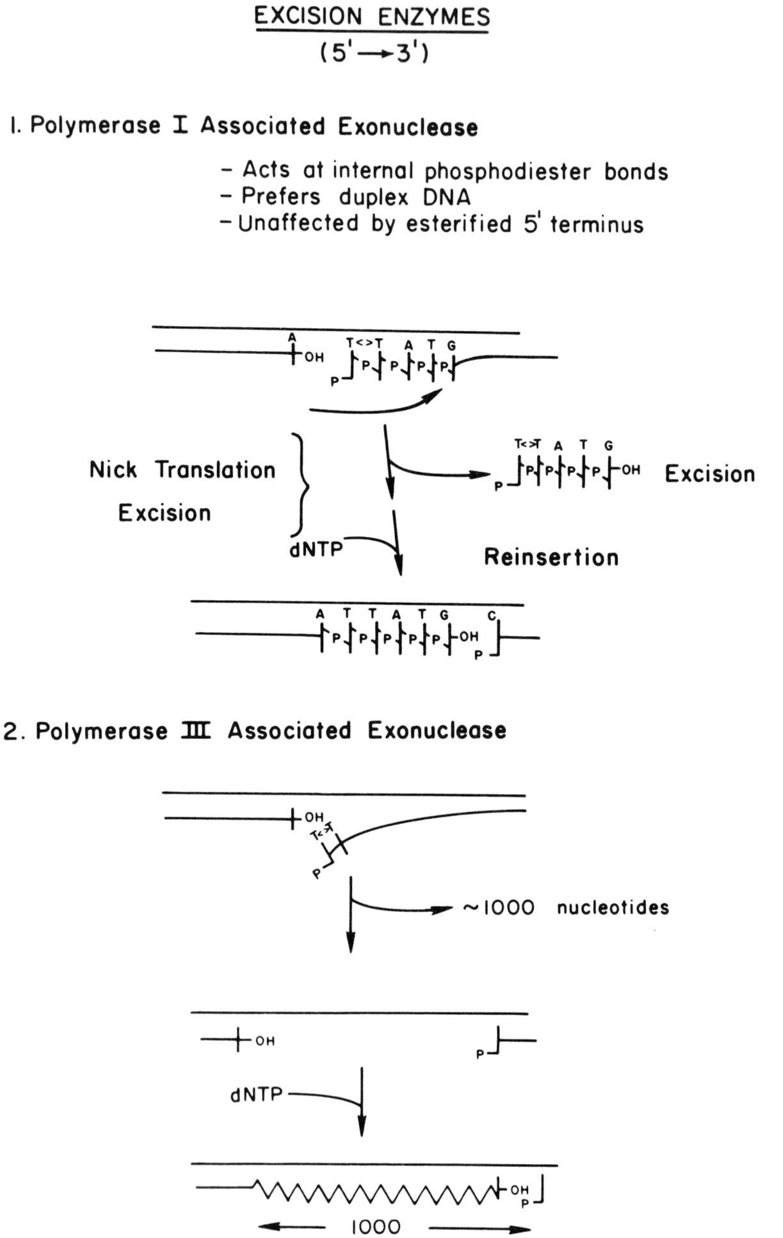

Fig. 4. Mechanism of DNA polymerase I-catalyzed dimer excision.

virtue of the stoichiometric nucleotide removal and reinsertion in a 5' → 3' direction. The ability of this associated exonuclease to function at internal rather terminal phosphodiester bonds makes it suitable for the removal of 5'-terminal damaged nucleotides. Kelly *et al*. (7) first demonstrated excision capabilities of the polymerase by its ability to remove pyrimidine dimers from DNAs incised by nonspecific endonucleases. Similarly, damaged DNA treated initially by the correctional endonucleases from either *M. luteus* or from *E. coli* followed by polymerase I loses the pyrimidine dimer containing fragments as an initial step during the excision reaction (8). This is a very efficient process because of the coordinated polymerization and excision by this multifunctional enzyme.

Unassociated Exonucleases

A variety of unique exonucleases have been isolated from *M. luteus, E. coli,* and from human placenta, which act specifically on single-stranded DNA regions but interestingly initiate hydrolysis from either the 3' or 5' terminus, generating small 5'-phosphorylated oligonucleotides (9 - 11). These single-stranded-specific exonucleases are also able to act on irradiated incised duplex DNAs in an efficient manner. The placental correctional exonuclease can, in addition, digest at a nick in an undamaged duplex DNA in both a 3' and 5' direction generating a gap approximately 40 nucleotides in length (11). The bacterial enzymes acting on incised irradiated DNA generate a gap approximately 10 nucleotides in length (12) presumably because their activities are limited to single-stranded-like regions of the incised irradiated DNAs.

Base - Nucleotide Incision Pathways (see Fig. 1)

In prokaryotes, nucleotide incision and base removal mechanisms potentially converge onto a common path. The initial removal of damage, catalyzed either by specific endonucleases acting at damaged sites or endonucleases acting at sites generated by DNA *N*-glycosylases lacking purines or pyrimidines are repairable by the efficient polymerase I excision-reinsertion pathway; strand restoration is catalyzed by nucleotide ligase. This most efficient pathway is mechanistically termed short patch repair (13). In polA mutants the unassociated exonucleases are capable of removing damaged oligonucleotides. By virtue of the presence of polymerases II or III in such mutants, it can be suggested that a gap expansion step is necessary to allow for the processive binding and reinsertion by these two polymerases (14). In either short or long patch repair, accommodation must be made for preserving the integrity of the complementary strands during the transient stage in which vulnerable gaps exist following excision just prior to reinsertion. Binding proteins by virtue of their ability to inhibit the action of many nucleases, which act at single-stranded sites, may as a consequence assume a protective role during repair (15, 16).

RECOGNITION OF PYR< >PYR CORRENDONUCLEASE

Structural Gene Control of Pyrimidine Dimer-Specific Correctional Endonucleases

The initial step in the excision repair process in *E. coli* is controlled by an endonuclease activity which appears to be absent in both *uvrA* and *uvrB* mutants (17). The distal relationship of these two structural genes seems paradoxal in terms of the control of what appears to be a single enzyme. A similar situation exists in *M. luteus*, although unmapped genetically. It is apparent that a similar enzymatic activity is controlled by at least two genes in this organism (13). It seemed advantageous to investigate this problem in *M. luteus* because of the greater concentration of such endonucleolytic activity in this organism. An analysis of a series of mutants and transformants of this organism indicates that the mutants which are obtained through nitrosoguanidine mutagenesis are uv sensitive and lack enzymatic activity (DB7). Transformation of such mutant cells with wild-type DNA produced a uv-sensitive transformant with some enzymatic activity (DB200). Further transformation of DB200 with wild-type DNA yields a uv-resistant revertant with full enzymatic activity (DB400). During purification of the enzymes on phosphocellulose chromatography wild-type levels of enzymes assayed with heavily irradiated DNAs were partially resolved (peaks I and II). The mutant DB7 possesses little or no activity; DB200 the first transformant, although still uv sensitive, has one peak of activity; DB400, the revertant, possesses wild-type levels of both peaks of activity. The third, fourth, and fifth peaks were used for reference for further identification of enzyme activities which recognize heavily irradiated DNA (Fig. 5). Their persistence in all of the mutants and transformants tested is diagnostic for their lack of involvement in ultraviolet damage repair process in these organisms.

An analysis of the same enzymatic activities at doses which induce no more than three to four pyrimidine dimers per genome is seen in Fig. 5 in which the presence of peaks III, IV, and V

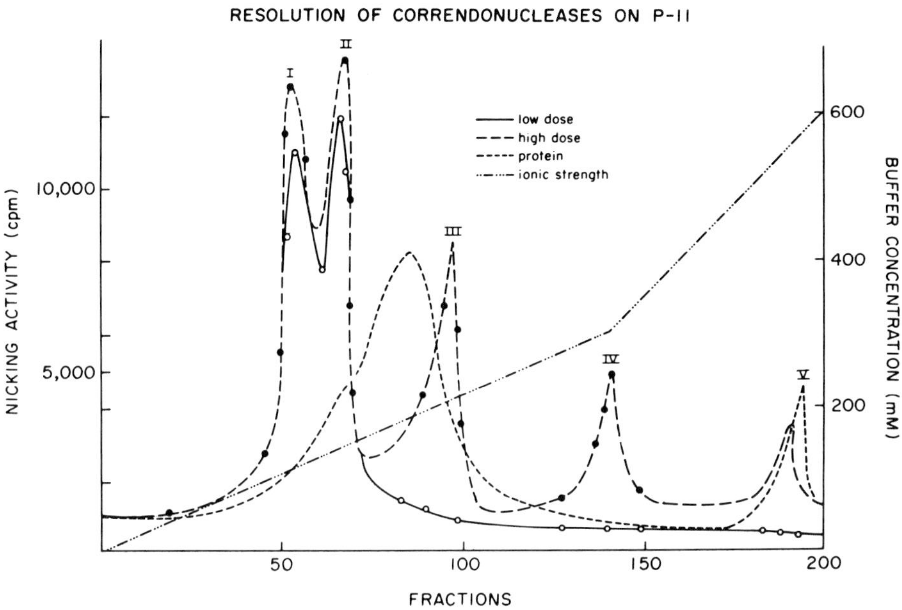

Fig. 5. Elution pattern of endonucleolytic activities from phosphocellulose (P-11). The ØX174 RFI incision assay was used to measure activity against native DNA (Δ-Δ), lightly irradiated (10 J/m²) DNA (O-O), and heavily-damaged (500 J/m²) DNA (O-O). The concentration of the eluting phosphate buffer was obtained from a measurement of the conductivity (----).

are not evident until DNAs have been more heavily irradiated. The important feature of this finding is that the first two peaks of activity are both present when either heavily or lightly irradiated DNAs are used as substrate. By taking advantage of the differences in DNA binding a resolution of the two enzymes is achieved by affinity chromatography. Final purification of these two peaks is achieved on isoelectric focusing.

An analysis of substrate saturation curves clearly show that both endonucleases reach saturation with no more than 2 to 4 pyrimidine dimers per genome. Peak III, IV, and V, however, required DNA substrates with 100 to 300-fold higher levels of modification before saturation was reached. These differences distinguish the damage specificity of peaks I and II from the damage dependency of the other three peaks of activity. Peak III has a wide range of specificity including irradiated DNA, alkylated-depurinated DNA, DNA treated with OsO_4 (which forms a thymine glycol derivative analogous to a gamma irradiation product) (19), and undamaged denatured DNA. This endonuclease probably recognizes the common secondary effects arising from the diverse forms of damaging agents.

That two endonucleases are specific for uv-irradiated DNA demanded a resolution of their substrate specificities. The ability of either peaks I or II to act specifically on pyrimidine pyrimidine dimers was examined. The action of photolyase on irradiated DNAs (photolyase monomerizes T< >T, C< >T and C< >C pyrimidine dimers) is to reverse the action of the two enzymes, implicating pyrimidine dimers as the substrate on the DNA for these enzymes.

The question of whether the two enzymes are specific for any of the three species of pyrimidines was examined by irradiating DNA with photosensitizers, in which thymine dimers are the major photoproduct and C< >T and C< >C are absent (20). It was found that both enzymes act on thymine-thymine dimers. Pyrimidine dimer specificity was further examined more closely by determining the effects of the separate endonucleases on excision of the respective dimers from irradiated DNA. This was accomplished by following the loss of pyrimidine dimers from DNA under conditions (shown in Table I), in which the endonuclease was subsequently treated with either exonuclease VII or polymerase I for excision. The data clearly suggest that neither enzyme I nor enzyme II show any unique pyrimidine dimer specificity.

The question remained as to why there are two separate endonucleases seeming to act on the same substrate. Furthermore, genetic analyses indicated that both proteins are required to confer uv resistance on cells.

The possibility that these two enzymes may be acting in different regions of the DNA molecule was pursued. The irradiated DNAs were incised by each endonuclease and the structure of the sites generated in each case was studied. Three different experimental approaches were utilized, all of which are diagnostic for the appearance of single-stranded-like regions at the site of incision. The first procedure utilized BAP, which is a heat stable and rather large protein. Because BAP is heat stable it is possible to examine the level of dephosphorylation under conditions of increasing temperature. The data in Fig. 6 show the extent of phosphomonoester group availability to BAP as a function of temperature. What is apparent is that the phosphomonoester sites generated by correndonuclease I are significantly more accessible to the action of BAP than is that site generated by correndonuclease II. These data are consistent with two unique sites being generated.

It was assumed that a distinction between these termini could be made by examining the susceptibility of these nicks to polynucleotide ligase. Ligase can catalyze the formation of a phosphodiester bond even though the 3' juxtaposed nucleotide is noncomplementary (21). Rates of ligation were followed with DNAs incised with either correndonuclease I, correndonuclease

TABLE I. Excision of Pyrimidine Dimer Species[a]

Treatment	Dimer content			Fraction excised		
	T<>T	C<>T	C<>C	T<>T	C<>T	C<>C
UV irradiation	8.5	8.05	3.15	-	-	-
$Endo^{uv}I$ + polI	3.91	3.94	1.42	54	51	55
$Endo^{uv}II$ + polI	3.49	3.22	1.10	59	60	65
$Endo^{uv}I$ + $Endo^{uv}II$ + polI	3.83	5.23	1.51	55	35	52
$Endo^{uv}I$ + exoVII	4.59	4.67	1.61	46	42	49
$Endo^{uv}II$ + exoVII	2.13	2.17	1.13	75	73	64
$Endo^{uv}I$ + $Endo^{uv}II$ + exoVII	2.38	5.6	1.57	72	30	50
$Endo^{uv}I$ or $Endo^{uv}II$ or polI or exoVII				0.5 to 4.0		

[a]Thymine or cytosine-labeled DNA (2 nmol) was incised with 2.5 BAP units of each correndonuclease (23) either individually or in combination. The incised DNA, after inactivation of the endonuclease at 70° for 5 min, was treated with DNA polymerase or human placental correxonuclease. The unexcised pyrimidine dimers remaining in the DNA were determined by its hydrolysis in 97% formic acid and paper chromatography of the hydrolysate in 1-butanol acetic acid; water (80:12:30) as discussed by Setlow and Carrier (24).

II, or pancreatic DNase I. The data in Table II demonstrates that distinctions exist between the two products of incision and that DNA incised with enzyme I is converted less rapidly to RFI in comparison to that generated by correndonuclease II. These data are in keeping with the suggestion that the site generated by correndonuclease I is more single-stranded-like than that generated by the second enzyme.

An additional approach to studying these structural differences, employed the cooperative binding properties of the *E. coli* binding protein to single-stranded DNA. In order to assess the effects of binding protein, use was made of an observation in which single-stranded DNA associated with binding protein is resistant to the action of exonuclease VII (22). In these experiments the incised irradiated DNA was preincubated with stoichiometric amounts of binding protein and sub-

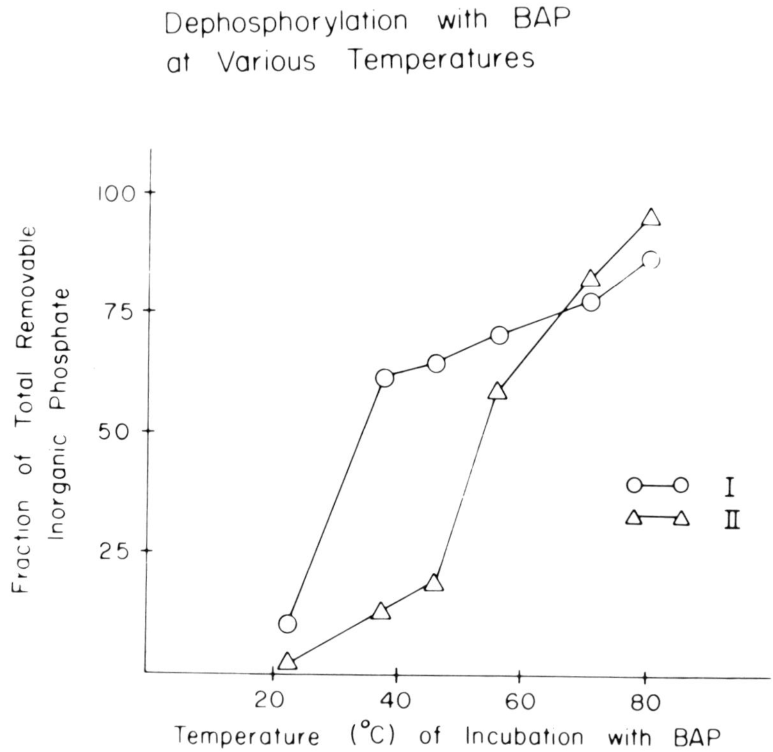

Fig. 6. Temperature dependence of dephosphorylation of bacterial alkaline phosphatase (BAP) of native uv-irradiated (1440 J/m²) ^{32}P-labeled Escherichia coli DNA after incision with Pyr<>Pyr correndonuclease I (O-O) or I (Δ-Δ). Five nanomoles of incised DNA In 50 mM Tris-HCl at pH 8.0 was dephosphorylated with 1.5 units of BAP. The values reported are relative to those obtained at 37 C with incised and heat-denatured DNA and were 39 and 43 pmol of phosphomonoester groups generated by Pyr<>Pyr correndonuclease I and II, respectively.

jected to hydrolysis by exonuclease VII. The data show that the site generated by enzyme I, when 25% hydrolyzed by exonuclease VII, causes complete shutoff of further hydrolysis. The site generated by correndonuclease II is quantitatively hydrolyzed, although at a reduced initial rate. These data support the suggestion that unique sites are generated by the resolved

TABLE II. Rejoining of Endonuclease Incised DNA[a]

Treatment	Radioactivity retained on the B-6 filters (cpm) -ligase (A)	+ligase (B)	Fraction rejoined $\frac{A-B}{A} \times 100$
Pancreatic DNase I incised DNA	9,876	3,437	65
UV endonuclease I treated DNA	8,267	7,870	5
UV endonuclease II treated DNA	8,103	5,275	35

[a]Twenty five picomoles of ØX174 RFI DNA containing five pyrimidine dimers/RFI molecule were incised at 37° for 30 min. with 1 unit of Pyr< >Pyr correndonuclease I or II. After inactivating the endonuclease at 55° for 60 min, the incised DNA was sealed at 30°C for 45 min, with 2 units of phage T_4 polynucleotide ligase in the presence of 66 μM ATP. The ligase-treated DNA, after denaturation and renaturation, was filtered through B-6 membrane filters and the radioactivity going through was taken as an index of ligase activity.

correndonucleases (18, 22, 23).

It can be inferred from the previous discussion that the correndonucleases may be functioning in different regions of the DNA. The data, though suggestive, are not sufficiently precise to come to definitive conclusions demanding another approach to this problem.

USE OF DNA FRAGMENTS OF DEFINED SEQUENCE AS SUBSTRATES

In order to examine endonucleolytic mechanisms with greater precision, a fragment of the lac operon promoter-operator (lac p-o) region labeled at the 5'-end with ^{32}P was chosen as a substrate for the enzymes. The sequence shown in Fig. 7 not only provides differing environmental sequences surrounding potential pyrimidine dimer sites for detailed analyses of endonucleolytic cleavage but also permits examination of the effects of dimer formation on repressor and RNA polymerase binding. Conversely, the protective effects of these proteins on such sites can also be examined. Two DNA fragments, 117 and

```
       1       10        20        30        40        50        60        70        80        90      99
117 5' AATTCTCACTCATTAGGCACCCCAGGCTTTACACTTTATGCTTCCGGCTCGTATAATGTGTGGAATTGTGAGCGGATAACAATTTCACACAGGAAACAG.......
       ........GAGTGAGTAATCCGTGGGGTCCGAAATGTGAAATACGAAGGCCGAGCATATTACACACCTTAACACTCGCCTATTGTTAAAGTGTGTCCTTTGTCTTAA 5' 168
               99'      90'       80'       70'       60'       50'       40'       30'       20'       10'      1'
```

Fig. 7. The lac operon - operator-promoter sequence. The 117 nucleotide-long fragment is one product of cleavage by the restriction endonuclease HaeIII and the 168 nucleotide-long fragment is the other product.

TABLE III. Frequency of Scission by uv Correndonuclease at Specific Sequences[a]

DNA Sequence with Dimer Region	*CPM's/Band: CPM's/Tract (%)*	*No. of Thymines in Region*	*Percent of Damage No. of Thymines*
5'... CCCC	0.1	0	0
GTTTCCTG	22.0	4	5.5
ATTG	14.0	2	7.0
GTTA	11.8	2	5.9
ATCCG	6.9	1	6.9
GCTCA	7.6	1	7.6
ATTCCA	15.9	2	7.95
ACACATTATACGC ...3'	25.3	2	12.65

[a]168 bp 5'-P^{32} end-labeled DNA was uv irradiated with 480 J/m^2 of 254 m light and incubated with saturating amounts of correndonuclease I + II. Cleavage products were electrophoresed on 20% polyacrylamide sequencing gels. Bands produced at each pyrimidine dimer region were excised from the gel and the radioactivity of each was determined by Cherenkov counting. The cpm's of each dimer region were divided by the total number of counts in the electrophoretic lane to produce the relative amounts of damage produced in each region. If this number is divided by the number of thymines present in that region, then it is revealed that the absolute quantum yield is predominantly dependent on the number of thymines present. The thymine dimer surrounded by a long sequence of alternating purines and pyrimidines, however, gives rise to a greater amount of damage and is termed a "hot spot" for damage.

168 base pairs in length, each containing one copy of the 95 base pair lac p-o region are employed in these studies. 5'-^{32}P phosphorylation results in labeling on only one strand of each fragment and the complementary strands of the known sequence are separately labeled after resolution of the HaeIII 117 and 168 bp fragments. Procedures for isolation and preparation of the fragments are described elsewhere (24, 25).

uv-irradiated DNA that is incubated with the pyrimidine dimer-specific endonuclease is cleaved at specific sites. To determine the position of cleavage, 5'-end-labeled 117 or 168

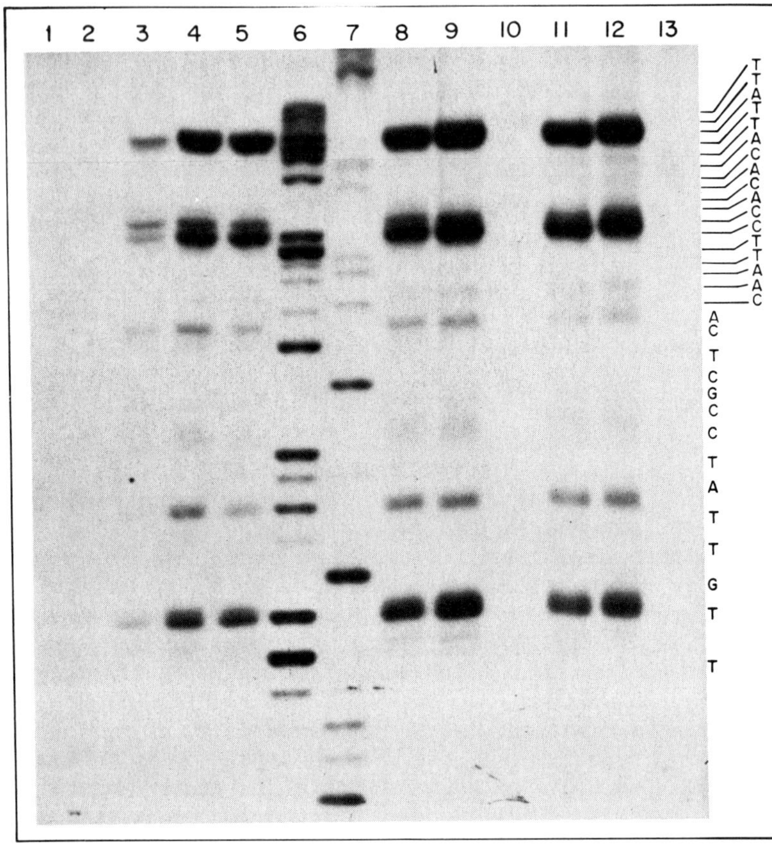

Fig. 8. Site-specific cleavage by purified Micrococcus luteus correndonuclease of DNA irradiated with varying doses of uv light. 168 base pair 5'-^{32}P end-labeled DNA sequence was irradiated with 254 nm light at varying doses, incubated with saturating amounts of purified M. luteus correndonuclease peaks I and II G-75 fraction, and loaded onto a denaturing 20% polyacrylamide sequencing gel alongside sequencing reactions of the Maxam and Gilbert protocol. The reactions contained: unirradiated DNA treatment with 5 µl (Lane 1) and 15 µl (Lane 2) of purified enzyme; DNA irradiated with 120 J/m^2 (Lane 3), 480 J/m^2 (Lane 4), 960 J/m^2 (Lane 5), 1440 J/m^2 (Lane 8), 4320 J/m^2 (Lane 9), 7200 J/m^2 (Lane 10), and incubated with 5 µl of purified enzyme; DNA irradiated at 10,080 J/m^2 (Lane 13) underwent the same manipulations as all other reactions but with no enzyme present. Lanes 6 and 7 contained C+T and A reactions of Maxam and Gilbert sequencing protocol. DNA, 60,000 Cherenkov cpm per tract, was irradiated in a 10 mM Tris-

base pair fragments are uv irradiated and, subsequently, incubated with endonuclease peaks I + II. The products are electrophoresed on high-resolution denaturing polyacrylamide gels. The adjacent tracts contain the products of DNA sequencing reactions as described by Maxam and Gilbert (26). Bands produced by sequencing result from migration of 5'-end-labeled fragments terminating in a 3'-phosphate, whereas endonucleolytic cleavage yields 5'-phosphorylated terminus.

The DNA fragments were irradiated with increasing 254 nm light, from 140 J/m^2 to 10,000 J/m^2, and treated with saturating enzyme concentrations. Since the enzyme was known to be at saturating levels, the different intensities of the bands reflect the absolute quantum yield based on nucleotide sequences recognized by the enzyme. Comparison of the mobility of the enzymatic cleavage products with those of the chemical sequencing reactions reveals that cleavage is taking place only at sites containing pyrimidine dimers (Fig. 8). No cleavage at possible dimers that might be formed near the staggered 5' EcoRI terminus of the fragment was apparent. This may be due to a requirement of the enzyme for a minimum length of duplex DNA or to a reduced activity on single-stranded DNA.

To investigate the distribution of damage more precisely, the enzymatically cleaved fragments separated by electrophoresis of the DNA substrate irradiated at 100 J/m^2 were excised from the gel and the radioactivity of each was determined by Cherenkov counting (Table III). It is evident that dimer formation is nonrandom and occurs to the extent greatest in those regions where there are at least two adjacent thymines. Furthermore,

(Fig. 8 cont'd.) HCl, pH 7.4, 0.1 mM EDTA buffer. Enzyme incubations were performed in 100 μl of 10 mM Tris-HCl (pH 7.4)/0.1 mM EDTA/50 mM NaCl buffer for 60 minutes at 37°C. Reactions were stopped by two extractions with phenol saturated with incubation buffer, following several other extractions. The DNA was precipitated by raising the salt concentration to 0.3 M NaOAX with 4 M NAOAC, pH 5.0, adding 10 μg of RNA carrier, and three times the volume of 95% ethanol. The mixture is kept in a dry ice/ethanol bath for 5 minutes, and then centrifuged for 5 minutes in an Eppendorf microcentrifuge. The supernatant is decanted, the pellet lyophillized and suspended in loading buffer, 7 M urea, 50 mM NaOH, 0.5 mM EDTA, and 0.02% bromophenol blue, xylene cyanol. Electrophoresis is performed for 11 hours at 1000 V.

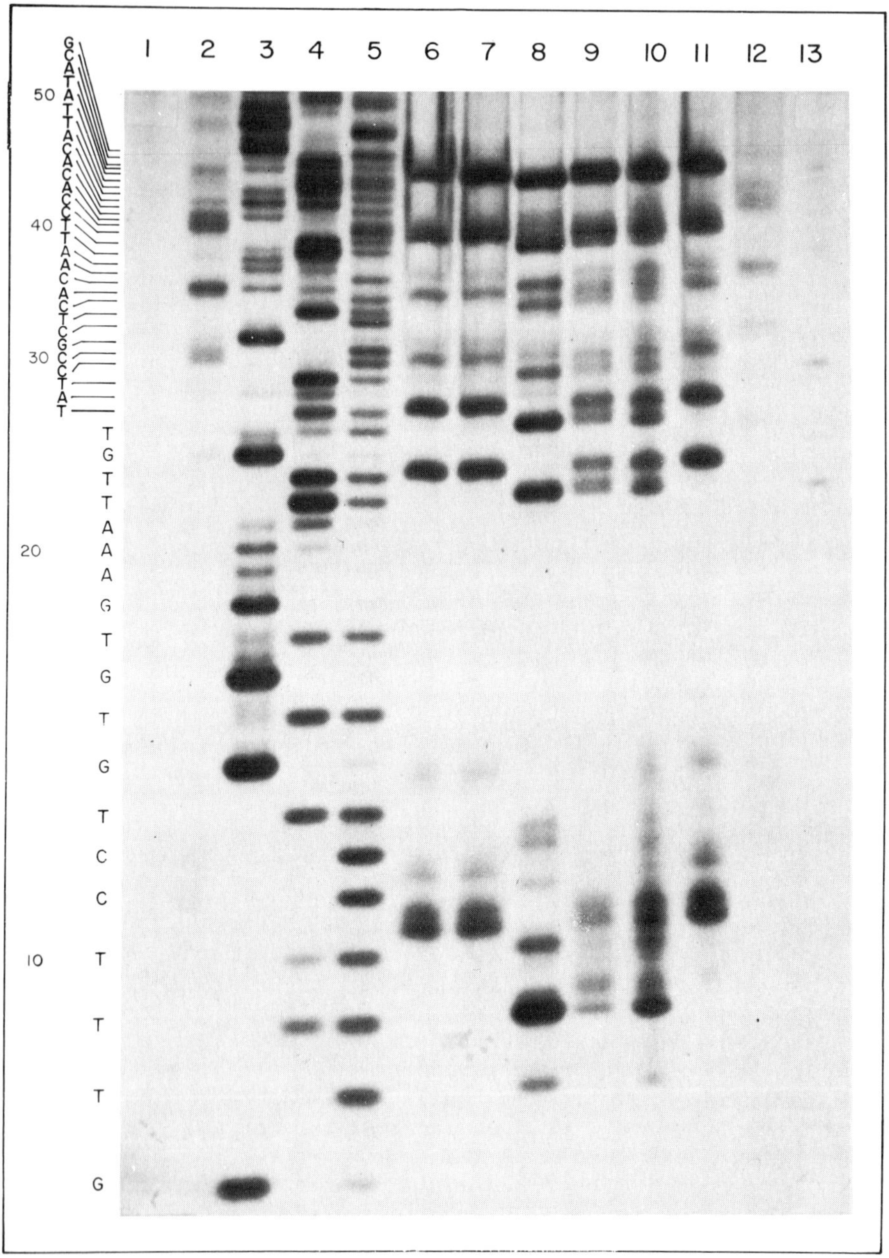

Fig. 9.

the extent of dimer formation can be precisely related to the number of thymines in each pyrimidine-rich region (see column 7 of Fig. 8). In the 117 base pair fragment, in a region containing 4 adjacent cytosines, irradiation produced negligible damage.

Interestingly, a nucleotide region of the 168 bp fragment which contains thymine pairs flanked by purines, the extent of damage is much greater than that arising at other isolated thymine dimer regions. This may be designated as a "hot spot" for damage which may result from the nonrandom sequences surrounding it which contain alternating purine-pyrimidine sequences not found in any other region of the substrate DNA.

A direct comparison of the migration patterns of the products of sequencing with the products of endonuclease hydrolysis should reveal the precise location of strand scission. Sequencing reactions selectively destroy a base with subsequent

Fig. 9. Heat and alkali lability of 3' terminus of enzymatically cleaved uv DNA. 168 bp 5'-P^{32}-labeled DNA was uv irradiated with 480 J/m^2 in 10 mM Tris/0.1 mM EDTA, pH 7.4, and subsequently incubated with saturating amounts of correndonuclease I + II in 10 mM Tris/0.1 mM EDTA/50 mM NaCl, pH 7.4, for 60 min at 37°C. DNA was layered in neutral NaP-glycerol buffer (Lane 6) or in 7 M urea/0.05 M NaOH (Lane 7) and electrophoresed. After enzymatic cleavage, some DNA was heated in 0.1 N NaOH at 90°C for 30 min (Lane 8) or at 37°C in 0.1 N NaOH for 30 min (Lane 9). DNA was also heated in neutral NaP buffer at 90°C for 30 min (Lane 10) or in NaP at 37°C for 30 min (Lane 11) and layered in NaP glycerol.

Lanes 3, 4, and 5 contain products of chemical sequencing reactions. Lane 3 contains DNA treated with dimethyl sulfate and hydrolyzed according to the Maxam and Gilbert protocol to produce cleavage at G's, and to a lesser extent at A's. Lane 4 contains DNA treated with neocarcinostatin to produce cleavage at T's and to a lesser extent at A's. Lane 5 contains products of DNA treated with hydrazine and piperidine according to Maxam and Gilbert, to cleave at positions of C's and T's.

Tracks 1 and 13 contain DNA that has been uv irradiated but not treated with enzyme and unirradiated DNA that has been incubated with correndonuclease. Tracs 2 and 12 contain DNA that has been uv irradiated and heated in 0.1 N NaOH at 90°C for 30 min (Lane 2) and in neutral NaP buffer at 90°C (Lane 12) but has not been treated with enzyme.

The electrophoretic gel contained 20% polyacrylamide with urea and was made according to the Maxam and Gilbert protocol. Electrophoresis was for 8 hr at 1000 V. Approximately 60,000 cpms were layered in each lane.

cleavage of the phosphodiester bond. The sugar is then hydrolyzed, generating a 3'-phosphoryl group. Cleavage of uv DNA by the endonucleases at a $\widehat{TT}$ dimer produces a DNA fragment which migrates parallel to a sequencing fragment in which the 3' dimeric thymine is destroyed. This would imply that the enzyme cleaves between the two thymines of the dimer. It is suggested, however, that the actual enzymatic cleavage product is a base-free deoxyribose originally associated with the 5'-thymine residue of the dimer and 5'-phosphorylated nucleotide.

The following observations support this view. Experiments were performed to determine the nature of the 3'-terminus of the incised DNA (Fig. 9). If enzymatically treated uv DNA was heated in 0.1 *N* NaOH at 90°C for 30 minutes prior to loading on gels, a shift in the migration pattern at the incised position occurred. A single band was produced at the position of a fragment one nucleotide shorter, that is, at the position of a fragment terminating in a 3'-phosphoryl group on the nucleotide 5' to the pyrimidine dimer. If, however, uv DNA was pretreated at 37°C in 0.1 *N* NaOH for 30 minutes after treatment with enzyme and prior to suspension in loading buffer, a mixture of fragments was produced. Two bands appeared for each thymine dimer, due to the original as well as the faster migrating fragment. This can be seen most clearly in positions 22 and 23 and 25 and 26, although shifting occurred at all positions in the tract. The same result was obtained if the endonuclease-treated DNA was heated in neutral sodium phosphate buffer at 90°C for 30 minutes. Incubation of incised DNA in neutral buffer at 37°C produced no shift in migration. It is proposed that the "alkali shift" is due to partial hydrolysis of the base-free sugar of the 3' dimeric thymine.

Further evidence for this suggestion is based on the well-characterized molecular mechanisms of depurination and hydrolysis of alkylated DNA which occurs during DNA sequencing (27). In the Maxam and Gilbert protocol, dimethyl sulfate is used to alkylate guanines and adenines. Heating at neutral pH depurinates the nucleotide, leaving a base-free sugar. Subsequent hydrolysis at high temperatures in 0.1 *N* NaOH cleaves the phosphodiester bond and destroys the sugar. Layering of depurinated DNA, which is only gently hydrolyzed, produces a fragment migrating in two positions. The faster migrating band corresponds to a fragment in which the damaged nucleotide has been completely removed, leaving a 3-phosphoryl group. The other band appears in a position of a fragment one nucleotide longer, but which actually arises from a DNA sequence terminating in a 3' base-free deoxyribose. If depurinated DNA is layered onto the gel in neutral phosphate-glycerol buffer, without heat denaturation, the DNA remains intact with only a few breaks. If, however, the standard loading mix of 7*M* urea and 50 m*M* NaOH is used with heating at 90°C for 15 seconds to effect denaturation, then fragments migrating at two positions occur. It is concluded that mild hydrolysis of depurinated sites takes place during this loading procedure in response to alkali and heat. uv DNA incubated with enzyme and loaded in NaP buffer produces only one band upon electrophoresis. If urea-NaOH plus heat denaturation is employed during the layering process, an additional more quickly migrating band is occasionally seen, resulting from hydrolysis of the base-free deoxyribose.

On the basis of these observations, it can be suggested that the endonucleases cleave the phosphodiester bond between the dimeric pyrimidines, leaving a 3'-deoxyribose and a 5'-phosphodiester bond. To complete strand scission, the glycosylic bond of the 5'-pyrimidine must also be broken. *N*-glycosylic bond hydrolysis may be a secondary consequence of phosphodiester bond hydrolysis. Once this bond is cleaved, then the deoxyribose of the 5'-pyrimidine becomes alkali or heat labile. Presumably also, this completes strand scission and permits excision to occur. If this second cleavage did not take place, then the DNA would migrate as an unbroken strand on the sequencing gel despite the break in the phosphodiester bond, due to the linkage of the pyrimidine dimer.

The enzyme preparation used in these experiments is composed of both correndonuclease peaks I and II as copurified on a phosphocellulose column. It is possible that the two enzymes, when separated on DNA cellulose, exhibit different activities. One enzyme may cleave the phosphodiester bond while the other labilizes the glycosylic bond. This is being investigated at the present time.

ACKNOWLEDGMENT

One of the authors (L.G.) was aided by Grants from the National Institutes of Health (5 RO1 GM 22846) and Department of Energy (EY-76-S-02-2814).

REFERENCES

1. Carrier, W. L. and Setlow, R. B. (1964). *Proc. Nat. Acad. Sci. U.S. 51*, 226.
2. Boyce, R. P. and Howard-Flanders, P. (1964). *Proc. Natl. Acad. Sci. U.S. 51*, 293.
3. Lindahl, T. (1974). *Proc. Natl. Acad. Sci. U.S. 71*, 3649.

4. Riazuddin, S. and Lindahl, T. (1977). "Institutional Congress on Environmental Mutagenesis." Edinburgh.
5. Weiss, B. and Milcarek, C. (1974). *In* "Molecular Enzymology" (L. Grossman and K. Moldave, eds.), Vol. 29, p. 180.
6. Weiss, B. (1976). *J. Biol. Chem. 251,* 1866.
7. Kelly, R. B., Cozarelli, N. R., Deutscher, M. P., Lehman, I. R., and Kornberg, A. (1970). *J. Biol. Chem. 245,* 39.
8. Hamilton, L., Mahler, I., and Grossman, L. (1974). *Biochemistry 13,* 1886.
9. Kaplan, J. C., Kushner, S. R., and Grossman, L. (1971). *Biochemistry 10,* 3315.
10. Chase, J. and Richardson, C. C. (1974). *J. Biol. Chem. 249,* 4553.
11. Doniger, J. and Grossman, L. (1976). *J. Biol. Chem. 251,* 4579.
12. Kushner, S. R., Kaplan, J. C., Ono, H., and Grossman, L. (1971). *Biochemistry 10,* 3325.
13. Cooper, P. K. and Hanawalt, P. C. (1972). *J. Mol. Biol. 67,* 1.
14. Gefter, M. L. (1974). *Progr. Nucleic Acid Res. 14,* 101.
15. Curtis, M. J. and Alberts, B. (1976). *J. Mol. Biol. 102,* 793.
16. Ruyechan, W. T. and Wetmur, J. G. (1976). *Biochemistry 15,* 5057.
17. Braun, A. and Grossman, L. (1974). *Proc. Natl. Acad. Sci. U.S. 71,* 1838.
18. Riazuddin, S., Grossman, L., and Mahler, I. (1977). *J. Biol. Chem. 252,* 6294.
19. Hariharan, P. V. and Cerutti, P. A. (1976). *Biochim. Biophys. Acta 447,* 375.
20. Lamola, A. A. (1969). *Photochem. Photobiol. 9,* 291.
21. Lehman, I. R. (1974). *Science 186,* 790.
22. Riazuddin, S. and Grossman, L. (1977). *J. Biol. Chem. 252,* 6287.
23. Riazuddin, S. and Grossman, L. (1977). *J. Biol. Chem. 252,* 6280.
24. Haseltine, W., Linden, C., D'Andrea, A., and Johnsrud, L. (1980). *Methods Enzymol. 65,* 235-248.
25. Johnsrud, L. (1978). *Proc. Natl. Acad. Sci. U.S. 75,* 5314-5318.
26. Maxam, A. and Gilbert, W. (1977). *Proc. Natl. Acad. Sci. U.S. 74,* 560.

GENETIC STRUCTURE OF A PLASMID REPLICON

Roberto Kolter
Manabu Inuzuka
David M. Stalker
Donald R. Helinski

Department of Biology, B-Ø22
University of California, San Diego
La Jolla, California 92093

INTRODUCTION

The replication of any DNA molecule can be conveniently divided into three stages: initiation, elongation, and termination. The work of several investigators has shown that, at least in bacteria, the elongation of DNA chains proceeds at a more or less constant rate (1). In addition, these studies indicated that the frequency of initiation of DNA replication is regulated during a cell cycle and it is this frequency of initiation that determines the copy number of a stably inherited replication element. For the most part, the biochemical mechanism of regulation of initiation of DNA synthesis remains undefined.

In an effort to explain the known properties of DNA synthesis in bacteria, a number of working hypotheses have been proposed for the regulation of DNA replication. In 1963, Jacob, Brenner, and Cuzin, drawing analogies from the results of their classical studies on the regulation of gene expression,

 ISBN 0-12-397580-8

proposed the replicon model for the regulation of DNA replication (2). This model defines a replicon as any DNA molecule with sufficient information to regulate its own synthesis. A replicon was proposed to have two specific components: a structural gene controlling the synthesis of a specific initiator and an operator of replication (a replicator or origin) upon which the initiator would act, followed by replication of any DNA contiguous to the replicator. The model thus assigns a positive regulatory role to the initiator such that replication is initiated only when the initiator substance is available. It was further proposed that attachment of the DNA molecule to a specific and limited number of membrane sites would limit the availability of initiator and thus regulate its copy number. In a contrasting model, Pritchard, Barth, and Collins proposed a negatively acting regulatory element for the control of DNA synthesis (3). This model postulated the existence of a replicon-specific repressor of DNA replication that can act at the origin. In the absence of the repressor, initiation of DNA synthesis occurs followed by replication of the DNA by the enzymatic machinery of the cell. The assumption is made that the repressor is synthesized during a short period soon after DNA synthesis begins and quickly reaches a level that establishes repression of further starts. Additional initiation events can not occur until the mass of the cell increases such that the concentration of repressor falls below a critical level. Direct evidence is lacking for either model and there are indications that the mechanism of regulation of DNA replication varies among replicons with the involvement of either or both negative and positive controlling elements depending on the replicon.

PLASMID DNA REPLICATION

Plasmids are extrachromosomal replicons that are maintained in a variety of bacteria (4). Their importance has become increasingly apparent as we better understand their role in the adaptation of microorganisms to the environment (5). In addition, as replicons, plasmids provide model systems for the studies of DNA replication (6). Their relatively small size makes physical and biochemical analysis possible. Different plasmids in *Escherichia coli* are known to replicate in a variety of modes (7) indicating that there may be at least several different mechanisms by which these DNA molecules are maintained in the cell. The genetic analysis of plasmid replication has not advanced rapidly due to the technical problems resulting from plasmid incompatibility and in some cases the maintenance of the plasmid in multiple copies. However, the recent development of *in vitro* DNA construction techniques utilizing site-specific endonucleases has provided a new and highly productive approach to the studies of the genetics of plasmid replication (8 - 10).

Plasmid R6K, a 38 kilobase pair (kb) resistance transfer plasmid naturally occurring in *E. coli* codes for resistance to streptomycin and ampicillin and is found in the cell in high copy number (11). Its relatively small size and relaxed mode of replication make it unique among the self-transmissible plasmids. The directionality of replication of both R6K (12) and a spontaneous deletion derivative of R6K, RSF1040 (13), has been determined. Replication initially proceeds from one of two origins toward an asymmetric terminus, followed by replication from the same origin in the other direction toward the terminus. These properties make R6K an attractive system for study. The physical and genetic map of R6K has been determined (14, 15) and is presented in Fig. 1.

THE R6K REPLICON

In an effort to better understand the role in replication of various segments of the circular map of R6K, the properties of this plasmid and *in vitro* constructed derivatives were analyzed both *in vivo* and *in vitro*. Using two non-self-replicating DNA fragments (a *Hin*dIII fragment of DNA carrying the *E. coli trp*E gene and a *Hae*II fragment of DNA carrying a gene for kanamycin resistance as selective markers), a series of deletion derivatives of R6K carrying the R6K replication region (14) were obtained after partial digestion of this replicon with *Hin*dIII and *Hae*II. The R6K sequence present in the two smallest plasmids isolated, designated pRK353 and pRK419, are shown in Fig. 2. Common to all derivatives is a contiguous DNA region, 2.1 kb in length mapping from 14.2 to 16.3 kb on the R6K coordinates presented in Fig. 1. This region consists of *Hin*dIII fragments 15 and 9 and 0.6 kb of *Hin*dIII fragment 4. All derivatives tested have the properties of relaxed replicons and exhibit incompatibility with R6K, indicating that the basic replication properties of the parental plasmid are still present. These results showed that, in the 38 kb plasmid, only 2.1 kb or less of the DNA contains sufficient information for the regulation of replication.

To analyze the requirements of R6K replication and to determine if any plasmid-encoded proteins are essential for R6K replication, an *in vitro* replication system was developed (16). The system for the replication of added covalently closed (CCC) R6K DNA contains CCC-R6K DNA, 0 - 40% ammonium sulfate fraction

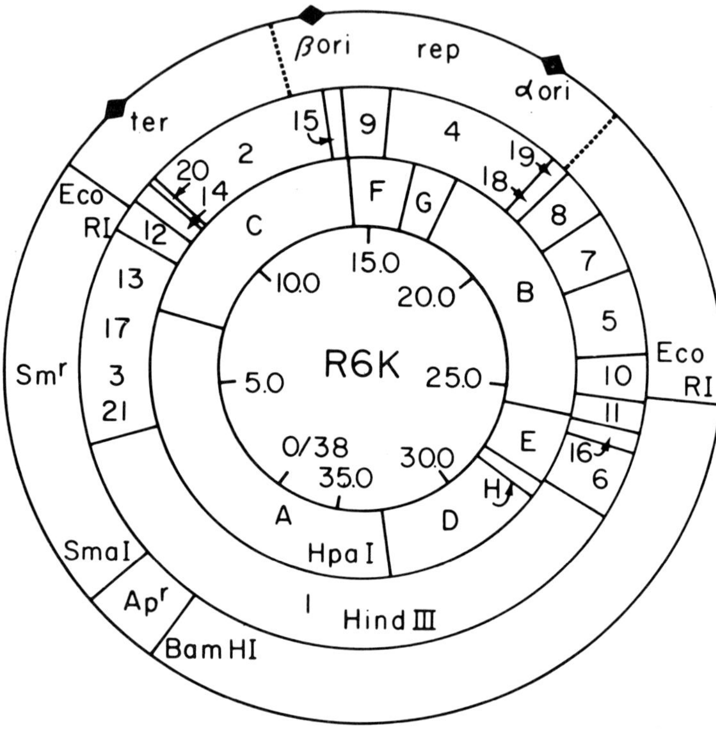

Fig. 1. Physical and genetic map of R6K. The recognition sites for BamHI, SmaI, EcoRI, HpaI, and HindIII are indicated relative to the kb map of R6K (innermost concentric circle). αori, βori, and ter are mapped on the basis of previously published data (12, 13). rep refers to the location of the replication region of the plasmid and encompasses the R6K region of pRK419 and pRK353 (see Fig. 2) (14).

from R6K$^+$ cell extract [AS(0-40)-R6K$^+$], 40 - 60% AS fraction from R$^-$ cell extract, four dNTPs, and four rNTPs. Approximately 30% of the added template CCC R6K DNA is initiated for replication and electron microscopic analysis of replicating molecules indicates that replication is initiated at an origin located in the replication region of R6K (see Fig. 1) (unpublished data, N. Inuzuka). The main product of the reaction is

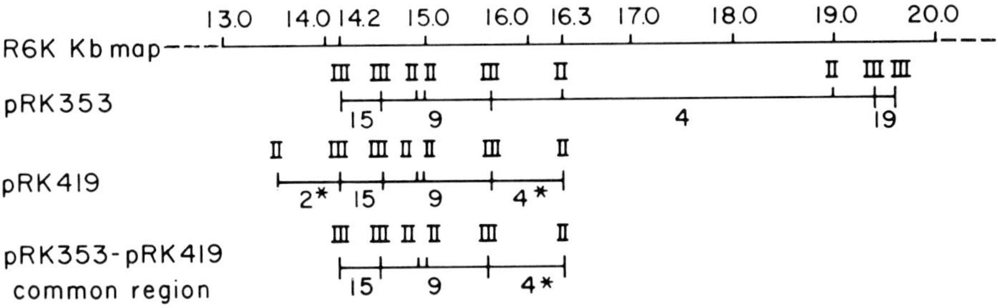

Fig. 2. The replication region of plasmid R6K. The area from 13 to 20 Kb of R6K has been enlarged. R6K was partially digested with HindIII and ligated to a HindIII non-self-replicating DNA fragment carrying trpE that was obtained from plasmid pDF38 (D. Figurski, unpublished results). trpE⁺ plasmids were screened and the smallest, pRK353, contained R6K HindIII fragments 15, 9, 4, and 19. Similarly, R6K was partially digested with HaeII and ligated to a HaeII non-self-replicating fragment carrying a kanamycin resistance gene. The smallest derivative obtained was pRK419 which contained 2.8 kb of R6K DNA in a region spanning 0.7 kb of HindIII fragment 2 (2), HindIII fragments 9 and 15, and 0.68 kb of HindIII fragment 4 (4*). The region common to both pRK353 and pRK419 covers from 14.2 to 16.3 kb. "II" and "III" refer to the recognition sites for HaeII and HindIII, respectively (14).*

CCC R6K DNA. The requirement of a plasmid-encoded protein for R6K DNA replication was directly demonstrated by an experiment in which a temperature-sensitive replication mutant of R6K derivative produced a heat-labile replication protein. Preincubation experiments in the presence and absence of deoxyribonucleotide triphosphates and/or AS(0-40) of $R6K^+$ have shown that an R6K encoded protein in this fraction acts positively to initiate replication and that it is required before or during primer RNA synthesis (Fig. 3) (17). This plasmid-encoded protein for initiation, designated π, has been partially purified from cell extracts and exhibits a molecular weight of 36,000. To date no other R6K encoded protein has been identified as being required for replication *in vitro*.

The 0 - 40% ammonium sulfate fraction prepared from cells harboring pRK353 or pRK419 is capable of supporting R6K DNA synthesis *in vitro*. These observations map the gene for π, *pir*, on the essential region of the plasmid. Figure 2 shows the position of *Hin*dIII and *Hae*II cleavage sites in this *pir* region. To define more precisely the location of the *pir* region, *Hin*dIII fragments 9 and 15 were cloned into a ColE1 derivative. The resulting plasmid was designated pRK496. It

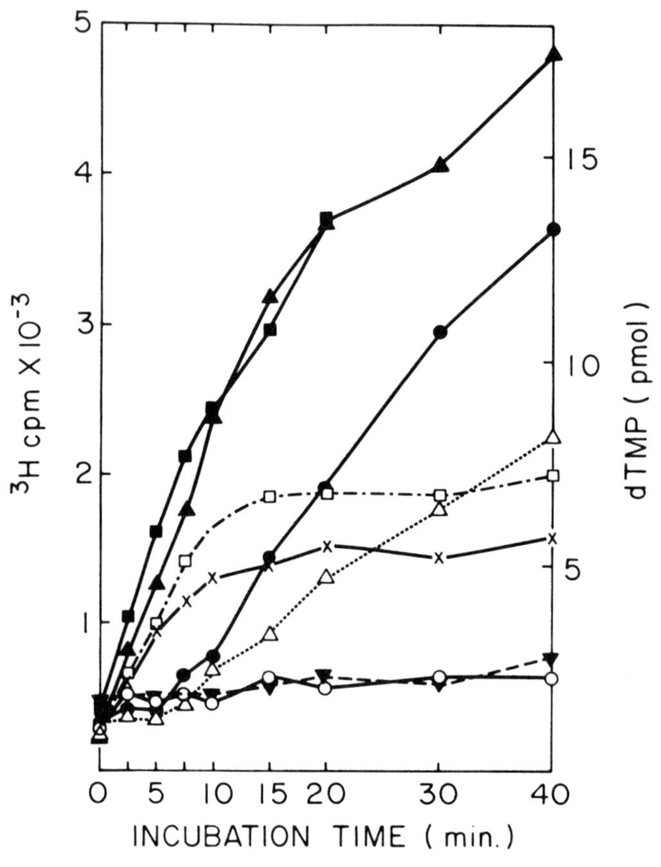

Fig. 3. Involvement of the π protein in the initiation process of R6K DNA replication. A standard reaction mixture (300 μl) containing an AS(0-40) fraction of KE357 (R6K) and CCC-R6K DNA was incubated with (○) or without (●) rifampicin (20 μg/ml) and 30 μl was withdrawn at indicated times. A reaction mixture without dNTPs was preincubated for 10 min (▲, X) or 20 min (■, □) and dNTPs containing ^{3}H dTTP was added at 0 time with (X, □) or without (▲, □) rifampicin. The reaction mixture also was preincubated for 10 min in the absence of dNTPs and the AS(0-40) fraction of KE357 and these components were added at 0 time with (▼) or without (△) rifampicin (17).

was possible to show that *Hin*dIII fragments 9 and 15 do not in themselves constitute a replicon. The presence of pRK496 in a cell, however, allowed temperature-sensitive replication mutants of pRK419 to replicate at the nonpermissive temperature (18). Furthermore, cell extracts prepared from strains that harbor pRK496 are active *in vitro*, mapping *pir* on *Hin*dIII fragments 9 and 15.

The activity of π in *trans* provides a functional *in vivo* assay for the target of π, presumably an origin of R6K replication. *Escherichia coli* cells carrying pRK496 or lysogens of *E. coli* carrying a λ*pir* specialized transducing phage that was constructed *in vitro* (18) were used to rescue a 420 bp *Mbo*I R6K DNA fragment carrying an origin of replication. This origin fragment was linked to a kanamycin resistance *Bgl*II fragment and designated pRK640 as is shown in Fig. 4. It has been observed that the *Hin*dIII site in the *Mbo*I fragment cannot be interrupted without loss of origin activity. This *Hin*dIII site represents the junction of R6K *Hin*dIII fragments 9 and 4 and, therefore, this origin fragment has 270 bp homologous to the helper region (*Hin*dIII fragments 9 and 15) that contains *pir*. The copy number of the origin containing plasmids (such as pRK640) was determined in λ*pir* lysogens and found to be approximately the same as that of the parent R6K (18). It has been found that cell extracts prepared from strains that harbor an elevated copy number of R6K DNA have an increased level of π activity in *in vitro* assays, suggesting that the control of π synthesis may be intimately involved in replication control (M. Inuzuka, unpublished data). If this is the case, the finding of constant copy number of the R6K derivatives regardless of the copy number of the *pir* gene suggests that the synthesis of π is auto-regulated.

The origin fragment has been sequenced (19) using the techniques developed by Maxam and Gilbert (20). The sequence is shown in Fig. 5. The base pairs are numbered in both directions from the *Hin*dIII site. The sequence has several major features: (a) the bases in the region of -30 to -20 resemble an RNA polymerase recognition sequence (21); (b) the nucleotide composition is 80% A+T from bases -15 to 92, a property common to several other replication origins (22 - 25); (c) bases 23 - 30 consist of a stretch of eight adenine residues that could represent a rho-independent RNA termination signal (21); (d) bases 49 - 86 closely resemble a "typical" *E. coli* promoter with an RNA polymerase recognition sequence followed by a Pribnow box (26); (e) bases 93 - 246 consist of seven 22 bp direct repeats; and (f) all reading frames for translation are interrupted by nonsense codons in both directions, indicating that

(a)

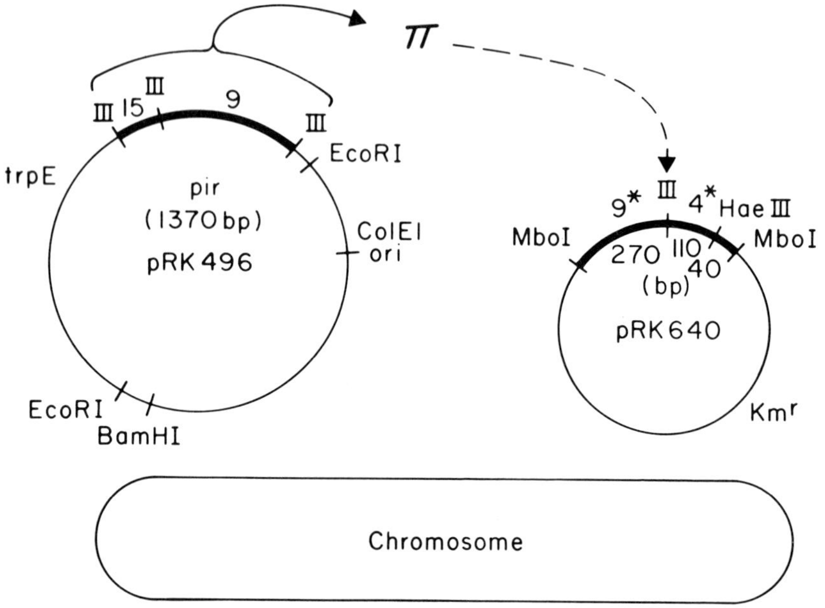
π
III
15
III
9
III
EcoRI
trpE
pir
(1370 bp)
pRK496
ColE1
ori
EcoRI
BamHI
9*
III
4*
Hae III
MboI
MboI
270
110
40
(bp)
pRK640
Km^r
Chromosome

(b)

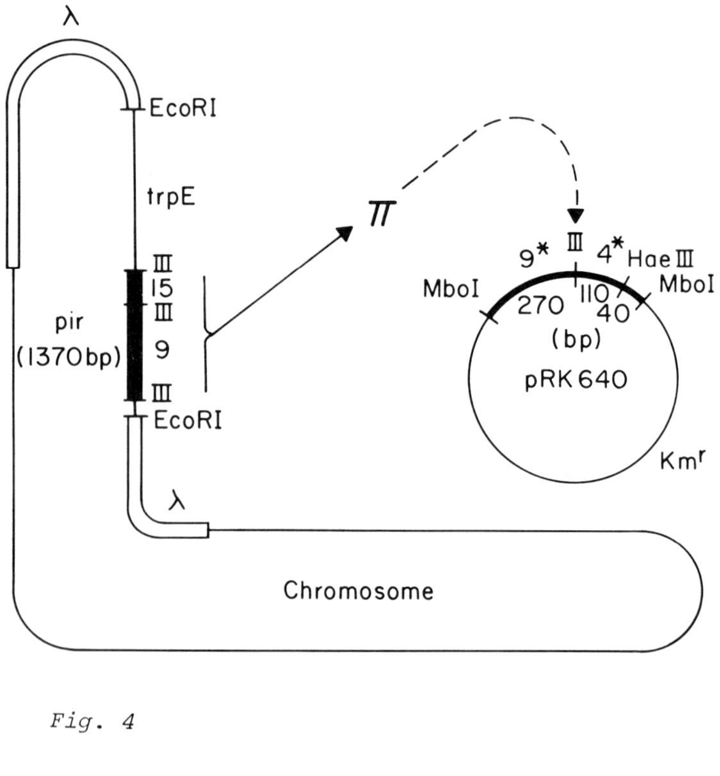
λ
EcoRI
trpE
π
III
15
III
pir
(1370bp)
9
III
EcoRI
λ
9*
III
4*
Hae III
MboI
MboI
270
110
40
(bp)
pRK640
Km^r
Chromosome

Fig. 4

no large polypeptide is coded for in this region. The sequence beyond the *Mbo*I/*Bgl*II site at the left end of the origin fragment includes another 22 bp repeat. This region contains one open reading frame and a possible start for π is located within the repeat (a GUG start).

Based on the above considerations, a working model was proposed for the recognition of initiation of replication (18). In this model, shown in Fig. 6, π recognizes the origin of replication near the *Hin*dIII site and functions there along with an RNA polymerase and other enzymes to initiate replication. A primer RNA is synthesized that crosses the *Hin*dIII site and terminates at the stretch of U's to allow an RNA to DNA transition. A second proposed property of π is its interaction with the direct repeats to control its own synthesis. Autogenous regulation involving direct repeats of nucleotide sequences has already been observed for the control of the repressor of phage lambda (27). This autogenous regulation of π would provide a constant level of the initiation protein during all stages of growth to maintain a constant copy number. The model assigns π a positive regulatory role and is therefore similar to the model proposed by Jacob, Brenner, and Cuzin in 1963 (2) with the added feature of an autogenously regulated initiator. It should be noted, however, that to date the data do not exclude the existence of a negatively acting (repressor) regulatory element to complement the action of the positive-acting π protein. Further analyses of the structure of the origin of replication of R6K and the *pir* control region and studies on the nature of the interactions between π and these sites are in progress to test the proposed model.

Fig. 4. Transcomplementation of a replication defective R6K derivative by a cloned pir gene. Strains carrying a segment of R6K containing the gene for π (protein of initiation of replication) either on a ColE1 plasmid or a lambda phage derivative were used to rescue a 420 bp MboI fragment of R6K that had been ligated to a BglII non-self-replicating fragment containing a kanamycin resistance gene and obtained from plasmid pIF11 (Figurski, Ito, and Helinski, manuscript in preparation). pRK640, isolated from the λpir strain, is unable to replicate in a strain that does not harbor the HindIII fragments 9 and 15 of R6K (28).

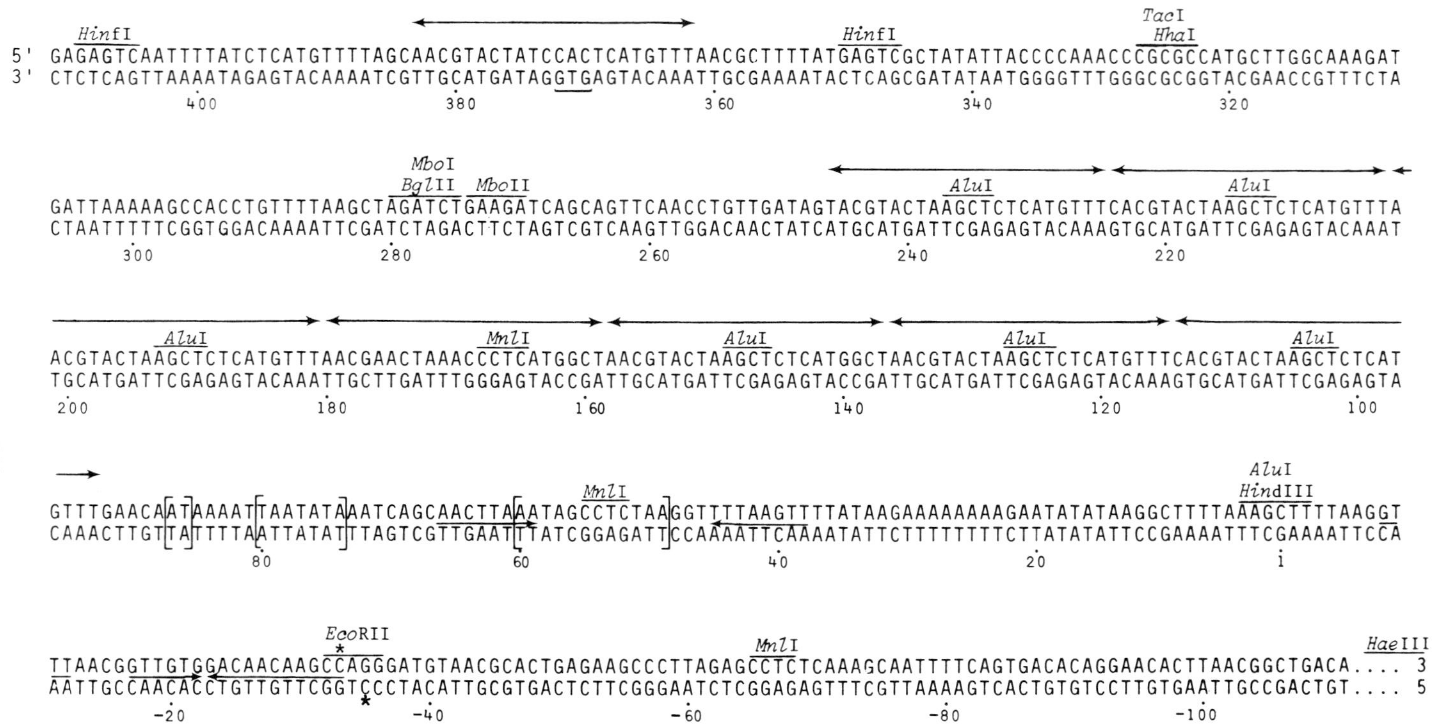

Fig. 5. Nucleotide sequence of the region of an origin of replication from plasmid R6K. Bases are numbered from the HindIII site and restriction enzyme sites are indicated. Arrows show the eight 22 base pair repeats. Regions of two-fold rotational symmetry are indicated by arrows between strands. The bracketed bases indicate a possible promoter for the pir gene (19).

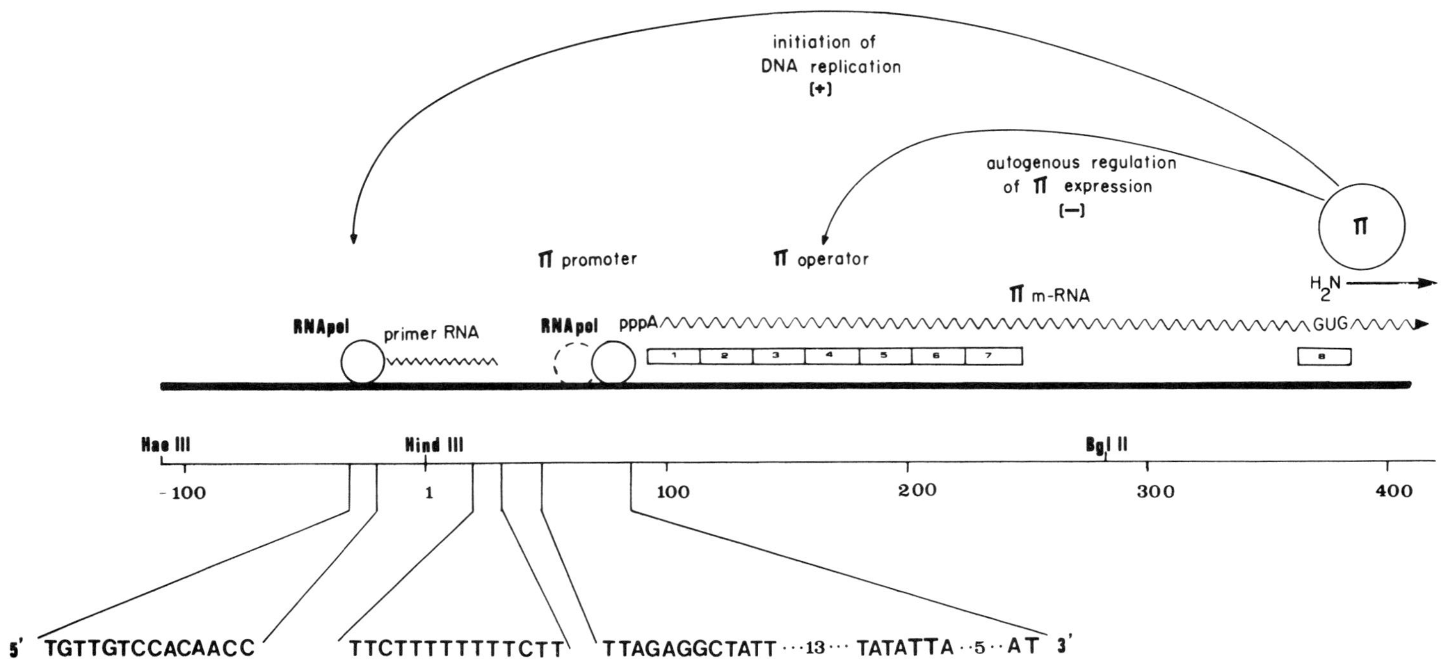

Fig. 6. Model for the regulation of R6K DNA replication. The relative orientation of the origin DNA has been reversed with respect to previous figures to have transcription and translation in the conventional left to right direction.

REFERENCES

1. Helmstetter, C., Cooper, S., Pieruci, O., and Revelas, E. (1968). On the bacterial life sequence. *Cold Spring Harbor Symp. Quant. Biol. 33,* 809-822.
2. Jacob, F., Brenner, S., and Cuzin, F. (1963). On the regulation of DNA replication in bacteria. *Cold Spring Harbor Symp. Quant. Biol. 28,* 329-348.
3. Pritchard, R. H., Barth, P. T., and Collins, J. (1969). Control of DNA synthesis in bacteria. *Symp. Soc. Gen. Microbiol. 19,* 263-298.
4. Falkow, S. (1975). "Infectious Multiple Drug Resistance." Pion Limited, London.
5. Reanney, D. C. (1976). Extrachromosomal elements as possible agents of adaptation and development. *Bacteriol. Rev. 40,* 552-590.
6. Helinski, D. R. (1976). Plasmid DNA replication. *Fed. Proc. 35,* 2026-2030.
7. Helinski, D. R., Lovett, M. A., Williams, P. H., Katz, L., Collins, J., Kupersztoch-Portnoy, Y., Sato, S., Leavitt, R. W., Sparks, R., Hershfield, V., Guiney, D., and Blair, D. G. (1975). Modes of plasmid DNA replication in *Escherichia coli. In* "DNA Synthesis and Its Regulation" (M. Goulian, P. Hanawalt, and C. F. Fox, eds.), pp. 514-536. Benjamin, Menlo Park.
8. Timmis, K., Cabello, F., and Cohen, S. N. (1975). Cloning isolation and characterization of replication regions of complex plasmid genomes. *Proc. Natl. Acad. Sci. U.S. 72,* 2242-2246.
9. Lovett, M. A. and Helinski, D. R. (1976). Method for the isolation of the replication region of a bacterial replicon: construction of a miniF'-Km plasmid. *J. Bacteriol. 127,* 982-987.
10. Figurski, D., Kolter, R., Meyer, R., Kahn, M., Eichenlaub, R., and Helinski, D. R. (1978). *In* "Microbiology 1978" (D. Schlessinger, ed.), pp. 205-110. American Society for Microbiology, Washington, D.C.
11. Kontomichalou, P., Mitani, M., and Clowes, R. C. (1970). Circular R-factor molecules, controlling penicillinase synthesis, replicating in *Escherichia coli* under either relaxed or stringent control. *J. Bacteriol. 104,* 34-44.
12. Lovett, M. A., Sparks, R. B., and Helinski, D. R. (1975). Bidirectional replication of plasmid R6K DNA in *Escherichia coli*: correspondence between origin of replication and position of single-strand break in relaxed complex. *Proc. Natl. Acad. Sci. U.S. 75,* 2905-2909.

13. Crosa, J. H., Luttropp, L. K., and Falkow, S. (1976). Mode of replication of the conjugative R-plasmid RSF1040 in *Escherichia coli*. *J. Bacteriol. 126,* 454-466.
14. Kolter, R. and Helinski, D. R. (1978). Construction of R6K plasmid derivatives *in vitro*: characterization of the R6K replication region. *Plasmid 1,* 571-580.
15. Crosa, J. H., Luttropp, L. K., and Falkow, S. (1978). Molecular cloning of replication and incompatibility regions from the R-plasmid R6K. *J. Mol. Biol. 124,* 443-468.
16. Inuzuka, M. and Helinski, D. R. (1978). Replication of antibiotic resistance plasmid R6K *in vitro*. *Biochemistry 17,* 2567-2573.
17. Inuzuka, M. and Helinski, D. R. (1978). Requirement of a plasmid encoded protein for the replication *in vitro* of plasmid R6K. *Proc. Natl. Acad. Sci. U.S. 75,* 5381-5385.
18. Kolter, R., Inuzuka, M., and Helinski, D. R. (1978). Trans-complementation dependent replication of a low molecular weight origin fragment from plasmid R6K. *Cell 15,* 1199-1208.
19. Stalker, D. M., Kolter, R., and Helinski, D. R. (1979). Nucleotide sequence of the region of an origin of replication of the antibiotic resistance plasmid R6K. *Proc. Natl. Acad. Sci. U.S. 76,* 1150-1154.
20. Maxam, A. M. and Gilbert, W. (1977). A new method for sequencing DNA. *Proc. Natl. Acad. Sci. U.S. 74,* 560-564.
21. Gilbert, W. (1976). Starting and stopping sequences for the RNA polymerase. *In* "RNA Polymerase" (R. Losick and M. Chamberlin, eds.), pp. 193-205. Cold Spring Harbor Laboratory, New York.
22. Sugimoto, K., Oka, A., Sugisaki, H., Takanami, M., Nishimura, A., Yasuda, Y., and Hirota, Y. (1979). Nucleotide sequence of *Escherichia coli* K-12 replication origin. *Proc. Natl. Acad. Sci. U.S. 76,* 575-579.
23. Meijer, M., Beck, E., Hansen, F. G., Bergmans, H. E. W., Messer, W., von Meyenburg, K., and Schaller, H. (1979). Nucleotide sequence of the origin of replication of the *Escherichia coli* K-12 chromosome. *Proc. Natl. Acad. Sci. U.S. 76,* 580-584.
24. Denniston-Thompson, K., Moore, D. D., Fruger, K. E., Furth, M. E., and Blattner, F. R. (1977). Physical structure of the replication origin of bacteriophage lambda. *Science 198,* 1051-1056.
25. Grosschedl, R. and Hobom, G. (1979). DNA Sequences and structural homologies of the replication origins of lambdoid bacteriophages. *Nature (London) 277,* 621-626.
26. Pribnow, D. (1975). Nucleotide sequence of an RNA polymerase binding site at an early T7 promoter. *Proc. Natl. Acad. Sci. U.S. 72,* 784-789.

27. Ptashne, M., Backman, K., Humayun, M., Jeffrey, A., Manier, R., Meyer, S., and Sauer, R. T. (1976). Autoregulation of a repressor in bacteriophage lambda. *Science 194,* 156-161.
28. Kolter, R., Inuzuka, M., Figurski, D., Thomas, C., Stalker, D., and Helinski, D. R. (1978). Plasmid DNA replication: RK2 and R6K encoded *trans*-acting factors and their site of action. *Cold Spring Harbor Symp. Quant. Biol. 43,* 91-97.

DEFINITION OF A DISEASE: EXPERIMENTAL ALLERGIC ENCEPHALITOMYELITIS

Fred C. Westall

Salk Institute
P.O. Box 1809
San Diego, California 92112

The work described in this manuscript spans a little over a decade and was initiated while I was in graduate school at UCSD. I was not one of Professor Kamen's students. However, his support of the UCSD peptide laboratory under the direction of Dr. Arthur Robinson and of my early career was extraordinary. I, like so many, owe Martin Kamen a tremendous debt.

INTRODUCTION

Six percent of central nervous system (CNS) myelin contains a substance identified as basic protein (see Fig. 1). Most mammalian basic proteins are single linear chains comprising 170 amino acids (1, 2). However, the basic protein of rat CNS myelin exists as two components (3), one (25%) approximately 18,500 MW and the other (75%) approximately 14,000 MW. The large rat basic protein closely resembles the single basic

 ISBN 0-12-397580-8

protein of other species (4) and the small rat basic protein is identical to the large except for a deletion of 40 amino acids toward the C-terminal end (see references 4 and 5 for review of known basic protein sequences).

For several decades this protein has been studied because of its involvement in various demyelinating diseases. When microgram doses of it in complete Freund's adjuvant are injected into mammals, a disease of the central nervous system, experimental allergic encephalomyelitis (EAE), can develop (6 - 8). Clinical signs ensue 2 - 3 weeks later and include paralysis, loss of balance, and incontinence of urine and feces followed by recovery or death within a few days. Lesions are confined to the CNS and consist characteristically of dense infiltrations of mononuclear leukocytic cells surrounding blood vessels. Within the protein are several regions which themselves are encephalitogenic. The encephalitogenicity of these regions depends upon the species being tested and the adjuvant used. Table I lists the regions which have been defined.

The contributions of the individual amino acids to the encephalitogenicity of two of the determinants have been ascertained. The so-called tryptophan region - Phe-Ser-Trp-Gly-Ala-Glu-Gly-Gln-Arg - has been studied thoroughly and is the only encephalitogenic region for guinea pigs. Our results obtained from testing various synthetic peptides indicate five amino acids are required (13, 19, 20): the glutamine, positively charged amino acid C-terminally adjacent to glutamine,

A S A Q K R P S Q R S K Y L A S A S T M
D H A R H G F L P R H R D T G I L D S L
G R F F G S D R G A P K R G S G K D G H
H A A R T T H Y G S L P Q K A Q G H R P
Q D E N P V V H F F K N I V T P R T P P
P S Q G K G R G L S L S R F S W G A E G
Q K R G F G Y G G R A S D Y K S A H K G
L K G H D A Q G T L S K I F K L G G R D
S R S G S P M A R R

Fig. 1. Amino acid sequence of bovine basic protein (2).

TABLE I. *Known Encephalitogenic Regions within the Myelin Basic Protein*

		Species in which disease is produced				
Sequence	Residues[a]	Monkey	Rabbit	Guinea Pig	Lewis rat	Hyperacute EAE in Lewis rats
1. Mid-peptide (9)	45-88	+	+	±	+	+[b]
2. Gly-Ser-Leu-Pro-Gln-Lys-Ser-Gln-Arg-Ser-Gln-Asp-Glu-Asn-Pro-Val-Val-His-Phe (10)	70-88	NT[c]	NT	+	+	+[b]
3. Phe-Lys-Leu-Gly-Gly-Arg-Asp-Ser-Arg-Ser-Gly-Ser-Pro-Met (11)	154-167	+	NT	NT	NT	NT
4. Phe-Lys-Leu-Gly-Gly-Arg-Asp-Ser-Arg (12)	154-162	NT	+	NT	+[d]	-
5. Phe-Ser-Trp-Gly-Ala-Glu-Gly-Gln-Arg (13)	113-121	-	+	+	-	-
6. Thr-Thr-His-Tyr-Gly-Ser-Leu-Pro-Gln-Lys (14)	66-75	NT	+	NT	-	-
7. Asp-Ser-Gly-Arg-Phe (15)	38-44[e]	NT	NT	NT	+	NT
8. Trp-Gly-Ala ....Ala-Arg-Arg (16)	115-170[f]	NT	NT	-	+	+
9. Phe-Lys-Asn......Glu-Gly-Gln (17)	88-120	NT	NT	NT	+	-
10. Gly-Ser-Lys-Tyr-Leu-Ala-Thr-Ala-Ser-Thr (18)	11-20	NT	NT	±	NT	NT

[a]Sequence numbers based on guinea-pig sequence (F. C. Westall, *Experientia*, submitted for publication).

[b]*B. pertussis* is required for disease induction.

[c]NT, not tested.

[d]While this sequence is not active in Lewis rats the Phe for Leu substituted analog is active if *B. pertussis* is used in the adjuvant (17).

[e]This region determined by overlapping peptides.

[f]Guinea pig sequence.

TABLE II. Encephalitogenicity of Peptides and Rabbit Myelin Basic Protein in Rabbits.

	Sequence
Peptides	
1a.	Phe-Lys-Leu-Gly-Gly-Arg-Asp-Ser-Arg[a]
1b.	Phe-Lys-Leu-Gly-Gly-Arg-Asp-Ser-Arg[b]
2.	Ser-Lys-Ile-Phe-Lys-Leu-Gly-Gly-Arg-Asp-Ser-Arg
3.	Phe-Lys-Leu-Ala-Gly-Arg-Asp-Ser-Arg
4.	Phe-Lys-Leu-Gly-Gly-Arg-Asp-Ala-Arg
5.	Phe-Lys-Leu-Gly-Gly-Arg-Asp-Ser-Ile
6.	Phe-Lys-Ala-Gly-Gly-Arg-Asp-Ser-Arg
7.	Phe-Ile-Leu-Gly-Gly-Arg-Asp-Ser-Arg
8.	Val-Lys-Leu-Gly-Gly-Arg-Asp-Ser-Arg
9.	Phe-Lys-Phe-Gly-Gly-Arg-Asp-Ser-Arg
Rabbit basic protein	

[a]Enzymatically derived.
[b]Synthetic sequence
[c]We were unable to find peripheral nerve lesions.

Dose (μg)	Dose (nmol)	EAE	Clinical Atypical EAE	Histological[c] CNS lesions
100	85.0	4/8	1/8	4/4
50	42.0	4/8	3/8	4/4
100	85.0	4/8	1/8	4/4
10	8.5	4/8	3/8	4/4
1	0.85	1/8	0/8	2/4
100	64.0	4/8	0/4	4/4
10	6.4	6/8	0/4	4/4
1	0.64	2/8	0/4	2/4
100	84.0	2/4	0/4	4/4
10	8.4	1/4	0/4	3/4
1	0.84	0/4	0/4	1/4
100	86.0	0/4	1/4	2/4
10	8.6	0/4	1/4	2/4
1	0.86	0/4	0/4	2/4
100	88.0	0/4	1/4	3/4
10	8.8	0/4	4/4	4/4
1	0.88	0/4	1/4	3/4
100	88.0	0/4	2/4	4/4
10	8.8	1/4	1/4	4/4
1	0.88	0/4	0/4	2/4
100	86.0	1/4	0/4	4/4
10	8.6	3/4	1/4	4/4
1	0.86	2/4	0/4	3/4
100	88.0	2/4	0/4	4/4
10	8.8	2/4	0/4	4/4
1	0.88	1/4	0/4	2/4
100	83.0	2/4	0/4	4/4
10	8.3	2/4	0/4	4/4
1	0.83	1/4	0/4	2/4
1000	53.0	6/9	0/9	9/9
320	17.0	5/9	0/9	9/9
100	5/3	3/10	0/9	8/10

the tryptophan located six residues from the glutamine toward the amino terminal end with the glycine C-terminally adjacent to it, and a hydrophobic region eight residues from the glutamine. Nagai *et al.* (21) has also suggested that the alanine adjacent to the glycine is important for disease induction. Nagai has shown (22) that when a synthetic fraction of the *Mycobacterium* is used in the adjuvant mixture in place of the whole bacterium the phenylalanyl and serinyl residues are not required for activity.

The sequence Phe-Lys-Leu-Gly-Gly-Arg-Asp-Ser-Arg is encephalitogenic in rabbits and probably also monkeys. As mentioned in Table I, this region containing a Phe for Leu substitution is encephalitogenic in Lewis rats when *B. pertussis* is included with the adjuvant.

It is obvious after examining Table II that the requirements for disease production in rabbits by the peptide Phe-Lys-Leu-Gly-Gly-Arg-Ser-Arg are not as stringent as those for the tryptophan region. However, the two C-terminal residues appear quite important for disease induction as is the case for tryptophan peptide. [The serine substitution for the glutamine in the tryptophan peptide is quite active (23).]

We have also observed in rabbits, using several of the peptides, an atypical form of EAE, beginning on day 9 rather than day 19, with considerable front leg paresis. We are not certain whether these peptides are actually producing two effects, EAE and a mild disease involving the front legs, or whether this effect is not just a slight variation of normal EAE. The so-called atypical form of EAE with front leg paresis is somewhat similar to the one observed in the monkey, if one compares the arm of the monkey with the front leg of the rabbit (24).

HYPERACUTE EAE

In Lewis rats a hyperacute variant of EAE can be induced using appropriate CNS immunogen with *Bordetella pertussis* vaccine as adjuvant with or without CFA (25 - 27). Hyperacute EAE (HEAE) differs from ordinary EAE clinically in its early onset and in its rapid and severe course, with high incidence of cerebral involvement and mortality. Histological lesions of HEAE are characterized by the presence in the CNS parenchyma of polymorphonuclear leukocytes, hemorrhage, edema, and fibrin. Because HEAE and ordinary EAE overlap histopathologically and immunologically they provide experimental evidence supporting the concept that certain human demyelinating diseases represent a spectrum of autoimmunity ranging from chronic multiple sclerosis to acute hemorrhagic necrotizing leukoencephalopathy, for which HEAE is a model (25 - 27).

With *B. pertussis* as adjuvant, the guinea pig basic protein is unique in its capacity to induce HEAE in Lewis rats when compared with basic proteins from seven other species (28). The unique antigenicity of the guinea pig basic protein leading to HEAE induction is determined by the primary amino acid sequence of a 14 amino acid peptide, residues 73-86 (Pro-Gln-Lys-Ser-Gln-Asp-Glu-Asn-Pro-Val-Val) (29). The corresponding region of the rat basic protein, which differs by only a single amino acid substitution at residue 79 (threonine/serine), induces ordinary EAE, but not HEAE. Thus the antigenic requirements for induction of two clinically distinct autoimmune syndromes are determined by one methyl group.

An attempt has also been made to determine whether other known encephalitogenic regions are capable of producing HEAE. There are at least three other encephalitogenic regions for Lewis rats which reside within residues 88-170: peptide 98-120, peptide 115-170 and an analog of peptide 154-162, Phe-Lys-Phe-Gly-Gly-Arg-Asp-Ser-Arg. This latter peptide requires *B. pertussis* to produce ordinary EAE (17). The other two peptides produce ordinary EAE with or without the *B. pertussis* (17, 30).

HEAE cannot be induced in either guinea pigs using 10^{10} or 20×10^{10} organisms of *B. pertussis* or in rabbits using 4×10^{10} organisms of *B. pertussis* and basic protein from the following species: guinea pig, Lewis rat, human, bovine, porcine, monkey, and rabbit. The only encephalitogenic region for Lewis rats which also produces HEAE in Lewis rats is not encephalitogenic in either rabbits or guinea pigs. Therefore, it seems highly probable that there is only one HEAE determinant for all species which are able to express HEAE (30).

CONFORMATIONAL CONSIDERATIONS

In studying both the C-terminal and tryptophan regions we made a series of analogs consisting of one amino acid substitutions. Comparing the encephalitogenicity of each peptide indicated the contribution to each amino acid of the peptide to the disease site. The contribution of the whole conformation of the peptide to disease activity was deemphasized. However, by comparing the disease-producing activity of larger peptides containing an encephalitogenic determinant, we have found significant clinical differences between EAE-inducing peptides and proteins.

Described below are examples of significant clinical differences in EAE disease induction apparently due to conformational differences between EAE-inducing peptides and proteins. These differences are found with the three major encephalitogenic regions; the C-terminal determinant, the tryptophan peptide, and the most potent region for Lewis rats, which is also responsible for hyperacute EAE.

1. *Peptide Phe-Lys-Leu-Gly-Gly-Arg-Asp-Ser-Arg (C-Terminal Determinant)*

While determining the requirements for disease induction in rabbits, several of the peptides containing this region appeared to produce atypical EAE with considerable front leg paresis and caused a general overall clinical weakness (see Table II). Furthermore, while we consistently observe EAE in rabbits on day 19 using basic protein, this atypical form on EAE began as early as day 9. It does not appear to be a more insidious disease, but just a slightly different form of EAE. Therefore, the peptide which contains only the encephalitogenic unit produces an altered form of EAE when compared to the whole basic protein, which contains within its sequence this unit.

2. *Tryptophan Peptide*

Previously we reported (31) the encephalitogenic dose response in guinea pigs of the tryptophan region. At low doses the peptide on a molar basis was an encephalitogenic as the whole basic protein. However, at high doses the peptide was decidedly less active than the whole protein (Table III). It is obvious that the peptide is much more susceptible to exopeptidases than the region within the protein. However, the high dose inhibition of encephalitogenicity is not the result of competitive inhibition by nonencephalitogenic fragments of the tryptophan peptide produced by *in vivo* exopeptidases (32). Moore (33) has recently shown that the three-dimensional structure of the tryptophan within the peptide is different from its structure within the whole protein.

3. *HEAE Determinant (Residues 73 - 86).*

As mentioned above a peptide comprising residues 70-88 of the guinea pig basic protein is known to be not only encephalitogenic but able to produce HEAE in rats. We have further narrowed this HEAE region using synthetic peptides to include residues 73-86.

The midpeptide (residues 45 - 88) of guinea pig basic protein is considerably less active on a molar basis in inducing

TABLE III. *Comparison of EAE Induction in Guinea Pigs Using the Tryptophan Peptide and Whole Basic Protein*

	Guinea pig basic protein in complete Freund's adjuvant				Tryptophan peptide in complete Freund's adjuvant				
nmoles	Clinical	% Clinical	Histological	% Histological	nmoles	Clinical	% Clinical	Histological	% Histological
17	56	90	62	100	8.50	11/42	26.1	25/42	59.5
5.3	57	91	62	100	2.80	17/51	34.0	29/51	56.8
1.7	44	70	62	100	0.85	30/51	58.8	34/51	66.7
0.53	36	58	54	87	0.280	6/26	23.1	18/26	69.2
0.17	28	45	47	75					
0.053	4	6	24	38					
0.0017	2	3	25	24					

TABLE IV. *Encephalitogenicity of Graded Doses of Guinea Pig Basic Protein and Peptides of Their Midregions*[a]

Immunogen	Dose (nmol)	Mean day of onset	Clinical EAE Total incidence of signs	Incidence of "cerebral" signs	Mortality	Mean day of death
Guinea pig basic protein	0.53	10	5/5	4/5	5/5	12
	0.17	10	5/5	4/5	5/5	15
	0.053	12	4/5	0/5	0/5	-
Guinea pig residues 45-88	0.62	11	10/10	6/10	4/10	17
	0.062-0.21	12	8/10	0/10	1/10	12
	0.021	-	0/5	0/5	0/5	-
Guinea pig residues 70-88[b]	0.16-2.3	9	33/33	22/33	27/33	12
	0.0046-0.037	13	13/14	4/14	7/14	15

[a]HEAE was distinguished from ordinary EAE clinically by a combination of early onset (usually 9 - 10 days), rapid and severe course and high incidence of mortality and cerebral signs (irritability, seizures, coma, and spasticity). Amino acid analyses were determined using a Beckman 121 analyzer after hydrolysis for 24 hr in 6 M HCl.

[b]70 ... 88

Gly-Ser-Leu-Pro-Gln-Lys-Ser-Gln-Arg- Ser -Gln-Asp-Glu-Asn-Pro-Val-Val-His-Phe

(1.0)(3.0)(1.0)(2.0)(4.0)(1.0) (1.0) (2.0) (2.0) (1.0)(1.0)

HEAE than either the smaller fragment (residues 70 - 88) or the complete basic protein (see Table IV). The greater potency of the intact guinea pig basic protein in comparison to its isolated midpeptide 45 - 88 could be explained by the additional encephalitogenic determinants known to reside in the complete protein. However, the lower potency of the midpeptide 45 - 88 in comparison to its smaller fragment 70 - 88 was unexpected.

This difference is not due to inhibitory sites within fragment 45 - 70 since peptide 45 - 70 when injected with peptide 70 - 88 has no effect on the antigenicity of the latter fragment.

In each of the above observations the clinical difference observed is most likely due to conformational differences between the basic protein and the peptide derived from it, i.e., the isolated peptides appear to have a different conformation from their corresponding shape within the whole protein.

SUPPRESSION OF EAE AND HEAE

There have been many nonimmunologic attempts at suppressing or preventing EAE (34 - 37). However, since EAE is an autoimmune disease, an immunologic approach has been sought. Most current methods of immunosuppression are nonspecific and have the recognized disadvantageous side effects of nonspecific immunosuppression; such maneuvers as lymphocyte depletion, conventional immunosuppressive drugs, x-irradiation, and treatment with anti-lymphocyte serum are effective in preventing EAE.

Specific immunotherapy also has been achieved. EAE can be suppressed by administration of (a) sera from recovered animals (38) or (b) the basic protein antigen or clinically modified forms (39, 40) in incomplete Freund's adjuvant or in saline prior to or after challenge with basic protein in Freund's complete adjuvant (41 - 45). Table V shows the results of suppressing ordinary EAE in Lewis rats with a basic protein from a variety of species all of whose sequences are known. Furthermore, using guinea pig basic protein for induction of HEAE and injecting doses of either guinea pig or bovine basic protein in saline starting on day 4 after challenge and ending on day 15, only guinea pig protein suppressed the variant of EAE (Table VI). By comparing the effective suppression and the amino acid sequences of these proteins one can conclude that both encephalitogenicity and suppression are produced by the same region. This conclusion is further supported by the fact that basic protein from a large number of different species essentially are as effective in suppressing EAE in guinea pigs. The major determinant for guinea pigs is essentially sequentially invariant.

TABLE V. Suppression of EAE in Lewis Rat

Challenge BP	Suppression BP	Dose[a] (μg)	Clinical (6 animals/group)
Rat (33 μg) in CFA	Rat	1500	0
		1000	0
		500	1
		250	1
		125	5
	Guinea pig	1500	0
		750	0
		375	2
		187	1
		94	4
	Bovine	2000	6
		1000	6
		500	6
		250	4
		125	6
	Rabbit	1000	6
		500	5
		250	6
		125	6
Bovine	Bovine	2000	0
		1000	0
		500	0
		250	0
		125	3
		0	6

[a]Daily injections day 4 - 15.

However, we have repeatedly failed in suppressing EAE with synthetic encephalitogenic determinants. Workers (46) have shown, however, that a synthetic random copolymer was capable of suppressing EAE.

TABLE VI. Suppression of Hyperacute EAE by Guinea Pig Basic Protein

Antigen and adjuvant	R_x	*Clin.*	*Onset*	*Death*
Guinea pig basic protein (33 μg) with Difco IFA and B. pertussis 10 million organisms	*Guinea pig basic protein 2000 μg in saline, daily day 4-15*	*0/6*	-	-
Guinea pig basic protein (33 μg) with Difco IFA and B. pertussis 10 million organisms	*Bovine basic protein 2000 μg in saline, daily day 4-15*	*6/6*	*8,8, 9,9, 10, 10*	*10, 10*

MECHANISTIC CONSIDERATIONS

Induction of ordinary EAE in guinea pigs and rabbits with defined encephalitogenic peptides is accompanied by demonstrable cellular immune responses (47 - 49), but in contrast to whole basic protein no circulating antibody has been detected (47, 50)., nor was antibody demonstrable in serum when EAE was induced in Lewis rats by the mid-peptide (51). However, Lennon and co-workers have shown (52, 53) that isolated lymph node cells produced antibody to basic protein in situations where no antibody was demonstrable in serum, and that antibody to basic protein was released in the brains of guinea pigs with EAE even with peptide encephalitogens (53), suggesting that antibody to the encephalitogenic determinants of the basic protein are absorbed from plasma by antigen exposed *in situ* in the CNS. Thus a potentially pathogenic role for antibody has not been excluded in ordinary EAE.

For hyperacute EAE most reports have been concerned with the morphology of the CNS lesions (25, 27, 54 - 56), but very few immunologic studies have been reported. Because pertussis vaccine preferentially induces IgE formation in the rat (57), and because of the resemblance of the histologic lesions to the Arthus reaction, it has long been suspected that IgE may play a role in the pathogenesis of the hyperacute form of EAE (55). Rapid passive transfer of histologic legions of hyperacute EAE by living immune lymphocytes can be achieved in certain defined experimental conditions (55). However, the rapidity of the passive transfer of hyperacute EAE with lymph node cells does not exclude a role for antibody since it has been shown that immune lymph node cells are continually releasing antibody to basic protein (52).

It generally has been assumed that each of the encephalitogenic regions by initiating immune processes produce cells which attack exposed *identical* regions within the basic protein bound in myelin resulting in the production of EAE. This is hard to believe for the following reasons: (1) While the guinea pig is susceptible to only one major encephalitogenic area (47), the rabbit and rat are susceptible to at least three, each located in distinctly separate areas of the protein; therefore, a sizeable portion of the protein would have to be exposed; (2) each species is susceptible to a different set of disease-producing regions. Thus the myelin of each species would be *required* to be of greatly different structure even though it contains essentially the same chemical components (58); (3) the structure of the basic protein *in vivo* is probably different *in vitro*. The molecule within the myelin does not react with antibasic antibody, while the isolated protein does (59). Furthermore, *in vivo* the protein is certainly bound to other substances, presumably phospholipids (60).

The tests for cellular immunity have shown that the requirements for encephalitogenicity can be dissociated from and appear more rigid than those for determining a cellular immune response. For example, it has been reported (49, 61 - 63) that using guinea pigs there are at least seven, but probably more, separate sequences within bovine basic protein encompassing determinants discernible in the macrophage migration inhibition test (for review see reference 64). The tryptophan region is the only major encephalitogenic region for guinea pig. *In vitro* immunoactive cells produced by the injection of the tryptophan region into guinea pigs respond to not only the tryptophan determinant but to other areas of the basic protein. Therefore, conceivably the cells which have been produced by the initiation of the immune response to the injection of a particular encephalitogenic determinant *in vivo* attack regions of basic protein impregnated within the myelin which are *not* exactly identical to the encephalitogenic determinant. Thus one exposed region of basic protein in the myelin might respond as a target immunologically for several different encephalitogenic determinants. The determination of whether a region within basic protein is encephalitogenic in a given species would therefore be determined primarily by the properties of the immune system, and, to a lesser degree, by the structure of the myelin.

If the immunologically active cells are attacking an exposed nonencephalitogenic region of the basic protein within the myelin then why is isolated myelin encephalitogenic? The basic protein within this myelin presumably would be of the same structure as *in vivo* myelin, i.e., the encephalitogenic determinants would be hidden. This, however, has been shown convincingly not to be true (59). While antibasic protein antibody does not react with intact myelin it will bind with isolated myelin. Thus the *in vivo* and *in vitro* structures of basic protein differ considerably. Therefore it is quite likely that encephalitogenicity results from the attack on regions of the bound basic protein which are different from the encephalitogenic regions. One exposed region of the protein *in vivo* might be the immunological target of cells which have been stimulated by several different encephalitogenic determinants.

In 1974 (65) we presented calculations which indicate that possibly the destruction of myelin in EAE and conceivably in human disease, for example, multiple sclerosis, is due to activated lymphocyte recognition of released soluble basic protein and not myelin-associated protein. This released basic protein accumulates during the proteins normal turnover which is probably initiated by nonenzymatic deamidation. Recently, Patterson (66) immunologically has detected myelin basic protein fragments in sera of normal rats. Knowing whether the immune system attacks released or membrane-bound basic protein does not assist in determining whether either properties of the immune system or the exposed regions within the basic protein are the deciding factor for determining encephalitogenicity. However, knowing the chemical requirements for disease induction in guinea pigs for the tryptophan peptide, one can observe a sequence similarity between it and the other major determinants, which are encephalitogenic in differing species.

(i) Phe-Ser-Trp-Gly-Ala-Glu-Gly-Gln-Arg
(ii) Phe-Lys-Leu-Gly-Gly-Arg-Asp-Ser-Arg
(iii) Phe-Ser-Trp-Gly-Ala-Glu-Gly-Gln-Arg
(iv) Thr-Thr-His-Tyr-Gly-Ser-Leu-Pro-Gln-Lys
(v) Pro-Gln-Lys-Ser-Gln-Arg-Ser-Gln-Asp-Glu-Asn-Pro-Val-Val

This similarity suggests that when the first four sequences mentioned above are encephalitogenic they fulfill the disease requirements of the same generalized structure rather than four separate such structures.*

**Peptide V is not only the most encephalitogenic region for the Lewis rat but is also responsible for hyperacute EAE (29). The corresponding synthetic peptide with Pro-Val-Val deleted is not encephalitogenic. Martenson et al. (67) has reported that removal of the Val-Val enzymatically results in a peptide possessing much reduced encephalitogenicity. If the serinyl residue is substituted by a threoninyl residue the ability of the peptide to produce hyperacute EAE is eliminated. Therefore, peptide V appears to possess similar characteristics to the other four peptides, i-iv, but in reverse direction. However, until it is determined which residues are important for encephalitogenicity within this region, it will not be known whether its potential sterochemical differences would prevent peptide V from being classed with the other four encephalitogens.*

For any given mammalian species there is certainly a generalized structure. For example:

tryptophan-glycine-alanine-nonspecific-	hydrogen bonding residue of general shape of glutamine	-positively charged residue
(tryptophan-glycine-alanine- x - x -	glutamine	-arginine)

for the guinea pigs, which will be observed as foreign by the antigen recognition system and results in an encephalitogenic response. From species to species the requirements for encephalitogenicity, i.e., this generalized sequence, are slightly varied, probably genetically, and these changes in this generalized sequence account for the difference in encephalitogenicity in different species.

Results from rabbits and Lewis rat further support this theory. Synthetic modifications of peptide (ii) indicate that the two C-terminal residues are very important for disease induction in rabbits. Variation of the other residues, however, produced peptides which were encephalitogenic to some degree. Thus the requirements for disease induction in rabbits are not as stringent as in the guinea pigs. Therefore it is not surprising that at least three (but probably four) of the regions are encephalitogenic in rabbits.

Of the four similar sequences (i-iv) mentioned above only a modified sequence of peptide ii (see footnote to Table I) and peptide 5 are encephalitogenic in Lewis rats.

Therefore the four (i - iv) similar sequences isolated from the basic protein vary in encephalitogenicity form species to species. When they are encephalitogenic they fulfill the requirement of the same generalized structure and not four separate such structures.

EAE RELEVANCE TO HUMAN DISEASE

EAE closely resembles several human virus and vaccine-induced encephalomyelitic diseases, especially rabbies encephalitomyelitis. This latter disease is quite prevalent in the countries such as India and Pakistan, which still use whole brain vaccine preparations.

For years EAE has been suggested as a model for multiple sclerosis. In this respect several groups (68 - 73) have initiated injections of basic protein into MS patients by protocols similar to the EAE suppression experiments described above. The Salk study has been by far the most extensive to date. This program has conclusively shown that the basic protein's actions in multiple sclerosis are not the same as in EAE. Even though EAE's relevance to multiple sclerosis is questioned, this experimental disease is certainly an excellent example for general autoimmunity and graft rejection.

ACKNOWLEDGMENT

Over the years many persons have made contributions to this program. These include: Arthur Robinson, Vanda Lennon, Joseph Kraut, S. J. Singer, Edwin Eylar, Jonas Salk, Juanita Caccam, Millie Thompson, Janet Clark, Francis Yurochko, and Chon Garcia.

REFERENCES

1. Carnegie, P. R. (1971). Amino acid sequence of the encephalitogenic basic protein from human myelin. *Biochem. J. 123,* 57-67.
2. Eylar, E. H., Brostoff, S., Hashim, G., Caccam, J., and Burnett, P. (1971). Basic Al protein of the myelin membrane. The complete amino acid sequence. *J. Biol. Chem. 246,* 5770-5784.
3. Martenson, R. E., Deibler, G. E., and Kies, M. W. (1970). Myelin basic proteins of rat central nervous system. Purification, encephalitogenic properties, and amino acid compositions. *Biochim. Biophys. Acta 200,* 353-362.
4. Dunkley, R. R. and Carnegie, P. R. (1974). Sequence of a rat myelin basic protein. *Biochem. J. 141,* 243-255.
5. Westall, F. C., Thompson, M., and Kalter, S. S. (1976). The proposed sequence of the encephalitogenic protein from chimpanzee brain. *Life Sci. 17,* 219-224.
6. Kies, M. W. and Alvord, E. C., Jr. (1959). *In* "Allergic Encephalomyelisis" (M. W. Kies and E. C. Alvord, Jr., eds.), pp. 239-299. Thomas, Springfield, Illinois.
7. Roboz-Einstein, E., Robertson, D. M., DiCaprio, J., and Moore, W. (1962). The isolation from bovine spinal cord of a homogeneous protein with encephalitogenic activity. *J. Neurochem. 9,* 353-361.
8. Paterson, P. Y. (1968). Experimental autoimmune (allergic) encephalomyelitis. *In* "Textbook of Immunopathology," 1 (P. A. Miescher and H. J. Muller-Eberhard, eds.), Vol. 1,

pp. 132-149. Grune and Stratton, New York.
9. Kibler, R. F., Shapira, R., McKneally, S., Jenkins, J., Selden, P., and Chou, F. (1969). Encephalitogenic protein: structure. *Science 164,* 577-580.
10. Westall, F. C., Thompson, M., and Lennon, V. A. (1976). Hyperacute EAE: unique antigenicity of a 19 amino acid peptide determined by a single methyl group. *Fed. Proc. 53,* No. 1271, 436.
11. Karkhanis, Y. D., Carlo, D. J., Brostoff, S. W., and Eylar, E. H. (1975). Isolation of an encephalitogenic peptide active in the monkey. *J. Biol. Chem. 250,* 1718-1722.
12. Westall, F. C. and Thompson, M. (1978). An encephalitogenic region for rabbits. *Immunochemistry 15,* 189-191.
13. Eylar, E. H., Caccam, J., Jackson, J. J., Westall, F. C., and Robinson, A. B. (1970). Experimental allergic encephalomyelitis: Synthesis of disease-inducing site of the basic protein. *Science 168,* 1220-1223.
14. Shapira, R., Chou, R. C-H., McKneally, S., Urban, E., and Kibler, R. F. (1971). Biological activity and synthesis of an encephalitogenic determinant. *Science 173,* 736-738.
15. Martenson, R. E., Levine, A., and Sowinski, R. (1975). The location of regions in guinea pig and bovine myelin basic protein which induce EAE in Lewis rats. *J. Immunol. 114,* 592-596.
16. Swanborg, R. A. (1975). Antigen-induced inhibition of experimental allergic encephalomyelitis. IV. Studies of the C-terminal end of the myelin basic protein molecule (1). *Immunol. Commun. 4,* 387-397.
17. Westall, F. C. and Thompson, M. (1977). Encephalitogenic regions for the Lewis rat within the myelin basic protein. *Immunol. Commun. 6,* 13-21.
18. Barton, M. A., McPherson, T. A., Lemieux, R. V., and Bain, G. O. (1972). Experimental allergic encephalomyelitis (EAE): Synthesis of the 7-11-20 and 12-20 peptides of human encephalitogenic protein and inhibition of histological EAE by the 12-20 peptide. *Can. J. Biochem. 50,* 689-702.
19. Westall, F. C., Robinson, A. B., Caccam, J., Jackson, J., and Eylar, E. H. (1971). Essential chemical requirements for induction of allergic encephalomyelitis. *Nature (London) 229,* 22-24.
20. Westall, F. C. (1971). Solid phase peptide synthesis as applied to experimental allergic encephalomyelitis. *In* "Multiple Sclerosis" (D. J. Field, T. M. Bell, and P. R. Carnegie, eds.), pp. 72-79. North-Holland, Amsterdam.
21. Nagai, Y., Nakano, K., Yasuda, T., Suzuki, K., Ohtani, S., and Tsumila, T. (1973). *In* "Fourth International Symposium

of Aetrology and Pathogenesis of Demyelinating Diseases" (H. Shiraki, T. Yonegawa, and Y. Kurowa, eds.), pp. 171-173. Kyoto, Japan.
22. Nagai, Y. (1976). Modulation of the immunologic response of myelin basic protein by extrinsic and intrinsic factors. *Neurology 26,* 45-46.
23. Westall, F. C. and Thompson, M. (1977). Further definition of the encephalitogenic region for guinea pigs. *Immunol. Commun. 6,* 23-31.
24. Rauch, H. C. and Einstein, E. R. (1978). Specific brain proteins: A biochemical and immunological review. *In* "Review of Neurosciences" (S. Ehrenpreis and I. J. Kopin, eds.), pp. 283-343. Raven, New York.
25. Levine, S. and Wenk, E. J. (1964). Allergic Encephalomyelitis: A hyperacute form. *Science 146,* 1681-1682.
26. Levine, S. and Wenk, E. J. (1965). A hyperacute form of allergic encephalomyelitis. *Amer. J. Pathol. 47,* 61-88.
27. Paterson, P. Y., Drobish, D. G., Hanson, M. A., and Jacobs, A. F. (1970). Induction of experimental allergic encephalomyelitis in Lewis rats. *Intern. Arch. Allergy 37,* 26-40.
28. Lennon, V. A., Westall, F. C., Thompson, M., and Ward, E. (1976). Antigen, host and adjuvant requirements for induction of HEAE. *Eur. J. Immun. 6,* 805-810.
29. Westall, F. C., Thompson, M., and Lennon, V. A. (1977). HEAE induced by a synthetic autoantigen. *Nature (London) 269,* 425-427.
30. Westall, F. C. (1977). HEAE: a single determinant. *Immunol. Commun. 6,* 227-237.
31. Young, J. D., Tsuchya, D., Geir, M., Geir, S., Westall, F. C., Thompson, M., Cyr, R., Ward, E., and Yurorhko, F. (1974). The encephalitogenic dose response in guinea pigs of the tryptophan region of myelin basic protein. *Immunol. Commun. 3,* 219-226.
32. Westall, F. C. (1978). High dose inhibition of the tryptophan peptide induced encephalitogenicity. *Immunol. Commun. 7,* 57-67.
33. Chapman, B. E. and Moore, W. J. (1976). Conformation of myelin basic protein in aqueous solution from nuclear magnetic resonance Spectroscopy. *Biochem. Biophys. Res. Commun. 73,* 758-766.
34. Good, R. A., Campbell, B., and Good, T. A. (1949). Prophylatic and therapeutic effects of para-amino benzoic acid and sodium salicylate on EAE. *Proc. Soc. Exp. Biol. Med. 72,* 341-347.
35. Arnason, B. G. and Richman, D. P. (1969). Effect of oral contraceptives on EAE. *Arch. Neurol. 21,* 103-108.
36. Saragea, M. and Vladutiu, A. (1969). The influence of nialamide on EAE. *Naturwissenschaften 52,* 564.

37. Kelemen, J. (1970). Research on experimental encephalopathies. *J. Stud. Cercet. Neurol. 15,* 85-92.
38. Paterson, P. Y. and Harwin, S. M. (1963). Suppression of allergic encephalomyelitis in rats by means of antibrain serum. *J. Exp. Med. 117,* 755-774.
39. Eylar, E. H. (1971). *In* "Multiple Sclerosis" (E. J. Field T. M. Bell, and P. R. Carnegie, eds.), pp. 90-104. North-Hollard, Amsterdam.
40. Einstein, E. R. (1971). *In* "Multiple Sclerosis" (E. J. Field, T. M. Bell, and P. R. Carnegie, eds.), pp. 105-110. North-Holland, Amsterdam.
41. Alvord, E. C. and Kies, M. W. (1965). Encephalitogen-induced inhibition of EAE: prevention, suppression and therapy. *Ann. N.Y. Acad. Sci. 122,* 333-345.
42. Einstein, E. R. and Rauch, H. (1968). Protective action of the encephalitogen and other basic proteins in EAE. *Immunochemistry 5,* 567-575.
43. Rauch, H., Ferraresi, A., Raffel, S., and Einstein, E. R. (1969). Inhibition of *in vitro* migration in EAE. *J. Immunol. 102,* 1431-1436.
44. Lisak, R. P., Falk, G. A., Heinze, R. G., Kies, M. W., and Alvord, E. C. (1970). Dissociation of antibody from disease suppression in the inhibition of EAE by MBP. *J. Immunol. 104,* 1435-1446.
45. Levine, S., Sowinski, R., and Kies, M. W. (1972). Treatment of EAE with encephalitogenic basic proteins. *Proc. Soc. Exp. Biol. Med. 139,* 506-510.
46. Teitelbaum, D., Meshorer, A., Hirsfeld, T., Arnon, R., and Sela, M. (1971). Suppression of EAE by a synthetic polypeptide. *Eur. J. Immunol. 1,* 242-248.
47. Lennon, V. A. (1972). Ceblular and humoral immune responses in experimental autoimmune encephalomyelitis. Ph.D. Thesis, University of Melbourne, Melbourne.
48. Bergstrand, H. and Kallen, B. (1972). Localization of antigenic determinants on bovine encephalitogenic protein. Studies in guinea pigs with macrophage migration inhibition. *Cell. Immunol. 3,* 660-671.
49. Bergstrand, H. (1972). Localization of antigenic determinants on bovine encephalitogenic protein. *Eur. J. Biochem. 27,* 126-135.
50. Lennon, V. A. and Carnegie, P. R. (1974). Immunological responses of guinea pigs to a synthetic autoantigen of brain. *Eur. J. Immunol. 4,* 60-62.
51. Dunkley, P. R., Coates, A. S., and Carnegie, P. R. (1973). Encephalitogenic activity of peptides from the smaller basic protein of rat myelin. *J. Immunol. 110,* 1699-1701.
52. Lennon, V. A. and Feldmann, M. (1972). The detection of autoantibody-forming cells. I. An assay for plaque-forming cells to the basic protein of myelin in guinea pigs.

Intern. Arch. Allergy 42, 627-640.

53. Lennon, V. A., Feldmann, M., and Crawford, M. (1972). The detection of autoantibody-forming cells. II. Cells in lymph nodes and central nervous system containing antibody to myelin basic protein. *Intern. Arch. Allergy 43,* 749-758.
54. Levine, S. and Wenk, E. J. (1965). Induction of experimental allergic encephalomyelitis in rats without the aid of adjuvant. *Ann. N.Y. Acad. Sci. 122,* 209-225.
55. Levine, S. and Sowinski, R. (1973). Hyperacute allergic encephalomyelitis. *Amer. J. Pathol. 73,* 247-257.
56. Hoenig, E. and Levine, S. (1974). Three localized forms of experimental allergic encephalomyelitis: An ultrastructural comparison. *J. Neuropathol. Exp. Neurol. 33,* 251-259.
57. Tada, T., Okumura, K., and Taniguchi, M. (1972). Cellular basis of IgE antibody formation in the rat. *In* "Proceedings of a Conference on the Biological Role of the Immunoglobulin E System" (K. Ishizaka and D. H. Dayton, eds.), pp. 89-102.
58. Wolfgram, F. and Kotorii, K. (1968). The composition of myelin proteins of the central nervous system. *J. Neurochem. 15,* 1281-1290.
59. Guarnieri, M., Himmelstein, J., and McKhann, G. M. (1974). Isolated myelin quantitatively absorbs antibody to basic protein. *Brain Res. 72,* 172-176.
60. Thompson, B. E., Kies, M. W., and Alvord, E. C., Jr. (1963). Isolation of an encephalitogenic phospholipid-protein complex by dialysis of myelin in organic solvents. *Biochem. Biophys. Res. Commun. 13,* 198-204.
61. Bergstrand, H. (1973). Localization of antigenic determinants on bovine encephalitogenic protein further studies with the macrophage migration inhibition assay in guinea pigs. *Immunochemistry 10,* 611-620.
62. Bergstrand, H. and Kallen, B. (1973). Is there a cross-reactivity between different parts of the bovine encephalitogenic protein in the macrophage migration inhibition assay? *Immunochemistry 10,* 471-476.
63. Bergstrand, H. and Kallen, B. (1975). Further evidence of an extensive cross-reactivity between the three main parts of the bovine encephalitogenic protein in the lymphoid cell transformation test and the macrophage migration inhibition test. *Acta Pathol. Microbiol. Scand. Sect. C83,* 165-172.
64. Bergstrand, H. (1977). Immunochemistry of encephalitogenic protein. *In* "Immunochemistry of Proteins" (M. Z. Atassi, ed.), Vol. 1, pp. 315-370. Plenum Press, New York.
65. Westall, F. C. (1974). Released MBP. *Immunochemistry 11,*

513-515.

66. Personal communication.
67. Martenson, R. E., Nomura, K., Levine, S., and Sowinski, R. (1977). EAE in Lewis rat: Further delineation of active sites in guinea pig and bovine myelin basic proteins. *J. Immunol. 118,* 1280.
68. Campbell, B., Vogel, P. J., Fishy, E., and Lorenz, R. (1973). Myelin basic protein administration in multiple sclerosis. *Arch. Neurol. 29,* 10-15.
69. Gonsette, R. R., Delmotte, P., and Demonty, L. (1977). Failure of basic protein therapy for multiple sclerosis. *J. Neurol. 216,* 27-31.
70. Salk, J., Westall, F. C., Romine, J., and Wiederholt, W. C. (1978). Studies with MBP in multiple sclerosis patients: 1. Rationale and preliminary report. Progress in MS-Research, Intern. Fed. of MS Soc., Sept., 1978.
71. Romine, J., Salk, J., Wiederholt, W. C., Westall, F. C., and Jablecki, C. K. (1978). Studies with MBP in multiple sclerosis patients: 2. Clinical Evaluation. Progress in MS-Research, Intern. Fed. of MS Soc., Sept., 1978.
72. Salk, J., Westall, F. C., Romine, J. S., and Widerholt, W. C. (1979). Preliminary phase studies of MBP in MS: 1. Rationale and immunologic observations. 31st Annual Meeting--American Academy of Neurology, April, 1979, Chicago, Illinois.
73. Romine, J. S., Salk, J., Wiederholt, W. C., Westall, F. C., and Jablecki, C. K. (1979). Preliminary phase studies of MBP in MS: 2. Clinical observations. 31st Annual Meeting--American Acadmey of Neurology, April, 1979, Chicago, Illinois.

CRYPTIC AMIDASE ACTIVE SITES CATALYZE DEAMIDATION IN PROTEINS

*H. Tonie Wright**

Department of Biochemical Sciences
Princeton University
Princeton, New Jersey 08544

Arthur B. Robinson

2251 Dick George Rd.
Cave Junction, Oregon

INTRODUCTION

One of us has proposed that the deamidation of asparagine and glutamine in proteins plays a role in regulating the rate of protein breakdown *in vivo* and so functions as a molecular clock (1, 2). This proposal was based upon the broad range of deamidation rates of peptides of different sequence, and the correlation of these peptide deamidation rates with those of the few proteins on which such rates have been measured.

We would like to propose here that there is a general acceleration of the rates of deamidation of asparagine and glutamine side chains when incorporated into polypeptide chains. Furthermore, as a result of specific sequence and structure ef-

**Present address: Department of Biochemistry, Medical College of Virginia, Richmond, Virginia 23298.*

 ISBN 0-12-397580-8

fects, the deamidation reaction can be further catalyzed intramolecularly. The magnitude of this catalysis will depend upon the number and kind of these unique sequence and structural arrangements which are realized in a peptide or protein. Finally, some of these special configurations resemble the active sites of proteases, which catalyze a hydrolysis reaction formally similar to deamidation, and so constitute cryptic active sites which can catalyze intramolecular deamidation.

DEAMIDATION RATES

In order to evaluate the acceleration of deamidation rates of asparagine and glutamine in peptides and proteins we must have a rate constant for this reaction in free asparagine and glutamine. There are two difficulties in taking the measured deamidation rates of asparagine and glutamine alone for this purpose. One is that free asparagine and glutamine have a positive and a negative charge which are lost when they are incorporated into a peptide chain, and which could be responsible for special effects upon the rate of deamidation. It is well known that glutamine readily cyclizes to pyrrolidone carboxylic acid due to the presence of the free amino group (3). The second is that the hydrolytic deamidation reaction is second order when measured for the free amino acids in either acid or base and first order for peptide or protein at neutrality.

We have eliminated the first of these difficulties by substituting acetamide as the standard against which to compare the intramolecular asparagine and glutamine deamidation rates. The chemical properties of the amide group of acetamide should closely resemble those of the side chains of asparagine and glutamine without the interference arising from the presence of charged groups. Furthermore, it is known that the polar effects of alkyl substituents on the hydrolysis rate of amides is nil, so that no influence of the rest of the asparagine or glutamine residue upon the rate of side chain deamidation would be expected.

In order to compare the second-order rate constant for deamidation of free acetamide with the first-order rate constant for deamidation of asparagine and glutamine in peptide chains, we have used the pseudo-first-order rate constant for the acetamide hydrolysis from the following expression of Bruice and Marquardt (4):

$$k_1' = k_{H^+}[a_{H^+}] + k_{OH^-}\frac{K_W}{[a_{H^+}]}$$

where a_{H+} is the hydrogen ion activity taken as the negative antilog of the pH and K_W is the dissociation constant of water. k_{H+} and k_{OH-} are the rate constants for the amide hydrolysis in dilute acid and base, respectively.

Using these conventions, we have tabulated some of the pertinent first-order rate constants for amide hydrolysis in Table I. The most rapidly deamidating of the measured peptides is one containing asparagine and is 1.8×10^4 faster than acetamide. The slowest deamidating of the measured peptides is a glutamine one which deamidates 34 times as fast as acetamide. The asparagine-containing peptides are from 220 to 18000 times more rapid in deamidation while the glutamine-containing peptides are from 34 to 1200 times more rapid than acetamide. The general conclusion is that both asparagine and glutamine deamidate more rapidly in peptides than in the free form, although the effect upon glutamine is less than that upon asparagine.

MECHANISM OF DEAMIDATION

In dilute base the rate of amide hydrolysis is first order in amide and first order in hydroxyl ion (5). In more concentrated base (5) and also in the case of anilide hydrolysis (8) the rate expression is second order in hydroxide ion.

Table I. First-Order Rate Constants for Amide Hydrolysis Reactions

Compound	k_1 (sec^{-1})	*Reference*
Acetamide	5×10^{-11}[a]	5,6
Gly-Ser-Asn-His-Gly	9.0×10^{-7}	7
Gly-Ile-Asn-Ala-Gly	1.1×10^{-8}	7
Gly-His-Gln-Ala-Gly	6.0×10^{-8}	7
Gly-Thr-Gln-Ala-Gly	1.7×10^{-9}	7

[a]*The second-order rate constant for alkaline hydrolysis was measured at 50°C (5) and that for acid hydrolysis at 65°C (6). Consequently, the rate constant at 37°C is lower than this value.*

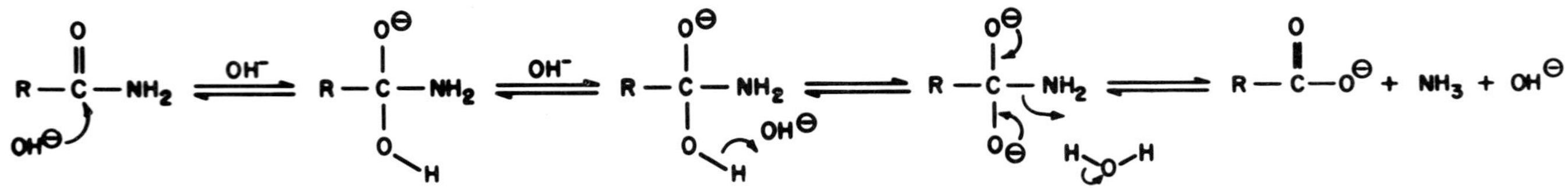

Fig. 1. General mechanism of specific base-catalyzed amide hydrolysis.

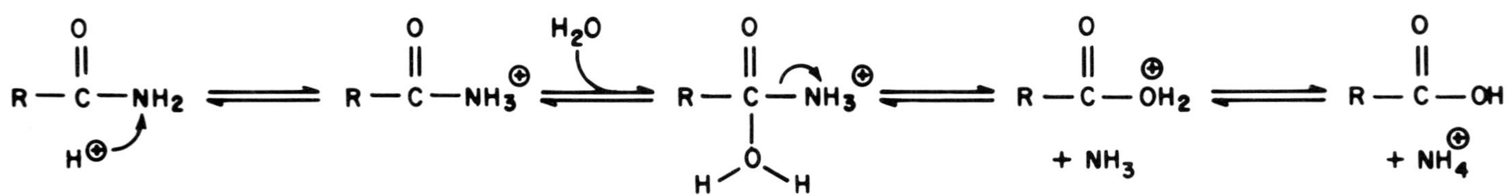

Fig. 2. General mechanism of specific acid-catalyzed hydrolysis of amides.

The most likely mechanism for alkaline hydrolysis (9) is shown in Fig. 1. It should be pointed out that the catalytic role of the second hydroxide ion can apparently be filled by a general base, at least in the case of anilides.

In dilute acid the rate of amide hydrolysis is first order in amide and first order in hydronium ion (6). The most probable mechanism consistent with this rate is shown in Fig. 2 (6).

In the basic hydrolysis of amides, the nucleophilic attack of the hydroxide ion on the carbonyl carbon is an efficient step whereas the base-catalyzed transfer of the proton to the leaving group is rate-limiting (9). In acid hydrolysis the protonation of the leaving group is rapid but the low concentration of strong nucleophiles causes attack at the carbonyl carbon to be rate-limiting. The resultant of these two effects is a pH-rate profile for amide hydrolysis which has a minimum around pH 6.0.

CATALYSIS OF DEAMIDATION

The largest body of data available on the deamidation of asparagine and glutamine in peptides is that for tetra- and pentapeptides (6). The peptides whose deamidation rates have been studied are, with one exception, pentapeptides with glycine at the termini and asparagine or glutamine at the third position flanked on each side by variable residues. The similar deamidation rates of peptides and proteins with the same central tripeptide sequence support the view that deamidation in these cases is affected primarily by the two residues immediately adjacent to the asparagine or glutamine. The absence of participation by the carboxyl terminal group has been confirmed in the case of one pentapeptide by blocking this group without effect upon the deamidation rate (9a).

The conditions under which these rate data were obtained are close to physiological and the concentrations of reactants sufficiently low that intermolecular reactions make little if any contribution. Therefore, the sources of enchanced deamidation rate of asparagine and glutamine in these peptides must be intramolecular.

It is somewhat surprising that no clear sequence-dependent pattern of deamidation rates emerges from the existing data. In general, those peptides with side chains which are charged or can function as general acids or bases deamidate more rapidly than those with functionally inert side chains. However, this generalization is a weak one and the many exceptions imply that peptide conformation is an important determinant of deamidation rates. We have explored by model building some of the possible conformations of the peptides whose deamidation rates have been studied. While this exercise is largely speculative, it has led us to plausible hypotheses for the general acceleration of asparagine and glutamine deamidation in peptides and proteins relative to the free amino acids. It has also suggested to us the possibility that the rapid deamidation of an asparagine in horse heart cytochrome *c* and rabbit muscle aldolase is brought about by a local conformation about each of these asparagines which resembles the active site of a protease.

Considering first the general acceleration of deamidation of asparagine and glutamine in peptides and proteins relative to acetamide, we propose that this arises from stabilization of the hypothetical oxyanion transition state by polypeptide chain —NH— groups. Such a role in transition state or intermediate stabilization has been ascribed to main chain —NH— groups at the active sites of the serine proteases, chymotrypsin, and subtilisin (10, 11). Although there is no data available to permit us to evaluate the magnitude of this effect upon the overall reaction rate, the structures of the serine proteases and their complexes with inhibitors and substrates support its role in catalysis.

The general rate enhancement of deamidation from transition state stabilization by main chain amido —NH— groups also suggests a structural reason for the generally higher rates of deamidation observed for asparagine relative to glutamine-containing peptides. Model building shows that the extra methylene group in the glutamine side chain prevents the side chain from reaching a position where main chain amido —NH— groups can make good hydrogen bonds to the amide oxygen. In contrast, the asparagine side chain can be oriented to form two and sometimes more hydrogen bonds to the peptide chain, depending upon the conformation of the peptide. If this is the case, we can conclude that the effect of such transition state stabilization is to increase the deamidation rate by an order of magnitude as observed in comparing glutaminyl and asparaginyl peptides.

DEAMIDATION IN PROTEINS

There are only a few well documented cases of deamidation in intact proteins although there are numerous reports, particularly of electrophoretic heterogeneity, which are very likely due to this phenomenon (see ref. 7, Table II). Deamidation in horse heart cytochrome *c* (11) and rabbit muscle adolase (13, 14, 15) have been carefully examined and the halflife of the labile residue in each case determined. There is also clear demonstration of deamidation in the mutant hemoglobins Wayne (16) and Providence (17), in the anti-tumor protein neocarzinostatin (18), bovine seminal ribonuclease (19), and in immunoglobulin chains, primarily myeloma proteins (20).

Although the sequences about the rapidly deamidating groups are different, there is a number of other aspects in which they are similar. In the well characterized cases the labile amide is always an asparagine. This residue occurs adjacent to or within one residue of a general acid group which is the side chain hydroxyl group of serine or threonine. In at least two cases, the strongly nucleophilic histidine side chain lies close to the rapidly deamidating residue. In three cases, the labile group is near the carboxyl terminus of the protein molecule. Despite these correlations, the considerable variation in sequence around the deamidating group and the disparate identities of the proteins which have been shown to deamidate, strongly imply that the conformation of the molecule is an important determinant of whether a particular residue will deamidate with anomalous rapidity.

While there is no reason to expect the mechanism of the deamidation reaction to be the same in all cases, the consistencies in the above cases support the following effects as contributory to intramolecular protein deamidation:

i. General acid catalysis by the Ser or Thr side chain hydroxyl groups, or less likely, by the histidine imidazole group.

ii. Nucleophilic catalysis by neutral imidazole of histidine or possibly by carboxylate group.

iii. Transition state stabilization of the oxyanion intermediate by main chain —NH— or other proton donors.

iv. Access to solvent which is necessary for the hydrolysis reaction.

v. Sufficient conformational flexibility in the protein structure to permit the reactive configuration to be attained.

In horse heart cytochrome *c*, whose structure is known (21), the carboxyl terminus is exposed to solvent on the surface of the molecule. While the residues in this terminal tetrapeptide as observed in the crystal structure are not in a configuration where catalysis of deamidation is likely to occur, relatively small deformations are necessary to bring them into such a configuration. Figure 3 shows the carboxyl terminus as observed in the crystal structure of horse heart cytochrome *c* and a hypothetical configuration for this region which we propose as the one in which Asn (102) deamidates. The conformational changes necessary to reach this configuration are confined to the last two residues with the exception of a small rotation in the side chain of Thr (101). The main chain dihedral angles of the three residues involved in our hypothetical deamidating sequence lie in acceptable ranges and are close in value to those of a 3_{10} helix.

In the proposed deamidation reaction, Thr (101) contributes a proton to the $-NH_2$ leaving group and the side chain carboxylate of Glu (103) attacks the amido carbon of Asn (102) (Fig. 4). Only the carboxylate side chain and not the terminal carboxylate group of the glutamic acid can approach the side chain amido group of the asparagine and function as a nucleo-

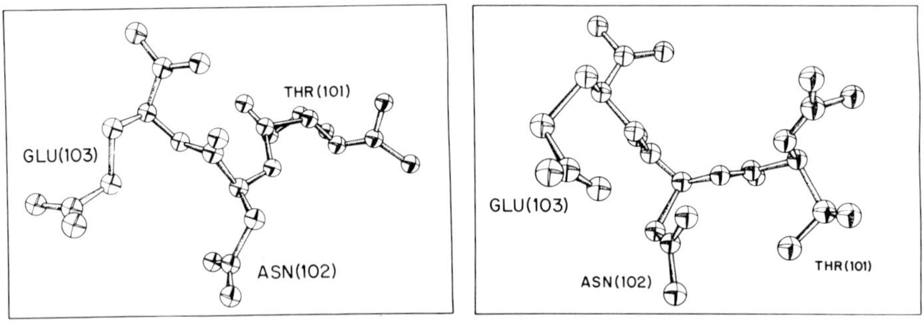

Fig. 3. (a) Conformation of the amino acid residues at the carboxyl terminus of horse heart cytochrome c. Coordinates for this figure were taken from the Princeton University Computer Graphics Laboratory files. (b) Hypothetical configuration of the amino acid residues at the carboxyl terminus of horse heart cytochrome c in which the side chain of Asn (102) is deamidated.

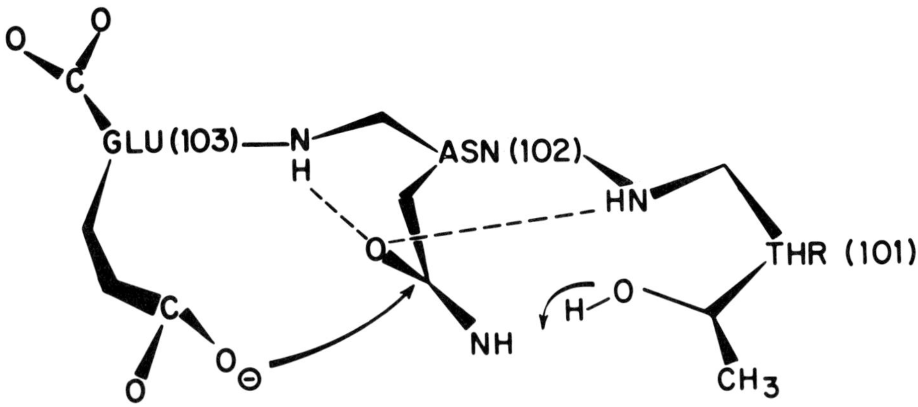

Fig. 4. Possible mechanism of deamidation of Asn (102) in horse heart cytochrome c.

phile in this mechanism. This distance is approximately 2.8 Å. An alternative mechanism in which either of the carboxylates of Glu (103) functions as a general base is also possible. We predict that if this terminal residue were not glutamic acid the deamidation rate of the adjacent asparagine would be diminished, the degree of diminution depending upon which mechanism is the dominant one. The $O_{\gamma 1}$ atom of Thr (101) can make a hydrogen bond of slightly less than 3.0 Å to the labile $-NH_2$ of the asparagine. The main chain amido —NH— groups of the three terminal amino acids are all directed toward the carboxyl oxygen of the asparagine side chain to which they approach within 3.0 Å. In these positions they can serve to stabilize the tetrahedral oxyanion which is the probable transition state of the hydrolysis reaction.

The observed deamidation rate of the carboxyl terminus tetrapeptide of cytochrome *c* is approximately 10^4 times that of acetamide. This is within the maximum likely rate enhancement effect attributable to intramolecular carboxylate nucleophile, general acid, and transition state stabilizing groups.

The deamidation of asparagine (358) in rabbit muscle aldolase is the only other case of deamidation in a protein which has been extensively characterized (15). The half-life of this amide is approximately 6 days, corresponding to a first-order rate constant of 7.2×10^{-7} sec^{-1}.

While there is no tertiary structure information available, the sequence around the rapidly deamidating Asn (358) presents two plausible mechanisms by which the deamidation could be catalyzed. In a reaction similar to the mechanism proposed for Asn (102) deamidation in horse heart cytochrome *c*, Ser (357)

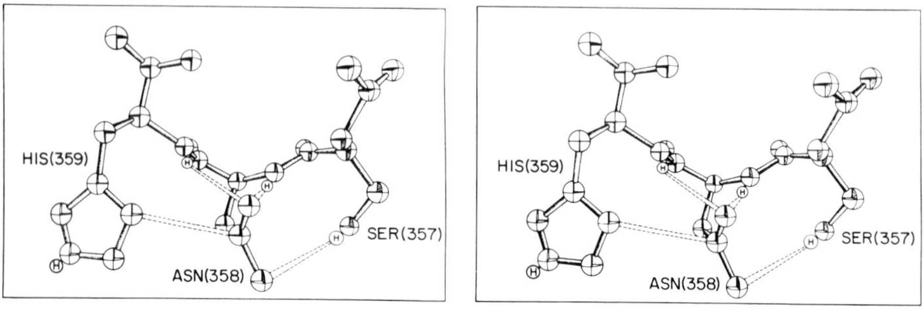

Fig. 5. Hypothetical configuration of the residues around Asn (358) in rabbit muscle aldolase which results in intramolecular catalyzed deamidation.

could act as a general acid in protonating the amido $-NH_2$ leaving group and histidine (359) could be the nucleophile which attacks the amido carbon. Such a hypothetical structure is shown in Fig. 5. Its similarity to the proposed cytochrome *c* deamidating conformation is obvious. On the other hand, it is possible, although somewhat less plausible, to build a model in which Ser (357) $O_\gamma H$ hydrogen bonds to $N_{\delta 1}$ of His (359). In such a conformation, which resembles part of the catalytic configuration of serine proteases, transfer of the proton from Ser (357) to His (359) would create a strongly nucleophilic seroxide group and a protonated His (359) to function as general acid. As in cytochrome *c*, main chain —NH— groups can hydrogen bond to the amido oxygen of Asn (358) and so stabilize it in its oxyanion form. If this latter mechanism occurs, rabbit muscle aldolase would contain a deamidating configuration resembling a neutral serine protease and cytochrome *c* one which resembles an acid protease.

The other documented cases of deamidation in proteins of known sequence have not been characterized with respect to their deamidation rates as have horse heart cytochrome *c* and rabbit muscle aldolase. However, they are similar in some respects to these two cases. The mutant hemoglobin, Wayne I, has been described in which the carboxyl terminal tripeptide of the α chain has been replaced by an octapeptide of sequence Thr-Ser-Asn-Thr-Val-Lys-Leu-Glu-Pro-Arg (16). This mutant exists in two forms, Wayne II being homologous with Wayne I, with the exception of the asparagine, which is aspartic acid in Wayne II. Since the mutation is a frameshift, the appearance of Wayne II is almost certainly the result of deamidation of the asparagine in Wayne I. It is noteworthy that this deamidating sequence is flanked by serine and threonine and occurs at a carboxyl terminus. The observation that Wayne hemoglobin is observed at low levels in the peripheral blood may be a result of rapid degradation resulting from conformational instability of the molecule caused by the deamidation reaction.

The cases of horse heart cytochrome *c*, rabbit muscle aldolase, and hemoglobin Wayne appear to be cases where the residues immediately adjacent in sequence to the rapidly deamidating asparagine catalyze the deamidation reaction. That this is the case for the first two is demonstrated by the similarity of deamidation rates in the protein and homologous model peptide (7).

While sequence proximity can guarantee spatial proximity of groups catalyzing deamidation, the conformation of a native protein could also concentrate such groups from sequentially remote parts of the molecule around an asparagine or glutamine residue as is observed in the active sites of enzymes. Hemoglobin Providence in which Lys β (82) is replaced by Asn is such a case (17). Hemoglobin Providence has been shown to be isomorphous to native deoxyhemoglobin in all respects except for the single replacement of Lys (82) by Asn (A. Arnone, personal communication). The mutant Asn at this position deamidates to aspartate. This reaction is almost certainly catalyzed by His β (143) acting either as a nucleophile or as a general acid. If His β (143) is a nucleophile, there is no good candidate for a general acid in this region of the molecule and such a role would have to be filled by solvent occupying the central cavity of the molecule. Alternatively, His β (143) may function as a general acid and solvent hydroxyl ions or possibly the 1-carboxylate group of bound 2,3-diphosphoglycerate could then serve as nucleophile. In normal human deoxyhemoglobin the NE2 atom of His β (143) lies 4.4 Å from C_γ of Lys (82). Minor movements of either or both of these residues would bring them into an acceptable orientation for the deamidation reaction (Fig. 6).

Neocarzinostatin (18) appears to be a similar case in which a disulfide bridge may bring the labile Asn (83) into proximity with Ser (39) as well as the nearby residues Thr (81) and Thr (85).

Finally, it has been shown recently that bovine seminal ribonuclease is rapidly deamidated (A. DiDonato and G. D'alessio, to be published). This molecule has considerable sequence homology with bovine pancreatic ribonuclease, which does not contain any rapidly deamidating amide groups. The positions and number of Asn and Gln residues is identical in the two molecules. However, unlike the bovine pancreatic enzyme, the bovine seminal ribonuclease exists as a dimer linked by two disulfide bridges between the corresponding residues Cys (30) and Cys (31) in each monomer (21). Several of the amino acid changes in bovine seminal RNase relative to bovine pancreatic RNase could account for the increase in deamidation rate of an Asn or Gln residue observed in the latter. It is also possible in this

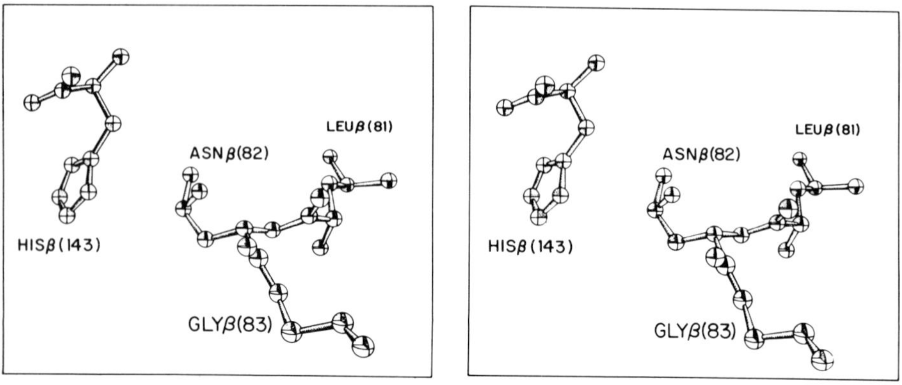

Fig. 6. Conformation of residues in the region of Asn β (82) in hemoglobin Providence. Coordinates were taken from the "Atlas of Macromolecular Structure on Microfiche," Richard J. Feldmann (1976) for human deoxyhemoglobin with Lys (82) replaced by Asn.

case that deamidation in one subunit is catalyzed by groups from the juxtaposed monomer which are brought into proximity with the labile residue as a result of dimer formation. Such an effect could expand the regulatory properties of oligomeric enzymes to include control of breakdown rate.

In vivo ROLE OF PROTEIN DEAMIDATION

The factors which we have listed in the preceding section as contributory to the deamidation of asparagine and glutamine residues in proteins have several other implications for this process. The flexibility in the protein structure which we invoke to bring the functional groups into the proper configuration for the deamidation reaction implies that such reactions are more likely to occur at chain termini and on the surface of the molecule. The three cases of horse heart cytochrome *c*, rabbit muscle aldolase and hemoglobin Wayne are consistent with this expectation. In hemoglobin Providence the rapidly deamidating Asp β (82) lies in the diphosphoglycerate binding pocket where some degree of side chain motility is likely, particularly in oxyhemoglobin. Those regions of protein structure which are most likely to have conformational flexibility are also most likely to be exposed to solvent, the participation of which is necessary in the hydrolytic reaction.

We expect that asparagine and glutamine residues in the interior of a protein molecule will deamidate more slowly than those near the surface. This follows from the decreased flexibility of the more tightly packed interior, the diminished access of solvent to a potential deamidating residue, and the lower probability of finding an effective, anionic nucleophile or general base in the interior of a protein molecule. The similarity in deamidation rates of peptides and proteins with homologous sequences is consistent with our expectation that the deamidating sites lie in exposed regions of the molecule where some conformational flexibility is more likely.

It has already been pointed out that the range of peptide deamidation rates in those which have been measured coincides with the range of *in vivo* turnover rates of proteins containing the same deamidating sequence (7). Since it is now clear that this range of rates represents various degrees of acceleration over those of free asparagine and glutamine, we can state that the *in vivo* rate of protein deamidation is a property of the protein and not of asparagine and glutamine themselves. Furthermore, in the cases of exceptionally rapid deamidation, the rate is so much higher than for the free amino acids and the configuration which catalyzes the deamidation so elaborate that it reinforces the earlier proposal that deamidation in these cases may have functional significance as a molecular clock.

If deamidation of asparagine and glutamine is a regulator of protein catabolism and if our proposed mechanisms for intramolecular protein deamidation are correct, deamidation provides an amplifier for the breakdown of proteins. This amplifier arises because the deamidation of an asparagine or a glutamine introduces a new carboxylate nucleophile or general base which can then catalyze deamidations of other asparagines or glutamines, particularly if changes in tertiary structure occur. Such an effect may account for the observation of Flatmark (12, 22) that the second asparagine to deamidate in cytochrome *c* does so faster than the first, once the first has occurred. It is quite likely, as previously suggested, that this acceleration of the second deamidation is the result of a tertiary structure change resulting from the first deamidation. Such a cascade process could rapidly change the nature of a protein, rendering it dysfunctional and susceptible to degradative reactions such as proteolysis. It is not beyond the realm of possibility that the genesis of new nucleophiles through deamidation could also create catalytic configurations which could split main chain peptide bonds as well as the side chain amides. It has been shown that Asn-Gly peptide bonds are labile in the presence of a strong nucleophile like hydroxylamine (23). Main chain peptide bonds adjacent to aspartate or asparagine residues are labile in relatively dilute acid (24) and Asp-Pro peptide bonds have been shown to hydrolyze under mild acid conditions by intramolecular, nucleophilic catalysis involving the aspartate carboxylate group (25). In proteins and peptides, a high local concentration of nucleophile could be realized through a conformation which brings neighboring nucleophilic side chain(s) into proximity to main chain peptide bonds.

Similar intramolecular deamidation reactions could affect the integrity of glycoproteins. In those glycoproteins where the carbohydrate is linked to an asparagine side chain through an *N*-glycosidic bond, deamidation of the asparagine residue would lead to loss of protein-bound carbohydrate. Such transformations could be important in some of the many processes involving the chemical nature of cell surface groups.

The recurrence of the general acid side chain groups of Ser and Thr suggests that either of these residues is a major contributor to the rate enhancement of the deamidation reaction in proteins. It is interesting in this regard that the immunoglobulins, which have frequently been reported to be heterogeneous as a result of amide hydrolysis (19), show a frequency of serine in the known sequences, which is roughly twice that observed in a sample of 205 proteins.* Perhaps more significant is the fact that asparagine occurs with an exceptionally high frequency (0.21) adjacent to asparagine and similarly glutamine adjacent to glutamine (0.24). Such a clustering of these amino acids could lead to a degradative cascade process such as that postulated above in which deamidation of one asparagine or glutamine residue then accelerates the deamidation of neighboring ones catalyzed by the newly introduced carboxylate groups.

The catalytic configurations which we have proposed as responsible for accelerated deamidation of asparagine and glutamine in proteins are attractive candidates for precursors of protease active sites. The presence of a general acid and/or basic group with a nucleophile has been demonstrated in a number of protease structures. Although we have suggested that in the deamidation reactions, serine or threonine is a general acid and histidine imidazole and carboxylate are nucleophiles, selective pressures upon the protein structure and the binding site of the enzyme could have led to a reversal of these functions. Such a reversal, succeeding the development of a substrate binding site and adaptive changes in protein conformation would lead to enhanced reactivity of these groups in their

**The frequency of serine in 205 proteins listed in reference 26 is 0.066 while that for 33 immunoglobulin chains (ref. 27) is 0.13.*

new roles. For example, the serine-histidine configuration around the rapidly deamidating asparagine in aldolase could be an antecedent of the serine proteases and the carboxyl terminus of horse heart cytochrome *c* an antecedent of the acid proteases.

Another implication of specific deamidation for evolution is that the deamidation reaction results in formation of an anionic carboxylate group. Such groups occur frequently as ligands of protein-bound metals. Deamidation of a primitive functional protein followed by capture of a stabilizing metal ion could open a new direction of functional adaptation for the molecule. It is also possible that a relationship exists between deamidating proteins and metal ions *in vivo* which modulates the function of either or both of them.

SUMMARY AND CONCLUSION

The deamidation of asparagine and glutamine in peptides and proteins is shown to take place at a rate which is two to four orders of magnitude greater than that of the free amino acids. This acceleration is the result of several intramolecular effects: (i) stabilization of the tetrahedral transition state by main chain amido hydrogens; (ii) nucleophilic catalysis by imidazole and carboxylate groups; and (iii) general acid catalysis by the hydroxyl groups of serine and threonine. The degree of rate enhancement will depend upon the accessibility of favorable stereochemical configurations in the molecule. Models of horse heart cytochrome *c* and rabbit muscle aldolase suggest that a configuration resembling a protease active site catalyzes a rapid deamidation in these molecules. Two mutant human hemoglobins, neocarzinostatin and bovine seminal ribonuclease are also likely to be deamidated by similar mechanisms. The presence of cryptic amidase active sites has a number of possible implications for protein catabolism, regulation, and evolution.

Asparagine and glutamine are the only naturally occurring amino acids which are chemically reactive under *in vivo* conditions on a time scale relevant to the lifetime of organisms. We suggest that the deamidation reactions of these residues in peptides and proteins *in vivo* is a widespread phenomenon, that it is intramolecularly catalyzed, and that it has been exploited by nature to control the rate of degradation of these molecules and quite likely the rates of other *in vivo* processes.

ACKNOWLEDGMENTS

H.T.W. was an Established Investigator of the American Heart Association. This work was supported by grants from the American Heart Association with funds contributed in part by the New Jersey Heart Association and from the N.I.H., GM23598. We would like to thank Dr. M. F. Perutz for a helpful discussion and Dr. A. Arnone for making unpublished results available to us.

REFERENCES

1. Robinson, A. B., McKerrow, P. H., and Cary, P. (1970). *Proc. Natl. Acad. Sci. U.S. 66,* 753-757.
2. Robinson, A. B. (1974). *Proc. Natl. Acad. Sci. U.S. 71,* 885-888.
3. Vickery, H. B., Pucher, G. W., Clark, H. E., Chibnall, A. C., and Westall, R. G. (1935). *Biochem. J. 29,* 2710-2720.
4. Bruice, T. C. and Marquardt, T.-H. (1962). *J. Amer. Chem. Soc. 84,* 365-370.
5. Packer, J., Thomson, A. L., and Vaughn, J. (1955). *J. Chem. Soc.,* pp. 2601-2603.
6. Bruylants, A., and Kezdy, F. (1960). *Rec. Chem. Prog. 21,* 213-240.
7. Robinson, A. B. and Rudd, C. J. (1974). *Current Topics Cell. Reg. 8,* 247-295.
8. Biechler, S. S. and Taft, R. W. (1957). *J. Amer. Chem. Soc. 79,* 4927-4935.
9. Jencks, W. P. (1969). "Catalysis in Chemistry and Enzymology," p. 524. McGraw-Hill, New York.
9a. Tsao, C., Sheets, A., and Robinson, A., unpublished observation.
10. Henderson, R. (1970). *J. Mol. Biol. 54,* 341-354.
11. Robertus, J. D., Kraut, J., Alden, R. A., and Birktoft, J. J. (1972). *Biochemistry 11,* 4293-4303.
12. Flatmark, T. (1966). *Acta Chem. Scand. 20,* 1487-1496.
13. Koida, M., Lai, C. Y., and Horecker, B. L. (1969). *Arch. Biochem. Biophys. 134,* 623-631.
14. Lai, C. Y., Chen, C., and Horecker, B. L. (1970). *Biochem. Biophys. Res. Commun. 40,* 461-468.
15. Midelfort, C. J. and Mehler, A. H. (1972). *Proc. Natl. Acad. Sci. U.S. 69,* 1816-1819.
16. Seid-Akhaven, N., Winter, W. P., Abramson, R. K., and Rucknagel, D. L. (1976). *Proc. Natl. Acad. Sci. U.S. 73,*

882-886.
17. Moo-Penn, W. F., Jue, D. L., Bechtel, K. C., Johnson, M. H., Schmidt, R. M., McCurdy, P. R., Fox, J., Bonaventura, J., Sullivan, B., and Bonaventura, C. (1976). *J. Biol. Chem. 251,* 7557-7562.
18. Maeda, H. and Kuromizu, K. (1977). *J. Biochem. (Tokyo) 81,* 25-35.
19. Reisfeld, R. A., Inman, J. K., Mage, R. G., and Appella, E. (1968). *Biochemistry 7,* 14-24.
20. Dickerson, R. E., Takano, T., Eisenberg, D., Kallai, O. B., Samson, L., Cooper, A., and Margoliash, E. (1971). *J. Biol. Chem. 246,* 1511-1535.
21. D'Alessio, G., Malorni, M. C., and Parente, A. (1975). *Biochemistry 14,* 1116-1122.
22. Robinson, A. B., McKerrow, J., and Legas, M. (1974). *Intern. J. Peptide Protein Res. 6,* 31-35.
23. Bronstein, P. and Balian, G. (1977). *Methods Enzymol. XLVII,* 132-145.
24. Schultz, J. (1967). *Methods Enzymol. XI,* 255-263.
25. Landon, M. (1977). *Methods Enzymol. XLVII,* 145-149.
26. "Handbook of Biochemistry," (1970). 2nd ed. C282-C287. The Chemical Rubber Co., Cleveland, Ohio.
27. Croft, L. R. (1975). "Handbook of Protein Sequences." Joynson-Bruvvers, Ltd., Oxford, England.

Appendix

The fifteen papers by Kamen included in this Appendix are representative of his broad scientific interest. They are of historical significance and are indicative of the importance of Kamen's work in modern science in a number of different areas.

Neutron-Induced Radioactivity of the Noble Metals

EDWIN MCMILLAN, MARTIN KAMEN

Radiation Laboratory, Department of Physics, University of California, Berkeley, California

AND

SAMUEL RUBEN*

Department of Chemistry, University of California, Berkeley, California

(Received July 19, 1937)

WE have started a study of the effects produced by neutrons on the elements in the neighborhood of platinum, and some of the results, although admittedly incomplete, seem sufficiently striking to merit publication at this time. Strong activities are induced in Ir, Pt and Aμ by slow neutrons, and in these metals and Hg by fast ($Li+H^2$) neutrons. Os, Tl, Pb and Bi give only comparatively weak activities with either type of neutrons. Our results so far on the strong activities, in order of increasing complexity, are given below.

MERCURY

This element has been reported to give a weak activity of 40 hr. half-life with slow neutrons,[1] and a strong activity of 43-min. half-life with fast neutrons.[2] We find the fast neutron 43-min. period to be indeed very strong, corresponding to an intensity with 10 μA of 5.5 Mev deuterons striking the Li target, after exposure to satura-

* Abraham Rosenberg Fellow.
[1] Anderson, Nature **137**, 457 (1936).
[2] Heyn, Nature **139**, 842 (1937).

Reprinted from *Physical Review*, **52**, 1 (1937).

tion, of 18 divisions per second on the electroscope.[3] Also a new period of 25 hrs., with a saturation intensity of 2 div./sec., was found. This was only weakly activated by the exposures used, and its properties have not been studied. The particles emitted by the 43-min. activity are negative (determined by a Thibaud trochoid apparatus, for the use of which we are indebted to Mr. Ernest Lyman). A chemical separation showed that this (as well as the 25-hr. period) is a mercury isotope. From this information we can place the 43-min. period in the system of isotopes, as being most probably Hg^{203}.

The 43-min. beta-particles are very soft; the absorption curve in aluminum shows a well-defined end point at 0.130 ± 0.005 g/cm^2, giving an upper limit of 0.42 Mev by Feather's formula. This low energy for such a short life does not fit at all on the usual Sargent curves. There is also a strong gamma-ray, with an absorption in Pb indicating a possible complex structure with energies in the range 70–250 Kev.

Gold

With slow neutrons, the well-known 2.7-day period[4] is produced with a saturation intensity of 860 div./sec. The absorption curve of the beta-rays has an end point at 0.31 ± 0.01 g/cm^2 of Al, corresponding to an upper limit of 0.78

[3] One division per second corresponds to about 10^4 beta-particles per second emitted from the target. All intensities given are calculated for an exposure to saturation with a 10 μA deuteron beam, although of course different intensities and times were actually used. For the Li neutron exposures, the samples were about 2 cm from the target, surrounded by a layer of B_2O_3 containing about 1/2 g/cm^2 of B. For the slow neutron exposures, they were about 5 cm from a Be target, enclosed in a large block of paraffin. The samples were all of about the same size ($\sim$0.1 g) but in some cases the beta-particles are so soft that even the thin sheets used do not allow them all to escape. No attempt has been made to correct for this.

[4] Fermi, Amaldi, D'Agostino, Rasetti, and Segrè, Proc. Roy. Soc. **146A**, 483 (1934).

Mev. The strong gamma-ray has an absorption in Pb indicating an energy of about 460 Kev. The beta-particles are all negative.

Fast neutrons on Au produce a much more complex activity. There is a 13-hr. period of saturation intensity 1 div./sec., and what appears at first to be a 3.3-day period of saturation intensity 6 div./sec. On following the decay of this, however, the logarithmic plot deviates from a straight line after about 10 days, and seems to be going into a slope of 4–5 days. This shows that the apparent 3.3-day period is a mixture, presumably of the 2.7-day period and a longer one in about equal amounts. We have shown that neither resonance nor thermal neutrons are responsible for producing the 2.7-day period here, since a sample surrounded by a thick sheet of gold is activated as strongly as one without this, and the boron layer is sufficient to stop all thermal neutrons.

Chemical separation shows that all these periods belong to gold isotopes, and they are also found to emit only negative electrons. This leads to some difficulty in placing the 13-hr. and 4–5-day periods in the system of isotopes, since only one space (Au^{196}) satisfies the necessary conditions for both of them. Au^{195} would be expected to emit positive electrons, unless a new stable Hg isotope (Hg^{195}) should exist. It seems possible that we have here another case of nuclear isomerism.[5]

Iridium

With slow neutrons, we find the known 19-hr.[4] and 2-month[6] periods, with saturation intensities 120 and 280 div./sec., respectively. We find also

[5] Snell, Phys. Rev. **51**, 1011 (1937), and Bothe and Gentner, Naturwiss. **25**, 284 (1937), find such a case in bromine.

[6] Fomin and Houtermans, Physik. Zeits. Sowjetunion **9**, 273 (1936).

a 1.5-min. period with intensity 28 div./sec. This is easily missed if the longer periods are activated strongly, and the observations delayed until several minutes after activation, but by using short exposures it is brought out prominently. It was observed in three different iridium samples, one of which is known to be of very high purity, and we feel that it cannot be due to a contamination. We have done no chemistry on iridium, but by analogy with other slow neutron processes all three of these periods should be attributed to the two isotopes Ir^{192} and Ir^{194}, so that we may have another case of isomerism here.

The 19-hr. beta-rays have an end point in Al of 0.99 ± 0.02 g/cm^2, corresponding to an energy of 2.1 Mev. The 2-month beta-particles are much softer.

With fast neutrons, a very long period (presumably 2 months) is produced with saturation intensity 10 div./sec., as well as a complex array of shorter periods that we have not succeeded in analyzing, using the present data. The 1.5-min. period is present with a saturation intensity of 0.2 div./sec., and the 19-hr. period is buried in the midst of a continuously curving logarithmic plot, so that we cannot be sure of its presence. It is certainly less intense relative to the 2-month period than with slow neutron activation, just as is the 1.5-min. period, so that we can provisionally assign the 2-month period to Ir^{192} and the other two to Ir^{194}. The other unresolved periods again offer difficulties, since all the emitted electrons are negative. No chemical separation has been made.

Platinum

Platinum with slow neutrons gives three periods, 31 min., 18 hrs., and 3.3 days, the saturation intensities being, respectively, 1.2, 0.8, and

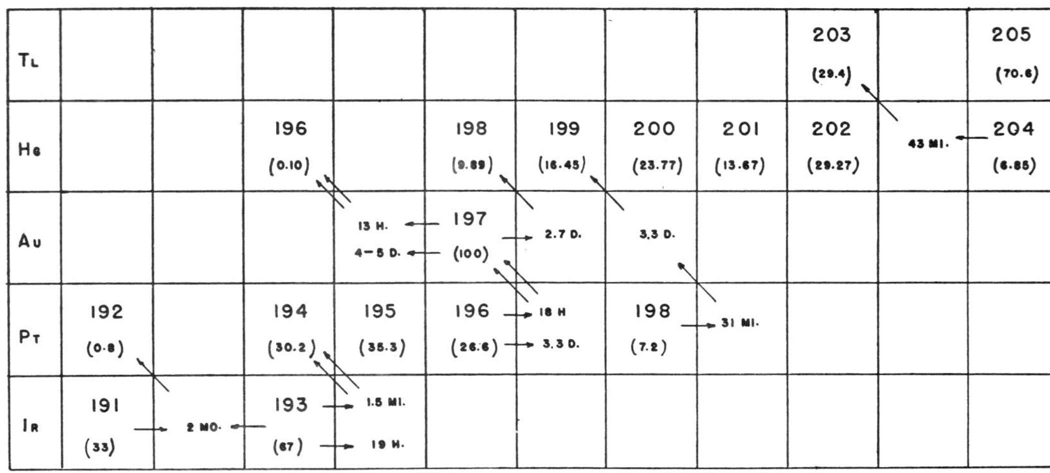

FIG. 1. Isotope chart, showing provisional assignment of identified activities. Figures in parentheses are percentage abundances of stable isotopes. Horizontal arrows indicate neutron capture or ejection, and diagonal arrows indicate β-particle emission.

1.2 div./sec. These might be assigned to the isotopes Pt^{193}, Pt^{197}, and Pt^{199}, but the observation that they all give negative beta-particles makes this unlikely, since Pt^{193} cannot form a known stable isotope by emitting a negative electron. A chemical separation helped to clear up this situation—the gold precipitate showed a 3.3-day activity. Reference to the isotope chart shows that one would expect Pt^{199} to form unstable Au^{199}. We made successive separations of gold from activated platinum to find which platinum period is its parent, and found that it does not come from the 18-hr. period, but most probably does come from the 31-min. period. When two separations $1\frac{1}{4}$ hr. apart were made, the second gold fraction showed about the right ratio of activity to the first to agree with this interpretation. The platinum fraction carries the 18-hr. activity (this incidentally shows that it is not due to Ir contamination), and also, even after repeated separations of gold, still has a period of about 3.3 days. Thus there seem to be two active bodies with the same period (within our present accuracy), one of which fits into the scheme very nicely, while the other (the Pt isotope) again causes trouble. The best that we can think of doing is to make the 18-hr. and 3.3-day platinum periods another pair of isomers. Au^{199} accounts for about 1/3 of the total 3.3-day activity; the agreement between the cross sections for the 31-min. and 3.3-day periods is spurious, since the latter has an extremely soft beta-radiation which can get out of only a fraction of the thickness of the sample.

Pt with fast neutrons gives, like Ir, what appears to be a very complex set of periods, ranging from about 30 min. to more than 5 days. One chemical separation made two days after activation showed that all the long period activity is isotopic with Pt, except for a trace of Au activity, which was much weaker relative to Pt than after slow neutron activation. (We should mention that any active Os would be lost in this and the other separations during the solution in aqua regia, so that we cannot exclude the possibility of its presence. With this exception, all the separations were made to distinguish an element from its two lower neighbors, and in some cases its upper neighbor.) The beta-particles from Pt activated with fast neutrons were found to be nearly all negative, but there was some evidence of a weak long period positron activity.

The results of this work so far do not seem to be capable of any simple explanation without the introduction of a fantastic number of isomeric nuclei, but other explanations may be suggested. Some of the arguments based on the nonexistence of certain stable nuclei may be wrong, or one might assume that some of the active bodies found are excited states of stable nuclei. A search was made using a linear amplifier for α-particles from Au, Ir, and Pt activated with both slow and fast neutrons, and none were found, so no explanation involving α-particle emission can be valid. It also seems strange that so little evidence for the formation of positron emitters is found, although there would appear to be ample chance for their production. We are continuing the work, and expect to publish a fuller account later.

In conclusion, we wish to express our gratitude to Professor Ernest O. Lawrence for his particular interest in this investigation. The financial support of the Research Corporation, the Chemical Foundation, and the Josiah Macy, Jr. Foundation is greatly appreciated.

Internal Targets in the Cyclotron

Robert R. Wilson and Martin D. Kamen
Radiation Laboratory, Department of Physics, University of California, Berkeley, California
(Received October 3, 1938)

The large circulating ion current in the cyclotron which ordinarily does not reach the target chamber has been utilized in the preparation in high concentration of extremely active samples of radio-iron and radiophosphorus. By adapting as internal target the probes used in demonstrating the presence of the circulating current, it has been possible to enhance greatly the yields of radioactive materials as compared with those obtained in the usual manner at the target chamber. The construction of the probe, the nature of the circulating current and the manner in which the target material is applied to the probe are discussed. The general applications of the internal target technique are indicated.

Introduction

It is customary in atomic bombardments with the cyclotron to draw the ion currents into a target chamber by means of a suitable deflector arrangement. As recently shown,[1] the ion beams obtained in this manner represent only a small fraction of the ion currents circulating in the vacuum chamber proper. By the insertion of probes, it is possible to demonstrate the presence of many hundreds of microamperes of high energy ions whereas the ion beams emergent in the target chamber are smaller by an order of magnitude. Indeed, one may install several probes, used as internal targets, all of which are effective simultaneously in intercepting some of this large current without seriously altering the intensity of the beam in the target chamber. While the activations possible with the ion beams in the target chamber are sufficient for most purposes, there has arisen recently the need for extremely powerful samples of radioactive phosphorus and iron. Accordingly, we have investigated bombardment techniques based on the presence of the large circulating currents which ordinarily are not utilized. This investigation has disclosed a technique which has proven very successful in resolving many problems connected with the preparation of very active samples of radioactive phosphorus and iron and which should be quite general in application.

Construction of Probes

Although the details of construction of probes must necessarily vary from one cyclotron to another, a description of the particular probes used by us may be of some interest. At Berkeley the diametral edges of the accelerating electrodes, or dees, run from north to south, and probes can be introduced directly into the region between dees through windows located at the extreme north and south ends of the cyclotron vacuum chamber. Fig. 1 illustrates the south probe which consists essentially of a piece of copper tubing *aa* bent as shown. This provides support and water cooling for the end of the probe *b* which is subjected to the large ion bombardment. The end is covered by a copper jacket, hard-soldered in place, to which various target materials can be applied. It should be pointed out that it is necessary to use hard solder in any part of the probe exposed to the large ion beam or to the high radiofrequency field. Soft solder because of its poor heat and electrical conductivity as well as its low fusion temperature usually melts at exposed surfaces and causes arcing. The jog in the tubing *c* is so placed because otherwise the deflected beam which comes past this region would be intercepted.

A novel feature is the sliding vacuum seal *d* by means of which the probe is moved in and out. This consists of a piece of brass tubing *e* encasing the copper tubing. A hole of diameter about $\frac{3}{4}$ that of the brass tube is cut in the center of a circular piece of Duprene rubber. Upon being stretched around the brass tubing, the rubber is distorted near the tubing and is bent out as shown, *f*. This is then clamped between the two pieces of brass *g* so cut as to maintain the bend in the rubber outward from the vacuum. Thus

[1] Wilson, Phys. Rev. 54, 240 (1938).

 Reprinted from *Physical Review*, 54, 1031–1036 (1938).

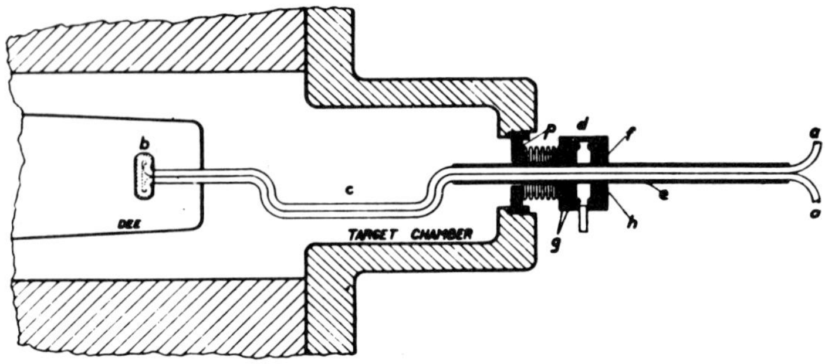

FIG. 1. The south probe.

the air forces the rubber against the tubing and so makes the seal. It is not necessary that the tubing be particularly straight, uniform, or scratch free for a good seal. A very small amount of stopcock grease on the tubing serves for lubrication. A second seal *h* is placed as shown behind the first seal and an airtight region left between the two. This provides for a means of evacuation behind the first seal in the event of a small leak. However, thus far this has proved unnecessary as the first seal acting alone holds even lower pressures than those used in the cyclotron. The probe can be moved in and out rapidly without disturbing the vacuum. Such a seal should find universal application wherever high speed pumping systems are employed, as it provides unlimited translational and rotational movements.

A sylphon connection between the sliding seal and the circular metal plate *p* which fits in place of a glass window provides for vertical motion of the end of the probe. The probe can be clamped at any position by means of a system of brass joints, clamps, and screw adjustments not shown in the diagram. The north probe is similar to the south probe, but there is no need for the jog in the copper tubing. A small spirit level is mounted on the outside of each probe in order to keep the target end vertical so as not to touch a dee.

At present, air must be let into the cyclotron chamber when a probe is installed. This could be avoided by an arrangement whereby the probe would first be installed in a small receptacle which could be roughly evacuated after installation and which could be connected to the cyclotron chamber by a vacuum gate. However, as the probe bombardments of the materials being activated at present extend over a period of several days each, no serious inconvenience is now caused by letting air into the cyclotron.

The Circulating Current

There are several methods of estimating the current of high energy ions striking the probe. The current can be measured directly by insulating the probe and filtering off the large radio-frequency current also picked up; the heating of the cooling water can be measured; or a rough estimate can be obtained from the activity induced in a substance whose radioactive yield is known. The first method has the disadvantages that low energy ions can be picked up and secondary emission can occur. In the second and third methods one must assume a value of the average energy of the ions. All three ways have been used here with consistent results, but the second, of measuring the water heating, is found to be the most convenient, and probably the most reliable. If Δt is the temperature rise of the water, F is its rate of flow, and V is the average energy of the incident ions in megavolts, then the current to the probe in microamperes is given by:

$$i = 4.18(F\Delta t/V).$$

The correct value of V to be used in the equation is uncertain because the ions do not all originate at the same place and because of wandering of ion paths due to inhomogeneities in the magnetic field. If for simplicity one assumes that the ions originate at the center and do not wander, the voltage will depend on the position of the probe as the square of the distance of the probe from the center. The values so assumed do not differ greatly from those obtained by dividing the product of V and i obtained from the heating measurement by the value of i obtained directly. Fig. 2 shows the current to the probe as measured by the heating plotted against the distance of the end of the probe from the center. As the probe is moved in from the edge of the dee, the current to it increases almost linearly with the distance for about the first ten centimeters, while the beam current decreases proportionally. When the beam current has been completely cut off, the current to the probe still increases, but not as rapidly as before. The manner in which the current increases as r becomes smaller can be ascribed to several causes: The ions do not originate at the same place; the ion paths wander because of magnetic field inhomogeneities; and ions are lost at lower energies by getting out of phase with the voltage.[2] Probably the most important of these is that the ions do not originate at the same place. This is so because of the length of the filament (about five cm) and because the electrons which form the ions can execute cycloidal motion in the crossed electric and magnetic field and wander several centimeters parallel to the gap between the dees—thus extending the region of ionization to about ten cm as the data indicate. Fig. 3 is also consistent with the occurrence of such a process. Here the current to two probes is shown; the north probe is fixed about five cm in from the north edge of the dee, and the current to each probe is plotted against the position of the south probe as the south probe is moved in. There is no shadow effect from the north probe, but as the south probe is moved in, it gradually robs the other probe of its current—taking most of it after being moved in ten or twelve centimeters.

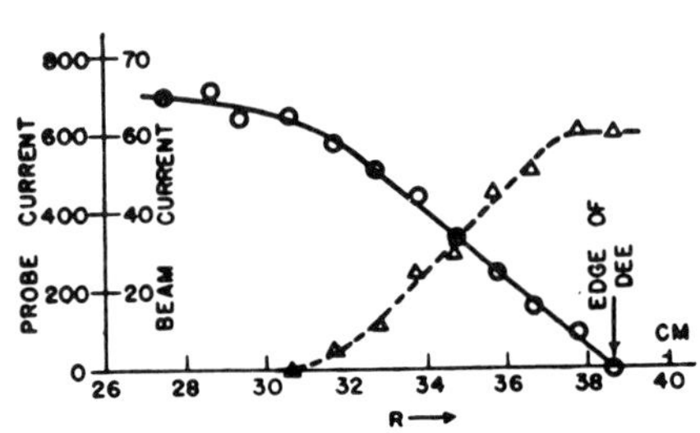

FIG. 2. The full curve is the current, in microamperes, to the south probe plotted against r, the distance of the end of the probe from the center of the cyclotron. The dotted curve represents the current, also in microamperes, that comes through the deflecting system to the target chamber plotted against r.

[2] Rose, Phys. Rev. 53, 392 (1938).

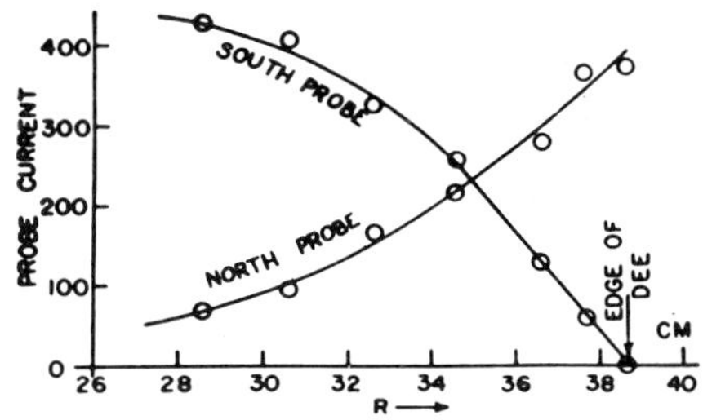

FIG. 3. The north probe is fixed in a position about 5 cm in from the periphery of a dee. The south probe is moved toward the center and the current in microamperes to each probe is plotted against the position of the end of the south probe.

The distribution of the beam falling upon the probe when it is located near the edge of a dee is shown in Figs. 4 and 5. These were obtained by passing the copper probe in front of a lead slit behind which was located an electroscope. Fig. 4 shows the vertical distribution, observed by placing the slit perpendicular to the leading edge of the probe. Because of the strong magnetic focusing most of the beam is confined within less than a centimeter. To obtain Fig. 5 which shows the horizontal distribution of ions on the probe, the slit was placed parallel to the edge. It is seen that the beam is spread over about one cm of the probe horizontally, whereas one would expect that it would spread over a distance corresponding to the change in radius of path an ion receives in being accelerated by twice the maximum voltage between dees. This turns out to be only about half of the observed spread, but numerical and graphical calculations show that the spreading out is just about what would be caused by the widening of successive paths because of the radially inhomogeneous magnetic field as sug-

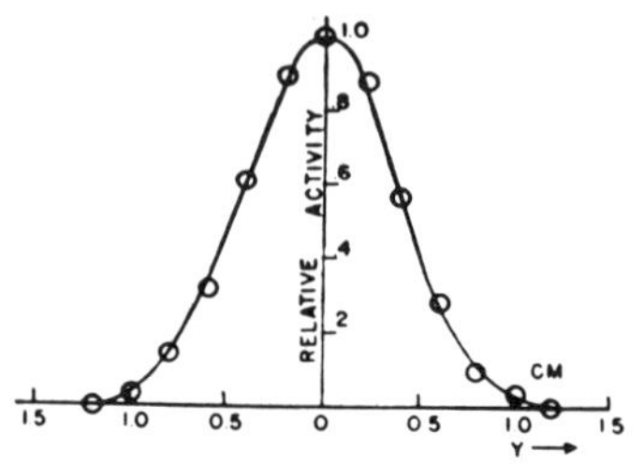

FIG. 4. The vertical distribution of the beam on the probe.

gested by Lawrence and Cooksey.[3] The maximum current to the probe as measured by the water-heating occurs at a slightly lower magnetic field strength than does the maximum of the beam current at the target chamber.

PRODUCTION OF RADIOACTIVE ISOTOPES BY PROBE BOMBARDMENT

The advantages inherent in probe bombardment when feasible are well exemplified in the two cases we have studied, radiophosphorus and radio-iron. These substances are produced by the deuteron bombardment of phosphorus and iron[4] according to the reactions

$$_{15}P^{31}+_{1}H^{2}=_{15}P^{32}+_{1}H^{1},$$
$$_{26}Fe^{58}+_{1}H^{2}=_{26}Fe^{59}+_{1}H^{1}.$$

A high energy beam (8 Mev) is necessary to produce the radio-iron (Fe^{59}) in appreciable quantities while the phosphorus reaction reaches good yields at 5.5 Mev. At present, the Berkeley cyclotron produces an 8-Mev beam, the intensity of which varies between 50 and 70 microamperes as measured at the target chamber. The activity required in phosphorus is of the order of 20 millicuries. To avoid serious limitations in therapeutic applications, the activity must be obtained with as little bulk of inactive material as possible (i.e., a high ratio of P^{32} to P^{31}).

To realize these yields with a 60-microampere beam at the target chamber, it is necessary to bombard the element itself as red phosphorus. An elaborate arrangement which has been found best consists of a knurled, water-cooled copper plate mounted with gaskets in an atmosphere of helium behind the thin aluminum window (5 cm in air equivalent stopping power) through which the beam emerges.[5] The phosphorus is pressed into the knurled grooves to a depth equal to the effective beam range in phosphorus and is then covered with thin gold-beaters foil (about 2 mm air equivalent stopping power). Thus, the phosphorus is in good thermal contact with a large, well-cooled surface. In this manner, it is found that thirty-five microampere-hours of bombardment at 8 Mev will yield one millicurie of activity. Hence 700 microampere-hours are required under these conditions to produce phosphorus samples strong enough for use in therapy. Moreover, since the effective portion of the beam is diffused over an area of approximately 10 cm^2, 1500–2000 milligrams of phosphorus are required to utilize the beam most effectively. Thus, one millicurie of activity is obtained for each 100 milligrams under the best conditions. For many purposes, it is desirable to obtain one millicurie per milligram of phosphorus. It can be appreciated readily that such a concentration of activity is far beyond the range of the target chamber bombardment.

In the case of iron, the problem is the production of enough activity even for "tracer" work. Because of the extreme rarity of Fe^{58}, the yield of Fe^{59} is extremely low, the best yields averaging one microcurie for 100 microampere-hours. Here again, a small bulk of material is required. It has been customary to bombard the iron simultaneously with the phosphorus by using an iron window (8 cm air equivalent stopping power) in place of the aluminum window. Unfortunately, it is quite difficult to prepare iron windows of this thinness completely hole-free, and those which are made do not last sufficiently long to produce strong iron samples. Moreover, the fact that only 8 cm of beam range is taken up in the iron reduces the activation by a factor of two.

All of these difficulties may be easily resolved by bombardment of ferrous phosphide on the probe. Ferrous phosphide as obtained from the City Chemical Corporation of New York is an easily-powdered, infusible mineral (melting point

[3] Lawrence and Cooksey, Phys. Rev. 50, 1131 (1936).
[4] Newson, Phys. Rev. 51, 624 (1937). Livingood, Fairbrother and Seaborg, Phys. Rev. 52, 135 (1937). Livingood and Seaborg, Phys. Rev. 54, 51 (1938).

[5] The final design of this chamber is due to Dr. F. N. D. Kurie, whose article is to appear shortly in the *Review of Scientific Instruments*.

about 1300°C) containing approximately 25 percent of phosphorus by weight. We have found that this material may be mounted satisfactorily in the following manner: A thin copper strip (2 cm ×1 cm) is tinned with "phos-copper," a hard solder alloy of copper and phosphorus. The copper is then placed on the part of the probe to be exposed and heated from below until the solder is molten. The ferrous phosphide powder is then sprinkled over the molten solder. On cooling, the mass is tamped free of excess phosphide. The residue on the surface represents about 100 milligrams of ferrous phosphide and is unusually hard and infusible. A surface of ferrous phosphide prepared in this way has withstood thousands of microampere-hours bombardment without the least deterioration. Tests have shown that the phosphide may be melted with an oxygen torch in air without noticeable decomposition. The presence of the phosphide in the vacuum chamber is not productive of gasiness and consequent unsteady operation.

The probe, having been prepared in this manner, is placed in position and adjusted. The adjustment is best described by reference to Fig. 2, an examination of which will indicate that 60–80 microamperes of current can be picked up on the probe without affecting the current at the target chamber. The final position of the probe will be governed both by the amount of current desired at the target chamber and the heating the target material on the probe can withstand safely. In the case of the phosphide, we have found that exposure to currents of 400 microamperes for several hours can be tolerated without serious deterioration. The procedure we consider best, however, is to insert the probe until it is taking out about 10 percent of the beam at the target chamber. Under these conditions the probe current as measured by the heating will vary between 150 and 200 microamperes. The length of time the probe is left in will vary from sample to sample depending on the half-life, yield desired, etc.

The chemical separation of iron and phosphorus from the other probe materials may be accomplished in a variety of ways depending on the final form in which the iron and phosphorus are desired. The following procedure may be recommended. The copper strip with its phosphide coating is dissolved in aqua regia with subsequent evaporation to dryness in excess HCl. The residue is then taken up in dilute HCl and the sulfides of copper and other heavy metals precipitated with H_2S. The filtrate is evaporated and the residue dissolved in a small amount of 6*N* HCl. The iron is extracted as iron chloride with ether saturated with 6*N* HCl.[6] The aqueous layer contains the phosphorus, cobalt and manganese as well as some nickel and zinc activities. The cobalt, manganese, nickel and zinc are present in microscopic quantities and it is advisable to add a minute amount of cobalt or manganese salts at this point to insure carriers for the cobalt and manganese activities. The solution should be evaporated, taken up in dilute nitric acid and the phosphorus precipitated several times as the phosphomolybdate. An alternative procedure consists in neutralizing the aqueous layer from the ether extraction with sodium hydroxide, the various "contaminants" such as cobalt and manganese then precipitating as phosphates. The amount of phosphorus lost in this way is negligible if the amount of cobalt added early in the procedure is very small. The phosphomolybdate precipitation is effective in recovering the phosphorus free of contaminants, but must be converted to sodium phosphate for use. This may best be accomplished by dissolving the phosphomolybdate in ammonia and precipitating the phosphorus as magnesium ammonium phosphate. The precipitate may be weighed and the calculated amount of NaOH added, the solution then being boiled until the ammonia no longer comes off. The solution is

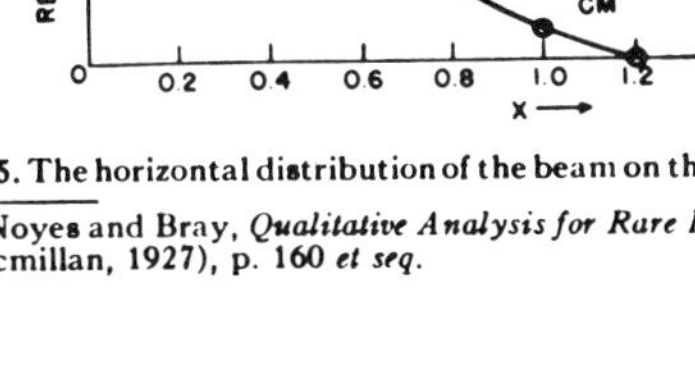

Fig. 5. The horizontal distribution of the beam on the probe.

[6] Noyes and Bray, *Qualitative Analysis for Rare Elements* (Macmillan, 1927), p. 160 *et seq.*

filtered from the magnesium hydroxide and contains the main portion of the phosphorus activity. The iron as extracted with ether must be washed with 6*N* HCl saturated with ether many times to ensure absence of contamination. It is advisable to have a small quantity of cobalt and phosphorus in the wash solution.

The yields obtained in this way show a remarkable enhancement compared to the yields observed in bombardments at the target chamber. In a day of operation, 10 millicuries of radiophosphorus have been obtained in 40 milligrams of phosphorus. Exposures of several days should enable one easily to realize many times this activity, thus meeting the desideratum of one millicurie per milligram. This represents a concentration factor of 100 over target chamber yields with the same total activity. For iron, we obtain one microcurie per day per 40 milligrams of inert iron. The largest activities observed with target chamber bombardment are of the order of three microcuries distributed over 90 milligrams of iron. Hence, an exposure of a week or more should result in greatly increased iron yields. In fact, we have found that in a two-week exposure, it is possible to obtain 20 microcuries per 40 milligrams of iron. In this case, pure iron metal was used on the probe. The yield of iron from ferrous phosphide we find to be half of that observed from pure iron, a result which is to be expected from the fact that only half the target material is iron in the ferrous phosphide bombardment. As by-products, large copper, cobalt, manganese, nickel and zinc activities are produced which may be of some utility.[7] It must be emphasized that all of these activities are produced without interfering with bombardments proceeding simultaneously in the target chamber, so that one cyclotron may satisfy the needs of both biologist and physicist without extremely long periods of operation.

[7] Several hundreds of millicuries of the short lived isotopes of the above-mentioned elements can be made while for the long period activities yields ranging from a few microcuries to many millicuries can be obtained.

The method of probe bombardment is capable of extension to a wide variety of elements. Beryllium and molybdenum have been tried and found to exhibit greatly enhanced activities. The neutron yields from the beryllium-deuteron reaction render intense neutron activations possible in reasonable periods of bombardment; moreover, a possible extension in the application of neutrons to therapy is indicated. The investigations of Doctors Segrè and Seaborg[8] on the activities in element 43 produced by deuteron bombardment of molybdenum have been expedited greatly by the extremely powerful samples of molybdenum prepared on the probe. It is proposed shortly to bombard sulfur and tellurium as sulfides and tellurides.[9] In short, any infusible substance which can be joined in good thermal contact with the probe material can be adapted to probe bombardment. It should be possible to prepare most elements in the periodic system either pure or as compounds, in such fashion as to satisfy probe requirements. Obviously the method of internal targets should find its most important application in the preparation of radio-isotopes which are long lived and difficult of activation, as well as in the demonstration of the existence of many radio-isotopes as yet undiscovered.

We wish to acknowledge our special indebtedness to Professor E. O. Lawrence, whose suggestion of the existence of a large beam initiated this research. The work has been greatly facilitated by the unstinting cooperation of our colleagues in the Radiation Laboratory and by the assistance of the W.P.A. The continued financial support of the Research Corporation has rendered this research possible.

[8] E. Segrè and G. T. Seaborg, Phys. Rev. **54**, 772 (1938).
[9] *Note added in proof:*—Since this article was written, tellurium has been bombarded successfully on the probes. By fusing tellurium into thin copper strips, Dr. J. G. Hamilton has obtained satisfactory telluride targets which, on exposure to the internal beam, show enhancement of the yield of radio-iodine (period 8 days) by a factor of 3 to 5, as expected.

Long-Lived Radioactive Carbon: C^{14}

S. RUBEN AND M. D. KAMEN
Radiation Laboratory and the Department of Chemistry, University of California, Berkeley, California

(Received December 2, 1940)

A long-lived activity has been obtained by deuteron bombardment of graphite. This activity has been shown by chemical methods to be isotopic with carbon. The same activity has been obtained by slow neutron irradiation of nitrogenous compounds (NH_4NO_3, NH_4NO_2, etc.). The radiations consist of low energy negative electrons with a range of 19±2 mg/cm² aluminum (145±15 kev upper energy limit). No soft or hard gamma-rays could be detected. The nuclear reactions must be C^{13} (*d*, *p*) C^{14} and N^{14} (*n*, *p*) C^{14}. A sample has been followed in a screen wall Geiger counter for 9 months and has shown no decay. From cross section and yield estimates the half-life of C^{14} is probably 10^3–10^5 years. The production, isolation, and detection of C^{14} for tracer work is described.

THERE are 5 known isotopes of carbon, stable C^{12} (~99 percent abundance), stable C^{13} (~1 percent abundance), C^{10} (8.8-second half-life),[1] C^{11} (21-minute half-life),[2] and C^{14} (half-life 10^3–10^5 years).[3] The production and properties of C^{14} will be described in this paper. In view of the possible application of C^{14} as a tracer,[4] a discussion of relative yields and specific activities obtainable is included together with a brief description of detection methods.

It is of interest to review briefly the history of C^{14}. In 1934 Kurie, using a Wilson cloud chamber, studied the fast neutron disintegration of nitrogen and observed in addition to the known[5] alpha-particle emission, another type of transmutation,[6] less frequent, in which singly charged heavy particles were emitted. Kurie suggested these were protons due to

$$n_0^1+N_7^{14}\rightarrow C_6^{14}+H_1^1+Q_1. \qquad (1)$$

Later Bonner and Brubaker,[7] irradiating nitrogen with slow neutrons, observed a few heavy tracks (~1.06-cm range) which they attributed to alpha-particles. Burcham and Goldhaber[8] demonstrated these tracks were protons and suggested they originated from reaction (1). Bonner and Brubaker[9] agreed with Burcham and Goldhaber and moreover estimated $Q_1=0.58\pm0.03$ Mev. From the known mass difference between neutron and proton, C^{14} was estimated to be unstable with respect to N^{14} by ~0.17 Mev so that it could be expected that C^{14} would be a beta-emitter of relatively long half-life. McMillan[10] found a long-lived soft radiation from metal scraped from inside the cyclotron vacuum chamber and suggested it might be due to C^{14} formed by the reaction

$$D_1^2+C_6^{13}\rightarrow C_6^{14}+H_1^1+Q_2. \qquad (2)$$

Unfortunately, the sample was accidentally lost before any chemistry was performed. However, the upper energy limit of the beta-radiation was measured by Libby and Lee[11] using a screen wall Geiger counter and found to be ~10 kev. (It seems quite likely this activity was due to H^3 which was discovered by Alvarez and Cornog.[12]) Shortly after, the writers had occasion to prepare N^{13} (by deuteron bombardment of graphite) for use as a tracer in nitrogen fixation experiments. The same graphite target was used in all bombardments and after ~1000 μ amp. hr. of 8-Mev

[1] Barkas, Creutz, Delsasso, Fox and White, Phys. Rev. **57**, 562 (1940).
[2] Cockcroft, Gilbert and Walton, Proc. Roy. Soc. (London) **A148**, 225 (1935).
[3] S. Ruben and M. D. Kamen, Phys. Rev. **57**, 549 (1940); **58**, 194 (1940).
[4] C^{11}, despite its short half-life has already been used rather extensively as a tracer, cf. G. T. Seaborg, Chem. Rev. **27**, 199 (1940).
[5] N. Feather, Proc. Roy. Soc. (London) **A136**, 709 (1932).
[6] F. N. D. Kurie, Phys. Rev. **45**, 904 (1934); **46**, 330 (1934).
[7] T. W. Bonner and W. M. Brubaker, Phys. Rev. **48**, 469 (1935).
[8] W. E. Burcham and M. Goldhaber, Proc. Camb. Phil. Soc. **32**, 632 (1936).
[9] T. W. Bonner and W. M. Brubaker, Phys. Rev. **49**, 223 (1936); **49**, 778 (1936).
[10] E. M. McMillan, Phys. Rev. **49**, 875A (1936).
[11] W. F. Libby and D. D. Lee, Phys. Rev. **55**, 245 (1939).
[12] L. W. Alvarez and R. Cornog, Phys. Rev. **56**, 613 (1939).

deuterons it was burned to CO_2 which was precipitated as $CaCO_3$. A fraction of this material, when mounted on the inner surface of a cylinder and counted in a screen wall counter of the type developed by Libby,[13] gave an effect of 16±2 counts/min. It was possible to show this effect was not due to natural contamination of the chemical reagents used. It seemed desirable, however, to obtain stronger samples before drawing any positive conclusions. It was apparent that a prolonged bombardment was necessary and, due to the heavy schedule under which the cyclotron was operating, this was not possible until recently. In the meantime B_2O_3 was bombarded in a gas-tight target chamber with 32-Mev alpha-particles accelerated in the 60″ cyclotron in a search for C^{14} produced by

$$_5B^{11}+_2He^4\rightarrow{}_6C^{14}+_1H^1+Q_3. \quad (3)$$

It had been observed that when B_2O_3 is bombarded with deuterons the major fraction of the newly formed C^{11} is expelled from the target as CO.[14,15] After ~10 μ amp. hr. of alpha-particle bombardment, the gas from the target chamber was introduced into an ionization chamber connected to an FP-54 electrometer tube and amplifier circuit.[16] No evidence for any long-lived activity was found.

At this time Pollard[17] and Bower and Burcham[18] presented additional evidence for the production and instability of C^{14}. Pollard investigated the protons from reactions (2) and (3). Using 3-Mev deuterons, the longest proton range was 81.1±2 cm yielding $Q_2=8.21$ Mev. Bombardment of boron with ThC alphas produced protons with a 97.6-cm range; Po alpha-particles gave a proton group of 42.1-cm range. These experiments yielded an average $Q_3=0.66$ Mev and using these values he proposed the mass of C^{14} to be 14.00780±0.00020 and thus unstable with respect to N^{14} by ~300 kev. His attempts t[o] detect an activity ascribable to C^{14} afte[r] ~6μ amp. hr. bombardment of graphite wit[h] 3-Mev deuterons were unsuccessful. Bower an[d] Burcham, using 0.8-Mev deuterons, found a 48 cm group of protons which they attributed t[o] reaction (2) with $Q_2=6.1$ Mev. Recently Bennett[,] Bonner, Hudspeth and Watt[19] have confirme[d] this observation, using targets enriched in C^{13}.

[13] W. F. Libby, Phys. Rev. **46**, 196 (1934).
[14] Yost, Shinohara and Ridenour, J. Chem. Phys. **3**, 133 (1935).
[15] Ruben, Kamen and Hassid, J. Am. Chem. Soc. **61**, 661 (1939); **62**, 3443 (1940).
[16] We are indebted to Dr. E. Segrè for the loan of the electrometer and for aid in many of the measurements made with it.
[17] E. Pollard, Phys. Rev. **56**, 1168 (1939).
[18] J. C. Bower and W. E. Burcham, Proc. Roy. Soc. (London) **A173**, 391 (1939).

Isolation of C^{14} After Prolonged Deutero[n] Bombardment

By resorting to the method of internal targets[20] it was finally found possible to produce appreci[-] able amounts of an activity which was prove[d] chemically to be isotopic with carbon. A prob[e] target, covered with a small quantity of Aquadag was inserted into the vacuum chamber an[d] subjected to a 40-μ amp. beam of 7–8 Me[v] deuterons for ~120 hr. The probe was frequentl[y] removed and examined and the Aquadag renewe[d] whenever necessary. The target prepared in thi[s] way proved unsatisfactory for prolonged bom[-] bardment and was finally removed and examine[d] for activity. For a second bombardment, th[e] graphite was electroplated with copper and har[d] soldered to the probe surface. The graphite wa[s] of a type prepared for use as an anode in a hig[h] power rectifier tube. This target proved mor[e] satisfactory and was able to withstand 500–600 [μ] amperes of ion current without appreciabl[e] deterioration. This probe was placed far into th[e] vacuum chamber to intercept high current[s] (~500 μ amperes) of 3–4-Mev deuterons unti[l] 13,000 μ amp. hr. had been accumulated.

Since high specific activities are necessary fo[r] many tracer experiments it is apparent that th[e] bombardment of carbon enriched in C^{13} i[s] desirable. Professor H. C. Urey and Dr. D. W[.] Stewart very kindly sent us 0.63 gram of ~2[…] percent C^{13} as amorphous carbon prepared b[y] the reduction of enriched NaCN with magnesiu[m] at ~400°C. A portion of this carbon was place[d]

[19] Bennett, Bonner, Hudspeth and Watt, Phys. Rev. **58** 478 (1940). Holloway and Moore (Phys. Rev. **57**, 1086A (1940)) were unable to detect the protons from $C^{13}(d, p)C^{14}$ They placed a lower limit of 1.2×10^{-28} cm² on the valu[e] for the cross section of this reaction at 1 Mev.
[20] R. R. Wilson and M. D. Kamen, Phys. Rev. **54**, 103[…] (1938).

in a steel mold and subjected to high pressure[21] (40 tons/in.^{2}). The material failed to cohere and examination under a microscope revealed a granular structure quite different from graphite found to press well. Since no equipment for very high temperature work was available, the carbon was sent to the Edgewater Works of the National Carbon and Carbide Corporation, at the suggestion of Dr. H. C. McPherson. In these laboratories, after considerable effort, Dr. McPherson succeeded in preparing several satisfactory targets. The procedure employed by Dr. McPherson involved dilution of the amorphous carbon with a suitable amount of pitch binder. The mixture was pressed into a mold to make a slab 2 mm thick and 1 cm square. The slab was baked at 1200°C and then "graphitized" (heated in vacuum) at 2800°C to drive off all volatile matter. The resultant material contained 9.3 percent C^{13}, a reduction by a factor of more than two from the starting material. (Such loss is unavoidable with small quantities of carbon). The slab was sprayed with molten copper (Schoop process) to provide a tenacious metallic backing for soldering the material to the water-cooled probe surface. It was necessary in some instances to repeat the spraying since the slabs showed a tendency to embrittle and disintegrate when reheated. The slabs which survived this treatment possessed about $\frac{1}{5}$ the heat conductivity of the best graphite obtainable and proved to be adequate as target material. One slab mounted with phos-copper solder on a probe withstood 10 hours' bombardment with 300 μ amperes of 3–4-Mev deuterons and showed no visible deterioration.

Chemical Identification

The graphite was burned in a stream of oxygen and the resultant gases passed over hot cupric oxide and finally absorbed in $Ca(OH)_2$. The precipitated $CaCO_3$ was filtered and carefully washed. The filter paper containing the precipitate, when placed over a thin-walled aluminum Geiger counter (0.1 mm wall thickness), failed to show any activity. A portion ($\frac{1}{3}$) of this sample, spread out in a thin film inside a cylinder which was introduced into a screen wall counter, was active (~400 counts/min.). In the counting arrangement used, the cylinder containing the sample could be slid to and fro over the counting region, thus enabling alternate readings of sample activity and background to be taken. The procedure was repeated using ordinary $CaCO_3$ prepared in the same way to test for activity in the reagents. The $CaCO_3$ precipitate obtained in this way showed no activity.

The active $CaCO_3$ was treated with acid and the CO_2 collected. After drying, this was introduced into a Geiger counter and gave an effect of ~2000 counts/min. The 5-fold increase in activity was due to increased solid angle and elimination of self-absorption. The active gas was absorbed in $Ca(OH)_2$ and reconverted into CO_2 several times with no loss in specific activity. However a preliminary measurement indicated the range of the radiations in aluminum to be approximately 12 mg/cm^2; which is close to the known value for S^{35}. The procedure used to this point did not exclude the possibility that part or all of the activity was due to $S^{35}O_2$. Carrier (inactive) SO_2 was mixed with the active gas and the mixture treated with $KMnO_4$ in sulfuric acid solution to oxidize the SO_2 to SO_4^{--}. There was no loss in gaseous activity. In another experiment the mixture of the gaseous activity and inactive SO_2 was absorbed in alkali and the SO_3^{--} oxidized to SO_4^{--} with I_3^-. The CO_2 was liberated by acidifying the solution and absorbed in $Ca(OH)_2$. The $CaCO_3$ was active, there being no loss in specific activity.

Production of C^{14} by Neutron Transmutation of Nitrogen

It was realized from the very beginning that deuteron bombardment of carbon was not the best method for preparing C^{14} in high specific activity suitable for tracer work. The transmutation of N^{14} by slow neutrons (reaction 1) offered many attractive features: (1) Possibility of obtaining very high specific activity since the target material is not carbon but nitrogen. (2) Production of C^{14} is achieved as a by-product of cyclotron operation and thus does not interfere with other research. (3) Any strong neutron source will suffice. It is not necessary to go to high bombarding voltages (3–4 Mev as is the case for

[21] We are indebted to the Department of Civil Engineering for cooperation in these experiments.

deuteron bombardment). (4) The nitrogenous material, also containing some hydrogen, can profitably replace water or paraffin as a shielding material.[22]

While the slow neutron experiments are by no means complete, the results obtained thus far conclusively indicate this to be the most practical method for C^{14} production.

Two five-gallon carboys filled with saturated solutions of ammonium nitrate were plaeed in the region opposite the deflector of the Berkeley medical cyclotron. The solutions were irradiated with neutrons during a period of ~6 months. Approximately 40,000 μamp. hr. of deuteron bombardment at 16 Mev of various targets (principally beryllium and phosphorus) took place in this time. A chemical analysis was performed in which carbon monoxide, carbon dioxide, methane, carbon, methanol, cyanide and formaldehyde were added as carriers and then separated and analyzed for radioactivity. After vigorous and prolonged shaking the gaseous compounds were pumped off. The gases were oxidized to carbon dioxide and absorbed in $Ca(OH)_2$. The $CaCO_3$ was very active, containing 10^5 counts/min. By the chemical methods described above the activity was shown to be isotopic with carbon.

Charcoal and other sediment was separated from the NH_4NO_3 solution by filtration. This proved to be completely inactive. Cyanide was precipitated as AgCN which also was inactive. Methanol and formaldehyde were distilled from an aliquot (~10 percent) of the NH_4NO_3 and the distillate divided into 2 portions. From one portion the insoluble hydrazone of formaldehyde prepared by the addition of excess 2,4 dinitrophenyl hydrazine was found to be inactive. The methanol from the other portion was volatilized and passed over hot cupric oxide. The resulting CO_2 was not active. For future experiments a more complete and elaborate chemical analysis has been devised.

In addition to NH_4NO_3 solutions, several samples of solid NH_4NO_3 (~1 lb.) as well as concentrated NH_4NO_2 solutions have been irradiated with neutrons. C^{14} was found only in the

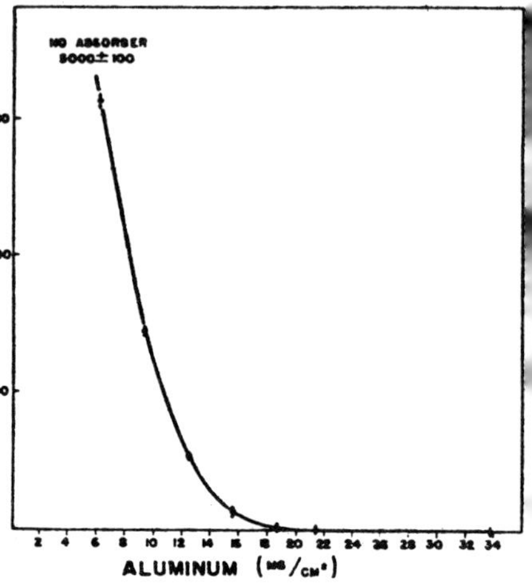

FIG. 1. Absorption of radiation from C^{14} in aluminum.

gaseous fraction. The solid NH_4NO_3 was irradiated in sealed soft glass containers. No volatile activity could be pumped off the solid even after addition of carrier gases. However, when the solid was dissolved in water the gaseous fraction contained 1000 counts/min. Evidently the gases are firmly trapped and are released only when the crystals are completely dissolved.

A large number of other nitrogen compounds are being irradiated not only to determine the best conditions for large scale[23] C^{14} production but also to investigate the chemistry of the recoiling carbon as a function of the chemical environment. Experiments are now in progress with urea, melamine, dicyandiamide, etc.[24]

PROPERTIES OF THE LONG-LIVED RADIATION

The absorption of the radiation in Al was measured by wrapping the screen wall of the Geiger counter with thin Al foils. With the first CaC^*O_3 sample (prepared by deuterons on carbon) which had an activity of only 40 counts/min. (background was 120/min.), the activity coming through 12 mg/cm² Al was 0±

[22] C. Lapointe and F. Rasetti, Phys. Rev. 58, 554 (1940), have found the total capture cross section of nitrogen for thermal neutrons to be 1.2×10^{-24} cm². Hydrogen has a cross section of 0.2×10^{-24} cm².

[23] It is a matter of common knowledge that NH_4NO_3 is thermodynamically unstable. However, it seems certain that its solution at room temperature offers no explosion hazard.

[24] We are indebted to the American Cyanamid Corporation for the gift of large quantities of these materials.

counts/min. However with a very thin layer ($\sim$0.1 mg/cm^2) of CaC^*O_3 (produced from NH_4NO_3) which had an activity of 5000 counts/min., 0.5 percent$\pm$0.1 percent of this effect was still observable at 13 mg/cm^2. The first estimate of the range was too low and it seems a more accurate value is 19$\pm$2 mg/cm^2. The absorption curve is shown in Fig. 1. Since the activity coming through 19 mg/cm^2 is <0.1 percent of the initial effect it is safe to conclude that γ-rays of $\sim$100 kev or more are absent. If the radiation consisted wholly of positrons the annihilation γ-rays should have given an effect of 100 counts/min. coming through the Al.[25] A comparison of the absorption in Cellophane and Al indicates the absence of soft γ-rays. 3.26 mg/cm^2 of Cellophane absorbed 69$\pm$4 percent while the same amount of Al absorbed 71$\pm$4 percent of the radiation.[26] Therefore it seems safe to conclude negative electrons are being emitted and the unstable isotope is C^{14}. If the upper energy limit of the beta-ray spectrum is taken to be 145,000$\pm$15,000 ev and the mass of N^{14} to be 14.007526,[27] the mass of C^{14} is calculated to be 14.00768. It is also possible, with the neutron-proton mass difference,[27] to determine $Q_1 = 0.64$ Mev. The estimate of Bonner and Brubaker[9] of 0.58$\pm$0.03 Mev is in agreement with this value.

The activity of a sample has been followed for 9 months and no decay observed. From this fact alone the half-life must be longer than 30 years. It is possible to make rough estimates of the half-life from the yields and cross sections. The $C^{13}(d, p)C^{14}$ experiments indicate 10^3–10^4 years while the $N^{14}(n, p)C^{14}$ data suggest 10^4–10^5 years. A lower limit of 1000 years is not unreasonable. Using the same technique employed in the H^3 work,[28] (${}_0n^1 + {}_3Li^6 \rightarrow {}_2He^4 + {}_1H^3$), Goldhaber concludes[29] the half-life to be longer than 300 years.

The upper energy limit of $\sim$150 kev coupled with the long half-life ($\sim 10^3$ years) is quite unexpected. It would have been supposed that the decay of C^{14} to N^{14} would be an allowed transition analogous to the He^6–Li^6 case. Dr. P. Morrison has calculated that on this basis, a maximum half-life value of the order of a day is to be expected. Since the actual half-life is quite certainly a factor of 10^5–10^6 greater it follows that C^{14}–N^{14} decay involves a forbidden transition. In consequence,[30] it must be supposed that configurations with high angular momentum make appreciable contributions to the ground state of C^{14}. A similar[31] situation has been found in the case of Be^{10}, a nucleus which might be expected to behave like He^6, i.e., decay with a short half-life. Actually, despite the high upper energy limit of the emitted electrons ($\sim$580 kev), Be^{10} seems to have a very long half-life (about 10^3–10^5 years).

It is of interest to note that because of their long half-lives the production of macroscopic amounts of C^{14} and Be^{10} is at hand and in the near future it may be possible to determine their nuclear spins, etc. If the half-life of C^{14} is assumed to be 10^4 years, then 0.1 microgram is present in the sample prepared from the transmutation of nitrogen.

In conclusion, a few remarks may be made regarding the measurement of C^{14}. For small samples (up to $\sim$50 mg), counting as gaseous CO_2 seems to be the simplest and most sensitive method. Carbon dioxide, with a trace of foreign gas (H_2O, C_2H_5OH, O_2, etc.) has been found[32] in this laboratory to be a quite satisfactory counter gas. However, since CO_2 requires relatively high counting voltages, the source of very steady high potential may be a limiting factor. For the measurement of large samples of low specific activity the use of thin carbonate (Li_2CO_3, etc.) layers inside a screen wall counter is the most efficient technique.

In many respects it is unfortunate that the half-life of C^{14} is so long, since, for the production of strong samples, long time exposures are required. On the other hand, it is possible to concentrate C^{14} by various isotope separation methods (chemical exchange, thermal diffusion, ultracentrifugation, etc.) and thus material of low

[25] This estimate is based on measurements with a positron emitter (Na^{22}) under the same conditions.
[26] A more detailed investigation of the spectrum by Dr. J. Halpern and Mr. R. C. Raymond, using a Wilson cloud chamber, is now in progress.
[27] Hahn, Flügge and Mattauch, Ber. 73, 1 (1940).
[28] R. D. O'Neal and M. Goldhaber, Phys. Rev. 57, 1086 (1940).
[29] Conference on Applied Nuclear Physics, Massachusetts Institute of Technology, 1940.

[30] We are indebted to Professor J. R. Oppenheimer and Dr. L. I. Schiff for valuable discussions on this point.
[31] E. M. McMillan and S. Ruben, unpublished.
[32] T. H. Norris, private communication.

specific activity (including the end products of tracer experiments), can be reconcentrated. A given sample of C^{14} (and C^{13}) may, therefore, be used repeatedly. When the use of C^{14} and C^{13} becomes widespread enough, it should be feasible for a commercial company to perform this task at a nominal cost.[33]

We cannot adequately express our gratitude to the members of the Chemistry and Physics Departments, as well as many others, for their help and encouragement. It is a pleasure to thank Professor E. O. Lawrence and the staff of the Radiation Laboratory for their assistance and cooperation. Financial assistance from the Research Corporation and the Rockefeller Foundation to the Radiation Laboratory is gratefully acknowledged.

---

[33] The Eastman Corporation, at the suggestion, and with the cooperation, of Professor H. C. Urey, has begun the concentration of N^{15} on a small scale. It is reasonable to believe that the interest and demand for carbon tracers will far exceed the supply of the radioactive isotopes and, therefore, the large scale concentration of C^{13} is of considerable importance.

Note added in proof.—The estimate for the C^{14} half-life was made by three different methods (1) the yield of C^{14} coupled with an extrapolation of Amaldi's curve (cf. Bethe's review) for the $C^{12}(d, p)C^{13}$ reaction and Pollard's value for the yield ratio $C^{13}(d, p)C^{14}/C^{12}(d, p)C^{13}$, (2) the $C^{13}(d, p)C^{14}$ cross section was estimated by comparison with the $B^{10}(d, n)C^{11}$ yield at bombarding energies of 7–8 Mev, (3) the C^{14} yield from the slow neutron irradiation of NH_4NO_3. Method (1) gave $2 \cdot 10^4$ years while (2) yielded $2 \cdot 10^3$ years. Schultz and Watson, Phys. Rev. 58, 1047 (1940), have found Pollard's $C^{13}(d, p)C^{14}/C^{12}(d-p)C^{13}$ yield ratio was high by a factor of 5–10. This seems to bring the first two C^{14} half-life estimates into agreement at 10^3 years. The $N^{14}(n, p)C^{14}$ experiments suggest a still longer half-life for C^{14}. Hence it seems reasonable to regard 10^3 years as a lower limit.

THE UTILIZATION OF GLYCINE IN THE BIOSYNTHESIS OF HEMOGLOBIN*

By MOISÉS GRINSTEIN, MARTIN D. KAMEN, AND CARL V. MOORE

(From the Department of Internal Medicine and the Mallinckrodt Institute of Radiology, Washington University School of Medicine, St. Louis)

(Received for publication, December 27, 1948)

Isotopic studies have demonstrated that glycine is used in the biosynthesis of hemoglobin. Glycine has been tagged in one of three ways: (*a*) with N^{15} in the amino group (1–3), (*b*) with C^{14} in the α-methylene carbon position (4), and (*c*) with C^{14} in the carboxyl group (5). When any of these three forms of glycine is fed to animals, the isotope is incorporated into hemoglobin, but interesting differences have been observed in the distribution of the isotope between the pigment and protein portions of the molecule. For instance, Shemin and Rittenberg, using glycine labeled with N^{15}, found greater concentrations of the heavy nitrogen in hemin than in red blood cell protein (2). Similarly, Altman and his associates observed that when the methylene carbon of glycine is tagged 7 to nearly 10 times as much C^{14} appear in hemoglobin protoporphyrin as in globin (4). On the other hand, recent observations in this laboratory have shown that when glycine containing C^{14} in the carboxyl group is administered the radioactive carbon is synthesized into globin but cannot be demonstrated in protoporphyrin (5).

The experiments to be described in this report were designed (1) to confirm the demonstration that the carboxyl carbon of glycine is used for the biosynthesis of globin but not protoporphyrin, (2) to determine whether globin within the red cell participates in protein interchange, and (3) to discover whether coproporphyrin I isolated from the urine and feces after the tagged glycine was fed would contain C^{14} even if the hemoglobin protoporphyrin did not.

Materials and Methods

Two healthy dogs and one rat were fed glycine tagged with C^{14} in the carboxyl group.[1]

Dog I, male, weighed 12 kilos. By removing 150 to 200 ml. of blood from the femoral artery on each of several days, this animal's packed

* Supported by a grant from Mr. John Mosby and one established in memory of Mr. Nathan Greenberg.

[1] The authors are indebted to Dr. R. B. Loftfield of the Massachusetts Institute of Technology for the synthesis of the labeled glycine.

erythrocyte volume was reduced from 51 to 29 per cent. 100 mg. of the labeled glycine were then given by stomach tube in divided doses on 3 successive days; the total radioactivity administered amounted to 5×10^7 counts per minute, assayed by a procedure described elsewhere (5, 6). During these 3 days, water was allowed *ad libitum* but all food was withheld. Thereafter, the stock ration of Purina dog chow was again fed. At intervals until the experiment was terminated on the 153rd day, 10 ml. samples of blood were obtained for analysis. Crystalline protoporphyrin IX dimethyl ester (7) and globin (8) were isolated for determination of radioactivity (5, 6).

Dog II, female, weighed 10 kilos. The packed red blood cell volume was reduced from 53 to 24 per cent by serial bleeding. The labeled glycine was again given by stomach tube in three divided doses within a period of 27 hours; the total dose was 50 mg. and the total radioactivity amounted to 1.7×10^7 counts per minute. Food was withheld during the period of administration. Blood was collected at intervals in 10 ml. samples as in the previous experiment for the same determinations, but observations were stopped at the end of 110 days. In addition, urine and feces were collected during the first 15 days following administration of the glycine and extracted for coproporphyrins (9). After purification and identification (10) coproporphyrin I was tested for radioactivity with a nucleometer.[2]

One white rat weighing 100 gm. was kept on a protein-free diet (2) for several days and then given by stomach tube, in four divided doses over 3 days, 200 γ of the labeled glycine (total radioactivity 2.1×10^5 counts per minute). 20 days later the animal was bled to death. From the 4.6 ml. of blood obtained, crystalline protoporphyrin IX dimethyl ester was isolated for determination of its radioactivity. Measurement was also made of the radioactivity in the remaining cell protein.

All samples were assayed without resorting to conversion to carbonate. Satisfactory checks were obtained in pilot experiments in which samples were assayed both as carbonate and as untreated material. All samples were measured in triplicate and values reported are averages, the spread of which is included in the errors assigned.

Results

No radioactivity could be demonstrated in the protoporphyrin isolated from the blood of any of the three animals. The activity in the globin

[2] The authors are indebted to Professor A. I. Lansing, Department of Anatomy, Washington University School of Medicine, for permission to use the nucleometer, a product of the Radiation Counter Laboratories, Chicago, Illinois. The nucleometer was calibrated with standard samples of labeled glycine assayed previously with the standard, thin-walled Geiger-Müller tube (5).

obtained from the dogs was at a maximum on the 1st day determinations were performed, the 23rd and 13th days, respectively (Fig. 1 and Table I). The concentration remained relatively constant for 77 days in Dog I and

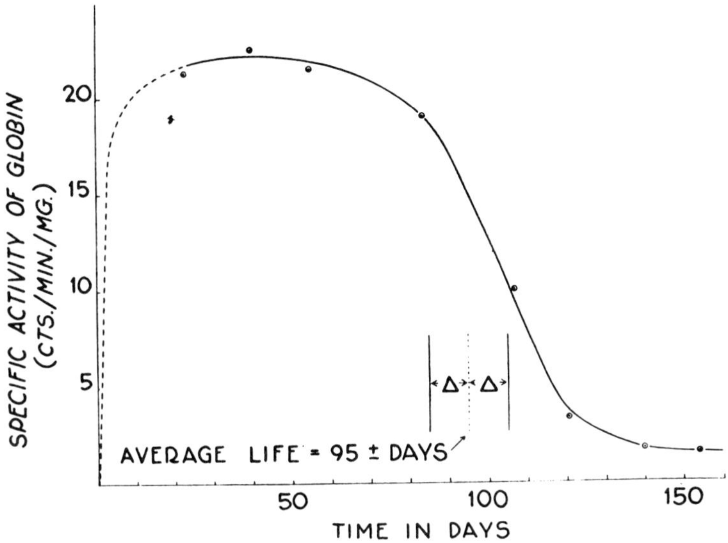

FIG. 1. Radioactivity in globin prepared from blood of Dog I after oral administration of glycine tagged with C^{14} in the carboxyl position. Δ represents half the period required for half the cells to die.

TABLE I

Radioactivity in Hemoglobin Protoporphyrin and Globin of Dog II after Oral Administration of Glycine Containing C^{14} in Carboxyl Group

Time	Radioactivity, counts per min. per mg. in	
	Protoporphyrin IX dimethyl ester	Globin
days		
13	<0.1	11.2 ± 0.6
33	<0.1	10.1 ± 0.5
47	<0.1	11.7 ± 0.4
66	<0.1	9.5 ± 0.5
80	<0.1	3.9 ± 0.1
93	<0.1	3.7 ± 0.2
110	<0.1	3.2 ± 0.2

until about the 66th day in Dog II. Thereafter, the radioactivity fell rather abruptly to low levels. In Fig. 1 are plotted the results obtained for Dog I. When mathematical analysis of this curve was made in the manner described by Shemin and Rittenberg (2) to estimate the average survival time of red blood cells of the dog, a figure of 95 ± 5 days was obtained. The period in which half of the red cells was destroyed extended from the 85th to the 105th day, as shown in Fig. 1. The average survival time of the red blood cells for Dog II, determined in the same manner, was 75 ± 5 days. The cell protein from 4.6 ml. of the rat's blood contained 1.5 ± 0.5 counts per minute per mg., a value reasonably close to that to be expected from the initial dose of glycine administered, in light of the experience with uptake into globin in the dogs.

From the urine and feces collected for 15 days following the last administration of glycine to Dog II, 310 γ of crystalline coproporphyrin I tetramethyl ester were isolated. No radioactivity was observed.

DISCUSSION

These results confirm the previous demonstration that in the biosynthesis of hemoglobin the carboxyl carbon of glycine is incorporated into globin but not into protoporphyrin. Since other work has shown that both the amino nitrogen (1–3) and the methylene carbon (4) of glycine are introduced into protoporphyrin during its formation, it would appear that only the $—CH_2 \cdot NH_2$ group of glycine becomes a part of protoporphyrin. If the whole of the glycine molecule is involved in the synthesis of protoporphyrin, CO_2 from the carboxyl group must be split off in the process.

On the other hand, since the amino nitrogen (2) and the methylene carbon (4) atoms as well as the carboxyl carbon are incorporated into globin, it seems likely that glycine as a whole is utilized for globin synthesis. The data available do not permit any conclusion as to whether CO_2 detached from glycine can be used for globin formation (11).

When the radioactivity of globin is plotted against time, the shape of the resulting curve is very similar to that obtained for protoporphyrin when the concentration of N^{15} is plotted against time (2). This result indicates that the globin of intact erythrocytes remains in the red cells without participating in protein interchange until the erythrocyte is destroyed.

The radioactivity of the globin fraction did not fall off with time to as low a value as might be expected on the basis that C^{14} from degraded globin was not reutilized for globin synthesis. Insufficient data are at hand to decide to what extent such reutilization occurred. It seems unlikely, however, that simple reutilization of glycine from degraded globin can be reconciled with results such as those shown in Fig. 1. Miller, Robscheit-Robbins, and Whipple have reported evidence indicating that globin contributes to the "protein pool" of the body, and that this "protein pool" in turn is used for the formation of new hemoglobin (12). When hemoglobin is injected intraperitoneally into dogs made both anemic and hypoproteinemic, new hemoglobin is formed. The conditions of these experiments, however, were quite different from those reported in the present paper. The two dogs used were not anemic at the conclusion of the experiment and had been fed a stock ration known to be nutritionally adequate.

The average times of survival of the red cells in Dogs I and II, as measured by the curves of radioactivity in globin, were 95 ± 5 days and 75 ± 5 days, respectively. Hawkins and Whipple estimated the life span of dog red cells to be about 124 days (13). These workers forced great numbers of new red cells into the circulation of four bile fistula dogs by blood destruction or blood withdrawal; as a consequence, the bile pigment output fell and did not rise again for 112 to 133 days. This interval was regarded as an indication of the survival time for the red cells. No conclusion is drawn regarding the difference between these figures and the ones calculated from the globin curves. More observations would be necessary to establish a more accurate average figure and the expected limits of variation.

Since available evidence indicates that coproporphyrin I is formed as a by-product during the synthesis of protoporphyrin IX (type III) for hemoglobin formation, and since hemoglobin protoporphyrin did not assimilate any radioactivity from the C^{14} of the carboxyl-tagged glycine, it is not surprising that the coproporphyrin I isolated from Dog II likewise possessed no radioactivity. Other experiments, furthermore, have shown that when coproporphyrin I is isolated from the feces of an animal fed glycine tagged with N^{15} the fecal coproporphyrin I as well as hemoglobin protoporphyrin IX (type III) contains heavy nitrogen in relatively high concentration (14). These results provide additional evidence in favor of the hypothesis that coproporphyrin I is a by-product of protoporphyrin IX formation.

SUMMARY

1. In the biosynthesis of hemoglobin, the carboxyl carbon is incorporated into globin but not into protoporphyrin.

2. Globin within the intact erythrocyte does not participate in protein interchange. No evidence was obtained to indicate that the C^{14} derived from the carboxyl carbon of glycine and synthesized into globin was reutilized for hemoglobin formation.

3. Coproporphyrin I isolated from the feces of one animal during a 15 day period, following the administration of carboxyl-tagged glycine, contained no radioactivity. Reference was made to other experiments in which coproporphyrin I was isolated from the feces of an animal fed N^{15}-tagged glycine; both the fecal coproporphyrin I and the hemoglobin proto-

porphyrin IX (type III) contained the isotope. These observations support the concept that coproporphyrin I is a by-product of protoporphyrin IX formation.

BIBLIOGRAPHY

1. Shemin, D., and Rittenberg, D., *J. Biol. Chem.*, **166**, 627 (1946).
2. Shemin, D., and Rittenberg, D., *J. Biol. Chem.*, **166**, 621 (1946).
3. London, I. M., Shemin, D., and Rittenberg, D., *J. Clin. Invest.*, **27**, 547 (1948).
4. Altman, K. I., Casarett, G. W., Masters, R. E., Noonan, T. R., and Salomon, K., *J. Biol. Chem.*, **176**, 319 (1948).
5. Grinstein, M., Kamen, M. D., and Moore, C. V., *J. Biol. Chem.*, **174**, 767 (1948).
6. Kamen, M. D., Radioactive tracers in biology, New York, 175 (1947).
7. Grinstein, M., *J. Biol. Chem.*, **167**, 515 (1947).
8. Anson, M. L., and Mirsky, A. E., *J. Gen. Physiol.*, **13**, 469 (1930).
9. Dobriner, K., *J. Biol. Chem.*, **120**, 115 (1937).
10. Grinstein, M., Schwartz, S., and Watson, C. J., *J. Biol. Chem.*, **157**, 323 (1945).
11. Olsen, N. S., Hemingway, A., and Nier, A. O., *J. Biol. Chem.*, **148**, 611 (1943).
12. Miller, L. L., Robscheit-Robbins, F. S., and Whipple, G. H., *J. Exp. Med.*, **81**, 405 (1945).
13. Hawkins, W. B., and Whipple, G. H., *Am. J. Physiol.*, **122**, 418 (1938).
14. Grinstein, M., Kamen, M. D., and Moore, C. V., *J. Lab. and Clin. Med.*, **33**, 1478 (1948).

Photoproduction of Molecular Hydrogen by *Rhodospirillum rubrum*

Howard Gest[1] and Martin D. Kamen

Mallinckrodt Institute of Radiology and *Department of Chemistry, Washington University*

THE PHOTOSYNTHETIC, nonsulfur purple bacterium, *Rhodospirillum rubrum* (strain SI)[2] will grow anaerobically upon illumination in a synthetic medium consisting of pure organic substrates, mineral salts including ammonium chloride, and a trace of biotin (*3*).[3] The organic substrate can be any of a large variety of compounds. Substrates more oxidized than carbohydrate are decomposed with a net production of CO_2 during growth. Aside from CO_2 and cell material, no other products have been observed in appreciable quantity. However, *R. rubrum* grown photosynthetically on certain oxidized substrates, with glutamate or aspartate instead of ammonia as a nitrogen source, exhibits in addition to CO_2 evolution a vigorous production of hydrogen.

In a preliminary survey, the only compounds so far found to be effective in evoking hydrogen production in growing cultures are malic, fumaric, and succinic acids. Of several dozen substrates tested at pH 6.6 with resting cells derived from hydrogen-producing cultures, only malic, fumaric, oxaloacetic, and pyruvic acids are effective. The magnitude of the hydrogen evolution from malic acid is of the order of one mole per mole of added substrate. Succinic acid has been tested with resting cells using pH values

[1] Predoctoral fellow of the American Cancer Society, recommended by the Committee on Growth of the National Research Council, 1947–1949. Present address, Department of Microbiology, Western Reserve University Medical School, Cleveland, Ohio.

[2] Original culture obtained through the courtesy of Prof. C. B. van Niel, Hopkins Marine Station, Pacific Grove, California.

[3] Hutner studied growth under aerobic conditions only. However, it appears that anaerobic growth requirements are similar.

 Reprinted from *Science*, **109**, No. 2840, 558–559 (1949).

varying from 5.7 to 8 because of the discrepancy in response of resting cells and of growing cells to this substance. Although extensive CO_2 evolution is observed over most of this range, no H_2 evolution is noted with succinic acid as substrate. This point is being investigated further.

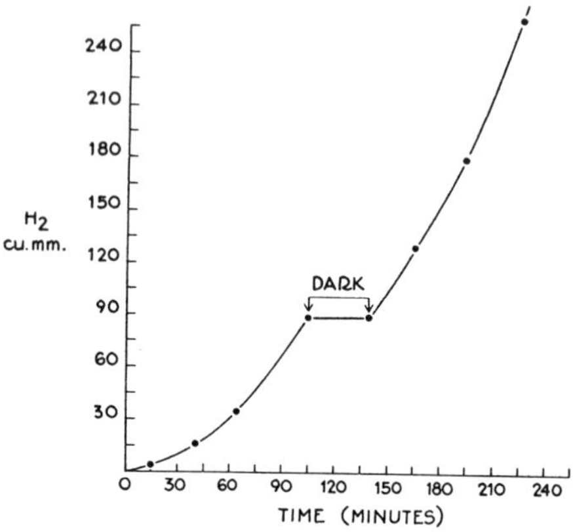

FIG. 1. Photochemical production of H_2 by *Rhodospirillum rubrum* (SI): 50 mm^3 of washed cells was suspended in 2 ml M/20 phosphate buffer (pH 6.6) and 5 mg of D,L-malate was tipped in at zero time. The center well contained 0.2 ml of 10% KOH; the gas space was filled with 100% helium.

When the gas phase in resting cell experiments is helium, argon, or hydrogen, a vigorous photoproduction of molecular H_2 is observed. A typical experiment is shown in Fig. 1, which demonstrates the results obtained with conditions as noted in the legend. In the presence of nitrogen there is no evolution of H_2. The possibility that a slight oxygen contamination of the nitrogen is responsible for the inhibition has been eliminated in three ways. First, a gas mixture consisting of 1 percent oxygen plus 99 percent helium fails to repress H_2 production. Second, nitrogen completely freed of oxygen (99.99 percent N_2) inhibits H_2 evolution. Third, a mixture of pure nitrogen (0.1 atmosphere) and helium (0.9 atmosphere) inhibits H_2 production. In addition it has been found that ammonia abolishes H_2 production under helium or hydrogen. Thus H_2 production is not observed if ammonium salts or molecular nitrogen are present. This effect of ammonia brings to mind nitrogen-fixing organisms, in which combined nitrogen inhibits production and activity of hydrogenase (*4*). The possibility exists that there may also be a hitherto unsuspected nitrogenase system in *R. rubrum*. This is now being investigated, using molecular nitrogen labeled isotopically.

The production of hydrogen during normal photosynthesis by either growing or resting cells of *R. rubrum* is unusual in that there is always present a large amount of CO_2 which can be reduced with hydrogen photochemically by these organisms, as can be demonstrated with resting cells.

Hydrogen evolution is dependent on exogenous substrate, is not inhibited at high light intensity, and does not require a prolonged adaptation period. Thus, photoproduction of hydrogen by *R. rubrum* differs from the only instance of photoproduction of hydrogen reported previously. Gaffron and Rubin (*2*) observed that in the green alga, *Scenedesmus,* an endogenous production of hydrogen of low magnitude occurred at low light intensity after long anaerobic incubation of these ordinarily aerobic organisms in the absence of CO_2, or in the presence of CO_2 when dinitrophenol was added. There is also to be remarked as a point of difference the unexpected effect of molecular nitrogen in inhibiting the H_2 evolution noted in *R. rubrum.*

Attempts to link hydrogen evolution to a formic hydrogenlyase system have failed. Growth in the presence of formate produces organisms which will evolve hydrogen in the dark with formate as substrate and with N_2 as the gas phase. However, no photoproduction of H_2 has been observed with formate. It may be recalled that fermentative production of H_2 by microorganisms is ordinarily inhibited by H_2 (*1*), whereas this is not the case with *R. rubrum.*

The data at hand do not permit of a decision as to the source of the photohydrogen. As a working hypothesis, it may be assumed that the substrate employed is the source, on the basis of the substrate specificity observed. The possibility of testing this hypothesis using labeled hydrogen is remote but is being considered.

In conclusion, it appears that *Rhodospirillum rubrum* under the influence of light and in the presence of a single organic substrate can liberate hydrogen as a photosynthetic product. This hydrogen evolution is linked to nitrogen metabolism and perhaps even to a nitrogenase system. It is hoped that further work now in progress with labeled nitrogen, hydrogen, and carbon will clarify the mechanisms involved in this production of molecular hydrogen during hydrogen transfer in photosynthesis.

References

1. CARDON, B. P. and BARKER, H. A. *Arch. Biochem.*, 1947 **12**, 165.
2. GAFFRON, H. and RUBIN, J. *J. gen. Physiol.*, 1942, **26** 219.
3. HUTNER, S. H. *J. Bact.*, 1946, **52**, 213.
4. LEE, S. B. and WILSON, P. W. *J. biol. Chem.*, 1943, **151** 377.

Evidence for a Nitrogenase System in the Photosynthetic Bacterium *Rhodospirillum rubrum*

Martin D. Kamen and Howard Gest[1]

Mallinckrodt Institute of Radiology and Department of Chemistry, Washington University

IT HAS BEEN DEMONSTRATED that the non-sulfur purple bacterium, *Rhodospirillum rubrum* (strain SI) will produce molecular hydrogen as a major photosynthetic product during illumination in certain media (*1*). This phenomenon does not occur when nitrogen is present either in the elementary form or as ammonium ion (*1*). These observations suggest the existence in these organisms of a nitrogenase or nitrogen-fixing system, hitherto unsuspected.

Evidence has been obtained using molecular nitrogen labeled isotopically with N^{15} which appears to provide unequivocal support for the existence of a light-stimulated nitrogenase system in *Rhodospirillum*. In one set of experiments, 3-day-old H_2-producing organisms were harvested by centrifugation, washed twice with *m*/20 phosphate buffer (pH 7.2) and suspended in a neutral medium containing $MgSO_4$, phosphate buffer, trace elements, biotin and D,L-malic acid. The bacterial suspension was divided into three 10-ml portions. Two portions were placed in 75-cc Warburg vessels on a vacuum line. The third portion was boiled and the dead organisms so obtained were placed in the third vessel on the vacuum line to provide a control. After evacuating and flushing the vessels and vacuum system several times with helium, the vessels were filled to 0.1 atmosphere with labeled N_2 (30% N^{15}) and 0.9 atmosphere helium. The nitrogen was generated from Eastman Kodak NH_4NO_3 (30% N^{15} in the NH_4 group) with alkaline NaOBr and freed of all combined labeled nitrogen (NH_3 or N oxides) by passage through two liquid air traps. Control experiments revealed no detectable NH_3 in the elementary nitrogen ($< 2 \times 10^{-3}$ mg NH_3 in a total of 6 mg as N_2). One vessel was illuminated, one was kept in the dark. After 6 days, the organisms were separated, and total Kjeldahl nitrogens obtained. These were analyzed for N^{15} content.[2] The organisms maintained in the light showed an N^{15} content in total cellular nitrogen of 3.14 atom percent excess; those in the dark, 0.189 atom percent excess; and the boiled controls 0.008 atom percent excess. In another set of experiments almost identical results were obtained with the additional observation that small amounts of ammonia (0.5 mg NH_4Cl/ml) inhibited N^{15} uptake from the labeled molecular nitrogen.

These observations appear to be the first reported in which there have been demonstrated (1) a nitrogenase system in photosynthetic bacteria, (2) a specific effect of molecular nitrogen on hydrogenase activity, and (3) a stimulating effect of light on turnover of molecular nitrogen. These observations have been confirmed in the laboratories of R. H. Burris and P. W. Wilson at the University of Wisconsin. We wish to express our appreciation for their wholehearted cooperation.

It has been ascertained that the N^{15} uptake observed is the result of a net fixation of nitrogen. An illuminated suspension of *R. rubrum* maintained for 30 days in an atmosphere of molecular nitrogen and hydrogen and in a medium devoid of all combined nitrogen except for a minimal quantity of yeast extract has shown a seven-fold increase in cellular nitrogen gained at the expense of molecular nitrogen.

Reference

1. GEST, H. and KAMEN, M. D. *Science*, **1949**, **109**, 558.

[1] Predoctoral fellow of the American Cancer Society, recommended by the Committee on Growth of the National Research Council, 1947–1949. Present address, Department of Microbiology, Western Reserve University Medical School, Cleveland, Ohio.

[2] We are indebted to Prof. D. Rittenberg and Mr. I. Sucher, Columbia University, and to Dr. Mildred Cohn, of Washington University, for the N^{15} analyses.

Reprinted from *Science*, **109**, No. 2840, 560 (1949).

HYDROGENASE ACTIVITY AND PHOTOASSIMILATION

MARTIN D. KAMEN

From the Department of Radiology, Washington University Medical School

ST. LOUIS, MISSOURI

THE UNIQUE ASPECTS of photosynthesis appear to derive from photo-activation and transfer of electrons involving the water molecule as ultimate source. It is assumed that the immediate result of the photochemical reaction is the production of two kinds of reaction systems. One presumably operates at a high oxidation level and catalyzes oxidation reactions which may be coupled energetically to biosynthesis. The other is assumed to operate at a low oxidation level supplying hydrogen for reduction of CO_2, the various intermediates in the CO_2 fixation cycle, and organic fragments involved in assimilation and synthesis.

Although little is known about the oxidizing components, it is possible to prepare relatively stable cell-free chloroplast homogenates which exhibit the partial reaction involving the photochemical production of oxygen, so that considerable optimism is warranted regarding the eventual elucidation of this fraction of the oxidizing systems. In the anaerobic photosyntheses carried on by bacteria, formation of oxygen is not observed because the oxidizing systems appear to use as substrates not water but an exogenous hydrogen donor. There appear to be no recorded attempts to ascertain whether extracts from photosynthetic bacteria can produce oxygen photochemically.

About the reduction systems even less is known than about the oxidation systems. It may be supposed that promising test systems are those organisms which carry on photosynthesis best in relatively strong reducing environments, i.e. the anaerobic photosynthetic bacteria. In this connection some recent developments in research on the metabolism of these bacteria appear sufficiently provocative to warrant presentation.

The organisms studied are representatives of the non-sulfur purple bacteria (*Athiorhodaceae*) which can grow photosynthetically only if supplied with an exogenous hydrogen donor, usually a single organic molecule, although molecular hydrogen itself can be utilized. The metabolic patterns involved have been most recently reviewed and systematized in a monograph by C. B. van Niel (1).

The phenomenon which forms the basis for this discussion, the production of molecular hydrogen as a characteristic product of anaerobic photoassimilation by growing or resting organisms on certain well-defined synthetic media, has come to the attention of investigators very recently, having been noted first by Dr. H. Gest in the summer of 1948 (2). The photoproduction of hydrogen in the presence of ample amounts of CO_2 appears to constitute the first clear cut demonstration of uncoupling of at least a part of the reducing system during photosynthesis. It will be convenient to describe and discuss this process first as it is observed to occur in the absence and then in the presence of exogenous N.

Of the many organisms tested, only two will be considered. One belongs to the genus *Rhodospirillum* and is a strain (S-1) of the species *rubrum* (1), referred to hereafter simply as *R*[1]. The other is a strain recently discovered by J. M. Siegel (3) and assigned to the genus *Rhodopseudomonas*, species *gelatinosa*, which appears to possess the unusual ability to attack ketones as well as secondary alcohols and which will be referred to as *G*.

Photoproduction of Molecular Hydrogen in the Absence of Exogenous N. All species of the *Athiorhodaceae* examined so far produce hydrogen when grown anaerobically under illumination in the presence of a simple medium consisting of a single organic substrate, mineral salts, carbonate and a trace of growth factor, such as biotin (2, 4, 5). Resting cells produce hydrogen when suspended in a phosphate buffer and a single organic substrate. Typical data are shown in figure 1. The substrate specificity varies to some extent depending upon the idiosyncrasies of the particular organism studied. As an example there is given in table 1 a comparison of some substrates and con-

---

[1] We are indebted to Prof. C. B. van Niel for supplying us with the original stock culture of *R. rubrum* (S-1).

 Reprinted from *Federation Proceedings*, **9**, No. 2, 543–549 (1950).

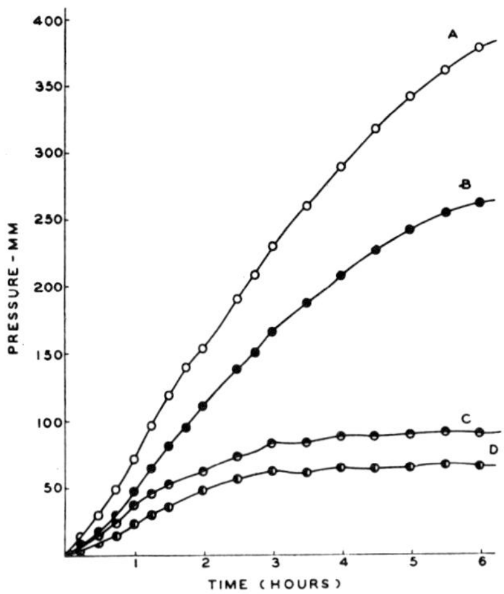

Fig. 1. PHOTOPRODUCTION of H_2 from succinate by *Rhodopseudomonas gelatinosa* (6). Seventy cu. mm. of washed cells suspended in 2 ml. of 0.05 M phosphate buffer (pH 6.7), 100% He; *Curve A*, 5 mg. succinate added at zero time, no KOH, evolution of $CO_2 + H_2$. *Curve B*, 5 mg. succinate added at zero time, 0.2 ml. of 10% KOH in center well, evolution of H_2. *Curve C*, endogenous control, no KOH. *Curve D*, endogenous control, 0.2 ml. of 10% KOH in center well.

TABLE 1. PHOTOPRODUCTION OF HYDROGEN IN RHODOSPIRILLUM RUBRUM (S-1) AND RHODOPSEUDOMONAS GELATINOSA (NOV. SPEC.) IN ABSENCE OF EXOGENOUS NITROGEN

SUBSTRATE	CONDITIONS	R. RUBRUM	GELATINOSA
Endogenous		−	+
Acetate	no KOH	?	+
	KOH	−	+
Fatty acids, (C_3, C_4)	no KOH		+
	KOH	−	−
Acetone	no KOH		+
	KOH		−
Isopropanol, ethanol	no KOH		−
	high HCO_3^-		+
C-4 Dicarboxylic[1]	no KOH	+	+
acids, acetoacetate and pyruvate	KOH	+	+
Succinate	no KOH	+	+
	KOH	−	+

[1] Malate, fumarate, oxalacetate.

ditions for photoproduction of H_2 in *R* and *G*. It can be seen that evolution of molecular hydrogen is associated with a considerable variety of substrates.

Correlation between effectiveness of substrate for hydrogen production and production of CO_2 during its metabolism can be established (6). Those substrates producing large quantities of CO_2 also produce H_2 despite the presence of KOH as an absorber, whereas substrates producing little CO_2 may not produce H_2 when KOH is present. This is particularly true with *G* which shows a relatively high light-stimulated endogenous production of CO_2 (figs. 1, 2). Quantitatively, it can be shown that the yield of photohydrogen is 60 per cent less when acetate is metabolized in the presence of KOH as a CO_2 absorber, compared to the yield with no KOH present (6). Absorption of CO_2 by KOH lowers the yield of photohydrogen 20 per cent with malate as substrate but does not affect the yield at all with oxalacetate as substrate (6). It may be remarked that the relative rates at which amounts of CO_2 are produced metabolically are in the

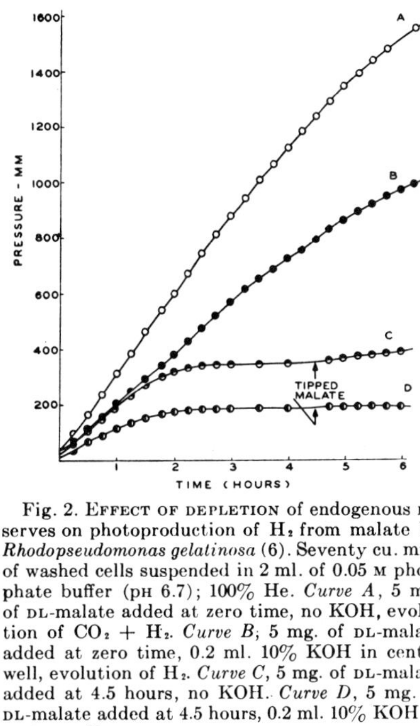

Fig. 2. EFFECT OF DEPLETION of endogenous reserves on photoproduction of H_2 from malate by *Rhodopseudomonas gelatinosa* (6). Seventy cu. mm. of washed cells suspended in 2 ml. of 0.05 M phosphate buffer (pH 6.7); 100% He. *Curve A*, 5 mg. of DL-malate added at zero time, no KOH, evolution of $CO_2 + H_2$. *Curve B*, 5 mg. of DL-malate added at zero time, 0.2 ml. 10% KOH in center well, evolution of H_2. *Curve C*, 5 mg. of DL-malate added at 4.5 hours, no KOH. *Curve D*, 5 mg. of DL-malate added at 4.5 hours, 0.2 ml. 10% KOH in center well.

order oxalacetate > malate > acetate. The mechanisms for the CO_2 effect remain obscure.

Another interesting feature of photohydrogen production in *G* is the dependence on active endogenous metabolism (fig. 2). This effect does not appear in the case of *R*.

The nature of the gas phase is important in expression of hydrogenase activity. Only the noble gases (helium, argon) are truly inert and, hence, utilizable. Molecular hydrogen does not affect photohydrogen production but its presence makes impossible identification of evolved gas by specific absorbers. The fact that hydrogen fails to suppress hydrogen production is in direct contrast to the typical behavior of fermentative hydrogenase systems (7). Oxygen in concentrations as high as 2 per cent does not interfere (5) but potentially it can be involved in an oxyhydrogen reaction as well as in respiration even though the latter process is apparently inhibited by light (8). Nitrogen gas exerts the most striking effect, namely, a complete inhibition of photohydrogen production (5). This will be discussed at more length in a later section.

The overall metabolism of the *Athiorhodaceae* resulting in the photosynthetic production of hydrogen is very much modified from what may be considered the conventional pattern (1) as exemplified by the general formulation:

$$(1)\quad CO_2 + 2H_2A \xrightarrow{h\nu} (CH_2O) + 2A + H_2O$$

('A' is used to designate an organic or inorganic residue.)

Thus, if one considers malate as a typical substrate producing cell material at the oxidation level of carbohydrate, one obtains

$$(2)\quad C_4H_6O_5 \xrightarrow{h\nu} (CH_2O)_3 + CO_2$$

as a special case of *equation 1*. Hydrogen is produced from malate according to the reaction,

$$(3)\quad C_4H_6O_5 + H_2O \xrightarrow{h\nu} (CH_2O)_2 + 2CO_2 + 2H_2$$

This equation as written does not mean that assimilation of malate carbon actually occurs in resting cell suspensions. It is only meant to imply that an organic residuum is formed at or close to the oxidation level of carbohydrate.

The net effect can be described as activation of a hydrogenase system with a concomitant inhibition of at least CO_2 assimilation. This result is the more unexpected because these organisms ordinarily effect a rapid reduction of CO_2 if supplied with molecular hydrogen as hydrogen donor.

This effect of the organic substrate in repressing hydrogen utilization and CO_2 reduction can be easily accommodated in the present schemes for the photometabolism of these bacteria as elaborated by van Niel (8), Gaffron (9) and others. Thus, it can be supposed that the primary reducing system HX is produced in equimolar amounts with an oxidizing system YOH by enzyme catalyzed photodegradation of water, i.e.

$$(4)\quad HOH + X + Y \xrightarrow[ENZ]{h\nu} HX + YOH$$

Photochemical oxidation of the organic substrate as a follow reaction will remove YOH but not HX. If the fragments derived from the organic material are available for assimilation neither the assimilation of CO_2 nor the functioning of a CO_2 fixation cycle remain obligatory, and utilization of HX need not occur. Under these conditions the excess HX can serve as substrate for a hydrogenase system whereby it is removed as molecular hydrogen, i.e.

$$(4a)\quad 4HOH + 4X + 4Y \xrightarrow{h\nu} 4HX + 4YOH$$

$$(4b)\quad C_4H_6O_5 + 4YOH \longrightarrow (CH_2O)_2 + 2CO_2 + 3H_2O + 4Y$$

$$(4c)\quad 4HX \longrightarrow 2H_2 + 4X$$

This system of equations is but one of an infinite number which can be written to satisfy the overall *equation 3*. Information on the quantum yield for photohydrogen production in *R* might settle the question of how many water molecules are actually involved in this process.

Preliminary data derived from a number of experiments support the notion that direct assimilation of the organic substrate, or fragments derived therefrom, can occur without participation in a CO_2 fixation cycle. Van Niel and Barker (10) noted that assimilation of C-14-labeled carbonate by cultures of *R* grown with unlabeled acetate as substrate was considerably less than would be expected on the basis that all acetate carbon was assimilated via carbonate. It has been found in our laboratory (11) using methyl-labeled acetate, that practically all the methyl carbon of acetate is assimilated as cellular carbon without a significant proportion of the labeled carbon appearing as carbonate. Experiments with carboxyl-labeled acetate have been performed by van Niel and Barker (12), as well as in our laboratory, and show considerably better equilibration of labeled acetate carboxyl with carbonate. This result would indicate that the two acetate carbons are not equivalent with regard to photoassimilation although they have been shown to be equivalent with regard to oxidation during dark respiration of labeled acetate (10–12). This in turn would support the notion that little or no cycling involving CO_2 occurs during light assimilation of organic material.

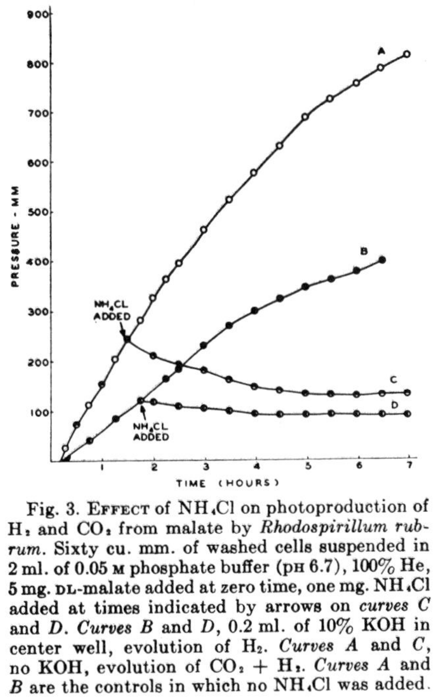

Fig. 3. EFFECT of NH_4Cl on photoproduction of H_2 and CO_2 from malate by *Rhodospirillum rubrum*. Sixty cu. mm. of washed cells suspended in 2 ml. of 0.05 M phosphate buffer (pH 6.7), 100% He, 5 mg. DL-malate added at zero time, one mg. NH_4Cl added at times indicated by arrows on *curves C* and *D*. *Curves B* and *D*, 0.2 ml. of 10% KOH in center well, evolution of H_2. *Curves A* and *C*, no KOH, evolution of CO_2 + H_2. *Curves A* and *B* are the controls in which no NH_4Cl was added.

It may be remarked that the experiments on light oxidation are complicated by the fact that so little acetate escapes assimilation. Experiments with labeled substrates more amenable than acetate in this respect (i.e. labeled malate, fumarate, succinate, etc.) are contemplated. In addition experiments are planned to investigate the ratio of labeled carbon in cells and evolved carbonate as a function not only of labeling position but also time and intensity of illumination.

Obviously, it is of fundamental importance to establish whether the fragments derived from photometabolism of the organic substrate are identical or in equilibrium with the endogenous C_2 compounds involved in the CO_2 fixation cycle. It is evident that no cycle need be involved in photoassimilation particularly if an active hydrogenase is present to remove reducing substances necessary for maintenance of a cycle under conditions required for photoassimilation.

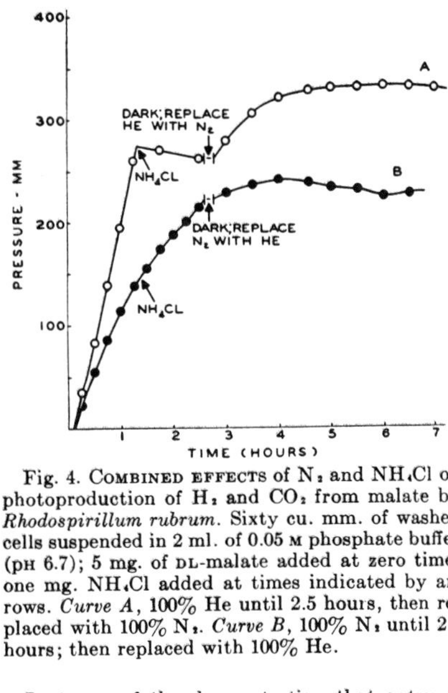

Fig. 4. COMBINED EFFECTS of N_2 and NH_4Cl on photoproduction of H_2 and CO_2 from malate by *Rhodospirillum rubrum*. Sixty cu. mm. of washed cells suspended in 2 ml. of 0.05 M phosphate buffer (pH 6.7); 5 mg. of DL-malate added at zero time; one mg. NH_4Cl added at times indicated by arrows. *Curve A*, 100% He until 2.5 hours, then replaced with 100% N_2. *Curve B*, 100% N_2 until 2.5 hours; then replaced with 100% He.

In terms of the demonstration that entry of CO_2 occurs only at a point involving production of glyceric acid phosphate (see other papers in this Symposium) the photo-oxidation of exogenous hydrogen donors could be imagined as resulting in direct formation of a C_3 compound which is closely related to, if not identical with glyceric acid phosphate, and which is assimilated directly. Thereby CO_2 fixation can be bypassed, and perhaps even turnover via the CO_2 cycle suppressed.

Photometabolism and Hydrogenase Activity in the Presence of Exogenous N. The photoproduction of hydrogen is completely inhibited when molecular nitrogen or ammonium salts are present (fig. 3–5). On the other hand certain amino acids, i.e. glutamic acid, aspartic acid, alanine and glycine do not suppress photohydrogen evolution (2, 5, 13). In fact initial experiments in which growth was observed concomitant with hydrogen production involved glutamate or aspartate as nitrogen sources. Amino acids which inhibit photohydrogen evolution, i.e. lysine, serine, appear to do so via deamination to free ammonia (13). It may be remarked that no species differences appear to exist as regards nitrogen and ammonia effects on photohydrogen production.

If *R* is suspended in the usual photohydrogen-evoking medium to which ammonia is added, it is found that ammonia is removed by the organisms (13). After complete removal of ammonia (<1–2 γ/ml.) photohydrogen production commences provided substrate is still present. The time required for inception of hydrogen production is roughly proportional to the amount of ammonia present. Ammonia differs from molecular nitrogen in that it also suppresses CO_2 production from malate as substrate (fig. 3) unless N_2 is the gas phase (fig. 4). On the other hand photo-utilization of molecular hydrogen with carbonate as substrate is not inhibited either by ammonia or nitrogen. Photo-utilization of molecular hydrogen has been observed with pyruvate or oxalacetate as substrates, but not with malate (6, 13). Studies on coupling of ammonia uptake with hydrogen uptake in the light in the presence of a variety of substrates are now being undertaken.

Before elaborating data available on the relation between nitrogen metabolism, photoassimilation, and photohydrogen production, it should be stated that *R* is now known to be a nitrogen-fixer (14, 5) as indeed appears to be the case for all the photosynthetic bacteria (15). This fact introduces what may be regarded, on the one hand, as an unexpected complication and, on the other hand, as an opportunity to study the coupling of nitrogen fixation to photosynthesis in organisms easy to culture and susceptible to a large variety of manipulations.

The results on inhibition of photohydrogen production by molecular nitrogen and free ammonia closely parallel previous studies of Wilson and his colleagues on the production of hydrogenase in the free-living nitrogen-fixing *Azotobacter* (16, 17). Thus glutamate and aspartate which fail to affect production of hydrogenase in *Azotobacter* also do not suppress photohydrogen production in *R*. Hydrogenase activity, insofar as hydrogen uptake is concerned, is not repressed by nitrogen or ammonia either in *Azotobacter* or in *R*. Furthermore ammonia which inhibits hydrogenase production in *Azotobacter* also inhibits photohydrogen production in *R*.

The one unique observation peculiar to the photosynthetic bacteria is that molecular nitrogen like ammonia inhibits photohydrogen production, whereas unlike ammonia it does not affect hydrogenase production in *Azotobacter*.

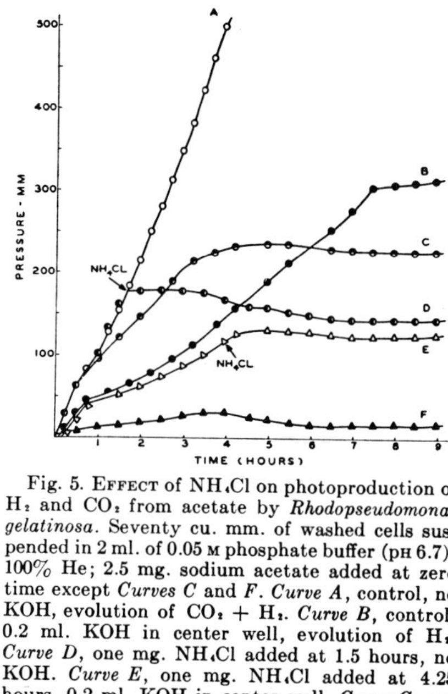

Fig. 5. Effect of NH_4Cl on photoproduction of H_2 and CO_2 from acetate by *Rhodopseudomonas gelatinosa*. Seventy cu. mm. of washed cells suspended in 2 ml. of 0.05 M phosphate buffer (pH 6.7); 100% He; 2.5 mg. sodium acetate added at zero time except *Curves C* and *F*. *Curve A*, control, no KOH, evolution of CO_2 + H_2. *Curve B*, control, 0.2 ml. KOH in center well, evolution of H_2. *Curve D*, one mg. NH_4Cl added at 1.5 hours, no KOH. *Curve E*, one mg. NH_4Cl added at 4.25 hours, 0.2 ml. KOH in center well. *Curve C*, endogenous control, no KOH. *Curve F*, endogenous control, no KOH, one mg. NH_4Cl added at zero time.

Another interesting difference between the hydrogen-producing hydrogenase in *R* and hydrogenase in another anaerobic nitrogen-fixing but non-photosynthetic organism, *Clostridium kluyveri* (18), is that hydrogenase in the latter organism is not affected by molecular nitrogen (19).

From the standpoint of photosynthesis, it should be recalled that the effect of ammonia and other forms of combined nitrogen in altering the reduction level of assimilated material in green algae has been known for some time and most recently reviewed by Spoehr and Milner (20) and by Myers (21). It is evident from the present studies that both molecular nitrogen and ammonia affect a return to the conventional type of metabolism described by *equation 2*. Thus, molecular N and ammonia appear to inactivate the hydrogenase concerned with light stimulated hydrogen production in the absence of exogenous N, although they have no effect on the hydrogenase involved in light stimulated uptake of hydrogen.

At present these observations can be understood on the basis of a number of hypotheses. First, it may be assumed that the hydrogenase systems involved in H_2 uptake differ from those involved in hydrogen evolution, and that molecular nitrogen is a specific inhibitor of the hydrogen-evolving enzyme systems. As a second hypothesis one might postulate a single hydrogenase system in competition with dehydrogenase systems for the photosynthetically-produced reducing substrate, HX. Addition of ammonia could activate and accelerate the dehydrogenase reactions by initiating processes like reductive amination and thus allow the dehydrogenases to compete successfully with the hydrogenases. Hence, repression of hydrogen production would be an indirect rather than direct result of the addition of ammonia.

The remarkable effect of molecular nitrogen appears to require activation of nitrogenase which in some way results in simultaneous activation of the dehydrogenases which are thereby enabled to compete successfully with the hydrogenase systems. The possibility of a direct reaction between molecular nitrogen and hydrogen or between molecular nitrogen and HX is not excluded, but available manometric data do not show either the uptake of nitrogen or the output of CO_2 to be expected on this basis (6).

It seems pertinent to note at this point that a hydrogenase activity leading to photohydrogen production was noted by Gaffron and Rubin (22) some time ago as a weak endogenous effect in the green alga, *Scenedesmus*, after anaerobic adaptation. This effect was best observed in the absence of CO_2 or in the presence of dinitrophenol, was not dependent on external substrates, but most important, was not inhibited by molecular nitrogen and inhibited by molecular hydrogen. The absence of the nitrogen inhibition for this hydrogenase is apparently related to the fact that the green algae, although possessing a hydrogenase system, do not fix nitrogen. It would appear to be of interest to extend these researches to other photosynthetic organisms which are known to contain a nitrogen-fixing system, such as certain of the blue-green algae (23, 24).

Prognosis for Analysis of Partial Reactions. Future progress in elucidating phenomena described in this paper depends largely on success in preparing cell-free extracts in which partial reactions involving light stimulated hydrogen transfer can be demonstrated. Cell-free preparations in which photohydrogen production is sensitive to nitrogen also are required. It is not difficult to obtain active dehydrogenase preparations from photosynthetic bacteria which are active in the oxidation or reduction of substrates like oxalacetate (25, 26) succinate (26) fumarate (26) and malate (26). An active and stable hydrogenase which accomplishes the classical reduction of methylene blue with molecular hydrogen can easily be prepared (27, 26). Cell-free juices of *R* have been prepared which in the presence of succinate show an uptake of oxygen which is reduced in magnitude by illumination (26). The small uptakes observed so far obviate attempts to specify whether this effect results from actual photohydrogen production or from competition between molecular oxygen and the photochemical oxidizing substance. Preparation of cell-free juices from the photosynthetic bacteria which exhibit light-stimulated hydrogen evolution or hydrogen uptake would go far to open up the photosynthetic proton transfer systems for investigations of the type which have already yielded so much information concerning biological oxidation in nonphotosynthetic systems. In addition, further studies are urgently needed to supplement those reported to date (28, 29) on the nature of the hydrogenase enzymes.

REFERENCES

1. van Niel, C. B. *Bact. Rev.* 8: 1, 1944.
2. Gest, H. and M. D. Kamen. *Science* 109: 558, 1949.
3. Siegel, J. M. and M. D. Kamen. *J. Bact.* In press.
4. Gest, H. and M. D. Kamen. *J. Bact.* 58: 239, 1949.
5. Gest, H., M. D. Kamen and H. M. Bregoff. *J. Biol. Chem.* 182: 153, 1950.

6. Siegel, J. M. and M. D. Kamen. Unpublished data.
7. Cardon, B. P. and H. A. Barker *Arch. Biochem*. 12 165, 1947.
8. van Niel, C. B. *Advances in Enzymol*. 1: 263, 1941.
9. Gaffron, H. *Biol. Rev*. 19: 1, 1944.
10. van Niel, C. B. and H. A. Barker. In *Photosynthesis in Plants*, edited by J. Franck and W. E. Loomis. Ames: Iowa State Press, 1949, p. 447.
11. Kamen, M. D., S. J. Ajl, S. Ranson and J. M. Siegel. Unpublished data.
12. van Niel, C. B. and H. A. Barker. Private communication.
13. Bregoff, H. M. and M. D. Kamen. Unpublished data.
14. Kamen, M. D. and H. Gest. *Science* 109: 560, 1949.
15. Lindstrom, E. S., R. H. Burris and P. W. Wilson. *J. Bact*. 58: 313, 1949.
16. Lee, S. B. and P. W. Wilson. *J. Biol. Chem*. 151: 377, 1943.
17. Lee, S. B., J. B. Wilson and P. W. Wilson. *J. Biol. Chem*. 144: 273, 1942.
18. Rosenblum, E. D. and P. W. Wilson. *J. Bact*. 57: 413, 1949.
19. Stadtman, E. R. and H. A. Barker. *J. Biol. Chem*. 181: 221, 1949.
20. Spoehr, H. A. and H. W. Milner. *Plant Physiol*. 24: 120, 1949.
21. Myers, J. See (10) p. 349.
22. Gaffron, H. and J. Rubin. *J. Gen. Physiol*. 26: 195, 1942.
23. Allison, F. E. and H. J. Morris. *Science* 71: 221, 1930.
24. Burris, R. H. and P. W. Wilson. *Botan. Gaz*. 108: 254, 1946.
25. Mehler, A. Quoted in (5).
26. Kamen, M. D. and S. J. Ajl. Unpublished data.
27. Gest, H. Private communication.
28. Hoberman, H. D. and D. Rittenberg. *J. Biol. Chem*. 147: 211, 1943.
29. Back, K. J. C., J. Lascelles and J. L. Still. *Australian J. Sci*. 9: 25, 1946.

THE MORTALITY OF BACTERIOPHAGE CONTAINING ASSIMILATED RADIOACTIVE PHOSPHORUS*

BY A. D. HERSHEY,‡ M. D. KAMEN, J. W. KENNEDY, AND H. GEST§

(From the Departments of Microbiology, Radiology, and Chemistry, Washington University, St. Louis)

(Received for publication, June 23, 1950)

When bacteria and bacteriophage are grown in a medium containing phosphate phosphorus of high specific radioactivity, the viral progeny are markedly unstable. Under suitable conditions one observes a progressive inactivation of the phage that is due solely to the effects of radioactive decay of the assimilated phosphorus. In the range of 30 to 300 atoms of P^{32} per million atoms of total phosphorus (about 10 to 100 curies per gm.), the course of inactivation of the phage can be used for the assay of assimilated P^{32}.

Two kinds of information are obtained in an assay of this kind. The rate of inactivation measures the specific activity of the phage phosphorus, and the form of the survivor curve tells something about the distribution of radioactivity within the phage population. This unique feature of the method may be illustrated in terms of the question which it was designed to answer; namely, whether all or only one of the progeny of a labelled phage particle contains isotopic phosphorus. Evidently the second alternative leads to the expectation that stable individuals will be found among the radioactive phage grown from non-radioactive seed in radioactive medium. We are attempting to apply the method to this and related problems of viral growth. In this paper we report the data on which our method is based, together with some observations having to do with the nature of the lethal reaction.

Materials and Methods

We have used phage T4 propagated on its host *Escherichia coli* strain B, and other closely related phages as noted below.

The tracer sample, either ordinary or "carrier-free" as received from the Oak Ridge National Laboratory, is concentrated by boiling at low pH, neutralized with NaOH,

---

* Supported by grants from the United States Public Health Service and the American Cancer Society. The radioactive phosphorus used in this investigation was supplied by the Oak Ridge National Laboratory on allocation from the Isotopes Division, United States Atomic Energy Commission.

‡ Present address: Carnegie Institution, Department of Genetics, Cold Spring Harbor, Long Island, New York.

§ Present address: Department of Microbiology, Western Reserve University, School of Medicine, Cleveland, Ohio.

Reprinted from *The Journal of General Physiology*, **34**, No. 3, 305-319 (1951).

centrifuged clear, and incorporated into a medium of the final composition per liter: 0.004 gm. or more phosphorus as orthophosphate, 10 gm. sodium lactate, 5.2 gm NaCl, 1.8 gm. K_2SO_4, 0.12 gm. $MgSO_4$, 1.0 gm. NH_4Cl, and 0.5 gm. bacto-peptone, at pH 6.8. The peptone contributes 0.0011 gm. of the phosphorus, at least 80 per cent of which is phosphate. The P^{32} content of the medium is assayed in absolute microcuries of beta radiation by means of a Lauritsen electroscope, using a secondary standard of uranium oxide calibrated in turn against a standard solution of P^{32} supplied by the National Bureau of Standards. The total phosphorus of the medium is determined by the method of Fiske and SubbaRow after digestion with sulfuric acid and hydrogen peroxide.

As used below "saline-df" means a diluting fluid containing 0.5 per cent NaCl and 0.001 M $MgSO_4$; "peptone-df" means saline-df containing 0.1 per cent peptone; and "gelatin-df" means saline-df containing 10 mg. per liter of gelatin (Adams, 1948, 1949).

Radioactive phage is prepared by either of two methods, the first being the one-step growth experiment of Ellis and Delbrück (1939) carried out as follows:—An 18 hour bacterial seed culture is prepared by aeration in the above medium containing 5 μg. P^{31} per ml., and diluted in saline-df to contain about 4×10^7 cells in 0.05 ml. This aliquot is used to seed 1 ml. of the radioactive culture medium, which is aerated for 2 hours at 37°C. There is at least a tenfold increase in cell substance during this time and the final bacterial titer is 10^8 cells per ml. This culture is called the "adsorption tube." The bacteria in the adsorption tube are infected by adding 0.05 ml. of a solution containing 1 mg. per ml. of *l*-tryptophan and sufficient washed phage in saline-df to infect the bacteria with an average of 5 particles per cell. After 5 minutes' further aeration, 90 per cent of the added phage has been adsorbed, and 0.05 ml. of antiphage rabbit serum (diluted 1:5 and dialyzed against saline-df) is added. The concentration of antiserum is chosen to neutralize 95 to 99 per cent of the unadsorbed phage in 5 minutes. Ten minutes after infection the adsorption mixture is diluted 10^3 times ("growth tube 3") or 10^4 times ("growth tube 4") with the radioactive culture medium. Growth of phage and lysis of bacteria take place in the growth tube. At 60 minutes after infection, NaCN to 0.01 M concentration is added to kill any uninfected bacteria, and to prevent any further growth of phage (Doermann, 1948). At 90 minutes after infection, dilutions from the growth tube of 10^5, 10^6, and 10^7-fold with respect to the adsorption tube ("titration tubes" 5, 6, and 7, respectively) are made in peptone-df. The titration tubes are kept below 5°C. for the ensuing period, during which the viable phage titer is measured daily by plaque counts. The yield of phage in these experiments is about 100 per bacterium.

The second, or "lysate" method of preparing radioactive phage consists in infecting the bacteria in the adsorption tube with a very small amount of phage, added together with 30 μg. *l*-tryptophan per ml. of medium, and allowing the culture to lyse without further dilution or addition. Lysis requires 6 hours or more under these conditions, and the cultures are usually left overnight. The next morning the lysate is centrifuged clear if necessary, and NaCN is added in 0.01 M concentration to the supernate. After 30 minutes longer at 37°C. the lysate is diluted serially in peptone-df, and the resulting "titration tubes," numbered to indicate the logarithm of the dilution, are stored at 5°C. for periodic titration. The principal feature of this method is that it is possible to infect the culture with one phage particle per thousand or more bacteria, so that the contributions from the seed phage of both total phosphorus and viable phage can be made as small as desired. The yield of phage in this method also is about 100 particles per bacterium.

We have observed the following precautions and controls to ensure that inactivation of phage in the titration tubes is due solely to the effects of assimilated P^{32}.

1. Phage grown by either method in a medium containing P^{32} with a large excess of added P^{31} proves to be stable, showing that the concentrations of antiserum and unassimilated P^{32} in the titration tubes are without effect on the phage. The safe concentration of antiserum has to be determined by trial. The concentration in titration tube 5 is in practice the critical factor limiting the amount to be added to the adsorption tube, and in the conduct of the experiments it has to be remembered that the temperature coefficient of inactivation of phage by antiserum is high (Hershey, 1941). The safe initial concentration of P^{32} (permitting 90 per cent survival) is 26 μc./ml. for 10 days, or 13 μc./ml. for 30 days (see below). These two sources of error are conveniently controlled within each experiment by checking that the titer of phage in titration tube 5 remains 10 times the titer in tube 6 throughout the period of observation.

2. In the one-step procedure, the results obtained using growth tube 3 and growth tube 4 are identical if the concentration of antiserum in tube 3 is not excessive. A delayed effect of antiserum on phage T2Hr1 was encountered, which caused phage from growth tube 3 to be somewhat less stable in the titration tubes than that from growth tube 4. In experiments of long duration with this phage, growth tube 4 had to be used. Growth tube 3 could be used in all experiments with phage T4, which is an advantage because of the lower concentration of P^{32} carried into the titration tubes.

3. In the one-step procedure, a blank run of seed phage plus antiserum without bacteria gives a base line titer of non-radioactive phage below which parallel observations on survival of radioactive phage cannot be carried. This allows a safety factor of 10 if 90 per cent of the seed phage is adsorbed in tubes with bacteria. A delayed effect of antiserum is again seen in the blank, causing progressive inactivation of seed phage in the titration tubes; in this case a convenient circumstance.

Additional controls serve to ensure that the specific radioactivity of the phage will be known from the assay made of the culture medium.

1. Decreasing the bacterial seed five-fold and increasing the period of growth before infection with phage by 1 hour does not affect the rate of death of the phage, showing that equilibration between phosphorus of bacteria and medium is virtually complete under the conditions used.

2. Agreement between the one-step growth method and the lysate method shows that the phosphorus introduced with the seed phage and antiserum is negligible.

3. Identical specific rates of death of phage grown in media with and without

peptone show that the peptone contains negligible amounts of preferentially assimilated phosphorus (see below).

4. Identical rates of death of phage grown in two media of the same specific radioactivity but with different amounts of total phosphorus (for example, 6 and 60 μg./ml.) show that the phosphorus contamination from the glassware is negligible. Results described below suggest that a total P content as low as 4 μg./ml. is satisfactory under our conditions, but we have not established the safe lower limit. This becomes important, of course, in experiments requiring high specific radioactivity.

The Relation between Specific Radioactivity of Viral Phosphorus and Survival of Phage

The presentation of data is made easier if we anticipate qualitatively what we have found; the results are compatible with the following assumptions: (1) The inactivation of a radioactive phage particle is the consequence of the disintegration of a single atom (not necessarily the first) of its assimilated P^{32}. (2) At least 999 per thousand of the phage particles are equally sensitive to this type of killing. (3) The P^{32} atoms are distributed at random among the phage particles. It should be observed that we are not concerned with the correctness of these assumptions, which we regard as plausible but essentially unprovable. The useful fact is that populations of the kind we are studying die out in an orderly fashion, which provides a technique for analyzing situations in which assumption (3) does not hold.

From the assumptions stated it follows that surviving phage containing assimilated P^{32} will die at the rate

$$-\frac{dS}{dt} = \alpha N^*\lambda S = 3.4 \times 10^{-6}\alpha N\lambda S A_0 e^{-\lambda t} \quad (1)$$

in which S is the titer of the surviving phage per ml.; α is the efficiency of killing per atomic disintegration; N^* is the number of radioactive and N the total number of phosphorus atoms per phage particle; $\lambda = 0.0485$ per day, the fractional rate of decay of P^{32}; and A_0 is the initial specific radioactivity of the phage phosphorus in curies per gram. The factor 3.4×10^{-6} comes from the definition of the curie and from λ and converts specific activity to the atomic ratio N^*/N.

For the conditions $S = S_0$ and specific activity $= A_0$ when $t = 0$, the solution of (1) is

$$\log_{10} S = \log_{10} S_0 - 1.48 \times 10^{-6}\alpha A_0 N(1 - e^{-\lambda t}) \quad (2)$$

It follows that a plot of logarithms of survivors against $(1\text{-}e^{-\lambda t})$ should form a straight line with origin log S_0 and slope proportional to the initial specific radioactivity of the phage phosphorus.

Typical data showing the relation between the specific radioactivity of the medium and survival of the phage T2Hr1 are shown in Fig. 1 and Table I. The phages were grown by the lysate method, excepting sample G which is a one-step growth experiment. The latter is taken from a larger experiment in which adequate samples of the medium could not be spared for determination

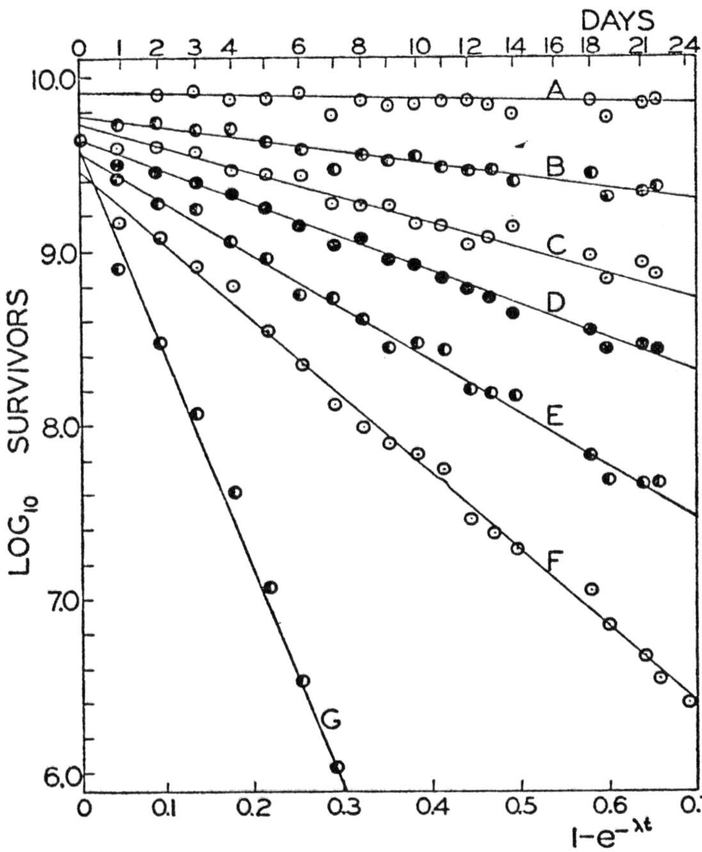

FIG. 1. Survival of phage T2Hr1 containing assimilated P^{32}. The origins of the curves have been adjusted to clarify the graph. See Table I for radioactivity measurements.

of total phosphorus; the specific activity given is approximate. With this exception it may be seen that the proportionality between rate of death of the phage and specific activity of the medium (constancy of αN) is fairly good, and that all the data fit the linear equation (2). The estimates of αN have been computed using equation (2) and the graphically measured slopes of the curves.

TABLE I

Relation between the Specific Radioactivity of the Culture Medium and the Specific Rate of Death of Phage T2Hr1

Curve (Fig. 1)	Initial radioactivity of medium		Specific death rate (α N)
	mc./ml.	*mc./mg. P*	$\times 10^4$
A	0.27	0.0	—
B	0.14	10.5	4.7
C	0.27	19	5.1
D	0.54	30	4.5
E	1.1	45	4.7
F	2.2	58	5.3
G	0.38	>93	<9.3

TABLE II

Survival of Phage Grown in Radioactive Media under Various Conditions

Curve (Fig. 2)	Method	Medium	[P]	A_0	Log S_0	αN	χ^2
			µg./ml.	*mc./mg.*	*per ml.*	$\times 10^4$	$\times 10^{-4}$
	Media containing depleted peptone						
C	1-step	Lactate-peptone	6.2	25	9.75	4.2	8
H	Lysate	"	6.2	25	9.79	4.2	33
B	1-step	"	15.6	29	9.47	4.3	5
G	Lysate	"	15.6	29	9.90	4.1	8
	Media containing whole peptone						
A	1-step	Lactate-peptone	15.6	29	9.32	4.3	7
F	Lysate	"	15.6	29	9.88	4.3	18
	Media without peptone						
D	Lysate	Glucose-lactate	5.8	31	10.06	4.3	11
E	Lysate	Lactate	5.8	31	8.93	4.4	17
	Media without P^{32}						
—	1-step	Lactate-peptone	5.0	0	9.78	(0.2)	8
—	Lysate	Lactate	1000	0	9.47	(0.4)	13

Figures in parentheses indicate per cent inactivation per day of non-radioactive phage. This datum and the accompanying χ^2 correspond to the equation $\log \frac{S_0}{S} = \frac{kt}{2.3}$. For radioactive phage, χ^2 measures the fit to equation (2). The χ^2 values are directly comparable for the two circumstances.

Similar results have been obtained with 25 samples of radioactive phage, grown in media prepared from five different samples of P^{32}. In these measurements the estimates of αN varied between 41,000 and 58,000, with a mean of

45,000. The data cited include parallel runs with the phages T2Hr1, T2L, T4, and an r mutant of T4, which did not differ significantly in radiosensitivity.

The most recent of these measurements are worth describing in further detail because they were designed to test a point of some technical importance, as well as to determine the accuracy of the methods finally worked out.

The point in question arose from the finding of Kozloff and Putnam (1950) that nutrient broth (peptone plus beef extract) contains small amounts of

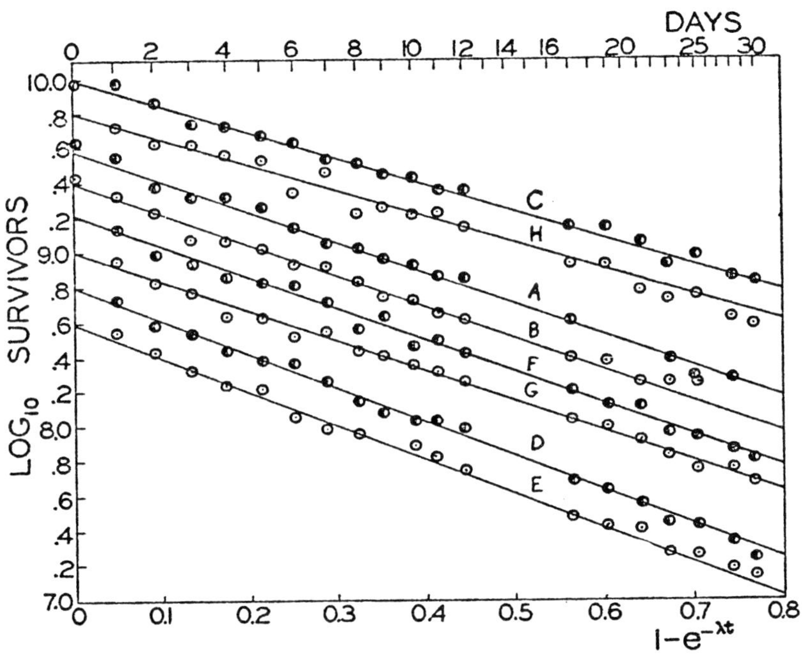

FIG. 2. Survival of radioactive phage T4 grown under various conditions. The lines have been fitted to the data for the first 12 days by the method of least squares. The origins of the curves are arbitrary. The conditions of growth and analysis of data are summarized in Table II.

preferentially assimilated organic phosphorus, causing bacteria and phage grown in broth with added P^{32} to contain phosphorus of lower specific radioactivity than that of the medium as a whole. Similar substances in our medium would introduce serious errors, particularly in the one-step growth procedure in which the culture is diluted with fresh medium during the period in which most of the phage phosphorus is being assimilated.

To test this possibility we have grown phage in media prepared as already described, and with the following variations: peptone omitted, 0.1 per cent glucose substituted for the peptone, and "depleted" peptone substituted for the whole material. The depleted peptone was prepared following Kozloff and Putnam by allowing bacteria to grow in 1 per cent aqueous peptone to a titer of about 5×10^8 per ml., centrifuging out the cells, and sterilizing the solution. Phage T4 was grown in these media by the one-step and lysate methods at two different concentrations of total phosphorus. The survival of phage was followed as already described, and the data were fitted to equation (2) by the method of least squares. The results are summarized in Table II and Fig. 2. Evidently none of the variations introduced had any effect on the specific rate of death of the phage, and the range of the estimated αN is less than 10 per cent of the mean.

We conclude that radioactive phage dies solely from the effects of the decay of assimilated radioactive phosphorus under the conditions of our experiments, that the specific activity of assimilated phosphorus is the same as that of the medium, and that $\alpha N = 43{,}000$ (the mean of the data of Table II) is a satisfactory estimate of the specific rate of death of phage T4.

The Phosphorus Content and Volume of the Phage Particle

In order to compute the quantum efficiency of the lethal reaction we wish to know as accurately as possible the phosphorus content of the phages we have used. Published analyses for T4 and related phages (Cohen, 1947; Kozloff and Putnam, 1949) vary upwards from 2.3×10^{-11} μg. per infectious unit. For a staphylococcal phage, Northrop (1938) found 3×10^{-11} μg. We have confirmed the lower estimate, and also have narrowed considerably, we believe, the margin of uncertainty.

In a first experiment phage was purified by centrifugation alone, but with special precautions to control inactivation of the phage during purification. Two lysates were prepared in a medium containing 1 per cent peptone, 0.5 per cent NaCl, 0.1 per cent glucose, and radioactive phosphate. Virtually all the phosphorus in this medium comes from the peptone. The lysates were passed through Selas porosity 02 porcelain filters. The phage was sedimented and washed three times with peptone-df, followed by two washings with saline alone. Eighty per cent or more of the phage present in the original lysates was recovered in active form in the three times washed concentrate. During the final saline washings, however, 30 to 40 per cent of the activity was lost which could not be found in the supernates. P^{32} was determined on samples of the concentrates after the third and fifth washings, total phosphorus was determined only after the final washing. The results are shown in Table III.

The phosphorus found corresponds to 2.5×10^{-11} μg. per phage particle, and the two phages T2Hr$^+$ and T2Hr1 do not differ significantly. For both phages, the five times washed concentrate contains a larger amount of phosphorus per lytic unit, evidently owing to the loss of activity of the phage during washing with saline.

In a second experiment the three phages T2H, T2L, and T4 were grown in a medium similar to the first but containing 0.5 per cent peptone and 15 μg./ml. of added phosphorus in the form of phosphate. The phages were purified by sedimentation and three

TABLE III

Phosphorus Content of Phage T2H and Its Mutant r1

	P^{32}	Total P	Phage titer	Specific activity	P/phage
	μc./ml.	*μg./ml.*	× *10^{11}/ml.*	*μc./μg.*	× *10^{-11} μg.*
Culture medium	27	28	—	0.96	—
T2Hr+					
3 times washed	8.0	—	2.9	—	2.5
5 " "	7.0	5.5	1.8	1.3	3.5
T2Hr1					
3 times washed	6.0	—	2.5	—	2.2
5 " "	4.0	3.6	1.2	1.1	3.0

The phosphorus per phage is computed from the activity measurement and from the mean specific activity 1.1.

TABLE IV

Phosphorus Content of Several Phages, and Fractionation by Specific Adsorption

Sample	Phage titer	P^{32}	P/phage
	× *10^{11} ml.*	*μc./ml.*	× 10^{-11} *μg.*
T4	2.3	5.4	2.3
T2L	1.1	3.6	3.3
T2H			
Unfractionated	0.68	1.9	2.8
Adsorbed on H (1)	(0.29)	0.79	2.7
Supernate (1)	0.39	1.2	3.1
Adsorbed on H (2)	(0.39)	1.1	2.8
Supernate (2)	0.29	0.90	3.1
Adsorbed on H/2 (3)	0.10	0.29	2.9
Supernate (3)	0.64	1.7	2.7

The phosphorus per phage is computed from the specific activity of the culture medium, 1.0 μc./μg., containing 31 μg./ml. total P.

Phage titers in parentheses are computed by difference.

washings with peptone-df. In addition, samples of the phage T2H were fractionated by adding bacteria (*E. coli* strain H) specifically adsorbing the phage, and separating into adsorbed and unadsorbed fractions by low speed centrifugation. Another sample was fractionated in the same way, except that a mutant strain of *E. coli* (H/2) not adsorbing the phage was used. In this case the phage found in the (unwashed) sediment is largely entrained supernate. All the samples were assayed for P^{32} and for phage. The results are given in Table IV.

The fractionation experiments show that phosphorus is removed from the preparations by adsorption to bacteria nearly in proportion to the removal of titratable phage. The preparations do not, therefore, contain more than a few per cent of the total phosphorus in the form of impurities not specifically adsorbed to bacteria. They also show that if any appreciable amount of phage not titratable by plaque count is present, it is adsorbed to bacteria in the same manner as the titratable phage.

A number of additional preparations of phage were subjected to the action of the enzyme desoxyribonuclease. For this purpose the phage was sedimented and resuspended in veronal buffer of pH 8 containing 0.003 M $MgSO_4$ and 10 μg. gelatin per ml. The suspension, containing not more than 10^{11} phage per ml., was divided into two parts, one of which received 20 μg./ml. of crystalline desoxyribonuclease (Worthington Biochemical Laboratory) or 200 μg./ml. of enzyme prepared according to McCarty (1946). (Either enzyme preparation at a concentration of 10 μg./ml. rapidly reduced the viscosity of 0.1 per cent sodium desoxyribonucleate under the conditions specified.) After 2 to 4 hours at 37°C., the phage was clarified if necessary by centrifugation, resedimented, and washed and resuspended in gelatin-df. These preparations were analyzed for nucleic acid in terms of the absorption of ultraviolet light at 260 mμ, and for total phosphorus by the colorimetric method. The analytical results for some of the better preparations are shown in Table V.

Evidently the enzyme has no effect either on the infectivity or nucleic acid content of the phage, which again shows a minimal phosphorus content of about 2.5×10^{-11} μg. per infective unit. The preparations show significant differences, however. Numerous preparations of T2H grown in synthetic medium showed an extinction coefficient of 1.0×10^{-11} cm.^{2} per infective unit, whereas the same phage grown in broth usually reads 30 per cent lower. The excessive readings are not due to nucleic acid impurities, because neither elaborate fractional sedimentation nor treatment with desoxyribonuclease has any effect. The variable factor is therefore the efficiency of plating (E) of the phage. Since the best preparations of T2H, T4, and T5 all have about the same nucleic acid content per infective unit, it seems likely that E is near unity in favorable cases. This is an assumption, however, and in what follows we retain E as an undetermined constant in all calculations. Thus the phosphorus content is 2.5×10^{-11} E μg. per particle, or $N = 5 \times 10^5 E$ atoms.

In other connections we need also an estimate of the volume of the phage particle. According to Taylor (1946), purified phage T2 contains 5 per cent dry weight of phosphorus. From this datum, a phosphorus content of 2.5×10^{-11} E μg. per particle, an assumed density of 1.15 and water content 38 per cent, the volume is $7.0 \times 10^{-16}\ E$ cm.³, which corresponds to a spherical diameter of 110 $E^{1/3}$ mμ.

TABLE V

Action of Desoxyribonuclease on Phage

Phage	Preparation No.	Titer	P/phage	E260/phage
		$\times 10^{11}$/ml.	$\times 10^{-11}$ μg.	$\times 10^{-12}$ cm.²
T4-S	1c	7.8	2.7	7.7
"	1e	7.8	2.7	7.4
"	2c	12	2.5	7.2
"	2e	12	2.5	7.0
"	3c	13	2.7	7.2
"	3e	12	3.0	7.4
"	4c	25	3.3	9.0
"	4e	26	3.2	8.5
T4-N	1c	0.69	3.3	9.3
"	1e	0.71	2.9	8.3
"	2c	0.95	2.7	7.7
"	2e	1.2	2.5	7.3
"	3	10	2.5	6.7
T2H-S	1	51	3.5	10
"	2	100	3.4	9.0
T2H-N	1	3.3	2.2	6.6
"	2	1.9	2.6	6.6
T5-S	1	4.4	—	5.0
"	2	3.1	—	6.0
"	3	3.1	—	5.0

T4-S, etc., mean phage grown in a synthetic glucose-ammonia medium.

T4-N, etc., mean phage grown in a medium containing 0.5 per cent peptone, 0.5 per cent NaCl, and 0.1 per cent glucose.

1e is phage incubated with desoxyribonuclease.

1c is the same phage incubated without enzyme, etc.

The titers given refer to independent assays of each sample after preparation for analysis.

*E*260/phage is the extinction coefficient per phage particle at 260 mμ wave length, partially corrected for scattering by subtracting the extinction at 330 mμ.

The Quantum Efficiency of Killing and the Cause of Death

From the estimates already made of $\alpha N = 4.3 \times 10^4$ and $N = 5.0 \times 10^5\ E$, where E is the efficiency of our plaque count of T4, the efficiency of killing of phage by decay of assimilated P^{32} is $0.086/E$ per atom disintegrating. A phage particle dies, therefore, after an average of about 11.6 disintegrations.

When an atom of P^{32} disintegrates, a beta particle with average kinetic energy 7.0×10^5 e.v. is emitted, the nucleus absorbs a recoil of the order 30 e.v., and the atom itself is left as sulfur. There is probably an additional local dissipation of energy coming from the rearrangement of electrons in the atom. Any of these effects can be thought of as the cause of death of the phage. The following argument seems to show that the beta radiation is not the cause of death.

A beta particle is supposed to produce its principal biological effects by ionizing atoms along its track. For the beta particles of the P^{32} spectrum the average initial specific ionization is 7.1 ion pairs per mm. of air. Beta particles arising from P^{32} randomly distributed within spherical phage particles of diameter 110 μ would traverse an average distance of 4.1×10^{-5} mm. on their way out, which is equivalent to a path in air of 0.04 mm. Thus there are on the average about 0.28 ion pairs produced within the phage by each beta particle emitted. There is, therefore, an average of 11.6 E times this number, or 3.3 E ionizations produced inside the phage particle before it dies. If these ionizations are the cause of death, the efficiency of killing is about 0.3 E per ionization.

The quantum efficiency of killing by beta radiation from P^{32} has been measured directly by mixing phage with radioactive phosphorus in 1 per cent peptone solution, and measuring the rate of death by periodic titration of the mixture. The rate of death under these conditions is expected to be

$$-\frac{dS}{dt} = 6.9 \times 10^{16} maVS = 6.9 \times 10^{16} m_0 aVSe^{-\lambda t} \tag{3}$$

from which

$$\log S = \log S_0 - \frac{6.9 \times 10^{16}}{2.3\lambda} m_0 aV(1 - e^{-\lambda t}) \tag{4}$$

where $m = m_0 e^{-\lambda t}$ is the radioactivity in millicuries per ml., a is the efficiency of killing per ionization within V, V is the volume of the phage particle, and 6.9×10^{16} is the rate of ionization per day per millicurie of radiation. We neglect in this computation any radiation absorbed by or passing through the walls of the tube.

The results of four experiments, in which the titration data are fitted to equation (4) by the method of least squares, are shown in Table VI. The data conform to the equation as shown by the values of χ^2, and the estimates of aV are in reasonable agreement with each other.

The experiments dated Apr. 3, 1950, in Table VI were done with the same phage and the same sample of P^{32} used in the measurements shown in Table II, and form the appropriate basis for comparison. From the constant $aV = 6.3 \times 10^{-18}$ ml. and the estimated volume of the phage particle given previously, we get an efficiency of killing by beta radiation of $0.009E$ per ionization.

This result is not in very good agreement with the data for killing by x-rays. Latarjet (1948) found, in substantial agreement with Luria and Exner (1941), a mean lethal dose for T2L of 3.4×10^4 r or 5.4×10^{16} ionizations per ml., as compared with our figure of 1.6×10^{17} ionizations per ml. According to our estimate of the volume of the phage particle, the x-ray data correspond to an efficiency of killing of 0.027 E per ionization. The discrepancy is unexplained.

Whichever estimate is used, it is clear that the efficiency of killing by beta radiation is too low to account on this basis for the deaths from assimilated P^{32}; according to our data 34 times too low. We conclude that the inactivation

TABLE VI

Inactivation of Phage by Beta Radiation from P^{32}

Date	Phage tested	Initial radio-activity	Log S_0	aV	χ^2
		mc./ml.		$\times 10^{-}$ *ml.*	$\times 10^{-4}$
10/12/49	T2Hr1	0.55	8.29	8.9	16
11/14/49	"	0.11	8.30	9.5	15
4/ 3/50	T4	0.68	9.25	6.3	18
4/ 3/50	"	0.34	9.24	6.3	12

The quantity aV is a specific death rate equivalent to a "sensitive volume;" $1/aV$ = mean lethal dose in ionizations per ml. All mixtures contain 1 per cent peptone. The phage-P^{32} mixtures in the experiments of Oct. 12 and Nov. 14 are identical except for intervening decay of P^{32}. Counts in the latter experiment have been corrected for spontaneous inactivation (2.3 per cent per day) of the control phage. There was less than 0.8 per cent per day spontaneous inactivation in the other experiments.

of phage containing assimilated P^{32} is rarely due to beta radiation, but is usually a direct consequence of the nuclear reaction, which kills with an efficiency of 1/11.6 or a little greater.

DISCUSSION

The finding that the nuclear reaction $P^{32} \rightarrow S^{32}$ kills radioactive phage with an efficiency of about 0.086 per atom transformed presumably means that at least this fraction of the phosphorus atoms of the phage particle is situated in vital structures, and therefore that the vital structures contain nucleic acid.

The relatively low efficiency found might be considered doubtful evidence for the conclusion stated if it were not that the efficiency of killing by ionizations produced in atoms in general is at least as low: about 0.009 for beta rays from P^{32} and about 0.027 for x-rays, both computed per ionization. These estimates, though precise as stated, are of somewhat doubtful appropriateness because of uncertainty about the quantum of action.

It is tempting to assume from our results that 9 per cent or more of the phosphorus atoms of the phage are strategic in the sense that disintegration of any one of them kills with high probability, and to relate this to the finding of Putnam and Kozloff (1950) that when a radioactive phage particle infects a bacterium, only 20 to 40 per cent of its phosphorus appears among the viral progeny. Thus one imagines a class of non-essential structures containing about 70 per cent of the phosphorus atoms and subject to metabolic turnover, and a second class containing about 30 per cent of the phosphorus atoms, whose disintegration kills with efficiency about 0.33, and which are conserved during replication of the structures. In this situation our method of assay would measure exclusively the radioactivity of the 30 per cent fraction, and would be competent to deal with the question whether the conserved atoms appear in one or in all of the progeny of a labelled parent.

Our present data exclude the possibility that the conserved atoms appear in only one of the phage progeny. When bacteria growing in radioactive medium are infected each with several non-radioactive phage particles, at least 99.9 per cent of the phage progeny contain essential phosphorous that is radioactive (curve G, Fig. 1). Thus if materially conserved essential structures containing phosphorus exist, it must be supposed that they are dispersed among the progeny during growth, a possibility already suggested by the phenomenon of genetic recombination, in which one sees a functional redistribution of functionally conserved structures during growth (Hershey and Rotman, 1949).

The test of the general hypothesis that materially conserved essential structures exist depends among other things on a test to determine whether the phage progeny as a whole contain all the essential phosphorus that their parents possessed. Our results suggest that this test may now be feasible.

SUMMARY

The bacteriophage T4 containing assimilated radioactive phosphorus is inactivated at a rate proportional to the specific radioactivity of the constituent phosphorus. The beta radiation from the phosphorus makes a negligible contribution to this effect. The inactivation is therefore a direct consequence of the nuclear reaction, which kills the phage with an efficiency of about 1/12. Several phages related to T4 behave similarly.

When radioactive phage is grown from a seed of non-radioactive phage, all of the phage progeny are subject to killing by radioactive decay.

The phage is killed by beta radiation from P^{32} with an efficiency of about 1/100 per ionization within the particle volume.

Bacteriophage T4 and its relatives contain about 500,000 atoms of phosphorus per infective particle. Virtually all this phosphorus is adsorbed to bacteria with the specificity characteristic of the infective particles, and none of it can be removed from the particles by the enzyme desoxyribonuclease. The phosphorus content per particle, together with the published data on analytical composition, indicates a particle diameter close to 110 mμ for the varieties of phage studied.

REFERENCES

Adams, M. H., 1948, *J. Gen. Physiol.*, **31,** 417.
Adams, M. H., 1949, *J. Gen. Physiol.*, **32,** 579.
Cohen, S. S., 1947, *Cold Spring Harbor Symp. Quant. Biol.*, **12,** 35.
Doermann, A. H., 1948, *Carnegie Institution of Washington Yearbook, No. 47*, 176.
Ellis, E. L., and Delbrück, M., 1939, *J. Gen. Physiol.*, **22,** 365.
Hershey, A. D., 1941, *J. Immunol.*, **41,** 299.
Hershey, A. D., and Rotman, R., 1949, *Genetics*, **34,** 44.
Kozloff, L. M., and Putnam, F. W., 1949, *J. Biol. Chem.*, **181,** 207.
Kozloff, L. M., and Putnam, F. W., 1950, *J. Biol. Chem.*, **182,** 229.
Latarjet, R., 1948, *J. Gen. Physiol.*, **31,** 529.
Luria, S. E., and Exner, F. M., 1941, *Proc. Nat. Acad. Sc.*, **27,** 370.
McCarty, M., 1946, *J. Gen. Physiol.*, **29,** 123.
Northrop, J. H., 1938, *J. Gen. Physiol.*, **21,** 335.
Putnam, F. W., and Kozloff, L. M., 1950, *J. Biol. Chem.*, **182,** 243.
Taylor, A. R., 1946, *J. Biol. Chem.*, **165,** 271.

SERENDIPIC ASPECTS OF RECENT NUTRITIONAL RESEARCH IN BACTERIAL PHOTOSYNTHESIS

Martin D. Kamen and Howard Gest
Mallinckrodt Institute of Radiology
Washington University Medical School, St. Louis
Department of Microbiology
Western Reserve Medical School, Cleveland

In 1887, Esmarch, desirous of obtaining some putrifying bacteria, immersed a mouse dead of septicemia in a beaker of distilled water and, three months later, examined the dried residue (7). Among the many microorganisms present, he was delighted to obtain, on isolation from nutrient gelatine, colonies of a brightly colored purple organism which appeared to be a pure specimen of a " true spirillum." Thus what was later to be known as *Rhodospirillum rubrum* made its first recorded appearance as an apparent heterotroph.

In those days, a pure culture of a spirillum (and a non-pathogenic one at that) was a comparative rarity, and Esmarch was delighted with his find. His description of the cavortings of the new organism is still as good as any. In his own words: " Die kurzen Spirillen zeigten sämmtlich sehr lebhafte Bewegung; theils schossen sie, wie vom Bogen geschnellt, in gerader Linie durch das Gesichtsfeld des Microskops, wobei die Windungen gänzlich verschwanden .... theils schroben sie sich etwas langsamer, und darum deutlicher zu beobachten, wie ein Pfropfenzieher in einem Weinkork in bohrender Bewegung durch die Flüssigkeit."

The photosynthetic character of *R. rubrum* became apparent soon after its discovery. To each succeeding generation of microbiologists interested in photosynthesis this organism, Katzenjammer-like in its hardihood and mischievousness, has appeared in different metabolic and physiologic moods. A milestone in the study of general metabolic patterns, physiology, and nutritional characteristics for this and related bacteria was reached with the appearance of van Niel's classic monograph on the *Athiorhodaceae* in 1944 (33).

One might have thought that this organism, now a sexagenarian,

Reprinted from "Phosphorus Metabolism," Vol. II
(William D. McElroy and Bentley Glass, editors).
The Johns Hopkins Press, Baltimore, 1952.

had little left in its bag of tricks. However, as we shall see, it remains as lively as ever. In view of the nature of the Institute sponsoring this Symposium it seems especially appropriate on this occasion to relate its most recent escapade, particularly as it provides an unusual example of the serendipic aspects of supposedly routine nutritional research.

We may begin by recalling the situation with regard to the role of phosphate in photosynthetic metabolism as it existed in 1947. At that time a number of theories were in existence (26, 29) involving the hypothesis that high-energy phosphate bonds, in particular adenosine triphosphate (ATP), could function as the reservoir for energy generated by light absorption. To test the relation of the light absorption act to the phosphate metabolism of the cell, we proposed to measure the rate of uptake and turnover of phosphate in various cell fractions as a function of illumination, using P^{32}-labeled inorganic phosphate to trace the flow of phosphate into the cells and thereafter through the various fractions. In addition to the well-known test organisms, the green algae *Chlorella* and *Scenedesmus*, we also had available various representatives of the photosynthetic bacteria, including a strain of *R. rubrum* kindly supplied us by Professor van Niel.

Almost immediately we encountered a number of phenomena which obviated unambiguous interpretation of the uptake and turnover experiments (11). The experimental procedures involved the usual technique of harvesting the cells and washing them to remove traces of nutrients and other substances from the occluded culture fluid. It was found that successive washes with saline solution, even at 0° C., removed internal phosphate gradually and continuously. Using cells uniformly labeled by growth in P^{32} labeled media, it was found that free exchange occurred between cellular phosphates and external phosphate, during washing with solutions containing inorganic phosphate. However, algae which had been washed exhaustively still showed normal photosynthetic and respiratory activity, in spite of considerable loss of cellular phosphate. Similar results were obtained with *R. rubrum*.

It was possible to show that the washing of *R. rubrum* suspensions with saline solution could result in loss of as much as 87 per cent of the acid-soluble phosphate of the cells without affecting their metabolic activity. When fresh medium (containing inorganic phosphate) was used as a wash liquid in the same experiments, the acid-soluble phosphate of the cells could be *increased* by some 50 per cent. Phenomena such as these completely obscured the distribution of labeled phosphate which might have obtained immediately after the test period.

The fact that much of the cellular phosphate could be removed without affecting the cellular respiration and photosynthetic activity suggested that the removable phosphate was "dispensable" in the cell economy. It was reasoned that storage of this excess phosphate might possibly be eliminated by growing the organisms in a medium with very low P content (5-10 γ per ml. as compared with the usual 200-500 γ per ml.). In fact, it was found that algae grown in low P medium contained about half the P per unit dry weight of that found in cells grown in the usual manner. Furthermore, a greater percentage of the phosphate in the "low-P" cells was in the acid-insoluble form. None of the acid-soluble P from such cells could be removed by washing with saline. On the other hand, the usual loss of P found after washing the "high-P" cells appeared to be at the expense of the acid-soluble P.

Thus, it appeared that there were two kinds of phosphate in the cell, one of which was "dispensable," removable by washing, or exchangeable with external phosphate, the other of which was firmly bound and non-dispensable.

With regard to this behavior of organisms grown at various concentrations of P, it should be remarked that similar observations were already available in the literature. Thus, Mann (27) observed that in growing cultures of *Aspergillus niger*, growth in 0.2-0.5 per cent KH_2PO_4 media resulted in mold with a P content of 1-2 per cent of its dry weight, in contrast with 0.3 per cent obtained when the medium contained only 0.02 per cent KH_2PO_4. Like observations had been recorded by Ketchum for *Nitzschia closterium* (22).

The phenomenon of "dispensable phosphate," as noted in algae and in the bacteria in our researches, has been most recently remarked in the diatom *Asterionella japonicum* (16) as well as in *Torulopsis utilis* (28). A similar phenomenon appears to be involved in sulfur uptake by *E. coli* (4).

Having removed the complication of excess soluble cell phosphate by proper culturing, we proceeded to study the uptake and turnover of phosphate in cellular metabolism as a function of light and dark exposure. It was possible to demonstrate unambiguously a light-stimulated uptake of phosphate and a light-stimulated turnover of ester-phosphate between acid-soluble and acid insoluble fractions of the cell. Although the fractionation of cellular P into only two gross fractions undoubtedly minimized the differential effect, it was possible to observe an increase by at least a factor of two in the turnover between the two fractions as the result of illumination (11). Recently, Vishniac (34) has shown that in spinach chloroplast suspensions incubated with diphosphopyridine nucleotide (DPN), rat liver mitochondria, ATP and labeled inorganic phosphate, there is a threefold increase in incorporation of radioactive phosphate in ATP in the light over that found in the dark. Simonis and Grube (31) also report a light-enhanced uptake of inorganic phosphate in the tri-chloroacetic acid-soluble fraction during short periods of CO_2 fixation by isolated leaves of *Helodea densa.* It must be emphasized that no experiments yet devised give information on the *direct* coupling of phosphorylation to initial hydrogen transfer in photosynthesis as it occurs in the cell. They do show, however, that thermal reactions initiated by the light-stimulated production of reducing substances can result in the synthesis of high-energy phosphate.

To come back to our story—the ease of cultivation, the hardy character and metabolic versatility of *R. rubrum* now suggested further researches on the role of phosphate in photosynthetic metabolism using this organism. By this time, the state of nutritional knowledge on photosynthetic bacteria had been elaborated far beyond that of the early days of Esmarch's mouse.

We may review briefly the nutritional requirements and metabolic patterns of *R. rubrum* and other members of the *Athiorhodaceae*, as elaborated so extensively by van Niel (33). Development of the organisms was found generally to be optimal under anaerobic conditions in the light, using a simple salt medium containing a single carbon compound as carbon source and small amounts of a source of growth factors, such as yeast extract. Hutner had shown (18) that biotin could, to a limited extent, replace the yeast complex. The reduction levels of the organic compound used and the products formed determined the necessity for the provision of exogenous carbonate to obtain growth. Thus, assuming that the over-all reduction level of the photosynthetic products was at the level of carbohydrate, it would be expected, and it was found, that carbonate was required for growth on substrates. More oxidized substrates manifested their photosynthetic utilization by exhibiting a net production of CO_2.

No gaseous products other than CO_2 had ever been noted in bacterial photosynthesis. This fact, together with the nutrition data on a wide variety of photosynthetic bacteria, had been primarily responsible for the well-known formulations of van Niel (32) expressed in a general photosynthetic equation of the type

$$CO_2 + 2H_2A \xrightarrow{h\nu} CH_2O + 2A + H_2O.$$

With the arrival of summer in 1948, it was necessary to aestivate because of the heat or continue the research in a more equable climate. It was decided to follow the latter course. One of us (HG) repaired to Professor van Niel's Laboratory, at Pacific Grove, California, with the primary object of accomplishing the apparently routine task of defining the minimal phosphate requirement for satisfactory growth of *R. rubrum.*

A modification of the most recently developed medium, that of Hutner, was chosen as a starting point. The medium contained in addition to the usual mineral salts, a small quantity of yeast extract, a carbon source such as malate or some other dicarboxylic acid, and, as nitrogen source, glutamate, rather than the classic nitrogen source, ammonia. This medium, while well defined in most respects, contained a rather high amount of phosphate, with the consequent production of cells too rich in dispensable P. Hence, the organism was to be grown in the usual anaerobic manner in glass-stoppered bottles using media of progressively lower P content. As controls, duplicate bottles were to be set up and kept in the dark.

Within a short time after these experiments were started, a copious evolution of gas was observed in the growing illuminated cultures but not in the dark cultures. It was quickly ascertained that this gas contained molecular hydrogen. Strenuous efforts to detect a bacterial contaminant which might account for the hydrogen production as a fermentative product failed. This, together with the failure to observe gas production in the dark controls, established the fact that *R. rubrum* was producing molecular hydrogen as a photosynthetic product, in addition to the usual CO_2 (12).

The finding that hydrogen was a normal product of photosynthesis under these conditions was most surprising, because these same organisms were known to carry out a vigorous photoreduction of carbon dioxide by molecular hydrogen; yet, in the presence of the organic substrate, both carbon dioxide and hydrogen were evolved.

These remarkable observations were further investigated upon resumption of work in the fall in St. Louis. Our first experience was failure to duplicate the observations made during the summer. We could obtain good growth with the same media, but no hydrogen production. After considerable investigation, we chanced on the observation that the production of hydrogen was critically dependent on the concentration of yeast extract. The yeast extract used in California had been a manually prepared liquid extract, whereas the commercial preparation we were now using was a concentrated powder.

It was found that over a restricted dilution range the production of hydrogen could be recaptured (13). In an attempt to identify the inhibiting agents in the yeast extract, we discovered that ammonium salts were very effective inhibitors of hydrogen production. This observation explained the failure to observe hydrogen production in all of the years previous to these investigations. Ammonium salts have always been the classic nitrogen source in photosynthetic research, whereas in our tests the nitrogen source was glutamic acid, which apparently could sustain growth without interfering with hydrogen production. The substitution of glutamate for ammonia was inadvertent and a holdover from Hutner's earlier experiments with strains incapable of using ammonia.

We now began attempts to elaborate the stoichiometry of the reaction. This was a critical feature of the investigation, because much depended on the magnitude of the hydrogen production, compared to the dissimilation of substrate, in assessing whether the effect was a primary or secondary metabolic reaction.

Manometric experiments were performed, using resting cells (i. e., cells not provided with an exogenous nitrogen source) to establish the magnitude of the photo-production of hydrogen. Again we found ourselves frustrated by this mischievous organism. Cells which had been producing H_2 uniformly in the bottles refused to produce hydrogen in the Warburg vessels. The only difference in the conditions employed was that in the manometric experiments a gas space was present for the pressure measurements. This gas space was filled with nitrogen gas, which we had assumed to be inert with respect to the metabolism of these organisms.

Continued failure to observe hydrogen production led us to assume that inhibition of hydrogen production might arise from the presence of small quantities of oxygen in the nitrogen gas used. This notion was predicated on the observations of Hoberman and Rittenberg (17) and others, who had shown that the hydrogenase of enteric bacteria was remarkably susceptible to traces of oxygen; inactivation was assumed to arise from oxidation of an iron component in the enzyme. It seemed likely that the functioning of the system responsible for the hydrogen production required a redox potential close to that of the hydogen electrode and that traces of oxygen might have raised the potential of the chemical environment into the region of inactivation of the enzyme.

During this discouraging period, the hydrogen metabolism of *R. rubrum* was compared with that of *Diplococcus glycinophilus* (3). The latter organism appeared to behave exactly opposite to *R. rubrum*, in that (a) it did not produce hydrogen during fermentation in stationary cultures in a closed system with small gas space, and (b) it did produce hydrogen in a vessel with a large gas space filled with nitrogen. Also, it had been noted that an atmosphere of 100 per cent hydrogen completely inhibited hydrogen formation by *D. glycinophilus*. Following a philosophy of desperation, it was decided that *R. rubrum* might show a positive production of gas in the presence of KOH under a hydrogen atmosphere, in direct contrast to the behavior of *D. glycinophilus*. This, in fact, was found. Unfortunately, it could not be conclusively established that the gas produced was hydrogen, when hydrogen itself was the gas phase employed.

The suspicion that molecular nitrogen might be the active inhibitor

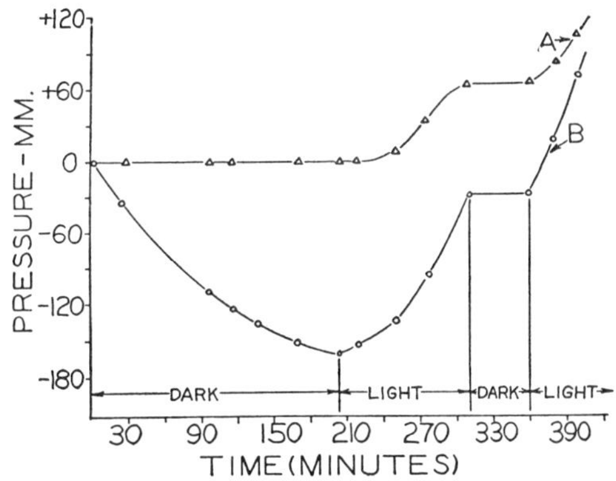

FIG. 1. Photoproduction of H_2 from Malate by *R. rubrum* in the Presence of Oxygen. After Gest, Kamen, and Bregoff (14).

The vessels contained 80 c.mm. of washed cells suspended in 2 ml. 0.05 M. phosphate buffer (pH 6.6). 5 mg. D,L-malate tipped in at zero time; 0.2 ml. 10% KOH in center well.

Curve A, 100% He; Curve B, 2.5% O_2-97.5% He.

in the manometric experiments was now entertained. The noble gases, argon and helium, were tried, and under these gases voluminous production of hydrogen occurred (12, 14). It could be shown that oxygen itself did not inhibit hydrogen production under illumination, because (a) some of the argon used in the manometric experiments contained traces of oxygen as a contaminant, and (b) cells allowed to assimilate malate in the dark with uptake of approximately half of the oxygen, present originally as a mixture of 2.5 per cent oxygen in 97.5 per cent helium, upon illumination immediately began to produce molecular hydrogen (Fig. 1). Thus, even long incubation under oxidative conditions did not abolish the production of hydrogen. Subsequent experiments, using nitrogen and helium mixtures, showed that as little as 5 per cent of nitrogen in 95 per cent helium sufficed to abolish completely the production of hydrogen, without noticeably affecting the production of CO_2 during the light dissimilation of malate (12, 14).

It was now impossible to escape the conclusion that nitrogen gas itself was not inert with respect to the metabolism of *R. rubrum.* It was shown then by experiments with N^{15}-labeled nitrogen as a sole nitrogen source that fixation of nitrogen gas occurred readily (21). In fact, it was found possible to obtain good growth of *R. rubrum* using the usual medium with molecular nitrogen as the sole nitrogen source (21, 14). Thus it was demonstrated that in addition to all of the other metabolic activities of this organism it was also a nitrogen fixer. This discovery was later elaborated by Wilson and his coworkers (24, 25) and shown to apply to practically all specimens of the photosynthetic bacteria.

Thus, within 4 years after the appearance of the monograph by van Niel which appeared to have embraced all the major aspects of the nutrition and physiology of *R. rubrum*, this organism continued to astound investigators with still another major metabolic trick. This serendipic episode in the history of nutrition research with *R. rubrum* quite effectively distracted our attention from further work on the phosphate metabolism of this organism, in favor of an intensive investigation of the phenomenon of photohydrogen production and nitrogen fixation in photosynthetic bacteria.

The present situation with regard to the metabolism of these organisms cannot be adequately summarized in the time allotted. We may describe in a few words the general metabolic pattern of *R. rubrum* as exhibiting both fermentative (23) and oxidative dark metabolism (33), in addition to anaerobic photosynthesis (33) and nitrogen fixation (21). It would appear to be as good an approximation to the Shmoo as we will see in microbiology. For the remainder of this period we will devote discussion to a summary of the facts as they are known at present concerning photohydrogen production in *R. rubrum* and in some related organisms (see refs. 1, 2, 5, 6, 9, 15, 19, 23) for general discussion and some recent findings).

Photoproduction of hydrogen can be observed in growing or resting cultures. Any of a wide variety of substrates appears to support hydrogen production (30). These include substances as reduced as alcohols and as oxidized as organic acids. No success has yet been attained in evoking photohydrogen production in strictly inorganic media (1). For the production of hydrogen with reduced substrates it appears that the presence of carbonate is essential (30). Photoreduction of CO_2 in resting cells does not occur as long as organic substrate is present to maintain photoorganic metabolism (20).

Concomitant with hydrogen production, there is observed production of carbon dioxide during the dissimilation of an organic substrate such as malate. However, the carbon dioxide production can be dissociated from the hydrogen production by various inhibitors, such as azide and cyanide in the usual range, viz., 10^{-3} to 10^{-4} molar. Using azide with malate as a substrate, inhibitions of 60 per cent to 100 per cent are obtained associated with inhibitions of from 10 per cent to 30 per cent in CO_2 production (20).

Ammonia and nitrogen gas, while inhibiting the production of hydrogen do not inhibit the uptake of hydrogen in photoreduction (20, 30). Photoreduction of CO_2 by hydrogen is not inhibited by the organic substrate in the presence of ammonia (20). Of the amino acids which may be used in growing cultures, glutamic and aspartic acids, proline, lysine, and arginine support hydrogen production, whereas alanine, serine, histidine, glycine, asparagine, glutamine, hydroxyproline, and to cystine inhibit it (2). In all cases where inhibition is found it is also possible to observe a steady-state concentration of ammonia, presumably resulting from deaminase activity of the cells. The inhibition by ammonia or N_2 is reversible (2). Removal of NH_4^+ by metabolism, or N_2 by flushing with helium, results in resumption of hydrogen production at rates identical with control cultures (Fig. 2).

The dissimilation of malate during photoproduction of hydrogen and CO_2 by resting cells proceeds in stages (2); in the first stage, the mol ratios of carbon dioxide and hydrogen evolved to malate dissimilated are 1:1:1. In later stages, when practically all the malate has been dissimilated, the gas production is maintained, resulting in some cases in an over-all production as much as 3 mols of hydrogen for each mol. of malate dissimilated (2). This ratio has also been observed in the case of lactate (9). These results are strong evidence for the primary nature of the phenomenon.

Exogenous phosphate is not required for the production of hydrogen in resting cells; phthallate or borate buffers are equally effective (1, 20). In *R. rubrum*, photoproduction of H_2 shows no marked pH

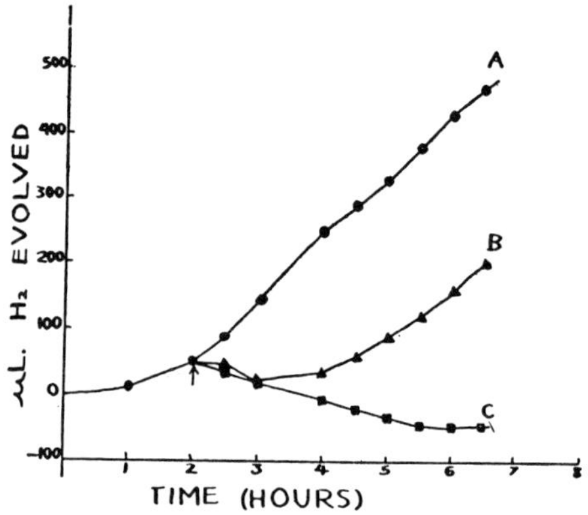

FIG. 2. Inhibition of Photohydrogen Production by NH_4Cl in Resting Cell Suspensions of *R. rubrum*. After Bregoff and Kamen (2).
Conditions as in Fig. 1. At 2 hrs. (arrow) NH_4Cl added
Curve A: control, no NH_4Cl.
Curve B: 4.6 µM. NH_4Cl added.
Curve C: 34.4 µM. NH_4Cl added.
Residual ammonia (6½ hrs.); Curve B, 0; C, 18.2 µM.

dependence over the range 5.8-7.2 (14). In *Chromatium*, which grows best at pH 8.0 or higher, H_2 production is observed at pH 8.0-9.0 (1). However, it appears that the photoreduction of CO_2 by hydrogen is more effective in *R. rubrum* at pH 7.5 than at more acid pH (20).

When molecular nitrogen is present, no uptake of nitrogen is observed in resting cells, despite the fact that in the same period of time large amounts of malate disappear without the production of the copious amounts of hydrogen observed under helium (2). In other words, the hydrogen metabolism of the organisms in the resting state is profoundly altered without an apparent net uptake of nitrogen.

Insufficient data have been accumulated to speculate profitably on the mechanism involved in the photoproduction of hydrogen (see refs. (9, 19) for discussion). It seems obvious that a renewed study of the production of hydrogen in heterotrophic organisms and, in particular, the isolation of hydrogenase and associated enzymes in a well defined state will be helpful. Preliminary experiments with cell-free hydrogenase preparations from *R. rubrum* indicate that the enzyme is very similar to the hydrogenase of the heterotrophic *E. coli* (10). In the former case, illumination has no apparent effect on the rate of reduction of acceptors by H_2. Although it cannot be stated that the phenomenon of photoproduction of hydrogen is directly connected with hydrogen transfer in the light absorption process, there is a sufficiently strong possibility that the production of hydrogen in quantities of such magnitude during illumination may be the result of an accumulation of the primary reduced hydrogen carrier consequent upon the light absorption process.

Experiments to elucidate directly the nature of the precursor of molecular hydrogen will be facilitated by discovery of a system in which equilibration between bound hydrogen and water hydrogen does not occur. Frenkel has shown (8) that in *Chlamydomonas* an endogenous production of hydrogen occurs which is stimulated by illumination. If this organism is incubated in a medium containing D_2O, then during the early stages of the reaction the gas produced is enriched in light hydrogen. Thus, exchange between bound hydrogen and water hydrogen appears to be slow enough to indicate the source of hydrogen as among the endogenous reserves of the cell. Unfortunately, exchange is rapid between water hydrogen, hydrogen gas, and cell hydrogen when *R. rubrum* is the organism involved (8). Experiments using carbon-labeled substrates and comparing the metabolic patterns and products produced under nitrogen or in the presence of other inhibitors are indicated. The intriguing relation between nitrogen metabolism and photosynthetic hydrogen transfer exhibited by these organisms remains a subject for future investigation.

There appears to be little doubt that *R. rubrum* and other photosynthetic bacteria afford as rich a field for investigation in the next half century as they have the first half of this century. We may repeat the sentiment of van Niel, who stated in the introduction to his monograph in 1944: " It is hoped that the publication of [these] results may make these bacteria more generally known to microbiologists, and thus lead to more intensive studies of the many fascinating problems which they still present."

REFERENCES

1. Bregoff, H. M., and Kamen, M. D., *J. Bact.*, **63**, 147 (1952).
2. Bregoff, H. M., and Kamen, M. D., *Arch. Biochem. Biophys.*, **36**, 202 (1952).
3. Cardon, B. P., and Barker, H. A., *Arch. Biochem.*, **12**, 165 (1947).
4. Cowie, D. B., Bolton, E. T., and Sands, M. K., *J. Bact.*, **60**, 233 (1950).
5. Cutinelli, C., Ehrensvaard, G., and Reio, L., *Ark. Kemi*, **2**, 357 (1950).
6. Cutinelli, C., Ehrensvaard, G., Reio, L., Saluste, E., and Stjernholm, R., *Ark. Kemi*, **3**, 315, 501 (1950).
7. Esmarch, E., *Centr. Bakt.*, **1**, 225 (1887).
8. Frenkel, A. W. (private communication).
9. Gest, H., *Bact. Rev.*, **15**, 183 (1951).
10. Gest, H., *J. Bact.*, **63**, 111 (1952).
11. Gest, H., and Kamen. M. D., *J. Biol. Chem.*, **176**, 299 (1948).
12. Gest, H., and Kamen, M. D., *Science*, **109**, 558 (1949).
13. Gest, H., and Kamen, M. D., *J. Bact.*, **58**, 239 (1949).
14. Gest, H., Kamen, M. D., and Bregoff, H. M., *J. Biol. Chem.*, **182**, 153 (1950).
15. Glover, J., Kamen, M. D., and Van Genderen, H., *Arch. Biochem. Biophys.*, **35**, 384 (1952).
16. Goldberg, E. D., Walker, T. J., and Whisenand, A., *Biol. Bull.*, **101**, 274 (1951).
17. Hoberman, H. D., and Rittenberg, D., *J. Biol. Chem.*, **147**, 211 (1943).
18. Hutner, S. H., *J. Bact.*, **52**, 213 (1946).
19. Kamen, M. D., *Federation Proc.*, 9, 543 (1950).
20. Kamen, M. D. (unpublished observations).
21. Kamen, M. D., and Gest, H.. *Science*, **109**, 560 (1949).
22. Ketchum, B. H., *J. Cell. Comp. Physiol.*, **13**, 373 (1939).
23. Kohlmiller, E. F., and Gest. H., *J. Bact.*, **61**, 269 (1951).
24. Lindstrom, E. S., Burris. R. H., and Wilson, P. W., *J. Bact.*, 58, 313 (1949).
25. Lindstrom, E. S., Tove, S. R., and Wilson, P. W., *Science*, **112**, 197 (1950).
26. Lipmann, F., *Adv. in Enzymol.*, **6**, 231 (1946).
27. Mann, T., *Biochem. J.*, **38**, 339 (1944).
28. Rautanen, N., and Miikulainen, P., *Acta. Chem. Scand.*, **5**, 89 (1951).
29. Ruben, S., *J. Am. Chem. Soc.*, **65**, 279 (1943).
30. Siegel, J. M., and Kamen, M. D., *J. Bact.*. **61**, 215 (1951).
31. Simonis, W., and Grube, K. H., *Z. Naturforsch.*, 7b, 194 (1952).
32. van Niel, C. B., *Adv. in Enzymol.*, **1**, 263 (1941).
33. van Niel, C. B., *Bact. Rev.*, **8**, 1 (1944).
34. Vishniac, W., *Federation Proc.*, **11**, 302 (1952).

A NEW SOLUBLE CYTOCHROME

Sir:

Rhodospirillum rubrum, a photosynthetic heterotrophic bacterium, contains large amounts of a haemprotein which may be obtained by treatment with warm trichloroacetic acid, by extraction with phosphate buffer from cell residues after acetone treatment and by sonic disruption of cell suspensions. The protein so obtained can be purified in good yield by a modified Keilin–Hartree procedure.[1] The same protein is also obtained with similar yields from another photoheterotrophe, *Rhodopseudomonas spheroides*. Although this haem protein (I) possesses many properties usually attributed to mammalian cytochrome-c (II) it is, in fact, a new cytochrome. Thus, although I, like II, can be reduced with DPNH via DPNH-cytochrome-c reductase (prepared either from *R. rubrum* or pig heart), is not auto-oxidized, can be reversibly reduced or oxidized by reagents such as the ferro-ferricyanide couple, hydrosulfite and ascorbate, and exhibits an absorption spectrum in the visible identical with that of reduced II, it differs from II in the following important particulars: (a) I is not oxidized in air in the presence of the cytochrome oxidase system, whether the enzyme is prepared from pig heart, pig kidney or rat kidney. Preparations of I, purified electrophoretically (see below), do not inhibit the oxidase, as evidenced by unabated activity of the enzyme in catalyzing oxidation of II in the presence of I. (b) I is not absorbed on the NH_4^+ form of Amberlite IRC-50 ion exchange resin, prepared according to the directions of Margoliash.[2] (c) I can be only partially oxidized by a ferro–ferricyanide oxidation–reduction buffer (20:1) under conditions which permit complete oxidation of II. (d) The cathodic mobility of I on ionophoresis in 1% ammonium acetate on Whatman no. 5 paper is less than that of II by at least one order of magnitude.

Preparations of I, as obtained by the Keilin–Hartree procedure, usually exhibit a purity of 10–50%, based on spectral absorption properties and the assumption that the extinction coefficients at the characteristic absorption maxima are closely similar to those for II. Electrophoretic treatment of such preparations results in a 2- or 3-fold purification because of removal of colorless impurities which migrate rapidly toward the cathode.

The conclusion that I is not identical with II is further strengthened by the observation that I can be oxidized via mammalian cytochrome oxidase, if

(1) L. P. Vernon, *Arch. Biochem. Biophys.*, **43**, 492 (1953).

 Reprinted from *Journal of the American Chemical Society*, 75, 6347 (1953).

some II is added to the reaction mixture (see Fig. 1). Complete oxidation in this manner cannot be attained, indicating in agreement with (c) above, that the oxidative–reduction potential of I is slightly positive with respect to II.

It is tempting, but premature, to discuss the many intriguing possibilities raised by the presence of this new cytochrome in photosynthetic organisms. We may note, however, that the existence of the bacterial cytochrome may provide the basis for a mechanism of photochemical H-transport and an explanation for the absence of a light-stimulated respiration in the photoheterotrophes. It is suggested that I be called tentatively "cytochrome-c2".[3] Further researches on the properties of this and other haem-proteins in the photosynthetic bacteria are proceeding.

The investigations at Washington University have been made possible by grants from the C. F. Kettering Foundation and the U. S. Public Health Service. We are indebted to Dr. S. Velick, Dept. of Biochemistry, Washington University Medical School, for aid in the experiments on electrophoretic purification of the bacterial cytochrome.

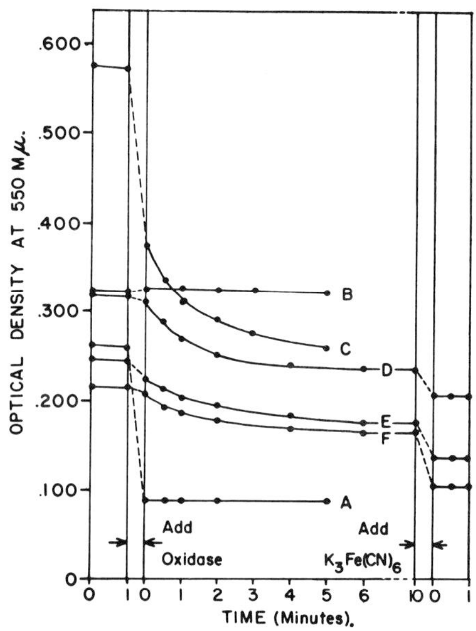

Fig. 1. Action of cytochrome oxidase on bacterial cytochrome: tests were performed with a Beckman DU spectrophotometer, using absorption cells in which were placed 1.0-ml. volumes of solutions containing 50 μmoles of phosphate buffer *p*H 7.4, 0.4 μmole of $AlCL_3$, 1 mg. of protein of a pig heart cytochrome oxidase preparation and cytochrome preparations as indicated. Reduced cytochromes were prepared with stoichiometric amounts of ascorbic acid. Oxidase was added to the test systems after an initial one minute period of observation as well as in equal amounts to the water blank. Mixing time was about 15 seconds and all values for optical density prior to mixing were corrected for the dilution: A, 0.01 μmole of reduced cytochrome c; B, 0.012 μmole of reduced bacterial cytochrome; C, 0.01 μmole of reduced cytochrome c plus 0.012 μmole of reduced bacterial cytochrome. D, E and F all contain 0.008 μmole of reduced bacterial cytochrome with 0.01, 0.004 and 0.002 μmole of oxidized cytochrome c, respectively.

(2) E. Margoliash, *Nature*, **170**, 1014 (1950).
(3) R. Hill and D. Keilin, private communication.

A.R.C. Unit of Microbiology, The University
Sheffield, England — S. R. Elsden
The Edw. Mallinckrodt Institute
of Radiology, Washington
University Medical School
Saint Louis, Missouri — M. D. Kamen, L. P. Vernon

Received December 3, 1953

EXISTENCE OF HAEM COMPOUNDS IN A PHOTOSYNTHETIC OBLIGATE ANAEROBE[1,2]

M. D. KAMEN AND L. P. VERNON

Edward Mallinckrodt Institute of Radiology, Washington University Medical School, Saint Louis, Missouri

Received for publication January 21, 1954

Recently, we have found that a strain of *Chlorobium limicola*, an obligate anaerobe and photoautotrophe, contains appreciable quantities of a cytochrome pigment. In addition, cell-free extracts prepared by sonic vibration of cell suspensions exhibit cytochrome oxidase, cytochrome photooxidase, and cytochrome reductase activity.

The organisms were obtained in a stab culture generously supplied by Dr. R. Y. Stanier and grown in liquid sulfide media of the composition described by Larsen (J. Bact., **64**, 187, 1952). Approximately 2.5 g wet weight of the organisms were suspended in 30 ml phosphate buffer (0.05 M, pH 7.3) and subjected to sonic vibration for 30 minutes at 5 C, using a Raytheon 9 kc magnetostriction oscillator. The thick turbid green suspension was freed of cell debris, unbroken cells, and sulfur particles by centrifugation at 5,000 G for 15 minutes at 0 C.

Microscopic examination of the supernatant fluid revealed no intact cells but considerable quantities of large cell fragments, indicating that disruption was much less complete than in comparable suspensions of the photoheterotrophes, *Rhodospirillum rubrum* and *Rhodopseudomonas spheroides*.

Centrifugation at 100,000 G (0 C) using the Spinco preparative centrifuge sedimented most of the green pigment and all of the particles. The green supernatant fluid exhibited a bacteriochlorophyll spectrum with broad absorption maxima at 420, 675, and 735 mμ. Oxidation of this supernate with ferricyanide and reduction with hydrosulfite resulted in spectra, the difference of which revealed traces of a cytochrome component. Similar treatment of the crude extract resulted in difference spectra which very definitely indicated the existence of a cytochrome with absorption maxima at approximately 415 and 550 mμ. Hence, the particles were subjected to analysis by the procedures already described

[1] This note is the eighteenth in a series of publications dealing with the metabolism of photosynthetic bacteria.

[2] This investigation was supported by the C. F. Kettering Foundation and, in part, by a research grant (E-292C) from the National Microbiological Institute, the National Institutes of Health, Public Health Service.

TABLE 1

Enzymatic activities affecting cytochrome-c in sonic extracts of Chlorobium limicola

Conditions: (a) for oxidase and photooxidase assays: total volume, 1.0 ml, containing 50 μM phosphate buffer (pH 7.4), 0.4 μM $AlCl_3$, 0.06 μM cytochrome-c (reduced with ascorbic acid); (b) for reductase assays: same as in (a) except that 0.04 μM cytochrome-c was present in oxidized form and 0.15 μM reduced diphosphopyridine nucleotide (DPNH). For all experiments, 0.1 ml *C. limicola* extract was added, containing 3.9 mg protein per ml, and having an optical density of 7.20 at 736 mμ (Beckman DU spectrophotometer). Photooxidase experiments involved illumination of test solution with intensity of 1,000 footcandles (well above light saturation) and comparison of resultant D^*_{550} with that for corresponding dark control. Data for *Rhodospirillum rubrum* are taken from a previous paper (Vernon and Kamen, Arch. Biochem. Biophys., **44**, 298, 1953).

ACTIVITY	CONDITIONS	D_{550}/MIN/MG PROTEIN*	
		C. limicola	*R. rubrum*
Dark oxidase	no treatment	−0.020	−0.065
	+ cyanide (10^{-3} M)	−0.005	−0.002
	boiled	−0.004	−0.003
Photooxidase	no treatment	−0.077	−0.140
	+ cyanide (10^{-3} M)	−0.043	−0.087
	boiled	−0.029	−0.002
DPNH-cytochrome-c	no treatment	+0.060	+0.093
	boiled	+0.001	+0.001

* Change in optical density at 550 mμ.

Reprinted from *Journal of Bacteriology*, **67**, No. 5, 617–618 (1954).

for isolation and purification of cytochrome in *R. rubrum* (Vernon, Arch. Biochem. Biophys., **43**, 492, 1953).

A soluble fraction containing a cytochrome pigment was obtained representing enrichment of cytochrome relative to bacteriochlorophyll by about 100-fold. The major contaminant appeared to be a bacteriochlorophyll. The Soret peak was found at 415 mμ and the α band of the reduced form at 553 mμ. This spectrum appeared to be quite similar to that found recently in another obligate anaerobe—in this case a sulfate reducer (Postgate, Biochem. J., *in press*, 1954).

By methods described previously (Vernon and Kamen, Arch. Biochem. Biophys., **44**, 298, 1953), it has been found also that the sonic extracts of *C. limicola* show cytochrome-c photooxidase as well as a weak dark cytochrome-c oxidase activity. The magnitudes of these activities, relative to those obtained with comparable quantities of *R. rubrum* extract, are shown in table 1.

It appears that this representative obligately anaerobic photoautotrophe possesses a cytochrome system similar to, but not identical with, that already established as present in typical facultative photoheterotrophes (Vernon and Kamen, *unpublished observations*). Further characterization of this system awaits availability of large quantities of *Chlorobium*.

COMPARATIVE STUDIES ON BACTERIAL CYTOCHROMES*

by

MARTIN D. KAMEN AND LEO P. VERNON

Edward Mallinckrodt Institute of Radiology, Washington University School of Medicine, St. Louis, Missouri, and

Department of Chemistry, Brigham Young University, Provo, Utah (U.S.A.)

INTRODUCTION

Interest in the occurrence and distribution of hematin compounds** in bacteria somewhat dormant since the first researches on this subject some 25 years ago, has revived markedly recently. It has been demonstrated that microorganisms, and in particular bacteria, provide excellent test material for the investigation of biological oxidation mechanisms, not only because they are well adapted for exploitation using such elegant methods as those developed by CHANCE and his co-workers[2,3], but also because they are found to be good source materials for the isolation and extraction of a large variety of hematin compounds and enzymes associated with such compounds[4,5,6,7,8].

In previous papers[7,8], it has been shown that a number of photosynthetic bacteria, including two strict anaerobes, contain cytochromes of the "*c*" type. Although these cytochromes possess a porphyrin moiety identical in character and in bonding to that of mammalian or yeast cytochrome *c*, their protein components differ with ensuing deviations in physico-chemical behaviour from that noted in the cytochrome *c* of mammalian tissue and yeast. As an example, (see[5,7]), the cytochrome *c* obtained from the photoheterotroph, ***Rhodospirillum rubrum***, differs from mammalian cytochrome *c* in that it is not absorbed on Amberlite resin IR-50 in the ammonia form, has a much smaller electrophoretic and ionophoretic mobility at pH 7, possesses an oxidation potential some 80 to 100 mV more oxidizing, and is not activated by mammalian cytochrome *c* oxidase. All of these changes are manifested without any appreciable shifts in the characteristic absorption spectra.

Enzymes associated with cytochrome *c*, such as the diphosphopyridine nucleotide (DPN) linked reductase and cytochrome *c* oxidase, are also found[9,8]. None of the oxidases obtained from these bacteria appear to be identical with those prepared from mammalian tissues. This observation appears to be fairly general for all bacterial oxidases[10].

* These researches have been made possible by the continued financial support of the C. F. Kettering Foundation. Much of the work was done at the Hopkins Marine Station, Pacific Grove, California, the facilities of which were most kindly placed at our disposal by Professor C. B. VAN NIEL and Professor L. R. BLINKS. Electrophoretic and centrifugal equipment needed for some of the purification procedures was made available by Dr. H. SCHACHMAN, Department of Biochemistry, University of California, Berkeley, California, who also was most generous in donating valuable time for helpful discussions. The technical assistance of Miss JEAN MILLER in Dr. SCHACHMAN's laboratory is also gratefully acknowledged.

** The term "hematin compound" is used as defined by LEMBERG AND LEGGE[1], p. 163.

References p. 22.

Reprinted from *Biochimica et Biophysica Acta*, **17**, 10–22 (1955).

A unique "photo-oxidase" for cytochrome *c* has been detected in all the known genera of photosynthetic bacteria and isolated from two of the facultative photoheterotrophs[9,8]. This enzyme exhibits a remarkable heat stability and resistance to cyanide. Together with the reductase and oxidase, it provides a rational basis at the enzyme level for a partial understanding of the known physiological traits of the photosynthetic bacteria, as they range from strict anaerobes to facultative heterotrophs[8].

In addition, a remarkable new type of hematin compound has been observed to occur in all the photosynthetic bacteria available in quantity as fresh mass cultures[7]. Yields obtained on isolation and purification from two facultative heterotrophs indicate that these hematin compounds occur in quantities comparable to those observed for the cytochromes of the "*c*" type. While too little is known about the chemistry and function of these compounds to propose a stable terminology, they have been labeled tentatively "pseudohemoglobins" because of their characteristic spectra, oxidation potential and auto-oxidizability, even though they are not true hemoglobins in that they fail to exhibit oxygen-carrying capacity and react to form hemochromogens characteristic of the cytochromes of "*c*" type[7].

All of these observations dictate the extension of studies, such as the foregoing to as many bacteria as possible, so that the metabolic pattern of hematin compounds in the photosynthetic bacteria can be assessed as influenced by the photochemical act in photosynthesis. In this paper, there are presented the results of researches on hematin compounds in all known genera of the facultative photoheterotrophs (non-sulfur purple bacteria, or *Athiorhodaceae*) as well as in two typical facultative denitrifiers.

METHODS

Cultures and other source materials

(a) *Photosynthetic bacteria*

Rhodospirillum (Rsp) rubrum. This bacterium was represented by the strain, S-1 (Van Niel), used in previous researches[7]. Its culture and manipulation have been described[11]. In addition, it was available in quantities of hundred of grams as a dry powder*, freed of fatty material and photoactive pigments by exhaustive extraction with various organic solvents. This powder, as well as those obtained from other bacteria described below, had been collected by VAN NIEL as part of a project devoted to the study of the bacterial photopigments (chlorophyll, carotenoids, etc.) some 20 years previous to the present researches. Although stored at room temperature the powders (with one exception) were found to be excellent sources of undenatured cytochrome *c* for each of the bacterial species involved. To further attest the remarkable stability of these bacterial cytochromes, it was found they had remained largely in the reduced form during storage through these two decades.

Rhodopseudomonas (Rps) spheroides. In addition to strain 241 (Van Niel) used in previous researches[7], there were also available large quantities of an aged dry powder, prepared from a strain labeled 242 (Van Niel) in the same manner as that described above for *Rsp. rubrum*.

Rhodopseudomonas (Rps) capsulatus (Van Niel). A 10-day culture of this bacterium, grown in 15 liters of a 5% yeast extract-mineral salt medium, was harvested by centrifugation and washed twice with tap water. The yield of fresh cells was 51 g wet weight.

Rhodopseudomonas palustris (Strain 5, Van Niel). 45 g dry powder, as described above for *Rsp. rubrum* were available.

(b) *Chemosynthetic bacteria*

Micrococcus (M) denitrificans. This strain was identical with that studied by KLUYVER AND VERHOEVEN[12]. It could be cultured either autotrophically, using a mineral medium with nitrous oxide, hydrogen and carbon dioxide as substrates, or heterotrophically under either anaerobic or aerobic conditions. The cultures examined for hematin compounds were grown heterotrophically, employing a medium made according to the following prescription: 10 g K_2HPO_4, 20 g KNO_3, 5 ml yeast autolysate, adjusted with conc. H_2SO_4 to pH 6.5–6.8 after making up 1 liter with tap

* See footnote on following page.

References p. 22.

water; after sterilization for 20 min at 120° C, added 12 ml sterile 80% glucose solution. For aerobic cultures, the nitrate was omitted and the medium aerated. It was possible to obtain readily yields of 130 g wet weight fresh cells from 18 liters of a 4–5 day culture.

Pseudomonas (Ps) denitrificans. The strain was that studied by SACKS AND BARKER[13]. It was available as 2.7 g of a dry cell preparation which had been stored frozen (—20° C) for several months. The powder had been prepared from cell suspensions grown anaerobically, with nitrate as H-acceptor.

Biochemical procedures

Extraction methods

The following procedures were used singly, or in combination:

1. Modified KEILIN-HARTREE method, beginning with suspension of cells in warm trichloracetic acid (TCA), as described previously[4]. This was the method of choice for all the isolations of cytochrome *c* components, as well as pseudohemoglobins, from the facultative photoheterotrophs (*i.e.*, *Rsp. rubrum, Rps. spheroides, Rps. capsulatus, Rps. palustris*).

2. Manual grinding in the cold with alumina, or glass beads. The procedure using alumina has been described previously[14]. The alumina powder was #1557 AB (levigated) obtained from Buehler, Ltd., 165 W. Wacker Drive, Chicago, Illinois. The glass beads were available from the Minnesota Mining and Manufacturing Company, St. Paul, Minnesota as "pavement-marking beads", grade 10, sieve size 60/80.

3. Sonic disruption at 0–5° C using a 9 kc Raytheon magnetostriction oscillator.

4. Disruption at 0–5° C in a Waring Blendor using glass beads, as described by LAMANNA AND MALLETTE[15]. This method proved to be the most reliable for general use.

For preparation of enzyme extracts for assay of oxidase and reductase activity, Method No. 4 was invariably employed, except for some facultative photoheterotrophs, which usually were amenable to sonic disruption.

In isolating cytochrome *c* from the bacterial powders, the first step was usually extraction in the cold (0–5° C) with 0.05 *M* phosphate buffer, pH 7.0. This step was repeated once and the combined extracts then fractionated, as in the KEILIN-HARTREE procedure. In all the powders prepared from the facultative photoheterotrophs, little or no hematin component resisted the cold extraction with phosphate. Both the pseudohemoglobin and cytochrome *c* components were usually recovered initially in the ammonium sulfate (AS) fraction between 25 and 50% saturation. On repeating the AS fractionation, the two components gradually separated, exhibiting more usual solubility relations, *i.e.*, the pseudohemoglobin was recovered in a fraction between 25 and 40% saturation, and the cytochrome *c* in a fraction between 50 and 60% saturation. From this point, further purification was achieved by the methods previously described[7]. The final purity was estimated on the basis of relative intensities of α-, β- and γ-bands in the visible and protein absorption in the ultra-violet.

Pyridine and cyanide hemochromogens were prepared as described previously[7].

Spectroscopic measurements

Spectra were determined using a Beckmann DU spectrophotometer. Preparations first were examined before treatment with oxidizing or reducing agents. The spectra of the reduced forms were determined after addition of sodium hydrosulfite. Spectra of oxidized forms were ascertained after addition of potassium ferricyanide, or where feasible, after vigorous aeration. In some cases where isolation was impractical, the presence of hematin compounds was detected by measuring the difference spectra of oxidized and reduced extracts.

Potential measurements

Ferro-ferricyanide oxidation-reduction buffers were used as described by DAVENPORT AND HILL[16]. All pH buffer systems used in these determinations were at a final concentration of 0.02 *M*. The following buffer systems were employed: pH 5, sodium acetate-acetic acid; pH 6, sodium monohydrogen phosphate-potassium dihydrogen phosphate; pH 8 to 9, "tris" (tris-hydroxy-methyl-amino-methane); pH 10–11, sodium tetraborate-sodium hydroxide.

Reagents

DPNH (reduced diphosphopyridine nucleotide) and cytochrome *c* (horse heart) were obtained from the Sigma Chemical Company, St. Louis, Missouri. Cytochrome *c* oxidase from pig heart was made by the Department of Biochemistry, University of California, Berkeley*. All other reagents

* The dried bacterial powders used in these researches were provided us by Professor C. B. VAN NIEL, to whom we are thus doubly indebted. We also wish to record the gift of a dried preparation of *Ps. denitrificans* from Dr. C. C. DELWICHE, Department of Plant Biochemistry, University of California, Berkeley, California. In addition, Dr. M. E. WINFIELD, in residence at the Hopkins Marine Station as a traveling Fellow of the Australian C.S.I.R.O., unselfishly made available to us

References p. 22.

were of C.P. reagent quality, as obtained from commercial sources, and were used without further purification, except for the potassium ferricyanide which was recrystallized twice from water.

Enzyme assays

Only cytochrome *c* oxidase and the DPNH-linked cytochrome *c* reductase activities were assayed in this series of researches. The methods employed have been described previously[8,9].

Preparation of bacterial enzymes

Oxidase and reductase activities in *M. denitrificans* were studied in extracts prepared as follows: 13 g wet weight of fresh washed cells from anaerobic cultures were blendorized with glass beads for 10 minutes (all operations being conducted at 0° C). The cell debris and beads were removed by centrifugation at 1200 *g* for 10 minutes, and discarded. The intact cells remaining in the supernatant liquid were removed by centrifugation at 20,000 *g* for one hour. The supernatant liquid (S) and the residue (H) were both tested for enzyme activity, most of which was found to remain in (S). This "soluble" material was stored at —20° C and used in experiments described below.

Terminology

The system proposed by SCARISBRICK[17] was adopted—thus, the hematin compound was named according to source and location of the α-band of the reduced form. This practice was extended to *Rsp. rubrum* cytochrome *c* as well as other bacterial cytochromes which previously had been referred to as "*c*" cytochromes[7].

RESULTS

Cytochrome c components of the facultative photoheterotrophs

1. *Rsp. rubrum* cytochrome-550. The properties of this compound, as prepared from fresh cell suspensions, have been described[7]. No changes were noted when this cytochrome was prepared from the aged bacterial powders. A yield of approximately 0.2 mg per g dry weight of fresh cells was noted. This yield was approximately 20% of that found using fresh cells.

The spectra characteristic of *Rsp. rubrum* cytochrome-550, as well as those of horse heart cytochrome *c*, are indicated in Table I which exhibits the characteristic spectroscopic data for the various cytochromes of the "*c*" type, obtained from the facultative photoheterotrophs.

2. *Rps. spheroides* (242, Van Niel) cytochrome-550. This cytochrome was obtained in rather poor yields from the powdered material prepared from the strain labeled 242 (Van Niel). Considerably less than 0.01 mg per g dry weight was recovered. However, this material appeared to be identical with the cytochrome *c* obtained in our previous researches using fresh cell suspensions of *Rps. spheroides* (see Table I). It is not known why this low yield occurred in this strain, inasmuch as the yields from the 241 strain were very similar to those noted for *Rsp. rubrum*.

3. *Rps. palustris* cytochrome-552. Yields closely comparable to those found for the *Rsp. rubrum* powder were also noted using powder prepared from *Rps. palustris*. The spectroscopic purity of the material, data for which are presented in Table I, was 25–30%.

4. *Rps. capsulatus* cytochrome-550. Forty g wet weight of fresh cells were employed as starting material. The final yield of cytochrome was approximately 0.5 mg per g dry weight. The purity estimated on the basis of spectroscopic data was 40–50%. The characteristic absorption maxima for this cytochrome, together with those for its cyanide and pyridine hemochromogens, are shown in Table I.

---

his cultures of *M. denitrificans* and *Rps. capsulatus* and also cooperated in helping us obtain mass cultures of these bacteria. A cytochrome *c* oxidase, prepared from pig heart by the usual KEILIN-HARTREE procedure, was obtained through the courtesy of Professor J. NEILANDS, University of California, Berkeley, California.

References p. 22.

TABLE I

WAVE LENGTHS OF ABSORPTION MAXIMA FOR "c" FERROCYTOCHROMES OF PHOTOSYNTHETIC BACTERIA (mμ)

*Compound**	*α-band***	*β-band***	*Soret band***
Rsp. rubrum-cytochrome-550	550	521	416
Rsp. rubrum-cytochrome pyridine hemochromogen	550	520	414
Rsp. rubrum-cytochrome cyanide hemochromogen	554	525	420
Rps. spheroides-cytochrome-550***	550	520	416–417
Rps. spheroides-cytochrome pyridine hemochromogen	550	520	414
Rps. spheroides-cytochrome cyanide hemochromogen	554	525	420
Rps. spheroides-cytochrome-550§	550	519–520	414–416
Rps. capsulatus-cytochrome-550	550	521	416
Rps. capsulatus-cytochrome pyridine hemochromogen	550	520	414
Rps. capsulatus-cytochrome cyanide hemochromogen	555	526	421
Rps. palustris-cytochrome-552	552	523	418
Chromatium D-cytochrome-552	552	525	417–418
Chlorobium limicola-cytochrome-553	553	520	415

* All these cytochromes upon oxidation show the characteristic fusion of the α- and β-bands into a broad band in the green with maximum between 530 and 535 mμ as well as the broadening of the Soret band with shifts toward the blue (maxima at 395–410 mμ).

** Corresponding values (mμ) for mammalian cytochrome *c*; α-band, 550; β-band, 520; Soret band, 416.

*** Isolated from strain 241 (Van Niel).

§ Isolated from strain 242 (Van Niel).

Cytochrome c components of denitrifiers

1. *M. denitrificans*-cytochrome-550. The yield of cytochrome *c* was not appreciably affected by whether the cells were grown under strictly anaerobic, or under aerobic conditions. The cytochrome *c* content appeared to be an order of magnitude less than that encountered in the photoheterotrophs. However, this posed no difficulties, as the amounts of this micro-organism obtainable were so great that quantities of cytochrome sufficient for purification procedures were readily obtained.

It proved necessary to disrupt these organisms by method 4 (see METHODS). Attempts to homogenize the cell suspensions by sonic vibration (with or without glass beads), by manual grinding with various abrasives and by phosphate extraction of acetone powders achieved only 5 to 10% disruption. Blendorizing with glass beads resulted in recovery of 80–90% of the original cell protein in the form of a cell-free extract.

The crude cytochrome resulting from the first AS fractionation was subjected to repeated cycles of AS fractionation until no further purification was obtainable by this means. Further purification by absorption on calcium phosphate gel[7] brought the sample to a spectroscopic purity of 15–20%. This material was then subjected to electrophoretic analysis, using ammonium acetate buffer (0.1 *N*, pH 7.0). Shortly after initiation of electrophoresis, two boundaries appeared both migrating toward the anode. Most of the colored component was concentrated in the slower moving boundary. It was noted that this direction of migration for the cytochrome was opposite to that observed for mammalian cytochrome *c* or *Rsp. rubrum*-cytochrome-550, both of which exhibit cathodic mobility at pH 7. Thus, it appeared that *M. denitrificans*-cytochrome-550 was negatively charged at neutral pH, indicating an isoelectric point at acid pH.

The anodic mobility was of an absolute magnitude (10^{-5} cm^2 sec^{-1} $volt^{-1}$) associated with cytochrome *c*.

By continuously displacing the buffer at the anode to compensate for the movement of the faster-moving colorless anodic impurities, it was possible to hold back most of the cytochrome in the cathode compartment. After 5–6 hours, the anode compartment showed little or no color, while the cathode and middle compartments were highly colored—indicating a good separation of the colorless impurities from the cytochrome component. The purest material (in the cathode compartment) exhibited a spectrospopic purity of approximately 50%. This material was used for establishing the nature of the visible absorption spectrum and the oxidation potential. The less pure material from the middle compartment—of spectroscopic purity approximately 35%—was used for preparation of the pyridine and cyanide hemochromogens. Spectroscopic data for wave lengths of absorption maxima are given in Table II.

TABLE II

WAVE LENGTHS (mμ) OF ABSORPTION MAXIMA FOR "*c*" CYTOCHROMES OF DENITRIFIERS

Compound	*α-band*	*β-band*	*Soret band*
(a) Reduced form			
M. denitrificans-cytochrome-550	550	522	416
M. denitrificans-cytochrome pyridine hemochromogen	550	520	414
M. denitrificans-cytochrome cyanide hemochromogen	554	525	420
Ps. denitrificans-cytochrome-552	552	525	418
Ps. denitrificans-cytochrome pyridine hemochromogen	551	520–521	415
Ps. denitrificans-cytochrome cyanide hemochromogen	556	527	422
(b) Oxidized form			
M. denitrificans-cytochrome-550	– (526–528) –		410–411
Ps. denitrificans-cytochrome-552	– (526–528) –		412–413

2. *Ps. denitrificans*-cytochrome-552. 2.7 g of a dry powder were used as starting material. It was found, as with *M. denitrificans*, that blendorizing with glass beads was most effective in extraction of the cytochrome component. However, all of the cytochrome initially precipitated in the first AS fraction (0–25% saturation) rather than at the more usual higher concentrations of AS. Resuspension of the precipitated cytochrome and repeated AS fractionation resulted in normal solubility behavior, *i.e.*, the major fraction of the cytochrome precipitated in the fraction 33–50% saturated AS. This material was refractionated a number of times with AS and finally reached a spectroscopic purity of 10–15%. This preparation was used for establishing the character of the visible absorption spectrum (see Table II for wave lengths of absorption maxima) and for determinations of oxidation potential.

Oxidation potentials of "c" cytochromes

The oxidation potentials at various values of pH for the bacterial cytochromes are shown in Table III. It was noted that the "*c*" cytochromes of the photochemically active bacteria had potentials much more oxidizing than those observed for cytochrome *c*, as

References p. 22.

isolated from yeast or mammalian tissue. (All the values given in Table III were determined in experiments using, as a standard, cytochrome *c* from horse heart, which always gave values for E_0' in good agreement with the accepted +0.265 v. at pH 7.0.) On the other hand, the nonphotochemical facultative anaerobes (denitrifiers) showed both "normal" and high values. Consideration of the E_0' values indicated that the cytochrome from *M. denitrificans* might be similar to mammalian cytochrome *c*, whereas the cytochrome from *Ps. denitrificans* appeared to resemble those obtained from the facultative photoheterotrophs.

TABLE III

OXIDATION POTENTIALS (E'_0) OF BACTERIAL CYTOCHROMES OF "*c*" TYPE (volts)

Compound	*pH*					
	5.0	*6.0*	*7.0*	*8.0*	*9.0*	*10.0*
Rps. capsulatus-cytochrome-550	0.33_2	0.33_0	0.32_9	0.32_4	0.31_8	0.31_1
Rps. palustris-cytochrome-552	0.31_0	0.32_7	0.30_7	0.32_0	0.30_4	—
Rsp. rubrum-cytochrome-550	0.38_0	0.35_8	0.33_8	0.30_5	0.30_4	—
M. denitrificans-cytochrome-550*	0.26_5	—	0.25_0	0.25_5	0.22_5	—
M. denitrificans-cytochrome-550**	—	—	0.24_7	—	—	—
Ps. denitrificans-cytochrome-552	—	0.32_0	0.32_0	0.32_0	—	—

* from anaerobic culture.
** from aerobic culture.

It should be noted that *M. denitrificans* cytochrome-550 is the first bacterial cytochrome of "*c*" type to be isolated which shows spectroscopic and electrochemical properties identical with those of mammalian cytochrome *c*. As remarked below, its enzymic properties are also similar to those of mammalian cytochrome *c*.

Enzymic properties of the bacterial cytochromes

Activation by mammalian cytochrome-c-oxidase

The responses of various bacterial cytochromes of the "*c*" type when incubated with a pig heart cytochrome *c* oxidase and air are shown in Fig. 1. It was found that none of

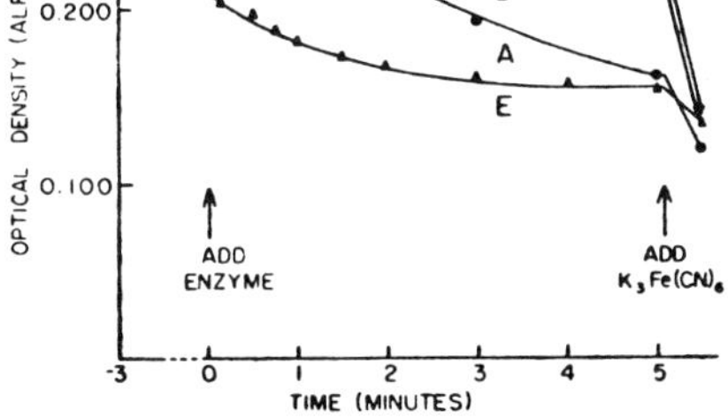

Fig. 1. Effect of pig heart cytochrome *c* oxidase on bacterial cytochromes. Test system: 0.1 *M* phosphate buffer pH 7, 0.1 ml; 0.1 % ascorbate-reduced horse heart cytochrome *c*, 0.2 ml, distilled water, 0.3 ml (Curve A). For other curves, water omitted and 0.5 ml each of appropriate cytochromes added as follows: Curve B, *Rsp. rubrum* ferrocytochrome-550; Curve C, *Rps. capsulatus* ferrocytochrome-550; Curve D, *Ps. denitrificans* ferrocytochrome-552; Curve E, *M. denitrificans* ferrocytochrome-550. Total volume of all test systems was 0.8 ml before addition of enzyme (0.2 ml). Experiment commenced, after 3 minute period (arrow). After 5 minutes, potassium ferricyanide added, as indicated. The rate of oxidation was determined by following the change in optical density at the wave length of maximal absorption for the α-band. This wave length was 550 mμ in all cases, except for the *Ps. denitrificans* pigment for which maximal absorption occurred at 552 mμ.

References p. 22.

the cytochromes of high potential tested were oxidized in the presence of the mammalian oxidase. This included not only the samples from photosynthetic bacteria (Curves B and C) but also that from *Ps. denitrificans* (Curve D). The cytochrome from *M. denitrificans*, which exhibited an oxidation potential similar to that of the horse heart cytochrome (Table III), was oxidized readily (Curve E).

Activation by bacterial cytochrome-c oxidases and reductases

It had been shown[8] that the oxidases and reductases derived from either *Rsp. rubrum* or *Rps. spheroides*, could activate cytochrome *c*, whether prepared from mammalian sources or from the bacteria themselves. A similar study of the enzymic activities associated with cytochrome *c*, in extracts of *M. denitrificans*, revealed that the oxidase and reductase could activate both horse heart cytochrome *c* and *M. denitrificans* cytochrome-550. In addition, it was ascertained that *Rsp. rubrum*-cytochrome-550 was oxidized in the presence of the oxidase from *M. denitrificans*.

The relative levels of oxidase and reductase activity in *M. denitrificans* were determined using both horse heart cytochrome *c* and *M. denitrificans*-cytochrome-550. With either cytochrome, closely identical rates were obtained. The oxidase levels were from 3–5 times greater than those for the reductase, after correction for a heat-resistant non-enzymic reducing component which appeared to be present in the bacterial extracts.

"Pseudohemoglobins"

In our previous studies[7], it was noted that both *Rsp. rubrum* and *Rps. spheroides* contained large quantities of a CO-binding heme protein tentatively identified as a "pseudohemoglobin". Evidence for the occurrence of this type of hematin compound was obtained for representative species of other photosynthetic bacteria[7]. In the present studies, "pseudohemoglobin" was recovered in good yield from the *Rsp. rubrum* powder, but none could be detected in the powders of the other facultative heterotrophs available. Examination of fresh cell cultures of *Rps. capsulatus* revealed the presence of a "pseudohemoglobin" apparently closely identical (spectroscopically) with those of the other facultative photoheterotrophs. As in the other "pseudohemoglobins", the "pseudohemoglobin" from *Rps. capsulatus* exhibited an absorption maximum in the reduced form at the same wave length as that for the α-band of the corresponding cytochrome *c* component (Table IV).

TABLE IV

SPECTROSCOPIC PROPERTIES OF "PSEUDOHEMOGLOBINS"

Source	*Absorption maxima (mμ)*	
	Reduced form	*Oxidized form*
Rsp. rubrum	423, 550	393, 490–500, 630
Rsp. rubrum, pyridine hemochromogen	415, 520, 550	406, 530
Rsp. rubrum, cyanide hemochromogen	420, 526, 555	404
Rsp. rubrum, CO-compound	416, 535, 560–565	—
Rps. spheroides (#241)	423, 550	395–400, 490–500, 630
Rps. capsulatus	425, 550	395–400, 495–500, 640
Chromatium	425, 552	400, 490, 630

As will be made apparent (see DISCUSSION), this "pseudohemoglobin" component

References p. 22.

may be of considerable importance for both the light and dark metabolism of the facultative photoheterotrophs, and possibly for the light metabolism of the strict photoanaerobes. It is desirable to adopt a name less likely to cause confusion than "pseudohemoglobin".

DISCUSSION

The results of the present researches re-enforce the previous findings[7,8] that all photosynthetic bacteria contain modified forms of cytochrome *c*, together with a full complement of associated enzymic activities, the levels of which can be correlated with the overall physiology of these bacteria. The presence of other heme proteins, such as cytochromes of the "*b*" type and the new "pseudohemoglobins", indicates that in the facultative photoheterotrophs, at least, a number of electron transport systems are available for coupling with cytochrome *c*[8].

Spectroscopic examinations by DUYSENS[19] of actively metabolizing cell suspensions of *Rsp. rubrum* have indicated that the major effect of illumination under normal physiological conditions is photo-oxidation of a cytochrome component with a difference spectrum (oxidized-reduced) very similar to that expected to arise owing to oxidation of a modified form of cytochrome *c*. More detailed observations by CHANCE AND SMITH[20] working with cell suspensions of *Rsp. rubrum*, incubated under a variety of conditions (anaerobic, aerobic, presence of respiratory inhibitors), have shown that the action spectrum is complex, appearing to arise from photo-oxidation of the whole cytochrome system, including a major contribution from a CO-binding pigment which has spectroscopic properties identical with those of the *Rsp. rubrum* "pseudohemoglobin"-550. It will be recalled that in the previous studies[8] it has been postulated that the shift from light to dark metabolism in the facultative photoheterotrophs involves a competitive interaction between two oxidase systems. One of these is light activated (photo-oxidase) and cyanide-insensitive. The other is light-insensitive and cyanide-sensitive. The steady state oxidation level of the bacterial cytochrome *c* is correlated with the relative levels of these oxidase activities in the various photoheterotrophs. CHANCE AND SMITH[20] suggest a similar competition mechanism. However, they propose the CO-binding pigment as the terminal oxidase in the dark metabolism and as the focal point for light mediation of the electron transport by cytochromes. Inasmuch as it has been shown that these heme proteins are capable of coupling enzymically or nonenzymically in a number of ways[8] it is not possible to deduce an unambiguous pathway for electron transport from the data available at present. In particular, the locus for introduction of the excitation energy absorbed through chlorophyll remains uncertain.

In connection with extension of this "cytochrome-mediation" hypothesis to photosynthesis in general, it is pertinent to examine data on (a) the distribution of hematin compounds in all photosynthetic systems, and (b) occurrence of enzymic activities associated with hematin compounds in these systems. Comparisons should also be made, where possible, with non-photosynthetic systems.

As regards (a), it may be stated with reasonable certainty that all photosynthetic systems examined to date contain a cytochrome *c*, which differs physically and chemically, and is distinguishable enzymically from the typical cytochrome *c* encountered in mammalian tissue or in yeast. It is less certain whether cytochromes of the "*b*" type are invariably associated with photosynthetic systems. The presence of the "pseudo-

References p. 22.

hemoglobins" has been established in all the photosynthetic bacteria available in fresh mass cultures. It is not known whether corresponding hematin compounds are present in green plant tissues. However, the occurrence of cytochrome "*f*", which may be considered a modified cytochrome *c*, appears to be quite general in green plants[16]. Thus, no data are in existence which contra-indicate the suggestion that a system of hematin compounds, modified from those encountered in mammalian tissues or in yeast, is a characteristic component of photosynthetic tissues.

Data on point (b) indicate that reductases and photo-oxidases for bacterial "*c*" cytochromes are found in all the photosynthetic bacteria, regardless of their ability to tolerate or utilize oxygen[8]. It is not clear what corresponds to a photo-oxidase in the green plant. However, data have been presented by LUNDEGÅRDH[21] which are claimed to provide evidence for photo-oxidation of cytochrome *f* during active photosynthesis in green plants and algae. The outstanding difference between bacterial and green plant photosynthesis is the production of molecular oxygen in the latter. This may be rationalized by supposing that the "photo-oxidase" in the green plant chloroplast is itself a catalase or is coupled with a catalase, whereas the bacterial photo-oxidase is coupled with a peroxidase.

In considering comparable phenomena in non-photosynthetic systems, it should be remarked that the photosynthetic bacteria are potentially facultative anaerobes in the sense that the photochemical "OH" is an oxygen substitute in electron transport, much as is sulfate or nitrate in the metabolism of non-photosynthetic facultative anaerobes. It may be noted that in the present researches cytochromes of "*c*" type have been isolated from two typical denitrifiers. Also, the existence of a cytochrome—apparently of the "*c*" type—in the strictly anaerobic sulfate reducer, *Desulfovibrio desulfuricans**, has been demonstrated by POSTGATE[22]. In these three instances, two-*Ps. denitrificans* and *Desulfovibrio*—yield cytochromes with spectra modified from those typical of the mammalian form. One—*M. denitrificans*—contains a cytochrome *c* spectroscopically identical with the mammalian compound.

All the cytochromes of "*c*" type from the photosynthetic bacteria, the oxidation potentials of which have been examined, appear to be "high-potential" cytochromes (Table III). This finding lends added significance to the observation by DAVENPORT AND HILL[16] that the chloroplast cytochrome (cytochrome *f*) also is a high potential cytochrome. It will be necessary to extend the present researches to the cytochromes of the photo-anaerobes such as *Chromatium* and *Chlorobium*—to test the generality of occurrence of high potential cytochromes in photosynthetic systems. A similar extension to non-photosynthetic bacteria—particularly the interesting cases of facultative anaerobes—has been effected in only two micro-organisms, the denitrifiers studied in the present researches. It is found that *Ps. denitrificans*- cytochrome-552 is a high potential cytochrome, whereas *M. denitrificans*-cytochrome-550 is not.

These observations can be correlated with the physiology of these two bacteria. *Ps. denitrificans*, when grown in the absence of oxygen in nitrate-glucose medium, contains much more of its cytochrome *c* than when grown aerobically[23]. It has been noted by SACKS AND BARKER[23] that cell-free extracts show characteristic absorption peaks at 552 and 522 mμ, in agreement with the values for the α- and β-bands of the reduced

* One of us (L.P.V.), has isolated an auto-oxidizable cytochrome, apparently of the "*c*" type, from a strain of *Desulfovobrio desulfuricans*. The absorption maxima of this hematin compound in the reduced form are at the same wave lengths as reported by POSTGATE[22].

References p. 22.

Ps. denitrificans-cytochrome-552 found in the present researches. In intact cell suspensions, it has been shown[23] that these peaks disappear on aeration and appear when oyxgen is removed. Further, addition of KNO_3 or KNO_2 (in the absence of oxygen) causes these bands to fade; after complete utilization of the KNO_2 the bands re-appear. Finally, KCN is effective in preventing this oxidation of cytochrome by KNO_2, while still permitting reduction by the extract (presumably through a DPN-linked reductase of the type found in the present researches in *M. denitrificans.*)

These observations provide a rational basis, if not conclusive proof, for supposing that *Ps. denitrificans*-cytochrome-552 mediates reduction of nitrate and nitrite, in a manner analogous to that postulated for the reduction of "OH" by *Rsp. rubrum*-cytochrome-550 in photo-metabolism[8].

On the other hand, it is observed in the present studies that no change in character or amount of cytochrome *c* occurs in *M. denitrificans*, whether grown aerobically or anaerobically. It appears that there is no adaptive response to nitrate, or the necessity to reduce nitrate, in *M. denitrificans* as there is in *Ps. denitrificans*. This raises the possibility that reduction of nitrate does not go through cytochrome *c* in *M. denitrificans*, whereas it appears to be a favored or required pathway in *Ps. denitrificans*.

Evidence has been presented by Postgate recently[24] which indicates participation of the cytochrome-*c*-like compound in sulfate reduction by *Desulfovibrio*. If this is true, it would appear that the oxidation potential of this cytochrome must be considerably less positive than most cytochromes of the "*c*" type, because any reasonable guess as to the oxidation potential of the sulfate-sulfite couple, or for reduction of sulfate to other lower valence forms of sulfur, indicates a value some 100 to 150 mV less oxidizing than the normal value for mammalian cytochrome *c*[25].

It is evident that more insight into the factors which govern the oxidation potentials in the various hematin compounds is required before a coherent picture of electron transport involving such compounds can be proposed. The availability of a much greater variety of cytochromes of the "*c*" type as well as of the new hematin compounds found in the present researches should facilitate further study.

It is particularly important to investigate the spectra in the near infra-red of the bacterial hematin compounds, and also of cytochrome *f*, because the nature of these spectra bears directly on the question of whether excitationen ergy from chlorophyll can be transferred directly to the hematin compounds. Thus, the existence of absorption bands in the bacterial cytochromes and other hematin compounds which lie just beyond the farthest emission frequencies of the bacterial chlorophylls (> 900 mμ) would indicate strongly that direct transfer by a process such as inductive resonance[26] could occur. In a similar fashion, the existence of absorption bands in cytochrome *f* in the region beyond 690 mμ would be of significance.

If it is demonstrated that light energy affects metabolism primarily through the cytochrome system, then it will be necessary to elaborate mechanisms to explain the eventual appearance of electrons at the potential of the hydrogen electrode. A scheme for achieving this by coupling of oxidized and reduced cytochrome systems has been suggested by Davenport and Hill[16]. It is evident there should be increasing attention to the necessity for adapting what is known about hematin chemistry to the problem of energy transfer in photosynthesis, even if such speculations run far ahead of the data available.

References p. 22.

SUMMARY

1. All photosynthetic bacteria appear to contain cytochromes of the "c" type together with associated enzymic activities, such as DPNH-linked reductase, oxidase and photo-oxidase. This is true of strict anaerobes as well as facultative bacteria. There also appears to be a general distribution of other hematin compounds, such as cytochromes of the "b" type, as well as of the new hematin compounds tentatively referred to as "pseudohemoglobins".

2. Two species of denitrifiers (facultative anaerobes) have also been examined and found to contain cytochromes of the "c" type. One of these species yields a cytochrome *c* which spectroscopically and enzymically resembles mammalian cytochrome *c*, whereas the other contains a cytochrome very similar in spectroscopic properties, oxidation potential and enzymic behavior to the cytochromes of "c" type found in photosynthetic bacteria.

3. The oxidation potentials of the various cytochromes of "c" type have been measured over a wide range of pH values. These results indicate that each microorganism examined has a cytochrome *c* unique to itself, but that all the "c" cytochromes isolated from the facultative photoheterotrophs have potentials much more oxidizing than those of cytochrome *c* as prepared from yeast or mammalian tissue.

4. These results are discussed and correlated with the overall physiology of the bacteria. Their significance for the general problem of transfer and storage of light energy in photosynthesis is indicated.

RÉSUMÉ

1. Toutes les bactéries photosynthétiques renferment des cytochromes de type "c" associés à des activités enzymatiques, telles que réductase liée au DPNH, oxydase et photo-oxydase. Ceci est vrai pour les anaérobies stricts aussi bien que pour les anaérobies facultatives. Il apparaît qu'il y a aussi une distribution générale d'autres composés hématiniques, tels que des cytochromes du type "b", aussi bien que des nouveaux composés hématiniques qu'on pourrait désigner par le terme de "pseudohémoglobines".

2. Deux espèces de bactéries dénitrifiantes (anaérobies facultatives) ont également été étudiées et se sont révélées contenir des cytochromes du type "c". L'une de ces espèces produit un cytochrome *c* qui ressemble du point de vue spectroscopique et du point de vue enzymatique au cytochrome *c* des mammifères, tandis que l'autre contient un cytochrome dont les propriétés spectroscopiques, le potentiel d'oxydo-réduction et le comportement enzymatique sont très voisins de ceux des cytochromes du type "c" trouvés chez les bactéries photosynthétiques.

3. Les potentiels d'oxydo-réduction des divers cytochromes du type "c" ont été déterminés dans un domaine étendu de pH. Les résultats montrent que chaque micro-organisme examiné possède un cytochrome *c* qui lui est propre, mais que tous les cytochromes "c" isolés de photohétérotrophes facultatifs ont des potentiels beaucoup plus oxydants que ceux des cytochromes *c* qu'on peut préparer à partir de la levure ou des tissus de mammifères.

4. Les résultats sont discutés en relation avec la physiologie globale des bactéries. Leur signification pour le problème général du transfert et du stockage de l'énergie lumineuse au cours de la photosynthèse est indiquée.

ZUSAMMENFASSUNG

1. Alle photosynthetischen Bakterien scheinen Cytochrome vom "c"-Typus, sowie damit verbundene Enzym-Aktivitäten, wie z.B. $DPNH_2$-Reduktase, Oxydase und Photooxydase zu enthalten. Das gilt sowohl für strenge Anaerobier, als auch für bedingt anaerobe Bakterien. Andererseits scheinen auch andere Hämatinverbindungen, wie Cytochrome des "b"-Typus, sowie die neuen Hämatinverbindungen, die versuchsweise als "Pseudohämoglobine" gekennzeichnet wurden, allgemein vorzukommen.

2. Zwei denitrifizierende Arten (bedingt anaerob) wurden ebenfalls geprüft und festgestellt, dass auch sie Cytochrome vom "c"-Typus enthalten. Eine dieser Bakterien liefert ein Cytochrom *c*, welches spektroskopisch und enzymatisch dem Cytochrom *c* aus Säugetiergeweben ähnelt; andere enthalten ein Cytochrom, welches in seinen spektroskopischen Eigenschaften, seinem Oxydationspotential und seiner Enzymwirkung denen ähnelt, welche in photosynthetischen Bakterien gefunden wurden.

3. Die Oxydationspotentiale verschiedener Cytochrome vom "c"-Typus wurden über einen weiten pH-Bereich gemessen. Die Ergebnisse lassen vermuten, dass jeder untersuchte Mikroorganismus ein artspezifisches Cytochrom *c* besitzt, dass aber andererseits die "c"-Cytochrome, die aus bedingten Photoheterotrophen isoliert wurden, ein viel höheres Oxydationspotential besitzen, als die aus Hefe oder aus Säugetiergeweben gewonnenen Cytochrome.

References p. 22.

4. Diese Ergebnisse werden diskutiert und mit der allgemeinen Physiologie der Bakterien in Beziehung gebracht. Ihre Bedeutung für das allgemeine Problem der Übertragung und Speicherung der Lichtenergie in der Photosynthese wird erwähnt.

REFERENCES

1 R. Lemberg and J. W. Legge, *Hematin Compounds and Bile Pigments*, Interscience Publ., Inc., N.Y., 1949.
2 B. Chance, *Nature*, 169 (1952) 215.
3 B. Chance, L. Smit and L. W. Castor, *Biochim. Biophys. Acta*, 12 (1953) 289.
4 L. P. Vernon, *Arch. Biochem. Biophys.*, 43 (1953) 492.
5 S. R. Elsden, M. D. Kamen and L. P. Vernon, *J. Am. Chem. Soc.*, 75 (1953) 6347.
6 M. D. Kamen and L. P. Vernon, *J. Bact.*, 67 (1954) 617.
7 L. P. Vernon and M. D. Kamen, *J. Biol. Chem.*, 211 (1954) 643.
8 M. D. Kamen and L. P. Vernon, *J. Biol. Chem.*, 211 (1954) 663.
9 L. P. Vernon and M. D. Kamen, *Arch. Biochem. Biophys.*, 44 (1953) 298.
10 L. Smith, *Bact. Rev.*, 18 (1954) 106.
11 H. Gest, M. D. Kamen and H. M. Bregoff, *J. Biol. Chem.*, 182 (1950) 153.
12 A. J. Kluyver and W. Verhoeven, *Antonie van Leeuwenhoek*, 20 (1954) 241.
13 L. C. Sacks and H. A. Barker, *J. Bact.*, 58 (1949) 11.
14 H. McIlwain, *J. Gen. Microbiol.*, 2 (1948) 288.
15 C. LaManna and M. F. Mallette, *J. Bact.*, 67 (1954) 503.
16 H. E. Davenport and R. Hill, *Proc. Roy. Soc. London*, Sec. B, 139 (1952) 327.
17 R. Scarisbrick, *Ann. Rep. Progress Chem.*, 44 (1947) 226.
18 L. A. Hyndman, R. H. Burris and P. W. Wilson, *J. Bact.*, 65 (1953) 522.
19 L. M. N. Duysens, *Nature*, 173 (1953) 692.
20 B. Chance and L. Smith, private communication.
21 H. Lundegårdh, *Physiol. Plant*, 7 (1954) 375.
22 J. Postgate, *Biochem. J.*, 56 (1954) xi.
23 L. E. Sacks, *Metabolic Studies on Denitrification by Pseudomonas denitrificans*, Ph.D. Thesis, Univ. of California, June (1948).
24 J. Postgate, *Biochem. J.*, 58 (1954) ix.
25 W. M. Latimer, *The Oxidation States of the Elements and Their Potentials in Aqueous Solutions*, Prentice-Hall, N.Y., 1952, 2nd ed., p. 74.
26 L. M. N. Duysens, *Transfer of Excitation Energy in Photosynthesis*, Ph.D. Thesis, Utrecht (1952).

Received November 16th, 1954

ON THE NEW HEME PROTEIN OF FACULTATIVE PHOTOHETEROTROPHS*

By ROBERT G. BARTSCH†‡ and MARTIN D. KAMEN‡

(*From the Edward Mallinckrodt Institute of Radiology, Washington University School of Medicine, St. Louis, Missouri*)

(Received for publication, June 24, 1957)

Vernon and Kamen (1, 2) demonstrated the existence of a new type of heme protein in TCA[1] extracts prepared from cell suspensions of the facultative photoheterotrophs, *Rhodospirillum rubrum* and *Rhodopseudomonas spheroides*. They noted that this new heme protein presented an anomaly in exhibiting spectrochemical properties of a myoglobin-like protein on the one hand and the thermal stability and hemochromogen reactions of a cytochrome *c* on the other. They also showed that the purified protein could be reduced by mammalian, DPNH-linked, cytochrome *c* reductase as prepared by Mahler *et al.* (3). This property, together with its autoxidizability and reactivity with carbon monoxide, suggested that it might function as a terminal oxidase.

Hill and Kamen[2] confirmed and extended these observations and ascribed to the new protein a structure in which the prosthetic heme group was bound by thio ether linkages to alkyl side chains as in cytochrome *c*, although possessing a free or loose linkage at the central iron as in myoglobin. They also showed that it had the peculiar property of reversibly binding *only* carbon monoxide and not at all any of the other usually effective reagents, such as azide, cyanide, hydrosulfide, oxygen, 4-methylimidazole, etc. They proposed that the new protein be called "RHP" (abbreviation for *Rhodospirillum*, or *Rhodopseudomonas*, heme protein). This name will be used in the present paper, although it appears that similar proteins exist in a number of bacterial species other than the facultative photoheterotrophs (1, 2, 4).

In none of the researches described was it possible to work with prepara-

* This work was supported by grants from the Charles F. Kettering Foundation, the National Science Foundation, and the Linde Air Products Company. One of us (M. D. K.) gratefully acknowledges tenure of a fellowship awarded by the John Simon Guggenheim Memorial Foundation during 1956.

† Research Associate, Linde Air Products Company.

‡ Present address, Graduate Department of Biochemistry, Brandeis University, Waltham, Massachusetts.

[1] The following abbreviations are used throughout: TCA, trichloroacetic acid; DPN, DPNH, oxidized and reduced diphosphopyridine nucleotide; Tris, tris(hydroxymethyl)aminomethane.

[2] R. Hill and M. D. Kamen, unpublished observations. See also footnote 8.

 Reprinted from *The Journal of Biological Chemistry*, **230**, No. 1, 41 (1958).

tions completely free from traces of the bacterial cytochrome *c*, nor could the purity of the samples used be determined precisely. Recently, we have developed new methods of extraction and purification. These, together with large scale production of *R. rubrum*, have made available sufficient crude proteins (approximately 250 to 500 mg.) so that highly purified samples, electrophoretically homogeneous and free from cytochrome *c*, could be obtained in amounts up to 50 mg. With this material we have investigated in more detail some of the important physicochemical properties of RHP.

Methods

Bacterial Culture—The medium[3] used for the photoheterotrophic growth of *R. rubrum* contained the following: malic acid 2.7 gm., L-glutamic acid 1.0 gm., sodium acetate·$3H_2O$ 1.0 gm., ammonium sulfate 1.0 gm., dipotassium phosphate 3.5 gm., magnesium chloride·$7H_2O$ 0.2 gm., calcium chloride 0.05 gm., nicotinic acid 1 mg., thiamine·HCl 0.5 mg., biotin 0.01 mg., neutral red indicator 0.1 mg., metal solution[4] 0.25 ml., tap water 1000 ml.

The acidic medium contained in Pyrex bottles (9 to 20 liter capacity) was autoclaved, cooled, and neutralized to neutral red end point by the addition of sterile 4 M sodium hydroxide solution, of which approximately 12 ml. per liter were required. Approximately 10 per cent (v/v) inoculum of a 24 to 48 hour-old culture was then added and any remaining space in the bottles was filled with sterile tap water. The culture bottles were finally sealed with rubber stoppers which were tied in place with strips of cloth.

The cultures were illuminated by a bank of 200 watt incandescent lamps which provided about 300 foot candles at the nearest surface of the culture bottles. Electric fans were used for cooling. After 7 to 10 days of incubation in the light, the bacteria were collected by centrifugation in a Sharples supercentrifuge. The cell mass was frozen and stored at $-15°$. After sufficient material was collected, it was thawed, suspended in an equal volume of water to make a thick cream, distributed into flasks, frozen, and dried by lyophilization. The dried bacterial mass was then fragmented into a powder with the aid of a high speed blender and the powder was stored at room temperature in tightly covered polyethylene containers. Usually 200 to 300 liters of culture were grown at one time

[3] The procedure used is essentially the same as one suggested to us by Dr. R. Y. Stanier.

[4] The metal stock solution contained per liter of final volume the following: disodium ethylenediaminetetraacetate (25 gm.), $ZnSO_4 \cdot 7H_2O$ (21 gm.), $FeSO_4 \cdot 7H_2O$ (28 gm.), $MnSO_4 \cdot 1H_2O$ (3 gm.), $CuSO_4 \cdot 5H_2O$ (0.8 gm.), $Co(NO_3)_2 \cdot 6H_2O$ (0.5 gm.), $Na_2B_4O_7 \cdot 10H_2O$ (0.34 gm.), $(NH_4)_6Mo_7O_{24} \cdot 4H_2O$ (0.4 gm.).

and from these mass cultures about 1 gm. of dry bacteria per liter of culture medium was routinely obtained.

Physical Measurements[5]—The sedimentation velocity constant s was determined in the Spinco analytical centrifuge with the procedure and calibrations described by Taylor (5). A synthetic boundary cell was used. The diffusion constant D was measured in a Klett electrophoresis apparatus by the procedure of Taylor and Lowry (6). All measurements of s and D were corrected for temperature and viscosity of medium (6) to give $s_{20,w}$ and $D_{20,w}$. The partial specific volume was found by a modification (6) of the original gradient tube procedure (7).

Electrometric titrations were performed by using the ferric-ferrous oxalate system, as described in the literature (8, 9).

The iron content of the protein was determined with 1,10-phenanthroline after wet ashing with nitric acid as described by Sandell (10). The well known procedure based on wet ashing with alkaline hydrogen peroxide (11) did not appear to be applicable to RHP. The reason for this is not known.

Spectra were obtained with either the Beckman DU spectrophotometer or the Cary spectrophotometers, model 11 or 14. All spectra shown in this report were recorded with model 14.

Isolation and Purification Procedures
General Remarks

The original method involving the classical Keilin-Hartree extraction with warm TCA of washed tissue was employed in early studies (1) before the existence of RHP was suspected and when a rapid extraction of cytochrome *c* only was desired. It was known, however, that TCA extraction was characterized by frequent low yields and excessive denaturation of both cytochrome *c* and RHP. Hence, it was quickly abandoned in favor of less drastic procedures, such as sonic disintegration combined with phosphate extraction in the cold, cold extraction of acetone powders with various buffers, mechanical disruption with blenders, etc. In all these procedures, RHP invariably was found associated with the bacterial cytochrome *c* in the soluble phase.

Further experience showed that, while the two heme proteins differed appreciably in electrophoretic and solubility characteristics, their separation from each other on a bulk scale was tedious. Hence, there appeared to be considerable advantage accruing to methods by which one or the

[5] We are indebted to Miss Carmelita Lowry, Department of Biochemistry, Washington University School of Medicine, for performing experiments on the electrophoretic, sedimentation, and diffusion properties of RHP. A Cary spectrophotometer, model 11, in the Department of Chemistry, was made available by Professor S. I. Weissman. A Cary spectrophotometer, model 14, in the Department of Microbiology, was placed at our disposal by Professor Arthur Kornberg.

other was extracted preferentially so as to yield material enriched in RHP relative to cytochrome *c*. The simplest procedure was found to be preliminary, exhaustive, cold extraction of the lyophilized bacterial powder at pH 4.0, a method employed frequently (12) for extraction of cytochrome from mammalian tissues. This procedure removed practically all of the cytochrome *c*, leaving RHP quantitatively in the residue.[6]

Although several procedures could be used for the extraction of RHP from the residual powder, the simplest one appeared to be incubation in a minimal volume of 1 per cent citric acid at 50° for 10 minutes, followed by cooling to room temperature, neutralization, and recovery of the liquid phase. Other extraction procedures at low temperatures, while more tedious, gave neither better yields nor preparations differing in degree of denaturation or other detectable properties from those obtained with warm citric acid. The use of this reagent was suggested by a number of successes attending its use, in place of TCA, in the isolation of cytochrome *c* from bacterial sources (13, 14).

Notwithstanding many attempts, it was not found possible to arrive at a precise procedure for isolation and purification of RHP. Many factors difficult to control appeared to affect ultimate recoveries, notably the previous history of the cultures, delays in procedure schedule occasioned by the need to accumulate material, etc. Once a certain degree of purity was achieved, it appeared that any of a number of procedures was effective in arriving at comparable degrees of purity and yield. With the exceptions as noted, the procedure given in the next section was found to be quite reproducible when adhered to.

Extraction and Purification of RHP

Procedure for RHP Optical Assay—The concentration of RHP in test solutions was monitored by measuring the reduced Soret absorption at 424 mμ for which ϵ (mg. per ml.) = 4. Because the reduced Soret absorption of the bacterial cytochrome *c* had its maximum at 415 mμ, and could contribute to light absorption at 424 mμ, this practice could be misleading when a crude preparation containing the cytochrome was examined. Hence, the ratio between optical densities at 415 and 424 mμ was used as a rough qualitative index for cytochrome *c* contamination of RHP, $R_{415/424}$ for the purest preparation being 0.84. The optical density ratio at 275 mμ (oxidized) to 424 mμ (reduced) ($R_{275/424}$) was used as a qualitative index of RHP purity. This ratio varied from more than 30 in certain crude extracts to 0.274 in the purest preparations reported here.

[6] One of us (M. D. K.) is indebted to Professor H. Theorell, whose hospitality and cooperation made possible researches sponsored by the John Simon Guggenheim Memorial Foundation during the course of which these observations were first noted.

In crude preparations both protein and nucleic acid contributed to the light absorption at 275 mμ; in purified RHP protein absorption was maximal at 275 mμ (see the spectral data below).

When $R_{275/424}$ was less than 0.3, the preparation appeared homogeneous upon electrophoresis at pH values from 5.8 to 7.8. No visible additional peaks could be seen unless the value of $R_{275/424}$ was ≥ 0.33, when a very small colorless component, migrating anodically and at a speed much less than the main RHP peak, appeared. This peak was less than 2 per cent of the total protein present as estimated by the relative areas of the peak patterns. It could be concluded that RHP with an $R_{275/424}$ less than 0.3 possessed a spectroscopic purity greater than 98 per cent. Crude preparations of RHP ($R_{275/424}$ greater than 2) usually showed at least one, and more often several, colorless components equal to, or greater in amount than, the RHP component and migrating faster anodically. These impurities could be correlated spectroscopically with nucleic acid-containing proteins or peptides or nucleic acid fragments. The small amount of residual protein contaminant remaining after removal of nucleic acid ($R_{275/424}$ = 0.33 to 0.6) was spectroscopically identical with RHP protein and may have been denatured RHP.

Extraction of RHP—Proportions for the complete procedure are described in terms of 100 gm. of starting bacterial powder.

100 gm. of dry *R. rubrum* were suspended in 1.5 liters of cold water and the pH was adjusted to 4 by the addition of 1 M sulfuric acid. The suspension was then stirred in the cold, the pH being checked occasionally and adjusted to 4 as necessary. After 1 hour the suspension was centrifuged, the supernatant solution containing mostly cytochrome *c* was set aside, and the residue was resuspended in 1 liter of cold water. The pH was again adjusted to 4; the suspension was stirred for 1 hour and centrifuged. The two supernatant solutions were combined and neutralized with 6 N sodium hydroxide and the flocculent precipitate which formed was centrifuged. If the second extract contained more than one-tenth as much cytochrome *c* as the first extract, the residue was washed again with 1 liter of water at pH 4 and this supernatant solution was discarded.

The residue from the cytochrome *c* extraction was suspended in 1 liter of 1 per cent (w/v) citric acid solution; the suspension was heated rapidly to 50–60° and then maintained at this temperature for 10 minutes. The suspension was then rapidly cooled to room temperature and the pH was adjusted to 7 by the addition of 6 N sodium hydroxide. After the preparation had stood in the cold (5°) overnight, the solids were centrifuged at 20,000 × *g* for 30 minutes in the Lourdes large capacity centrifuge. The nearly clear yellow-brown supernatant solution was set aside. The gummy dark brown residue was resuspended in 500 ml. of 10 per cent (w/v) ammonium sulfate solution and stirred for 1 hour and the suspension was centrifuged. A clear extract was obtained only with prolonged centrifugation if the salt was omitted from the wash solution. The initial extract was combined with this second yellow-brown supernatant solution which contained about one-tenth as much RHP as the first. The solid residue was discarded.

For each 100 ml. of combined crude RHP extract, 20 gm. of solid ammonium sulfate were stirred into solution and the suspension was stirred in the cold for 1 hour. The precipitate (P-1) was centrifuged and then dissolved in a minimal amount of M/15 phosphate buffer, pH 7, and the solution was dialyzed free of salt and lyophilized.

To the supernatant solution were added 40 gm. of solid ammonium sulfate for each initial 100 ml. of extract and the suspension was left overnight in the cold. The crude RHP precipitate (P-2) was centrifuged, dissolved in M/15 phosphate, pH 7, dialyzed, and lyophilized. The supernatant solution contained a small amount of cytochrome *c* which was readily precipitated when one-tenth volume of cold 10 per cent (w/v) TCA was added. This precipitate was added to the crude cytochrome *c* obtained above.

Isoelectric Precipitation and Extraction of RHP—The crude dry RHP (P-2) was dissolved in 20 ml. of cold water for each gm. of powder to give a clear brown solution. To this solution 1 M citric acid was added to reduce the pH from an initial value of approximately 6 to about 5, at which pH the first trace of precipitate appeared. This reddish brown precipitate which contained traces of bacterial pigments such as carotenoids was centrifuged and discarded. The clear supernatant solution was adjusted to pH 4.6 by the further addition of 1 M citric acid, the suspension was stirred for 1 hour in the cold, and then the relatively small gray-brown precipitate was centrifuged. This precipitate (P-3) was taken up in 0.1 M citrate-Tris buffer, pH 5.0, and the solution was clarified by centrifugation and set aside. The residue which contained some crude RHP was dissolved in M/15 phosphate, pH 7, dialyzed, lyophilized, and set aside to be reworked with a later batch of crude starting material.

The pH of the supernatant solution from the precipitation at pH 4.6 was adjusted to pH 3.6 and the suspension was stirred for 1 hour in the cold. The precipitate (P-4) was centrifuged. The precipitate was washed twice to remove soluble nucleic acid-containing material by suspending it as well as possible in cold 0.1 M citrate-Tris buffer, pH 3.6, equal to one-half the volume of the first supernatant solution of pH 3.6, stirring in the cold for 20 minutes, and then centrifuging. The two wash solutions were added to the supernatant solution at pH 3.6 and the small amount of heme compounds present was precipitated by the addition of 60 gm. of solid ammonium sulfate for each 100 ml. of solution. The precipitate was centrifuged, dissolved in M/15 phosphate, pH 7, dialyzed, lyophilized, and set aside to be reworked as crude RHP.

The washed RHP precipitate (P-4) was washed successively with 0.1 M citrate-Tris buffer of pH 3.8, 4.0, 4.25, 4.5, 4.75, and 5.0. At each step the supernatant wash solutions were checked for RHP concentration and relative purity. Any step that appeared to aid the purification was repeated until it was obvious that further repetition was valueless. At the lower pH values, nucleic acid was usually dissolved preferentially and the reverse held above pH 4. However, this finding was not invariably the case. In different preparations the best RHP fraction appeared between pH 3.5 and pH 4.75. Finally, the residue was suspended in water, neutralized by the addition of M Tris, clarified by centrifugation, dialyzed, lyophilized, and set aside to be reprocessed as crude RHP.

Supernatant wash solutions of comparable purity were combined and the RHP was precipitated by the addition of M citric acid to adjust the pH to 3.5. Each precipitate was dissolved in 0.1 M citrate-Tris, pH 5.0, dialyzed, and lyophilized.

It was noted that the cytochrome *c* initially present in the crude RHP tended to concentrate in the precipitates which were insoluble at pH 5.0. After repetition of the precipitation and extraction process, the RHP was free of any spectroscopically detectable contamination by the cytochrome *c*. By repeating the isoelectric precipitation and extraction process once or twice on a given preparation, and by reworking the various tailing fractions, several hundred mg. of the purest material reported here were prepared. Best results were obtained when the RHP from the first precipitation-extraction step was stored for several weeks in the cold, either in solution or as a dried powder, before the process was repeated. Apparently the nucleic acid-like material that was the chief contaminant at that stage was degraded to such an extent as to become soluble at pH 3.6 to 4.

Purified RHP was ordinarily stored in the cold, preferably at −15°, although no spectroscopic change was noted in dry material stored at room temperature for as long as 30 days. In Table I are given results selected from typical runs.

Alternative procedures in which are employed either continuous paper electrophoresis, batch paper electrophoresis, or percolation through carboxymethylcellulose columns have been examined.[7] All of these methods have effected good purifications of RHP. Continuous paper

[7] We are indebted to Dr. G. Zweig of the Charles F. Kettering Foundation, Yellow Springs, Ohio, for his cooperation in testing various continuous electrophoresis procedures using facilities at the Kettering Foundation laboratories, kindly made available to us by Dr. H. A. Tanner.

electrophoresis at pH 7 has appeared well suited to the processing of crude RHP, especially in regard to removal of cytochrome *c* contamination. To bring samples of intermediate purity to a degree of purity shown in Step 5 the isoelectric extraction procedure is most convenient. Purification beyond Step 5 is readily achieved invariably by paper electrophoresis at pH 7.

It is conceivable that enzymic degradation of nucleic acid could be combined with the precipitation and extraction steps to effect savings in

TABLE I

*RHP Extraction and Purification**

Step No.	Fraction	$R_{275/260}$	$R_{275/424}$	$R_{415/424}$	Dry weight	RHP or cytochrome *c*†
	Extraction of cytochrome *c* and RHP					
					mg.	*mg.*
1	Cytochrome *c*	1.17	1.23	1.55	1050	256
2	RHP (P-1)	0.89	5.9	1.02	567	27
3	" (P-2)	0.57	25.5	0.98	2660	190
	Purification of RHP (P-2)					
					mg.	
4	Combined pH 4.25, 4.5, 4.75 washings	0.75	3.6	0.92	438	141
5	Combined pH 4.25, 4.5, 4.75 washings (from Step 4)	0.75	1.9	0.87	161	96

* Based on 100 gm. of starting *R. rubrum* powder.

† Based on Soret absorption peaks; ϵ (mg. per ml.) = 4.0 for RHP; and mg. per ml. = 10.9 for cytochrome *c* (calculated from ϵ (micromoles per ml.) = 142 (Paléus and Neilands (15)).

time and total number of operations, but this possibility remains to be investigated.

Results

General Physical and Chemical Properties of RHP[8]—RHP is soluble in water or buffers above pH 5. Below pH 5 it exhibits minimal solubility, in accord with its isoelectric point which lies close to this pH (see below). It possesses a thermal stability much like that of bacterial cytochrome *c*. Thus, it can be boiled for 1 to 2 minutes at 100° with little irreversible

[8] Many of the observations described in this section were made originally in previously published researches (1, 2), or in unpublished observations by R. Hill and M. D. Kamen, and confirmed in the present studies.

denaturation. Heating for 10 to 15 minutes at 50–60° in 1 per cent citric acid is without effect. However, prolonged boiling at neutral pH will result in irreversible denaturation, as evidenced by a typical acid hematin spectrum with absorption maxima at 400 and 530 mμ.

RHP is stable in the cold in the range pH 2 to 11. RHP at pH > 11 appears to undergo a transition to a reversibly denatured form which upon reduction exhibits a hemochromogen spectrum which is identical with that of the bacterial cytochrome *c* at neutral pH (see below). The oxidized form is spectroscopically similar to that of an alkali hematin, with absorption maxima at approximately 415 mμ and approximately 540 mμ. When the pH is brought back to 7, RHP reappears in its original spectral form. Treatment with glacial acetic acid results in a compound with a spectrum like that obtained by prolonged boiling.

RHP can be oxidized and reduced reversibly with reagents commonly employed for this purpose in chemical manipulation of hematin compounds. It is reduced completely at neutral pH by hydrogen-palladium, ferrous oxalate, dithionite, and hydrogen sulfide. Reduction by ascorbate is incomplete even when the ascorbate is present in 100-fold molar excess. RHP is reduced enzymically by DPNH when incubated with the mammalian DPNH-linked cytochrome *c* reductase (1) but not by DPNH alone. The specific cytochrome b_5 reductase of liver microsomes (16) fails to catalyze reduction with DPNH.[9]

RHP was found to be rapidly autoxidizable. No precise determination of the oxidation kinetics was attempted, but a semiquantitative estimate was made on the basis of the following observations. RHP (18 mg.) dissolved in 4 ml. of M/15 phosphate buffer at pH 7.0 was reduced with hydrogen-palladium. The solution was filtered free of the reductant under helium and then exposed to air and shaken vigorously at room temperature. The disappearance of the characteristic α-band at 550 to 560 mμ (see below) was observed visually with a Hartridge reversion spectroscope. It could be estimated that the half life of the reduced form was approximately 3 seconds.

If RHP is brought to pH > 11 in the presence of lauryl sulfate and then subjected to several cycles of reduction with dithionite and oxidation with air, it is observed that the heme moiety disappears. This behavior is not observed with the bacterial cytochrome *c* and is much like the well known degradation of hemoglobin and similar hematin compounds when subjected to oxidative denaturation (17). Presumably the products of degradation of RHP are bile salts, as in the case of myoglobin.

RHP does not bind reversibly many reagents which form addition compounds with hemoglobin or myoglobin. Thus, it fails to react at

[9] P. Strittmatter, private communication (1957).

neutral pH to form stable addition compounds, either in the oxidized or reduced form, with hydrogen sulfide, oxygen, nitric oxide, sodium azide, sodium cyanide, and 4-methylimidazole. On the other hand, reduced RHP reversibly binds carbon monoxide, the resultant compound showing a typical hemochromogen spectrum (see below). The carbon monoxide compound is extremely sensitive to light at pH 7, but appears quite stable at pH $\leq$ 4.0. This anomalous behavior is unique to RHP, as far as we are aware. The half life of the carbon monoxide compound at pH 7 and at light intensities in the range of 100 to 200 foot candles is of the order of 10^{-1} second.

If RHP is partially denatured by treatment with lauryl sulfate in dilute alkali, it appears to be able to form a complex with 4-methylimidazole, as assayed by the appearance of a typical hemochromogen spectrum. However, reaction with cyanide or pyridine requires complete denaturation by heating in strong alkali (6 N sodium hydroxide). When this is done, the derivative hemochromogens are identical with those obtained in the same manner from pure beef heart cytochrome *c* or bacterial cytochrome *c* (1, 2).

The prosthetic heme of RHP cannot be split off, as in hemoglobin, by treatment with cold dilute acid-acetone mixtures. Reductive cleavage with sodium amalgam (18) produces a mixture of mesoporphyn and chlorin spectroscopically absolutely identical with that obtained from pure cytochrome *c*, either mammalian or bacterial. No other method, such as the silver-salt cleavage (19), or strong acid cleavage (20), has been attempted.

Analytic Data

Electrophoretic Measurements—The combined results of all electrophoretic determinations are presented in Table II. Unfortunately, insufficient amounts of the purest preparation were available to determine mobilities at the various pH values shown, so less pure samples ($R_{275/424}$ = 0.35 to 0.50) were used for pH 5.8 and pH 7.8. In these cases, however, impurity peaks which were visible were colorless and only one colored peak could be seen. The colored component was used in determining mobilities at these pH values.

Another difficulty arose in the inadvertent use of buffers with different ionic strengths. While corrections for ionic strength effects are uncertain, an attempt was made to reduce all mobility μ values to a constant ionic strength of τ = 0.2. This was done by using the two values at pH 6.0 and 6.8 to establish a linear relation of mobility and pH. Then, the value at pH 5.8 was corrected downward to fall on the same line. By this procedure, it was possible to obtain a factor by which to increase the observed mobility at pH 7.8 so as to arrive at a mobility for RHP at pH 7.8 and ionic strength of $\tau = 0.2$. The predicted value was -7.5×10^{-5} e.s.u. The value determined by extrapolation of the line based on the experimental points at pH 6.0 and 6.8 was -7.35×10^{-5} e.s.u.

Another procedure was based on correction of mobilities, assuming that these depended on the square root of the ionic strength, all values being normalized to $\tau = 0.2$.

By using either of these procedures, it was found by extrapolation to $\mu = 0$ that the isoelectric point was at pH 5.0 to 5.1. This value must be considered as tentative until sufficient pure material is available to make possible enough determinations under proper conditions, but it is reasonable in view of the observed tendency of RHP to precipitate at pH < 5.

TABLE II

Electrophoretic Mobilities (μ) *of RHP*

pH	Ionic strength*	μ (cm.^{2} sec.$^{-1}$ volt^{-1}) $\times 10^5$ (observed)	μ (cm.^{2} sec.$^{-1}$ volt^{-1}) $\times 10^5$ † (corrected)
5.8	0.1156	−2.54	−1.6
6.0	0.2000	−2.2	−2.2
6.8	0.2000	−4.39	−4.39
7.8	0.2888	−5.1	−7.5

* Phosphate buffers of the following composition were used: pH 5.8, 78 ml. of M Na_2HPO_4 and 922 ml. of M KH_2PO_4 per 10 liters; pH 6.8, 500 ml. each of M Na_2HPO_4 and M KH_2PO_4 per 10 liters; pH 7.8, 944 ml. of M Na_2HPO_4 and 56 ml. of M KH_2PO_4 per 10 liters.

† See the text.

As noted from the sign of the mobility coefficient, RHP is an acidic protein, migrating anodically at pH > 5.

Sedimentation and Diffusion Constants—The sedimentation velocity constant of RHP was determined with a sample of RHP approximately 98 per cent pure ($R_{275/424} = 0.33$) which displayed only one major colored peak and a barely detectable minor colorless peak in the schlieren picture. The average value obtained from several runs (concentration of RHP, 0.9 per cent (w/v)) in phosphate buffer at pH 7.0 (ionic strength (τ) = 0.2) and 20° was 2.66 Svedberg units (S). Corrected to pure water solvent, the value ($s_{20,w}$) was 2.76 S. This value could be estimated to have a standard deviation of approximately ±2 per cent. Systematic errors noted occasionally (5) in this procedure could increase the estimated error to ±5 per cent.

The diffusion constant for the same RHP preparation was determined by measuring the rate of area broadening of the peak in the electrophoretic pattern at pH 6.8 (6). Corrected to 20° and pure water solvent, the value ($D_{20,w}$) was 8.65 × 10^{-7} c.g.s. unit, with an uncertainty estimated from previous experience (5) of ±5 per cent.

Partial Specific Volume—The density of the protein solution, ρ, and the density of the solvent, ρ_0, for a given concentration of RHP in water containing a trace of Tris buffer sufficient to maintain the pH at 7 and at 20° (±0.01°) were determined by the gradient tube method (6). The partial specific volume, $\bar{v}_{20}$, then was calculated according to the formula

$$\bar{v} = \left(1 - \frac{100(\rho - \rho_0)}{c}\right) \Big/ \rho_0$$

The value obtained was 0.731.

TABLE III
Spectral Properties of RHP

	Extinction coefficients of absorption peaks at					
	275 mμ*	390 mμ*	424 mμ	500 mμ*	550 mμ	640 mμ*
ϵ, *mg. per ml.*	1.09	3.63	4.00	0.452	0.464	0.132
E *μmoles per ml.*	30.7	101.6	112.0	12.7	13.0	3.70

Characteristic ratios of extinction coefficients as follows: $\epsilon_{275}/\epsilon_{424} = 0.274$, $\epsilon_{424}/\epsilon_{550} = 8.62$, $\epsilon_{275}/\epsilon_{640} = 8.3$.

* Oxidized.

Molecular Weight (M)—By substituting the values for the various parameters determined as described previously in the well known formula

$$M = \frac{RTS_{20,w}}{D_{20,w}(1 - \bar{v}\rho)}$$

the value of 27,740 was determined for the molecular weight of RHP. The systematic errors could be estimated as ±5 per cent; hence a value of 28,000 ± 1000 was taken for calculation of millimolar extinction coefficients of Table III. The frictional ratio (21) was calculated as 1.33.

Determination of the heme content as the pyridine hemochromogen according to the procedure of Drabkin (22) gave a value of M = 25,000, assuming one heme per mole.

The agreement between these two independent procedures can be considered satisfactory.

The iron content of the same highly purified RHP preparation used for the ultracentrifugal and electrophoretic and gradient measurements was found to be 0.18 per cent. This indicated a molecular weight of 31,000, assuming one iron per mole. Because the iron analysis is likely to be low rather than high, leading to a high estimate of the molecular weight, it was assumed that the molecular weight was close to 28,000 and that all iron was accounted for as heme iron, of which there was 1 mole per mole of protein.

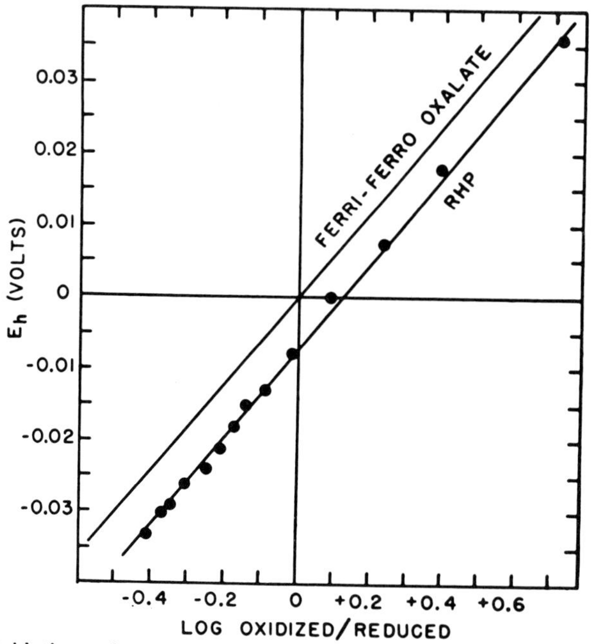

FIG. 1. Oxidation-reduction equilibrium of RHP with the iron-oxalate system at pH 7 and 30°. The reaction mixture contained initially 5×10^{-3} M potassium phosphate, pH 7, 0.5 M potassium oxalate, pH 7, 10^{-3} M ferric-ammonium sulfate, and 0.36 mg. of RHP (approximately 1.3×10^{-2} μmole) in 2.0 ml. and was titrated anaerobically by the addition of increments of 0.02 M ferrous-ammonium sulfate.

Electrochemical Potential—The oxidation-reduction equilibria between RHP and ferric-ferrous oxalate at pH 7 were measured in the manner described by Velick and Strittmatter (9), except that the equilibrium concentrations of RHP were determined from optical densities at 550 and 640 mμ rather than at the Soret absorption peak. The calculated results were identical at the two wave lengths.

The reversibility of the equilibrium was indicated by the ready reoxidation of ferrous oxalate-reduced RHP by the ferric salt. In Fig. 1 the results of an electrometric titration are represented in the manner described by Hill (8). It is apparent that the ferric-ferrous oxalate and RHP curves are

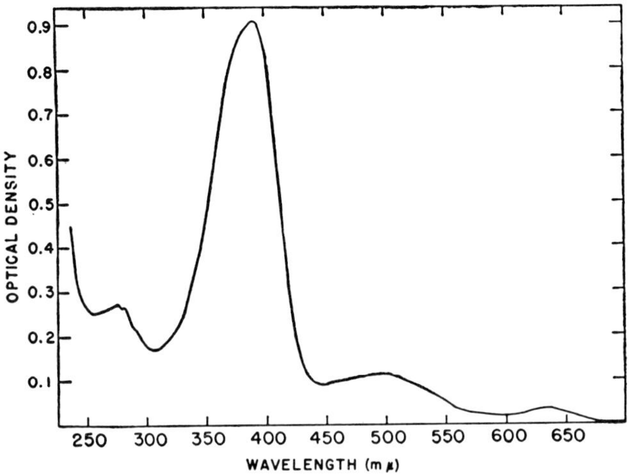

FIG. 2. Spectrum of oxidized RHP (0.25 mg. per ml. of RHP, approximately 0.93×10^{-2} μmole per ml.) in M/15 phosphate, pH 7.

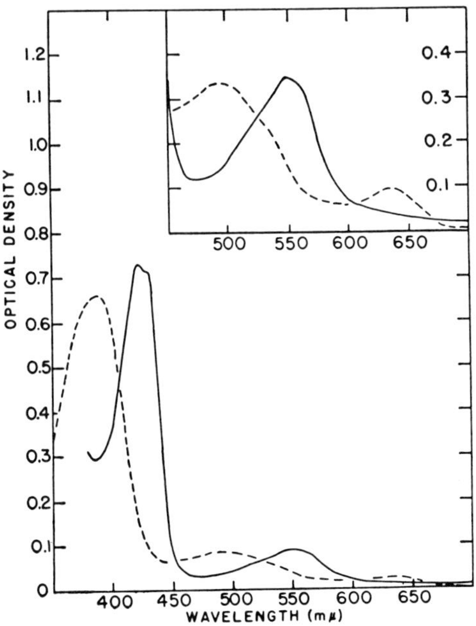

FIG. 3. Spectrum of oxidized (broken line) and reduced (solid line) RHP at pH 7. For the main curves 0.186 mg. per ml. of RHP (approximately 0.66×10^{-2} μmole per ml.) and for the inset curves 0.745 mg. per ml. of RHP (approximately 2.7×10^{-2} μmole per ml.) were used in M/15 phosphate, pH 7.

almost coincident and that for RHP

$$E_0' = E_h \text{ (0.5 reduced, pH 7)} = -0.008 \text{ volt}$$

Spectroscopic Observations

The spectroscopic properties of RHP exemplify the novel character of this heme protein as compared with those commonly encountered. It exhibits a spectrum such as one might expect from a myoglobin-like protein in which the prosthetic group possesses saturated side chains.

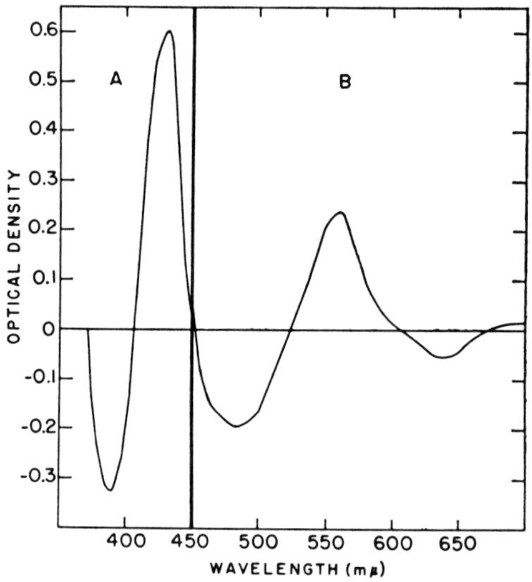

FIG. 4. Difference spectrum of RHP at pH 7, reduced minus oxidized. In *A* 0.186 mg. per ml. of RHP (approximately 0.66×10^{-2} μmole per ml.) and in *B* 0.745 mg. per ml. of RHP (approximately 2.7×10^{-2} μmole per ml.) were used in M/15 phosphate, pH 7.

Thus, the reduced compound shows a typical sharp Soret band and a rather diffuse α-band, as in myoglobin, but shifted 10 to 20 mμ toward the blue. The complex spectrochemical behavior of RHP is shown in Figs. 2 through 8, wherein are presented direct tracings of RHP spectra obtained with the Cary model 14 recording spectrophotometer. In Table III are summarized data on location and intensity of absorption peaks of the RHP sample used in obtaining these spectra.

The spectrum of oxidized RHP at pH 7, reproduced in Fig. 2, shows the protein absorption band with a maximum at 275 mμ and a shoulder at 280 mμ, the Soret band at 390 mμ, another band at 500 mμ, and the hematin band at 640 mμ.

The reduced RHP spectrum is shown in Fig. 3. There is a broad α-band extending from 550 to 560 mμ. The Soret band is sharp with a maximum at 424 mμ and a shoulder at approximately 431 mμ. The significance of this shoulder is not known. The spectrum of the oxidized

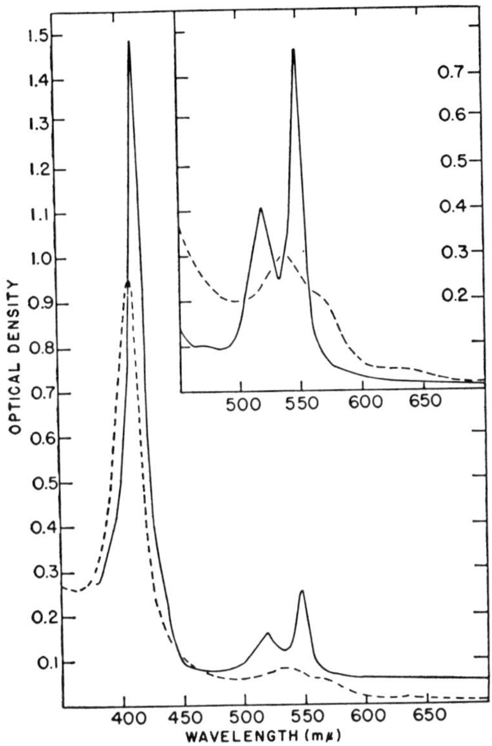

FIG. 5. Spectrum of oxidized (broken line) and reduced (solid line) RHP at pH 11.8. For the main curves 0.186 mg. per ml. of RHP (approximately 0.66×10^{-2} μmole per ml.) and for the inset curves 0.745 mg. per ml. of RHP (approximately 2.7×10^{-2} μmole per ml.) were used in 0.1 M trisodium phosphate buffer adjusted to pH 11.8.

form is included for comparison, demonstrating the shift in absorption maxima toward the red for both the Soret and α-bands, as well as the loss of the hematin band that accompanies reduction of RHP. The reduced pigment possesses no β absorption band. It may be that the unsymmetric nature of the reduced α-band conceals fine structure that could be revealed by examination at low temperature. Alternatively, the presence of traces of contaminating cytochrome *c* may be betrayed by the slightly skewed α-band and the hint of a slightly high absorption in the regions of 520 to 525 mμ.

The RHP difference spectrum at pH 7 is given in Fig. 4. Minimal absorption peaks, corresponding to oxidized components, fall at 388, 485, and 640 mμ and maxima, corresponding to reduced components, occur at 432 and 510 mμ. This difference spectrum resembles closely that of Chance and Smith (23), based on differential spectrophotometry of cultures *in vivo*,

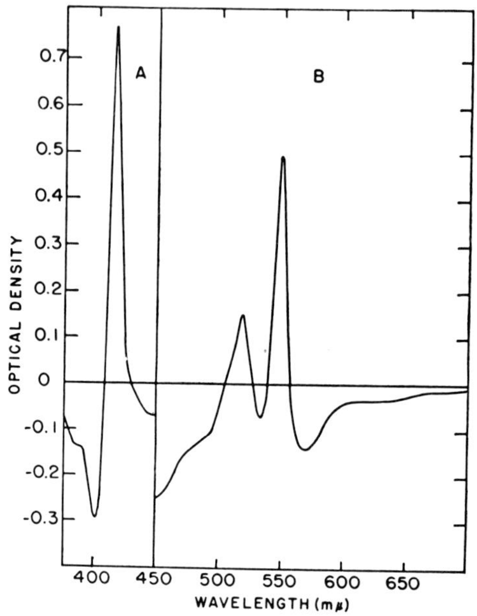

FIG. 6. Difference spectrum of RHP at pH 11.8, reduced minus oxidized. In *A* 0.186 mg. per ml. of RHP (approximately 0.66 × 10^{-2} μmole per ml.) and in *B* 0.745 mg. per ml. of RHP (approximately 2.7 × 10^{-2} μmole per ml.) were used in 0.1 M trisodium phosphate, pH 11.8.

in which a heme compound showing a difference Soret absorption peak at 430 mμ has been suggested as identical with RHP.

The oxidized and reduced hemochromogen type spectra of RHP at pH 11.8 are presented in Fig. 5. Major absorption maxima for the oxidized spectrum occur at 407 and 537 mμ and for the reduced spectrum at 413, 518, and 549 mμ. In the reduced spectrum the values of optical densities at 413, 518, and 549 mμ are 1.48, 0.11, and 0.205, respectively. These values give ratios for Soret to α and β absorption not identical to, but somewhat higher than, those found for the hemochromogen-like spectrum of *R. rubrum* cytochrome *c* (1).

Fig. 6 represents RHP difference spectra (reduced minus oxidized) at pH 11.8, of which the structure and orders of magnitude are very similar to those of cytochrome *c*. Minimal absorption peaks occur at 402, 450, 534, and 569 mμ and maxima occur at 415, 532, and 549 mμ.

The absorption spectrum of the reduced RHP-carbon monoxide complex is given in Fig. 7. Formation of the complex appears to shift the Soret absorption band from 424 to 415 mμ, to intensify the absorption (ε (mg.

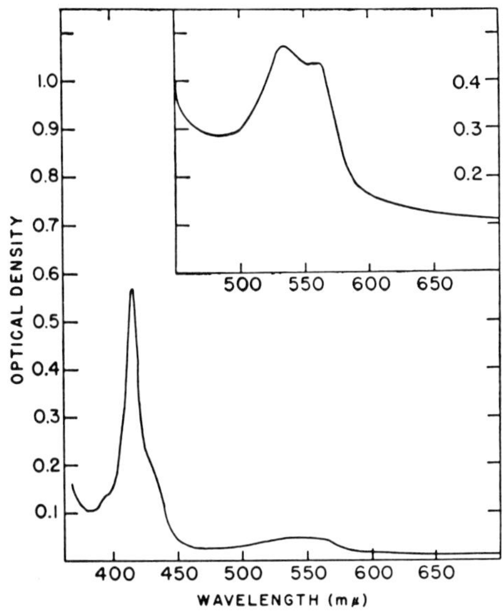

FIG. 7. Spectrum of the carbon monoxide-reduced RHP compound at pH 4.75. For the main curve 0.0745 mg. per ml. of RHP (approximately 0.27×10^{-2} μmole per ml.) and for the inset curve 0.745 mg. per ml. of RHP (approximately 2.7×10^{-2} μmole per ml.) were used in 0.1 M citrate-Tris buffer, pH 4.75, with pure carbon monoxide in the gas phase.

per ml.) changes from 4 to 7.65), and almost to obliterate the shoulder seen in the absence of carbon monoxide. However, the remnant of that shoulder may be responsible for the inflection near the base of the carbon monoxide Soret peak. The α-peak of the carbon monoxide complex is a composite of two peaks with maxima at 534 and 560 mμ. When the RHP-carbon monoxide complex is examined at pH 4 or less with the Hartridge reversion spectroscope, two sharp bands are seen at 535 and 560 mμ, whereas at pH 7 only one broad band over the range 540 to 560 mμ is noted. At the lower pH much of the protein is insoluble, making it difficult to obtain a satisfactory spectrum. For the spectrum in Fig. 7, pH 4.75

was chosen as the lowest pH value which would permit an indication of the splitting of the α-band without undue protein precipitation; however, as is indicated by the inset tracing, sufficient protein precipitation occurred at the protein concentration used for the α-band spectrum to move the base line upward from zero to approximately 0.1 optical density.

Finally, in Fig. 8 is presented the difference spectrum for reduced RHP-carbon monoxide minus reduced RHP at pH 4.75. Minimal absorption occurs in the Soret region at 396 and 432 mμ and in the yellow-green at 407, 550, and 586 mμ. Maxima occur in the Soret region at 413 mμ and in the yellow-green region at 477, 531, and 565 mμ.

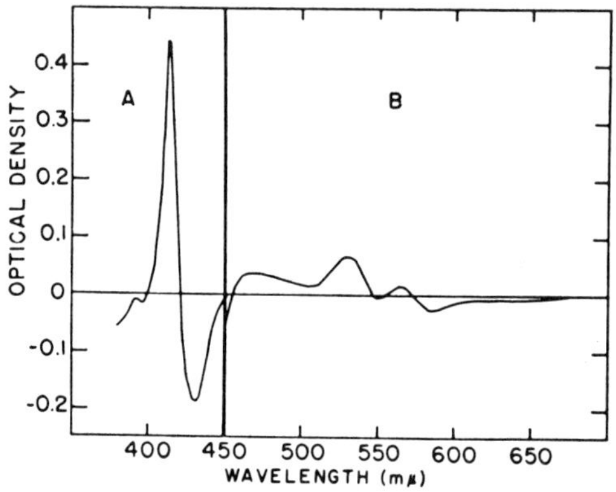

FIG. 8. Difference spectrum of the carbon monoxide-reduced RHP compound at pH 4.75 and carbon monoxide plus reduced minus reduced. In *A* 0.0745 mg. per ml. of RHP (approximately 0.27×10^{-2} μmole per ml.) and in *B* 0.745 mg. per ml. of RHP (approximately 2.7×10^{-2} μmole per ml.) were used in 0.1 M citrate-Tris buffer, pH 4.75, with pure carbon monoxide as the gas phase in the experimental cuvette.

DISCUSSION

All of the spectroscopic and chemical observations on RHP described in this report can be rationalized in so far as the prosthetic group and its relation to the apoprotein are concerned by assuming that RHP represents a variant of cytochrome *c* (note (24) for brief summary of present concepts of cytochrome *c* structure) in which one (or both) of the extraplanar protein links to the central iron atom is loose or missing. Such a structure would account for the myoglobin-like spectra and at the same time explain the characteristic shift in absorption maxima of some 10 mμ toward the blue, as compared with myoglobin. In this respect, the compound can be thought of as a myoglobin type of protein with a meso- or hematoheme prosthetic group like those reported by Hill and Holden (25). RHP appears to share with cytochrome *c* and cytochrome *f* (Davenport and Hill (25)) the property of being the only naturally occurring heme protein compound containing these stable linkages between side chains and protein.

The failure to observe fission of the heme from protein by the classical cold acid-acetone treatment effective for myoglobin shows that the prosthetic group is linked stably to the protein through the heme side chains. The fact that reductive cleavage with sodium gives both qualitatively and quantitatively identical mixtures of mesoheme and chlorin strongly supports the notion that the binding of the heme in RHP is identical with that in cytochrome *c*. It is known that hematoheme is also reduced to mesoheme under the conditions of the cleavage procedure (18). However, it would not be expected that quantitative agreement in nature and yield of products from RHP compared to cytochrome *c* would result if the linkage in RHP were, say, an ether linkage to a hydroxyamino acid rather than a thio ether linkage to cysteine. The proof of the identity of the heme and its binding in RHP and cytochrome *c* must await accumulation of sufficient pure material to enable isolation of iron peptides after tryptic digestion, as done in the cytochrome *c* studies. From such peptides it may be possible to ascertain the nature of the linkage between the heme side chains and the moiety involved in the protein.

It is of interest that reduced RHP exhibits a true cytochrome *c* spectrum when brought to pH > 11. This behavior is opposite to that of cytochrome *c* at high pH. In other words, RHP can be "changed" into a hemochromogen almost identical with cytochrome *c* spectroscopically by going to high alkaline pH. The change is strictly reversible, however, for on lowering the pH to 7 the original acid hematin spectrum is restored.

While RHP can be changed in this fashion to exhibit a spectrum similar to that of cytochrome *c* at neutral pH, it is not possible to effect a similar change in cytochrome *c* whereby it resembles RHP other than superficially. If cytochrome *c* is brought to pH 4 in the presence of lauryl sulfate, it exhibits a characteristic acid hematin type spectrum, a fact well known for some time (26). However, the resultant spectrum, while resembling RHP in general appearance, is quite different quantitatively. The oxidized form of the acidified cytochrome *c* has maxima at 405 and 630 mμ, the ratio of the Soret to red bands being 12.5. The corresponding Soret and α-band maxima for oxidized RHP either at pH 4 or 7 are at 395 and 640 mμ, the ratio being 23.6.

These changes in RHP at high pH may result from a reversible twisting of the protein in which a group hindering chelation of the iron to a nitrogenous residue in the protein moves aside and a nitrogen atom from the protein moves in, thus forming a cytochrome *c*-like hemochromogen.

RHP is unusual among heme proteins in that, although it exhibits an acid hematin spectrum, it is incapable of forming stable addition complexes with any of the usual reagents (cyanide, azide, 4-methylimidazole, oxygen, hydrogen sulfide, nitric oxide). This fact would argue for some form of steric hindrance, as postulated to explain the spectrochemical behavior during reversible alkaline denaturation. However, more than this is needed to understand the remarkable reaction of reduced RHP with carbon monoxide to form a typical hemochromogen addition complex. Furthermore, the RHP-carbon monoxide complex appears to be the first instance reported of a light-sensitive heme-carbon monoxide compound in which the photodissociation rate depends on pH. Thus, the remarkable chemical properties of RHP require further elaboration before any attempt can be made to rationalize them in terms of structure.

The possibility that RHP is an artifact of the extraction procedure may be dismissed as most unlikely on the following grounds: (1) RHP is localized in bacterial chromatophores along with cytochrome *c*, cytochrome *b*, and all the photoactive pigments. It can be seen in these chromatophores directly when they are treated with acetone to remove the photoactive pigments. (2) RHP is obtained in the same yields relative to cytochrome *c*, regardless of the manner of extraction, except when TCA is used. With TCA, yields of RHP are too erratic from one culture to the next to determine a relative yield. (3) Isolated RHP is not formed from bacterial cytochrome *c*, or any other components of the chromatophores, by treatment with reagents used in the extraction procedures. (4) Difference spectra of actively metabolizing cells determined under a variety of conditions (anaerobic with and without carbon monoxide and other inhibitors, aerobic with and without inhibitors) can all be correlated with known spectroscopic characteristics of a mixture of RHP and cytochrome *c* (23). (5) RHP is one of the most abundant proteins in the bacteria. It can be obtained in yields as high as 3 mg. per gm. dry weight, thus constituting 0.6 per cent of all the cellular protein. This amount is comparable to that of the cytochrome *c* component in *R. rubrum* on a molar basis. The relative amounts of RHP and cytochrome *c* are in accord with those estimated visually in the experiment with acetone-treated chromatophores.

The identity of RHP with the terminal oxidase of *R. rubrum* is suggested by the following considerations. First, the difference spectra (reduced minus oxidized and carbon monoxide reduced minus reduced) are identical with those for a component in the oxidation chain of the intact cell suspensions actively metabolizing under a variety of conditions (23). The major portion of the differences in absorption in the visible light region when *R. rubrum* suspensions are shifted from light anaerobic to dark anaerobic, or dark anaerobic to dark aerobic conditions (23), can be accounted for by assuming RHP as the active reagent. Similarly, the shift in Soret absorption seen when the cells are subjected to carbon monoxide in the dark (23) is precisely that to be expected for RHP. Secondly, RHP is reversibly oxidized and reduced readily by DPNH-linked cytochrome *c* reductases, such as those found in bacterial or mammalian tissue, although it is not coupled with DPNH-linked cytochrome b_5 reductase as found in mammalian liver microsomes. Thirdly, RHP is rapidly oxidized in air.

The rate of autoxidation is of an order of magnitude consistent with the respiratory rate observed in the weakly aerobic photoheterotrophically grown cells from which RHP is isolated. In a typical manometric experiment, 200 c.mm. cells suspended in 2 ml. of buffer will take up 60 c.mm. (approximately 2.5 μmoles) of oxygen per hour in dark aerobic metabolism of a substrate such as acetate. This quantity of cells corresponds to at least 90 γ of RHP or 3×10^{-3} μmole of RHP. The minimal rate of oxidation, assuming first order kinetics and the observed half life of RHP of 3 seconds, is approximately 1.5×10^{-3} μmole in 3 seconds or approximately 2 μmoles per hour. This corresponds to a rate of oxygen uptake of at least 0.50 μmole per hour.

If RHP is an integral portion of the respiratory chain in *R. rubrum*, functioning as a terminal oxidase, it may be possible to use it for monitoring oxidase and reductase activities in cell-free extracts, thereby leading perhaps to isolation and characterization of other components in the electron transport systems supporting both light and dark metabolism. It is also interesting in this regard that RHP is a major constituent of the chromatophores, which have been demonstrated to support light-catalyzed anaerobic phosphorylation (27). The possibility of RHP as a prototype for bacterial oxidases in general remains to be explored.

Spectrophotometry of metabolizing cell suspensions has revealed the presence in a variety of bacteria of moieties with spectroscopic properties like RHP which appear to function as terminal oxidases (4). However, at least one strict anaerobe, the photosynthetic bacterium *Chromatium*, also contains an RHP type of pigment (1, 2). The isolation of this compound from *Chromatium* has been accomplished on a mg. basis recently.[10]

Further research on these heme proteins directed toward clarification of their chemistry and structure should help to solve the problem of how they function in the richly variegated metabolism of the photosynthetic bacteria and by extension how analogous proteins act in other photosynthetic tissues, as well as related non-photosynthetic systems, such as facultative anaerobes which are chemosynthetic.

SUMMARY

1. The isolation and purification of a new heme protein from the facultative photoheterotroph, *Rhodospirillum rubrum*, are described.

[10] M. D. Kamen, unpublished observations.

2. The new protein, termed a "pseudohemoglobin" in previous studies and renamed "*Rhodospirillum* heme protein" (RHP) in these studies, is characterized on the basis of its spectroscopic and chemical properties as an autoxidizable variant of cytochrome *c*, in which one of the extraplanar protein bonds to the central heme iron is missing or hindered.

3. RHP is a soluble acidic protein, with an isoelectric point at pH 5.0 to 5.1 and with one heme per molecular weight of 25,000. Measurements of sedimentation velocity, diffusion, and partial specific volume lead to a calculated molecular weight of 28,000. Total iron analysis yields a somewhat higher figure. There appears to be no iron in excess of that accounted for as heme.

4. A function as a terminal oxidase for *R. rubrum* is suggested for RHP and evidence for this suggestion is presented. The possible significance of RHP as a prototype for bacterial oxidases in general is noted. The circumstance of its presence in the strict anaerobe *Chromatium* is confirmed.

BIBLIOGRAPHY

1. Vernon, L. P., and Kamen, M. D., *J. Biol. Chem.*, **211**, 643 (1954).
2. Kamen, M. D., and Vernon, L. P., *Biochim. et biophys. acta*, **17**, 10 (1955).
3. Mahler, H. R., Sarkar, N. K., Vernon, L. P., and Alberty, R. A., *J. Biol. Chem.*, **199**, 585 (1952).
4. Smith, L., *Bact. Rev.*, **18**, 106 (1954).
5. Taylor, J. F., *Arch. Biochem. and Biophys.*, **36**, 357 (1952).
6. Taylor, J. F., and Lowry, C., *Biochim. et biophys. acta*, **20**, 109 (1956).
7. Linderstrøm-Lang, K., and Lanz, H., Jr., *Compt.-rend. trav. Lab. Carlsberg, Série chim.*, **21**, 315 (1938).
8. Hill, R., *Nature*, **174**, 501 (1954).
9. Velick, S. F., and Strittmatter, P., *J. Biol. Chem.*, **221**, 265 (1956).
10. Sandell, E. B., Colorimetric determination of traces of metals, New York (1944).
11. Drabkin, D. L., *J. Biol. Chem.*, **140**, 387 (1941).
12. Paléus, S., *Acta chem. Scand.*, **8**, 971 (1954).
13. Verhoeven, W., and Takeda, Y., in McElroy, W. D., and Glass, B., Inorganic nitrogen metabolism, Baltimore, 159 (1956).
14. Kamen, M. D., and Takeda, Y., *Biochim. et biophys. acta*, **21**, 518 (1956).
15. Paléus, S., and Neilands, J. B., *Acta chem. Scand.*, **4**, 1024 (1950).
16. Strittmatter, P., and Velick, S. F., *J. Biol. Chem.*, **221**, 277 (1956).
17. Lemberg, R., and Legge, J. W., Hematin compounds and bile pigments, New York, chapter 10 (1949).
18. Davenport, H. E., *Nature*, **169**, 75 (1952).
19. Paul, K. G., *Acta chem. Scand.*, **5**, 389 (1951).
20. Theorell, H., *Enzymologia*, **4**, 102 (1937).
21. Svedberg, T., and Pedersen, K. O., The ultracentrifuge, Oxford (1940).
22. Drabkin, D. L., *J. Biol. Chem.*, **146**, 605 (1942).
23. Chance, B., and Smith, L., *Nature*, **175**, 803 (1955).
24. Ehrenberg, A., and Theorell, H., *Nature*, **175**, 158 (1955).
25. Hill, R., and Holden, H. F., *Biochem. J.*, **20**, 1326 (1926). Davenport, H. E., and Hill, R., *Proc. Roy. Soc. London, Series B*, **139**, 327 (1952).
26. Keilin, D., and Hartree, E. F., *Nature*, **145**, 934 (1940).
27. Frenkel, A. W., *J. Biol. Chem.*, **222**, 823 (1956).

Early History of Carbon-14

Discovery of this supremely important tracer was expected in the physical sense but not in the chemical sense.

Martin D. Kamen

When, how, and why was carbon-14 discovered? As T. S. Kuhn has remarked (*1*), discovery is seldom a single event that can be attributed wholly to a particular individual, time, or place. He notes that some discoveries, such as those of the neutrino, radio waves, and missing isotopes or elements, are predictable and present few problems, as far as establishment of priority is concerned. Others, such as the discoveries of oxygen, x-rays, and the electron, are unpredictable. These put the historian in a "bind" when he tries to decide when, how, who, and where the discovery was made. Much more rarely does he have a basis for an answer to the question "Why?"

I propose in this account of the "prenatal" history of carbon-14 to provide the answers to my leading questions (*2*). These make a story which is a fragment of the whole record. That record must be constructed by future historians who seek to probe the events of a period in which there has been an unparalleled impact of intellectual curiosity and scientific creativity on the structure of society.

The tremendous outburst of technology in the past half century, the result of the rise of nuclear science, has crowned man's quest for the philospher's stone so successfully as to be hardly credible even to the most optimistic alchemist. Tracer methodology, an offspring of nuclear science, has provided essential support for the ever-widening and deepening knowledge of structure and function in biological systems, expressed as the dynamic science of molecular biology.

These developments have profound, but unknown, implications for the future of our social structures. They obviously bring with them an unexampled load of grist for the mills of cultural historians, social scientists, and philosophers. Perhaps the novelists will dig into the record of these exciting times for fresh insights into the age-old drives of mankind.

Carbon-14, the long-lived carbon isotope, is the most important single tool made available by tracer methodology, because carbon occupies the central position in the chemistry of biological systems. Thus it plays, and will continue to play, an essential role in the elucidation of biochemical mechanisms, knowledge of which is essential in the further development of molecular biology. Obviously, the circumstances surrounding its discovery are valid objects of interest for the historian (*3*).

Initial Phases, 1934–36

In the early 1930's, nuclear physics, immersed in the great traditions of the Cambridge school led by Ernest Rutherford, was concerned primarily with observations of processes associated with the scattering of elementary nuclear particles by various atomic nuclei. Reports in those times show

The author is professor of chemistry at the School of Science and Engineering, University of California, San Diego. **This article is adapted from a paper which he presented at a meeting of the American Chemical Society in Los Angeles in April 1963, when he received the Society's 1963 Award for Nuclear Applications in Chemistry. This paper is also being printed in the May issue of the *Journal of Chemical Education*.**

Reprinted from *Science*, **140**, 584–590 (1963).

painstaking determinations of range-energy relations for the fundamental projectiles (protons, deuterons, alpha particles). The energies used did not exceed approximately 10 Mev, because of the limitations set by the relatively primitive accelerators and by the radiation characteristics of the naturally radioactive materials that were available. The rationale for such work, which often involved tedious attention to detail and much labor, was that if enough precise facts were put together, accurate binding energies for nuclei could be deduced. From these energies, it was reasoned, there could be derived a solid basis for further attack on the problem of the nature of nuclear forces.

By 1933, such data—binding energies, angular distributions in scattering experiments, and so on—had demonstrated that nuclear forces could be described as analogous to saturation exchange forces like those postulated previously for chemical bonding. The so-called "alpha-particle" model of the nucleus already contained the seeds of what was to be the full-fledged modern "shell" theory of nuclei, to be developed later by Maria Mayer, Eugene Feenberg, and others.

As to my part in this, I was a young, eager student and had just begun doctoral research, using the Wilson cloud chamber to study the angular distribution of neutrons scattered in collisions with protons and other nuclei. These researches were part of a general program initiated in the laboratory of W. D. Harkins in the chemistry department at the University of Chicago (*4*). My decision to work in this field was largely a result of the influence of D. M. Gans, Harkins' associate and an assistant professor in the department (*5*).

Most significantly for this history, similar work was also under way at Yale, where F. N. D. Kurie, investigating neutron-induced disintegration of light elements, had obtained certain anomalous results for the angular distributions of protons in collisions with neutrons. In 1934 he proposed a radical interpretation (*6*) of certain events he noted in the cloud chamber. When nitrogen was exposed to fast neutrons, for instance, he noted that in some cases the ejected nucleus produced a very long, thin track. This he ascribed to a proton, rather than to an alpha particle. Thus, he supposed that the usual reaction $N^{14}(n,He^4)B^{11}$, was accompanied by a less frequent but readily observable reaction, $N^{14}(n,H^1)C^{14}$. (As far as I am aware, this is the first suggestion in the literature that C^{14} might exist.) Kurie also suggested, however, that the tracks he was observing might arise from H^2, or even H^3, and thus that the reactions $N^{14}(n,H^2)C^{13}$ and $N^{14}(n,H^3)C^{12}$ were also possibilities. In fact, he felt the reactions with emission of H^2 and H^3 were the more likely because they resulted in nuclei of known stability.

What was radical about Kurie's suggestion was the idea that something other than an alpha particle could emerge in a disintegration of a nucleus such as N^{14}. But the physicists at the time assumed from their everyday experience that the alpha particle was much the most likely nucleon to be formed in such a nuclear reaction. This belief found a ready basis in the relatively great stability of the alpha particle, which was considered to exist as an entity in all nuclei because of its relatively enormous binding energy per nucleon, and because invariably in natural radioactivity it was the only heavy nucleon ejected.

In the meantime, T. W. Bonner and W. M. Brubaker (*7*) published observations on the energies of recoils induced by neutrons in inelastic collisions with nitrogen nuclei. Assuming the usual reaction, $N^{14}(n,He^4)B^{11}$, they calculated from the mass values given by Hans Bethe (*8*) that *Q*, the heat of reaction, was about 1.5 Mev. Most significantly, however, Bonner and Brubaker (*6, 7*)

and W. Chadwick and M. Goldhaber (*9*) independently reported that the distintegration of N^{14} occurred also with *slow* neutrons.

This, it turned out, was the crucial observation in the "prenatal" history of C^{14}, because it prompted a re-evaluation of the assumptions on which analyses of the nitrogen disintegration tracks were based.

Chadwick and Goldhaber had detected disintegrations in an ionization chamber connected to a linear amplifier and oscillograph. From the size of the oscillograph deflections they had deduced that *Q* was about 0.5 Mev. Bonner and Brubaker had used a cloud chamber and observed a group of tracks with a sharply defined range of 1.06 centimeters (in air, at normal temperature and pressure) which they thought to be alpha particles. On this basis, they had calculated a value for *Q* of 2.33 Mev. The discrepancy in the two values for *Q* was far beyond any experimental error. W. E. Burcham and Goldhaber (*10*) then were inspired to suggest that both sets of data were referred to the wrong reaction, and that if the disintegration were assumed to take place with proton emission—that is, $N^{14}(n,H^1)C^{14}$—then the value for *Q* taken from the range observed by Bonner and Brubaker became 0.58 Mev, in good agreement with the value of *Q* deduced from the ionization measurements. Bonner and Brubaker not only concurred but advanced further evidence in support of this suggestion by noting that when they measured *Q* for a related reaction, $B^{11}(H^2,n)C^{12}$, they found a value of 13.5 Mev, which, taken together with *Q* values known for some other reactions, enabled them to construct a reaction cycle in which the $N^{14}(n,He^4)B^{11}$ reaction turned out to be endergonic ($Q = -0.28$ Mev), rather than exergonic ($Q = 2.33$ Mev). Thus the alpha-particle emission could not have been induced by slow neutrons, at least if one accepted the rather marginal negative value for *Q*.

Burcham and Goldhaber proceeded to use the visualization technique of nuclear emulsion, by means of which they could distinguish more certainly between protons and alpha particles. They showed conclusively (*10*) that alpha particles were not emitted in the slow-neutron disintegration of N^{14} and proposed that the recoil particles be assigned definitely to the reaction, with C^{14} as a product. They could not exclude experimentally the possibility that H^2 or H^3 particles, rather than protons, might account for the recoil tracks observed. On the other hand, they remarked that reactions that could produce such particles were unlikely on energetic grounds. Thus, Kurie's original intuitive suggestion, based on the appearance of certain unusually thin, long tracks in his cloud-chamber experiments of 1933–34, was raised to the status of a practical certainly by 1936.

I think it is safe to say that in the *physical* sense the discovery of C^{14} had been established by the observations of Burcham and Goldhaber. But the discovery of C^{14} in the *chemical* sense—which I must say has turned out to be the more important—was delayed by many obstacles. If I may refer again to the remarks of Kuhn (*1*), the discovery of C^{14} in a physical sense belongs in the "expected" category but its discovery in the chemical sense does not.

Intermediate Phases, 1936–38

Meanwhile, "back at the ranch," I was plugging away collecting pictures of proton recoils produced by collisions with neutrons from a pathetically weak mesothorium-beryllium source of a few millicuries equivalent. By the end of 1936 I had obtained results from an analysis of 730 tracks, which showed a marked asymmetry in the angular distribution of protons in the energy range from ~ 0.1 to 3 Mev (*11*). This result confirmed results obtained by Kurie a few years earlier (*12*). At the time, Kurie's results had been in contradiction to theoretical expectations. Before his and my anomalous findings could be taken seriously, it was necessary that greater numbers of events be analyzed—an objective quite unrealizable with the feeble neutron sources then available. It had taken 3 years of constant labor to produce and analyze a few hundred proton recoils—in fact, 10 man-hours per track had been required. This statistic shows how slender were the means available for research in nuclear physics in 1936—which, after all, is not so long ago.

It was natural in the mid-1930's to look toward Berkeley, where Ernest O. Lawrence was assembling a group of young physicists drawn by the fact that the cyclotron, a much stronger particle source than any then existing, was in operation.

Thus, in January 1937 I found myself at the Radiation Laboratory, along with Kurie, who had arrived there a short time before. Because of our mutual interest in neutron scattering, we quickly began a collaboration in which we proposed to investigate not only the various apparent anomalies in the neutron-proton interactions but neutron-nuclear interactions in general. The neutron fluxes available with the cyclotron, even in its primitive 1937 form (an accelerator 27 inches in diameter), were already greater by four or five orders of magnitude than any with which we had been familiar.

Unfortunately, the neutron flux obtainable with the new cyclotron was not a well-collimated beam with minimal energy spread. The cyclotron contained innumerable scattering sites from which great numbers of neutrons of undefined energies emerged in a welter of direction. To achieve some measure of order in the neutron beam, Lawrence and his group set to work on a major program, with responsibility vested largely in Arthur Snell. The procedure was to lead the deuteron beam away from the main vacuum accelerating chamber so that it would hit a beryllium target at some distance from the scattering bulk of the cyclotron. The emergent neutrons from the $Be^9(H^2,n)B^{10}$ reaction were to be piped through a hole in the water shield around the cyclotron. It was hoped that, with judicious use of auxiliary absorbers, this effort, called "snouting," would produce a relatively well collimated, intense beam of neutrons with energies only slightly lower than the maximum obtainable in the reaction.

Kurie and I found that when a cloud chamber, filled with hydrogen as target, was placed to either side of the "snout" hole, there were relatively few recoil pictures, whereas the yield was enormous when the chamber was placed directly in front of the hole. A typical single expansion, in front of the snout hole, with an exciting deuteron beam of a few microamperes incident on the beryllium target, exhibited more recoil tracks than either Kurie or I had seen in all our previous experience with the strongest naturally radioactive sources obtainable—for example, hundreds of millicurie equivalents of polonium or radium mixed with beryllium.

In the light of the earlier findings on the production of C^{11}, we decided to investigate the neutron disintegration of N^{11}. The slow-neutron disintegration of N^{14}, with its characteristic recoil product—a proton with a range of 1.06 centimeters in air, as dictated by the Q value of 0.58 Mev—manifested itself on almost every other expansion in the cloud chamber as a short, stubby track with a knob at the starting end. This track was the track of the proton, which, because of its relatively low mass, took up practically all the energy of the reaction as kinetic energy; the knobby stub was the C^{11} recoil product, which, because of its relatively high mass, possessed very little of the kinetic energy released in the reaction. These characteristic short, stubby tracks provided a convenient and accurate internal monitor for calibrating the stopping power of the mixture of nitrogen gas and water vapor in the cloud chamber. In this way we solved a very bothersome technical problem—that of determining just what value for stopping power we should use in calculating recoil energies. Thus, by early 1937 tracks of C^{11} were being used in the calibration of cloud-chamber experiments, and its existence was well established.

However, nothing was known about its physical characteristics. One could attempt to make some predictions. First, it seemed certain that C^{14} was radioactive and that it must emit negative beta rays in the transformation to N^{14}. This conclusion followed from the observation that in no known element of low atomic number were there neighboring isobaric pairs in which more than one member of the pair was stable. Examples of pairs with an unstable member are He^{6}-Li^{6}, Be^{7}-Li^{7}, Be^{10}-B^{10}, and C^{11}-B^{11}. So, one might expect that for the isobaric pair C^{14}-N^{14} the stability of N^{14} requires that C^{14} be unstable and that it decay to N^{14} by transformation of a neutron into a proton, with emission of a negative beta particle and a neutrino. Second,—and here a great uncertainty arose—one would expect the rate of decay to be fairly high, as it is in one known analogous case (He^{6}-Li^{6}) in which a nucleus with two excess neutrons decays to a nucleus with equal numbers of neutrons and protons. For this pair as well as for another similar pair, Be^{10}-B^{10}, the expected decay ratio was in quite good agreement with the ratio actually observed. Thus, He^{6}, with an upper energy limit for beta rays of 3.5 Mev and a possible spin difference of no more than 1 unit, could be expected to decay with a period of 10^{-1} to 10^{1} seconds. Its half-life is, in fact, 0.8 second. The decay of Be^{10} with its enormously long half-life—of the order 10^{6} to 10^{7} years (*13*)—represented a highly forbidden transition, despite a fairly high beta-ray energy maximum of about 0.5 Mev. In this case, a high degree of forbiddenness could be rationalized on the basis of the shell model available—that is, a large spin difference was a possibility. The C^{14}-N^{14} pair was analogous to He^{6}-Li^{6}, with a maximum spin difference of 1. The lower energy limit to be expected depended on the difference in mass between C^{14} and N^{14}. Thus, for the reaction with slow neutrons,

$$(n - H^{1}) = (C^{14} - N^{14}) + Q.$$

From then-accepted values for the masses and a value for Q of +0.58, the maximum energy for the emergent beta particles was estimated to be about 0.3 Mev. On this basis, a tentative estimate of a few hours or days for the half-life of C^{14} was made by P. Morrison and J. R. Oppenheimer. Thus it appeared that C^{14} was probably quite short-lived. If attempts to isolate it were unsuccessful, it could be assumed that the half-life was too short for the isotope to be isolated, rather than too long—that is, $\tau_{1/2}$ was expected to be on the order of seconds, not centuries.

None of these arguments were wholly convincing, for it was recognized that the state of beta-decay theory was quite unsatisfactory; nevertheless, they were sufficiently impressive to discourage us from making a determined effort to isolate C^{14} at that time. E. M. McMillen had noted the presence of some very long lived activities in old cyclotron targets and bits of metal scraped from various parts of the acceleration chamber. With what now is seen to have been extraordinary intuition he surmised that these activities could represent radioactive species of beryllium and carbon. In an abstract (*13*) submitted to a meeting of the Physical Society in 1936 he described two activities residual in an old beryllium target. The major activity was a very soft component, with a decay rate consistent with a half-life of 10 years or longer. This component, he thought, might be assigned tentatively to Be^{10}. He suggested that the other activity, which was weaker but harder and decayed with a half-life of a few months, was attributable to a radioactive isotope of carbon—in particular, to C^{14}. In later years (*14*) he published a final summary of work done with S. Ruben in the years 1938–40, in which the activity assigned to Be^{10} was reassigned with some certainty to H^3. The other activity obviously could not have been attributable to C^{14}, because, as we know now, the half-life of C^{14} is about 5700 years. In an effort to produce C^{14} by the $N^{14}(n,H^1)C^{14}$ reaction, McMillen exposed solid ammonium nitrate to the intense neutron flux of the 37-inch cyclotron for several months in 1938–39, but this experiment ended when the bottle containing the salt was inadvertently knocked off the magnet coil tank and smashed.

However, a new—and what was to prove decisive—factor entered late in 1937. As we all know, the ancient alchemists ascribed to the "philosopher's stone" two magical powers—the ability to transmute elements and the ability to banish disease. The old alchemical treatises bear constant witness to the state of mind which supported, through centuries, what were tedious, disappointing, and often hazardous—even fatal—searches. Even after chemistry was well established and alchemy was in some disrepute, Robert Boyle could still make his well-known statement (*15*), "There may be some agent found out so subtile and powerful, at least in respect of those particular compounded corpuscles, as to be able to resolve them into those more simple ones, whereof they consist."

With the discoveries of radioactivity by Becquerel and x-rays by Röntgen at the end of the 19th century and the experiments of Rutherford early in this century, the prophecy of Boyle and the hopes of the alchemists had been realized. Lawrence saw even greater potentialities in the exploitation of the immensely more powerful agents available from artificial radioactivity and man-made accelerators and was eager to see them applied. For this reason he had interested his brother, John Lawrence, a physician and worker in medical research, in the possible use of the neutron as a therapeutic agent and had assigned P. C. Aebersold, then a graduate student, to help him establish the physical and radiological procedures involved. A small beginning had already been made in utilization of the radioactive isotopes of phosphorous, sodium, and iodine as tracers in biological research. To facilitate the development of such researches, both at Berkeley and elsewhere, I was asked to assume responsibility as staff chemist to develop procedures whereby target materials could be bombarded, processed, and delivered in forms suitable for direct application in biological systems. The problems involved were so challenging and urgent that from the early months of 1937 until well into 1940 most of my energies were channeled into this activity.

Almost immediately I found it necessary to devise a dependable procedure for rapid preparation of C^{11}, with half-life of 21 minutes, for use by S. Ruben, W. Z. Hassid, and I. L. Chaikoff, who planned to investigate carbohydrate metabolism by means of C^{11}-labeled sugars prepared photosynthetically from $C^{11}O_2$. In a short time it became apparent (*16*) that bombardment of B_2O_3 powder by deuterons in a special external target chamber, designed by Kurie, produced as $C^{11}O$ or $C^{11}O_2$ almost all of the C^{11} made in the $B^{10}(H^2,n)C^{11}$ reaction. The success of this procedure, which eliminated the exposure to hazardous radiation incidental to manipulation of target material, expedited the research to such an extent that Ruben was encouraged to initiate research on photosynthesis itself—an undertaking into which I was soon drawn

full time, outside of my duties as radiochemist.

In the meantime, Ruben observed, in May 1938, that a number of graphite targets, which I had been using to make N^{13} by the $C^{12}(H^2,n)$ reaction for some research on nitrogen fixation (*17*), contained no residual long-lived activity isotopic with carbon. He concluded that any C^{14} formed by the $C^{13}(H^2,H^1)$ reaction, if it had a half-life up to 200 years, could have been detected. This finding underscored the probability that C^{14}, if it existed, was a short-lived isotope.

My constant exposure to cyclotron targets kept me in a steady state of radioactive contamination which rendered me *persona non grata* around assay equipment. I recall an experience during collaboration with Philip Abelson in late 1937, when we were attempting to use an apparatus consisting of an ionization chamber connected to an FP–54 Pliotron tube amplifier. We were harassed by the occurrence of an eccentric and irregular background drift of variable magnitude. Finally, Abelson noted that the effect was correlated with my movements toward and away from the apparatus. While I stood in a corner he systematically stripped me and established that the disturbance originated from the front of my pants. Likewise, in my collaboration with Ruben, it was necessary to keep me away from the counting equipment. Eventually, Ruben assumed sole responsibility for assay and I concentrated on the production of tracer isotopes.

Ruben deserves almost all the credit for the growth of interest in tracer methodology which occurred at Berkeley in the years 1937–38. His unique combination of skill, energy, wide-ranging interest, and quick grasp of essentials when confronted with new and unfamiliar areas of science provided a focus for the efforts of an ever-increasing number of able investigators.

By the middle of 1938 the demands for cyclotron time on the part of biologists and clinicians was so great that round-the-clock operation of the cyclotron—by that time a 37-inch accelerator—was mandatory. In addition, the pressure to build bigger accelerators had impelled Lawrence to seek additional subsidies from biologically oriented foundations and other organizations likely to support such projects. As an argument, Lawrence cited preliminary results of neutron therapy and of the use of radioactive isotopes, such as Na^{24}, P^{32}, and Fe^{59}, in medical diagnosis. A 60-inch cyclotron was almost ready for operational tests, and further development depended critically on financial support from private organizations. The demonstration that energies and ion currents far in excess of the limits earlier thought to be reasonable might ultimately be obtainable lent added urgency.

The researches of R. R. Wilson in this period were of considerable importance in pointing the direction for further development of cyclotron design. Wilson showed that very large beam currents at energies somewhat lower than those in the external target area circulated inside the accelerating electrodes (*18*). He realized that if a means could be devised to use these internal beams, bombardments of much greater magnitude than those possible with external targets could be effected. Moreover, he was able to show that a large fraction of the internal circulating ion current could be intercepted by appropriately designed "probes" without appreciable diminution of external beams. However, the problems involved in the achievement of acceptable internal targets were formidable, and no immediate efforts were made to exploit the internal ion currents. For example, the external beams of 7 to 8 Mev and 50 to 100 microamperes required dissipation of 300 to 800 watts—a power input which, even with well-cooled targets of good heat conductance, required sequestration of target material from the cyclotron vacuum by aluminum or other metal foil windows. Internal beams with ion currents of as much as several milliamperes and energies only slightly less than 6 to 7 Mev generated power inputs of an order of magnitude greater than those associated with the external beams.

However, the cyclotron was increasingly unable to meet the demands for radioisotopes, as well as for bombardment time for clinical trials; even the needs for essential nuclear physical research could not be satisfied. This led Wilson and me to try to devise acceptable internal targets for radioisotope production—an enterprise in which we succeeded late in 1938. This research was to prove the turning point in the sequence of events which led to the production of C^{14}. In fact, as we remarked at that time (*19*), "Obviously, the method of internal targets should find its most important application in the preparation of radio-isotopes which are long-lived and difficult of activation, as well as in the demonstration of the existence of many radio-isotopes as yet undiscovered."

Final Phase, 1938–40

The search for funds proceeded with some success through 1938 and into 1939, and thus both cyclotrons were in operation on a full schedule when the discovery of nuclear fission burst on the world in January 1939. In the meantime, Lawrence's hopes of producing an even greater machine had been raised by the successful operation of the 60-inch cyclotron. Citing the expanded collection of clinical data and the enormously developing tracer program made possible by the Berkeley cyclotrons, he pressed for more support from the organizations that had provided it before.

In our own researches on photosynthesis, C^{11} had by this time been exploited to its ultimate limits (*19–21*).

There had been many difficulties. We had found it essential to try to make an ultracentrifugal determination of the molecular weight of the labeled intermediates produced during photosynthesis in the presence of $C^{11}O_2$. The necessary apparatus was at Stanford University, 50 miles away from the Berkeley cyclotrons along a heavily traveled road. We calculated that it would be possible to produce the $C^{11}O_2$, incubate algae with it, extract the resultant intermediates, and drive with these to Stanford. There would be insufficient time, however, to make a proper centrifuge run. We considered a number of possible courses of action, such as arranging a police escort for the motor trip to Stanford or posting one of us at Stanford to have the counting apparatus ready. One night a brilliant solution occurred to Sam Ruben: he woke me by phone at 2 A.M. to suggest carrier pigeons!

Fortunately, this problem was finally solved by the discovery that an apparatus identical with that at Stanford was available at the Shell Oil Company research laboratories, only 10 minutes' drive from the cyclotron. Even so, we found it impossible to obtain the precision needed to establish with certainty the average size of the early intermediates in photosynthesis.

In a gloomy conference late in September 1939, Ruben and I reached the conclusion that without a long-lived isotope of carbon our researches were at an end. I suggested one last desperate try to produce C^{11} by means of the internal-target technique. The difficulty was that, in view of the general pessimism about the probability of isolating a long-lived isotope, we could not arrange to use any of the internal targets, which were being constantly used to produce P^{32} and Fe^{59}.

It was with some amazement therefore, that I found myself shortly after this conference being told by Lawrence that both cyclotrons must be diverted forthwith to a full-time effort to determine definitely whether long-lived isotopes of hydrogen, carbon, nitrogen, and oxygen did or did not exist. The reason for this was soon evident. During Lawrence's most recent efforts to

Table 1. Plan for production reactions for possible long-lived carbon isotope, September 1939. Target chemistry for all but the reaction in the bottom row is as follows: Collect C^*O in gas, burn residue to CO_2, with carrier C. Target chemistry for the reaction in the bottom row: Aspirate CO_2-free air through solution, burn emergent gases, and trap C^*O_2.

Reaction†	Target material
Atmosphere: 3% CH_4 in O_2	
$Be^9(He^4,n)C^{12*}$	BeO
$B^{10,11}(He^4,H^1)C^{13*,14}$	B_2O_3
$B^{11}(H^1,\gamma)C^{12*}$	B_2O_3
$B^{11}(H^2,n)C^{12*}$	B_2O_3
$C^{12,13}(H^2,H^1)C^{13*,14}$	Graphite
$N^{14,15}(n,H^1)C^{14,15}$	Ammonium nitrate (nitrite)

† Energies available: protons (4–8 Mev), deuterons (8–10 Mev), alphas (16–32 Mev).

achieve increased and continued subsidies for cyclotron development, some question had been raised as to the real value of radioactive isotopes in biological research, relative to the rare stable isotopes such as H^2, C^{13}, N^{15}, and O^{18}. Thus, while these very useful stable isotopes existed as tracers for the elements of primary importance in biology, there were no comparable radioactive isotopes with reasonably long lifetimes available; no radioactive hydrogen isotope existed (it was thought that tritium was stable relative to He^3), the carbon isotopes C^{10} and C^{11} had half-lives of 8 seconds and 21 minutes, respectively, and only very short-lived N^{13} and O^{15} were known for nitrogen and oxygen.

Lawrence asked me to organize a complete and systematic campaign to determine whether or not long-lived isotopes existed for any of the elements in the first row of the periodic table—especially hydrogen, carbon, nitrogen, and oxygen. From experience gained in previous years it was a simple matter to draw up a comprehensive plan which included protocols for every nuclear reaction obtainable with the projectiles available, including the choice of target materials and target chemistry. It was natural to concentrate first on carbon. An abridged plan taken from my notebook, as drawn up in September 1939, is shown in Table 1.

Similar plans were devised for nitrogen and oxygen activities, based on bombardments of B_2O_3, graphite, BN, and $(NH_4)_2F_2$ targets. Every possibility was assumed—even that there might be long-lived isomers of stable nuclei (for example, C^{12*}, C^{13*}) or of short-lived nuclei (for example, N^{13*}, O^{15*}).

The first trials involved bombardment of B_2O_3 with 16-Mev alpha particles in the 37-inch cyclotron, from 27 to 29 September. A 5 microampere-hour exposure yielded no long-lived activity in the gas space, as assayed by direct introduction of the target gases into an ionization-chamber and pliotron apparatus. The residual powder was burned with a small bit of filter paper as a carbon carrier, but the resultant gases were also inactive. It could be concluded that the 16-Mev alpha particles were inadequate to produce significant quantities of N^{13*}, N^{14*}, or C^{13*} or of the isotope C^{14} by means of the (He^4,n) or (He^4,H^1) reactions. Next, B_2O_3 was bombarded for 34 microampere-hours with 16-Mev deuterons in the 60-inch cyclotron. In this experiment, E. Segré collaborated with me and examined the activities produced in the gas phase, which were introduced again directly into an ionization chamber detector. He found the expected C^{11} in enormous quantities and also an activity with a half-life of 112 minutes. But we traced this to F^{18} produced by the $(H^2,2n)$ reaction with O^{18}. By 16 October 1939 I had reached the conclusion that none of the alpha-particle-induced reactions were feasible: it appeared that at least 5×10^5 microampere-hours of 32-Mev alpha particles would be needed to produce one microcurie of radioactivity with $\tau_{1/2} \sim 1$–3 hours.

On 17 October, I tried exposing methane to 20 microampere-hours of 16-Mev deuterons in the 60-inch cyclotron and obtained the expected N^{13} (several hundreds of millicuries), and about 0.01 microcurie of F^{18} from a small amount of contaminant oxygen, but again no long-lived activity.

In the meantime, I had begun continuous exposure of a graphite probe target introduced through the north port of the 37-inch cyclotron. This probe was allowed to collect stray deuterons in the internal cyclotron beam for nearly a month, throughout January 1940. The probe was inserted so as to intercept practically all deuterons during night operation and was retracted to allow normal operation in the daytime. I undertook the night bombardments, aided occasionally by others, who needed the copious supply of neutrons produced for further studies on the uranium fission reactions. The probe target was not designed to withstand intense bombardment. I had merely smeared colloidal graphite on the water-cooled copper surface and had counted on replacing, during frequent inspections, whatever graphite was found to be blasted off. The weather was unusually violent, even for January, and on most nights there were heavy drenching rains and wind storms. The noise of the rain on the tin roof of the laboratory, accompanied by cannonades from high-voltage discharges from the cyclotron, created an appropriate fanfare for the birth of C^{14}, which was fated to occur during this bombardment.

This experiment had been regarded as the one most likely to produce C^{14}. It was performed in a spirit of mixed desperation and resignation, and it involved a considerable degree of hazard from radiation exposure, as it was necessary to examine the intensely radioactive probe nightly to insure that some graphite still clung to the target surface. Occasionally I found the irradiated graphite almost on the verge of flaking off and had to cement it back on with more graphite.

On 15 February, during a particularly violent storm, I terminated this bombardment, which had involved exposure of the graphite to 5700 microampere-hours of 7- to 8-Mev deuterons. Shortly before dawn I left the graphite, which looked like bits of gravel, in a weighing bottle on Ruben's desk. On the way home to get some sleep for the first time in several days I must have presented a sorry spectacle—unshaven, red-eyed, and dazed—for I was intercepted and questioned by police looking for an escaped convict. Fortunately I failed to pass muster and was released, to continue stumbling onward toward sleep.

On awakening some hours later I phoned Ruben, who had found the sample, burned it to CO_2, precipitated it as $CaCO_3$, and found some activity when he examined the precipitate inside a screen-wall counter of a type designed by W. F. Libby (*20*) to permit assay inside the sensitive volume of a Geiger-Müller tube. (No activity could be detected with the usual thin-walled tubes.) The effect was very small—about four times the counting background—but reproducible. In some excitement I hurried back to the laboratory, prepared a new probe target—this time one made up of graphite solidly bonded to copper—and then joined Ruben to press on with identification of the activity. By Tuesday afternoon, 27 February, we had disposed of the last uncertainty—that the activity might have arisen from S^{35} produced by the (H^2,H^1) reaction on the surface as a possible contaminant of the graphite used—but had only about one-eighth of the original activity left. We wrote a preliminary account for publication as a letter, and later as an abstract, in the *Physical Review* (*21*), and on Wednesday evening we motored to Lawrence's home to acquaint him with the result. Lawrence was resting in an attempt to banish a cold before his appearance the next night, Thursday, 29 February, to receive the Nobel prize in physics. His pleasure was unbounded but it revived nagging doubts in our own minds, about the reality of the activity we were ascribing to C^{14}. After all, it was an activity only half that of the counting background when the sample was counted as solid $CaCO_3$, even though it had persisted through repeated cycles of precipitations with $CaCO_3$ and acidification to CO_2. It was a comfort to realize that, of all the elements in the periodic system, only carbon possessed an oxide which could be liberated repeatedly from acid solutions under oxidizing conditions—a fact which made it clear that the activity observed was, in fact, isotopic with carbon.

The new probe target, with its improved mode of bonding, withstood much more intense probe beams; in 1 week a probe inserted in the south port of the 37-inch cyclotron withstood 13,500 microampere-hours of bombardment with 3- to 4-Mev deuterons. The activity obtained was 10 to 20 times the counting background and sufficient to show that the isotope was indeed C^{14}; that the beta-energy maximum lay in the neighborhood of 120 kilovolts; and that the half-life was certainly greater than 20 years. I was able to make an estimate for $\tau_{1/2}$ on the basis of considerations as follows. The first probe sample, after 5700 microampere-hours, had shown a total activity of ~100 disintegrations per second. From a knowledge of the cross section at 3 to 4 Mev for the reaction $C^{12}(H^2,H^1)C^{13}$, and from the assumption that the cross-section for $C^{13}(H^2,H^1)C^{14}$ was similar, it could be calculated that the ratio of C^{14} nuclei produced to the number of deuterons stopped in the graphite would have been about 1/60,000 if 100 percent of the carbon were C^{13}. Hence $1/(6 \times 10^6)$ was assumed for the graphite, which had the normal isotope content of 1 percent C^{13}. The total number of deuterons was $\sim 1.2 \times 10^{20}$ (5700 μamp-hr), hence the number of C^{14} nuclei produced ($N_{C^{14}}$) was $1.2 \times 10^{20}/(6 \times 10^6)$, or 2×10^{13}. Since $N_{C^{14}}$ and $dN_{C^{14}}/dt$ (100 disintegrations per second) were known, it was simple to deduce the $\tau_{1/2}$ as $\sim 4 \times 10^3$ years. This value was remarkably close to the true disintegration half-life of ~5700 years, determined many years later, but the agreement was quite accidental, as this 1940 estimate was uncertain by as much as an order of magnitude in either direction. However, it did shed light on the reason for the negative results of previous years: the half-life of C^{14} was too great to permit production of the isotope in significant quantities until the internal-target technique had been developed to the point where 5000 microampere-hours and more of bombardment was possible. Moreover, it was clear that C^{14} had an enormously long half-life—a result most surprising on theoretical grounds, as I remarked previously.

All of these investigations were based on the assumption that the (H^2,H^1) reaction with C^{13} would be much the most likely to succeed. A reaction of this type exhibited the largest cross-section among those excited by charged particles and had the advantages that the target chemistry was simple and that dilution of isotopic material was minimal. The reaction of slow neutrons with N^{14}, which had led to the initial postulation of the existence of C^{14}, was regarded as a possible, but not promising, process for producing C^{14}. It may come as a surprise to many readers that this impression prevailed in 1940, in view of the fact that the $N^{14}(n,H^1)C^{14}$ reaction is the method of choice now.

There were many cogent reasons for neglecting the slow-neutron reaction. (i) The neutrons produced in the cyclotron were not primaries but were secondary particles with ranges up to many meters in dense media, so that only a fraction could be captured by N^{14} nuclei, even with the best possible geometry. On the other hand, all deuterons produced as primaries with very small ranges could be absorbed in a minimal amount of target. (ii) Cross sections for slow-neutron capture were high only for the (n,γ) process, with which the (n,H^1) process was expected to compete poorly. (iii) The recoil C^{14} nuclei produced would not be expected to reach equilibrium in chemical species sufficiently uniform to permit simple, efficient extraction. Nevertheless, two carboys, each containing 5 gallons of saturated ammonium nitrate solution, were placed near a good source of neutrons—the deflector region of the 60-inch cyclotron—in January. Prior to exposure, the solutions were acidified with dilute nitric acid, aspirated with CO_2-free air to remove contaminant carbonate (which would dilute excessively any radioactive carbonate formed —not that we expected to see any!), and sealed tightly. The success with the graphite probes had completely distracted my attention from these carboys, and I was busily engaged in an attempt to improve the (H^2,H^1) yields by fabricating C^{13}-enriched graphite probes when an angry deputation from the 60-inch cyclotron paid me a visit and demanded that the carboys, which had sprung leaks and were proving an intolerable nuisance, be removed. The deflector region was in constant need of adjustment, and the cyclotron crew was weary of the constant pushing and pulling required to move the box and get at the deflector controls, especially since the box was wet with acid.

So, with no great enthusiasm, I went over to the 60-inch cyclotron with a cart and moved the box to Ruben's laboratory in the ramshackle hut affectionately labeled the "Rat House." Ruben and I decided to make a gesture and aspirate some air, freed of CO_2 by passage through soda lime, through the carboys in the hope that some C^{14} might be entrained and removed as $C^{14}O_2$. A copious precipitate of $CaCO_3$ which formed in the $Ca(OH)_2$ trap after passage of the effluent gases through a combustion train did not encourage us, since it indicated that large quantities of CO_2 had diffused into the ammonium nitrate solutions and that any activity formed was likely to have been lost because of excessive dilution. To our astonishment, we found that a small fraction of this precipitate was so active it completely paralyzed the screen wall counter! In a short time we ascertained that we had several microcuries of C^{14}—a quantity greater by two or three orders of magnitude than any we had seen from the probe bombardments. Needless to say, our interest in the $C^{13}(H^2,H^1)C^{14}$ reaction vanished, never to return.

Where had I gone astray in assessing the slow-neutron process? All the assumptions about poor cross sections and complex target chemistry were eminently sound and eminently wrong! As it developed later, the (n,H^1) process for producing N^{14} was favored heavily over the (n,γ) process—possibly the only exception to the general rule. It is remarkable that the only other instance of appreciable yield from the (n,H^1) process with slow neutrons also involves production of an important radioactive tracer (22)—S^{35} from the (n,H^1) reaction on Cl^{35}. Moreover, as we found in further experimentation (*23*), and as was later confirmed by Yankwich, Norris, and Rollefson (*24*), well over 80 percent of the recoil C^{14} found its way into the volatile oxides of carbon (CO and CO_2) and so made possible a simple, practically quantitative recovery of the C^{14} from any amount of bulk solution. When all these facts were known there was some talk of forming a syndicate to build a battery of cyclotrons designed solely for the production of C^{14}.

In the meantime, Lawrence authorized construction of a special set of stainless steel cans with aspirator inlets and outlets to be mounted as a permanent shield around the 60-inch cyclotron. This setup was expected to produce hundreds of microcuries of C^{14} each month. After these cans had been in place for a month, however, worry over the possibility, however remote, that ammonium nitrate solutions were an explosion hazard induced Lawrence to order them removed. A few years later, in 1944, these cans served as a source of C^{14} in the first tracer researches with C^{14} to be reported in a scientific publication—research which I carried out in collaboration with Barker (*25*).

The story of the birth of C^{14} ends here. But, of course, there were many chapters to come. One of these was the expansion of C^{14}-production from the microcurie to the curie level which followed the development of nuclear reactors, and the resultant proliferation of C^{14} as a tracer isotope in every area of biological research. Another chapter (which provides a thread from the early history up to 1940) concerns something that may prove of great significance in the future of nuclear theory—the anomalously low rate of C^{14} decay. All through the period from 1938 to 1940 the search for C^{14} was conducted in the belief that the half-life of C^{14} was of the order of seconds, or less. When it was seen, by July of 1940, that the half-life of the material we had produced was certainly of the order of years—in fact, millennia—doubt was expressed that the observed activity was isotopic with carbon. Even the testimony of algae, which absorbed the activity photosynthetically, and the chemical behavior of the activity failed to still these doubts at the time. The question raised—Why is C^{14} so long-lived?—remains with us today.

Explanations have been offered seriatim; (i) that there is a change in parity; (ii) that C^{14} is wholly $^1S_0$ and N^{14}, purely ^3D, hence, that a transition between them is ΔL forbidden; (iii) that there is a fortuitous cancellation in the matrix element for decay (*26*). The third suggestion is favored at present (*27*) but is not wholly acceptable. The first two are definitely excluded on experimental as well as theoretical grounds (*26*, *27*). It is apparent that the answer to the riddle of C^{14} decay will be an important part of the future history of nuclear theory. In conclusion, I quote, as a moral for our story, the words of the late psychiatrist E. Winzholz: "We never let our theories interfere with our practice."

References and Notes

1. T. S. Kuhn, *Science* **136**, 760 (1962).
2. A brief account, concerned mainly with technical aspects, appeared previously; see M. D. Kamen, *Isotopic Traces in Biology* (Academic Press, New York, ed. 3, 1957), p. 300.
3. I wish to record a deeply felt personal indebtedness to two former associates, Dr. Franz N. D. Kurie and Dr. Samuel Ruben. The interruption of Kurie's career by a debilitating illness a few years ago deprived nuclear physics of one of its foremost investigators. The untimely death of Ruben at the age of 29, while he was engaged in research on chemical warfare in 1943, was an unmitigated catastrophe for modern biochemistry.

4. W. D. Harkins, D. M. Gans, H. W. Newson, *Phys. Rev.* **44**, 529 (1933).

5. I am happy to record my great debt to Dr. Gans and also to Dr. Henry W. Newson, my immediate predecessor in the research.

6. F. N. D. Kurie, *Phys. Rev.* **45**, 904 (1934); **46**, 330 (1934).
7. T. W. Bonner and W. M. Brubaker, *ibid.* **48**, 469 (1935).
8. H. Bethe, *ibid.* **47**, 633 (1935).
9. W. Chadwick and M. Goldhaber, *Nature* **135**, 63 (1935); *Proc. Cambridge Phil. Soc.* **31**, 612 (1935).
10. W. E. Burcham and M. Goldhaber, *ibid.* **32**, 632 (1936).
11. W. D. Harkins, M. D. Kamen, D. M. Gans, H. W. Newson, *Phys. Rev.* **50**, 980 (1936).
12. F. N. D. Kurie, *ibid.* **49**, 461 (1933).
13. E. M. McMillan, *ibid.* **49**, 875 (1936).
14. ——— and S. Ruben, *ibid.* **70**, 123 (1946).
15. See the scholarly article by F. Paneth in the introductory lecture of his *Radioelements as Indicators* (Baker Lectures) (McGraw-Hill, New York, 1928).
16. S. Ruben, W. Z. Hassid, M. D. Kamen, *J. Am. Chem. Soc.* **61**, 661 (1939).
17. See S. Ruben, W. Z. Hassid, M. D. Kamen, *Science* **91**, 578 (1940), for an account of the tracer experiments involved.
18. R. R. Wilson, *Phys. Rev.* **54**, 240 (1938).
19. ——— and M. D. Kamen, *ibid.* **54**, 1031 (1938).
20. W. F. Libby, *ibid.* **46**, 196 (1934).
21. S. Ruben and M. D. Kamen, *ibid.* **57**, 549 (1940); see also M. D. Kamen and S. Ruben, *ibid.* **58**, 194 (1940).
22. M. D. Kamen, *ibid.* **60**, 537 (1941); **62**, 303 (1942).
23. S. Ruben and M. D. Kamen, *ibid.* **59**, 349 (1941).
24. P. E. Yankwich, G. K. Rollefson, T. H. Norris, *J. Chem. Phys.* **14**, 131 (1946).
25. H. A. Barker and M. D. Kamen, *Proc. Natl. Acad. Sci. U.S.* **31**, 219 (1945).
26. See D. R. Inglis, *Rev. Mod. Phys.* **25**, 390 (1953).
27. R. Sherr, J. B. Gerhart, H. Horie, W. F. Hornyak, *Phys. Rev.* **100**, 945 (1955).

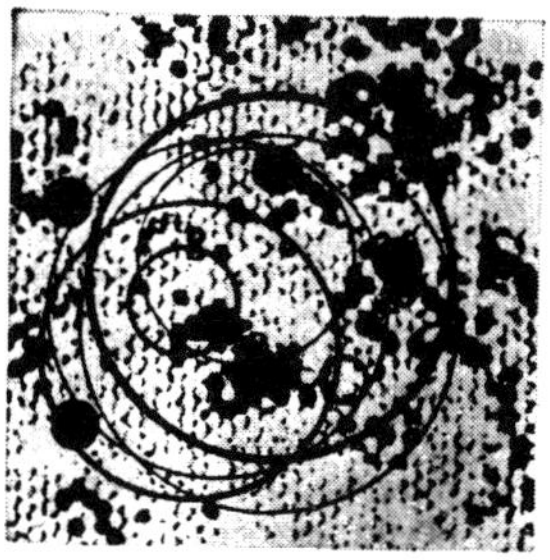

Toward a comparative biochemistry of the cytochromes

Martin D. Kamen※

I. Introduction

The ultimate objective of biochemical research is an understanding of the molecular mechanisms which lead to expression of biological function in organisms as they arise, grow and evolve. A necessary approach in this effort is that of "comparative biochemistry," in which variations in biochemical entities, such as proteins, enzymes, nucleic acids, and other metabolic determinants, are examined and compared as they relate to identical or similar functions from system to system. The range of such comparisons may be restricted, involving a small group of biological systems serving a limited function. Examples[A-7] are studies of the primary structures of fibrinopeptides, globins, lactalbumins, immunoglobulins, and histones··· all of which have relevance only to function in mammalian eucaryotes. Comparative biochemistry in its most general sense includes such studies on bioorganic compounds encompassing given functions as they are expressed over the whole biological range, from bacteria to whales.

Development of a comparative biochemistry at the molecular level requires availability of bioorganic molecules, such as specific enzymes and proteins, which are the active agents in well-defined functions which can be traced from organism to organism, and with molecular properties which render them amenable to structure analyses practicable with present or foreseeable biotechnology. In this article, I propose to discuss the cytochrome systems as appropriate test objects for purposes of comparative biochemical elucidation of structure relationships to function, and to outline the present status and future prospects for researches looking toward a general comparative biochemistry.

II. The Cytochrome Systems

Everyone is familiar with the ubiquitous occurrence and essential importance of the cytochrome system in mammalian respiration. All eucaryotes live aerobically and the energy transformations which support aerobic life are based on electron transport coupled to phosphorylation involving in large part the chain of cytochromes, first elucidated and studied mostly in mitochondria[A-1,4]. Bacterial (procaryotic) organisms exhibit aerobic systems which are analogous, but not identical, to those seen in eucaryotes[A-5,6,8]. These include electron transport systems which support strictly anaerobic metabolism (as in the chemosynthetic sulfate-reducing bacteria), facultative metabolic patterns (as in anaerobic denitrifying processes of facultative anaerobic bacteria), and electron transport coupled to photophosphorylation in the essentially anaerobic photometabolism of plant and algal chloroplasts,* and in the strictly anaerobic

---

※ Department of Chemistry, University of California, San Diego, LaJolla, California, 92037, U.S.A.

This article was written on Sabbatical leave as Visiting Professor at the University of Osaka Institute for Protein Research (auspices, Japan Society for Promotion of Science, 1972) and during tenure of a John Simon Guggenheim Fellowship (1972-73). Grants-in-aid from the National Institutes of Health (U.S.A.)···HD-01262 and GM-18528···and the National Science Foundation (U.S.A.)···GB-7033 X, GB-25050, and GB-36019···supported researches which provided some of the data presented. The writer is happy to acknowledge helpful suggestions provided in discussions with Drs. T. E. Meyer and F. R. Salemme.

* see next page.

Reprinted from *Protein, Nucleic Acid and Enzyme*, **18**, No. 8, 753-773 (1973).

bacterial photosyntheses[A-3,5,6,8].

Thus, except for the obligate fermentative anaerobes (e.g. clostridia) and certain facultative microorganisms (e.g. streptococci), cytochrome systems occur as essential parts of energy metabolism in all life processes, ranging from obligate anaerobes through facultative anaerobes to strict aerobes. Hence, it is to be expected that a great diversity of structural proteins will be found among cytochrome systems, evolved in response to metabolic demands of a bewildering variety of living organisms.

This richness in structural variation and functional behavior is a source of present difficulties[A-9] in arriving at an adequate classification of cytochromes. Expedience dictates that provisional categories be established based on the relatively limited range of properties found in the traditional eucaryotic mitochondrial system. Cytochromes, in general, are defined[A-9] as "heme proteins" which function in electron transfer as reversible redox systems based on the valence change (ferrous ⇄ ferric) of the iron atom at the center of the prosthetic heme group. The term "heme"··· the prosthetic group originally defined as the characteristic heme of hemoglobin, protoporphyrin IX···is now understood as applicable to any tetrapyrrolic chelate of iron, including dihydro- ("chlorin") type and tetrahydroporphin-type tetrapyrroles. (This provision anticipates occurrence of cytochromes in which the usual porphyrin chelate is replaced by derivative types, as in certain "cytochromes *D*," *vide* below.)

The present categories of cytochromes are based on the mode of binding of side chains of the prosthetic heme group and its magnetic spin state, most readily established by spectroscopic observation. Nomenclature of cytochromes includes specification, in nm, of the absorption ("α") band maximum for the reduced form, e.g. "horse heart cytochrome *c*-550." While most cytochromes show absorption spectra characteristic of low-spin metal complexes ("hemochromes" and "hemichromes"), some are also found which reveal high-spin character. In these cases, the letter designations employed (such as "*A*", "*B*", "*C*") are supplemented by a primed suffix (as "*B'*", "*C'*", *vide* below). Classes are denoted by the letters *A*, *B*, *C* and *D*, this order being an accident of history[A-1,9].

Class *A* refers to cytochromes with a non-covalently bound formyl porphyrin···"heme *a*" ··· and comprises a number of oxidases, including the mitochondrial cytochrome *a* of mammalian and fungal origin, as well as some bacterial forms. None of the latter have ever been characterized adequately[A-5,8] and their relation to the mammalian eucaryotic forms is unknown. However, generally the bacterial oxidases function well only with their characteristic substrate cytochromes *c*, and not with the mitochondrial cytochromes *c*[A-6]. Thus, the cytochromes *a* of *Azotobacter* species, free-living strictly aerobic nitrogen-fixing bacteria, are among the most active known, but show little or no activity with the mitochondrial eucaryotic cytochromes *c*[A-6] while being functional with the bacterial c-type cytochromes[B-1,2]. On the other hand, the oxidase of the facultative denitrifying bacterium, *Micrococcus denitrificans*[B-3,4] not only exhibits good activity with its own bacterial c-type cytochrome but also apparently significant activity with the mammalian form (however, *vide* Section IV). Conversey, the bacterial cytochrome *c* appears to be marginally active in the mitochondrial system[A-6,B-4]. The heme *a* oxidases are all functional as particulate systems, but can be solubilized in some cases by detergent treatment (*vide*, Section IV).

The second class···cytochromes *B*··· includes all heme proteins with non-covalently bound "heme *b*," which is identical to the classical heme···iron protoporphyrin IX···of hemoglobin and myoglobin. In the mitochondrial system, the b-type cytochromes are membrane-bound, and function as electron carriers intermediate in redox potential between the dehydrogenases and the terminal portion which includes cytochromes *c* (*vide* below) and the oxidase. Similar b-type cytochromes participate in analogous aerobic electron transport systems of non-mitochondrial character, such as are found in bacteria. Generally, b-type cyto-

* The interesting question of a possible respiratory system in another category of procaryotes, the blue-green algae, is outside the scope of material to be discussed in this article, *vide* e.g. ref. B-111).

chromes are difficult to solubilize and, when this is achieved, often aggregate.

The large variation in b-type forms is reflected in varying functions in other kinds of metabolic systems, of which a few may be enumerated. The lactate dehydrogenase of yeast is a flavocytochrome *b*, called "b_2"[A-9]. Recently, a spermidine dehydrogenase with FAD and heme prosthetic groups has been reported in the bacterium, *Serratia marcescens*[B-5]. Various ill-defined cytochromes of the b-type are membrane-bound in chloroplasts ("b_6", "b_{559}", "b_{564}") where they serve as electron carriers in photometabolism. In facultative bacteria they function often in modified electron transport which does not include the complete cytochrome chain (as in *E. coli* cytochrome b_{562} which has been solubilized and for which the primary structure has been determined[B-6]). A subgroup of b-type cytochromes, of uncertain function, exists in plant microsomes ("b_3") and liver microsomes ("b_5"). Primary structures of five mammalian microsomal cytochromes b_5 have been reported[A-7].

A particularly interesting group of heme proteins which might be included in the cytochrome *B* class is that called "P-450," functional[A-5] in bacterial oxidation of terpenes and other hydrocarbons (exhibited by certain pseudomonads), and in hydroxylation of steroids in mammalian systems[B-7~9]. These are high-spin protoheme proteins,* hence could be called "cytochromes *b′*". Another high-spin protoheme form, unrelated to P-450, has been seen in the aerobic *Azotobacter vinelandii*, the photoanaerobic *Chromatium* (strain D), and in an aerobe, *Flexibacter roseolus* (T. E. Meyer, R. G. Bartsch (unpublished)).

The third class…by far the best understood and most thoroughly studied…is cytochromes *C*, characterized as all cytochromes in which the heme group (of classical type to the present) is bound covalently. Binding, presumably occurring in all cases but not proven generally, involves condensation of one or more cysteines of the protein to the vinyl side chains of the heme to form a thioether linkage(s). If the heme subsequently is cleaved chemically to establish its identity as a pyridine hemochrome, it appears as a mesoheme, rather that a protoheme, because the vinyl side chain(s) has been transformed by the addition reaction into saturated adducts containing the cysteinyl residues of the protein. Functionally, it is the substrate for terminal oxidation in coupled aerobic electron transport.

The great variety of proteins in this class can be listed briefly as follows:

I(A). "Mitochondrial-type" cytochrome *c*

These are readily solubilized, monomeric, monoheme proteins (~100–110 residues of low-spin type with $E_{m,7} \simeq 250$ mV). The heme is bound to at least one cysteine and usually to both cysteines when they are present, e.g. residues 14 and 17 in mammalian proteins, with concomitant ligand binding to His 18 and Met 80, or homologous side chains. They are highly stable in evolution[A-7], as seen from the primary structures of well over 40 eucaryotic cytochromes *c* which have been sequenced[A-7]. In some trypanosomes and euglenoid algae, examples are found in which only one cysteine is present, the second being replaced by alanine[B-10]. The usually invariant region of primary structure (residues 70 to 80) which has figured so prominently in speculations about the redox mechanisms of mitochondrial cytochrome *c*[B-11] is altered in these cases also (as in the replacement of Tyr 74 by phenylalanine). Only a few examples have been isolated from bacteria in which their presence, in general, has been inferred only from spectral data. Demonstrations of mitochondrial character require isolation, purification and study of interaction with the mitochondrial transport system (*vide* Section V). The mitochondrial cytochromes *c* can be solubilized readily, without detergent treatment usually, and shown to retain function as substrates for the oxidase[B-12~14]. Their definition functionally follows from these observations. Tertiary structures for a few of these cytochromes have been determined recently[B-11] (M. Kakudo (priv. comm.)). It has been proposed that a general definition of cytochrome *c* be based on the demonstration that

* The rigorous demonstration of protoheme character, e.g., extraction by acid-acetone and characterization of the derivative pyridine hemochrome, has been achieved recently (R. Sato (private communication)).

any given example tested must react as substrate with the corresponding oxidase[A-6] (see further discussion below and in Section IV).

I(B). Cytochromes c_1

These are bound forms of mitochondrial cytochromes c[B-15,16]. The primary sequence of the heme-bearing peptide has been published[B-16], and isolation procedures for purification of the protein, based on detergent treatment, have been described[B-17,18]. It is the terminal protein in the portion of the electron transport chain of mitochondria occurring just prior to the site of cytochrome c involvement (*vide* Section V).

I(C). Cytochrome c_2

These are low-spin, readily solubilized c-type cytochromes occurring in non-sulfur photosynthetic bacteria which show considerable degrees of homology to class I(A). There is an even closer resemblance (*vide*, Section IV) in tertiary structure than might be inferred from the primary sequences between the one example determined…that of *Rhodospirillum rubrum* cytochrome c-550[B-19]…and that of horse heart mitochondrial cytochrome c[B-11]. $E_{m,7}$ ranges between 320 and 350 mV at physiological pH. Its function, in soluble or bound form, is to provide an electron carrier at the high potential terminus of the electron-transfer chain linked to photophosphorylation. It is a poor substrate for the mitochondrial oxidase and is not functional in respiration in the organism[A-5]. Its evolutionary position seems close to that of class I(A)[A-7].

II. Cytochromes c_3

These are low-spin, low potential cytochromes ($E_{m,7} \simeq -150$ to -250 mV) which are multiheme (3-4 hemes) and predominantly monomeric with molecular weights in the range 9,000—14,000[B-20~22]. Thus, per heme residue the protein weight can be as small as ~3000., One extraplanar ligand is histidine, but the other is not known. Major criteria for inclusion in this class are the low redox potential and characteristic spectra[B-22]. Evolutionary history, as revealed by sequence determinations on three examples[B-23]…indicates at least one gene duplication with many insertions and deletions. There is no homology to class I(A). They are found in the non-photosynthetic obligately anaerobic sulfate-reducing bacterial genus, *Desulfovibrio*, in purple photosynthetic nonsulfur bacteria and in a blue-green alga[B-22].*

In view of the demonstrated genetic heterogeneity of the sulfate-reducing bacteria, based on nucleotide base ratios of DNA isolated from them[B-28], and the general distribution noted[B-22], it is unlikely that this sub-class, based solely on present criteria, particularly low redox potential, will retain the status of a homogeneous grouping.

III. Cytochromes c′

These are monoheme, easily solubilized high-spin c-type cytochromes ($E_{m,7} \simeq 0$—20 mV) with spectral properties resembling myoglobin or alkaline peroxidase. They are found in purple photosynthetic bacteria, in at least one facultative anaerobic denitrifier[A-5], and in the aerobic N_2-fixing *Azotobacter vinelandii*[B-29], and occur mostly in dimeric form. ~125-130 residues per heme are present and sequence determinations of six proteins have been determined completely, or nearly completely[B-30]. No homology is exhibited to class I(A). The heme group is bound to the carboxyl terminal portion rather than close to the N-terminus, as in classes I(A) or I(C). Anomalous binding of ligands[B-31] occurs; thus, only CO in the reduced form, and NO in either reduced or oxidized forms, act as ligands. The fifth ligand (extraplanar) appears to be histidine, based on sequence analysis, but the sixth ligand is unknown (it could by H_2O or carboxyl…at neutral pH).

These proteins have a history of uncertain and varying nomenclature, having been termed "yellow pigment," "*R. rubrum* heme protein (RHP)," "cytochromoid c," and "cryptocytochrome"[B-32]. Their remarkable magnetic and spectral properties[B-32], including a partially reversible pH-induced transition from high-spin to low-spin state in alkali[B-33], make them especially interesting test systems

* Its reputed occurrence in the green sulfur photosynthetic bacterium, *Chloropseudomonas ethylica*[B-24], appears to be in error as this organism has been shown to be a mixture of *Chlorobium* species and a sulfate-reducing bacterial symbiont[B-25] and the associated cytochrome c_3 is not found in the pure *Chlorobium* component[B-26] (*vide*, also, ref. B-27)).

for physico-chemical studies.

IV. "Spheroides Heme Proteins"

While found so far only in two photosynthetic bacteria[B-34], this cytochrome variant appears to be a form of high-spin, monomeric, cytochrome ($E_{m,7} \simeq -22$ mV) definitely different from III, with spectral properties like myoglobin. It is monomeric, with molecular size ~12,000, and binds CO, N_3^-, CN^-, etc.

V. *Pseudomonas* Cytochrome *c*-551

These are low-spin, readily solubilized, monoheme monomers[A-5] possessing ~80—85 residues with $E_{m,7} \simeq 260$—290 mV at pH 7.0. There appears to be a distant relationship but little homology[B-35] to class I(A) as well as tenuous homology to VI(A) (*vide* below). Considerable homology exists among the six examples sequenced so far[A-7,B-30,35]. They function as the characteristic c-type cytochrome substrate for the oxidative (anaerobic) electron transport system leading to the modified *Pseudomonas* oxidase (nitrite reductrase, *vide* below and Section IV), and possibly to an a-type oxidase io these bacteria[A-8]. In close analogy to class I, heme binding to Cys 12 and 15 and to ligands His 16 and Met 51, or ligands in homologous positions, is found. They have been isolated from the facultative pseudomonads, in cultures grown under anaerobic or semianaerobic, but not under aerobic, conditions. On the other hand, an example is known in the strict aerobe, *Azotobacter vinelandii*.

VI(A). Cytochrome c_5* (for "*Cytochrome c_4*" *vide* below)

These[B-2,30,34,34a] are low-spin, readily solubilized, dimeric monoheme proteins ($E_{m,7} \simeq$ 300 mV, ~85—90 residues). Heme is bound, as in class I, to Cys 19 and 22 and ligands His 23 or to ligands in homologous positions. Sequence determinations[B-30] are complete, or nearly complete for five pseudomonads (similar distribution as in class V) and *Azotobacter vinelandii*.

VI(B). *Chlorobium* Cytochrome *c*-555[B-37]

This protein appears best placed in a class by itself although it exhibits a considerable homology[B-38] to VI(A). It is a low-spin, readily solubilized, monomeric and monoheme protein (~80—90 residues, $E_{m,7} \simeq 150$ mV) with reduced α-peak red-shifted and asymmetric. It is found only in the green sulfur photoanaerobes (*Chloraceae*). It is thought to function both at the high potential (oxidizing) end of the photophosphorylation chain and possibly as a high-potential redox pool. It has no respiratory function (*vide*, however, Section V).

VII. Flavocytochromes *c*

This subclass of soluble, low-spin hemeflavoproteins, is a c-type analogue of cytochrome b_2, found in photosynthetic bacteria, but with function undetermined and flavin prosthetic group covalently bound[B-39]. Variable properties are exhibited depending on the source. In *Chromatium* (purple sulfur photoanaerobe) the molecular weight of 72,000 includes two non-identical subunits, one being di-heme in character (MW 21,000), the other with a covalently bound flavin, MW ~45,000. $E_{m,7}$ for the protein is ~10 mV. In *Chlorobium thiosulfatophilum* (green sulfur photoanaerobe), MW is 50,000 and there are monoheme and monoflavin subunits ($E_{m,7} \simeq 100$ mV). The flavin prosthetic group reacts to form charge transfer adducts with CN^-, SO_3^{2-}, $S_2O_3^{2-}$, Coenzyme A or reduced glutathione [B-39].

VIII(A). Cytochromes "*f*"

These are soluble, low-spin proteins which are monoheme monomers of MW ~10,000—12,000 when isolated from algae[B-40]. They appear in aggregated forms when derived from higher plants[B-41,42] and the minimal size of the monomer in the case of spinach is ~35,000[B-43]. The reduced α-peak is red-shifted and asymmetric as in class VI(A). Functionally, they act at the high potential end of "System I" in chloroplast photophosphorylation and have no respiratory function (*vide* Section V). A so-called "split-α cytochrome *c*-553"···present in small amounts in the photoanaerobe *Chromatium*[B-44]···has been suggested as a variant cytochrome "*f*", but it does not seem to function in the photophosphorylation mechanism of the bacterium.

This subclass may be expected to be heterogeneous. It must be remembered that these

* Another c-type cytochrome ("*c*-553"), found in *Desulfovibrio* sp.[B-36], resembles cytochromes c_5 in primary structure, has 79 residues, is readily solubilized, is low-spin and a monoheme monomer. Its potential and function are unknown.

cytochromes arise from both eucaryotic (green plants and algae) and procaryotic (blue-green algae) sources, and represent a long evolutionary history culminating in highly specific and efficient photosynthetic systems for conversion of solar energy to biochemically useful free energy. Two sequences recently completed[B-30)]...those from the red alga, *Porphyra tenera*, and the green alga, *Euglena gracilis* ...show good homology, and also some degree of homology to VI(A) and VI(B). In view of the considerations above, it will be interesting to see the sequences of cytochromes *f* derived from blue-green algae, on the one hand, and from higher plant sources on the other, as well as from a wider selection of algae.

VIII(B)

This listing includes various other forms insufficiently characterized, so far[A-5,B-30)], e. g. "purple sulfur bacterial cytochrome *c*-553," "*palustris* cytochrome *c*-556," "*T. pfennigii* (purple sulfur) cytochrome *c*-552, 545," etc. There are important bound forms functional in photometabolism, notably "purple-sulfur bacterial cytochrome *c*-556, 552" which when solubilized with cholate, cannot be resolved into separate heme units, although evidence is clear that the 556 heme is low-spin with $E_{m,7} \simeq 350$ mV and the 552 heme is low-spin with $E_{m,7} \simeq 10$ mV[B-45)]. Other important cytochromes not included in this scheme are "cytochrome *cd*" (the nitrite reductase of *Ps. aeruginosa*, *vide* below), a "*Chlorobium* cytochrome *c*-551" of size 60,000 with $E_{m,7} \simeq 135$ mV[B-34)], "cytochromes c_4"[B-1,2,46)] and others which are being found from time to time.

The fourth accepted class, cytochrome *D*, with wide bacterial distribution, mostly among aerobic forms, is characterized by noncovalent binding of a heme derivative of chlorin (dihydro-porphyrin) type···termed "heme *d*." The structure of this prosthetic group has been determined for only one case, that of a strain of *Aerobacter aerogenes*[B-47)]. A distinct spectrochemical feature is the marked red displacement of the α-band maximum to 620—630 nm, as contrasted with the other cytochromes which exhibit such maxima at 590—605 nm (cytochromes *a*), 555—565 nm (cytochromes *b*) and 550—558 nm (cytochromes *c*). Cytochromes *d* were discovered originally by spectroscopic means in *E. coli* and *Shigella dysenteriae*[B-48,49], and seen later in many other microorganisms[B50)]. A particularly important soluble variant in this group, known as cytochrome "*cd*," and found in pseudomonads and denitrifying bacteria, contains both a covalently bound heme *c* and non-covalently linked heme *d*. In the case of *Pseudomonas aeruginosa*, it has been purified and characterized extensively[B-51,52)]. It can function as an oxidase[B-52)] for its substrate cytochrome *Ps. aeruginosa c*-551[B-53)], but is physiologically a nitrite reductase[B-51)]. A preparation active in both modes from the facultative denitrifier, *Micrococcus denitrificans*, has been isolated in highly purified state and compared with that of the pseudomonad[B-54)]. The heme *d* groups appear to be identical to each other and different from that reported for the *Aerobacter* sp.[B-47)]. In *M. denitrificans*, the enzyme only appears in anaerobic culture[B-54)] and is considerably larger (MW 120,000) than that of *Ps. aeruginosa* (MW 80,000).

III. The Cytochromes as Test Systems

The foregoing discussion, however brief, should indicate clearly that of all the components in the cytochrome systems, those grouped in the category of cytochromes *C* are best adapted as test systems for comparative biochemical purposes. They present the most complete body of knowledge at the structural and functional level over the largest range of living organisms available. This fact has not gone unnoticed, as even a casual survey of the literature will make evident (*vide* A-7)). Nevertheless, the basic data needed for the achievement of a complete comparative biochemistry···a full complement of primary and tertiary structures over the whole range of cytochrome *C* proteins, together with analyses of variations in active sites which are likely to occur when the proteins are functional (in solution, or membrane bound and in interaction with associated redox enzyme systems) are still to be assembled. To date, approaches have been concerned on the one hand with constructions of phylogenetic trees[A-7,B-55)] based on the extensive library of primary structures for the mammalian, fungal and plant proteins (with only a passing attempt, however, to include bacterial data) and, on the other, with functional assessment of evolutionary history by examination of cross reactions between cytochromes *C* and selected redox enzyme systems[A-6].

The hurdles to be overcome in accumulating reliable data at both structural and functional levels are most formidable. Consider first the acquisition of sequence data. Once the choice of protein has been made, based on general distribution, molecular weight, yields available, and minimal variations in isolation procedures caused by source materials, there can occur numerous types of error. These can be technical, such as assignment of wrong positions to block sequences, deletions owing to missed peptides, uncertainty about amide locations, transpositions, etc., or they can arise from failure to characterize adequately the source material. Undoubtedly, examples of all these kinds of errors lurk in the literature ! (*vide*, ref. B-35 for a full discussion.)

The eucaryotic cytochromes *C* are a class of proteins for which errors in primary structure data[A-7] are minimal or absent, so that they can be utilized with some sense of surety for derivation of evolutionary histories. The corresponding data for procaryotic cytochromes *C* are rapidly approaching the same highstandard of reliability, so the necessary correlations with tertiary structure, when available, with the vast body of biochemical data can be attempted, and thus the full biologi cal range of these proteins exploited.

Another requirement, determinations of tertiary structure, depends at present on production of adequate crystals···requiring success in solubilization of starting material. The present highly developed technology, based on X-ray diffraction analyses, can treat successfully, in reasonable time, structures of proteins up to moderate sizes (maximum MW ~50,000) but for the large functional proteins, such as dehydrogenases and synthetases, hope must be placed in the existence of subunits of manageable size. Again, the cytochromes *C* offer the best opportunities, from the standpoint of ease of solubilization, molecular size and stability. Two high resolution structures are available in both oxidized and reduced form···that of horse heart mitochondrial cytochrome *c*[B-11] and of a procaryotic analogue from the cytochromes c_2, *R. rubrum* cytochrome *c*-550[B-19]. Other structures, both from eucaryotic[B-56] and procaryotic sources[B-57] (F. R. Salemme (priv. comm.)) are in prospect.

Finally, there is the question of functional evaluation. In the cytochromes *C*, problems in this area are present in no small degree. A major difficulty is the establishment of a given protein in the free state as the physiological substrate for the terminal oxidase. In some cases, a constellation of soluble forms is found, as in the photosynthetic bacterium, *Rhodopseudomonas palustris*, which exhibits perhaps half-a-dozen readily prepared different soluble c-type cytochromes in more than trace amounts[B-58~60]. In other cases, as in the photoanaerobe *Chromatium vinosum* the functional c-type cytochrome (cytochrome *c*-552, 556) is not solubilized readily, and those forms found in readily soluble state appear to bear no relation to this functional membrane bound cytochrome[B-45]. Again, variations in the nutritional history, or in culture conditions, of an organism can lead to uncertainties As an example[A-6], the fungus, *Aspergillus oryzae*, when grown in surface culture produces copious quantities of a cytochrome *c* which, however, fails to interact with the standard eucaryotic fungal cytochrome *c* oxidase from yeast (*Saccharomyces oviformis*). However, strangely enough, when grown under submerged conditions, an active form is obtained.

Other sources of difficulty arise from the possible formation, or presence, of deamidated and noncovalently-bound polymerized modifications of the native protein[A-4,B-61,62]. These can produce unsuspected inhibitory effects in reactions with associated redox systems.

While such phenomena create difficulties, they are not insurmountable hindrances to continued progress, although of course interesting and worthy of investigation in their own right. The richness of material among the cytochromes *C* clearly guarantees that there should be enough proteins available, which can be freed of uncertainties both with regard to reliability and extent of structural data and definition of function.

IV. Some Approaches to a Comparative Biochemistry of Cytochromes *c*

Approaches to a comparative biochemistry of cytochromes *c* have been concerned mainly with delineation of events in the evolutionary history of these proteins. One approach, extensively documented[A7,B-63], has been based on examination of variations in primary structure. Attempts have been made, using highly sophisticated computer techniques[B-64], to trace the evolutionary development of organisms dependent on redox processes involving cytochromes, and to infer structures of possible precursors, as well as a common precursor for animals and plants. However, these studies do not have as their immediate objective the elaboration of structure-function relationships.

Another approach, confined to the limited group of mitochondrial cytochromes *c*, has been based on immunochemical properties as useful in placement of organisms in an evolutionary sequence[B-65]. Difficulties with this approach, using as criteria immunochemical cross-reactions with antisera, are obvious enough…e.g. antigenic sites may be redundant portions of the protein unrelated to function, intermediate structural forms which are the active components may not be present in appreciable concentrations in solution, important features of structure may not be available to particular surface antigens, etc. Very probably, immunochemical approaches can become most significant when sufficient data are available on tertiary structures and physiological modifications therein, correlated with activation through binding to membrane active sites.

The third approach to be considered is that based on comparative biochemical properties exhibited by cross-interactions between cytochromes *c* and associated selected redox enzyme systems. This approach has also been extensively documented[A-6]. Some difficulties arise in the choice of the assay redox enzymes. Among those available, one may mention the particulate mammalian mitochondrial cytochrome *a* (usually prepared from beef heart or a yeast), various[A-6,B-66] soluble cytochrome peroxidases from yeast and aerobic bacterial sources (thiobacilli), a soluble sulfite reductase from the obligate aerobic *Thiobacillus novellus*[B-67], and the soluble, well-characterized *Ps. aeruginosa* nitrite reductase[A-6]. With the exception of the *Pseudomonas* preparation, none are available in readily standardized and completely characterized form.

The mammalian oxidase preparations have only relatively recently begun to achieve some degree of standardization and reliability as regards purity. They are usually particulate preparations which always present uncertainties as to presence of contaminants and variable occluded ionic constituents. A considerable difficulty is complete removal of traces of the physiological substrates, the mitochondrial cytochromes *c*, which remain bound in the particulate form of the preparation. These can couple non-enzymatically with the foreign test cytochromes *c* and lead to false conclusions regarding the cross-reactivity of the test protein with the mitochondrial system.

A few examples can be cited. When beef heart oxidase is incubated with the *Ps. aeruginosa* cytochrome *c*-551, a reaction occurs if a small amount of beef heart cytochrome *c* is present[A-6]. If cytochrome *f* from algal[B-68] or plant[B-41,88] sources is incubated with beef heart oxidase, a reaction is seen on addition of beef heart cytochrome *c*, even though a potential difference of as much as 100 mV exists ($E_{m,7}$ of cytochrome $\simeq$ 350—400 mV, $E_{m,7}$ cytochrome $c \simeq 260$ mV) to limit the extent of non-enzymic reduction of the beef heart cytochrome *c*. In early experiments with mixtures of *R. rubrum* cytochrome *c*-550 and mammalian oxidase[B-69], a similar effect on adding the mammalian cytochrome was reported. Obviously, presence of any oxidizing contaminant which is autoxidizable and capable of nonenzymic redox coupling with the test cytochrome *c* can mislead the observer.

Other complications arise because of the complex kinetic behavior of the oxidase reaction, stemming in part from the fact that the reaction system is heterogeneous, the polar substrate, cytochrome *c*, being required to interact at the phase-boundary between solution and the particulate membrane system.

Detergent solubilization to obviate such difficulties may introduce new uncontrolled variables in the kinetic behavior of the oxidase system[A-7,B-70]. A final comment may be inserted to the effect that ion-binding to substrate or oxidase can produce large changes in kinetic responses of the oxidase system, particularly anion binding[B-71]. It is essential, then, that functional tests of this kind be exhaustively monitored and interpreted with caution.

Despite these difficulties, an impressive correlation of cross-reactivities with reasonable inferences about evolutionary age of organisms over a large range of cytochromes *c* has been assembled[A-6].

V. New Approaches and Prospects

The recent successes in production of high resolution X-ray structural studies of horse heart mitochondrial cytochrome *c*[B-11] and of the analogous c-type cytochrome c_2 from the photosynthetic non-sulfur purple bacterium, *R. rubrum*[B-19], provide new strategies whereby one may relate structure to function in more direct fashion than previously possible. The approaches can be of two kinds. In the first which has been documented extensively already[B-11,72], a single structure is explored in depth, comparing[A-4] changes in function with variations in composition obtained either by chemical or by biological means. The second is a comparative approach in which a single well-defined functional redox system is challenged by substrate c-type cytochromes for which the high-resolution tertiary structure is known, and which represent metabolic patterns removed in evolutionary time from that represented by the test system. Combinations of both these strategies can also be imagined.

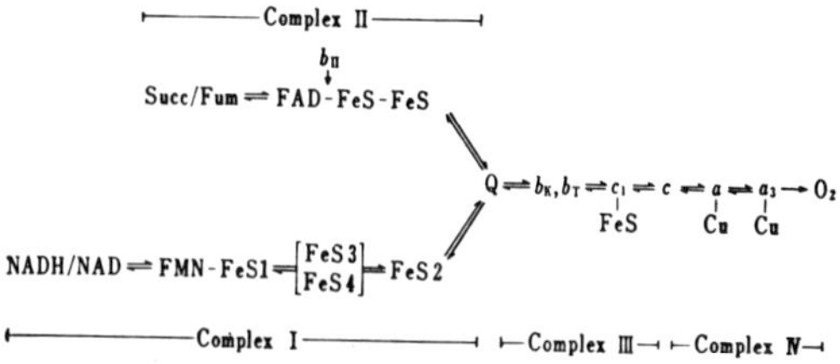

Fig. 1. Representation of eucaryotic mitochondrial electron transport system, after Y. Hatefi[B-73].

Test cytochromes *c* are inserted at site shown filled by mitochondrial cytochrome *c*.

It is already possible to make a beginning in applying these strategies. Thus, previous studies on reactivity of cytochromes *c*, employing proteins of widely variable purity and tested with incompletely characterized redox systems, can be improved and extended by exploiting the present availability of highly purified cytochromes *c*, derived from the whole biological range of organisms available, and of well-defined and essentially pure enzyme systems. Undoubtedly the best among the latter is the reconstituted bovine heart mitochondrial electron transport system[B-73] which can be fractionated into four functional complexes and recombined to produce the original, fully functional redox chain (Fig. 1). In this system, the mitochondrial cytochrome *c* provides a natural bridge as a specific substrate, between two of these complexes…one of which ("Complex III") carries electrons from ubiquinone (UQ) to mitochondrial cytochrome *c*[B-74], and the other[B-75] from mitochondrial cytochrome *c* to oxygen ("Complex IV"). By combination of a third fraction ("Complex I") which connects NADH to UQ, electron transport is bridged from NADH to oxygen. This test system is available in solubilized form and can be manipulated and controlled in a variety of ways quite impossible with the particulate preparaqions used hitherto. Moreover, it can be denuded completely of traces of the natural substrate, cytochrome *c*.

An assumption, based on reasoning from notions in evolutionary biochemistry, is that the mammalian mitochondrial system represents the most highly specialized redox process among the eucaryotes and that the structures of its substrate cytochromes *c* are the ultimate forms developed to serve the unique requirement of the eucaryotic aerobes…namely, reduction of oxygen to water coupled to conversion and storage of biochemical free energy by means of oxidative phosphorylation. Thus, all cytochromes *c* can be evaluated from the standpoint of their reactivity in this system as a criterion for the degree to which they approach in structure the reference mammalian mitochondrial cytochrome *c*.

In applying this strategy, a formidable set of requirements arises. The techniques represented include three major activities. First, there are those of modern microbiology for production of pure cultures of microorganisms, isolation of cytochromes therefrom, and purification and characterization of these heme proteins. Secondly, enzymological facilities must be readily available for production of fresh, fully characterized and functional mitochondrial electron transport complexes, together with the assay instrumentation involved. Third, a complete X-ray diffraction structure analysis laboratory, including all auxiliary facilities ranging from crystal preparation laboratories to computers, must be provided.

To bring all these activities into a harmonious and fruitful collaboration requires circumstances not found often. Nevertheless, it must be done to achieve maximal progress from such a strategy. We have been fortunate in having an opportunity at San Diego to exploit a happy juxtaposition of laboratories at the campus of the University of California and at the Scripps Clinic and Research Foundation. I will devote the rest of this article to a description of our collaborative activities as examples of the manner in which one may expect to see progress implemented in the future toward elaboration of comparative biochemistry at the molecular level.

In the writer's laboratory, we have studied for some twenty years the various forms of cytochromes *c* derived from bacterial and algal sources. These efforts have provided a laboratory of the type fulfilling the first prerequisite···that of the modern microbiological and biochemical facility. In roughly the same period, a research group headed by Prof. Y. Hatefi, who has been a leading investigator in the successful fractionation and reconstitution of the mitochondrial electron transport system, has been assembled and provides the second requirement··· that of a fully equipped enzymological facility to prepare the test redox system. The third requirement···that of the X-ray structure facility···is filled by the research group and laboratory headed by Prof. J. Kraut. These three laboratories have begun a collaboration of the type I have described.

Of course, there are other ways to achieve the desired results. It is not absolutely necessary to have all of these facilities in locations side by side. Collaboration does appear essential, however, as it is quite impossible for any single research group to include all the needed technolcgies of modern biology.

The choice of the test redox system is clear, as already discussed. The substrate cytochromes *c* to be tested should be chosen with the following consideration in mind: (1) each protein should be available in a high degree of purity, free of contaminant and degraded (deamidated or polymerized) forms, and characterized as fully as possible with respect to absorbed ionic constituents; (2) the selection should include cytochromes *c* which represent a diversity of function, ranging from those completely non-functional in the mitochondrial system to those with function closely analogous or very similar; (3) crystalline preparations suitable for X-ray structure analyses must be provided; (4) reliable primary sequence structures should be available not only to aid in tertiary structure determination, but also to provide preliminary data on structure parameters (e.g. distribution and number of basic and acidic residues) needed in interpretation of reaction rates; and (5) tertiary structures should be known, or their determination should be in reasonable prospect.

A preliminary selection, based on available preparations in our laboratory, has been made. These are: (1) horse heart mitochondrial cytochrome *c* (the reference protein, termed "I" henceforth), (2) *R. rubrum* cytochrome *c*-550, (3) *Euglena gracilis* cytochrome *c*-558, (4) *Chlorobium thiosulfatophilum c*-555, (5) *Porphyra tenera* cytochrome *f*, (6) *R. rubrum* cytochrome *c'*, and (7) *Rhodopseudomonas palustris* cytochrome *c'*. Each of these six test proteins (2) to (7) may be discussed briefly in turn.

R. rubrum cytochrome *c*-550···a representative of the cytochrome c_2 subclass (IC)···was the first bacterial form to be isolated and characterized[B-69,76] and is the first for which

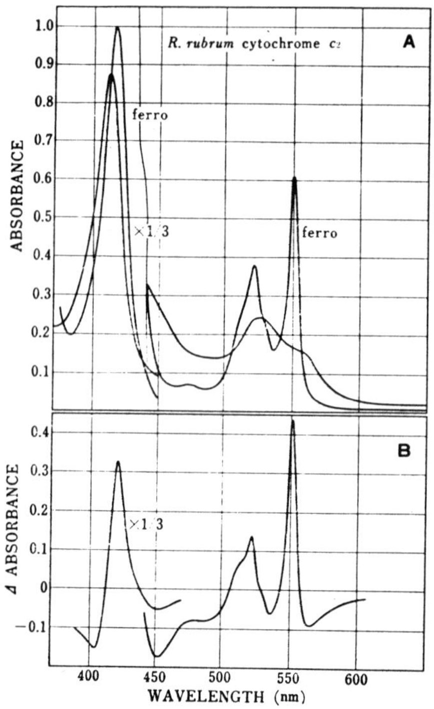

Fig. 2. Absorption spectra of *R. rubrum* cytochrome *c*-550, after R. G. Bartsch[B-59].
Upper portion shows reduced and oxidized spectra, as indicated. Lower portion shows difference spectrum.

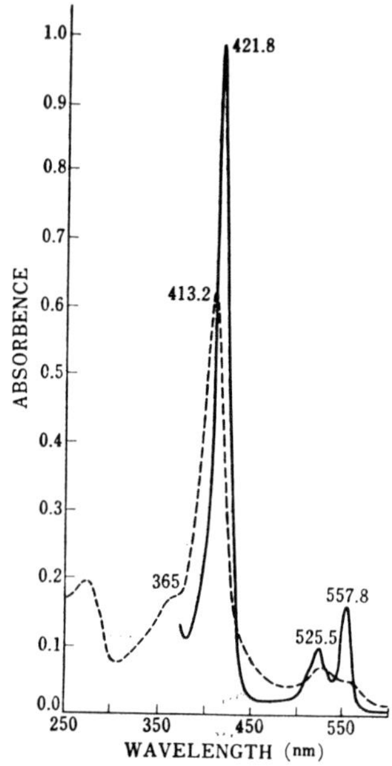

Fig. 3. Absorption spectra of *E. gracilis* cytochrome *c*-558, after T. E. Meyer[B-34,84].
oxidized, ··············; reduced, ————

both the primary structure[B-77] and a high resolution tertiary structure[B-19] have been obtained. As with other members of its class, it differs from the mammalian form in having a longer polypeptide chain (112 residues as compared to 103—104) with a freely available amino terminus, a more positive redox potential ($E_{m,7}$=310 mV) which is variable with pH in the range from 5 to 8[B-78], and an isoelectric point at acidic pH[B-79]. Its physicochemical characteristics have been the subject of much study and it is probably the most thoroughly documented of the bacterial proteins[B-58,59]. Its function is non-respiratory, being instead a substrate for the strictly anaerobic electron transport coupled to photometabolism in bacterial photosynthesis, wherein it is located at the oxidizing end of the photophosphorylation chain[A-5]. In wild-type *R. rubrum*, it is present largely in reduced form and is not changed in its steady-state level of reduction by aeration, only shifting to a more oxidized state when the organism is illuminated with actinic light and is in active photosynthesis, or photometabolism[A-5, B-80]. Its participation in a high-potential pool has also been inferred[B-81]. Its spectral characteristics are quite similar to those of mitochondrial cytochrome *c*, so its absorption spectra are shown in Fig. 2[B-59] as examples of "typical" cytochrome *c* spectra.

E. gracilis cytochrome *c*-558 is the functional substrate cytochrome *c* for the respiratory system of the parent alga[B-82,83]. It is a basic monoheme protein, like (I). Recently, it has been prepared[B-84] in a high degree of purity and in amounts sufficient to produce crystals for X-ray studies. It is atypical in that it contains only one cysteine residue, while otherwise approaching closely in total residue number that of the mammalian forms. In another modified mitochondrial system…

that of the trypanosome *Crithidia oncopelti*, a mosquito parasite···the functional cytochrome *c*, for which the primary sequence is available[B-10], shows the same anomaly, the cysteine which normally binds heme at residue position 14 being replaced by alanine. As seen in Fig. 3, this substitution, the same in the *Euglena* as in the *Crithidia* protein (G. Pettigrew (priv. comm.)), shifts the reduced α-peak maximum 8 nm to the red, as expected for an electron-withdrawing substituent on a heme pyrrole β position. Thus, we have mitochondrial cytochromes *c*, in which the so-called "invariant" cysteine residue at position 14, or a homologous position, is missing and the heme can be bound by only one covalent linkage in the normal thioether fashion. This preparation provides an opportunity to determine how loss of one covalent linkage affects mitochondrial function dependent on electron transport between Complexes III and IV. The sequence for the *E. gracilis* cytochrome *c*-558 will be available shortly (G. Pettigrew (priv. comm.)). The redox potential $E_{m,7}$ is reported as 307 mV [B-83] which is somewhat higher than that of (I). It does not vary[B-83] over the range of pH 5—11. This pH invariance of E_m is of interest in regard to some speculations on function of Tyr 67 in cytochromes *c* (*vide* below).

C. thiosulfatophilum *c*-555 is derived from a green sulfur photoanaerobe in which no respiratory function exists. It is a basic monoheme protein, like (I), but its molecular weight is less (~10,000, ~92 residues) and its redox potential is more reducing (E_m varying from 145 mV at pH 6.0 to 114 mV at pH 8.0) while its spectral absorption properties are atypical (Fig. 4) compared to the mammalian form[B-37]. In this latter respect ···low α/Soret ratio and asymmetric α-peak ···it seems to resemble the cytochromes c_5. Its primary sequence shows some homology to these cytochromes, as well as algal cytochromes "*f*"[B-30,38]. This cytochrome *c* appears to function like *R. rubrum* cytochrome *c*-550, at the high positive redox potential terminal portion of the coupled photophosphorylation chain in the anaerobic photometabolism of the parent organism[B-22,85]. Thus, it provides another example of a basic cytochrome *c* of size and monoheme type similar to the mammalian protein, with an overall positive charge at functional pH, and yet with a function far removed from that of the mitochondrial cytochromes *c*.

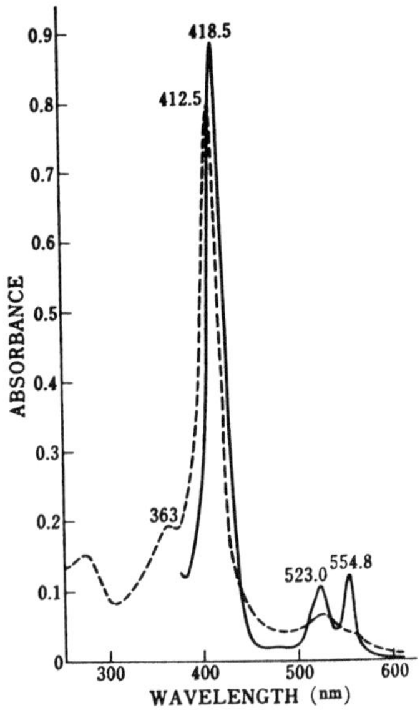

Fig. 4. Absorption spectra of *C. thiosulfatophilum* *c*-555, after T. E. Meyer, *et al.* [B-37]. oxidized, ··············; reduced, ————

Porphyra tenera cytochrome *f* (cytochrome *c*-553) is typical of the algal chloroplast c-type cytochromes[B-40] which function in the anaerobic photophosphorylation process linked to "System I" in eucaryotic photosyntheses. It is a monoheme protein, molecular weight ~10,000, with an elevated Soret/α absorption ratio, and an $E_{m,7}$ of ~ 350 mV. It is readily prepared in crystalline form[B-86,87]. Its primary structure is known[B-30]. This c-type cytochrome represents an ultimate evolutionary form also, as does the reference protein (I). As a component of the highly evolved chloroplast organelles, it functions in an anaerobic mode uniquely for conversion and conservation of radiant energy in a biochemically useful form. This function is diametrically opposed to that of the mitochondrial oxidase in that it is obligately anaerobic. Oxygen not only is unutilized in this system, rather

Table I. Reduction of Cytochromes *c* by Complex I-III[B-68] (Reproduced courtesy of *Biochem. Biophys. Res. Comm.*)

Electron Acceptor	Specific Activity[a)]	% Inhibition	
		Antimycin A (1 μM)	Polylysine (5 μM)
Cytochrome *c*	24.6	100	>98
〃 c_2	13.6	> 98	84
〃 *c*-558	16.7	100	>98
〃 *c*-555	0.45	100	73
〃 *f*	0.39	None	Activation[b)]
〃 *c'* (*R. palustris*)	0.0	—	—
〃 *c'* (*R. rubrum*)	0.0	—	—

a) μmoles cytochrome reduced × min^{-1} × mg^{-1} complex I-III protein at 38°C. The extinction coefficient used for all cytochromes was that of cytochrome *c*, i.e., ΔA (reduced minus oxidized) = 18,500 *l* × $mole^{-1}$. The wavelengths used for measuring the reduction of the cytochromes, in order of their listing in the Table, were in nm: 550, 550, 558, 555, 554, 550 and 550.

b) At this concentration, polylysine activated cytochrome *f* reduction sevenfold. Maximum activation at 10 μM polylysine was 13 fold.

it is excluded! In an early classic experiment, Hill and Scarisbrick[B-88] showed that cytochrome *f* of higher plants could couple with mitochondrial cytochrome *c*, thereby providing a means of reaction with cytochrome *c* oxidase, whereas alone, it did not react directly with the oxidase.

R. rubrum cytochrome *c'* is a dimeric monoheme protein with typical high-spin spectra and properties, but atypical in that it has a highly restricted range of ligand-binding characteristics[B-31,89]. Its function, as an electron carrier in the anaerobic photophosphorylation process of bacterial chromatophores, is expressed only in the bound state when it appears to mimic a b-type heme protein[B-90, A-5]. Its molecular weight is close to 28,000 in the free form (14 kD per heme) and it exhibits $E_{m,7}$ of ∼−10 mV[B-33]. Similar properties are found for the cytochrome *c'* of *Rps. palustris*[B-60,90a], which, however, is isolated as a monoheme monomer[B-91].

When these heme proteins, representing a functional range of c-type cytochromes of great diversity, are challenged[B-68] as substrates for the mitochondrial cytochrome *c* reductase system (Complexes I+III), they exhibit reactivities shown in Table I. Thus, relative to (I), *R. rubrum* cytochrome *c*-550 shows∼60% activity and appears as a reasonably good substrate for the reductase. This fact strongly implies that, structurally, it possesses many features in common with those of the mitochondrial protein, as far as the reductase specificity is concerned. This conclusion is further strengthened by the fact that Antimycin A has an identical inhibitory effect on the cytochrome c_2 reduction, as it does in the cytochrome *c* reference reaction. Here we see an advantage of the use of the well-characterized solubilized reductase system in that Antimycin A is known to exert its inhibitory effects only at the point between cytochromes *b* and c_1 of the Complex III; thus, we can be assured that electron transport is taking place through the same pathway for both cytochromes *c* and c_2.

Table II. Oxidation of Cytochromes *c* (plus Ascorbate-TMPD) by Cytochrome *c* Oxidase[B-68] (Reproduced courtesy of *Biochem. Biophys. Res. Comm.*)

Cytochrome	Specific Activity[a)]	% Inhibition by Polylysine (5 μM)
Cytochrome *c* (11 μM)	9.7	100
〃 c_2 (11 μM)	0.7	72
〃 c_2 (100 μM)	2.7	90
〃 *c*-555 (11 μM)	10.7	100
〃 *c*-558 (11 μM)	19.8	100
〃 *f* (64 μM)	nil	—

The assay mixture (1.8 m*l*) consisted of 10 mM potassium phosphate, pH 7.0, 22 μM tetramethyl-*p*-phenylenediamine (TMPD), 1.1 mM sodium ascorbate, and 0.16 mg cytochrome oxidase.

a) μmoles cytochrome oxidized × min^{-1} × mg^{-1} cytochrome oxidase protein at 30°C.

Other important points emerge when it is seen that the polycation, polylysine (average MW≃15,000), inhibits the reduction of both cytochromes *c* and c_2 (Table I) as noted previously in mitochondrial preparations[B-14], and that when the two cytochromes *c* are tested with the Complex IV oxidase (Table II), results closely identical with those obtained previously with the less well-characterized particulate oxidase preparations[A-6, B-69] are found: that is, cytochrome c_2 at the same concentration as cytochrome *c* is oxidized ∼ 5—8% as rapidly and this activity

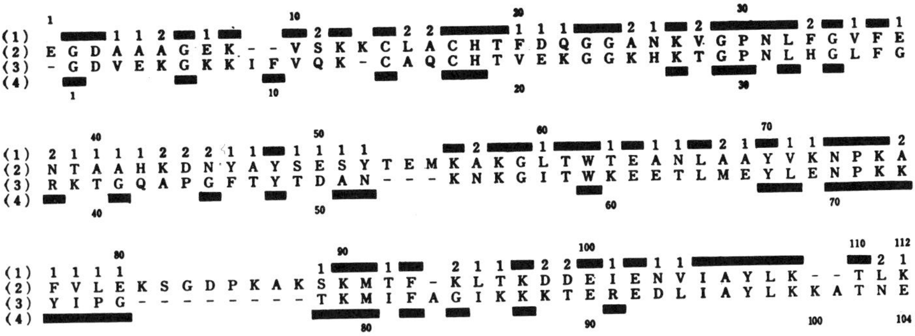

Fig. 5. Primary structures and homology of horse heart mitochondrial cytochrome *c* and *R. rubrum* cytochrome *c*-550, after F.R. Salemme, *et al.*[B-19,77].

(1) Homologous residues indicated by bars, (2) *R. rubrum* cytochrome *c*-550 sequence (112 residues), (3) Horse heart cytochrome *c* sequence (104 residues), (4) Invariant residues in most eucaryotic cytochromes *c*, indicated by bars (exceptions occur as discussed in text).

Numbers between bars in line (1) indicate nucleotide base changes; viz., there are 43 identical residues, 38 single base changes, 18 double base changes.

abbreviations A : Alanine, R : Arginine, N : Asparagine, D : Aspartic acid, C : Cysteine, Q : Glutamine, E : Glutamic Acid, G : Glycine, H : Histidine, I : Isoleucine, L : Leucine, K : Lysine, M : Methionine, F : Phenylalanine, P : Proline, S : Serin, T : Threonine, W : Tryptophan, Y : Tyrosine, V : Valine

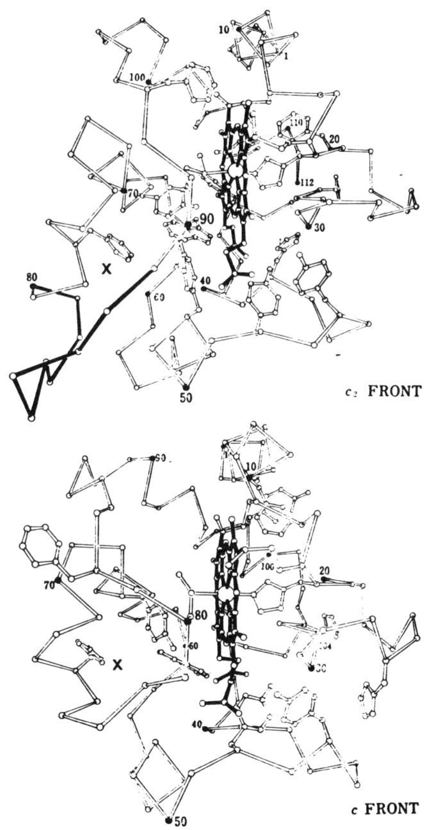

Fig. 6. Tertiary structures of beef heart mitochondrial cytochrome *c* and *R. rubrum* cytochrome *c*-550, after F.R. Salemme, *et al.*[B-19] and R.E. Dickerson, *et al.*[B-11].

These are two-dimensional projections showing the heme cleft and "front" views. Salient features of interest (*vide* text discussion) are (1) the concentration of lysine residues on the front of the molecule (*vide* also Fig. 7), and (2) the homologous positions of the mitochondrial cytochrome *c* Tyr 74 and the *R. rubrum* cytochrome *c*-550 Phe 77, seen on left sides, denoted by "X" in lower left quadrants.

is also inhibited by polylysine as expected[B-70] and also found with (I).

These results provide an excellent example of the advantages derived from the comparative approach as can be appreciated from the following considerations. It has been suggested for several reasons that the entry of electrons mediated by the mitochondrial reductase occurs at a binding site on the substrate cytochrome *c* independent of that involved in electron loss mediated by the mitochondrial oxidase, e.g.: (1) an apparently invariant region[A-7] of primary structure (residues 70—80) exists in some 30 mitochondrial cytochromes *c* which could be rationalized as a site for reductase binding in that it contains an invariant tyrosine residue (no. 74) in contact with an internal tyrosine (no. 67) near the ligand Met (no. 80), constituting (with the invariant Try 59) through a free-radical mechanism[B-93] a path of entry for the electron coming from cytochrome c_1[B-11,72]; (2) antibodies can be prepared against cytochrome *c* which block oxidation by the oxidase but not reduction by the reductase[B-92].

This reasoning is brought into question by the results shown in Table I because in *R. rubrum* cytochrome c_2, which is quite functional in the reductase system, a phenylalanine (no. 77) is homologous with this so-called invariant tyrosine 74, not only in the primary structure (Fig. 5) but even more certainly in the tertiary structure (Fig. 6). In addition the whole "invariant" sequence (residues 70—80 of cytochrome *c*) is interrupted by an insertion of 8 residues in *R. rubrum* cytochrome c_2. These facts render the suggestion of reductase binding at the protein surface covered by the sequence from residues no. 70 to no. 80 dubious, to say the least. It seems even more questionable when it is found that there are at least two known cases of *bona fide* mitochondrial cytochromes *c* in which tyrosine 74 is not invariantly present but is replaced by phenylalanine…namely in the trypanosomal *C. oncopelti* cytochrome *c*-558 [B-10] and in the fungal *L. humicola* *c*[B-94]. Thus, there are at least three instances in which the "essential" tyrosine of the electron transfer mechanism proposed for cytochrome *c*-reductase interaction is missing and yet in which the reductase reaction occurs readily.

The dissatisfaction with the tyrosine no. 74 mechanism and correlative assignment of a bonding site increases further when it is seen that polylysine inhibits both oxidase (Table II) and reductase (Table I). It is reasonable to suppose that the oxidase inhibition arises from competition between the polylysine and the polylysine-like surface of cytochrome *c*[B-11,72], as also in cytochrome c_2[B-19], for

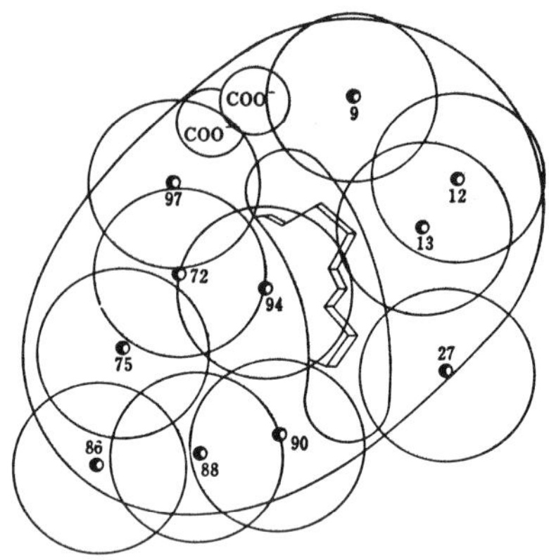

Fig. 7. Lysine residue distributions around heme crevice of *R. rubrum* cytochrome *c*-550, after F. R. Salemme, *et al.*[B-19].

binding to a homologous counter-ion structural feature on the oxidase (Fig. 7). Hence, as proposed from the structural studies in horse heart cytochrome *c*[B-11,72] and from studies on polycation activation of electron transport[B-95], binding of the oxidase takes place on the front of the molecule over the heme cleft. This idea is supported further by results of chemical modification experiments. Thus, progressive acetylation of the external lysines gradually diminishes oxidase activity[B-96,97] and trinitrophenylation of lysine no. 13, lying at the top of the heme cleft, is particularly effective in inhibition of the oxidase reaction[B-98]. The fact that polylysine also inhibits the reductase forces attention to the possibility that the binding sites for the reductase may also involve a portion of the protein surface near the heme cleft. This consideration does not rule out the possibility that a portion of the reductase binding may involve the surface area including the residues nos. 70—80; however, the presence of tyrosine no. 74 as crucial appears to be ruled out.

Another fact recently brought to light is that in *E. gracilis* cytochrome *c*-558, the crucial Tyr 67, an obligate member of the suggested electron transport mechanism through the left side of the molecule, is replaced by a phenylalanine (G. Pettigrew (priv. comm.)). This finding would appear to completely invalidate the mechanism basing entry of electrons through the region of invariance in the mammalian mitochondrial cytochromes *c*.

An interesting recent result from our laboratory (unpublished) is that polyaspartate, also as a mixture of polymers of varying chain lengths with average molecular weight ~24,000, inhibits the reductase reaction. This result can be rationalized by the supposition that binding of the polyanion to the highly positively charged region occurs over the heme cleft, thus rendering the protein inaccessible to the reductase, a notion consistent with binding of the reductase to the front part of the protein.

The results reported[B-92] on preferential inhibition of oxidase activity by antibodies directed against cytochrome *c* without parallel inhibition of the reductase prove only that binding of the antibody near the cleft of the heme can be achieved so that entry through the cleft via the reductase is possible even though exit via the oxidase is not. Inasmuch as two regions of clustered lysines exist, separated by the cleft and on either side of it, there is no apparent difficulty in visualizing binding of reductase on one side of the cleft while the anti-oxidase activity antibody blankets the other side. However, as remarked below, there may be no necessity to have simultaneous binding of both oxidase and reductase to the same substrate molecule of cytochrome *c* in the actual *in vivo* case.

Proceeding further with results of these preliminary comparative biochemical trials, one may note that *E. gracilis* cytochrome *c*-558, missing one thioether linkage, nevertheless functions well in the reductase reaction (Table I), and, astonishingly, almost three times better than the reference (I) in the oxidase reaction (Table II). This result indicates that control of the orientation of the heme group relative to the natural redox reactions is effective without the necessity of more than one covalent linkage at the top corners of the prosthetic group, and that the overall rigidity of the heme group can be achieved mainly owing to the multiplicity of residue interactions. These include the hydrogen bonding network which directs the heme orientation into its unnatural position (with polar propionate side-chains buried in the non-polar environment of the cleft) and the extensive hydrophobic bonding interactions arising between the heme group, and the numerous non-polar side-chains of the internal residues concentrated in the cleft. As mentioned previously, the recent finding (G. Pettigrew (priv. comm.)) that Tyr 67 is replaced by phenylalanine proves that reductase activity does not require electron entry via Tyr 67 to Met 80. As will be noted below, this result has some implications for a suggested mechanism whereby pH variation of E_m is controlled.

The results with cytochrome *f*…its very low reactivity with the reductase (Table I) and its complete lack of reactivity with the oxidase (Table II)…are consistent with the assumption that it represents, as expected, an evolutionary form of c-type cytochrome far removed in surface structural features from mitochondrial cytochrome *c* even though it retains the common methionyl-iron ligand feature as well as the thioether binding, so far common to all cytochromes *c*. The few primary structures known for cytochromes *f*[B-30]…those of *P. tenera*[B-30], *E. gracilis*[B-30], and *Monochrysis lutheri* *c*-553[B-99] … show little obvious homology to mitochondrial cytochromes *c*. Also, a relatively small number of lysyl residues (no more than 5—6 per molecule) is found, suggesting that the polylysine-like binding feature of the mitochondrial cytochrome *c* is absent from the surface structures of cytochromes *f*. Interestingly enough, polylysine activates, rather than inhibits, the low residual reactivity of *P. tenera* cytochrome *f* with the reductase (Table I). (One may speculate wildly that there may be binding of the polylysine to the cytochrome *f*, thus presenting an increased surface positive charge to the reductase and leading accidentally to reductase binding !) However, the expectation of heterogeneity in the cytochrome *f* subclass, commented upon previously (Section III), inhibits general statements about cytochrome *f* structure. As an example, the cytochrome *f* of the cellulosic-walled blue-green alga, *Oscillatoria agardhii* cytochrome *c*-553, appears (K. Sletten (priv. comm.)) to contain 12 lysine residues per molecule (~109 residues total). This number is quite sufficient to provide the necessary surface lysines for reductase or oxi-

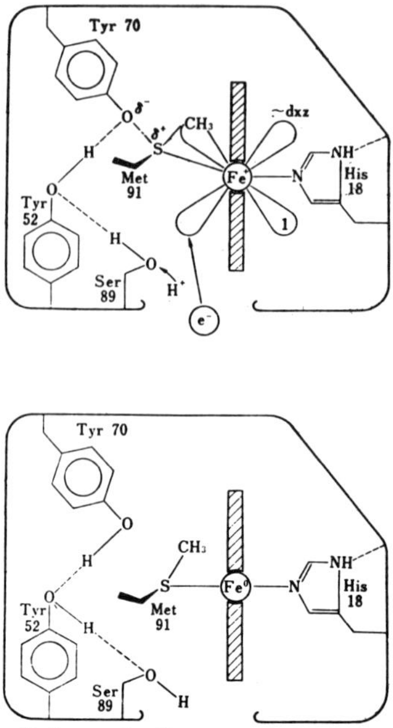

Fig. 8. Hydrogen-bonding network (schematic) in *R. rubrum* cytochrome *c*-550, after F. R. Salemme, *et al.*[B-19] (*vide*, discussion in text). (a) oxidized conformation, (b) reduced conformation

dase binding. It will be interesting to observe the reactivity of this cytochrome *f* with the mitochondrial system.

Before leaving the cytochromes *f*, it may be of interest to note that polylysine and other polycations inhibit photosystem I reactions in chloroplasts, but not those of photosystem II[B-100]. In particular, cytochrome *f* photooxidation in spinach chloroplasts is found to be decreased but not that of the active chlorophyll pigment, "P-700." The evidence available[B-101] indicates that polylysine binds the plastocyanin (copper protein) component in System I. This state of affairs is reminiscent of the situation in mitochondria in which polylysine inhibits electron transfer from cytochrome *c* to a copper-containing system, Complex IV···the heme a-Cu oxidase moiety of the mitochondrial electron chain··· as described above.

The responses of the cytochromes *c′* are as expected···they react not at all with the reductase. (Being rapidly auto-oxidizable they could not be assayed for oxidase activity.) However, if challenged with a soluble NADH-dehydrogenase derived from Complex I[B-68], they show a significant reactivity indicating ···as seen very often···the decrease in specificity of the mitochondrial redox enzymes upon partial denaturation or modification consequent on removal from the membrane[B-102].

A final comment on the potentialities of the comparative approach can be provided based on a recent ingenious suggestion[B-19] relating protonation to redox potential in cytochrome c_2. It is noted that the redox potential, E_m, of *R. rubrum* cytochrome *c*-550 depends on pH, dropping from a value of 370 mV at pH 5.0 to ~290 mV at pH 8.0 [B-33]. This dependence of pH is not seen with (I) in this pH range. Inspection of the cytochrome c_2 structure indicates a displacement of the bond between the sulfur atom of the methionyl (no. 91) ligand and the central iron atom some 0.3 Å away from its expected axial position in a rigorously symmetric octahedral iron chelate. This displacement can be attributed to a charge-pair interaction between the hydroxyl oxygen of a nearby internal tyrosine residue (no. 70) and the partially positive sulfur of the methionine ligand (no. 91) resulting from delocalization of the positive charge on the ferri-iron. The hydroxyl of Tyr no. 70 bears a negative charge···resulting from partial displacement of its proton toward another tyrosine (no. 52) in contact via a hydrogen bond to an external serine residue (no. 89). Protonation of the serine residue by lowering of the external pH can displace protons to the Tyr 52, thence to the Tyr 70, thus destabilizing the ferri-state bent-bond configuration and restoring the orientation of the S-Fe bond to the symmetrical perpendicular orientation favored in the reduced protein. This series of events would tend to stabilize the reduced form as the degree of protonation of the Ser 89 increases, thus raising the redox potential (shifting the equilibrium $e^- + Fe^{3+} \rightleftarrows Fe^{2+}$ to the right) with decreasing pH. This is precisely the behavior seen. In the case of mitochondrial cytochrome *c*, the homologous residues present (Tyr 67 in cytochrome *c* for Tyr 70 in c_2, Asn 52 for Tyr

52, and Thr 78 for Ser 89) comprise a similar network, except that the Asn 52 (truly invariant so far in all mitochondrial cytochromes *c*) may break the protonation path from the exterior of the molecule to the binding Met 80 ligand during the extensive conformation changes described as occurring when (I) is reduced or oxidized[B-72]. Such conformation changes do not occur in *R. rubrum* cytochrome c_2 (the oxidized and reduced crystals are isomorphic[B-19]) so that the protonation mechanism described is not so subject to uncertainties (possibly associated with movements of the residue at position 52). It may be noted that the Thr 78 residue in (I) is observed as invariantly hydrogen-bonded to Tyr 67 in both reduced and oxidized states[B-72], so that a protonation pathway is always available. Nevertheless, the E_m of cytochrome *c* shows no change over the pH range in which that of cytochrome c_2 varies. Thus, it appears that considerations based solely on the static structures of (I) may be inadequate without inclusion of perturbations arising from dynamic effects on structure, such as seem to be occurring in (I) but not in cytochrome c_2. The recent observation that in *E. gracilis* cytochrome *c*-558, the crucial Tyr 67 is replaced by a phenylalanine residue (G. Pettigrew (priv. comm.)) supports the protonation mechanism presented[B-19] in that one would predict invariance of E_m with change in pH in this cytochrome *c*, because the hydrogen-bonding network from Thr 78 to Met 80 is wholly broken by the insertion of the Phe 67 for the Tyr 67. As remarked previously, pH invariance of E_m is seen over the whole physiological range of this cytochrome *c* (*vide*, above).

It should be evident from all the foregoing discussion that a rich harvest of knowledge is in prospect as the approaches outlined are exploited. It is outside the scope of this review to indicate more than a few random examples of future researches.

Thus, it has been shown that a particular absorption peak far upfield from the proton envelope of the proton in NMR spectra of cytochrome *c* is associated with the methionyl-80 methyl protons in the central ligand field [B-103,104]. This peak can be used to monitor conformation changes, similarly to the well known charge-transfer absorption peak in ferricytochrome *c* at 695 nm. Analogous studies on *R. rubrum* cytochrome *c*-550 can be performed to elucidate the expected conformation changes of the ferri form with protonation described above.

The indications that reductase and oxidase systems can bind to sites on mitochondrial cytochrome *c* which possess overlapping structural features at the front of the molecule over the heme crevice (*vide*, above) provide a basis for speculation that more than one cytochrome *c* is inserted at the junction of the reductase and oxidase systems. Thus, if two molecules were present in contact, one attached to reductase and the other to oxidase, a simple mechanism would be available for electron transfer. It is likely that sufficiently rapid (~millisec) non-enzymatic electron exchange occurs between two molecules of cytochrome *c* in contact, because the aforementioned double NMR experiments on mixtures of ferri- and ferrocytochrome *c*[B-103,104] show that even in solution such exchange occurs with average lifetimes of ~50 millisec. Fast redox reactions in contact processes between heme proteins and redox reagents are well-documented. Thus, a bulky group like ferricyanide ion (at 15 mM) exhibits an extraordinary rate of 0.1 sec^{-1} in oxidation of *Chromatium* cytochrome *c′*[B-105], a molecule in which direct access to the heme group appears impossible for charged ligands[B-31]. At the protein concentration of ~10^{-5} M used, this reaction rate corresponds to a second-order rate constant of ~10^{10} M^{-1} sec^{-1}. Again, in the reaction between ferricyanide ion and (I), a second order rate constant of ~10^7 M^{-1} sec^{-1} is found (M. Cusanovich (priv. comm.)). Such enormous reaction rates might not be expected for even bulkier reagents, such as in inter-cytochrome electron transfer reactions, but impressive enough rates are seen, nevertheless, as reported[B-106] for the reaction between *Ps. aeruginosa* cytochrome *c*-551 and horse heart cytochrome *c*, which exhibits a second order rate constant of 1.8×10^3 M^{-1} sec^{-1} at pH 7.0, 4.5°C and ionic strength 0.2.

Another phenomenon worth mention is the apparent biological specificity seen in electron-transfer processes between homologous proteins as compared with the same types of proteins derived from differing tissue sources. Thus, it has been reported[B-107] that the second-order rate constant for the reaction of azurin from *Ps. aeruginosa* with its homologous cytochrome *c*-551 is greater than $10^6 \mathrm{M}^{-1} \mathrm{sec}^{-1}$ whereas the corresponding constant for the reaction with horse heart (I) is of the order $10^3 \mathrm{M}^{-1} \mathrm{sec}^{-1}$.

No data exist which unambiguously require just one cytochrome *c* molecule for each respiratory mitochondrial chain. In fact, those data which are available suggest that as many as two per chain is a possible stoichiometry (Y. Hatefi (priv. comm.)). The numerous speculations, based on the notion of a single cytochrome *c* per chain, involving rotations, preferential diffusion, etc., thus may be unnecessary.

Further experiments comparing results of chemical modifications of *R. rubrum* cytochrome c_2 with the numerous analogous researches on (I), (*vide*, e.g.,[A-2,4,B-96-98]) readily cometo mind. A particular example is the demonstration that selective cleavage of (I) at Met 65[B-108] leading to formation of two large polypeptide fragments, can be exploited to show that the polypeptide chain, including residues 1—65, containing the heme moiety, and the chain comprising residues 66—104, can be recombined to yield a protein in which function in the succinoxidase system is essentially fully restored as well as the original structural features of the native mitochondrial form, despite the fact that the two chains are not joined covalently[B-109]. It will be of much interest to establish whether a similar result can be achieved with *R. rubrum* cytochrome *c*-550, in which the analogous available methionine (Met 55) exists. In this manner, one may determine how crucial the insertion of eight extra residues in the mitochondrial cytochrome *c* region of residues 70—80 is with respect to proper refolding of the primary structure. This is another example of the great advantage conferred on comparative biochemical studies by availability of a well-characterized tertiary structure for a c-type cytochrome modified in such a manner as to affect functionality in the well-defined mitochondrial system. Inasmuch as several other bacterial cytochromes *c* may yield tertiary structures in the near future…e.g. *Micrococcus denitrificans* cytochrome *c*-550[B-57] and *Pseudomonas aeruginosa* cytochrome *c*-551 (R. E. Dickerson (priv. comm.))…it can be expected that other biologically modified forms of cytochromes *c* will be at hand to extend potentialities of the present studies.

Some other experiments which suggest themselves involve plant mitochondrial cytochromes *c* all of which exhibit[A-7,B-110] relatively low lysine contents (~12 compared with ~19 of the animal mitochondrial species), and which can be subjected to progressive acetylations, similar to the experiments done on the yeast and beef heart cytochromes[B-97,98] to probe essentiality of lysine residues in the oxidase and reductase binding reactions. The effect of polylysine on the interactions between Complex IV and *P. tenera* cytochrome *f* can be tested to see if a reaction can be observed where none is found ordinarily, paralleling the activation of the reductase reaction by polylysine.

These few very randomly chosen examples give some impression of future prospects. It is hoped that they enable the reader to share in some degree the optimism of investigators now engaged in active research on structure-function relationships in the cytochromes *c*.

It has been a unique honor to receive and respond to the invitation of the Editor to inaugurate a new practice…and hopefully a tradition…of English language articles in this outstanding Japanese biochemical publication, all the more so as the subject chosen is one to which the contribution of Japanese biochemists and microbiologists has been, and continues to be, basic and massive. The task of preparation has been lightened to a considerable degree by the opportunity to meet with many colleagues in Japan who have been most cooperative and helpful in providing new, and in many instances, unpublished data. I may mention among others, Professors T. Horio, R. Sato, M. Kakudo, and K. Narita at the Institute for Protein Research, Osaka; Prof. H. Matsubara, and Drs. T. Yamanaka, K. Wada and Y. Orii at Osaka University; Dr. S. Taniguchi and Prof. S. Hino, Hiroshima University; Prof. Huzisige, Okayama University; Prof. K. Shibata, Institute for Physical Chemical Research, Saitama;

Profs. T. Mori and H. Iwasaki, Nagoya University, and Profs. O. Hayaishi and S. Sano, Kyoto University. Finally, it is a privilege to recall the well-known pioneering efforts of Profs. K. Okunuki, H. Tamiya, E. Yakushiji, Y. Ogura and F. Egami which played so important a role in the establishment of cytochromology as a biochemical discipline not only in Japan but internationally.

Bibliography

A. General References and Source Materials

1) D. Keilin : *The History of Cell Respiration and Cytochromes*, Cambridge Univ. Press, London, New York (1966)
2) *Symposium on Structure and Function of Cytochromes*, Osaka, 1967 (ed. K. Okunuki, M. D. Kamen, I. Sekuzu), Univ. of Tokyo Press (1968)
3) M. D. Kamen, K. M. Dus, T. Flatmark, H. De Klerk: *Treatise on Electron and Coupled Energy Transfer in Biological Systems* (ed. T. E. King, M. Klingenberg), Chapter 5, Dekker, New York (1970)
4) E. Margoliash, A. Schejter : *Advances in Protein Chemistry*, Vol. **21**, 114 *et seq*, Academic Press, New York (1966)
5) M. D. Kamen, T. Horio : *Annual Reviews of Biochemistry*, **39**, 673 *et seq.* (1970) ; *Annual Reviews of Microbiology*, **24**, 399 *et seq.* (1970)
6) T. Yamanaka : *Advances in Biophysics*, **3**, 227 *et seq.* Academic Press, New York (1972); *Evolution of the Cytochrome c Molecule and the Energy Acquiring System*, in Supplement, this Journal (1972)
7) M. O. Dayhoff : *Atlas of Protein Sequence and Structure*, Vol. **5**, Nat'l. Biomedical Res. Foundation, Washington, D. C. (1972)
8) N. S. Gel'man, M. A. Lukoyanova, D. Ostrovskii : *Respiration and Phosphorylation of Bacteria* (transl. from Russian), Plenum Press, New York (1967)
9) *Comprehensive Biochemistry* (ed. M. Florkin, E. H. Stotz), **13**, Chapter 5, Elsevier Publ. Co., Amsterdam (1965)

B. Specific References (These are selected as appropriate to the scope of the discussion presented. A complete bibliography can be obtained by reference to the general sources given above (A).)

1) A. Tissieres : *Biochem. J.*, **64**, 582 (1956)
2) R. T. Swank, R. H. Burris : *Biochim Biophys. Acta*, **180**, 473 (1969)
3) M. D. Kamen, L. P. Vernon : *ibid.*, **17**, 10 (1955)
4) P. B. Scholes, G. McLain, L. Smith : *Biochemistry*, **10**, 2072 (1971)
5) C. W. Tabor, P. D. Kellogg. *J. Biol. Chem.*, **245**, 5429 (1970)
6) E. Itagaki, L. P. Hager : *Biochem. Biophys. Res. Comm.*, **32**, 1013 (1968)
7) I. C. Gunsalus, P. J. Chapman, J. F. Kuo : *ibid.*, **18**, 924 (1965)
8) C. J. Sih : *Science*, **163**, 1297 (1969)
9) T. Omura, R. Sato : *J. Biol. Chem.*, **239**, 2379 (1969)
10) G. W. Pettigrew : *FEBS Letters*, **22**, 64 (1972)
11) R. E. Dickerson, T. Takano, O. B. Kallai, L. Samson : *Structure and Function of Oxidation-Reduction Enzymes*" (Wenner-Gren Symposium, 1970), (ed. Å. Åkesson, A. Ehrenberg), p. 69, Pergamon Press, Oxford (1972); R. E. Dickerson : *Scientific American*, April, p. 58 (1972)
12) C. L. Tsou : *Biochem. J.*, **50**, 493 (1952)
13) D. Keilin, E. F. Hartree : *Nature* **176**, 200 (1955)
14) L. Smith, K. Minnaert : *Biochim. Biophys. Acta*, **105**, 1 (1965)
15) E. Yakushiji, K. Okunuki : *Proc. Imp. Acad. (Tokyo)*, **16**, 299 (1940)
16) K. Wada, H. Matsubara, K. Okunuki : in ref. A-2, p. 309
17) Y. Orii, K. Okunuki : *Ann. Report Biol. Works, Fac. Sci., Osaka Univ.*, **17**, 1 (1969)
18) C. A. Lu, L. Yu, T. E. King : *J. Biol. Chem.*, **247**, 1012 (1972)
19) F. R. Salemme, S. T. Freer, Ng. H. Xuong, R. A. Alden, J. Kraut : *J. Biol. Chem.* (1973) (in press)
20) J. Postgate : *Biochem. J.*, **56**, xi (1954)
21) M. Ishimoto, J. Koyama, T. Omura, Y. Nagai : *J. Biochem., Japan*, **41**, 537 (1954)
22) T. E. Meyer, R. G. Bartsch, M. D. Kamen : *Biochim. Biophys. Acta*, **245**, 453 (1971)
23) R. P. Ambler, M. Bruschi, J. LeGall : *FEBS Letters*, **5**, 115 (1969); **18**, **347**, 351 (1971)
24) J. M. Olson, E. K. Shaw, *Photosynthetica*, **3**, 288 (1969)
25) B. H. Gray, C. F. Fowler, N. A. Nugent, N. Rigopoulos, R. C. Fuller : *J. Bacteriol.* (1972) (in press)
26) B. H. Gray, C. F. Fowler, N. A. Nugent, R. C. Fuller: *Biochem. Biophys. Res. Comm.*, (1972) (in press)
27) Y. Shioi, K. Takamiya, M. Nishimura : *J. Biochem., Japan*, **71**, 285 (1972)
28) M. Mandel, E. R. Leadbetter, N. Pfennig, H. G. Trüper : *Int'l. J. System Bacteriol.*, **21**, 222 (1971)
29) T. Yamanaka, S. Imai : *Biochem. Biophys. Res. Comm.*, **46**, 150 (1972)
30) R. P. Ambler (1973) (in preparation)
31) S. Taniguchi, M. D. Kamen : *Biochim. Biophys. Acta*, **74**, 438 (1963)
32) M. D. Kamen : *Acta Chem. Scand.*, **17**, S 41 (1963)
33) T. Horio, M. D. Kamen : *Biochim. Biophys. Acta*, **48**, 266 (1961)
34) T. E. Meyer : Doctoral Dissertation, University of California, San Diego (1971)
34a) R. P. Ambler, E. Taylor : *Biochem. Soc. Trans., London*, **1**, 111 (1973)
35) R. P. Ambler, M. Wynn : *Biochem. J.* (1973) (in press)
36) J. Le Gall, M. Bruschi-Heriaud : in ref. A-2, p. 467
37) T. E. Meyer, R. G. Bartsch, M. A. Cusanovich, J. H. Mathewson : *Biochim. Biophys. Acta*, **153**, 854 (1968)
38) J. Van Beenmen, R. P. Ambler : *Ant. V. Leewenhoek* (1973) (in press)
39) R. G. Bartsch, T. E. Meyer, A. B. Robinson : in ref. A-2, p. 443
40) Y. Sugimura, F. Toda, T. Murata, E. Yakushiji : in ref. A-2, p. 452
41) H. E. Davenport, R. Hill : *Proc. Roy. Soc. (London) Ser. B*, **139**, 327 (1952)
42) G. Forti, M. L. Bertoli, G. Zanetti : *Biochim. Biophys. Acta*, **109**, 33 (1965)

43) N. Nelson, E. Racker : *J. Biol. Chem.*, **247**, 3848(1972)
44) M. A. Cusanovich, R. G. Bartsch : *Biochim. Biophys. Acta*, **189**, 245 (1969)
45) S. J. Kennel, M. D. Kamen : *ibid.*, **253**, 153 (1971)
46) R. P. Ambler, S. Murray : *Biochem. Soc. Trans., London*, **1**, 107 (1973)
47) J. Barrett : *Biochem. J.*, **64**, 626 (1956)
48) H. Yaoi, H. Tamiya (1928) : quoted by D. Keilin(ref. B-39)
49) D. Keilin : *Nature*, **132**, 783 (1933)
50) A. Fujita, T. Kodama : *Biochem. Z.*, **273**, 186 (1934)
51) T. Yamanaka : *Ann. Rep. Sci. Works, Fac. Sci. (Osaka Univ.)* **11**, 77 (1963)
52) T. Yamanaka : *The Biochemistry of Copper* (ed. J. Peisach, P. Aisen, W. E. Blumberg), p. 275, Academic Press, N. Y. (1966)
53) T. Horio, T. Higashi M. Sasagewa, K. Kusai, M. Nakai, K. Okunuki : *Biochem. J. (Japan)*, **77**, 194 (1960)
54) N. Newton : *Biochim. Biophys. Acta*, **185**, 316 (1969)
55) E. Margoliash, W. M. Fitch, R. E. Dickerson : *Structure, Function and Evolution of Proteins, Brookhaven Sympos. Biol.*, No. **21**, 239 (1968) ; W. M. Fitch, E. Margoliash : *Evolutionary Biology* (ed. T. Dobzhansky, M. K. Hecht, W. C. Steere), Vol. **4**, p. 67, Appleton, Century, Crofts Press, N. Y. (1970)
56) T. Ashida, T. Ueki, T. Tsukihara, A. Sugihara, T. Takano, M. Kakudo : *J. Biochem., Japan*, **70**, 913 (1971)
57) R. Timkovich, R. E. Dickerson (1973) (in preparation)
58) R. G. Bartsch : *Methods in Enzymology* (ed. A. San Pietro), **23**, 344 (1971)
59) R. G. Bartsch : *Bacterial Photosynthesis* (ed. H. Gest, A. San Pietro, L. P. Vernon), p. 475, Antioch Press (1963)
60) R. G. Bartsch, T. Horio, M. D. Kamen : *Biochim. Biophyys. Acta* (1973) (in press)
61) E. Margoliash : *Brookhaven Sympos. Biol.*, **15**, 266 (1962)
62) E. Margoliash, J. Lustgarten : *J. Biol. Chem.*, **237**, 3397 (1962)
63) E. L. Smith : in ref. A-2, p. 282
64) T. H. Jukes, C. R. Cantor : *Mammalian Protein Metabolism* (ed. H. N. Munro), p. 21, Academic Press, N. Y. (1969)
65) E. Margoliash, M. Reichlin, A. Nisonoff : in ref. A-2, p. 269
66) T. Yamanaka : *Biochim. Biophys. Acta*, **275**, 74(1972)
67) T. Yamanaka, S. Takenami, N. Akiyama, K. Okunuki : *J. Biochem., Japan*, **70**, 349 (1971)
68) K. A. Davis, Y. Hatefi, F. R. Salemme, M. D. Kamen: *Biochem. Biophys. Res. Comm.*, **49**, 1329 (1972)
69) S. R. Elsden, M. D. Kamen, L. P. Vernon : *J. Am. Chem. Soc.*, **75**, 6347 (1953)
70) Various papers in ref. A-2, p. 35 *et seq.*
71) G. H. Barlow, E. Margoliash : *J. Biol. Chem.*, **241**, 1473 (1966)
72) T. Takano, O. B. Kallai, R. Swanson, R. E. Dickerson (1973) (in press) ; *Cold Spring Harbor Symp. Quant. Biol.* (1971)
73) Y. Hatefi : in ref. A-9 (1966), p. 199
74) Y. Hatefi, A. G. Haavik, P. Jurtschuk : *Biochim. Biochim. Acta*, **52**, 106 (1961)
75) L. R. Fowler, S. H. Richardson, Y. Hatefi : *ibid.* **64**, 170 (1962)
76) L. P. Vernon : *Arch. Biochem. Biophys.*, **43**, 492 (1953)
77) K. Dus, K. Sletten, M. D. Kamen : *J. Biol. Chem.*, **243**, 5507 (1968)
78) L. P. Vernon, M. D. Kamen : *J. Biol. Chem.* **211**, 643 (1954)
79) K. Sletten, M. D. Kamen : in ref. A-2, p. 422
80) L. Smith, M. Baltscheffsky : *J. Biol. Chem.*, **234**, 1575 (1959)
81) T. Horio, M. D. Kamen : *Biochemistry*, **1**, 144(1962)
82) J. Gross, J. Wolken : *Science*, **132**, 357 (1960)
83) F. Perini, M. D. Kamen, J. A. Schiff : *Biochim. Biophys. Acta*, **88**, 74 (1964)
84) T. E. Meyer, M. A. Cusanovich : *ibid.*, **267**, 383(1972)
85) C. Sybesma, *Photochem. Photobiol.*, **6**, 261 (1967)
86) S. Katoh : *Plant Cell Physiol.* **1**, 29 (1959)
87) E. Yakushiji, Y. Sugemura, I. Sekuzu, I. Morikawa, K. Okunuki : *Nature*, **185**, 105 (1960)
88) R. Hill, R. Scarisbrick : *New Phytol.*, **50**, 98 (1951)
89) M. D. Kamen : *Hemes and Heme Proteins*, p. 521, Academic Press, N. Y. (1966)
90) T. Kakuno, R. G. Bartsch, K. Nishikawa, T. Horio : *J. Biochem, Japan*, **70**, 79 (1971)
90a) H. deKlerk, R. G. Bartsch, M. D. Kamen : *Biochim. Biophys. Acta*, **97**, 275 (1965)
91) K. Dus, H. deKlerk, R. G. Bartsch, T. Horio, M. D. Kamen, *Proc. Nat. Acad. Sci. U. S. A.*, **57**, 367 (1967)
92) L. Smith, H. C. Davies, M. Reichlin, E. Margoliash : *J. Biol. Chem.*, **248**, 237 (1973)
93) M. E. Winfield : *J. Mol. Biol.*, **12**, 600 (1965)
94) W. T. Morgan, C. P. Hensley, J. P. Riehm : *J. Biol. Chem.*, **247**, 6555 (1972)
95) E. E. Jacobs, E. C. Andrews, H. Wohlrab, W. Cunningham : in ref. A-2, p. 114
96) K. Wada, K. Okunuki : *J. Biochem., Japan* **64**, 607 (1968)
97) K. Wada, K. Okunuki : *ibid.*, **66**, 263 (1969)
98) K. Wada, K. Okunuki : *ibid.*, **66**, 249 (1969)
99) M. V. Laycock (1973) (unpublished) ; M. V. Laycock, J. S. Craigie : *Can. J. Biochem.*, **49**, 641 (1971)
100) J. Brand, T. Baszynski, F. L. Crane, D. W. Krogmann : *Biochem. Biophys. Res. Comm.*, **45**, 533 (1971); *J. Biol. Chem.*, **247**, 2814 (1972)
101) J. Brand, A. San Pietro, B. C. Mayne : *Arch. Biochem. Biophys.*, **152**, 426 (1972)
102) Y. Hatefi, B. Stempel : J. Biol. Chem., **244**, 2350(1969)
103) R. K. Gupta, A. G. Redfield : *Science*, **169**, 1204(1970)
104) A. G. Redfield, R. J. Gupta : *Cold Spring Harbor Symp. Quant. Biol.*, **36**, 405 (1971)
105) Q. H. Gibson, M. D. Kamen : *J. Biol. Chem.*, **241**, 1969 (1965)
106) R. A. Morton, J. Overnell, H. A. Harbury : *ibid.*, **245**, 4653 (1970)
107) E. Antonini, A. Finazzi-Agno, L. Avigliano, P. Guerrieri, G. Rotilio, B. Mondovi : *J. Biol. Chem.*, **245**, 4847 (1970)
108) K. Ando, H. Matsubara, K. Okunuki : *Biochim. Biophys. Acta*, **118**, 256 (1966)
109) G. Corradin, H. A. Harbury : *Proc. Nat. Acad. Sci. U. S. A.*, **68**, 3036 (1971)
110) Y. Morita, S. Ida, S. Yagi (1973) (in press)
111) D. S. Hoare, L. O. Ingram, E. L. Thurston, R. Walkup : *Arch. Mikrobiol.*, **78**, 310 (1971)

Index

D

E

F

G

H

I

K

L

M

N

O

P